T0075086

Linux with Operating System Concepts

Linux with Operating System Concepts

Second Edition

Richard Fox

CRC Press is an imprint of the
Taylor & Francis Group, an **informa** business
A CHAPMAN & HALL BOOK

Second Edition published 2022
by CRC Press
6000 Broken Sound Parkway NW, Suite 300, Boca Raton, FL 33487-2742

and by CRC Press
2 Park Square, Milton Park, Abingdon, Oxon, OX14 4RN

First edition published by CRC Press 2015

CRC Press is an imprint of Taylor & Francis Group, LLC

Library of Congress Cataloging-in-Publication Data
Names: Fox, Richard, 1964- author.
Title: Linux with operating system concepts / Richard Fox.
Description: Second edition. | Boca Raton : CRC Press, 2022. | Includes bibliographical references and index.
Identifiers: LCCN 2021031731 | ISBN 9781032063454 (paperback) | ISBN 9781032066707 (hardback) | ISBN 9781003203322 (ebook)
Subjects: LCSH: Linux. | Operating systems (Computers)
Classification: LCC QA76.774.L46 F69 2021 | DDC 005.4/32—dc23
LC record available at https://lccn.loc.gov/2021031731

ISBN: 978-1-032-06670-7 (hbk)
ISBN: 978-1-032-06345-4 (pbk)
ISBN: 978-1-003-20332-2 (ebk)

DOI: 10.1201/9781003203322

Typeset in Times
by codeMantra

Access the Support Material: https://www.routledge.com/9781032063454

With all my love to Cheri Klink, Sherre Kozloff and Laura Smith.

Contents

Preface

The Unix/Linux textbook market is full of books for people who are looking to acquire hands-on knowledge of Unix/Linux, whether as a user or a system administrator. There are almost no books that serve as textbooks for an academic class. Why not? There are plenty of college courses that cover or include Unix/Linux. We tend to see conceptual operating system texts which include perhaps a chapter or two on Unix/Linux or Unix/Linux books that cover almost no operating system concepts. Most Unix/Linux books either focus on how to use Unix/Linux or how to administer Unix/Linux.

This book is unique. It explores operating system concepts while introducing Linux-specific content. It is divided roughly into two parts: an introduction to Linux for users and an introduction to Linux for system administrators. It covers Linux and operating system concepts not as a how-to, hands-on guide but as a true textbook, complete with definitions, concepts, chapter reviews and a collection of ancillary material to help support the instructor and student if used in a class.

If you are reading this book, I hope you find it helpful and learn a great deal. My intention is not to cover all things Linux but to explore Linux and offer background for why things are set up the way they are.

HOW TO USE THIS TEXTBOOK

In this edition of the textbook, Chapters 1–6 cover Linux from a user perspective, while Chapters 7–12 cover Linux from an administrator's perspective. Throughout the book are operating system concepts related to topics of the chapter. This textbook is envisioned to be used in a 1- or 2-semester course on Linux. A 2-semester course should be able to cover all 12 chapters, dividing the semesters between user and administrator topics. Such a course could be split between lectures (roughly 50% of the time) and labs. See the lab manual which accompanies this text (available online).

A 1-semester junior or senior-level IT or computer science course should be able to cover a majority of the book. An IT course might cut out some of the operating system concepts and use the lab manual, while a computer science course might include portions of the lab manual and eliminate some of the less needed topics like `sed` (Chapter 5), `systemd` initialization (Chapter 9), open-source software installation (Chapter 11) and system troubleshooting (Chapter 12).

A lower-level 1-semester course that is using this text as an introductory to Linux should cover Chapters 1–4 in detail and then select relevant topics to fill out the course, for instance including some portions of user accounts from Chapter 7, file system topics from Chapter 8, network configuration topics from Chapter 10 and software installation from Chapter 11.

Below is a list of topics found in the chapters that can be considered optional. Omitting some of these topics will not reduce the readability of the remainder of the text.

- Chapter 1: All content outside of Section 1.5, on Linux installation, can be omitted as it mostly explores a history of operating systems and Linux development.
- Chapter 2: Removing Sections 2.7.1 and 2.7.2 should not be an issue; Section 2.6 on `vi` can be skipped if students will not be using `vi`.
- Chapter 3: Sections 3.2 and 3.8 on storage terminology and devices are not essential; Section 3.9 on file compression (although 3.9.4 might be useful) can be skipped; Section 3.7.1 on File Browser searching is only needed if students are not to learn the `find` program.
- Chapter 4: Forms of process management (Section 4.2) and memory and virtual memory (Section 4.7.1) can be skimmed over if the course is an IT course.
- Chapter 5: Sections 5.5 and 5.6 (`sed` and `awk`) can be omitted if these two programs are not going to be used, although some parts of `awk` should still be covered as it is referenced later in the textbook.

- Chapter 6: Section 6.3.5 on the `expr` instruction can be skipped; depending on how much scripting the student is to learn, Sections 6.7–6.9 on arrays, string pattern matching and functions may be omitted.
- Chapter 7: Section 7.5 on PAM and strong passwords may not be needed; Section 7.8 on SELinux could be reduced to just an introduction; Section 7.9 is not necessary.
- Chapter 8: Sections 8.2.1 and 8.2.2 provide background for better understanding inodes but 8.2.2 can be eliminated and 8.2.1 reduced; if the course will not cover topics like partitioning, then Section 8.5 can be skipped.
- Chapter 9: If the course is an IT course, coverage of booting (Section 9.2) may not be warranted; if the course is a computer science course, detailed information about `systemd's` initialization process may not be desired (Section 9.3); some content from Section 9.6 can be eliminated unless the specific service is going to be examined in detail.
- Chapter 10: Depending on the student's background, Section 10.2 may not be needed; Section 10.4.2 is only needed if the course covers DHCP; based on the version of Linux used, either cover Sections 10.6.1–10.6.3 or Section 10.6.4.
- Chapter 11: Sections 11.2 and 11.8 provide background on software but can be skipped; depending on whether the course covers Red Hat or Debian-based Linux, subsections within Sections 11.3–11.5 can be skipped; Section 11.7 can be omitted if the students will not use `gcc`.
- Chapter 12: Coverage of RAID (12.2.2) and encryption (12.2.3) may be skipped if the topics are beyond the scope of the course and similarly the content on operating system issues (12.4.1) may be omitted.
- Chapter 13: This chapter, available online, covers Apache and Squid installation and configuration; it can be skipped entirely if the course will not cover either of these software titles.

NEW TO THIS EDITION

The first edition of this book was written specifically for Red Hat 6 and correspondingly, CentOS 6. At the time of its publication, Red Hat 7 had come out, but Red Hat 7 was viewed as an in-between step as Red Hat was moving from `init` to `systemd`. With both Red Hat 6 and Red Hat 7 reaching the end of their lives, this edition has updated all content to move on to `systemd` Linux distributions, concentrating on Red Hat 8. While preparing this manuscript, CentOS announced that they would not support CentOS 8 beyond 2021 and would instead concentrate on CentOS Stream. At the time of this writing, CentOS 8 and Stream are nearly identical.

So, the first major change between the first edition of this textbook and this new edition is updating to Red Hat 8/CentOS Stream. In an attempt to broaden the appeal of this text, this edition has a good deal of content on Debian/Ubuntu versions of Linux when those versions differ from Red Hat. The information on Debian/Ubuntu is not complete, but attempts have been made to note the differences when space permits.

Chapters 1–7 in the first edition are now Chapters 1–6. Material from Chapter 5 has been moved (`vi` is now in Chapter 2, network software is now in Chapter 10, compression is now covered in Chapter 3, and encryption topics are now in Chapter 12) or removed. Most of these chapters have been rewritten to improve their clarity with improved examples and more tables and figures. Some of the removed content has been uploaded to the textbook's companion website as supplemental reading material.

Chapter 8 from the first edition has been removed. Content on the Linux kernel is now described in Chapters 1, 4 and 9. Installation of Linux has been moved to Chapter 1. Other content, such as virtual memory, has also been moved. With this chapter removed, Chapters 7–12 are primarily what Chapters 9–14 had been. All content in this part of the book has been updated to `systemd`-versions of Linux (primarily Red Hat) and include such new topics as the new top-level directory layout, the xfs file system (briefly), the NetworkManager service, `firewalld` and `ufw` as new front-ends to `iptables`, the `journald` service and Cockpit (covered briefly). These chapters have also been substantially rewritten and new examples, figures and tables added.

Some material from the former Chapters 9–14 are being moved online via supplemental readings, including for instance the Red Hat 6/init initialization process. The former appendix (binary numbers) is also being removed and made available online.

AVAILABLE ONLINE SUPPLEMENTS

There is a zipped file available for instructors who adopt this book and a zipped file available to everyone. This latter file is available via the textbook's companion website at https://www.routledge.com/9781032063454. If you are an instructor, contact your CRC Press/Taylor & Francis book rep. The contents of these files are listed below.

Instructor-Only Resources

- Instructor's manual complete with answers to chapter review questions
- Testbank

Other Available Material

- PowerPoint notes
- Glossary of terms from chapter reviews (consolidated into one file)
- Select answers to some chapter review questions
- Complete lab manual (assuming CentOS Stream can be adapted for other Linux distributions)
- Supplementary readings (reference to first edition chapters)
    - The fetch-execute cycle and details on the CPU (previously from Chapter 1)
    - Comparing Bash to Csh (previously from Chapters 2 and 7)
    - A brief introduction to emacs (previously from Chapter 5)
    - `vi` and `emacs` cheatsheets
    - System V/Upstart initialization process (previously from Chapter 11)
    - Some example network scripts (previously from Chapter 12)
    - A look at disaster planning and recovery (previously from Chapter 14)
    - Review of binary numbers (previously from the Appendix)
    - Apache/Squid installation (previously Chapter 15, available only online)
    - Perl scripting

Acknowledgments and Contributions

First, I am indebted to Randi Cohen for her encouragement and patience in my writing and completing this text. I would also like to thank Stan Wakefield for connecting me with Randi and CRC Press/Taylor & Francis Group. I would like to thank Olivia Snowden, a recent graduate from Northern Kentucky University (NKU), for so kindly volunteering to proofread the entire manuscript. She caught many errors and typos and helped me improve the book considerably. I would also like to express my thanks to colleague Dr. Wei Hao (NKU) and to Dr. Jim Furstenberg (Ferris State University) for proofreading select chapters and providing both corrections and insightful comments. Dr. Yi Hu (NKU) has maintained a list of errata from the first edition and has given me some great ideas for this new edition. I would also like to thank Micah Sidebottom, a student who, while taking our Linux course in fall 2020, pointed out several inconsistencies in the first edition that I have hopefully corrected. I am also indebted to several former colleagues and associates for their insight into networking (Professor Scot Cunningham) and Linux (Dr. Xiannong Meng, Bucknell University and Professor Peter Bartol, San Diego State University).

I would also like to thank everyone in the open-source community who contribute their time and expertise to better all of our computing lives. Without all of the efforts put into Linux, this book would obviously not exist!

On a personal note, I would like to thank Cheri Klink for all of her love and support.

Author

Richard Fox is a professor of computer science at Northern Kentucky University (NKU). He primarily teaches artificial intelligence, computer architecture, computer systems, concepts of programming languages, object-oriented programming and Unix systems. He has also taught data structures, IT fundamentals, web development and web server administration, among other courses. Dr. Fox, who has been at NKU since 2001, is currently the undergraduate program director of Computer Science and chair of NKU's University Curriculum Committee. Prior to NKU, Dr. Fox taught for 9 years at the University of Texas – Pan American. He has received two Teaching Excellence awards, from the University of Texas – Pan American in 2000 and from NKU in 2012, and an award for Outstanding Service from NKU in 2016.

Dr. Fox received a Ph.D. in Computer and Information Sciences from The Ohio State University in 1992. He also has an M.S. in Computer and Information Sciences from Ohio State (1988) and a B.S. in Computer Science from the University of Missouri Rolla (now Missouri University of Science and Technology) from 1986.

Dr. Fox has published two other books with CRC Press/Taylor & Francis Group: an introduction to information technology text (in its second edition) and a book on Internet infrastructure (coauthored by colleague Dr. Wei Hao). He has also authored or coauthored over 45 peer-reviewed research articles primarily in the area of artificial intelligence.

Richard Fox grew up in St. Louis, Missouri, and now lives in Cincinnati, Ohio. He is a big science fiction fan and progressive rock fan. As you will see in reading this text, his favorite composer is Frank Zappa.

1 Linux

What, Why, Who and When, and How

This chapter's learning objectives are to be able to

- Describe what operating systems are and how and why we use them
- Compare the more popular versions of Linux
- Explain the term open source software and the role the open-source community has had in the development of Linux
- Enumerate reasons why IT personnel should learn Linux
- Identify the roles of the most significant individuals in the development of Linux
- Install Linux
- Use Linux from the GUI to start applications and open terminal windows

1.1 INTRODUCTION

As you are reading this book, you must want to learn about Linux. Why should you learn about Linux? From a user's point of view, Linux offers a different approach than other operating systems. It is an open-source product meaning that you can obtain it for free. Similarly, most software for Linux is open source. With open-source software, you can also enhance the code (if you have the capability of doing so). From a system administrator's perspective, Linux, like Unix, lets you dive deeply into the operating system (OS) and have more control than you can in Windows or MacOS. Learning Linux teaches you not only how to use it but more about operating systems and computers.

The intention of *open-source software* is to make the software available in its source code format. Those who are skilled at coding can then enhance or alter the software or develop new software that uses some portions of the existing software. Although open-source software was originally synonymous with the development of Unix, Linux and related operating systems, the open-source community has produced a number of highly useful applications software. You may have used open-source software yourself. Table 1.1 lists the top ten open-source products as rated by TechRadar as of April 2021. All of the software listed in Table 1.1 can run in Linux, Windows and MacOS unless noted.

One drawback of open-source software is that it does not come with *commercial support.* When we buy software, we are not purchasing the software itself but purchasing the right to use the software. With that purchase almost always comes a guarantee that the software developers will provide timely feedback on issues identified with the software. Bugs will be fixed for us. Security holes will be resolved and patches released quickly. Manuals (whether printed or online) show users how to use the software.

Most commercial software comes with a guarantee of this support; open-source software products are supported by those who developed the titles but that support is not guaranteed. This is an issue that organizations might have when considering the adoption of open-source software. Even with support available, open source might lag behind commercial software with respect to bug and security fixes and improved features.

DOI: 10.1201/9781003203322-1

TABLE 1.1
Some of the Best Open-Source Products

Title	Type of Software	Website
LibreOffice	Productivity software/office suite	libreoffice.org
VLC media player	Plays movies, music, live streams (also available for Android, iOS)	videoland.org/vlc
GIMP	Photo and image editor	gimp.org
Shotcut	Video editor (Windows only).	shotcut.org
Brave	Web browser which supports privacy in browsing (also Android, iOS)	brave.com
Audacity	Audio editor for music and spoken word	audacityteam.org
KeePass	Password generator, credential storage tool (Windows although has been unofficially ported to other platforms)	keepass.info
Thunderbird	Email manager	thunderbird.net
FileZilla	FTP client	filezilla-project.org
Linux	Operating system (can be dual booted from Windows; can be installed in a virtual machine in Windows and MacOS)	Varies by distribution

But while this is a concern, we can also see that the open-source community can operate at a speed greater than that of commercial software companies. One interesting example demonstrating the support from the open-source community comes from the OpenSSL project. OpenSSL is a popular software product used to support secure communication over computer networks, with the capability of creating and utilizing public- and private- key encryption.

A security flaw was discovered in OpenSSL on April 1, 2014. The bug, called *HeartBleed*, would allow clever hackers to break the encryption generated by OpenSSL making websites protected by OpenSSL insecure. Six days after discovery, OpenSSL announced the security flaw and released a patch to solve the problem. One of the biggest security flaws discovered in open source software was patched in *less than a week*. Such agility is found among the open-source community. Many security flaws found in commercial software, including Windows, have taken far longer to fix. Note that while the security flaw was patched, it often took weeks, months or even years for administrators of the various websites that used OpenSSL to apply the patch and thus remove the security flaw.

Linux is free and supported. Is that the only reason to use Linux? Actually, no. It is perhaps the most visible reason but for anyone in IT, it is not the most important. If you are a Windows user, you may have used any number of Windows GUI-based tools to administer your operating system. With these tools, you can create new accounts, change permissions on files and directories, schedule tasks, view events (recorded in logs), modify disk partitions and modify any number of hardware settings, to name only a few of the tasks that these tools support. Nearly every one of these tasks uses a GUI (although some of these operations can be handled through PowerShell commands).

In Linux, every type of administrative operation is handled from the command line. This sounds like a detriment; who wants to type cryptic commands when those operations can be input through a GUI? But, in fact, using the command line gives the user greater flexibility in issuing commands which in turn provides the administrator *greater control* over the operating system. There are some OS settings that cannot be. modified in Windows. Just about everything in Linux is modifiable, whether via the command line or, when necessary, by modifying Linux source code; fortunately, we won't have to resort to that step in this textbook!

Greater control through the command line also leads to, in many cases, *easier control*. This seems unlikely in that using a GUI should always be easier. And yet if you have experience with Windows, you know that some settings are hidden behind numerous operations. Click here, select this, open that, select yet another link, change tabs, fill in this form, select OK and then confirm. Making multiple changes may require repeating the same operations with minor modifications, over and over. We will learn that Linux offers very convenient ways to recall a previous command line instruction, make minor modifications to it and execute the revised version. Alternatively, we can write shell scripts to perform such operations.

If you are a student studying Linux or hold an IT position, another aspect of Linux that you might not gain in Windows or MacOS is learning about operating systems. Windows and MacOS tend to shelter many OS concepts from you. You may learn to control processes using a tool like the Task Manager, learn about disk partitions and file systems, or learn about virtual memory because of heavy disk usage. But in Linux, you may need to delve into these details. You may want to change process priorities, launch some processes in the background and change effective user ID of other processes. You may need to manually mount and unmount partitions. You may need to examine memory usage statistics to see how frequently your computer is swapping between main and virtual memory. Learning Linux from the command line forces you to understand, at least to some extent, what the OS does and how.

In learning Linux, you will expand your knowledge of computers in general. And, well, using Linux is cool (at least that's what many people think!).

In this textbook, we look at Linux from two different perspectives. Early on, we learn Linux as a user. We explore how to enter commands from the command line prompt and navigate through the Linux file space. We learn how to launch and manage processes. We learn some of the powerful tools available in Linux including regular expressions and shell scripting. Over the course of these early chapters, we learn dozens of Linux commands and useful features that make command line entry easier. Midway through the text, we shift focus to learning Linux as a system administrator. In this set of material, we look at creating and managing user accounts, managing files and file systems, the Linux boot and initialization process, controlling services, configuring network access and installing software. Throughout the book, we introduce OS concepts.

To get us started in this chapter, we focus on five questions. *What is Linux*, and more generally what are operating systems? We define operating systems and examine the components that make up operating systems before we turn to Linux specifically and its components. *Why should we use Linux*? We gave a brief answer to this question but delve more into the significance of Linux as an OS of choice. *Who developed Linux and when*? We look at the history of Linux and the important players, starting with some earlier operating systems that led to or factored into the development of Linux. Finally, *how do we use Linux*? Although that is a topic for the entire book, in which chapter we explore how to install several different versions of Linux. We also introduce different forms of Linux interfaces, concentrating on how to open a terminal window for command line input, which we will then use throughout most of the book.

SECTION ACTIVITIES

1. How many operating systems do you have experience with? Count not only desktop/laptop computers but any servers, mainframes, supercomputers, tablets and smartphones. If you have experience with more than one, what similarities do you find between those that you know? What differences?
2. Make a list of your own reasons for learning Linux. How do they compare to the reasons covered in this section?
3. Read about Heartbleed at https://heartbleed.com/. Were you aware of it when the problem was first announced in 2014?

1.2 WHAT IS LINUX?

We start with something more general, what is an *operating system*? A computer's OS is a collection of programs that, as a whole, support our use of the computer through managing hardware resources and running processes. Operating systems usually comprise many different programs. The heart of the OS is a single program called the kernel. The *kernel* is loaded into memory when a computer is booted, and it remains resident in memory until the computer is shut down. The kernel's role is to handle process and resource management, among other tasks. The kernel calls upon other OS components to accomplish some of its tasks. Some of these components are loaded and run as needed. Users may also call upon some of these other OS components, again loaded and run as needed. Although we asked specifically what Linux is, we will concentrate first on operating systems in general. We then shift to Linux specifically later in this section.

1.2.1 Early Operating Systems

The earliest electronic computers, developed in the mid-1940s through the early to the mid-1950s were one-of-a-kind devices, created as much to explore how to build computers as to be useful computational devices. These computers had no operating systems at all. For a "user" to use a computer, that user would write their program code and submit it. Code might have been entered into the computer by making connections of various components through cables, setting switches on the computer's console and pressing the start button. Such programs used no external resources. If a program required a resource like access to input from magnetic tape or punch cards, the instructions to perform such input had to be included in the program itself. Otherwise, the computer would not know how to access the tape drive or punch card reader.

Around 1958, programmers began shifting from low-level machine languages to more sophisticated, high-level languages. Among the first were FORTRAN and COBOL. A program written in one of these languages could not be directly executed. Instead, the program had to be translated from the high-level language into machine language using a separate program called a *compiler*.

The programmer had several distinct steps to run their program. First, they would mount and load the compiler. Next, they would input their FORTRAN or COBOL program into the running compiler. The compiler would execute, outputting the executable version of the program onto magnetic tape. Now, the programmer would unmount the compiler and mount the tape containing their executable program. The programmer would next load and run the executable program. The program's input would likely come from another tape or punch cards. Remember, every one of these steps requires that the programmer implement the steps by additional program code. That's a lot of work to run a program!

To simplify this process, some of the tasks were captured into a program called the *resident monitor*. This program would be loaded into memory after the computer booted. The program would stay in memory until the computer was shut down, thus the word *resident*. The term *monitor* described the role of this program: to monitor a running program's requests for access to the system resources available. The system resources were generally limited to magnetic tape and punch card reader (printers usually were separate devices that would print data from magnetic tape).

The first resident monitor predates FORTRAN and was released in 1955. By the early 1960s, the tasks of the monitor had grown to the point that people were referring to these programs as operating systems. Over the course of decades, operating systems have grown in complexity and size. Table 1.2 examines some of the earliest operating systems/resident monitors.

Today, the OS contains the kernel (what had been the resident monitor) and supporting software including device drivers, utilities, shells, services and servers. Nearly all of our modern computers have and require operating systems. Without an operating system, most users would be unable to use their computer.

TABLE 1.2
Early Resident Monitor/Operating Systems of Note

System	Platform	Notable Features
MIT's Tape Director (1955)	UNIVAC 1103	Mounting and access to files on tape.
General Motors Operating System (unnamed, 1955)	IBM 701	First batch operating system.
GM-NAA I/O (1956) SHARE OS (1959)	IBM 704	Next-generation follow-ups to the General Motors OS, executing scheduled jobs in a batch mode but sharing routines that were common across processes.
Atlas Supervisor (1957)	Atlas Computer	Managed resources including virtual memory.
BESYS (Bell Labs Systems, 1957)	IBM 704 (and later 7090 and 7094)	Batch processing and tape management, use of punch cards, program libraries, core dumping.
IBSYS (1960)	IBM 7090, 7094	Batch processing, job control cards to control operations.
Compatible Time-Sharing System (CTSS, from MIT, 1961)	IBM 7094	Time sharing (multitasking).
Master Control Program MCP, 1961)	Burroughs mainframes	Multiprocessors and virtual memory; the first OS written in a high-level language.
Bolt, Beranek and Newman (BBN) Time-Sharing System (1962)	PDP-1	Time sharing, supported by both the operating system and specialized hardware.
OS/360 (announced 1964, released 1966)	IBM 360 mainframes	Batch processing with multiprogramming and a separate process scheduler.

Computers without an OS are known as *bare machines*. You might find a bare machine if you either construct your own computer hardware and run it without installing an OS or delete the existing operating system. The only reason to use a bare machine is to experiment with hardware and to gauge hardware efficiency irrespective of OS or application software load.

1.2.2 The Operating System Kernel

We identified several types of software that make up the OS in the last subsection. Let's take a closer look at each. The kernel, as already noted, is an expanded version of the resident monitor. It is the kernel that is responsible for most of the important tasks of the OS. We highlight several of these tasks in Table 1.3. The tasks are listed in alphabetical order rather than in order of importance. Process management is likely the most important role of any OS.

There are three different types of kernel: monolithic, microkernel and hybrid. A *monolithic* kernel is one in which the kernel is a single program which operates solely within the computer's *privileged mode* and in its own address (memory) space. Communication between the user side (running applications) and the kernel side is handled through *system calls*. A system call invokes a specific portion of the OS kernel. Upon receiving a system call, the kernel changes from *user mode* to privileged mode. The kernel then ensures that the user program's request is legitimate, meaning that the user program (or user) has an appropriate level of access for the call to be carried out.

Early operating systems used a monolithic kernel. But, as these operating systems governed computers with limited amounts of memory and few system resources, the kernel was not asked to do a great deal and so the kernel was fairly small (in comparison to later OS kernels). As computer capabilities grew, operating systems became larger and monolithic kernels became more and more complex.

In response to the complexity of the monolithic kernel, the *microkernel* was developed starting in the 1980s. The microkernel is based on a smaller kernel that operates within privileged mode and in its own address space. In order to support the expected range of functions, the OS

TABLE 1.3
Roles of the OS Kernel

Role	Meaning	Example
Auditing and accounting	Keep track of users logged in and the resources allocated to them; log events.	Auditing and logging programs.
Device management	Ability to add, remove and interface with peripheral devices.	Mounting disk drives, adding devices via USB ports.
File management	Ability to open, close, create, save, rename, move and copy files and directories.	Windows File Explorer or similar commands available in MS-DOS and PowerShell.
Interprocess communication	Information sharing between running processes.	Shared data in memory or message passing.
Interrupt handling	Dealing with device interrupts and error situations.	I/O interrupts, interprocess interrupts, timer interrupts.
Memory management	Manage memory allocation and deallocation; protect areas of memory from memory violation; move code and data into and out of memory as needed.	Virtual memory (demand paging, demand segmentation), contiguous memory allocation and compaction, memory overlays.
Process management	Start new processes; monitor processes as they run to initiate the handling of requests; detect and initiate the process of handling errors; switch between processes; remove processes from the computer when they terminate.	Single tasking, batch processing, cooperative multitasking, pre-emptive multitasking; multithreading.
Protection	Ensure a process can only access resources available to that process or that process' owner.	Access control lists, user accounts and a login process, user mode versus privileged mode.
Resource management	Grant a process access to a resource (e.g., a file); maintain mutually exclusive access; maintain process liveness.	Synchronization mechanisms, deadlock handling mechanisms.
Scheduling	Determine the order with which processes will run.	First-come first-served, priority, round-robin.
Security	Extend protection across a computer network.	User accounts and a login process, encryption.
User interface	Provide a means for the user to interface with the OS, running applications and hardware.	Graphical-User interface (GUI), command line interface (CLI), menu-driven interface.

is extended by a number of subsystems referred to as *servers* (not to be confused with the use of the word server as used in webserver, file server, print server and so forth). Servers operate based on requests from user applications in the computer's user mode and the user section of memory's address space. Communication between application software and kernel is much like with the monolithic kernel, but the kernel then calls upon servers to handle many of the operations. Thus, the microkernel involves a far greater amount of system calls as the servers are separate from the kernel.

Among the server components are file system servers, network servers, display servers, user interface servers and servers that can directly communicate with device drivers. Remaining in the kernel are the process scheduler, memory manager (including virtual memory) and interprocess communication between OS components. One of the most prominent operating systems to use the microkernel is Mach, Carnegie Mellon University's implementation of Unix.

The *hybrid* kernel compromises between the two extremes where the kernel is kept small but server-like components are added on. Here, the server components run in kernel mode but often in the user's address space. In this way, processes can call upon the servers more easily than with the microkernel approach and thus bypass some of the time-consuming system calls. The smaller kernel might handle such tasks as interrupt handling and process and thread scheduling. The servers might handle virtual memory, interprocess communication, the user interface, I/O (device drivers) and process management.

Windows NT and later versions of Windows, including Windows 7, 8 and 10, use forms of hybrid kernels. Mac OS X and iOS (used on Apple mobile devices) combine the microkernel of Mach with components from FreeBSD and NetBSD. The resulting kernel for Mac OS and iOS is known as XNU (X is Not Unix) and is another hybrid kernel.

Those who work with either microkernels or hybrid kernels cite that the monolithic kernel is too complicated and large. As the monolithic kernel is essentially one large piece of code, making a minor modification to one portion of the code might have completely unexpected impacts on other portions. Errors may arise that are very hard to identify and locate. This could lead to an unstable OS which yields errors and system crashes for reasons that have little to do with the modified code. There are concerns that the monolithic kernel will be inefficient because of its size.

Given the problems with a monolithic kernel, why would anyone want to produce one? In order to improve on the efficiency and reduce the impact of errors of the monolithic kernel, the modern monolithic kernel includes numerous *modules*. Each module implements one or more of the core kernel's responsibilities. Thus, the kernel can itself be partially modularized. Modules are loaded either at kernel initialization time or on demand. The efficiency of a monolithic kernel's execution is in part based on the number of modules loaded. Loading fewer modules leads to a more efficiently executing kernel. Through modules, the Linux kernel can be kept relatively small.

Figure 1.1 illustrates the difference, at a rudimentary level, between the monolithic kernel and microkernel. On the top, the monolithic kernel is large and handles all kernel operations. System calls are limited to just the applications software invoking the kernel. On the bottom, the microkernel is smaller and simpler, but system calls occur between the kernel and servers as well as between the application software and servers.

As seen in Figure 1.1, a running application invokes a part of the kernel through a system call. The *system call* is a function invocation where the function is not part of the application that invoked it but part of the kernel. In this way, a programmer writing an application has a ready-made interface by which to call upon the kernel.

In Linux, a system call is not directly intercepted by the kernel but instead handled by a *wrapper function*. Wrapper functions are part of one or more Linux libraries such as `glibc`. The role of the wrapper function is to place the arguments of the function call into appropriate hardware

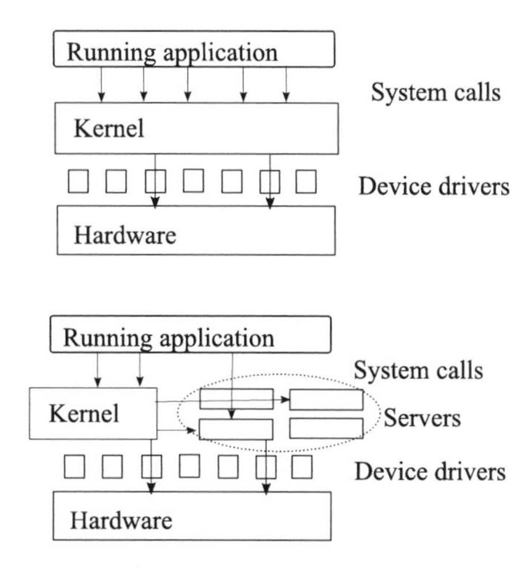

FIGURE 1.1 Monolithic kernel (top) compared to microkernel (bottom).

registers where the kernel expects them and to switch processor mode from user mode to privileged mode. The kernel operates under privileged mode while the user's process operates under user mode. If an error arises during the execution of the system call, it is the wrapper's responsibility to intercept the error from the kernel and modify this into an error number, stored in the special variable `errno`. This variable can then be referenced by the application or the user from the command line. Some wrappers are more complicated and perform various forms of preprocessing and/or postprocessing.

Table 1.4 provides a list of some of the many system calls available in Linux. Note that system calls will vary by both distribution and between versions. The system calls in the table, listed in alphabetical order, are those found in most versions of Linux although the name might vary slightly. The table describes the listed system calls and provides a basic categorization of the type of activity it oversees. Keep in mind that any implementation of Linux will likely have several hundred system calls available.

If you look over the list of system calls you might recognize some of the names by their type of operation (e.g., `open`, `close`, `read` and `write`). Others have names similar to that of Linux instructions like `chmod`, `kill`, `stat` and `mount`/`umount`. If you are already familiar with Linux, the system call `pipe` should sound familiar.

TABLE 1.4
Some Linux System Calls

Name	Category	Use
`access`	File management	Test user permissions on a file.
`chdir`	File management	Change the current working directory.
`chmod`	File management	Change file permissions.
`close`	File management	Close a file descriptor.
`creat`	File management	Create a file or device.
`exec`, `execv`, `execve` (and others)	Process management	Create a child process.
`exit`	Process management	Terminate current process.
`fork`, `vfork`	Process management	Create a child process, create a child process while blocking the parent process.
`fstat`, `stat`	File management	Get file status.
`getpid`, `getppid`	Process management	Get the ID of a process, process' parent.
`getpriority`, `setpriority`, `nice`	Process management	Get or set the process' priority, change process priority.
`ioctl`	Device management	Send control signal to I/O device.
`lseek`	File management	Move pointer within file to new location.
`kill`	Interprocess communication	Send signal to a process (may result in exiting the process).
`mlock`, `munlock`	Memory management	Lock/unlock page in memory.
`mount`, `umount`	Device management	Mount or unmount file system.
`open`	File management	Open a file/device, assign to a file descriptor.
`pipe`	Interprocess communication	Create interprocess channel.
`poll`	Process management	Cause process to wait for event from file descriptor.
`read`	File management	Read from a file descriptor.
`stat`	File management	Get file status information.
`truncate`	File management	Set a file to a specified length.
`write`	File management	Write to a file descriptor.

Before we leave this topic, we have two additional comments. First, the *file descriptor* is a designator (number) that the kernel assigns to an open file or resource. It is through the designator that the kernel will control access to that open file or resource. Thus, when we want to read data from an open file, the operation to the kernel becomes "read from this file descriptor". Second, the terms `exec` and `fork` are used to express what happens when a process is created. In Linux, all processes are generated, or spawned, by an existing process (with the exception of the first process that runs during system initialization). The system calls of `execve` and `fork/vfork` are used so that the current process can generate a new process. We explore the exec and fork system calls in more detail in Chapter 4.

1.2.3 Other Operating System Components

We wrap up our introduction to the OS with a look at other OS components, first with the *device driver*. Referring back to Figure 1.1, you can see that device drivers are positioned between the kernel and the hardware devices. Each driver is a program written specifically to facilitate communication between the kernel and the device. The reason for device drivers is that I/O devices vary dramatically in terms of type of operation, quantity of data transmitted and speed of transmission. It is too much to expect the CPU to be able to communicate directly with the great variety of I/O devices. Instead, the CPU sends generic instructions to the device which the kernel intercepts and converts into actions by calling upon the corresponding driver.

Many device drivers are already preinstalled in most operating systems. The device driver merely needs to be enabled to be used. For less common devices, or devices that were not available when the OS was implemented, the drivers must be installed. These drivers might come from the manufacturer on optical disc provided with the device but more commonly are downloaded over the Internet and installed.

Some OSs, most notably Unix/Linux, Windows and Mac OS, also employ operating system *services*. Services run in the *background* meaning that they run without user interaction. A service runs on demand when it is needed. When unneeded, it uses few system resources. There are a great many services in Linux supporting such tasks as process scheduling, management of hardware and the network, logging and file system handling. We explore services in Chapter 9.

Another component of many OSs is, like the device drivers and services, a collection of programs, referred to as OS *utilities*; also possibly referred to as *tools*. Each utility offers some functionality in improving system performance. One class of utility pertains to files and file systems. There are disk defragmenters to make disk access more efficient, disk recoverers to scavenge deleted files from disk, file system cleaners to remove unused or outdated files, backup and archive utilities and encryption/decryption utilities. Another form of utility is an anti-malware program which not only searches your hard disk for computer viruses but also examines website URLs to see if the website should be avoided as untrustworthy.

Most utilities run at the request of the user. This makes the utility different from other OS components which run as needed or are invoked by application software or requests from other OS components. And unlike the kernel and OS services, utilities run in the foreground i.e., they present the user with an interface. Windows has a great number of utilities, some of which are part of the Windows OS and some are third-party add-ons. Linux has fewer utilities because some of the same tasks are handled by Linux services. Some utilities are not part of Linux at all and must be installed as third-party products.

The last "add-on" component to the OS is tailorable user interface. In Windows and Mac OS, the user can tailor the desktop with files and shortcut icons. In Linux, we more commonly use the shell to personalize our interface. The *shell* is an environment in which the user enters commands. The shell permits definitions and instructions to be recalled later. Definitions include shell scripts and aliases. We explore the Linux shell in detail in Chapter 2.

We noted earlier that the system call wrapper functions, among other tasks, cause the computer to switch modes. The Linux OS has a clear differentiation of roles between user mode and

TABLE 1.5
Uses and Platforms Where We Find Linux

Uses/Platforms	Explanation
Arts	Preferred choice for animation by many movie studios.
Cloud	Used to support cloud computing and other cloud services.
Gaming	Often used in gaming development.
Governments	Used extensively by dozens of governments including the US, China and India.
Mobile device	Android and Raspberry Pi, among others.
Networking	Found in firewall and router hardware devices.
Smart devices	Found in many smart TVs and home automation devices.
Webserver	Linux is used to run more webservers than any other OS.

privileged mode. User mode is applied when running all applications software, Linux services, the GUI, common software libraries and system calls. Privileged mode consists of the OS kernel, including the execution of system calls and device drivers.

1.2.4 So, What Is Linux?

We posed this question to start this section but then dedicated most of the space to describing OSs in general. What is Linux? Succinctly, Linux is a family of OSs, loosely based on Unix, and largely developed by and supported by the open-source community. Linux is an OS that uses a monolithic kernel supported by modules that can be loaded and unloaded. Linux is an OS whose interface with applications software is made through system calls.

Linux is also one of the few portable operating systems available. The term *portable* means that the OS is not written for one or a few forms of hardware but could be run on many different platforms of hardware. Linux is also one of the more efficient OSs. It is small enough to run on nearly any type of computer device but powerful enough to run large-scale applications such as webservers. It is modern enough to use features found in most other OSs while also having its own cutting-edge facilities such as its security handling (SELinux). It is similar enough to Unix so that users of Unix will have no problem learning Linux, and its GUI is relatable enough for a common user to be able to use it. As Linux is available in open source, a programmer can modify how Linux runs by updating and recompiling Linux source code.

There are many uses for Linux. We share some in Table 1.5. Move on to Section 1.3 for more reasons for using Linux.

SECTION ACTIVITIES

1. How important is it for a user to understand the operating system concepts explored in this section? For instance, do you need to know what a kernel is? What services are? Think about your own knowledge of the OS. Does it make it easier for you to use computers?
2. In this section, we saw, briefly, the development of early OSs until 1966. You might wonder about Windows and MacOS. As personal computers were not introduced until the 1970s, it's no surprise we didn't see them included in the list. IBM PC and compatible computers started off using MS-DOS, with Windows being introduced in 1985, 1 year after the first Macintosh came out. Research the development of both MacOS and Windows. You can find a great many websites that describe their individual histories or the history of all operating systems combined.

1.3 WHY USE LINUX?

What makes Linux different from other operating systems? We noted earlier in this chapter that it is open source and provides the user more control. There are other open-source operating systems, albeit not that many, and none have been as successful as Linux (Android by Google is perhaps the most popular alternative but is itself based on a modified version of the Linux kernel). There are other OSs that provide the user with the degree of control as Linux although the one that equals Linux is Unix, on which Linux is based.

It's worth reiterating that open source does not *just* mean free. With Linux, open source comes with an entire community of programmers. These programmers work, usually in large groups, on related projects. Some are involved in enhancing existing Linux versions. Others work on applications that run within one or more versions of Linux. And those same people or others are involved in providing the support that comes with commercial products.

In addition to the open-source nature of Linux, it was initially built to require few resources. We might refer to Linux as a *lightweight* OS compared to that of Mac OS and Windows because of the approach taken in implementing Linux. Linux should boot and initialize faster than other operating systems. A computer running Linux should require less memory and should be able to function with a slower processor. Although these statements are generally true, some versions of Linux require more resources than others.

Another advantage of Linux is portability. The Unix OS was developed to be run on any platform of mainframe computer, and later versions were released for workstations, personal computers and servers. Linux is even more portable. Versions of Linux can run on just about every classification of computer: supercomputer, mainframe, server, personal computer and laptop, tablet, smartphone and embedded devices. This level of portability does not exist with other OSs.

Another reason to choose Linux is that it is less attack-prone than Windows. Hackers more commonly target Windows-based computers when attempting to breach computers, inflict viruses or perform other forms of attacks because there are far more Windows computers than Linux computers (excluding servers and supercomputers). You might think that Linux would be open to attacks of all kinds because its source code is readily available. In fact, the open-source community takes great efforts to ensure a lack of security flaws in the OS before a Linux version is released. That is not to say that they are 100% successful. But because of the efforts taken combined with the lower incidents of attack, some companies and even whole governments are moving to Linux as the preferred OS because it is, at least currently, more secure.

Linux has become the OS of choice for many computer programmers and system administrators. It is also an OS of choice for many computer users, hobbyists and companies. But in spite of its increased popularity, the number of desktop/laptop computers running Linux remains in the minority when compared to Windows and Mac OS. It is estimated that no more than 1.93% of the world's desktop computers are currently running Linux. Figure 1.2 highlights some of the more interesting statistics about Linux usage, as of spring 2021.

SECTION ACTIVITIES

1. What open-source software have you used? Make a list. It's probably more than ten titles!
2. The statistics in Figure 1.2 provide an interesting dichotomy: Linux is the operating system of choice for supercomputers and most servers but few desktops. Why do you suppose this is the case? Write down three reasons. Now research the topic and see if your reasons are those cited by others.

*Linux runs on **all** of the world's fastest supercomputers.*

***Every** major space program in the world uses Linux-based computers to run their software.*

***96.3%** of the world's top 1 million servers run on Linux computers (less than 2% use Windows).*

*Of the world's top 25 most visited websites, 23 (**92%**) run on Linux computers.*

***95%** of the world's top 1 million Internet domains run on Linux computers.*

***90%** of all public cloud facilities run on Linux computers, and all of Amazon Web Services runs on Linux computers.*

***90%** of all Hollywood special effects are created on Linux computers; special effects found in such movies as Titanic and Lord of the Rings were generated on Linux computers.*

***85%** of all smartphones are Linux-based.*

***83.1%** of all software developers prefer to work in Linux while **54.1%** actual use Linux (as of 2019).*

Countries whose governments prefer to use Linux computers include Austria, India, Cuba, Russia, Turkey, French and Dutch police use Linux, and both the White House and the US military have migrated their computers to Linux.

Linux is most popular in the states of California and Utah.

FIGURE 1.2 Interesting statistics about Linux usage.

1.4 WHO DEVELOPED LINUX AND WHEN?

We can start the history of Linux with one person and one event: Linus Torvalds, a Finnish computer science student, who in 1991 was dissatisfied with the operating system that came with his operating system textbook. He set about creating his own OS which he eventually named Linux. But rather than starting with Torvalds in 1991, we start 27 years earlier.

1.4.1 The Birth and Development of Unix

In 1964, a group of software engineers from AT&T Bell Labs, General Electric and MIT began developing a new and novel operating system called MULTICS (Multiplexed Information and Computer Service). MULTICS was a multiuser, time-sharing (multitasking) OS that pioneered a number of innovative features. While the development of MULTICS seemed promising, AT&T withdrew from the project in 1969, shortly before its release.

Ken Thompson and Dennis Ritchie were two of the AT&T members involved in the early development of MULTICS. They did not want to see their efforts end at that time. AT&T was using a batch processing OS that they felt was unsatisfactory. They worked on their own version of a MULTICS-like OS with help from other AT&T employees, including Rudd Canaday. The result was a new OS, written for the PDP-7 mainframe and written in that computer's assembly language. They called it UNICS, short for Un-multiplexed Information and Computing Service, likely a joke name spoofing the name MULTICS.

Further development targeted the PDP-11 computer but rather than reimplement the OS in PDP-11 assembly language, Thompson and Ritchie wanted to implement their OS in a neutral language so that the OS could be ported more easily to other computers. Ritchie, along with another AT&T employee, Brian Kernighan, created a new programming language called C. It was in C that UNICS was rewritten and eventually released in 1973. During this time, the OS was renamed from UNICS to Unix. Other employees contributed significantly to Unix as it was developed.

A look at Unix features shows that it was at least partially derived out of the experience that Thompson and Ritchie gained in their work on MULTICS. Table 1.6 shows some of the features either introduced in MULTICS and used or enhanced in Unix, or introduced in Unix. Compare these features to the brief descriptions of other early 1960s OS from Table 1.2, and you will see that first MULTICS and then Unix popularized many features that had yet to be developed.

In spite of Unix being referred to as platform-independent, the 1973 version still contained some code specific to the PDP-11. It was not until 1978 that Unix was successfully ported to another platform. When distributed by AT&T, Unix was provided in source code format (for a small fee). Although AT&T released many versions of Unix, they were all referred to as Research Unix. But because AT&T offered no support in the versions they distributed, it was up to other organizations to modify the Unix source code as needed to run on different platforms, fix errors and add desired features. This led to a splintering of Unix into many versions.

One such version was provided to the University of California Berkeley's computer science department. At Berkeley, programmers developed their own OS, using some of the Unix code as components. In 1975, Ken Thompson became a visiting professor at UC Berkeley. Under his

TABLE 1.6
Unix Features (From Richard Fox, 2021, *Information Technology: An Introduction for Today's Digital World*, 2nd edition)

Feature	Explanation	Comments
Command line interpreter	Beyond a CLI, Unix's CLI provides flexibility with operations to redirect input and output, define constructs like aliases, and the ability to write shell scripts.	Popularized from Multics
Commands as programs	Rather than having the kernel handle all operations, many basic commands are provided as their own programs such as file system commands, text editing and processing commands, and process management commands.	First
Hierarchical file system	Directories can have their own subdirectories and administrators to better organize files.	Popularized from Multics
Interprocess communication	Allows synchronized access to shared data among multitasking processes.	
Multitasking/multiuser OS	Runs multiple processes of multiple logged-in users at a time.	
Portability	Unix could be installed on a large variety of computer types.	First
Regular expressions	Ability to search files for strings through command-line programs like `grep`, `awk` and `sed` and the programming language `Perl`.	Popularized from Multics
Shells	The CLI was one component of a shell; shells introduced convenient shortcut features to assist the user.	First
System calls	Written in C, Unix made many kernel operations available as C functions known as system calls, which could be invoked by user programs.	First
Top-level directory structure	Separate directories for OS components and user files simplify the process of searching for a file.	First
Treating devices as files	Nearly everything in Unix is either a process or a "file"; this allows file commands to be executed on devices either from the command line or from a script.	First

guidance, a new version of Unix was produced. This became known as the Berkeley Standard Distribution (BSD) Unix. Unlike AT&T's Unix, which was proprietary, BSD Unix was freely distributed under the BSD License (which imposed only minimal restrictions on the use and distribution of this version of Unix). BSD versions 4.2 and 4.3 became leading versions of Unix, distributed to other organizations. BSD led to other versions such as SunOS, Darwin (which would become the basis for the Mac OS), FreeBSD and NetBSD. Other versions splintered from Unix to become IBM's AIX, HP's HP-UX, Microsoft's Xenix and Sun's Solaris. By 1990, there were dozens of versions of Unix, many of which were proprietary.

1.4.2 GNU

In the early 1980s, Richard Stallman, a researcher at MIT's artificial intelligence laboratory, wanted to create his own Unix-like operating system. He wanted to make this OS available in source code format for free and with no licensing restrictions so that other programmers could modify the OS as they desired just as long as any modifications would similarly be made available in source code for the community to use. He called upon many members of the Unix user and programming community for help. He dubbed the project GNU for GNU Not Unix, a recursive definition (recursion is a programming tool commonly used in artificial intelligence).

Stallman's group began work on the GNU OS at the same time BSD was becoming popular. But while BSD became successful, a GNU kernel was never completed. Stallman succeeded, though, in his version of a free software movement. His vision was that software should be free. He would explain free as in "free speech" not "free beer". He opposed proprietary software and saw the software field as one where all programmers could or should contribute to the development of free software. Software, he felt, were based on ideas and ideas could not be owned.

Stallman created the GNU General Public License (GNU GPL). The GPL requires that software published under the GPL must be free for anyone to use for any purpose: free to be studied, free to be changed, free to be redistributed and free to be improved. The proviso is that anything created by GPL software would also be published under the GPL so that further distribution of such software would also be available as source code allowing others the same freedoms.

GPL version 1 was released in February 1989. Its main points are that any software distributed under the GPL can be copied and distributed in the source code format (as long as it is distributed exactly as provided) and distributed in any medium, and as long as the redistribution maintains the exact same notices as the original. The same stipulations apply to modified code: such code must contain the original notices as well as notices that this version has been modified, including the files changed and the dates of those changes, and that the new distribution be carried under the GPL. The GPL also permits distribution of the code in its compiled, executable form (known as object code). There are ten clauses in GPLv1 including two that clearly indicate that the software comes with no warranty because it is free.

GPLv2 was released in June 1991 adding a restriction that works covered by the GPL can only be redistributed if they can satisfy all of the license's obligations. The inclusion of this clause is to discourage any form of patent infringement lawsuit being claimed over a GPL-distributed product. GPLv3 was released in June 2007, which added more detail and language to protect GPL-covered items (which at this point went beyond software).

Although we will not examine the GPL itself, Figure 1.3 shows the first version's Preamble, which is worth reading. It states, right from the start, that the GPL is there to express our freedom rather than the restrictions found in most software licenses. It is for this reason that Stallman called the GPL a *copyleft* instead of a copyright. The entire GPLv1 license, along with all newer versions, is available at https://www.gnu.org/licenses/ with version 1 at https://www.gnu.org/licenses/old-licenses/gpl-1.0.en.html.

```
                          Preamble

  The license agreements of most software companies try to keep users
at the mercy of those companies.  By contrast, our General Public
License is intended to guarantee your freedom to share and change free
software--to make sure the software is free for all its users.  The
General Public License applies to the Free Software Foundation's
software and to any other program whose authors commit to using it.
You can use it for your programs, too.

  When we speak of free software, we are referring to freedom, not
price.  Specifically, the General Public License is designed to make
sure that you have the freedom to give away or sell copies of free
software, that you receive source code or can get it if you want it,
that you can change the software or use pieces of it in new free
programs; and that you know you can do these things.

  To protect your rights, we need to make restrictions that forbid
anyone to deny you these rights or to ask you to surrender the rights.
These restrictions translate to certain responsibilities for you if you
distribute copies of the software, or if you modify it.

  For example, if you distribute copies of a such a program, whether
gratis or for a fee, you must give the recipients all the rights that
you have.  You must make sure that they, too, receive or can get the
source code.  And you must tell them their rights.

  We protect your rights with two steps: (1) copyright the software, and
(2) offer you this license which gives you legal permission to copy,
distribute and/or modify the software.

  Also, for each author's protection and ours, we want to make certain
that everyone understands that there is no warranty for this free
software.  If the software is modified by someone else and passed on, we
want its recipients to know that what they have is not the original, so
that any problems introduced by others will not reflect on the original
authors' reputations.

  The precise terms and conditions for copying, distribution and
modification follow.
```

FIGURE 1.3 The GPLv1 preamble.

1.4.3 Enter Linus Torvalds

In 1991, Linus Torvalds, a computer science student at the University of Helsinki, was unhappy with the toy OS that came with his operating system textbook. Torvalds began to develop his own small OS on his Intel 386-based computer, written in Intel 80386 assembly language. He developed only a few components: a small multithreading kernel, a `bash` interpreter and the `gcc` compiler. He posted on the USENET newsgroup, `comp.os.minix`, to attract other programmers who wanted to help develop an OS that was different from and better than MINIX. He posted the source code for his fledgling OS on an FTP server. Over the next few months, members of the GNU community contributed GNU components including the facility to compile code using `make` as well as `compress` and `sed`, among other programs.

As the OS took shape, he published it under a version of the GPL (GPLv2). Over the next several months, the community of programmers working on Linux grew. A new USENET newsgroup was established for the community, `alt.os.linux` (which would be renamed `comp.os.linux` a couple of months later). Enhancements to Linux during this time included a number of device drivers, a more fleshed-out kernel, the adoption of the POSIX standard and an implementation of the X Window System. By 1994, Linux had grown to over 175,000 lines of code, written in C. This version was distributed as Linux version 1.0.0. Aside from versions released for Intel processors, Linux was successfully ported to the DEC Alpha, Sun SPARC and MIPS processors.

In 1992, the first well-known splintering of Linux took place when Canadian software engineer Peter MacDonald developed Softlanding Linux System (SLS). Whereas Torvald's Linux was a kernel, device drivers and full utilities like gcc, SLS included a full windowing system (based on X Windows) and a number of applications software. Computer science student Patrick Volkerding (at Moorhead State University in Minnesota) downloaded SLS and modified it to fix some bugs. He distributed his version to fellow students who urged him to publish his version. His version became known as Slackware.

A German software company, Software und System Entwicklung (Software and System Development, or SUSE for short) adopted Slackware as a platform, translating it into German and enhancing it. This led to three different versions from the same branch of the Linux "tree". All three versions continued to be updated although SLS was largely ignored in favor of Slackware. But it was SUSE that found the most popularity in the end and has had several spinoff versions under various names including SUSE, SuSE, SuSELinux and openSUSE. In comparing SUSE to other versions of Linux, we find it comes closest to Red Hat Linux, and as SUSE/openSUSE is of limited popularity, we will not examine it any further.

The SLS/Slackware/SUSE version of Linux leads us to the term *distribution*, or *distro* for short. More and more distributions took place in the 1990s and into the 2000s resulting in a large number of Linux versions. There are far too many distros to track them all here. The remainder of this subsection looks at some of the more significant developments of the distros.

In 1994, software developer Marc Ewing developed his own version of Linux that he named Red Hat Linux. Released late in 1994, his business was bought by entrepreneur Bob Young. The new company was called Red Hat Software and largely revolved around a commercial version of Linux (both the OS and application software). Unlike the versions of Linux being developed by the open-source community, the intention of Red Hat was to include support so that companies could trust that bugs would be fixed quickly. Red Hat Software was later renamed Red Hat, Inc., which was purchased by IBM in 2018.

Two spinoffs of Red Hat are supported by Red Hat, Inc., Fedora and CentOS. Both of these versions are open-source versions of what is now called Red Hat Enterprise Linux (RHEL). There are a few differences between these three distributions but mostly they are built on the same kernel with most of the same services, applications and features. Similarities and differences between the three distributions are described in Table 1.7. Note that the most recent version is of spring 2021.

Debian was first announced in 1993 with the intent to be the first release of a Linux outside of Torvald's. The project was headed by programmer Ian Murdock, and the name was a combination of his girlfriend (Debra) and his first names. Although the first release of Debian (version 0.01) was in 1993, the first major release did not take place until 1996, thus it trailed both SLS/Slackware and

TABLE 1.7
Comparisons between RHEL, Fedora and CentOS

	RHEL	Fedora	CentOS
Initial release	2000	2003	2004
License type	Commercial	GPL	GPL
Maintained by	Red Hat, Inc/IBM	Open-source community	Open-source community
Processor targets	x86, ARM, IBM Z, IBM power systems	x86, ARM, aarch64	x86, ARM64, Power8
Computer types	Servers, mainframes, workstations, supercomputers	Servers, desktops, cloud	Servers, desktops, workstations, supercomputers
User interface	GNOME	GNOME	GNOME
Release targets	2–4 minor releases per year, major releases every 5 or so years	Every 6 months	Same as RHEL
Most recent version	RHEL version 8 (Oct 2020)	Version 33 (Oct 2020)	Stream (Dec 2020)

Red Hat. Murdock handed off control of Debian to Bruce Perens in 1996 and then leadership was established by elections every year (starting in 1999).

Unlike Red Hat which had both commercial and open-source versions, Debian was intended to be open source from day one. The Debian development community took their time with each version, planning to release a new version only every 2 years. Three different versions of Debian are available at any time: the current stable version, a testing version of the intended next release and an unstable version which includes libraries and packages that may not install correctly or at all (that is, these items are untested).

The deliberative and slow nature of new releases under Debian was a point of contention with some developers and so Ubuntu spun off from Debian with new releases targeted for every 6 months. The first version of Ubuntu came out in 2004. Ubuntu attempts to be a more secure version of Linux and as such, pioneered the idea that rather than users having access to the system administrator account (*root*), the first user account would automatically receive access to administrator commands through the `sudo` command. We explore this later in the text. Other Linux distributions later adopted this practice.

Two other notable Debian descendants are Knoppix and Linux Mint. Knoppix is noteworthy as a form of Linux that can be booted and run entirely off of an optical disc. This feature, known as a *live boot*, allows the user to use Linux without installing it on their computer. The disadvantage is that you start with a new version of the OS every time you launch it. Any previously created accounts or changes made to the OS or environment are lost. Files must be saved externally, for instance via a USB-based drive. Linux Mint is based on a Ubuntu-derivative called Kubuntu, which was first released in 2006. As Linux Mint was further developed, it became more based on Debian than Ubuntu and is now called Linux Mint Debian Edition (LMDE). Two differences between Linux Mint and Ubuntu are that Linux Mint supposedly uses less memory when running and it has more preinstalled applications than Ubuntu.

In reading through the distributions listed above you might question why there are so many. What are their differences? All of the spinoff distributions are based around the same Linux kernel. Each version's kernel shares many of the same modules with the other versions. The versions share many of the same C libraries, system calls, shells and X Windows interface (although the specific look of the desktop will differ). Many of the same applications run on all of the Linux distributions.

There are perhaps four significant differences between these main distributions. The most significant difference in terms of usage is the package manager program used to perform software installation and upgrades. A second difference is the quantity of available applications software as some distributions have fewer titles available or support fewer third-party applications. There is also a philosophical difference behind the developer's plans for major version releases. Finally, there is a different philosophy taken regarding whether the root password is made available.

There are two primary package managers used in today's Linux distributions: `dnf` (or the slightly older `yum`) and `apt`. These are built on top of other package manager programs: `rpm` (Red Hat Package Manager which `dnf/yum` uses) and `dpkg` (Debian Package Manager which `apt` calls upon). Amazingly, other than the specific desktop design, this is the single most visible difference between the various distributions.

In terms of available software, we find most distributions come with a `Firefox` web browser (usually built in) and can run the `emacs` editor (not built in). They also all have the ability to run an office suite (built in or not), usually `LibreOffice`. However, some have far more applications than others. For instance, Ubuntu is a popular choice for gaming enthusiasts.

The development philosophy is less visible but no less impactful. The Debian community takes a slow but thorough approach when creating and vetting a new version. The target of every 2 years seems extreme to some Linux users and so they favor the more rapid development and deployment of Ubuntu. With Red Hat distributions, Red Hat, Inc. drives most of the development but as new versions are made available, members of the Red Hat community take the source code and use it to develop a new release of CentOS. So, in fact, CentOS will have mostly the same code and features as RHEL.

TABLE 1.8
Comparing Linux Distributions

	CentOS	Debian	Fedora	Linux Mint	RHEL	Ubuntu
Initial release	2004	1996	2003	2006	1995	2004
Parent (if any)	Red Hat	N/A	Red Hat	Ubuntu	N/A	Debian
Package manager	rpm, dnf/yum	apt, dpkg	rpm, dnf/yum	apt, dpkg	rpm, dnf/yum	apt, dpkg
Approximate amount of software packages	6000–7000	90,000	65,000	30,000	10,000–12,000	90,000+
Target release frequency	Coupled with RHEL	Every 2–3 years	Yearly	Every 2–3 years	Major releases every 5–10 years	Every 6 months
Degree of open source	Completely	Completely	Completely	Mostly	None	Mostly
Most recent version	8 and Stream	10.8	33	20	8	20.04
Most recent release date	2020	2021	2020	2021	2020	2020

In previous releases of RHEL, a related version of CentOS was released shortly afterward. Starting in 2020 however, a new approach is being taken. CentOS is now releasing versions under the name Stream. The idea is that rather than waiting for a new, completed version of RHEL, CentOS will release versions earlier that contain features being proposed for the next RHEL release.

Finally, there is a distinction between RHEL, and the rest of the Linux distributions described here in that RHEL is not a free OS. Instead, a commercial purchase must be made in order to receive RHEL support. The other versions of Linux are free and can be found in source code format. Within the different distributions, some software may be proprietary. Early versions of Linux Mint had proprietary software (which has supposedly been removed).

We review the most significant differences between these most notable distributions in Table 1.8. In the table, we cite which of the original distributions it evolved from, the choice of package manager, the quantity of available software (estimated from various website reports) and the development philosophy. We also note the most recent version and its distribution date. The Linux distributions are listed in alphabetical order (left-to-right) rather than by year of initial or most recent release.

There is no official count of the total number of Linux distributions available or in use today, but the estimate is between 600 and 1000. Some of these distributions vary more dramatically than what was explained above. Nearly all of the distributions are available for free (RHEL being one of the notable exceptions), and these can come in either source code or as an executable installation program (usually as an ISO image). The history is so complicated that mapping the distros requires a very large chart (see the Linux distro tree at https://5thlobe.com/reference/gnulinux-distro-chart/).

We should point out that Linux is not the only open-source OS. Three notable other open-source OSs, all of which are Unix-like, are FreeBSD, Dragonfly BSD and GhostBSD. Another open-source OS is XNU (which stands for X is Not Unix), which is the basis for the iOS operating systems found on such devices as the Apple iPad and Apple TV. ReactOS was developed in 1996 as a clone for Microsoft Windows 95 and has the look and feel of more recent Windows OSs. And then there's FreeDOS which is primarily used to run old MS-DOS games and other legacy software. There are others, but none of these operating systems has achieved anything close to the popularity or name recognition of Linux.

1.4.4 The Open-Source Community

The earliest open-source projects were the further development of Unix, GNU and then Linux. The first application software that the open-source community contributed to was the Netscape web browser, which the original developers offered to the community for input. It was at this point that the term *open source* was first used.

As work continued on both Linux and Netscape, the open-source software community fell into a dispute. Torvalds did not require that modified versions of software be made freely available. He felt that if a person modified a piece of open-source code, the modified code could continue to be freely distributed as source code, or it could be freely distributed but as executable code, or it could even be sold for profit. But some members of the open-source community eschewed commercial software and felt that anything developed out of the open-source community must remain open source and freely available.

This rift caused the community to splinter into two groups: the Open-Source Initiative (OSI) and the Free Software Foundation (FSF). OSI was founded in 1998 by members of the open-source community developing versions of Linux including Bruce Perens from Debian. FSF was founded by Richard Stallman in 1985. The difference between the two groups is that OSI is willing to accept copyrights on some software restricting freedoms that were established under the GPL while the FSF generally feels that anything created from a GPL-produced product must also be made available in open source under the GPL. Although this rift has driven some projects to splinter, it has not impacted the productivity of the open-source community and the number of titles being produced.

The shift in perspective from commercial software development to open source not only has impacted software developers' perspective of open source but the corporate world. Early on, open-source software was primarily used by hobbyist and people in the open-source community. Open-source software was largely shunned by companies. There was a stigma attached to open source products that they would have errors that could potentially harm an organization because the software would not work when it was needed. This could result in a loss of productivity and revenue and cause the organization to receive a bad reputation. Companies preferred the guarantee that proprietary software would be fully supported such that errors would be fixed in a timely manner.

Today, open-source software competes regularly with proprietary software in part because the open-source community has been found to be as or more responsive than many of the companies producing proprietary software. A comparison between open-source software and proprietary software products often shows that open-source software has an equal level of security, is of equal (or even higher) quality, has a greater degree of interoperability (available on multiple platforms), and provides a reasonable amount of support. The support might be available for free, or for a fee but the fee would in most cases still be less than the expense of the proprietary software and its support.

We also find today that the open-source community exists well beyond individual developers wanting to make contributions to Linux or other projects. For instance, Microsoft, who would have opposed the open-source initiative years earlier, is now a steady contributor with such products as .net development tools, the Visual Studio Code editor, PowerShell, and tools for building and training machine learning systems. Their open-source contributions are not limited to Windows; they also offer versions of software for MacOS, Linux, Unix and in the cloud via Microsoft Azure. Other companies that contribute to the open-source community include Adobe, Facebook, Google, IBM, Intel, LinkedIn, Netflix, Oracle and Twitter.

Why are people willing to contribute their time to the open-source community? It is especially puzzling because contributions are often made by programmers who develop software for a livelihood. If they are willing to freely contribute, then they are in essence producing for free something that they could earn a salary for creating instead. It is possible that the software they are helping to produce might compete against software that they are paid to produce or maintain. And yet being involved in the open-source community could help their careers. Table 1.9 lists several reasons for why developers freely contribute to open source projects as a means to further their career.

Participation in open-source development also gives the programmer an opportunity to give back to the open-source community, or even a larger audience, all computer users. Mozilla's Firefox, for instance, is such a success that a contribution to its development can be thought of as a contribution to much of humanity.

Contributions are not limited to just software developers. There are a lot of open-source programmers who program for a hobby and not as part of their job. Their motivations may be similar

TABLE 1.9
Some Reasons to Contribute to Open-Source Software Development

Improve coding skills.	Demonstrate a wider range of skills than what the developer uses at their current position.
Learn new technologies.	Learn different software approaches.
Gain experience with developers from other regions of the world.	Learn programming concepts not already known.
Improve on coding habits.	Make contacts with other developers.
Participate in a large group project.	Learn about or develop tools that can be applied in the development career.

to those who want to further their career: learning, working with other developers and producing something useful. But in their case, the impact may be more of a personal nature over improving their career.

Contributions can also be made by non-programmers through software testing. Most open-source software consists of very complex programs. Testing is a crucial aspect of software development, and it is typical that testers are not the same group as the developers. Testing can be performed at different levels. Software testers commonly run the software on a number of test cases, that is, groups of input. They then compare the output to the expectations and write up reports where the output diverges from expectation. In other cases, testers are users who compile lists of problems that they detect when using the software. For instance, if the software crashes when opening a file or if a particular feature operates too slowly, these cases are noted and sent back to the developers. Aside from testing, there is also a need for documentation writing.

The open-source ideology has extended beyond software development to other community-involved pursuits. Wikipedia is one of the best examples of this. The online encyclopedia receives contributions from people all over the world who want to share knowledge. Unlike open source software which requires some expertise as a programmer, Wikipedia only requires knowledge of a particular topic and the willingness to contribute. With Wikimedia, knowledge-based contributions extend to other media, namely images (whether drawings or photographic). We see artists sharing their products through the Internet using various forms of social media such as through YouTube and Bandcamp.

SECTION ACTIVITIES

1. We often hear of basement inventors who go on to make millions of dollars on their inventions (of course we mostly only hear about the successes, how many other inventors never profited from their inventions?) Torvalds made no attempt to profit from Linux. Yet, as a software developer he has made an immense success for himself and is estimated to be worth well over $100 million. Research his personal history to learn how he turned Linux into a fortune without actually making much money off of Linux itself.
2. In this section, we provided a URL that shows all of the Linux distributions. Use that link to examine this amazing figure. Try to estimate how many Linux distributions exist.
3. You have likely accessed and used other products of the open-source community like Wikipedia, Wikimedia and YouTube. Come up with a list of other websites where information is freely shared that you have used.

1.5 HOW DO YOU USE LINUX?

In this section, we look at the steps to install Linux. We examine several different Linux distributions. In each case, we make the assumption that the installation will be performed onto a virtual machine. If you want to try any of these on your own computer, it is best to download and install a virtual machine client program. VirtualBox from Oracle and VMWare Workstation are both open-source products. For complete information on installing Linux into a VM, see lab 0 in the lab manual that accompanies this textbook.

To install Linux, we need to get ahold of the Linux installation program, which is in the form of an `iso` file. We download the needed `iso` file directly onto hard disk from which to perform the installation. The versions of Linux we examine here are free (and available in open-source format for those interested in looking at or modifying the code) where the `iso` can be found on the Linux projects' websites.

We perform a default installation in most cases although discuss some of the options. Later in the textbook, we look at the steps required to perform a nonstandard partitioning of the Linux computer's file space. This section ends with a brief look at each of the Linux distributions we installed including their basic desktop setup and how to open and use a terminal window to enter commands.

1.5.1 Installing Debian Linux

The Debian installation image can be obtained from `www.debian.org/distrib`. For users who have slow Internet connections, a small image is available that can be saved onto an optical disc. Otherwise, the full image can be saved to optical disc or hard disk. A cloud image is available if you have an account in one of the commercial clouds although it is likely that your cloud provider already has a Debian image available so that you can skip installation and just create a copy.

Upon starting our VM software, we create a new VM. When asked for the type of OS, we select `Linux` and then more specifically `Debian`. We are asked for a location of the installation image and we select the downloaded `iso` file. Figure 1.4 shows the first four screens that we see during installation. First, we select the type of installation. We have selected `Graphical install` to ensure that our Linux installation comes with a GUI. Next, we are asked for our language (the default is English) and location. The location is used to set the time. We are next asked for the keyboard configuration (not shown in the figure, we selected `American English`). At this point, installation begins. A bar indicates the amount of installation completed. Messages appear during installation describing what is currently being installed. The next step is to specify the root password. Instructions explain the importance of selecting a strong password. We are asked to enter the password twice in this step.

As installation continues, the next step is to create an initial user account. We want to have a non-root account available so that when we log in, we log in as a normal user instead of root. Logging in as root through the GUI can lead to mistakes that could damage components of the OS or accidental deletion of files. It is best to switch to root sparingly and only from the command line. For the new user account, we are asked for the user's full name. Debian creates an account for this user. We are then asked to enter the new user account's initial password, twice, much like we did for root.

At this point of installation, we are asked to select how we want the disk space partitioned. The options are to have a guided approach or to perform partitioning manually. There are three forms of guided approaches, each uses the entire hard disk available (which in the case of a VM is not our computer's full hard disk, just the section reserved for the VM based on the size specified when we created the VM). One guided approach uses a standard partitioning, one uses a logical volume manager (LVM, covered in Chapter 8), and one uses an LVM but employs encryption. Note that if we were installing Linux on a dual boot computer, we would have to perform a manual installation so that we do not wipe out the existing operating system and files.

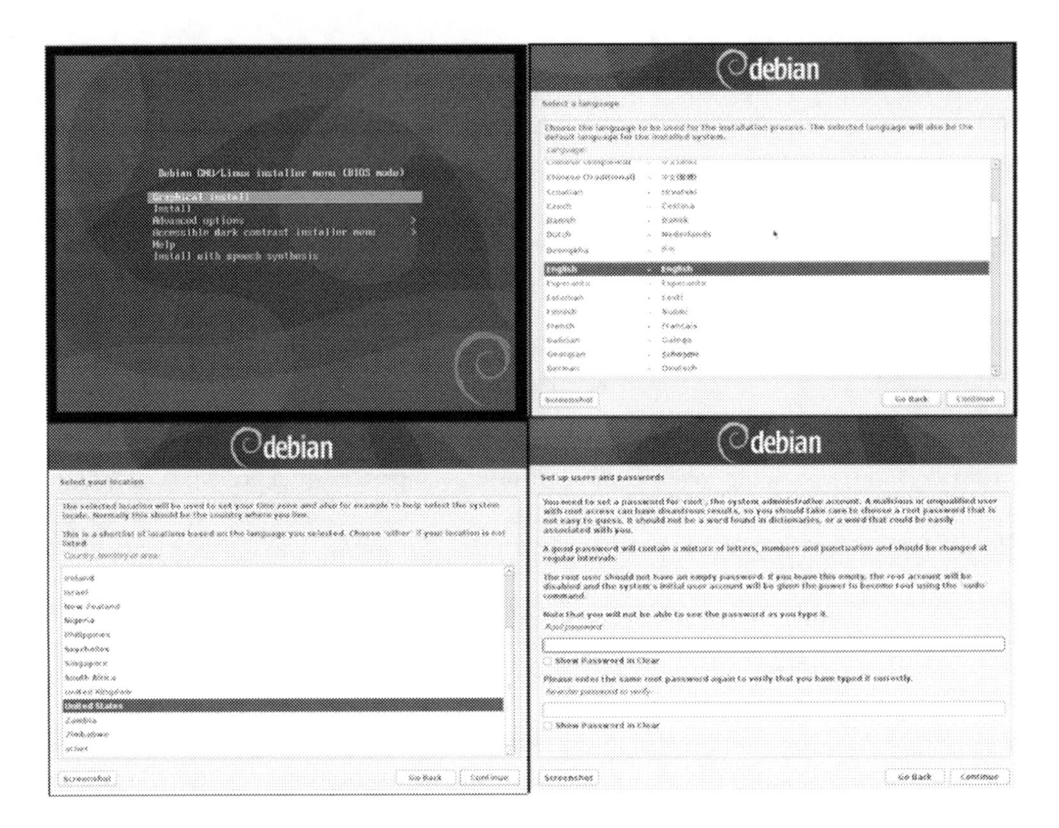

FIGURE 1.4 Four steps during Debian installation.

In Figure 1.5, we see the partition choices and, having selected `Guided - use entire disk`, we see three options. It is preferable to separate `/home` (the partition containing the user directories), `/var` (the partition containing various software data) and `/tmp` (the partition used by software to save temporary information) from the rest of the operating system. We next see in Figure 1.5 the recommended partitioning with sizes. Note that our disk space was limited to 32 GB as we were creating this Linux computer in a VM with limited hard disk space).

With partitioning specified, installation continues with the core packages. We are later asked if the GRUB boot loader should be installed in the master boot record. Our only choices are `no` or `yes`. Selecting `no` will require that we install our own boot loader later. We select `yes` and are asked where to place GRUB. This is shown in the lower right of Figure 1.5. The notation `/dev/sda` represents our hard disk. The rest of the installation is handled with no user interaction.

Upon completion, the VM should boot to our Debian installation and we will be shown the login screen like that shown in the left half of Figure 1.6. Here, there is only one user, the user we had established during installation. Having selected this user, we are asked for the user's password to complete the login process. Notice in the upper right (shown in full in the left side of the figure but not in the right side of the figure) are controls to change accessibility settings, network connection, the volume and "power settings" (log out, shut down, reboot).

Upon logging in, we are presented a mostly blank desktop. Along the upper left is a menu labeled Activities. Selecting this menu brings up a list of icons along the left margin, representing from top-to-bottom, the Firefox web browser, the Evolution Email program, the RhythmBox music program, LibreOffice Writer, the Linux File Browser, the Software program (software store), Debian Help and applications software. Upon selecting the application software icon, the installed titles are displayed in the desktop. On the right side of Figure 1.6, roughly half of the installed titles appear. A search bar appears at the top for easy searching (useful if we had more software titles installed). You will find that this basic setup is similar to both CentOS and Ubuntu.

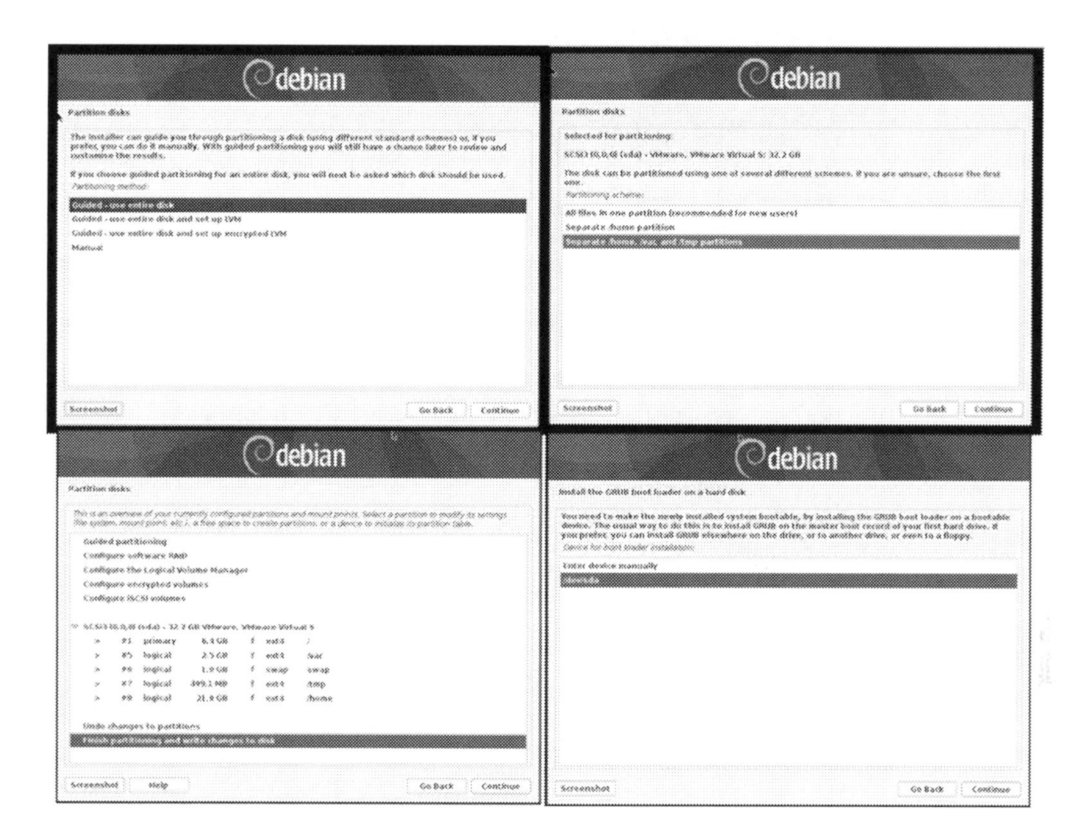

FIGURE 1.5 Partitioning the Debian installation.

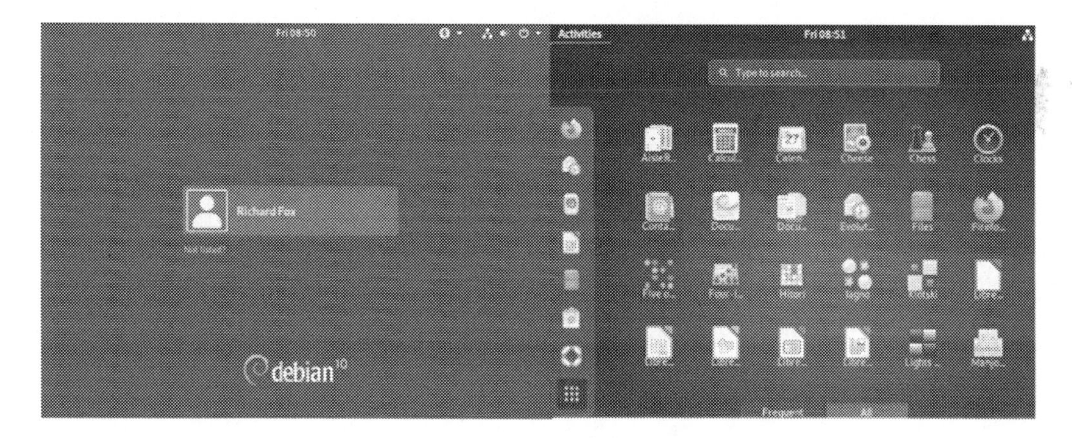

FIGURE 1.6 Debian login screen and desktop.

1.5.2 Installing CentOS Linux

At the time of this writing, there are two different "current" versions of CentOS: CentOS Linux version 8 and CentOS Stream. In all previous releases of CentOS, the current version followed from the most recent release of RHEL. Thus, CentOS 8 is what you would expect to install. However, starting in 2020, CentOS announced that rather than releasing new CentOS versions based on the most recent RHEL version, CentOS Stream would offer versions of Red Hat that were, at least in some cases, ahead of the next RHEL version. This means that CentOS Stream releases may include features that are proposed but not yet available in the most recent RHEL releases. Because of this shift in philosophy, CentOS has announced that CentOS 8's end of life will occur at the end of 2021. By the time you read this, it is likely that CentOS will have moved entirely on to Stream releases. In this section and throughout this textbook, we focus on CentOS Stream rather than CentOS 8. With that said, as of August 2021 the two releases are virtually identical.

We can obtain installation ISOs for both CentOS 8 and CentOS Stream at the `centos.org` website. Specifically, we find versions of CentOS 8 at `www.centos.org/centos-linux` and versions of CentOS Stream at `www.centos.org/centos-stream`. Download the `iso` of the version you prefer (to match the textbook, Stream is recommended). Upon downloading the `iso`, create a new VM and select the `iso` as the installation image.

The first step in installing CentOS is to select whether we want to install the OS, test the media and install the OS or troubleshoot. The troubleshooting selection is for a partially or fully installed OS that is not functioning. As we have just downloaded the `iso` and stored it on disk, there is no need to test the media and so we select Install (which should be the first choice). Like with Debian, we are asked for the language and keyboard layout, both of which default to English. Next, we will be presented with the `Installation Summary` window. We see an example of this in Figure 1.7. From the summary window, we are presented with choices that specify the remainder of the installation. We step through these choices one at a time.

We must respond to selections which appear with an orange triangle and an exclamation mark in it with a description listed in orange font. If all selections are in black font then the `Begin`

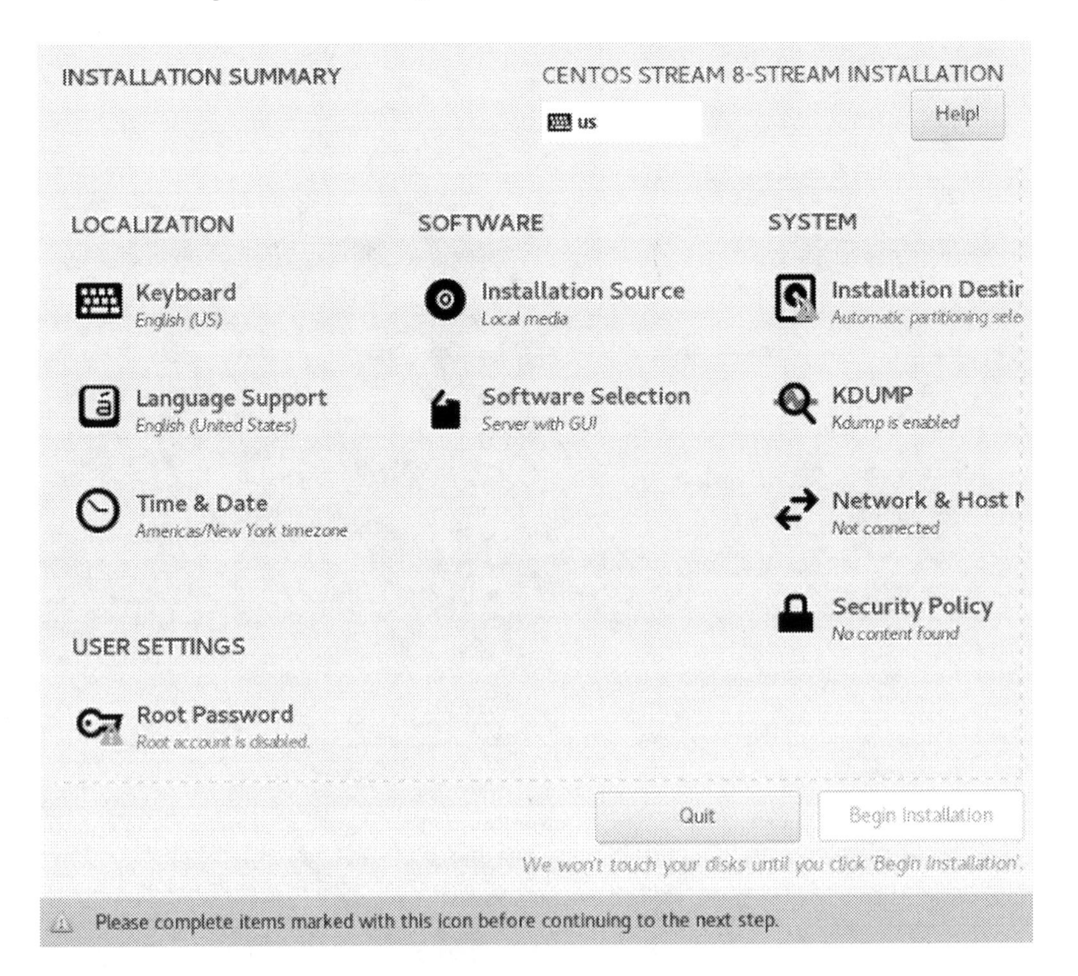

FIGURE 1.7 CentOS installation summary.

Installation button is accessible. In Figure 1.7, Installation Destination and Root Password are the only two selections with the orange triangle. We can select any other item to override the default but do not have to.

Selecting any item will take us to a new window so that we can specify our choice(s). For each window, there will be a Done button in the upper left of the window. Selecting it returns us to the Installation Summary window.

As we had previously selected Keyboard and Language Support, we can skip those. Time & Date is already set for us and we would only select this if we needed to alter the default. We will also not need to select Installation Source as this was specified when we started to create the new VM by specifying the location of the iso. We will not need to modify KDump or Security Policy. So, for us, we will concentrate on Software Selection, Installation Destination, Network & Host Name, and Root Password.

The Software Selection defaults to Server with GUI (note that earlier versions of CentOS defaulted to Minimal Install, which provides a text-based only version of Linux). Although Server with GUI is fine, if you do not intend to run a server (like a webserver) from your Linux VM, it may not be the best choice. The list of choices runs along the left pane, as shown in Figure 1.8. For us, we will select Workstation.

Upon selecting one of these base environments, a list of Additional software is provided in the right pane. The software listed is specific to the base environment selected. Among the items available for a workstation are GNOME applications (GNOME is the GUI desktop), Internet applications, Office suite, legacy UNIX compatibility software and development tools. If we were going to use our Linux computer to write code, we would want the last of these, perhaps the last two. For us, we will choose GNOME applications only. We select Done to return to the Installation Summary window.

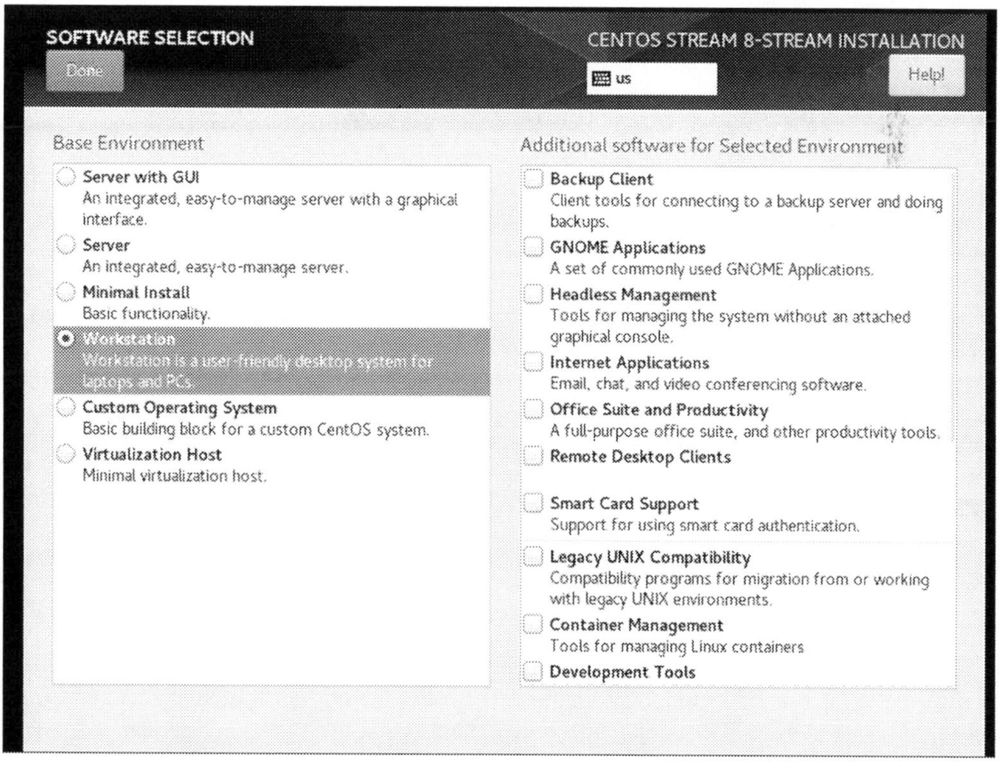

FIGURE 1.8 Base environment and software selection choices.

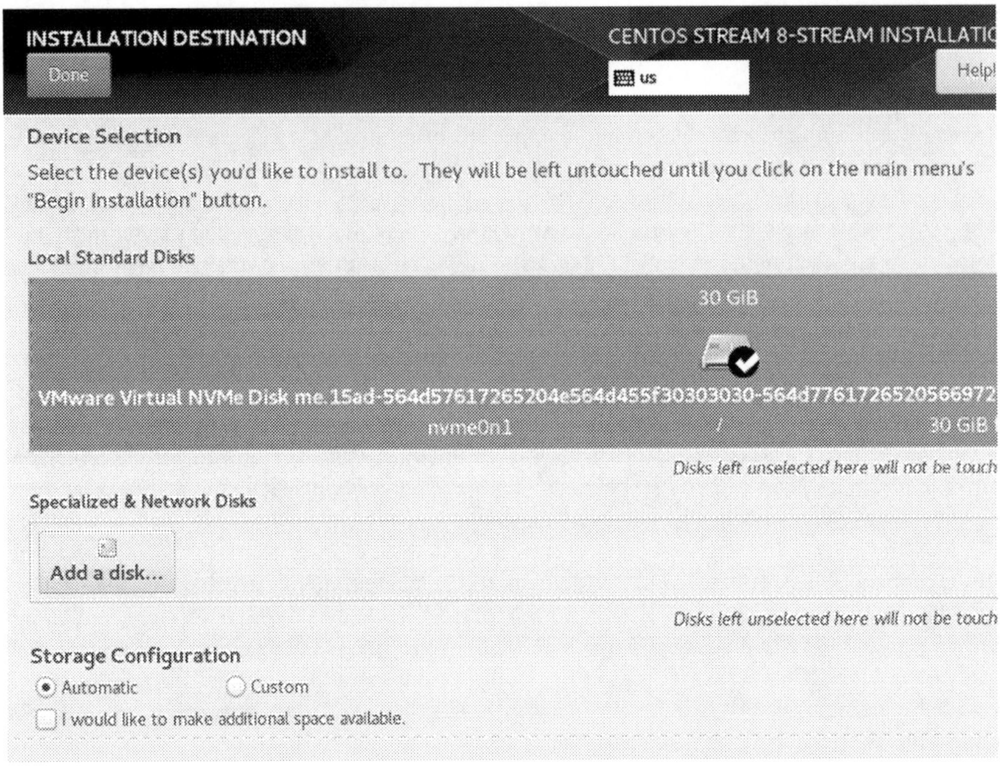

FIGURE 1.9 Installation destination window to specify partitioning.

Next is Installation Destination. By default, Automatic partitioning is selected. Should we wish to change this, select Installation Destination. The Installation Destination window is shown in Figure 1.9. Here, we would select the disk(s) onto which to install. There is only one choice in this case, labeled as VMWare Virtual NVMe Disk me followed by a lengthy identifier. To select this, click on it twice. The first time highlights it, and the second time adds a checkmark. Without the checkmark, the disk is not selected. Note that as we are installing into a VM, we have a small hard disk size.

Selecting Done returns us to the Installation Summary. If we had chosen Custom, selecting Done would take us to a Manual Partitioning Window, as shown in Figure 1.10. We cover this step for completeness but suggest that you use the Automatic partitioning.

From the Manual Partitioning Window, we can select between Standard Partitioning, LVM and LVM Thin Partitioning (the default is LVM). The Standard Partition selection allows us to specify the exact number, type and size of partitions. We omit detail for a standard partitioning until we cover this in Chapter 8.

The LVM (also covered in Chapter 8) allows our partitions to grow as needed and is the more reasonable choice, at least at this point. Thin Provisioning makes more efficient use of disk space at the expense of added run-time overhead. LVM will be sufficient for our needs.

One issue with using an LVM is that we need to separate out the boot partition. We need to create one separate partition, for /boot. Near the bottom left of the Manual Partitioning window are buttons labeled + and -. Select the + to add a new partition. From the Add A New Mount Point window (shown at the bottom of Figure 1.10), select /boot for the mount point and specify a desired capacity. For CentOS Stream, it is recommended that this partition be 1 GB, so we enter 1024 and select Add Mount Point. After being returned to the Manual Partitioning window, select Done. We will be presented with a summary of the changes. Select Accept Changes

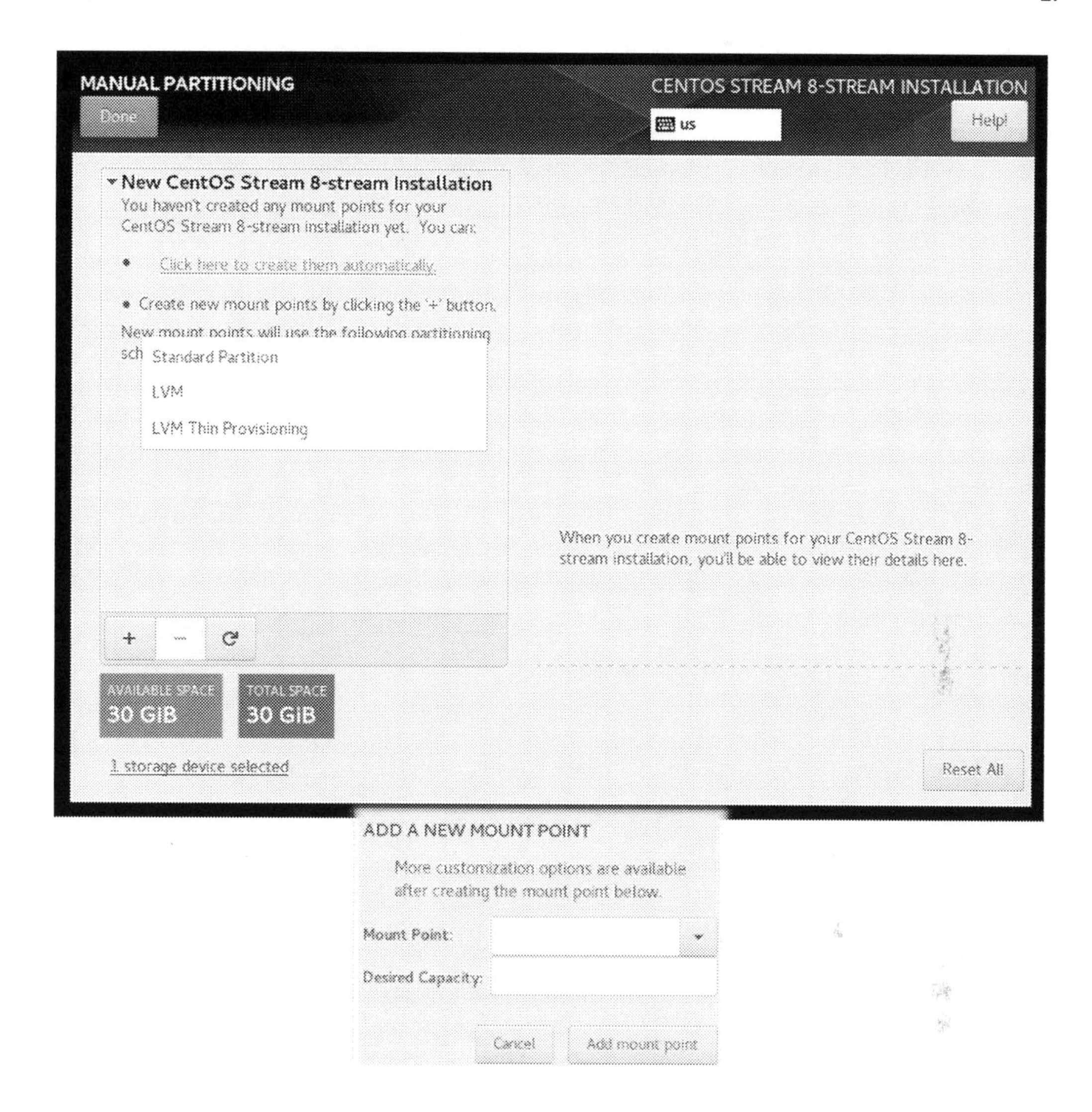

FIGURE 1.10 Manual partitioning.

and we will be returned to the `Installation Summary` window to continue. Note that should we choose, we can return to `Installation Destination`, select `Automatic` and `Done` to continue.

For `Network & Host Name`, the default is to disable the network interface. We might want to change this to enabled. We can also change the default hostname, which is initially set to `localhost.localdomain`. We can also add other interface devices. The default device is an Ethernet connection. Select `Network & Host Name` and from this window, `enable` the network and select `Done`.

Selecting `Root Password` allows us to specify and confirm the root password. A bar indicates how strong the password is. We would prefer to use a strong password. Use a password for root that you will not forget! Upon returning from the `Root Password` window, the `Installation Summary` has another entry that we can select, `User Creation`. Selecting this entry lets us specify an initial user account. As noted in the subsection on installing Debian Linux, it is preferable to have an initial user account so that we can log in as a normal user rather than root.

From the Create User window, we are asked for the user's full name. Typing in the name causes CentOS to automatically create a username using the format first initial last name. We might prefer to alter the default name. For instance, if we enter Frank Zappa as the user's name, we are given the username of fzappa but we might prefer something like zappaf or zappaf1. We are also asked to specify the user's initial password and confirm it.

Two checkboxes allow us to require a password for this account (selected by default) and to make this user an administrator (and thus give this user account access to all administrator programs, this is not selected by default and we want to leave it unselected). The Advanced... button lets us tailor this account manually by changing the home directory, change UID/GID and group membership. Do not bother changing the initial user's default values. We explore such options in Chapter 7. Selecting Done returns us to the Installation Summary where we are finally ready to perform the installation. We select Begin Installation.

Installation takes some time. Upon completion of installation, we are given a new option, Reboot. We must reboot to start using Linux. Upon booting (and in all future instances of booting or rebooting our CentOS VM), the first screen shown asks us which version we want to boot to. By default, CentOS installs with two kernels: the normal kernel and the rescue kernel. The rescue kernel is only needed if our OS develops some faults and we can't successfully boot to it. The normal kernel is selected automatically if we do not make a selection within a few seconds.

After rebooting, we are brought to an Initial Setup window with one choice, to view and accept the License Information. The license agreement is displayed (its brief!); select I accept the license agreement and Done. Upon returning to the Initial Setup window, select Finish Configuration in the lower right of the window. The computer then takes us to a login screen very much like that of Debian (see the left side of Figure 1.6).

The first time a user logs into his or her account, CentOS runs a user setup program. The first window is a Welcome screen whereby the user selects their language (followed by keyboard language). These will default to whatever language we selected during installation (English in our case). Next, the user can select whether to permit location services to run or not (privacy). Finally, the user can select which online accounts the computer should connect to. This step can be skipped. After this initial user setup is completed, the user is presented with a Ready to Go window, which is a tutorial for showing us how to use GNOME. Closing this window brings us to the CentOS GNOME Desktop. We can return to the GNOME Help window any time we like as explained below.

The GNOME Desktop in CentOS is very similar to that of Debian (see the right side of Figure 1.6). The Activities menu brings up our list of favorites, just as it did in Debian, although the initial favorites may differ. One of the items in this list looks like a life preserver. Selecting it returns us to the GNOME Help window. Selecting the last entry expands all installed software. It is likely that the CentOS Stream installation has fewer titles than Debian's installation. One item is Utilities. Selecting this displays the various CentOS utility programs. Two are of particular interest to us, System Monitor (which is similar to Windows Task Manager and a program we explore in Chapter 4) and Terminal. Right click on Terminal and select Add to Favorites. This places the terminal window icon on the activities list for easy access. We can drag this icon to the top of the list for even more convenient access.

Another of the software selections is called Settings. This presents a GUI much like Windows 10's Settings tool, although there are fewer selections available. We see some of the selections in the left pane of Figure 1.11 where we have selected Network. The available network settings are shown in the right pane. Note that while we can use this GUI to modify settings, we will find other ways to modify settings as well, including Cockpit (which we explore next) and through programs launched from the command line.

As Linux users, we would likely not need administrative tools beyond the settings available through the Settings tool. As a system administrator, we will need to be able to change any number of OS settings, most of which are not available through the Settings tool. In older versions of CentOS/RHEL, there were some administrator GUI tools available, including those to manage user and group

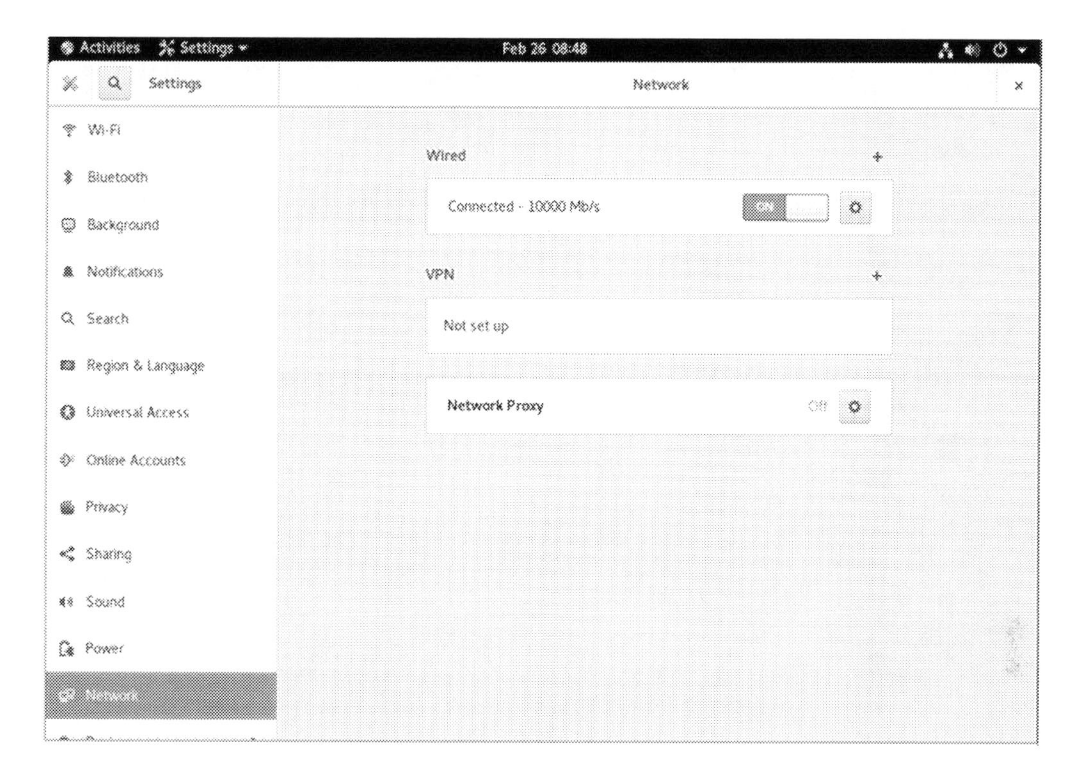

FIGURE 1.11 Changing settings in CentOS.

accounts, control `KDump`, start and stop services and alter the firewall. With Red Hat 8 (CentOS 8, CentOS Stream, RHEL 8), most of the GUI tools no longer exist and instead we manage these tasks either through command line or through the new *Cockpit* interface. Our access to Cockpit is through a web browser using our computer's IP address followed by `:9090`, as in `10.11.12.13:9090`.

As Cockpit loads, we are asked to log in. If we log in under our user account, our access will be restricted to mostly the same items as we saw in Settings. If we log in as root, we are given additional capabilities including user account management. Figure 1.12 shows the Cockpit interface with System selected to show performance information. The controls along the left pane allow us to view log files and examine currently used and available storage, network settings, current user accounts, available software and other items.

1.5.3 Installing Ubuntu Linux

Having started the Ubuntu installation, we are presented with some installation screens like those shown in Figure 1.13. The last two screens of the figure show some of the information that we are presented with during installation. We are asked a few installation questions, similar to that of CentOS: language and keyboard layout, time zone, type of software installation (our choices here are normal, minimal or other), installation type and initial user account information. For the installation type, if there is already a version of Ubuntu installed, we are asked whether to erase it and install the new one or install the new version alongside the existing version. Other choices are to erase the disk, encrypt the installation and use an LVM. We can also select `Something else` to specify our own partitioning. We skip these details.

The initial user account selection differs from that in CentOS and Debian in one significant way. In CentOS and Debian, we are asked to provide a root password so that we can have user accounts and a separate administrator account. In Ubuntu, the root account's password is randomly generated and not given to the user. In place of having root-level access, the first user account is given full

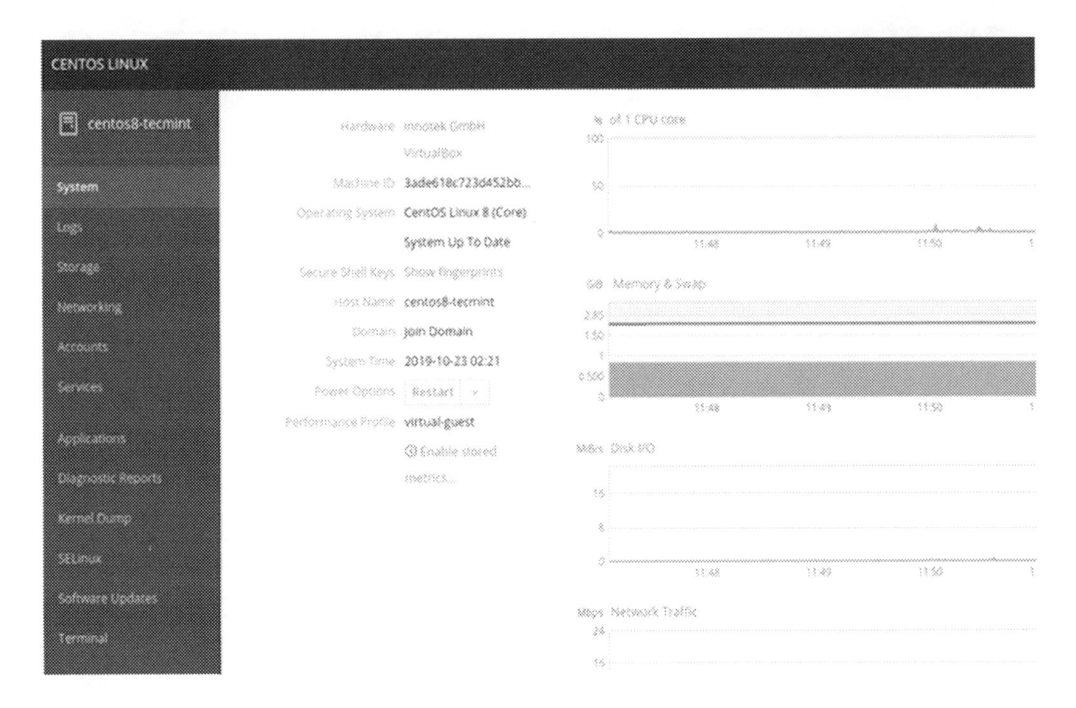

FIGURE 1.12 `Cockpit.`

administrator privileges through `sudo`. The `sudo` program, covered in Chapter 7, allows a user to run programs under different access rights. In the case of Ubuntu, the initial user account is given access to all system administrator programs so that this user never has to log in as root.

This change in approach to installing Linux originated with Ubuntu and is found in several Linux distributions, mimicking to some extent Windows 10 installation for home users where the first user account is given full administrator privileges. To specify the initial user, we are asked for the user's name, the username, the computer's name and a password. We are not, however, asked if this user should be made an administrator because this choice is made automatically. After installation, we are asked to reboot.

After rebooting, we are presented a login screen which will be similar to that of Debian and CentOS except for the color scheme. After logging in, the user is presented with a desktop very similar to that of Debian and CentOS except that there are two desktop icons already created (the user's home directory and a trash can). The Ubuntu desktop is shown in Figure 1.14.

1.5.4 Installing Linux Mint

Linux Mint is the easiest of all installations as there is almost no user interaction. Upon creating a new VM and selecting the Linux Mint `iso`, we are presented with a menu of installation choices much like in Debian (the upper left window of Figure 1.4). Installation proceeds from this point with no interruption. Unlike the other Linux installations, Mint installs with a generic user account and no initial password. This initial account, like in Ubuntu, has full administrator privileges as the root password is unknown.

After installation, the desktop appears, as shown in the left half of Figure 1.15. Here, we see two desktop icons for the computer and for the user home directory, and a list of icons along the bottom left. The leftmost icon brings up a menu similar in style to the Windows 10 Start Button menu. Upon selecting this start button, the menu, which we see in the right half of Figure 1.15, appears. The other icons along the lower left, running from left to run, display the desktop (the default view), open a web browser, open a terminal window and launch the file browser GUI.

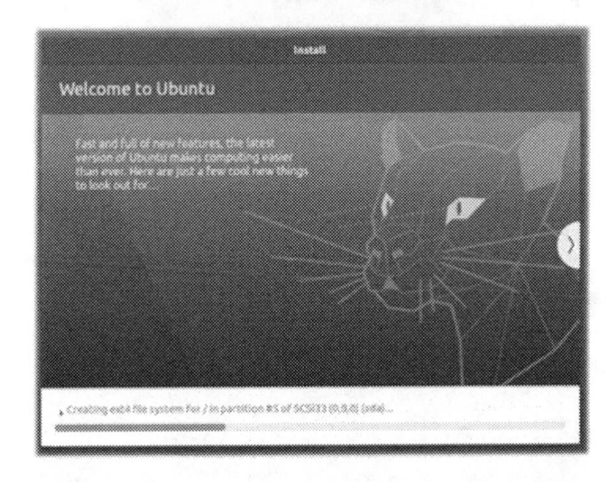

FIGURE 1.13 Ubuntu installation.

FIGURE 1.14 Ubuntu desktop.

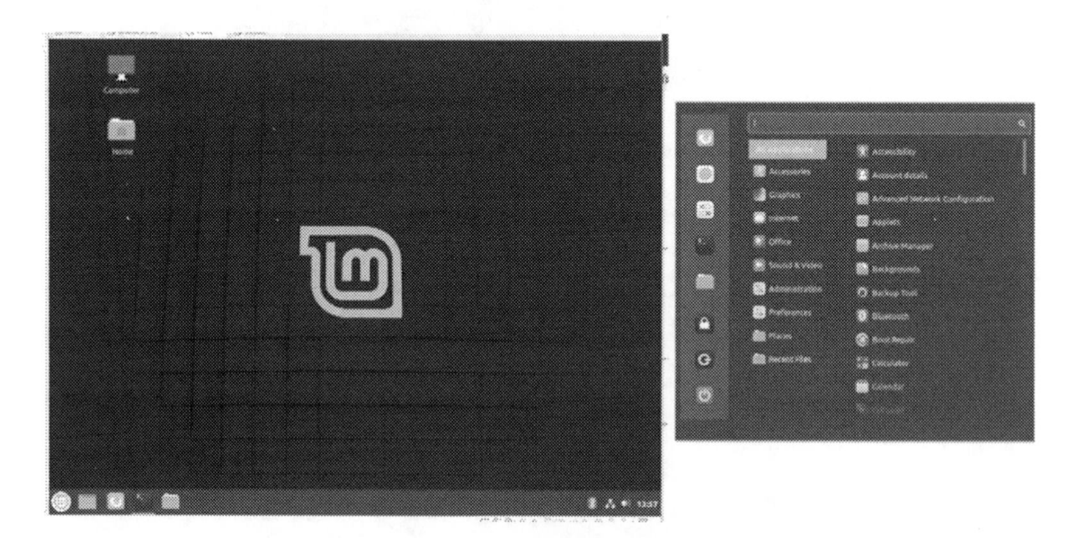

FIGURE 1.15 Linux Mint desktop.

1.5.5 An Introduction to the Shell and Command Line

No matter which version of Linux you may have installed, they all have the capability of opening a terminal window. A terminal window is a GUI component which itself runs a shell and an interpreter. The window has limited capabilities such as minimize, maximize, resize, zoom in and out, and cut, copy and paste text from within it. The shell running inside the window provides the user with a command line prompt.

From the prompt, the user inputs Linux instructions. Any instruction entered is then handled by the shell's interpreter. Most Linux users have a default shell of Bash, so commands are run by the Bash interpreter. As we will explore in Chapter 2, we can open shell sessions from within a shell and in doing so, change the type of shell (e.g., from `bash` to `tcsh`). In Chapter 7, we cover how to change the default shell of a user.

Opening a terminal window from the desktop is handled differently depending on the distribution. In Debian, CentOS and Ubuntu, select `Activities` and from the favorites list, select the `Terminal Window` (which looks like window with a black background and a white prompt). The CentOS Stream Workstation installation does not have a terminal window icon in this menu and so we have to add it as discussed previously to make it a part of our favorites list. In Debian, the terminal is called `XTerm`.

We see an example of an open terminal window in CentOS in Figure 1.16. Notice at the top of the desktop we have an added menu alongside `Activities` called `Terminal`. From this menu, we can close our terminal window (by selecting `Quit`; we can also close the terminal window by clicking on the `x` in the upper-right hand corner of the terminal window, or typing `exit` from the command prompt) but we can also open additional terminal windows (`New Window`). The terminal window itself has menus for controlling copying and pasting, zooming, etc. The `File` menu also has a selection to open a new window, open a tab, close the window or close the current tab.

The terminal window gives us the ability to enter Linux commands from the command line prompt in the window. Upon pressing `<enter>`, the interpreter executes the command for us. We see some very simple interactions in Figure 1.17 where we first ask who we are logged in as (`whoami`), the current time and date (`date`), where we are located in the file space (`pwd`) and who else is logged in (`who`). These are among the simplest instructions available in Linux. We spend most of the rest of this textbook looking at dozens of other commands.

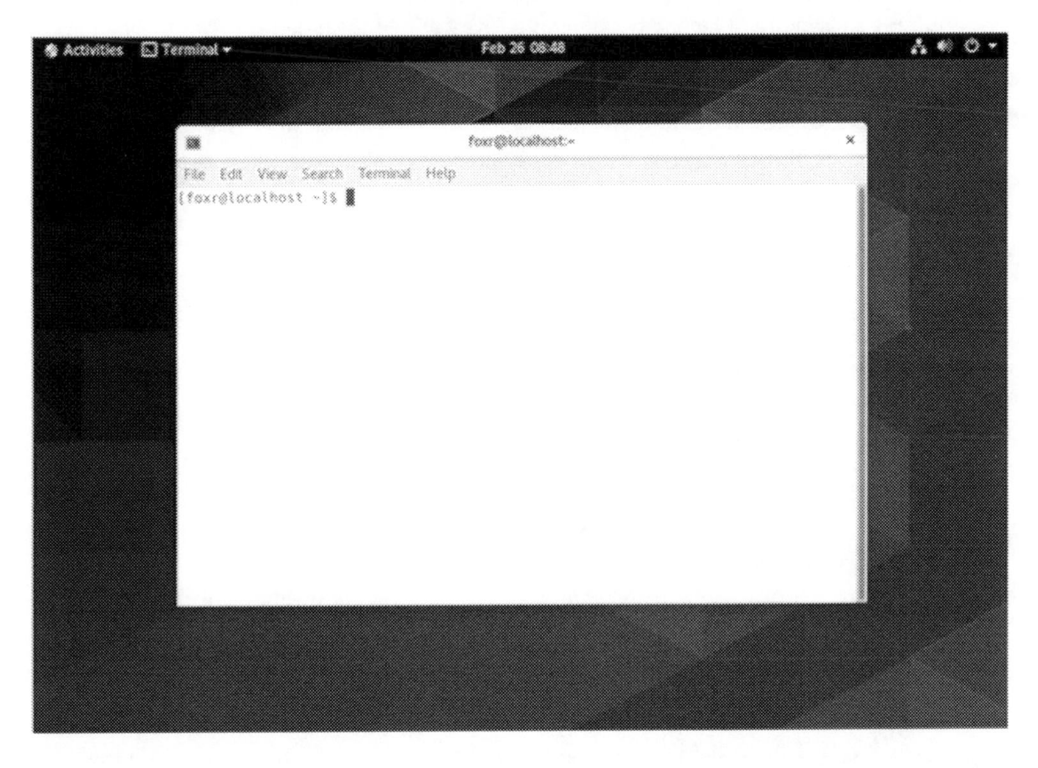

FIGURE 1.16 A terminal window.

```
[foxr@localhost ~]$ whoami
foxr
[foxr@localhost ~]$ date
Fri Mar 12 13:10:58 EST 2021
[foxr@localhost ~]$ pwd
/home/foxr
[foxr@localhost ~]$ who
foxr      tty1      2021-03-12 13:11 (tty1)
zappaf    pts/2     2021-03-10 08:44 (10.11.12.13)
```

FIGURE 1.17 Some simple interaction through a terminal window.

One additional instruction to learn is `passwd`, to change our password. It is likely that a user is given an initial password from the system administrator and is told to change that password at their first login. In our case, this is not necessary because we created our own initial password when we installed Linux.

The `passwd` command, when issued with no parameter, attempts to change our password. We are asked for our current password. As we type in the password, it is not displayed on the screen. We are then asked for the new password and again, as we type it is not displayed on the screen. If the password satisfies whatever strong password requirements are set up in the system, we are asked to confirm the password. As we will explore in Chapter 7 when we look at user accounts, root can also use password to change other users' passwords or change attributes about passwords.

SECTION ACTIVITIES

1. Of the four versions of Linux we examined in this installation section, which would you pick to use? Would you select the one easiest to install? The one that looked the most user-friendly? The one with the most available preinstalled software? The one that has the most games available?
2. If you are taking a Linux class, you will probably do some labs using Linux. If this is not the case, download and install an open-source virtual machine client like VirtualBox and then download and install one of the Linux distributions covered in this section.

1.6 CHAPTER REVIEW

Concepts and terms introduced in this chapter:

- Bare machine – a computer with no operating system installed to ensure the fastest possible hardware execution.
- BSD Unix – one of the most significant spinoffs of Unix, developed by Ken Thompson and faculty and students at UC Berkeley; widely distributed in the 1980s and 1990s.
- Commercial support – the ability to ask for (and expect) assistance with software when purchased; support includes timely bug fixes and help when you do not know how to accomplish some task that the software should be able to perform.
- Device driver – a piece of software tailored for a specific hardware device allowing the CPU to communicate with that hardware device.
- CentOS – a distribution of Red Hat Linux that is open source, lightweight and a popular choice for servers.
- Cockpit – a recent addition to Linux to give the user and system administrator a uniform means of viewing the system and changing settings; Cockpit runs through a web browser.

- Copyleft – a term coined by Richard Stallman about the GPL to express that the license supports freedom rather than restrictions on software usage.
- Debian – one of the earliest distributions of Linux; popular because of its stable releases.
- Fedora – a distribution of Red Hat Linux that is open source and supported entirely by the open source community.
- File descriptor – a means for the OS kernel to refer to a resource such as an open file.
- Free software foundation – the name originally given to the open-source community but after a splintering is the group that feels that all software should be free.
- GNU – a project to develop a Unix-like operating system that was entirely free with no licensing restrictions; the project was headed by Richard Stallman but never completed; members of the GNU project contributed time and already implemented code to the development of Linux.
- GPL – the GNU Public License, a copyleft which supports users' (and developers') rights rather than restrictions on using a piece of software.
- HeartBleed – a flaw found in the encryption technology of OpenSSL which was patched less than a week after the flaw was discovered, showing the responsiveness of the open-source community.
- Hybrid kernel – a type of OS kernel which combines aspects of microkernels and monolithic kernels.
- ISO – the abbreviation references the International Organization of Standards where ISO 9660 specifies optical disc storage; in this context, an ISO is the format used to store Linux installation programs.
- Kernel module – in order to make a Linux kernel more efficient, components of the kernel, called modules, can be loaded into and out of memory.
- Knoppix – a distribution of Linux descended from Ubuntu; Knoppix was the first Linux distribution that featured live booting.
- Lightweight – the idea that an OS (or other piece of software) can run on limited resources; many Linux distributions are considered lightweight particularly when compared to OSs like Windows.
- Linux distribution – a specific version of Linux; there are hundreds of different distros that are all based on the same Linux kernel but have different features, services, available applications software and GUI components.
- Linux kernel – a monolithic kernel with modules to make it more efficient; most Linux distributions are based on the same kernel.
- Linux Mint – a Linux distribution based on Debian Linux known for being lightweight.
- Live boot – the ability to boot from and run the OS off of an optical disc (or USB flash drive) rather than one that is installed on hard disk.
- Microkernel – a form of OS kernel which hands off some of its duties to lesser programs called servers.
- Monolithic kernel – a form of OS kernel which handles all of its tasks internally and thus makes the kernel more complex; Linux, while based on a monolithic kernel, hands off some of its tasks to modules and services (not to be confused with servers).
- MULTICS – an early 1960s operating system that introduced many new OS innovations and was a basis for many of Unix's original features.
- Open-source community – people who freely contribute their time and effort to developing and supporting open-source software.
- Open-source initiative (OSI) – a splinter of the open-source community who believe that some software can be commercially developed and distributed as opposed to the *Free Software Foundation*; the OSI makes up the larger fraction of the open-source community.

- Open-source software – a software product that is made available in its source code format so that developers can download and modify the code; as it is available in source code format it is freely available and usually also available in an easy-to-install executable format as well for non-developers.
- Operating System (OS) – a piece of software responsible for monitoring the resources of a computer and for acting as an intermediary between the user, the application software and computer hardware.
- OS kernel – the core program of any OS; the kernel is loaded into memory at the time the computer boots and remains in memory while the computer runs; the kernel is responsible for handling the most important tasks of the OS including process management, resource management, memory management, interprocess communication and interrupt handling.
- OS utility – a standalone program that is typically invoked by the user to help manage the computer system; unlike the kernel and OS services, a utility runs in the foreground and often presents to the user a GUI-based tool to interface with.
- Portability – the aspect of software where the program is written to run on multiple platforms; Unix was the first OS developed to be platform-independent.
- Privileged mode – most processors run in one of two modes: user mode to run user programs and privileged (or supervisor) mode to run system software.
- Red Hat Enterprise Linux (RHEL) – the version of Red Hat that is only available as a commercial product but comes with Red Hat, Inc. commercial support; CentOS is an open source version related to RHEL.
- Red Hat Linux – one of the earliest distributions of Linux and one of the most popular; Red Hat is somewhat unique because one of its spinoffs, Red Hat Enterprise Linux, is a commercial product.
- Resident monitor – before OSs were created, many mainframes loaded and ran a resident monitor to support simple user interactions such as loading and running programs off of magnetic tape.
- Server – in the context of this chapter, a server is a piece of the OS separate from the kernel and invoked by the kernel so that the kernel can be smaller in size; this strategy is used in support of microkernels.
- Service – an OS program separate from the kernel used to support some functionality of the OS; we find services used in OSs like Linux and Windows.
- Settings – a GUI tool available in most modern Linux distributions so that the user can change personalized settings; similar in style to modern Windows operating systems Settings tool.
- SLS/Slackware/SUSE Linux – SLS was the earliest Linux spinoff distribution which was the basis for the improved Slackware which itself was used to create SUSE Linux.
- System call – an invocation by a user program or OS component to call upon a function of the kernel.
- Ubuntu – a spinoff of Debian Linux which became popular because of its user-friendly approach and a large number of software products.
- Unix – developed in the late 1960s by AT&T Bell Labs and modeled somewhat on MULTICS, Unix became a popular, portable and powerful operating system; Linux' look is very similar to that of Unix.
- User mode – the default mode for processors so that user processes can only make requests of the OS; the processor must switch modes to privileged mode to invoke the OS and access system resources.
- Wrapper function – a system call invokes a wrapper function which itself invokes the OS kernel; the wrapper function is responsible for handling pre- and post-invocation tasks such as reporting an error to the user.

REVIEW QUESTIONS

1. Why is commercial support desired when purchasing software? What are some of the guarantees that come with commercial support?
2. In what way is the HeartBleed bug a good indicator that open-source software does not need commercial support?
3. Are all Linux distributions free?
4. One argument for using Linux over an OS like Windows is greater control. How does Linux offer greater control?
5. Which of the following is true of the earliest mainframe computers: they were bare machines, they had resident monitors, they had operating systems or they were virtual machines?
6. Explain the steps by which a FORTRAN programmer would compile and run their FORTRAN program on a mainframe computer of 1958.
7. At what point in time would a resident monitor be loaded into memory? How long would it remain in memory?
8. What resident monitor/OS was the first to provide time sharing (multitasking)?
9. Which role of the OS kernel allows multiple running processes to share data in memory?
10. Virtual memory would be considered a feature handled by which of the OS kernel's roles?
11. Synchronization and deadlock handling are tasks handled by which of the OS kernel's roles?
12. Upon receiving a system call from a user process, the kernel switches the processor from _____ mode to _____ mode.
13. Linux contains which type of kernel: microkernel, monolithic kernel, hybrid kernel?
14. Which of the forms of kernel is the most challenging to debug because of its complexity: microkernel, monolithic kernel, hybrid kernel?
15. Examine the system calls in Table 1.4 to answer the following questions.
    a. Which system call(s) would be involved in creating a new file?
    b. Which system call(s) would be involved in starting a new process?
16. True/false: All device drivers are preinstalled into the OS.
17. Linux services run in the foreground or background?
18. List three types of OS utilities.
19. What members of AT&T Bell Labs worked on MULTICS and later started developing Unix?
20. What was the original name for Unix?
21. Who was Richard Stallman and what project did he begin?
22. Read Figure 1.3 (the GPL Preamble). Instead of restrictions on software, it talks about users' _____.
23. In what year did Linus Torvalds begin working on Linux? In what year was the first full implementation made available?
24. Between SLS/Slackware, Debian and Red Hat, which appeared first?
25. List three popular distributions based on Debian Linux.
26. List three popular distributions based on Red Hat Linux.
27. In what way is RHEL different from CentOS?
28. How does CentOS Stream differ from previous CentOS releases?
29. What is the main difference between Debian-based Linux distributions and Red Hat-based Linux distributions?
30. List three reasons for why a developer might contribute to open-source software.
31. What is an ISO file?
32. True/false: When installing Linux, you must specify how the disk drive is to be partitioned.
33. When installing CentOS Linux, what two choices are you required to specify from the Installation Summary window? At what point does the Begin Installation button become accessible?

34. When installing a recent version of Debian, CentOS and Ubuntu, there is a similar Desktop. What is the name of the only menu that appears when the desktop first appears?
35. What is GNOME?
36. As a user, what can you do through the Cockpit interface?
37. What is the name of the one menu available in current versions of Debian and CentOS Linux?
38. List five types of settings that a user can change.
39. One difference between installing one of Debian or CentOS Linux from Ubuntu is that in Debian and CentOS you are asked for the root password. How does Ubuntu provide root-level access if the user does not know the root password?

2 Bash

This chapter's learning objects are to be able to

- Explain the role of the shell
- Enter commands via the command line prompt
- Apply history, command-line editing, brace expansion, tilde expansion, wildcards and redirection
- Define and use aliases and variables from the command line prompt
- Utilize forms of help
- Personalize a shell using `.bashrc` or `.bash_profile`
- Use the `vi` editor

2.1 INTRODUCTION

A *shell* is a user interface. There are three different types of user interfaces. Although the graphical-user interface (GUI) is popular today, many operating systems also provide a command-line interface (CLI). A third interface is a text-based menu interface which is largely no longer in use because the GUI offers the same functionality but is far easier to use.

In Linux, we are offered both a GUI and a CLI. Many Linux users prefer the CLI, for reasons enumerated in Figure 2.1. Another reason to learn the CLI is that many Linux distributions do not offer equivalent GUI tools for many of the Linux commands.

New users often complain about having to use the CLI because of the large number of Linux commands that the user must know to be successful. The learning curve is steep. But to fully learn Linux, we need to use the CLI early and often. In order to simplify CLI usage, Linux shells have a number of features making interaction more convenient. Here, we explore the Bash shell and its many features.

A shell runs an interpreter. The *interpreter* is a program which accepts user input and determines how to execute that input. The original Unix shell, the Thompson shell (named after the program's author, Ken Thompson), had a built-in interpreter called `sh` (the executable program). Thompson built the shell based on the shell available in the MULTICS operating system.

Thompson's intention of providing a shell was to combine a command line prompt, an interpreter and an environment. The command-line prompt gives the user the ability to input commands. The

1. Control: Linux commands often have numerous options that provide both flexibility and power over the defaults when issuing commands by GUI.

2. Speed: many programs issued at the command line are text-based and therefore start faster and run faster than GUI-based programs.

3. Resources: as with speed, a text-based program uses fewer system resources.

4. Remote access: when remotely logging into a Linux computer, we are usually only presented with a text-based interface.

5. Reduced health risk: this one sounds odd but excessive mouse usage is one of the most harmful aspects of using a computer as it strains numerous muscles in our hand and wrist, while typing is less of a strain (although poor hand placement can also damage our wrist).

FIGURE 2.1 Advantages of using the command-line over the GUI.

DOI: 10.1201/9781003203322-2

interpreter's role is to parse an input, convert that input into an executable statement and execute it. The environment's role is to capture definitions and previous commands of the current session to be recalled by the user to simplify the user's interaction with the computer.

Features of the Thompson shell included redirection operators so that input and output could be redirected from default locations to other locations (e.g., input from keyboard instead of file, output to file instead of monitor) and pipes (redirection of program output to become input for another program). `sh` also gave the user the ability to refer to file names using *wildcards*. For instance, all files whose names start with an `a` can be denoted using `a*`. Developed on top of the Thompson shell was the Mashey Shell (also called `sh`). This shell added environment variables such as the `PATH` variable as well as an expanded set of shell language instructions to make shell scripts more useful.

The Thompson shell was written in 1971. By 1977, Stephen Bourne had enhanced it into the Bourne Shell. Bourne added several new features including a fully defined scripting language so that the interpreter could be used to interpret commands from the command line or commands placed into a file, called a script. He also introduced the *here document* for input. This permits the user to enter unlimited amounts of text from the keyboard while the program is running. This was somewhat novel in the 1970s when most computers were batch processing (whereby input had to be submitted to the program by means of input files). The Bourne Shell was distributed with Unix 7 starting in 1979 and became a standardized shell for most Unix distributions.

Two criticisms of the Bourne Shell were that it was "user unfriendly" and that the scripting language's syntax was based on ALGOL, a language that was rarely used in the United States. Bill Joy developed the C-shell (`csh`) in 1978 as his answer to the Bourne shell. `csh` introduced its own scripting language based on syntax from the C programming language. It also included features not found in the Bourne Shell. These features included a history list and the ability to recall instructions from the history list, command line editing features, tilde (~) notation to indicate the user's home directory, the ability to define and recall aliases of instructions and escape completion (in most modern shells, we now use tab completion). As we find these (or variations of these) features in Bash, we will describe them more fully in this chapter.

The TC-Shell (`tcsh`) is backward compatible with `csh` but adds features including the use of the up and down arrows on the keyboard to step through the history list. It also provides more controls for recalling an instruction from the history list. `tcsh` also saw an expanded scripting language over `csh`. Distributed in 1983, it became one of the most popular of the shells. Also released in 1983 was the Korn shell (`ksh`), which shared many of the same features. TC-Shell remained the most popular shell until 1989, when the Bourne Again Shell (Bash) was released.

Since the release of Bash, we have seen many other shells offer variations of features found in either `bash` or `tcsh`. These include `zsh` released in 1990 as a merger of `tcsh`, `bash` and `ksh`, and the Almquist Shell (`ash`), released in the early 1990s.

Bash is the default shell for Linux users, although this can be altered at the user's request. Bash again expanded the scripting language keeping both the ALGOL notation of the Bourne Shell while later adopting C-like syntax similar to `csh` for many types of commands. Bash introduced yet other features including *brace expansion* and features that could utilize regular expressions. As we move through this chapter, we will concentrate on the Bash shell. We will look more fully at Bash shell scripting in Chapter 6.

Modern shells can be tailored by the user. The user is able to define aliases as shortcuts to commands and define their own environment variables and functions that can be utilized from the command line or script. The user can set up script files that are automatically executed upon login or when a new shell is opened. Definitions (whether aliases, variables or functions) can be placed in these script files so that the environment is already set up as the user likes upon startup.

SECTION ACTIVITIES

1. Have you had any prior experience with command line entry? Aside from interaction with an operating system, you may have done this if you have entered programming instructions in an environment like Python or Ruby. If you have, did you enjoy the experience?
2. Does Windows have shells and scripts? If you are a Windows user, open PowerShell (you can find this under Windows PowerShell in the Start menu). This tool operates in many ways like the Linux shells although has a more complex list of instructions. It has a prompt, it accepts instructions that run commands, it has many features similar to Linux shells that you will explore in this chapter like help screens, history and the ability to define aliases. In PowerShell, type `Get-Verb` to list the available commands. Every one of these "verb commands" has accompanying nouns so that a command is issued as verb-noun such as `Get-Date` or `Test-Path`. Type `Get-Help Test` to view the list of Test nouns. NOTE: PowerShell is case insensitive so you can type `get-help test` instead. After you play around with PowerShell briefly, type `Exit` to close the window.
3. Do a web search on `Linux shell` and view the different shells to see how they differ.

2.2 ENTERING LINUX COMMANDS

Upon logging into Linux through the GUI, we can run programs from the GUI, but instead let's open a terminal window. In the terminal window, a shell runs based on our default shell program. We will assume its Bash (note that we can refer to Bash as the shell or `bash` as the program, we will largely refer to Bash throughout the text).

The shell offers us a prompt, awaiting our commands. The prompt might look like `[zappaf@localhost ~]$`. This prompt informs us of several pieces of information: who we are, where we are, where we are and who we are. You are not seeing double, as we listed both questions twice!

The first "who we are" is our username (zappaf in this case). The first "where we are" is the machine we are logged into. Localhost means that the shell we have opened is on *this* computer. The name Localhost indicates that we have not named our computer, so Bash uses the default name of `localhost`. Had we remotely logged into another computer, we would see a different value for the computer's name.

The second "where we are" is provided by ~ (tilde). This notation indicates that we are in the user's home directory, which is likely `/home/zappaf` for user zapapf. The second "who we are" is provided by the `$`, representing that this account is a normal user account. If we had logged in as root (system administrator), the prompt would be presented as #. The reason for the difference is to warn a user that they are currently logged in as root in case the user enters a command that could do harm to the computer. As we will probably know who we are and which machine we are using, the most useful piece of information in the prompt is the current working directory.

The Bash interpreter awaits our command. To use the interpreter, we type in a command and press the `<enter>` key. Whatever was typed is now interpreted. The command, however, may include a number of the features we named in the introduction. So, the interpreter goes through several steps to execute the command. We will look at these steps near the end of this chapter. In this section, we will concentrate on commands.

2.2.1 SIMPLE LINUX COMMANDS

Although the prompt tells us a lot of information, there are several simple Linux commands that can tell us who we are, where we are and what can be found at our current location. These commands are listed in Table 2.1.

TABLE 2.1
Simple Linux Commands

Command	Meaning	Usage
cd	Change working directory.	cd .. (go up one level) cd /home/foxr (move to foxr's home directory)
echo	Output a message, including values stored in variables.	echo Hello world! echo Hello $USER
hostname	Display the computer's host name.	
ls	List the items in the current working (or specified) directory(ies).	ls (list current working directory) ls /home/foxr (list contents of foxr's home directory)
pwd	Print the current working directory.	
vi, emacs	Run the vi or emacs text editor.	may be followed by a filename as in vi newfile1.txt
who	Print all logged in users.	
whoami	Print your username.	

Most of the commands in Table 2.1 are simple to use. They require no or few parameters and have no or few options. We explore the meaning behind parameters and options in the next subsection. From Table 2.1 though, we see that cd, ls and vi/emacs may include parameters. The echo instruction expects parameters (although using echo without a parameter does not result in an error).

If you are currently logged into a Linux computer while you read this, try the instructions hostname, pwd and whoami. You will see that these respond with information that you can also find in your prompt. If you are logged in as root, whoami returns root. Note that if the user prompt tells you that you are currently at ~, pwd would still return the full directory path, such as /home/foxr. Additional, simple, Linux commands are listed in Table 2.2. The passwd and su command may include a parameter of a username but is not required. The others do not take parameters.

We can enter multiple commands at once by separating each command with a semicolon (;). Figure 2.2 shows a short interaction between the user entering commands and the results. First, three different instructions from Tables 2.1 and 2.2 are entered. Then, the three commands are entered together as a single instruction. Note that the responses will vary depending on the specifics of your computer and who is logged in.

TABLE 2.2
Additional Simple Linux Commands

Command	Meaning
arch	Output the computer's architecture (processor type).
bash	Start a new bash session.
exit	Exit the current bash session and if this is the "outermost" session, exit the current window.
passwd	Change your password; you are prompted to input the current password followed by a new password twice; if the password entered is not strong enough, you are warned (and depending on settings your new password may be rejected).
su	Switch user to the specified user; if no username is provided, switch to root; unless you are currently root you are queried to enter the user's password.
uname	Output information about your operating system.

```
[foxr@localhost ~]$ uname
Linux
[foxr@localhost ~]$ arch
x86_64
[foxr@localhost ~]$ who
foxr      tty7      2013-10-11 09:42 (:0)
foxr      pts/0     2013-10-11 15:14 (:0)


[foxr@localhost ~]$ uname; arch; who
Linux
x86_64
foxr      tty7      2013-10-11 09:42 (:0)
foxr      pts/0     2013-10-11 15:14 (:0)
```

FIGURE 2.2 Simple command-line interaction.

2.2.2 Commands with Options and Parameters

As we explore Linux commands in this and the next several chapters, we will find most Linux commands have a large number of options and may require parameters. The basic format of a Linux command is `command [options] [parameter(s)]`.

Options typically follow a hyphen (-) and are usually a single letter or sometimes a digit. With many instructions, some of those same options are available in a longer format. For instance, some instructions use `-f` for force and have a variation of `--force`.

Options vary by instruction but sometimes the same option is used in the same way and has the same meaning across multiple instructions. For instance, copy (`cp`), move (`mv`) and delete (`rm`) all have `-f`/`--force` for *force* and `-i`/`--interactive` to perform the operation in *interactive* mode. Similarly, `-r` (or `--recursive`) performs a *recursive* operation for both `cp` and `rm`. Many Linux instructions use `-a` for *all* and `-h`/`--help` for *help*.

Parameters are sometimes required and sometimes optional. Consider the two `ls` commands shown in Table 2.1. The first, without a parameter, displays the contents of the current directory. The second displays the contents of the specified directory. The `ls` command, like many other commands, can include any number of parameters. The instruction `ls /home/foxr /home/zappaf /etc/sysconfig /usr/bin /var/log` displays the contents of all five of the specified directories.

Depending on the instruction, parameters may be usernames, as used in commands like `passwd` and `su`, processes indicated by their ID, as used in an instruction like `kill`, or files or directories. There are other forms of parameters which we will learn about later in this book.

Let's focus on the `ls` command. We use `ls` to list the contents of one or more directories or details of one or more files. Perhaps the most popular option for `ls` is `-l` (lower case "L"), which results in `ls` responding with a long listing. The *long listing* shows detail of the files and directories. The command `ls -l file1.txt` provides the long listing for the named file. Table 2.3 explains the contents displayed in a long listing. Figure 2.3 provides an example of a long listing so that you can see specifically what it looks like.

The entries shown in Figure 2.3 are excerpted from the full long listing of the directory `/etc`. The file types and the permissions are combined together to form a string of ten characters. We see a variety of different types of permissions among these items. Directories, denoted with a first character of `d`, either have `rwxr-xr-x` or `r-xr-xr-x` permissions. Regular files, denoted with a first character of `-`, are commonly `rw-r--r--` in this directory but `.pwd.lock` has permissions of `rw-------`. Notice that there is a symbolic link shown in the figure as well. Its type is denoted with `l`, and its name includes both the name of the link (`grub.conf`) and the location of the linked file (`../boot/grub/grub.conf`). We explore permissions and directory notation (e.g., `../grub`) in Chapter 3.

TABLE 2.3
The Contents of a Long Listing

Item	Meaning	Example
File type	In Linux, "files" include true files, directories, symbolic links, devices and other items.	- Regular file d Directory l Symbolic link
Permissions	Also called *mode*, the file's access permissions.	`rw-r-----` owner has read/write access, group members have read access, other users have no access.
Hard links	Number of hard links that point at this item.	Often 1 for files and 2 for directories but can be larger.

User, Group	The user who owns the file and the group that the file belongs to; for most user files, the group is the user's private group.	`foxr foxr`
		File owned by foxr and in his private group.
		`foxr students`
		With a different group it opens up access rights to all users in the `students` group.

Size	Size of object in bytes.	A numeric value, 0 for empty.
Last	Modification date/time (if not modified then creation date/time).	Jan 19 09:31.
Name	Name of the item.	If a symbolic link the item's name is followed by `->linked item`, see Figure 2.3 for an example.

```
total 2264
drwxr-xr-x. 120 root root  12288 Dec 17 03:16 .
dr-xr-xr-x.  28 root root   4096 Aug  7 07:46 ..
drwxr-xr-x.   3 root root   4096 Jan  8  2012 abrt
drwxr-xr-x.   4 root root   4096 Dec  1  2011 acpi
-rw-r--r--.   1 root root     45 Aug  7  2019 adjtime
drwxr-xr-x.   2 root root   4096 Aug  6  2012 akonadi
-rw-r--r--.   1 root root   1512 Jan 12  2010 aliases
-rw-r--r--.   1 root root  12288 Dec  1  2011 aliases.db

lrwxrwxrwx.   1 root root     22 Dec  1  2011 grub.conf -> ../boot/grub/grub.conf
----------.   1 root root    795 Sep 17 08:32 gshadow
----------.   1 root root    803 Sep 17 08:32 gshadow-

-rw-------.   1 root root      0 Dec  1  2011 .pwd.lock
-rw-r--r--.   1 root root    220 Oct 13  2008 quotagrpadmins
-rw-r--r--.   1 root root    259 Jul 19  2011 quotatab
```

FIGURE 2.3 Example long listing showing files, directories and links.

All of the long listing permissions end with a period. The period indicates that the item has an SELinux context. *SELinux* is security-enhanced Linux and provides a greater degree of protection over what is available through ordinary Linux permissions. An SELinux *context* means that SELinux is being applied to the item. We cover SELinux in Chapter 7.

We saw earlier that `ls` permits multiple parameters, separated by spaces. For instance, we could obtain the long listing of multiple files using an instruction like `ls -l file1.txt file2.txt file3.txt file4.txt file5.txt`. In this example, all of the files listed end with `.txt`. We will see an easier way to specify such a list of files through wildcards, a feature we cover in Section 2.4, and expand on wildcards in Chapter 3.

The default for `ls` is to display visible items. The term *visible* indicates items whose names do not start with period (.). Files whose names start with a period are considered *hidden*. For instance, `file1.txt` is a visible file and `.file1` is a hidden file. To view both visible and hidden items, use the option `-a` (for "list all"). This option not only displays all of the contents of the directory(ies) but also two special directories indicated by `.` (the current directory) and `..` (the parent directory).

Instructions with multiple options can usually receive the options in any combination or order. To provide a long listing including hidden files, we can use `ls -al`, `ls -la`, `ls -a -l` or `ls -l -a`. While we can separate the options as shown in the last two cases, it is more common to combine them using one hyphen to save typing. The output from Figure 2.3 was produced using the instruction `ls -al /etc`. Thus, the output shows a long listing of all items, which is why we see the hidden items (., .., `.pwd.lock`). Had we issued the command `ls -l` without the `-a`, we would not see the entries for the current and parent directories or the `.pwd.lock` file.

We have now explored two of the commonly applied options for `ls`. There are dozens of others. Table 2.4 describes some of those that you might find useful. We have used the word *link* now both in the text and in Table 2.3. There are two types of links in Linux, hard links and symbolic links. We explore these in Chapter 3 (and later in Chapter 8) and omit any further detail for now.

Through options, we are able to take greater control of Linux by specifying more precisely how an instruction should execute. On the flipside of having greater control, understanding the options and remembering the great variety of them can be a challenge. Fortunately, we can view all of the options of a command easily through the command's *man page* (discussed in the next section) or by issuing the command followed by the option `--help` (not all Linux commands have this option).

TABLE 2.4
Useful Options for `ls` (Other Than `-a` and `-l`)

Option	Meaning
`-A`	Same as `-a` except that `.` and `..` are not shown.
`-B`	Ignore items whose names end with ~; the tilde indicates a backup file.
`-C`	List entries in columns; fits more items per screen should the directory contain many items.
`-d`	List directories by name but not their content.
`-F`	Append listings with item classifications; ends directory names with /, ends executable files with *; ends symbolic links with @.
`-g`	Same as `-l` except that owner is not shown.
`-G`	Same as `-l` except that group owner is not shown.
`-h`	When used with `-l` modifies file sizes to be "human readable".
`-i`	Include inode number; inodes are discussed in Chapter 8.
`-L`	Dereference links; that is, display information about item being linked to rather than the link itself.
`-r`	List items in reverse alphabetical order.
`-R`	Recursive listing (list contents of all subdirectories).
`-s`	When used with `-l`, outputs sizes in blocks rather than bytes.
`-S`	Sort files by size rather than name.
`-t`	Sort files by modification time rather than name.
`-X`	Sort files by extension name rather than name.
`-1`	(the number '1') List files one per line (do not use columns).

SECTION ACTIVITIES

1. Of the instructions listed in Tables 2.1 and 2.2, which do you feel will be the ones you use most often? Write them down. At the end of Chapter 2, reconsider this and see if your answer has changed. Do the same at the end of Chapter 3.
2. From a Bash prompt, type `man ls` and look at all of the options listed. Compare these to those listed in Table 2.4. Are there options not listed that should be? Are any of the options listed in Table 2.4 ones that you feel you will never use?
3. What is the benefit of having hidden files? Windows also has hidden objects. Research on the Internet "hidden files" to see why some files are purposefully hidden. Does this seem of value to you? Have you ever made hidden files visible in Windows? If so, why?

2.3 FORMS OF LINUX HELP

Using the Linux CLI will initially be a challenge for most users as most users are used to using a GUI for most or all of their interactions with a computer. Remembering names of commands is challenging enough but when we add the variety of options and parameters it makes command entry that much more complicated. Fortunately, Linux comes with several forms of support that we can take advantage of. The most used form is the command's man (manual) page. We explore that form of help in the first subsection. The second subsection offers other forms of help which may or may not be as useful.

2.3.1 `man` PAGES

The `man` command expects the name of a Linux command as its argument. The response of the `man` command is to display the *manual page* for that command, called a *man page* for short. The man page is displayed within the `vi` text editor (but without the ability to edit the man page). Movement and search commands available in `vi` can be used to move through the man page. We explore `vi` at the end of this chapter so for now we will concentrate on what the man page gives us and just a few basic movement commands.

The man page fills the entire terminal window. At the bottom of the window is a colon (:) which serves as a prompt. From this prompt, we can enter one of several movements or search commands.

The man page itself is divided into sections. The specific sections will vary based on the command although there are several sections that will always appear. We describe many of these sections in Table 2.5. Figure 2.4 shows excerpts of the `ls` command's man page.

As mentioned, the man page is opened in a `vi` editor but in a view/search-only mode. There are many convenient commands we can use to move through the man page. Most of these commands are single characters or a combination of a few characters. Some of the commands used to move around or search in `vi` are described in Table 2.6. Once you learn `vi` (see Section 2.6), you will find other commands that will work.

Most Linux commands have accompanying man pages (although applications software and some services do not). One group of commands is known as *Bash Built-ins*. These commands are built into Bash instead of being Linux commands; they include `alias`, `cd`, `exit`, `history` and `jobs`. The Bash built-in commands share the same man page. Consulting any of these commands' man pages displays all of the built-in commands with brief descriptions of each. As this man page presents dozens of commands, we may need to scroll down or search for the command we are interested in. Some of these Bash built-in commands also have their own man pages like `bash`, `kill` and `pwd`.

TABLE 2.5
Sections Found in man Pages

Field	Meaning	Examples
NAME	Command and its name.	`ls` – list directory contents `pwd` – print name of current/working directory
SYNOPSIS	Syntax (format) for the command; might contain multiple formats.	`ls [OPTION]... [FILE]...` `su [options] [-] [user [argument...]]`
DESCRIPTION	Explanation of the command; may include options or these may be in a separate section.	

OPTIONS	Complete list of options (if separate from DESCRIPTION section) along with their meaning and syntax/usage.	`-a, --all`
		do not ignore entries starting with .
		`-A, --almost-all`
		do not list implied . and ..

CONFIGURATION FILE	Many Linux commands/services have configuration files; this section describes the name/location of the file and valid entries; we explore services and configuration files in Chapter 9.	The `su` command lists files `/etc/defaults` and `/etc/login.defs` and configuration directives like `FAIL_DELAY`, and `ENV_PATH`

EXIT CODE	Some commands will exit with a code indicating success or failure.	Excerpts from `passwd` command:
		`0 success`
		`1 passwd/libuser operation failed`
		`2 unknown user`
		`252 unknown user name`

FILES	Files that this command uses or are related to the command.	The `su` command lists multiple files including `/etc/pam.d/su` and `/etc/default/su`
SEE ALSO	Other man pages (or websites) to consult.	The `su` command lists `setpriv`, `login.defs`, `shells`, `pam` and `runuser`
EXAMPLES	Some instructions are complicated enough to include example usage of the instruction.	The `tar` instruction used to have an example section but examples have been moved to the description section; the `ip` command includes a few examples.
AUTHOR, REPORTING BUGS, COPYRIGHT	Author name, email address to send bug reports to, the copyright of the instruction's executable code.	

2.3.2 Other Forms of Command-Line Help

Reading man pages is something of a learning process in itself. An introductory user might be overwhelmed by the content of even a short man page. Many of the commands have a separate *help page* available. If, in searching a man page for a command you find `-h/--help`, then you can view the command's help page in the future. The help page is often limited to one screen's worth of content, give or take a few lines.

For instance, `rmdir --help` displays the help page shown in Figure 2.5. `rmdir` is used to delete directories. The help page provides the synopsis (here, called `Usage`), a brief description, and common options. The help page may also display other content found in a man page such as exit codes, examples or other useful information.

```
LS(1)                          User Commands                          LS(1)

NAME
       ls - list directory contents

SYNOPSIS
       ls [OPTION]... [FILE]...

DESCRIPTION
       List  information  about  the  FILEs (the current directory by default).
       Sort entries alphabetically if none of -cftuvSUX nor  --sort  is  speci-
       fied.

       Mandatory arguments to long options are mandatory for short options too.

       -a, --all
              do not ignore entries starting with .

       -A, --almost-all
              do not list implied . and ..

       --author
              with -l, print the author of each file
```

```
   Exit status:
       0      if OK,

       1      if minor problems (e.g., cannot access subdirectory),

       2      if serious trouble (e.g., cannot access command-line argument).

AUTHOR
       Written by Richard M. Stallman and David MacKenzie.

REPORTING BUGS
       GNU coreutils online help: <https://www.gnu.org/software/coreutils/>
       Report ls translation bugs to <https://translationproject.org/team/>

COPYRIGHT
       Copyright © 2018 Free Software Foundation, Inc.  License GPLv3+: GNU GPL
       version 3 or later <https://gnu.org/licenses/gpl.html>.
       This is free software: you are  free  to  change  and  redistribute  it.
       There is NO WARRANTY, to the extent permitted by law.

SEE ALSO
       Full documentation at: <https://www.gnu.org/software/coreutils/ls>
       or available locally via: info '(coreutils) ls invocation'
 Manual page ls(1) line 216/240 99% (press h for help or q to quit)
```

FIGURE 2.4 Two excerpts from the `ls` command's `man` page.

If a command does not respond to `-h`/`--help`, another possibility is to add `--usage` to the command. Some commands will automatically display the usage if the command is entered incorrectly. This might be the case if, for instance, the wrong number of parameters are submitted or an illegal option is used. The `vmstat` command (display virtual memory information) does not require options but if options are included, `-b` is not one of them. Figure 2.6 shows the result of typing `vmstat -b`. Here, we are told that our command had an illegal option (`b`) and displays the correct usage of the command.

Yet another form of help is through the command `help`. The `help` command is used only for the built-in Bash commands. When used, it provides a help page which is often no longer than one page, like the `-h`/`--help` options. The `help` command includes an option, `-d`, to shorten the output to just the description of the command. Figure 2.7 provides the output for the command `help echo`. The output shows us a reduced set of information available in the man page. We get a usage/synopsis type statement and a description of some of the more useful or popular options, as well as

TABLE 2.6
`vi` Keystrokes That Work with `man`

Movement	Keystroke
Down one line	`<enter>`, `<down arrow>`, `e`, `j`
Down one page	`<space>`, `<page down>`, `f`, `z`
Up one line	`<up arrow>`, `k`
Up one screen	`<page up>`, `b`, `w`
Down half a page	`d`
Up half a page	`u`
Move to specific line	`#G` (# is a line number), e.g., `10G` moves to line 10
Move to top of man page	`1G`
Move to end of man page	`G`
Obtain help	`h`, `H`
Redisplay screen	`R`
Search forward	*/pattern* as in `/EXAMPLES`, additional / will move to next found instance of pattern
Search backward	*?pattern*, additional ? moves to the previous found instance
Exit	`q`

```
Usage: rmdir [OPTION]... DIRECTORY...
Remove the DIRECTORY(ies), if they are empty.

      --ignore-fail-on-non-empty
                  ignore each failure that is solely because a directory
                    is non-empty
  -p, --parents   remove DIRECTORY and its ancestors; e.g., 'rmdir -p a/b/c' is
                    similar to 'rmdir a/b/c a/b a'
  -v, --verbose   output a diagnostic for every directory processed
      --help     display this help and exit
      --version  output version information and exit

GNU coreutils online help: <http://www.gnu.org/software/coreutils/>
For complete documentation, run: info coreutils 'rmdir invocation'
```

FIGURE 2.5 The output of `rmdir --help`.

a brief exit status section. We also receive useful information on how to control the output of `echo` using escape characters.

To view just a command's description (as shown in the man page's NAME entry), use the command `whatis`. The command is provided one parameter, the name of the command in question. For instance, `whatis ls` responds with two lines of output, as shown below.

```
ls (1)   - list directory contents
ls (1p)  - list directory contents
```

The reason for two responses is that `ls` appears in two sets of man pages. `whatis` has its own set of options including `-w` to search for commands by specifying the command name with wildcards as in `ls` * to list all `ls` commands, and `-r` to permit regular expression characters when expression the command name. Both of these options allow us to search for a command whose name we may not remember exactly. We can also control which set of man pages are consulted (if our system has multiple sets available) using options `-s`, `-m` and `-M`.

The `man` command is essential to learning how to use Linux commands. But what if we want to learn about commands whose names we can't remember? Using `whatis` requires that we know

```
vmstat: invalid option -- 'b'

Usage:
 vmstat [options] [delay [count]]

Options:
 -a, --active             active/inactive memory
 -f, --forks              number of forks since boot
 -m, --slabs              slabinfo
 -n, --one-header         do not redisplay header
 -s, --stats              event counter statistics
 -d, --disk               disk statistics
 -D, --disk-sum           summarize disk statistics
 -p, --partition <dev>    partition specific statistics
 -S, --unit <char>        define display unit
 -w, --wide               wide output
 -t, --timestamp          show timestamp

 -h, --help     display this help and exit
 -V, --version  output version information and exit

For more details see vmstat(8).
```

FIGURE 2.6 Displaying proper usage of an illegally entered command.

```
echo: echo [-neE] [arg ...]
    Write arguments to the standard output.

    Display the ARGs on the standard output followed by a newline.

    Options:
      -n        do not append a newline
      -e        enable interpretation of the following backslash escapes
      -E        explicitly suppress interpretation of backslash escapes

    `echo' interprets the following backslash-escaped characters:
      \a        alert (bell)
      \b        backspace
      \c        suppress further output
      \e        escape character
      \f        form feed
      \n        new line
      \r        carriage return
      \t        horizontal tab
      \v        vertical tab
      \\        backslash
      \0nnn     the character whose ASCII code is NNN (octal).  NNN can be
        0 to 3 octal digits
      \xHH      the eight-bit character whose value is HH (hexadecimal).  HH
        can be one or two hex digits

    Exit Status:
    Returns success unless a write error occurs.
```

FIGURE 2.7 Example help page.

some part of the command name. We could also search all of the Linux commands by viewing the contents of the directories that contain the Linux commands (primarily `/usr/bin` and `/usr/sbin`). For instance, we might view the contents of `/usr/bin` to see if a command name looks familiar. As there are hundreds of such commands stored in these directories, this approach may be very time consuming and still may not yield the result we want.

Fortunately, there is another command available that lets us search for commands based not on their names but their descriptions. The command is `apropos`, and it expects a string as a parameter. It then searches through all of the man page descriptions for matches to this string. The string should be quoted for an exact match, otherwise keyword matching is performed against any of the words listed in the string specified.

Let's take a look at using `apropos` to find the instruction used to delete a directory (`rmdir`). We type `apropos directory`. Unfortunately, we receive a list of 121 responses! We next try `apropos delete` and receive a list of 41 commands. This is better. We try `apropos remove` and receive a list of 63 commands. Note that the number of commands will vary based on which specific version of Linux you are using.

Can we combine our keywords? Yes. We next try `apropos remove delete directory`. This leaves us with over 200 response! What went wrong? By listing multiple keywords, `apropos` searches all descriptions for *any* of the keywords. The three keywords are found together or separately in over 200 different instructions' man pages.

We use quote marks around our list of keywords to find an exact match for the group of keywords. However, as it must be an exact match, we may find no commands that fit our exact phrase. So, we try `apropos "remove directory"`. We get just two responses, `File::Path` and `rm`. Unfortunately, neither response is the correct command (`File::Path` is used to create or remote directory trees and `rm` deletes files).

With a little more effort we can locate the correct command. The notation .* is a regular expression that says "match anything". We want to use a search string of the form "remove"+*anything*+"directory". We use the command `apropos "remove .* directory"`. Now `apropos` returns all commands whose descriptions include the words "remove" and "directory", in that order, but possibly with additional words between them. This command returns three instructions, `remove`, `rmdir`, `unlink`. Success! We could also issue the command `apropos "remove .* directory" "delete .* directory"`. Here, the command searches for either set of strings and finds the same three instructions but also `mdeltree`. It also lists `rmdir` twice because both strings match part of `rmdir`'s description.

The `apropos` command takes some getting used to. And for now, we don't know enough about regular expressions to use it to its fullest. We examine regular expressions in Chapter 5. But before we leave `apropos` behind, let's consider one additional example. We want to issue a command to view the current virtual memory usage. We don't know the name of the command. We try `apropos virtual memory` and receive a list of 149 commands. We use `apropos "virtual memory"` and receive the output as shown in Figure 2.8. Fortunately, we are able to find the instruction we want (`vmstat`) without having to resort to regular expressions.

We will continue to explore Linux commands in every chapter of this text. Remember to use the various forms of help presented here, especially `man`. For the remainder of this chapter though, we move forward with more features of the Bash shell.

```
[rfox@localhost ~]$ apropos "virtual memory"
mremap (2)           - remap a virtual memory address
vfork (3p)           - create a new process; share virtual memory
vmstat (8)           - Report virtual memory statistics
```

FIGURE 2.8 Result of apropos command.

SECTION ACTIVITIES

1. In a Bash shell, type `man tar` and read the man page. What is `tar`? How many synopses are there? From just reading the man page, do you understand how to use the instruction? Now try the same with `man ls`. How much easier (if at all) is this man page to understand? Do you think the reason for the difference is that you have had some experience using `ls` but not `tar`, or is it that ls is just easier to use than `tar`?
2. Select three instructions introduced in this chapter (for instance, see those in Tables 2.1 and 2.2) and see if any have help pages or respond to the option `--help` as in `ls --help`. Which ones respond to `help` and which ones respond to `--help`?
3. Use `apropos` and try to identify Linux commands for each of these strings. For multi-word strings, try without quotes and then with quotes.
   a. mounting (or mount)
   b. domain socket
   c. user account

2.4 BASH FEATURES

As previously noted, aside from interpreting commands and permitting the execution of scripts, shells also provide the ability to define entities. Starting with `csh` in Unix, features were introduced into the shell that allowed the user to interact with the environment more easily. In this section, we cover features found in Bash. We do not cover them in the order introduced but instead in a relative order of usefulness.

2.4.1 Recalling Commands through the History List

Every time we enter a command, that command is stored in a *history list*. To recall the history list, type `history`. This will provide us a list of every command that has been entered. The history is limited in size, so as it grows, earlier commands are discarded. By default, the history list will store up to 1000 commands. An example of a short history list is shown in Figure 2.9.

The history list can be useful if we don't remember what we were doing or if we entered a command and want to re-execute it. Combined with command-line editing (a feature we explore in a few subsections), we can also recall a previous instruction, edit it and launch the revised version.

```
1  pwd
2  cd /home/zappaf
3  ls -al
4  cat file1.txt
5  cd ~
6  ls -l
7  vi script1
8  ./script1
9  rm script1
10 cd ~underwoodr
11 ls
12 history
```

FIGURE 2.9 Sample history list.

Notice each command in the history list in Figure 2.7 is preceded by a number. We can use this number to recall that instruction. The ways to recall an instruction are explained in Table 2.7 along with examples that reference the list from Figure 2.9. For short, the exclamation mark (!) is often referred to as "bang" so for instance "bang 9" (written as `!9`) recalls the `rm script1` instruction.

When using ! to recall an instruction from the history list, the instruction is re-executed. Another role of the history list is to allow us to modify an instruction on the history list. For this, we must recall it on the command line. We review command line editing in a few subsections. The approaches to recall and print instructions on the history list are shown in Table 2.8.

The `history` command has a number of options available. The command `history` by itself lists the entire history list up to the limit set by the environment variable `HISTSIZE` (which defaults to 1000). The options are shown in Table 2.9. Examples in the third column refer back to the sample history list in Figure 2.9.

2.4.2 Shell Variables

A shell is not just an environment to submit commands with a viewable and recallable history. The shell remembers items defined in script files and at the command line. Specifically, a shell can contain variables, aliases and functions. In this and the next subsection, we explore variables and aliases. We hold off on functions until we look at shell scripting in Chapter 6.

A *variable* is merely a name that references something stored in memory. Variables are used in nearly every programming language. Variables, aside from referencing memory, have associated with them a *type*. The type dictates what can be stored in that memory location and what kind of operations can be applied to the value stored there. Bash variables can only store strings but strings that are storing integer numbers can be treated as integers with special syntax. Floating point values are treated as strings only.

TABLE 2.7
Recalling Instructions from the History List

Method	Meaning	Examples Using Figure 2.9
`!!`	Recall last instruction.	Would re-execute `history`.
`!`*n*	Recall instruction *n* from the history list.	`!7` to recall `vi script1` `!6` to recall `ls -l`.
`!`*str*	Recall the latest instruction that uniquely starts with *str*.	`!c` to recall instruction 10. `!ca` to recall instruction 4. `!ls -a` to recall instruction 3. `!l`, `!ls` both recall instruction 11`!ls -` recalls instruction 6.
`control+s` *str* `control+r` *str*	Search forward/backward through the history list for *str*; each `control+r` moves us further back to the previous *str*; *str* does not have to start the command.	`control+r s` – recalls history. `control+r` – recalls `ls` `control+r` – recalls `rm script1`. `control+r` – recalls `./script1`. `control+r` – recalls `vi script1`. `control+r` – recalls `ls -l`.
command `!^`	Issue command but use the first parameter/option from the last instruction.	After instruction 3, typing `ls -l !^` recalls `ls -a file1.txt`.
command `!$`	Issue command but use the last parameter/option from the last instruction.	After instruction 3, typing `ls -l !$` recalls `ls -l file1.txt`.
command `!*`	Issue command but use all of the parameters/options from the last instruction.	After instruction 3, typing `ls -l !$` recalls `ls -al file1.txt`.

TABLE 2.8
Recalling a Previous Instruction to Edit or Display

Keystroke(s)	Explanation
`<up arrow>` or `control+p`	Retrieve last instruction; issuing it multiple times steps backward through the history list.
`<down arrow>` or `control+n`	Move forward to the next instruction in the history list (only useful after having moved backward through the history list with up arrow/`control+p`).
`!!:p`	Output last instruction from history.
`!:`*n*	Output instruction *n* from the history list.
`!:`*str*	Output last instruction which starts with *str* from the history list.
`control+<`	Move to beginning of history list.
`control+>`	Move to end of history list.

TABLE 2.9
Options for the `history` Command

Option	Meaning	Example
`-c`	Clear the history list.	
`-d` *n*	Delete entry *n* from the list, causes the remainder of the list to "move up".	`history -d 10` causes `ls` to become #10 and `history` to become #11 with `history -d 10` becoming the latest on the history list.
`-r` *filename* `-w` *filename* `-a` *filename* `-n` *filename*	Load the history list saved in *filename* and use it in place of current history list; save history list to *filename*; append current history to *filename*; read the lines in the history file that are not in the current history list and place them in the current history.	
`-p` *arg* [*arg*...]	Display the corresponding item from the history list but do not execute it; multiple *args* are allowable, separated by spaces.	`history -p "!4"` will display `cat file1.txt`

`-s` *arg* [*arg*...]	Save the commands, denoted by *arg(s)*, from the history list as a single entry.	`history -s "!7" "!8" "!9"` adds a new entry consisting of
		`vi script1`
		`./script1`
		`rm script1`

Bash variables can be defined from the command line and then used later in the same shell session or from within a script and used throughout that script. In order to extend the variable's usage beyond the shell or script which defined it, we need to *export* the variable. We explore this at the end of this subsection. We generally refer to variables defined and exported from shell scripts as *environment variables* because they are variables that are available for us to use in the environment.

To view the environment variables currently defined, use the `env` instruction. There could be dozens or more environment variables. In Red Hat 8, there are 45 variables defined by default in a Bash shell while Debian 11 defines 40 and Ubuntu 20 defines 48. We do not look at them all but some of the more significant ones are described in Table 2.10.

Aside from `env`, we can view environment variable values using the `echo` instruction. The instruction is `echo` *`string`* where *string* is output. To output the value of a variable, precede the variable name with a `$` as in `$USER` or `$OLDPWD`. The instruction `echo $USER $SHELL $PWD $OLDPWD $MAIL` outputs all five of the variables' values, each item separated by a space but output on a single line.

TABLE 2.10
Some of the Bash Environment Variables

Environment Variable	Usage	Example Value
DESKTOP_ SESSION	Name of desktop GUI.	gnome
HISTSIZE	Size of history list.	1000
LANG	The specified language and character encoding used by this user.	en_US.UTF-8
MAIL	Location of user's email storage.	/var/spool/mail/foxr
OLDPWD	Current working directory prior to this one; see PWD.	/home/zappaf
PATH	List of directories to search to find executable program.	/home/zappaf/bin: /usr/bin: /usr/sbin: /usr/local/bin:...
PS1	Defines the user's prompt.	[\u@\h \W]\$
PWD	Current working directory; content of this variable is moved to OLDPWD when changing directory.	/home/foxr
SHELL	User's default shell.	/bin/bash
USER	Displays the user's name based on who they currently are, this is the response when using whoami.	zappaf
USERNAME	Same as USER except that if you su to another user, USERNAME stays as your original value.	zappaf

Notice that the environment variables from Table 2.10, and those output by the env command, are all capitalized. As Linux variables are case sensitive, if we refer to a variable without proper capitalization, we will not get the value expected. A variable that has no value outputs nothing. For instance, echo $user $shell will merely output a blank line.

To store a value into variable, we use an *assignment statement*. The form of assignment statement in Bash is *VAR=VALUE* with no blank spaces around the equal sign. We may define our own variable or redefine an existing one. We do not want to change an environment variable unless we clearly know what we are doing. If we are defining our own variable, the name can consist of letters, digits and underscores (_) as long as it begins with a letter or underscore.

The *VALUE* on the right-hand side (RHS) of the assignment statement can comprise a number of different things such as literal values and values stored in variables (we explore other types of items that can appear on the RHS in Chapter 6). But there is a restriction. If the RHS has any blank spaces in it, we must embed the entire RHS in quote marks. We can use single or double quote marks although we generally use double quote marks as explained later in this subsection.

One of the most significant environment variables is PATH. This variable defines the directories that the Bash interpreter will search when looking for a command. Recall in Linux that commands are implemented as programs. These programs are distributed throughout various directories, but most are found in either /usr/bin or /usr/sbin. The reason we have the PATH variable is to save us from having to either change directory to the location of the program's executable before running it or referencing the directory when entering the command. Thus, we can type ls rather than /usr/bin/ls.

Without the PATH variable, we would not only have to type more to enter a command but also remember the location of the various commands. Instead, Bash searches every directory in our PATH variable, one at a time, for the command we have issued. In Red Hat 8, a user's PATH is preset through various startup scripts.

As an example, user zappaf might have for his PATH variable the following.

```
/home/zappaf/.local/bin:/home/zappaf/bin:/usr/local/bin:
/usr/bin:/usr/sbin
```

The entries here are of two directories within zappaf's home directory named .local/bin and bin followed by /usr/local/bin, /usr/bin and finally /usr/sbin. The notation .local indicates that this directory is hidden. If a directory does not exist, we do not receive an error but instead the Bash interpreter will move on to the next directory in the PATH variable.

Another environment variable worth knowing is PS1. This variable defines the user's prompt. If you refer back to Table 2.10 you will find a cryptic value stored there. The *char* notation indicates a type of information that the Bash interpreter will fill in to create the prompt. For instance, \u is the user's username and \h is the hostname of the computer. Anything that does not appear with a \ is treated literally so part of our prompt includes the characters [and] plus a blank space. Table 2.11 shows what other values can be stored in the PS1 variable that are used by Bash to control the prompt's output.

We can redefine our PS1 variable should we prefer a different prompt. Consider changing it using the assignment statement PS1="[\t - \!]\$". Instead of the prompt as we saw throughout the chapter, it might now look like [21:49 - 1042]$. In this case, we see the time is 21:49 (9:49 pm), and the next instruction we enter will be placed in the history list at number 1042. There are also options for changing the color of the prompt and formatting the time and date. We place the RHS of the above assignment statement in quote marks because it contains blank spaces.

Aside from PS1, there are likely environment variables to define other prompts. PS2, if defined, is the prompt to use when a command is so long that it needs to be continued on another line. PS3 will be the prompt to use when inside a select loop. PS4 defines the prompt to use when in debug mode. Finally, PROMPT_COMMAND, if set to non-null, executes the specified command immediately before evaluating PS1.

Let's focus more attention on both assignment statements and output statements. The top portion of Figure 2.10 shows two assignment statements whereby we set the variables FIRST and LAST. The two echo instructions then are used to output those variables' values.

The two echo statements in Figure 2.10 output six things, each on a single line. The first echo statement outputs Hello, a space, the value stored in FIRST, a space, the value stored in LAST and an exclamation mark. What would happen if we omit the $? The second echo statement in Figure 2.10 shows us the result. We get the same output except that instead of outputting the values

TABLE 2.11
Other Usable Characters for Our PS Variables

Character	Meaning	Character	Meaning
\d	Date.	\v	Bash version.
\h	Hostname up to the first period (.).	\V	Bash version+patch version.
\H	Full hostname.	\w	Current working directory.
\j	Number of jobs in current shell	\W	Last part of current working directory (e.g., foxr instead of /home/foxr).
\s	Name of shell.	\!	History list number of this instruction.
\t	Time (24-hour format).	\#	Command number of this instruction.
\T	Time (12-hour format).	\$	Dollar sign.
\@	Time using am/pm.	*nnn*	ASCII character equal to the octal value of *nnn*.
\u	Username.	\n, \\	Newline, backslash.

```
[foxr@localhost ~]$ FIRST=Frank
[foxr@localhost ~]$ LAST=Zappa
[foxr@localhost ~]$ echo Hello $FIRST $LAST!
Hello Frank Zappa!
[foxr@localhost ~]$ echo Hello FIRST LAST!
Hello FIRST LAST!
```

FIGURE 2.10 Sample assignment and echo statements.

```
[foxr@localhost ~]$ echo $FIRST
Frank
[foxr@localhost ~]$ echo $LAST
Zappa
[foxr@localhost ~]$ echo -n $FIRST
Frank[foxr@localhost ~]$ echo -n $LAST
Zappa[foxr@localhost ~]$
```

FIGURE 2.11 Additional examples of echo output.

stored in FIRST and LAST, we literally get the strings FIRST and LAST. The $ is required to recall the value stored in a variable, otherwise echo treats the item as a string.

Let's look at another interaction, shown in Figure 2.11. Assuming FIRST and LAST store Frank and Zappa, respectively, the first two echo statements output their values. Although echo does not have many options, one is -n. With this option we inform echo not to output a new line at the end of the output. We see the result of this after the third and fourth echo statements. Notice how the user's prompt appears immediately after each of these outputs.

The output shown in Figure 2.11 shows why we might not want to use -n as it tends to provide hard-to-read output. We will likely only use echo -n from within shell scripts when we want several echo statements to have their outputs appear on the same line. As we would not see a prompt appear after the output (unless the script ended), it would not look messy. Two other options available with echo are -e and -E. These options enable and disable backslash escapes respectively. We explore the role of backslashes shortly.

Let's return to the assignment statement. Assuming FIRST and LAST have the values of Frank and Last, respectively, we can use these in other assignment statements. Let's create FULL_NAME, a variable that will store $FIRST and $LAST with a blank space between them. This is shown at the top of Figure 2.12. Unlike the two assignment statements at the top of Figure 2.10, neither of which provided output, here we receive a cryptic error message! The output says bash: Zappa: command not found.

Why did the Bash interpreter think we wanted to execute an instruction called Zappa? The difference between this assignment statement and the ones from Figure 12.10 is the blank space on the RHS. The Bash interpreter reaches the blank space found between $FIRST $LAST and thinks that $LAST is in fact a part of a different operation. Thus, the interpreter does two things.

```
[foxr@localhost ~]$ FULL_NAME=$FIRST $LAST
bash:  Zappa:  command not found...

[foxr@localhost ~]$ $FULL_NAME='$FIRST $LAST'
[foxr@localhost ~]$ echo $FULL_NAME
$FIRST $LAST

[foxr@localhost ~]$ $FULL_NAME="$FIRST $LAST"
[foxr@localhost ~]$ echo $FULL_NAME
Frank Zappa
```

FIGURE 2.12 Using single versus double quote marks.

First, it assigns to FULL_NAME the value stored in FIRST. The interpreter then attempts to execute $LAST. As LAST is a variable storing Zappa, $LAST is Zappa, so the interpreter tries to execute Zappa as if it were an instruction. As Zappa is not an instruction, we receive the error.

To resolve the problem, we must tell Bash that the two items on the RHS of the assignment statement are to be treated as one item. We do so by placing quote marks around the entire RHS. Interestingly, we get a different behavior depending on whether we use single quote marks, '$FIRST $LAST', or double quote marks, "$FIRST $LAST".

We see two more interactions in Figure 2.12. In the middle of the figure, we use single quote marks around the RHS of the assignment statement. When we output $FULL_NAME, we get $FIRST $LAST instead of the values stored there. The single quote marks inform the Bash interpreter to use the items literally and not apply the $ to retrieve the values from the variables.

We try again using double quotes. With double quotes, the Bash interpreter interprets whatever appears in the quote marks so that $FIRST and $LAST provide the values they store. So, FULL_NAME is assigned the value stored in $FIRST, a space, and the value stored in $LAST.

Consider the two pairs of instructions in Figure 2.13. The first and second pairs are nearly identical, only differing in the types of quote marks used. Both assignment statements result in the same value stored in X and thus the same output. This is because there are no items in the quote marks for Bash to interpret.

Let's consider another example. Assume FIRST=Frank and LAST=pwd. Here, we are storing the name of a Linux command in a variable. FULL_NAME=$FIRST $LAST does not yield an error. Instead, this command assigns FULL_NAME to store the value stored in FIRST and then the Bash interpreter executes what is stored in LAST, and since LAST stores the value pwd, the Bash interpreter executes pwd. We receive output from this assignment statement which is the output of the pwd command.

Let's return to echo to look at the -e option. Before we do so, we motivate why we might want to use -e. We have stored in the variable FULL_NAME a user's full name and in the variable AMOUNT some numeric amount. We enter the instruction echo $USERNAME owes $$AMOUNT. The notation $$AMOUNT is meant to convey "output a $ and then output the value in the variable AMOUNT". The notation is odd in that there are two dollar signs together, but it is syntactically correct. The output we receive, however, is not what we expect! The output is Frank Zappa owes 4896AMOUNT. What happened?

Unfortunately, $$ has its own unique meaning so that Bash interprets this in an unexpected way. Instead of outputting a $ followed by the value in AMOUNT, we get the process ID of the running Bash interpreter. We have to "trick" the Bash interpreter into not treating $$ to mean "return the PID". How? Here, we can apply the *escape character*, \.

The escape character is used to tell the Bash interpreter to treat the character that follows literally and not to interpret it. We specify \$$AMOUNT to indicate that the first $ is meant to be treated as a dollar sign. The second $ tells Bash to return the value stored in the variable AMOUNT rather than the word AMOUNT. In order for echo to treat \$ as expected, we add the option -e. The revised instruction is echo -e $USERNAME owes \$$AMOUNT. Assuming AMOUNT stores 123, we get output of Frank Zappa owes $123.

```
[foxr@localhost ~]$ X="Hello Linux User!"
[foxr@localhost ~]$ echo $X
Hello Linux User!

[foxr@localhost ~]$ X='Hello Linux User!'
[foxr@localhost ~]$ echo $X
Hello Linux User!
```

FIGURE 2.13 Single versus double quotes often will not change behavior.

TABLE 2.12
Other Escape Characters of Note

Escape Character	Meaning
`\\`	Output a \
`\b`	Backspace (back cursor up 1 position)
`\n`	Newline (start a new line)
`\t`	Tab
`\!  \$  \&  \;  \'  \"`	!, $, &, ;, ' and " respectively

Notice that we could have avoided the problem caused by `$$` by adding a space between the two dollar signs as in `echo $ $AMOUNT`. As the first dollar sign clearly is not modifying the second dollar sign, the Bash interpreter handles this as we would expect.

There are several other escape characters of note that we might want to use in our `echo` statements. Some of these are shown in Table 2.12. The most used of these are to insert a new line (`\n`) and a tab (`\t`) although there may be times when we need to output one of the special characters like a backslash, exclamation mark, dollar sign or other.

2.4.3 Aliases

The history list provides one type of command line shortcut. Another very useful shortcut is the alias. The *alias* is used to define an instruction in a shortened form. As with a variable, an alias is defined using an assignment of a name to a value, but in this case the value is a Linux command. We can then have that command executed by issuing the alias rather than the command itself. This saves us time.

To define an alias, use an assignment statement preceded by the word `alias`. The instruction then expects an assignment statement of the form *`name=command`* where *name* is the alias and *command* is the Linux command. The command can but does not have to include options and arguments.

The names we use for our aliases are not as restrictive as with variables since just about any set of characters can be used. If a command on the RHS includes any blank spaces, the command must be enclosed in single or double quote marks. The difference between the quote marks is the same as with assignment statements, so, for the most part, we will want to use double quote marks. It is a good habit to always use the quote marks whether there are spaces in the command or not.

Aliases are used by users for several reasons. Long commands can be shortened to save typing. Complicated instructions, once defined in an alias, do not have to be remembered. Common typos can be "repaired" through aliases so that if we have a command that we commonly mistype, the alias can be used in its place. Some commands are dangerous and by redefining those commands using an alias, we can define them to operate in safer modes. To illustrate each of these ideas, we look at some examples shown in Figure 2.14.

```
alias ..="cd .."
--------------------------------------------
alias ~="cd ~"
alias ~="cd /home/zappaf"
--------------------------------------------
alias mountcd="mount /dev/cdrom /cd iso9660
     ro,user,noauto"
--------------------------------------------
alias lss=less
alias sl=ls
--------------------------------------------
alias rm='rm -i'
```

FIGURE 2.14 Example `alias` statements.

To reduce the amount of typing, if there is a common instruction that a user issues often, an alias can be defined. For instance, we see the command `cd ..` from Figure 2.14 to change to the parent directory (we talk about this in Chapter 3). It is less typing to enter `..` in place of `cd ..`. A related alias is to define ~ for `cd ~` so that we can type ~ to return to our home directory. Figure 2.14 shows two versions of this alias, one where we show the full path to the directory (`/home/zappaf`) and one where we use the shortcut ~ notation. We make the assumption that the user is zappaf.

The `mount` command shown in Figure 2.14 mounts the optical drive. It is a command that might be hard to remember and certainly a lot to type. Figure 2.14 shows how we can define the alias `mountcd` so that we do not have to recall or enter the lengthier command. Notice that the command includes a number of options: `ro` (read-only), `user` (user is allowed to mount the drive) and `noauto` (do not automatically mount at boot time). Should we use `mountcd`, it is applied as defined, with all of these options in place. Note that because of a lack of space, this command is shown in the figure on two lines but would be entered on a single line.

At this point, you may not be very comfortable with command-line input. As you become more familiar, chances are you will start typing commands at a faster pace which will invariably lead to typos. Simple typos might include misspelling a command such as by accidentally typing `sl` for `ls` or `lss` for `less`. If you find yourself with frequent typos, you might create aliases whereby each of those typos is converted into the instruction you meant. We see these two examples as the next two entries in Figure 2.14. These two aliases did not require quote marks because there are no blank spaces on the RHS of the assignment.

Let's assume the user has defined `sl` as an alias. How does this work? The user types `sl ~` intending to perform `ls ~` (i.e., a listing of their home directory). The Bash interpreter first identifies that `sl` is an alias and replaces it with `ls`. Now the instruction is `ls ~`. Next, the interpreter replaces ~ with the user's home directory (the value stored in `$HOME`). Assuming the user is zappaf, the instruction is now `ls /home/zappaf`. The instruction is then executed. Thus, the alias may not be a complete instruction but can be a portion of an instruction that we can add options and parameters to once the alias has been replaced.

Dangerous commands include those that manipulate the file system without first asking the user for confirmation. As an example, the command `rm *` will remove all files in the current directory. The `rm` (remove/delete) command has two modes: force mode (the default) and interactive mode. In interactive mode, the command pauses before any deletion to ask the user for permission.

Out of carelessness, we might delete something we want, so it's safer to always use the `-i` option. However, if we are in a hurry, we might not think about it or we might be too lazy to add the `-i` option. The last alias in Figure 2.14 defines the alias `rm` to be `rm -i` so that any use of the `rm` instruction uses the interactive mode. Entering the instruction `rm *` becomes `rm -i *` as the `rm` alias is replaced by the command `rm -i`.

As with environment variables, some aliases are predefined for us in various startup scripts. We can view all already defined aliases by just typing `alias` by itself. We can remove any already defined alias by using `unalias` as in `unalias sl` to remove the alias for `sl`.

2.4.4 Command-Line Editing

To support the user in entering commands, the Bash interpreter accepts a number of special keystrokes that, when entered move the cursor or copy, cut or paste portions of whatever is already on the command line. We give this feature the name *command-line editing*.

The keystrokes are based on keystrokes used in the `emacs` text editor. Learning these keystrokes helps you learn `emacs` (or vice versa, learning `emacs` first will help you perform command-line editing). The keystrokes are combinations of the control key or the escape key and another key on the keyboard. We have already seen two such keystrokes as `control+p` and `control+n` are used to move through the history list (as mentioned in Table 2.8).

Table 2.13 shows many of the keystrokes available. In the table, `control+key` is indicated as `c+key` and `escape+key` is indicated as `m+key`. The use of `m` for escape is because Unix users often referred to the escape key as "meta". The control and escape keys are used differently in that the control key is held down and then the other key is pressed, while the escape key is pressed first (and released) followed by the other key. So, for instance, `c+p` means "press the control key and while holding it type p" while `m+p` means "type escape and then type p". The table does not include keystrokes that we already know such as the backspace key, delete key, space bar and arrow keys.

Let's consider an example. Our Linux system has four users named dukeg, marst, underwoodi and underwoodr. Each of these users has a home directory under `/home`. We wish to view the contents of each directory as a long listing. The command we would issue would be `ls -l /home/username` as in `ls -l /home/dukeg`. Rather than entering four separate commands from scratch, we will use command line editing to recall and edit the previous instruction. We enter the first command, `ls -l /home/dukeg` <enter>, in its entirety This displays the contents of dukeg's directory.

To view the next user's home directory, marst, we recall the last instruction and use keystrokes as described in Table 2.14. Remember that `c+key` stands for `control+key` and `m+key` stands for `<escape>` followed by *key*. Having performed the steps in the upper part of the table, we repeat the steps except that we type `underwoodi` instead of `marst`. For the last user, after `control+p`, the steps to change `underwoodi` to `underwoodr` are shown in the bottom part of the table (the last three instructions in table 2.14).

TABLE 2.13
Command-Line Editing Keystrokes

Keystroke	Meaning
c+a	Move cursor to beginning of line.
c+e	Move cursor to end of line.
c+n (also up arrow)	Move to next instruction in history list.
c+p (also down arrow)	Move to previous instruction in history list.
c+f (also right arrow)	Move cursor one character to the right.
c+b (also left arrow)	Move cursor one character to the left.
c+d (also delete key)	Delete character at cursor.
c+k	Delete (kill) all characters from cursor to end of line.
c+u	Delete everything from the command line.
c+w	Delete all characters from cursor to previous white space.
c+y	Yank, return all deleted characters (aside from c+d) to cursor position.
m+y	Rotate the kill ring (those previously deleted) and yank the new top; only permissible if last command was c+y or m+y.
c+_	Undo last keystroke (you can continue to use this to undo multiple keystrokes).
m+f	Move cursor to space after current word.
m+b	Move cursor to beginning of current word.
m+d	Delete the remainder of the word.
m+del (delete key)	Delete from cursor to beginning of word (does not include white space making this different from c+d).
m+u	Uppercase the word where the cursor is.
m+l	Lowercase the word where the cursor is.
m+c	Capitalize the letter where the cursor is
c+t	Transpose the character at the cursor position and the character after that.
c+l (lower case "L")	Keep the command on the command line but clear the screen.

TABLE 2.14
Keystrokes to Edit Example Command

Keystroke	Explanation
`c+p`	Move back one instruction in the history list; now the command line has the previous command, placing `ls -l /home/dukeg` on the command line.
`m+b`	Move back to the beginning of the current word; the cursor is now at the `d` in `ls -l /home/dukeg`.
`c+k`	Kills (deletes) the rest of the line leaving `ls -l /home/` on the command line.
`marst`	The instruction now appears as `ls -l /home/marst`.
`<enter>`	Causes the instruction to be entered and executed.
`<backspace>`	Backup over the `i` after `underwood` leaving `ls -l /home/underwood`.
`r`	The command is now `ls -l /home/underwoodr`.
`<enter>`	Causes the instruction to be entered and executed.

Note that we could view all of the directories with a single instruction such as `ls -l /home/dukeg /home/marst /home/underwoodi /home/underwoodr` <enter> but this is nearly as much typing as entering four separate instructions. With command-line editing, after typing the first instruction, we only needed to enter a few keystrokes and their usernames. We will see another approach we could have used near the end of this section. We will visit another example of command-line editing in Chapter 3.

2.4.5 Redirection

In Linux, there are three predefined locations that Linux commands can interact with. These are standard input (abbreviated as STDIN), standard output (STDOUT) and standard error (where to send error messages, abbreviated as STDERR). Typically, a Linux command will receive its input from STDIN and send its output to STDOUT. By default, STDIN is the keyboard, and both STDOUT and STDERR are the terminal window in which the instruction was entered.

These three locations are predefined file descriptors. A *file descriptor* is a means for representing a file as it is used in a running process. In Linux, file descriptors are denoted as integer numbers with 0, 1 and 2 reserved for STDIN, STDOUT and STDERR, respectively. A *redirection* is an alteration from one or more of STDIN, STDOUT and STDERR to other locations. Bash provides us with five redirection operators so that the specified Linux command receives input from and/or sends output to a location of our own specification.

There are two forms of output redirection, two forms of input redirection and a form to link the output of one command to be the input of another command. Output redirection sends output to a file rather than to the terminal window from which the command was issued. One form writes to the specified file while the other appends to the file. The difference is that if the file previously existed, the first approach overwrites the file. If the file doesn't already exist, writing to and appending to the file do the same thing.

One form of input redirection accepts input from a text file while the other redirects input from keyboard. This latter approach uses the previously mentioned *here document* in that Bash drops us into an editor to enter the input.

The last category of redirection is known as a *pipe*. The Linux pipe sends the output of one program to be used as the input to another. This allows us to chain commands together without having to use temporary files as an intermediate.

To specify redirection, the syntax is typically `command operator item`. The *item* varies based on the redirection/operator. Output redirection operators are > to write to and >> to append to. The *item* is a file name.

Input redirection uses either the operator < to input from file where *item* is a file, or the operator << to input from keyboard. In this case, *item* is either omitted or is a string. Redirection from keyboard causes Bash to drop us into the here document. If *item* is omitted then the Bash interpreter accepts input until we type `control+d`. If *item* is a string, then the interpreter accepts input until we enter the string on a line by itself.

The pipe operator is |, and *item* is another command. Table 2.15 provides numerous examples along with explanations. Note that `cat` is the concatenate command. It is used to combine multiple files together and output the result.

Figure 2.15 provides an example of keyboard input using the `cat` command. Using << drops us into the here document where the prompt is simply > (which is the prompt defined by our `PS2` variable). We now enter content including the `<enter>` key. Because we included the string `done` in our command, input continues until we type `done <enter>` on a line by itself. At that point, the entire set of input is sent to the `cat` command, which simply outputs the result. Notice that `done` does not appear in the output.

TABLE 2.15
Redirection Examples

Command	Explanation	Comments
`cat *.txt > newfile.txt`	Combine the contents of all files in the current directory whose file names end with `.txt` and store this combined set of text to the new file `newfile.txt`.	This overwrites `newfile.txt` if it already exists; files are concatenated with no extra blank lines to separate their content.
`cat *.txt >> newfile.txt`	Same as above but appends to `newfile.txt`.	Same result if the file didn't already exist but if `newfile.txt` exists, the new content is added at the end of the file.
`cat < *.txt`	Send all of the files whose names end with `.txt` to the `cat` command to be output.	As `cat` accepts files as parameters, this instruction is no different from `cat *.txt`; we discuss this further in the text.
`cat << quit`	Accept all input from the keyboard until the user types `quit <enter>`, sending that input to `cat`.	Displays whatever we have typed in after typing `quit <enter>`.
`cat *.txt | sort`	Take the contents of all of the files whose names end with `.txt` but before displaying them with `cat`, `sort` the content.	Effectively sorts all of the content rather than outputting the files line by line.

```
[foxr@localhost ~]$  cat << done
> pear
> banana
> cherry
> strawberry
> done
pear
banana
cherry
strawberry
[foxr@localhost ~]$
```

FIGURE 2.15 Example of redirection from the here document.

In looking at the example in Figure 2.15, the content that we entered is merely displayed back to the same terminal window. We might wish to instead redirect the output of cat to a file to save the entered content. We can do so by adding another redirection operator to our command. In this case, we want to send the output of cat to a file that we will call fruit. Our command becomes cat << done > fruit. This notation is more complex but perfectly acceptable.

If we had specified redirection using << without done, the input would end with control+d. But doing so, in this case, generates an error. The reason for the error is that Bash can't handle the notation << > (or << >>), as in cat << > fruit.

We can see in the above example that we can use multiple redirection operators in a single instruction. We can similarly use pipes with <<, >, >>, and we can use multiple pipes in one instruction. The instruction cat file1.txt file2.txt file3.txt | sort > newfile.txt takes the contents of the three .txt files and sends them to newfile.txt but sorts the output first, so that the lines in the new file appear in alphabetical order.

One last comment to end this subsection. The < redirection operator indicates that we are redirecting STDIN to come from a file. Most Linux instructions receive input from file by default. We noted that the third entry in Table 2.15, cat < *.txt does the same thing as cat *.txt. There are not many occasions where we will want to use < with Linux commands. However, there is a value with < which is to send input from a file to a script. The input instruction is read, which by default receives input from the keyboard. Using < forces our scripts to input from file instead of keyboard whenever we have a read instruction. We visit the use of < with scripts in Chapter 6.

As we progress through this textbook and learn more and more Linux instructions, we will find more situations where redirection is useful. At times in fact, we may need to use multiple pipes, possibly combined with > or >>. But to wrap up this subsection, we look at a set of simple examples. Assume we want to save the output of our ls -l commands from earlier where we applied command-line editing (see Table 2.14) to a file called user_entries.txt. Figure 2.16 demonstrates this. Notice that the first command uses > while the remaining three use >>. If we were using command-line editing, we would have to not only change the username but also change > to >> when we modify the first instruction into the second. The third and fourth instructions would not require a change to the redirection operator as it will have already been changed.

All five forms of redirection can be useful, but the pipe is probably the one you will use the most. We will hold off on further examples of the pipe until we reach Chapter 3.

2.4.6 Other Useful Bash Features

We wrap up this section by looking at a few additional features that make command-line entry easier. We have already mentioned two of these, the tilde (~) and the *. The use of ~ is known as *tilde expansion*. The Bash interpreter replaces ~ with our home directory as defined in the environment variable $HOME. Alternatively, ~*user* is replaced by *user*'s home directory.

We use tilde expansion to shorten instructions. Most user directories are stored under /home. For instance, zappaf's directory is /home/zappaf. By using ~, we shorten any reference by eliminating /home/*username* (where *username* would be our username). By using ~*username*, for someone else's username, we eliminate /home/. Thus, cd ~ takes us to our home directory, doing the same thing as cd /home/*username*. cd ~zappaf takes us to /home/zappaf.

```
[foxr@localhost ~]$ ls -l /home/dukeg > user_entries.txt
[foxr@localhost ~]$ ls -l /home/marst >> user_entries.txt
[foxr@localhost ~]$ ls -l /home/underwoodi >> user_entries.txt
[foxr@localhost ~]$ ls -l /home/underwoodr >> user_entries.txt
```

FIGURE 2.16 Using > and >>.

When we viewed example redirection commands in Table 2.15, we used the *. This is one form of *wildcard* operator available in Bash. There are several wildcard characters available that denote different groupings of characters. With * the wildcard is replaced by everything found. Others can replace single characters.

The use of a wildcard is known as *filename expansion*, or informally is sometimes called *globbing*. Although `ls -l *.txt` looks like we are saying "perform a long listing on all items in the current directory whose names end with `.txt`", in reality the Bash interpreter operates differently. Before executing `ls -l`, the Bash interpreter literally replaces `*.txt` with all files in the current directory that match this notation. If this directory contains `bar.txt`, `file1.txt`, `file2.txt` and `somefile.txt`, then `ls -l *.txt` is expanded into the instruction `ls -l bar.txt file1.txt file2.txt somefile.txt`. Thus, the instruction expands from the version with a wildcard to the full list of matching files (or directories). There are several different wildcards available but for now we only consider *. We view the others in Chapter 3.

The other forms of shortcut are new, some being more useful than others. One that we might find useful when we have a lot of files or directories to reference in an instruction is called *brace expansion*. We can group files or subdirectories that share a path in common and list the different files or subdirectories in curly brackets ({, }), as in `/directory/{item1,item2}`. In this case, both `item1` and `item2` are to be found in the directory named `directory`. This notation would expand to `/directory/item1 /directory/item2`.

Let's look at a specific and lengthier example. Earlier, we used command-line editing to view the contents of several user directories. Let's assume that aside from those four users' directories, `marst` has a subdirectory called `stuff` and `underwoodr` has two subdirectories named `music` and `lyrics`. We want to view the contents of all of these directories using `ls -l`.

Normally, we would issue a single instruction rather than separate commands like we had previously, but the single command involves a lot of typing. With brace expansion, it reduces the typing. Figure 2.17 shows this revised version of the command along with the list of directories that will be output. The command should appear on one line but does not fit in the width so it has been divided into two lines. The notation `.` as found in the braces (twice) indicates the home directory for marst and the home directory for underwoodr.

Last, but certainly not least, is a feature called *tab completion*. Introduced in `csh` as escape completion, this feature allows the user to enter a partial file or directory name, press <`tab`> and the Bash interpreter will attempt to complete the name. The portion of the name entered must uniquely identify the item. If not, the interpreter responds with a beep. Pressing <`tab`> twice causes the interpreter to list all of the matches so that the user can see what is available and type more characters.

Let's consider an example. We are currently in the directory `/etc`. We want to view the file `/home/underwoodr/music/chords/inca_roads2.txt` using the `cat` command. Let's further assume that one other user, underwoodi, exists (we choose this name because it is similar). underwoodr has two subdirectories called `misc` and `music`, and under `music` is a subdirectory called `chords` which itself contains two files named `inca_roads1.txt` and `inca_roads2.txt`.

```
[foxr@localhost ~]$ ls -l /home/{dukeg,marst/
   {.,stuff},underwoodi,underwoodr/{.,lyrics,music}}

/home/dukeg
/home/marst
/home/mars/stuff
/home/underwoodi
/home/underwoodr
/home/underwoodr/lyrics
/home/underwoodr/music
```

FIGURE 2.17 Using brace expansion.

We type cat /h<tab>. As home is the only directory at the top-level that starts with an h, the Bash interpreter completes this to cat /home/. We type u<tab>, but this cannot be completed fully. Instead, the Bash interpreter completes as much as possible and emits a beep, giving us cat /home/underwood on the command line. We type r<tab>, and this is completed to cat /home/underwoodr/.

We continue by typing m<tab> and get a beep again indicating a non-unique name. We type <tab> twice in a row and Bash responds, on a separate line, with misc/ music/ while the command line prompt keeps us at the end of cat /home/underwoodr/m. We type u<tab> and now the command is completed to cat /home/underwoodr/music/. We type c<tab> and this completes to cat /home/underwoodr/music/chords/.

We type i<tab> and two things happen. First, the command is completed to cat /home/underwoodr/music/chords/inca_roads and we get a beep. Again, the Bash interpreter was not able to complete the full file name because it is not unique but completed as much as it could. We type <tab> twice and find two matching files: inca_roads1.txt inca_roads2.txt. We type 2<tab> and the command becomes cat /home/underwoodr/music/chords/inca_roads2.txt. We press <enter> and the command executes.

Tab completion, as seen in this example, can complete directory names and filenames. If there is only one match, it is completed. If only part of the file/directory name can be completed, it is, but then we receive a beep. If there are multiple matches, pressing <tab> twice will result in one of three outputs. If the number of matches is small, it will display them all. If the number of matches is sufficiently large that it will scroll beyond one screen's worth, we are shown all that will fit in one screen and prompted to display more. If there are more than a few dozen responses, we are queried as to whether we want to display them all. For instance, issuing cat /usr/lib/systemd/system/s<tab><tab> causes the Bash interpreter to respond with a query like Display all 131 possibilities? (y or n).

Tab completion might be the best of the Bash features in terms of saving typing. Try to remember to use it because it is easy to use. Other features like command-line editing and redirection will take practice.

SECTION ACTIVITIES

1. Between history, defining/using variables, defining/using aliases, redirection, command-line editing, wildcards, brace expansion, tab completion and tilde expansion, which do you think you will use the most often and the least often? After using Linux for a few months, return to this question and see if in fact you were right.
2. In a Bash shell, type history. Experiment with recalling several instructions from your history list using !*n*, !*str* and !!. Do not use !! if your last instruction was history because that recalls history and the result looks almost identical.
3. Using history combined with command-line editing gives the user the ability to issue instructions and then change them or correct mistakes. How useful would this feature be if the keystrokes were easy to remember? What can you to do help remember not only what the keystrokes are but to use them? The author has to remind his students frequently to use command-line editing. It is a feature that most students either ignore or forget about! Tab completion is also often forgotten.
4. List two instructions that you have learned to this point that you feel you might want to define an alias for. Try to define such an alias and use it.

2.5 TAILORING OUR ENVIRONMENT

A shell contains an environment. This environment consists of all of the entities defined during this session. Unfortunately, ending the session or starting a new session means that those entities are no longer defined. Typing `bash` from within a session opens a new session. From a Bash session, typing `exit` closes the session. We are either returned to the previous session (if one existed), or the window closes.

Consider the user interaction in Figure 2.18. The figure demonstrates what happens to items defined as we move from one session to another. When we start a new shell from within a shell, we refer to the new shell as a *subshell* (we can also refer to the shells as *nested shells*). In effect, we now have two Bash shell sessions running although only the subshell is currently active. We will refer to the two shell sessions in the figure as the "Outer session" and the "Inner session".

In Figure 2.18, we start in a session labeled "Outer session", and define `X` to store `1`. The `echo` statement confirms its value. The command `bash` starts a new Bash session. We now have two sessions. The active session is labeled "Inner session". We output `X` and find it has no value. Why not? `X` has not been defined in this session. We assign `X` to `2` and output it. The `exit` instruction causes "Inner session" to end. The previous Bash session ("Outer session") resumes. In this session, `X` had been defined as `1` and still exists. This session's version of `X` was not visible in the inner session but is again visible because we have left the inner session.

Why would we want to have multiple sessions? The above example does not illustrate the value of the subshell. Imagine that we are in the middle of an important task when something else pops up on us. We want to tackle that new task without interfering with our current shell's history list, aliases or variables. We start a subshell, take care of the new task, exit the new shell and resume without having introduced anything new in the first, outer shell.

In modern windows-based operating systems, we could easily open a new terminal window which has its own shell. This would allow us to accomplish the new task without impacting the other shell. But if we are working in a text-based environment only, we would have to create a subshell.

Can we have an item defined in one shell persist in subshells? Yes, by using the `export` built-in Bash command. The `export` command typically is followed by one or more variables that we want to export. We can also export functions using the option `-f`. We can remove the `export` property from an item using the option `-n`. The option `-p` displays all exported items. Figure 2.19 is similar to Figure 2.18 except that we demonstrate exporting `X` from a shell to subshells.

The session in 2.19 is more complicated so let's step through it, line by line. First, we set `X` to `1`. We then create a subshell and output `X`. As the outer session's `X` was not exported, the subshell has no value of `X`. We `exit` the subshell and in the outer shell we export `X`. Now `X` (storing 1) will be accessible in a subshell.

```
[zappaf@localhost ~]$  X=1              }  Outer session
[zappaf@localhost ~]$  echo $X
1
[zappaf@localhost ~]$  bash             }
[zappaf@localhost ~]$  echo $X
                                           Inner session
[zappaf@localhost ~]$  X=2
[zappaf@localhost ~]$  echo $X
2
[zappaf@localhost ~]$  exit
exit
[zappaf@localhost ~]$  echo $X          }  Outer session resumed
1
```

FIGURE 2.18 Creating and exiting an inner session.

```
[zappaf@localhost ~]$  X=1
[zappaf@localhost ~]$  bash
[zappaf@localhost ~]$  echo $X

[zappaf@localhost ~]$  exit
exit
[zappaf@localhost ~]$  export X
[zappaf@localhost ~]$  bash
[zappaf@localhost ~]$  echo $X
1
[zappaf@localhost ~]$  X=2
[zappaf@localhost ~]$  bash
[zappaf@localhost ~]$  echo $X
2
[zappaf@localhost ~]$  exit
exit
[zappaf@localhost ~]$  export -n X
[zappaf@localhost ~]$  bash
[zappaf@localhost ~]$  echo $X

[zappaf@localhost ~]$  exit
exit
[zappaf@localhost ~]$  echo $X
2
[zappaf@localhost ~]$  exit
exit
[zappaf@localhost ~]$  echo $X
1
```

FIGURE 2.19 Using `export`.

We start a new subshell and output `X` and get the value `1`, as expected. Next, we set `X` to `2`. Notice that we now have two values of `X`. In the outer shell, `X` is still `1`. `X` as a variable is still being exported. We type `bash` to enter yet another subshell. We output `X` and get the value `2`. We `exit` out of this subshell to resume the other subshell. Now, we alter `X`'s property to not be exported. We enter another subshell and output `X` and get no value. We exit this inner subshell and the resumed subshell still has `X=2`. We `exit` this subshell and resume the outer shell where `X` is `1`.

Although we might find a use in defining our own variables, it is the environment variables that we want to persist. Typing `export -p` displays all of the exported variables. You might find dozens of such variables defined, including those we explored earlier in the chapter like `USER`, `USERNAME`, `PATH`, `PWD` and `OLDPWD`. As noted earlier, we can export both variables and functions. We cannot export aliases however. If we want an alias to persist across all of our shell sessions, we have to define it differently. This leads us to shell scripts.

A *script* is a file that contains instructions to be executed. With Bash, there is a complete language available so that scripts are not limited to Linux commands but can include input, output, loops, selection statements and so forth. Bash scripts can also include Linux instructions, including `alias` instructions (the `alias` instruction is actually a Bash built-in).

There are several significant scripts already available to users and the system administrator. As a user, we might want to update one or more of these scripts to tailor our environment. We might add alias definitions, variables that we export, functions that we export or just code that we want to run. As a system administrator, we would use these scripts so that user environments are established using variables, aliases and functions that we want available.

There are four script files of note, two for users and two for system administrators. These are `/etc/profile`, `/etc/bashrc`, `~/.bash_profile` and `~/.bashrc`. The two in `/etc` are available for system administrators to modify, and the two in our home directory (~) are for us, the user, to modify. The two profile files execute for both interactive and non-interactive shell sessions. A non-interactive shell session is one where a shell is needed to run a script but where there is no user interaction. When we log in to Linux and open a window, we are running an interactive shell. In such a case, both the two profile scripts and the two bashrc scripts will execute.

Let's step through these four files and see what each does and the order that they run. The first to execute is /etc/profile. This script first defines a function called pathmunge. It is this function that is responsible for constructing the PATH variable. The script then begins defining a number of environment variables including USER, LOGNAME, MAIL and possibly EUID or UID. With the user's ID (EUID/UID) established, pathmunge is called to begin building the PATH variable. The HOSTNAME, HISTSIZE and HISTCONTROL environment variables are defined. All defined environment variables are exported. A umask instruction is executed (we study umask in Chapter 7).

The script continues by executing all of the scripts stored in the directory /etc/profile.d. There are potentially dozens of files in this directory that include both .sh (Bash) startup scripts and .csh (C-shell) startup scripts. For instance, there might be lang.sh and lang.csh both of which establish the LANG environment variable, and less.sh and less.csh which establish the LESSOPEN environment variable.

The /etc/bashrc file sets up further environment variables including PS1 and PROMPT_COMMAND. It also adds to the PATH variable begun in /etc/profile. It has some overlapping code as /etc/profile in case /etc/profile did not execute. As a system administrator, we would want to place system-wide functions and aliases in this file.

Now, the user's ~/.bash_profile executes. This file, by default, has only one executable instruction, an if-statement which tests whether the user has a .bashrc file in the user's home directory, and if so, runs it. Users are free to edit their own .bashrc file, adding their own instructions to it.

The ~/.bashrc file tests to see if /etc/bashrc exists and if so, executes it (although by default it will have already been executed by /etc/profile). The ~/.bashrc file concludes by adding local bin directories ($HOME/.local/bin and $HOME/bin) to the PATH variable established by /etc/profile and /etc/bashrc, and then exports the revised PATH variable. At the bottom of the ~./bashrc file is a comment whereby we can add our own aliases and functions.

A system administrator might wish to define an alias for rm to always invoke rm -i. This alias would be placed in /etc/bashrc so that it would be defined by all users. A user might want to define aliases for such commands as cd .., cd ~ and ls -al and place these in their ~/.bashrc file.

One additional file of note is ~/.bash_logout. This file is invoked when a bash session closes. A user might define code here to run before any session ends to clean up the user's environment. As an example, the user might define a function in this or one of the other scripts to locate and delete empty files. The bash_logout script then can contain the instruction to invoke this function.

Changing a script does not cause the change to be executed. For instance, if we edit ~/.bashrc and add an alias, that alias is not currently defined. Instead, it would become available only when we opened a new Bash session (either opening a new window, creating a subshell or logging out and later logging in). However, we can have a script execute by issuing the source command. Having modified ~/.bashrc, we can execute source ~/.bashrc. This causes the interpreter to read the file, executing the statements therein.

SECTION ACTIVITIES

1. List three aliases you might want to define in your .bashrc file.
2. You are currently working on something important and don't want to interrupt the flow of instructions but want to experiment with a set of Linux commands. Which do you think you would do? Try to justify your answer in terms of both your own convenience and what might happen if you make a mistake.

- open a subshell/inner shell to experiment with something and then exit to resume your work
- open a new window to experiment with and close the window when done
- experiment in your current shell

3. Explore the four files discussed in this section (`/etc/bashrc`, `/etc/profile`, `~/.bashrc` and `~/.bash_profile`). You can view each using the `cat` command (or the command `less`, which pauses at the end of each window's worth of output). How much of these files' contents do you understand?

Note that for users who are using other shells there are similar files such as `/etc/cshrc` and `~/.cshrc` (for C-shell users). If such scripts are not available, the system administrator may wish to write them.

2.6 vi

`vi`, or the newer `vim` (`vi` improved), is the default text editor found in Linux. Users will typically use this text editor to create text files (including shell scripts), and system administrators will typically use `vi` to create and edit scripts and modify configuration files. `vi` can be a challenge to learn. As with command-line editing, the only way to truly master `vi` is through practice. In this section, we take a look at `vi`.

2.6.1 vi Commands

Much of `vi` involves understanding the keystrokes used as commands to move the cursor and perform editing tasks. Some users prefer `emacs`. Both can be challenging to learn. We cover `vi` here. You can explore `emacs` through one of the online supplemental readings.

The first thing to know about `vi` is that the interface operates in one of three modes. The main mode is known as the *command* mode. In this mode, keystrokes entered operate as commands rather than characters entered into the document being edited. For instance, typing `G` moves the cursor to the last line of the document and typing `k` moves the cursor up one line rather than entering the letter `G` or `k` into the document.

The other two modes are insert and replace. In *insert*, any characters entered are placed at the current cursor position while in *replace* any character entered replaces the character at the current cursor position. To move from insert or replace mode back to command mode, type the escape (`<esc>`) key. Keystrokes are roughly divided into one of five categories: mode commands, cursor movement commands, editing commands, file commands and miscellany.

For *mode* commands, we already saw that `<esc>` is used to move from insert/replace mode to command mode. Other mode commands move from command mode into one of insert or replace. To enter insert mode, use one of `I`, `i`, `a`, `A`, `O` or `o`. Each of these places the cursor in a different location. Respectively, these place the cursor at the beginning of the line, at the current cursor position, immediately after the current cursor position, at the end of the current line, in a blank line immediately above the cursor or in a blank line immediately below the cursor. These insert-mode commands are all illustrated in Figure 2.20.

To enter replace mode, there are two commands, `r` and `R`. With `r`, we enter replace mode to enter one character only and as soon as we have entered the new character (which replaces the characters at the cursor position), we immediately move back to command mode. With this mode, we do not type `<esc>` to return to command mode. With `R`, we move to replace mode and stay there. As we type, anything we enter replaces (overwrites) what is currently in the document. We remain in replace mode until we press `<esc>`.

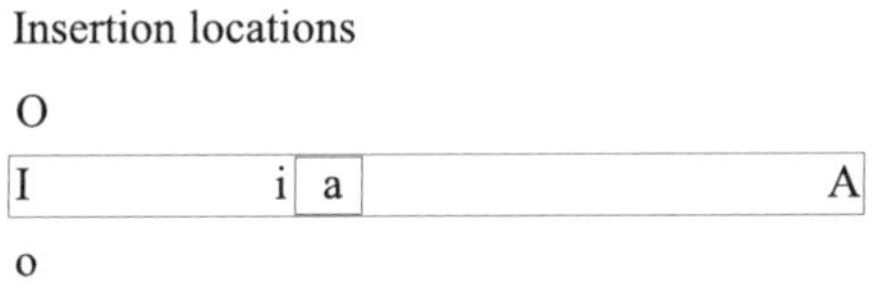

Insertion locations

O

I	i	a	A

o

FIGURE 2.20 `vi` insertion locations.

The next set of commands define *cursor movement*. Many of the keystrokes used for moving about in `vi` are the same as what we saw earlier to move about a man page. We can move one character up, left, down and right using `k`, `h`, `j` and `l` (lower case "L"), respectively. Keep in mind that the arrow keys will likely also work but in some cases may not be mapped correctly so it's important to learn these four commands.

We can move to the beginning or end of the current line using `0` and `$`, respectively. We can move forward or backward by one word using `W` and `B` while `w` and `b` move us forward or backward to the next/previous word or punctuation mark. For instance, if we are at the beginning of a hyphenated word, `W` takes us to the start of the next word while `w` takes us to the hyphen. Similarly, `E` and `e` take us to the end of the current word or to the next punctuation mark.

We can move to line *n* with *n*`G` where *n* is a number such as `1G` (line 1) or `10G` (line 10). Typing `G` by itself moves us to the last line of the document. We can move up or down through half and whole screens. These commands are `control+u` and `control+f` to go back or forward by one full screen, and `control+b` and `control+d` to go forward or backward by one-half screen. The keystrokes `H`, `M` and `L` move the cursor to the top, middle and bottom of the current screen, respectively. All of these cursor movement commands are summarized in Figure 2.21.

Editing commands allow us to cut, copy and paste. There are numerous forms of cut (deletion) commands. `dd` deletes the current line, `dw` the current word, `d)` deletes all content from the cursor to the end of the sentence, `D` deletes to the end of the line, `db` deletes the previous word and `x` deletes the current character. We can specify an integer number for `dd` and `dw` to delete multiple consecutive lines or words. With `dd`, it appears as *n*`dd` as in `5dd`, and with `dw`, it appears as `d`*n*`w` as in `d5w`. With `dd` and `dw`, whatever is deleted is moved to a buffer so that it can be recalled.

If we have moved an item into the buffer, it can be pasted elsewhere. Pasting is handled through either `P` or `p` to paste the contents of the buffer before the cursor and after the cursor respectively. To copy content, use `yy` or `yw` to copy the current line or current word to the buffer.

If we do not specify otherwise, the general buffer is used for any cut or copy action. Adding a letter `a` to `z` lets us select one of 26 additional buffers. For instance, `a6yy` copies the six lines from the cursor forward into buffer `a`, and after moving the cursor, typing `ap` pastes those six lines after the cursor.

There are a few other editing commands. `J` joins two consecutive lines into one. `xp` transposes the current character with the next character. `u` undoes the last change made while `U` undoes all changes made to the current line.

File commands start with a colon (:). To save the file, use `:w`. To perform save as (change the file name), use `:w` *`filename`* <enter>. To open a new file, use `:r` *`filename`* <enter>. To exit, use `:q` but if the file has been changed, use either `:w` followed by `:q`, or `:wq` combined (to save and exit), while `:q!` discards any changes to the file before exiting.

The last category is the vaguely named *miscellany*. These commands did not fit within another category but mostly have nothing to do with each other. We can mark a location in the document and then move there later. To mark a location, use `m`*`char`*, where *char* indicates a label for the mark, as in `ma`. To move to a marked location, use `'`*`char`* as in `'a`. We can mark numerous locations in a document for easy movement.

We can search forward or backward for a string, a regular expression or a single character. To search forward from the cursor for *string*, use `/`*`string`* <enter> and to search backward, use `?`*`string`* <enter>. Once an instance of *string* has been found, we can continue to the search

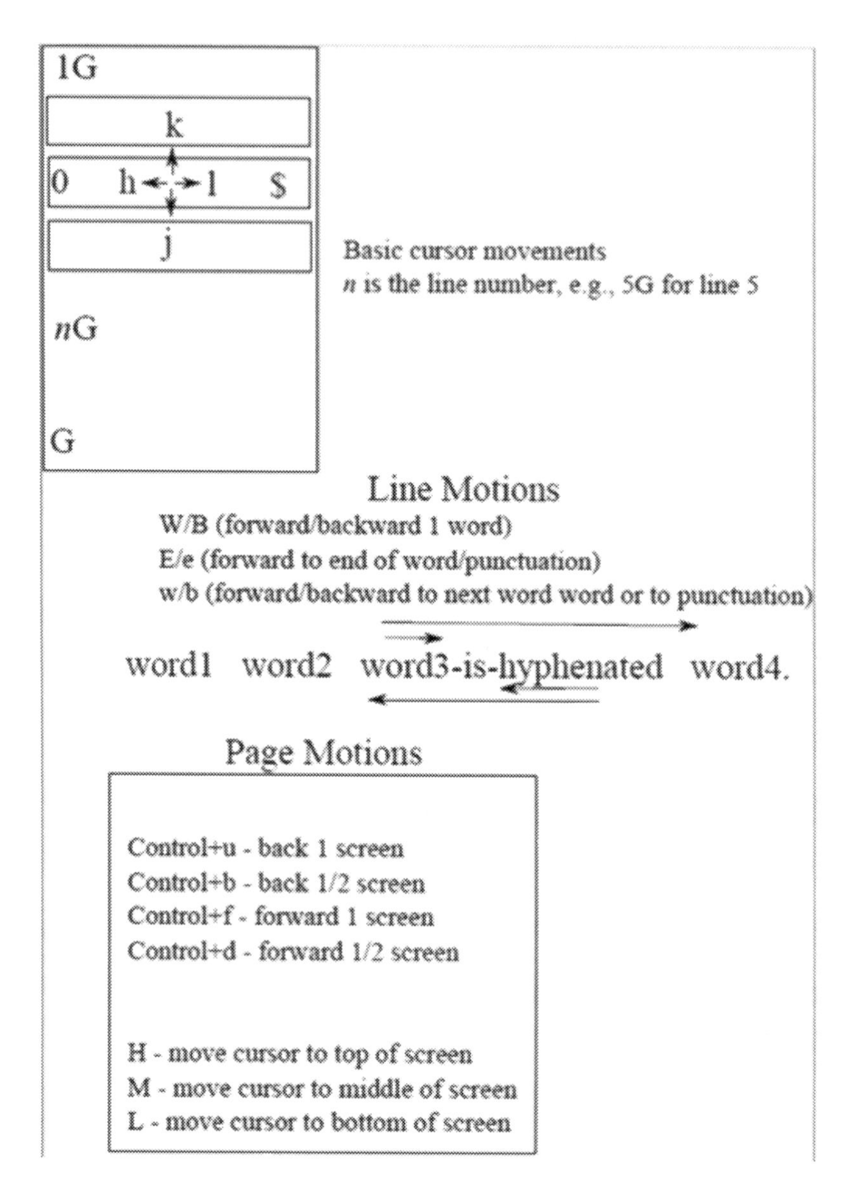

FIGURE 2.21 `vi` cursor movement commands.

forward using `/<enter>` and backward using `?<enter>`. The string may be a literal string or contain a regular expression.

To search for a single character, use `f`*`char`*, `F`*`char`*, `t`*`char`* and `T`*`char`* where `f` and `t` search forward for the character *char*, positioning the cursor on the character or immediately before the character respectively, and `F` and `T` search backward on the line for the character *char*, positioning the cursor on the character or immediately before the character respectively. The command `;` is used to repeat the last `f`, `F`, `t` or `T` command.

Another form of search is to find the next (or previous) parenthesis. If the cursor is resting on an open or close paren, the `%` key moves the cursor to the corresponding close or open paren. For instance, if we have the text `the value of x (heretofore unknown) is not a concern` and the cursor is on `)`, pressing `%` will move it to the `(`.

2.6.2 An Example to Illustrate How to Use vi

Let's put all this together and see how to create, edit and save a document. This little exercise will help you get started in understanding vi even though only practice will help you memorize the many keystrokes of vi. If you are going to be a common vi user, download the vi cheatsheet which is part of the supplemental readings on the textbook's website.

We start from the command line by typing vi<enter>. Notice we have not yet named the file. We could also type vi *filename*<enter> which starts vi in a newly created file named *filename*.

Type i to enter insert mode. Type the following two lines, pressing <enter> and <esc> where indicated.

```
Information is not knowledge,<enter>
knowledge is not music.<esc>
```

Upon pressing <esc>, we return to command mode. Type 1G and the cursor move to the start of the document. Move down one row with j (or down arrow). Type o to enter insert mode with a new line below the cursor. Now type Wisdom is not truth,<esc> in that blank line We now have three lines in our file.

We realize we have made a mistake on the previous line. Type k to move up to the previous line and 0 to move to the beginning of the line. The k in knowledge should be uppercased. Type rK. The r puts us in single character replace mode and K replaces k. We are back in command mode. Type 3w. The 3 indicates that we want to do the w command three times. With each w, we move forward to the next word. This positions the cursor at the beginning of "music.". This should be changed into "wisdom,". Type R and type wisdom,<esc>. We have successfully replaced the wrong word with the correct word. Notice that we entered full replace mode here because music has five letters while wisdom has six.

Type G to move to the end of the file and type o to again enter insert mode in a blank line. Now type the additional lines shown below.

```
Love is not music,<enter>
Truth is not beauty,<enter>
Beauty is not love,<enter>
Music is the best!<esc>
```

If you read through these lines, you might notice a pattern that each line ends with a word that the next line starts with, except that there is a mistake. The line starting with Love follows the line ending with beauty. We need to move the line starting with Love to appear below the line starting with Beauty. As we are currently at the bottom of the file, we want to move back three lines. Do so typing 4G. The cursor can appear anywhere in the line starting with Love. Type dd to delete the entire line. The cursor automatically appears in the following line, which starts with Truth. We want to move one more line down, so type j (or 6G). To paste the cut line below this current line, type p.

Note that we could also delete the two lines starting with Truth and Beauty and paste them above the line starting with Love. To accomplish this, we would position the cursor anywhere in the fifth line (5G) and type 2dd to delete two lines. We would move the cursor up one line (k) and then type P.

The text is complete. These seven lines are spoken words which make up a small part of the Frank Zappa song Packard Goose. The spoken word part continues but this is the only part we will reference. Let's save our work to a file. Because vi was invoked without a file name, the contents are currently unsaved. Using :w <enter> will cause vi to respond with No file name error message and the file would remain unsaved. Instead, type :w goose <enter>. Type :q to quit.

To further your understanding of vi, we will continue with this example. Type vi <enter> to restart vi. Once in vi, type :r goose <enter>. This step could have been avoided by starting vi with vi goose <enter> but by using :r you can experience another command.

Move the cursor to the end of the file using G$ (G for last line, $ for end of line) and then back up to the beginning of the current word using b. Fully capitalize best to BEST. This can be accomplished in several ways. We can replace each character one at a time using rBrErSrT (each r enters replace-one-character mode) but this is tedious. It is easier to enter full replace mode and then type over the four characters, using RBEST<esc>.

Another way to fully uppercase this word is to use the *uppercase* command. Here's how we will accomplish this. Mark the text to change using a mode called *visual*. While on the b type v and then move to the right using l (lower case L) until the cursor is over t. Now, to capitalize the selected text, type gU. The cursor moves back to the beginning of the marked area having uppercased the entire marked set of characters.

We will make one additional change to the document. Although song lyrics lines often end with commas, each new line has started with a capital letter. Replace all of the commas with periods. The search and replace operation is based on the same syntax as a program called sed, which we will learn about in Chapter 5. The syntax is quite strange. The command, in its entirety, is :%s/,/./ <enter>.

Let's break down this command. The : gives us access to a prompt to enter the rest of the command. %s indicates a search and replace command. Between the slashes, we put the string to search for and the string to replace it with. In this case, both strings are individual characters: a comma and a period. Issue this above command and all of the commas will be replaced with periods.

Notice that u is used to undo the last change. But in this case, if we type u it undoes the entire search and replace change. Why? Because this was the last command we entered even though it impacted many lines. If you typed u, you can redo the last command undone by typing control+r. Type :wq to save and exit vi.

SECTION ACTIVITIES

1. Open vi and follow the instructions covered to create and modify the goose file. When done, reflect on how challenging it was to use vi. How successful will you be in remembering the vi keystrokes?
2. Compare using vi to a WYSIWYG (what you see is what you get, or GUI-based) word processor. Under what circumstance might you use vi instead of a WYSIWYG? Note that there are WYSIWYG word processors available in Linux such as LibreOffice Writer.
3. Download the supplemental online document on emacs and compare the keystrokes to control emacs to those of vi. Which seem easier? Notice that the command-line editing keystrokes all come from emacs. Learning command-line editing will help you learn emacs should you choose to use emacs instead of or in addition to vi.

2.7 INTERPRETERS

We wrap up this chapter by considering the interpreter. As described in Chapter 1, an interpreter is a program whose task is to execute instructions. To execute an instruction, the instruction must be converted into an executable form. In Linux, this means parsing the instruction entered on the command line into its component parts and then executing the program that the command is implemented by on the arguments of the rest of the instruction.

For instance, `rm *.txt` executes the `rm` program on the list of files supplied as parameters. Before executing `rm`, the Bash interpreter checks to see if `rm` is aliased. Assuming it is, then the alias `rm` is replaced with the instruction, which is likely `rm -i`. Next, the Bash interpreter performs filename expansion, exchanging `*.txt` for the list of all `.txt` files in the current directory. Now, the instruction can be executed. In this section, we look at interpreters in general before focusing on the Bash interpreter.

2.7.1 Interpreters in Programming Languages

We generally categorize high-level programming languages as interpreted and compiled languages. A compiled language, which includes languages like C, C++, COBOL, FORTRAN, Ada, Java and Pascal, requires that the program written in that language be translated *all at once* from its source code into the executable code of machine language. In most of these languages, the entire program must be completely written before translation can take place. C, C++ and FORTRAN are exceptions in that pieces of code can be compiled separately so that some parts of a program are already compiled when we compile the remainder.

The *compiler* is a program which performs this translation process, converting an entire program into an executable program. The compiled approach to programming is commonly used for large-scale applications as will be described below.

The interpreted approach to programming utilizes an environment in which the programmer enters instructions one at a time. Each instruction, upon being entered, is parsed, converted into executable code and executed. The parsing, converting and executing steps are all handled by the *interpreter*. This approach to programming greatly differs from the compiled approach because it allows the programmer to experiment with instructions while writing a program. Thus, a program can be written one instruction at a time. This certainly has advantages for the programmer who might wish to test out instructions without having to write an entire program around those instructions.

Having to have a full program written to compile it might be thought of as a disadvantage when writing programs. Another disadvantage with compiled programs is that the high-level language program must be compiled to run on a specific platform. If, for instance, we compile a C program for a Windows computer then that executable program would not run in a Linux environment nor could it run on an IBM mainframe. Instead, the programmer would separately compile the C program for Windows, for Linux, for IBM mainframe and so forth. Each compilation results in an executable program that can run on the one platform it was compiled for.

While the translation time is roughly the same between compiling any single instruction and interpreting any single instruction, we desire to have applications compiled before we make them available to users. The reason for this is that the entire translation process can be time consuming (seconds, minutes, even hours for enormous programs). If a large program is being interpreted, then the user is sitting through this translation process rather than seeing the program execute immediately, as would be the case if the program were already compiled. So, we compile the program and release the executable code.

From a programmer's standpoint, the compiler will find syntax errors for us so that we can correct them. If a program with syntax errors was given to a user to run in an interpreted environment, the user would see the errors and not be able to run the program, perhaps growing frustrated. Of course, most interpreted programs are thoroughly tested before being released but in general we prefer to have our software compiled.

There are many situations where interpreting can come in handy. These situations primarily occur when the amount of code to be executed is small enough that the time to interpret the code is not noticeable. This happens, for instance, with scripts that are executed on a webserver (server-side scripting) to produce a web page dynamically or in a web browser to provide some type of user interaction. As the scripts only contain a few to a few dozen instructions, the time to interpret and execute the code is not a concern.

There are numerous interpreted programming languages available although, historically, most programming languages have been compiled. One of the earliest interpreted languages was LISP, a language developed for artificial intelligence research. It was decided early on that a LISP compiler could be developed allowing large programs to be compiled and thus executed efficiently after compilation. This makes LISP both an interpreted language (the default) and a compiled language (when the programmer uses the compiler) giving the programmer the ability to both experiment with the language while developing a program and compile the program to produce efficient code.

In the 1990s, a new scheme was developed whereby a program could be partially compiled into a format called byte code. The *byte code* is a platform-independent but low-level code that can be quickly interpreted. Having compiled a program into byte code, one would run it using an interpreter.

The advantage of the byte code approach is that the byte code does not need to be compiled for every platform it is to run on. Instead, byte code being platform-independent only requires compiling it once. And because byte code is low-level, the time it takes to interpret (translate and execute) is only slightly more than the time it takes to execute already compiled code.

Java was the first programming language to use this approach where the interpreter is built into software known as the Java Virtual Machine (JVM). Nearly all web browsers contain a JVM so that Java code can run inside the web browser. Notice that JVM contains the term "virtual machine". The idea behind the JVM is that it is a VM capable of running the Java environment, unlike the more generic usage of VM mentioned in Chapter 1. More recent languages like Python and Ruby are interpreted but can be compiled into a byte code format for quicker execution.

2.7.2 Interpreters in Shells

When it comes to operating system usage, using an interpreter does not accrue the same penalty of slowness as with programming languages. There are several reasons for this. First, the user is using the operating system in an interactive way so that the user needs to see the result of one operation before moving on to the next. The result of one operation might influence the user's next instruction. Thus, having a pre-compiled list of steps would make little or no sense. Writing instructions to be executed in order will only work if the user is absolutely sure of the instructions to be executed.

Additionally, as the user is entering commands from the command line, it will probably take the user longer to enter a command than it will for the interpreter to translate and execute the command. As the interpreter runs in a shell which is text-based, it means the operating system is spending more of its time running text-based programs than GUI-based programs. GUI programs are nearly always going to be more resource intensive than text-based programs. Thus, the command entered can be interpreted and executed in less time than it would take the operating system to launch the equivalent GUI-based program.

The other primary advantage to having an interpreter available in the operating system is that it gives the user the ability to write their own interpreted programs or scripts. A script, at its most primitive, is merely a list of operating system commands that the interpreter is to execute one at a time. Most scripting languages today contain instructions similar to those found in any high-level language. For instance, the Bash scripting language has input and output statements, assignment statements, loops, selection statements, arrays, functions and function calls. Thus, scripting languages are as, or nearly as, powerful as compiled languages. This gives the user a lot more flexibility when writing scripts to automate activities or perform applications.

The other advantage to interpreted code when interacting with the operating system is that the interpreter runs within an environment. Within that environment, the user can define entities that

- Input is broken into individual tokens
  - a token is a known symbol (e.g., <<, |, ~, *, !, etc) or
  - a word separated by spaces where the word is a command or
  - a word separated by spaces where the word is a filename, directory name or a combination
  - an option combining a hyphen (usually) with one or more characters (e.g., `-l`, `-al`).
- Quotes (if any appear) are handled.
- Any alias is replaced the RHS of the alias definition.
- All of the commands, keywords, and operators are broken into individual commands (if there are multiple instructions separated by semicolons).
- Brace expansion unfolds into individually listed items.
- ~ is replaced by the appropriate home directory.
- `$Variables` are replaced by their values.
- Commands appearing in `$( )` or `` ` ` `` are executed with the results inserted into the remaining command (we explore this in chapter 6).
- Arithmetic operations (if any) are executed.
- Redirections (including pipes) are performed.
- Wildcards are replaced by a list of matching file/directory names.

FIGURE 2.22 The bash interpreter's steps in interpreting a command.

continue to persist during the session that the environment is active. In Linux, these definitions can be functions, variables and aliases. Once defined, the user can call upon any of them again. Only if the item is redefined or the session ends will the definition(s) be lost (unless, as described earlier, the definitions are placed in one of the startup scripts or exported). Similarly, commands are saved in the environment to be recalled, edited and re-executed.

2.7.3 The Bash Interpreter

The Bash interpreter, in executing a command, goes through a series of steps. The first step is of course for the user to type a command on the command line and press `<enter>`. Upon entering the command, the interpreter takes over. The interpreter reads the input. Now, the interpreter performs various steps depending on what features the command utilized. These steps are described in Figure 2.22.

With all of the modifications made in Figure 2.22 by the interpreter to the instruction, the command is executed, and upon completion, the output is displayed (if no output redirection was called for). Alternatively, if an error arose, an error message is displayed. In fact, in Linux, we could potentially see both output and error messages.

There are a few items described in Figure 2.22 that we have not covered in this chapter such as `$( )` and `` ` ` ``. We explore these and arithmetic operations in Chapter 6. We examine further wildcard characters in Chapter 3.

SECTION ACTIVITIES

1. Do you know any programming languages? If so, are they compiled, interpreted or both? You might need to research this.
2. Given that you can experiment as you go along in an interpreted environment, which format do you prefer to use: compiled or interpreted?
3. Of the steps of the Bash interpreter covered in this last subsection, which of these types of features have you used and which haven't you used?

2.8 CHAPTER REVIEW

Concepts and terms introduced in this chapter

- Alias – a substitute string for a Linux command.
- Bash – a popular shell in Linux whose acronym stands for Bourne Again Shell; Bash comes with its own interpreter and an active session with variables, aliases, history and features of command-line editing, tab completion and so forth; syntax is somewhat based on the ALGOL programming language.
- Bash built-in – commands that are part of Bash instead of Linux and share a man page.
- Bash session – the shell provides an environment for the user so that the user's interactions are retained, creating a session; the session contains a history of past commands and any definitions entered such as aliases, variables or functions.
- Bourne shell – predecessor to Bash; contains a few of the features found in Bash.
- Byte code – languages whereby programs are compiled into an intermediate form to be later executed by an interpreter; byte code is platform-independent so that the source code does not need to be compiled to multiple platforms.
- Command-line editing – keystrokes that allow the user to edit/modify a command on the command line prompt in a Bash session; keystrokes based on those found in `emacs`.
- Compiler – a program that converts a high-level language program (source code) into an executable program (machine language); many but not all high-level languages use a compiler (e.g., C, C++, Java, FORTRAN, COBOL, Ada, PL/I).
- C-Shell – a popular Unix and Linux shell that predates Bash; syntax is similar to that of the C programming language.
- Dot (hidden) files – files whose names start with a period; when using the `ls` command, these files only appear if you use the `-a` (all) or the `-A` (all but `.` and `..`) option.
- Environment variable – variable defined in a script that can be used by other scripts and the user; examples include `HOME` (the user's home directory), `PWD` (the current working directory), `OLDPWD` (the previous working directory), `PS1` (the user's command-line prompt definition) and `USER` (the user's username).
- Filename expansion – replacing wildcards with matching files as performed by the interpreter before executing the command; also called *globbing*.
- File type – in Linux, "file" extends to many types of entities including directories, physical devices and symbolic links; the first character of a long listing indicates the file type with `-` used for regular file and `d` for directory.
- Here document – Linux commands expect input from file but input can be redirected from keyboard using << which inserts the user into the "here document", an editor to enter input.
- History list – Bash maintains the list of the most recent instructions entered by the user (usually limited to the last 1000) during the current session; instructions can be recalled from the history list and displayed, edited or re-executed.
- Home directory – the user's home directory, typically stored at `/home/`*`username`* and can be denoted using the tilde (~) character; see tilde expansion.
- Interpreter – a program that interprets code, that is, receives an instruction from the user, translates it into an executable form and executes it; Linux shells contain an interpreter so that the user can enter commands in the shell via a command line.
- Java virtual machine (JVM) – a program which contains a Java byte-code interpreter; nearly all web browsers contain a JVM so that compiled Java code can be executed without having to compile a Java program for every platform.
- Long listing – an option (`-l`) in the `ls` (list) command to display files and their properties (permissions, size, last modification date, etc).

- Man pages – help information available for nearly every Linux instruction; view a man page using `man command` as in `man ls`.
- Options – specifiers to modify how a Linux command operates; most commands have options; see the command's man page.
- Parameters – the entity(ies) that the command will execute on; most Linux commands have files/directories as parameters but some operate on processes, users or other system entities; some commands require parameters, some have optional parameters; view each command's man page for details.
- PATH variable – one of the Bash environment variables which stores a list of all directories to search through to locate an executable program; without the `PATH` variable, you would have to specify the full path to the executable program to run a command like `/usr/bin/ls` instead of `ls`.
- Permissions – access rights to a file/directory; covered in Chapter 3.
- Pipe – a form of redirection in which the output of one Linux command is used as the input to another; the pipe operator is | which is placed between two commands as in `ls -l *.txt | wc -l`.
- Prompt – the command line presents a prompt by which the user enters commands whose format is specified by the variable `PS1`.
- PS1 variable – by default, this variable stores `[\u@\h \W]\$` so that the user's prompt is of the form `[foxr@localhost ~]$` indicating the username (`foxr`), the hostname (`localhost`), the current working directory (~, or foxr's home directory) and the user's status (normal user, `$`).
- Redirection – overriding the default input (STDIN) and output (STDOUT) of a command using operators <, <<, >, >> and | (the pipe).
- Shell – an environment that includes an interpreter, a command-line prompt, a history and previously defined entities (variables, functions, aliases).
- STDERR – default output location for the running program to send error messages.
- STDIN – default input source for the running program.
- STDOUT – default output for the running program.
- TC-Shell – a variation of C-shell with updated features, many of which are found in Bash.
- Tilde (~) expansion – the ~ represents the user's home directory; Bash expands ~ to the proper location (e.g., `/home/foxr`); when used as *~username*, bash expands this to `/home/`*`username`*.
- Wildcard – characters used as substitutions for groups of items; in Linux these are typically used to represent collections of files; for instance `*.txt` is substituted for all `.txt` files in the current directory; see Filename expansion.

Linux commands covered in this chapter

- = – used in an assignment statement of the form *VAR=VALUE*.
- alias – a command used as a substitution for another, perhaps longer, command; to define an alias use `alias` *`name=instruction`* where *instruction* must be placed in quote marks if it includes a blank space; once defined, *name* can then be used in place of the instruction from the command line.
- apropos – find instructions whose descriptions match the given string.
- arch – output the architecture (processor type) of the computer.
- bash – start a new Bash session; from within a Bash session, using `bash` launches a subshell or an inner session; see also `exit` and `export`.
- cat – concatenate file(s), by default sending output to the terminal window.
- cd – change working directory to directory specified.
- echo – output the parameters supplied, including literal values and values stored in variables if variable names are preceded by `$`; for more details, see Chapter 6.

- exit – leave current Bash session; if this is not a subshell, `exit` will also close the terminal window.
- export – make a variable available in other subshells or to other scripts.
- help – display help page for built-in Bash commands (generally not available for Linux commands that are not Bash built-ins but many Linux commands have a -h/`--help` option).
- history – output the last part of the history list (the number of instructions is based on the environment variable `HISTSIZE`).
- hostname – output the computer's host name.
- ls – list the contents of the specified file(s)/directory(ies), or if no directory is specified, the current working directory; option `-l` provides a long listing and option `-a` outputs hidden items.
- man – display the given instruction's man(ual) page.
- passwd – allow the user to change passwords, or if specified with a username allow root to change the specified user's password.
- pwd – print the current working directory.
- source – execute the given script by the current interpreter.
- unalias – remove an alias from the environment.
- uname – print the operating system name.
- vi – launch the vi/vim editor.
- who – list all logged in users and their login point.
- whoami – output the current user's username.

Files of note introduced in this chapter

- /etc/bashrc – script which executes for all user-interactive shell sessions; used by the system administrator to set up aliases and functions for all users.
- /etc/profile – script which executes for all user-interactive and non-interactive shell sessions; used by the system administrator to set up an initial environment for all users.
- ~/.bash_profile – script which executes for this user's interactive and non-interactive shell sessions; used by this user to tailor their environment.
- ~/.bashrc – script which executes for this user's interactive shell sessions; used by this user to define additional items like aliases, functions and variables.

REVIEW QUESTIONS

1. Research Bash and TC-shell (`tcsh`) and list three differences between them.
2. Research Bash and Ash and list three differences between them.
3. A user prompt appears as `[fox@zappa duke]$`. Who is the user? What is the name of the machine? In what directory is the user located?
4. Under what circumstance would your prompt end in # instead of `$`?
5. How does `whoami` differ from `who`?
6. How do the commands `arch`, `hostname` and `uname` differ?
7. True/false: Both the `su` and `passwd` commands can be issued with or without a username as in `passwd` and `passwd foxr`.
8. What character is used to separate multiple Linux commands entered on one line (i.e., entered at one time rather than separately)?
9. True/false: All Linux instructions have options and parameters.
10. When using `ls -l`, under what circumstance will the file's name look like `name1 -> name2`?

11. You want to perform a long listing on all contents of the current directory. Which of the following is the proper way to specify this (there can be multiple answers)?

`ls -a -l`	`ls -a l`
`ls -al`	`ls -a-l`
`ls -a,-l`	`ls -l -a`
`ls -la`	`ls -l a`
`ls -l,-a`	`ls -a ls -l`

12. Match the following `ls` options with their usage.

a. `-A`	i. Output contents in reverse order
b. `-B`	ii. List items one per line (one column)
c. `-C`	iii. Sort by file size rather than name
d. `-h`	iv. List entries in columns
e. `-L`	v. Show all hidden items but not `.` and `..`
f. `-r`	vi. Show sizes in "human readable" format
g. `-R`	vii. Ignore items whose names end with ~ (backup files)
h. `-S`	viii. Perform a recursive listing
i. `-1`	ix. Display information about item pointed to rather than the link

13. True/false: The command `ls /home/foxr /home/zappaf` is erroneous because `ls` expects only a single item.
14. True/false: All Linux instructions have man pages using the same fields (e.g., `SYNOPSIS`, `OPTIONS`, `EXAMPLES`, `SEE ALSO`).
15. What do you expect to find under the `EXIT CODE` section of a man page?
16. Under what circumstance will a Linux command have a `CONFIGURATION FILE` section? What do you find in that section?
17. True/false: The man page for a Linux command will include definitions for all of the command's options.
18. True/false: While not all Linux commands have a help page, they all respond to the option `-h`/`--help` to display help information.
19. You can't remember the name of the Linux command to compress a file. Provide a Linux command that should help you use to locate the command's name.
20. From the history list in Figure 2.9, how can you recall the instruction `cat file1.txt`. There are multiple answers.
21. From the history list in Figure 2.9, which command will be recalled with !v? With !.? With !c? With !~?
22. Which shell variable is used to display the response to `whoami`?
23. You are currently at `/home/foxr`. You type `cd /etc`. What two environment variables are updated after executing the `cd` command?
24. Write assignment statements to do each of the following:
    a. Store your first name in the variable `FIRST`
    b. Store your middle initial (without a period) in the variable `MIDDLE`
    c. Store your last name in the variable `LAST`
    d. Store in the variable `FULL` your name in the form `Frank V. Zappa`, note that while `MIDDLE` does not include a `.` we want one included here.
25. Assume `X` stores `5`. We execute the following five instructions. What is the output of the `echo` statement?

```
A=X
B=$X
C='$X'
D="$X"
echo $A $B $C $D
```

26. What will your prompt display if we define `PS1` as `"<\!: \t> \w \$"`

27. What is wrong with the following alias statement? Fix it. alias ~=cd ~
28. Is this legal? alias ...="cd ../.." (note: ../.. moves you up two directories)
29. What does the alias instruction do if there is no assignment statement with it (i.e., alias by itself)? What does the unalias instruction do?
30. Match the following command-line editing function with the keystroke that provides it
    a. Move the cursor to the end of the line i. c+k
    b. Move back one character ii. c+-
    c. Move forward one word iii. c+e
    d. Delete one character iv. m+f
    e. Retrieve the last instruction entered v. c+b
    f. Undo the last keystroke vi. c+d
    g. Delete all characters from here to the end of the line vii. c+p
31. Your previous instruction was cd /home/foxr/stuff/music. This directory does not exist. You realize that you meant to type this same command but for zappaf instead of foxr. Using command-line editing, fix the instruction. Specify in a step-by-step manner how you will accomplish this. Try not to use the arrow keys at all in your answer. NOTE: there are multiple answers.
32. When navigating through a man page, you use keys from ____ but when doing command-line editing, you use keys from ____ (fill in the blanks with emacs and vi).
33. Explain what the following instruction does. cat << eof > stuff.txt
34. Explain what the following instruction does. cat *.txt | sort > stuff.txt
35. Under what circumstance might you use the redirection operator <?
36. True/false: Bash permits you to use no more than one pipe per instruction.
37. True/false: The following instruction is illegal. cat << >> stuff.txt.
38. To what directory does ~dukeg reference? What about ~?
39. List all of the directories that will be listed with the following instruction.
    ls /home/{abcd/{temp1,temp2},defg,hijk/{sub1,sub2,sub3/{a,b}},lmno}
40. The current directory has these six files:
    abc.txt abd txt.ghi mno.txt mno .txt
    a. What is listed with the command ls?
    b. What is listed with the command ls *?
    c. What is listed with the command ls *.txt?
    d. What is listed with the command ls a*?
    e. What is listed with the command ls a* m*?
41. Explain what filename expansion is.
42. The current directory has these six files:
    abc.txt abd txt.ghi mno.txt mno .txt
    a. You type cat a<tab>. What happens?
    b. You type cat a<tab><tab>. What happens?
    c. You type cat abc<tab>. What happens?
    d. You type cat .<tab>. What happens?
43. You have typed the following sequence of commands.

```
X=1
bash
echo $X
```

    What is output?

44. You have typed the following sequence of commands.

```
X=1
export X
bash
X=2
exit
echo $X
```

What is output?

45. You have typed the following sequence of commands.

```
X=1
export X
bash
echo $X
exit
export -n X
bash
echo $X
```

What is output with both `echo` statements?

46. In what order will these files execute?

```
~/.bash_profile
~/.bashrc
/etc/bashrc
/etc/profile
```

47. As a user, you want to define an alias. In which of these files would you insert the alias?

`~/.bash_profile, ~/.bashrc, /etc/bashrc, /etc/profile`

48. Explain the role of the following characters as used in Bash.

```
a. ~
b. !
c. -
d. *
```

NOTE: There are no review questions for Sections 2.6 and 2.7.

3 Linux File Commands

This chapter's learning objectives are to be able to:

- Utilize Linux file commands `cat`, `cd`, `cmp`, `comm`, `cp`, `diff`, `head`, `join`, `less`, `ln`, `ls`, `more`, `mkdir`, `mv`, `paste`, `pwd`, `rm`, `rmdir`, `sort`, `tail` and `wc`
- Specify both absolute and relative paths
- Apply wildcards in file commands
- Describe the role of and create hard and symbolic links
- Use `find` to locate files
- Explain the Linux permissions scheme and use commands `chgrp`, `chmod`, `chown`
- Compare the types of storage devices
- Describe lossy versus lossless file compression and utilize file compression programs

3.1 INTRODUCTION

The operating system is there to manage our computers for us. This leaves us free to run applications without concern for how those applications are run. Aside from starting processes and managing resources (e.g., connecting to a printer or network, adding hard disks), the primary interaction that a user has with an operating system is in performing file operations. You no doubt have already had experience in creating, copying, moving and deleting files in your computer usage. Other operations include creating and moving directories, searching for files and viewing file and directory properties like their sizes. Because file operations are such a large part of what we do when interacting with an operating system, early personal computer operating systems usually had a name containing the phrase "disk operating system" (e.g., MS-DOS, Apple DOS, Quick DOS).

Linux offers a rich and complex set of file operations. Some of these involve file/directory manipulation: creating files/directories, deleting files/directories, copying files/directories, viewing directory contents, viewing file contents (for text files), moving/renaming files/directories, modifying file/directory permissions, searching directories for files and searching files for content. Administrator operations include creating and manipulating partitions, searching for and repairing disk issues, placing and altering disk quotas on users, mounting/unmounting partitions and backing up the files.

We divide file commands roughly into two categories: those used by users and those used by system administrators. In this chapter, we explore the user commands. We return to file commands, and a deeper examination of files, in Chapter 8 when the focus is on system administration.

We might also break file commands into those that manipulate files at a logical level and those that manipulate files at a physical level. Our file space is stored on storage media (hard disks, optical disks, solid-state storage, magnetic tape). We think of a file as a singular object but it is often stored in smaller units (e.g., blocks). We issue commands from the command line or GUI that operate on logical constructs: files and directories. The operating system translates our commands into physical actions that operate on blocks and partitions.

It should be noted that users are not actually capable of *commanding* the operating system. Instead, users make requests. Recall from Chapter 1 that the computer operates in one of two modes: user mode and privileged mode. When we run programs and we enter commands from the command line, we do so in user mode. The operating system receives each request and switches to privileged mode. In this mode, the operating system determines if we, as a user, have adequate access rights to perform the operation. If so, then the OS executes the command. This adds a layer

DOI: 10.1201/9781003203322-3

of protection on our computer systems so that, for instance, a user cannot alter or delete a file that the user does not have permission to access.

This chapter explores files and the file space from a user's perspective. We start with some terminology including the notion of a hierarchical file space. We then introduce numerous Linux commands. We view these in several sections based on the category of instruction. Later in the chapter, we focus on Linux permissions. We also look at how to perform various types of searches for files. The chapter ends with a look at storage devices.

SECTION ACTIVITIES

1. When you save files on your home computer (no matter the platform), do you save them in a My Documents-like folder set up for you by the operating system, do you create your own set of folders or do you save files on the desktop? Have you created subdirectories to better organize your files?
2. Do you save all of your files on your internal hard disk (or solid-state drive)? Do you have any external drives that you connect and disconnect to your computer? If so, are these hard disks or are they USB drives or both? Do you use cloud storage?

3.2 STORAGE TERMINOLOGY

We will define the *file space* of our computer as the collection of storage devices which, combined, make up the locations of all files and directories accessible to the users of the system. When users view the file space, we see files and directories. This is a *logical* view of the file space. With various tools, we can view the file space as devices, partitions, mountings and the breakdown of the files into their disk blocks. This is the *physical* view of the file space. It is important to understand that operating systems have commands that show us both perspectives but that we generally view (and prefer to view) the file space logically. We have been careful to this point as to not refer to our file space as a *file system* because that term has a different meaning, one which we explore in Chapter 8.

The *file* represents the smallest logical unit within the file space. That is, our file space stores files. Files can be categorized in multiple ways. We might view files as either programs or data files. Within this categorization, programs might be executables stored as binary files or scripts stored in text. Data files might be stored as raw text, formatted text using some form of tag (e.g., XML) or formatted binary. Files themselves are stored on disk as a collection of disk blocks. Blocks of a single file are not necessarily stored *contiguously* on disk but instead scattered across one or more disk surfaces.

Files have properties. Among them is a filename. Naming rules differ by operating systems by, for instance, restricting length, restricting characters that can be used to name a file and treating names case sensitively or insensitively. In Linux, filenames are *case sensitive* and limited to 255 characters.

Filenames can use only a subset of characters. It is recommended that names include only letters, digits and underscores. Spaces can be used but this can create a challenge when specifying a name from the command line (because a blank space indicates to Bash that the next item is a different item like a different file). Other punctuation is also available but discouraged and includes the comma (`,`), vertical bar (`|`), colon (`:`), ampersand (`&`) and semicolon (`;`). Because many of these characters are used in Bash for other purposes, using some of these characters may cause problems when issuing commands from the command line, unless we explicitly specify the character using the escape character (`\`), as in `\&`. We explore other file properties in this chapter and still others in Chapter 8.

Directories are a way to organize files. When directories can contain subdirectories, it creates a *hierarchical* file space. This is the case with Linux where there is virtually no limit to the number of subdirectories that can be created. Linux comes with a pre-established set of directories often referred to as the *top-level directories*. We explore some of these in this chapter but look at them more thoroughly in Chapter 8.

When interacting with the operating system, we are placed into a directory called the *current working directory*. If we need to access a file in another directory, or we wish to change to a new directory, we do so by specifying a directory path. A *path* is itself expressed hierarchically showing the directories we are moving upward or downward through to the destination directory.

There are two types of paths: absolute and relative. An *absolute path* always starts from the root level of the Linux directory hierarchy, /. A *relative path* describes the change from where we are to where we want to go. Moving downward requires listing the next subdirectory by name. Moving upward requires specifying the parent directory, denoted as `..` (two consecutive periods with no space between them). A path can contain several subdirectories as we move downward, or several parent directories as we move upward. We separate each directory in our path with a forward slash (/). We noted above that file names have a length of 255 characters. Linux restricts paths to no more than 4096 total characters.

Directories are grouped together into a partition. The *partition* is a physical division of the collection of storage devices into independent units. A partition can be operated upon without impacting other partitions. For instance, to back up content we would unmount the specific partition containing that content. While unmounted, no one can access it. But other partitions are unaffected so that users can continue to access those that are still mounted.

Partitioning comes with a disadvantage. We create our partitions at system installation time and when we create the partitions we are required to specify each partition's size. Once partitioned, this size is a physical limitation and thus the specified size limits the amount of growth of the file space of that partition. Repartitioning our storage devices to change size is possible but can be risky as it could destroy data from another partition. We will see in Chapter 8 that most partitions are named after the top-level directory they contain, but this is not always the case.

An alternative approach to physical partitioning is to use a logical volume manager (LVM). The LVM is software which logically partitions our disk space. The advantage of the LVM is that it does not restrict any partition's growth until we reach the total size of our storage devices. There are two disadvantages of the LVM. First, because the LVM is software it adds overhead to disk access. Second, one partition is used to store the boot loader program and Linux kernel (this partition is usually called `/boot`), and as the LVM is part of the Linux OS, it is not accessible until the Linux kernel has been loaded and is running. Thus, the `/boot` partition, which will contain the top-level directory `/boot`, must be separate from the LVM. Notice that the name of this partition is the same as the directory.

Depending on the specific distribution, the default installation for Linux usually provides three separate partitions: `/boot`, `LVM` and `swap`. As just noted, `/boot` must be separate from the `LVM`. The `LVM` will contain logical partitions of such content as the operating system, the user home directories (the `/home` partition containing the `/home` directory) and the variable data (the `/var` partition containing the `/var` directory). The `swap` partition is also separate from the `LVM` because the swap space (virtual memory) is treated differently from the rest of the file space and not directly accessible by users.

When installing Linux, we could choose to create our own partitioning. If we do so, we might create partitions for the operating system, /, and for each of `/home` and `/var`, and possibly `/usr` (which store applications software). We explore partitions, partitioning and LVM in Chapter 8 and omit further details from this chapter.

We have introduced files, directories and partitions. There is an additional component in the Linux file space called the inode. An *inode* is a data structure which stores file information (metadata) and pointers to the various file blocks on disk. This leads us to another file space component, the *link*. The link is a pointer which points from a file in a directory to its inode which contains the pointers to point at the file's physical blocks.

There are two types of links. The *hard link* points directly at the file's inode. The *symbolic* (or *soft*) *link* points at a hard link. This explanation might be hard to grasp for now, and like other

```
drwx------. 3  root   root   4096  Dec 23 01:33  accounts
-rw-rw----. 2  root   root    258  Dec  4 09:21  config
-rw-r--r--. 2 sales  sales  18532  Feb 19 11:41  customers
lrwxrwxrwx. 1  root   root    577  Mar  1 06:04  files ->
    /usr/local/somesoftware/data.txt
lrwxrwxrwx. 1 sales  sales   1853  Dec 19 12:01  data2 -> data
```

FIGURE 3.1 Sample long listing demonstrating hard and symbolic links.

content, we will hold off on exploring it further until Chapter 8. But before leaving this topic, there are mechanisms to determine if a link is a hard link or symbolic link. We will use `ls -l`.

Recall from Chapter 2 that the long listing of an item provides several pieces of information including file permissions, a hard link count and the file's name. For a symbolic link, the first character of the permission (the file's type) is `l` (lower case L) and the name is indicated as *`link`* `->` *`file`* where *link* is the symbolic link (the name) and *file* is the item being pointed at. Typically, the file is stored in another directory so *file* is listed as a full path.

The hard link's type is not `l` but instead the type of file being pointed to (e.g., `-` for a regular file, `d` for a directory, etc.). The name is the link's name. There is no ->, nor is there a separate file name. The hard link count, though, is greater than one. When a file has a single pointer pointing to it, its hard link count is one. A file or directory whose link count is greater than one indicates more than one link to that item. All directories have a link count of at least two.

Figure 3.1 illustrates the long listing for three hard links and two symbolic links. The first three in the listing are hard links, indicated by no `l` as the file type (the first character in the listing) and a hard link count greater than one. The symbolic links (the last two in the figure) show both the link's name and the name of the item being pointed to. The first symbolic link points to a file in another directory while the other points to a file in the current directory.

We noted above that an inode is a data structure that contains information about the file it represents as well as pointers to the disk blocks that physically store the item. An inode is dedicated to *every* file, directory and symbolic link in the file space. The information it stores includes the file's type, owner, group, permissions and last access/create/modification date. We will largely ignore inodes in this chapter and return to them in Chapter 8.

We wrap up this introduction with a look at the typical Linux *top-level directory* structure. Note that the specific top-level directories will differ by Linux distribution. All Linux distros contain at a minimum `/boot`, `/dev`, `/etc`, `/home`, `/proc`, `/var` and `/usr`. One version is shown in Figure 3.2. Aside from the top-level directory structure, some directories have a pre-established internal structure. Shown in the figure is that `/etc` has its own subdirectories (although only a few of the actual subdirectories are shown), `/home` has one subdirectory per user, `/proc` has subdirectories for every running process, `/run` has subdirectories for users, `/usr` has subdirectories to store the various software in logically named places and `/var` has subdirectories dependent on the installed software.

SECTION ACTIVITIES

1. What's the longest file name you've ever used? Have you used characters aside from letters, digits and periods? Have you used upper and lowercase letters? Have you used blank spaces in your file names (or directory names)? Do you use extensions? When you view files, do you see the extensions or hide them?
2. Windows, by default, has a single partition for the internal hard disk called C: and gives you only a few top-level directories. Linux divides the space into several partitions and as noted in Figure 3.2, many more top-level directories. Which approach do you prefer?

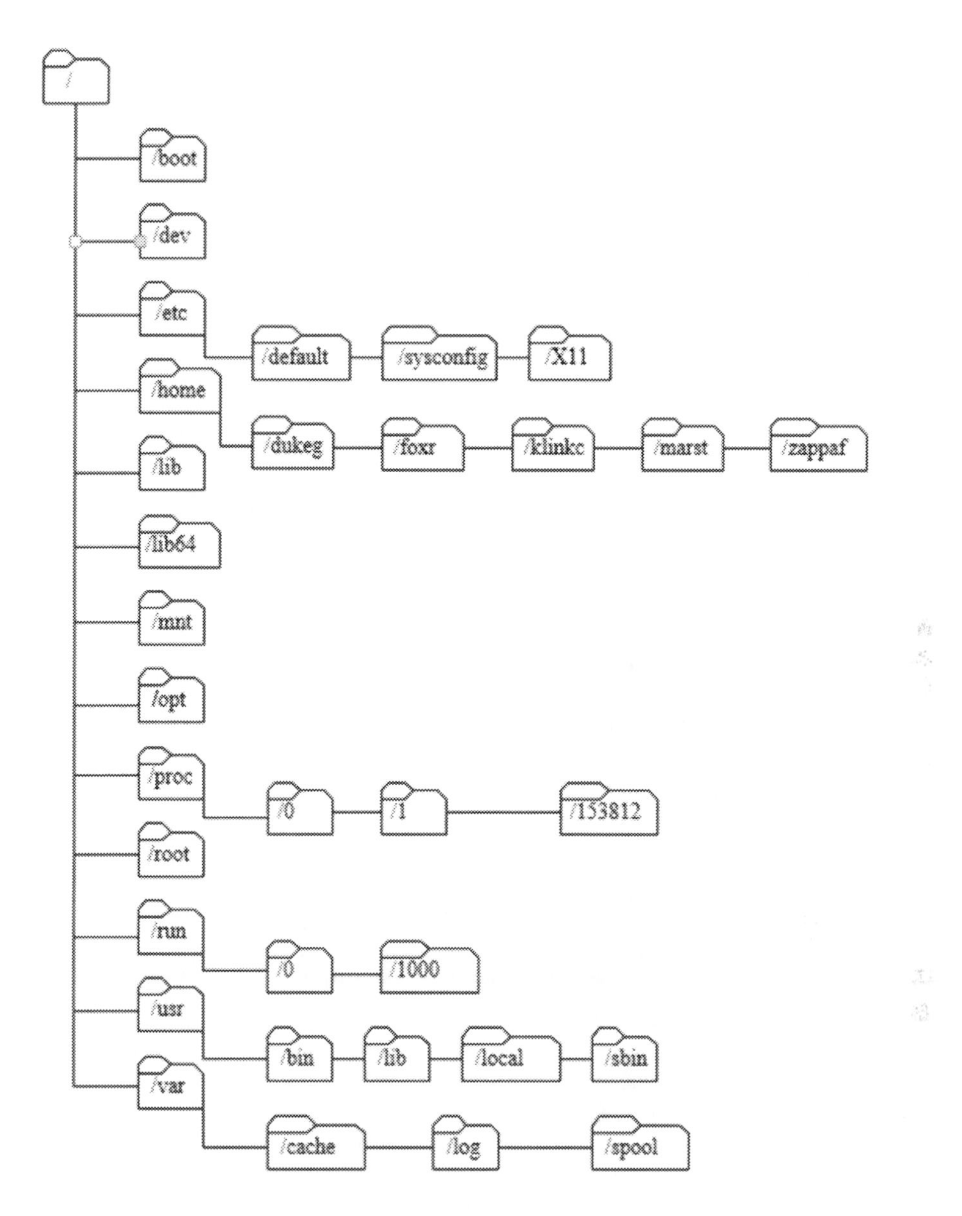

FIGURE 3.2 Sample top-level directory structure.

3.3 FILENAME SPECIFICATION

In this section, we focus on specifying paths and files as parameters in Linux instructions. We limit our look at instructions to four instructions that we explored in Chapter 2: `cat`, `cd`, `ls`, and `pwd`. In the first two subsections, refer to Figure 3.3 as needed. In this figure, we see two top-level directories, `etc` and `home` with `home` having two subdirectories, `foxr` and `zappaf`. The items in normal font are directories while the three items in italics are files.

3.3.1 The Path

A *path* is a description of how to reach a particular location in the file space. A path is required if the specified item is not in the current working directory. For executable programs, we can omit the

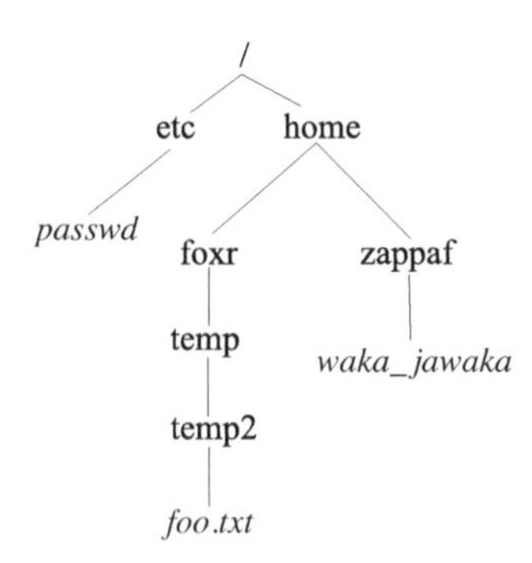

FIGURE 3.3 Example file space.

path if the directory containing the program is stored in our PATH variable. Paths can be expressed as absolute and relative. To determine where we are, we use pwd (print working directory), and to change directory, we use cd.

An absolute path always begins at the root level of the file space (/) and specifies each directory from the root level down to the location of interest, separating each directory in the path by a forward slash (/). The length of an absolute path is determined by the number of directories/subdirectories that exist between the root level and the target file or directory.

As an example of specifying a file with its path, the passwd file is stored in the top-level directory etc. The absolute path for this file is /etc/passwd. In Figure 3.3, we see that foxr's home directory has a subdirectory (temp) which has a subdirectory (temp2) which has a file (foo.txt). The absolute path for this file is /home/foxr/temp/temp2/foo.txt.

When we specify a directory in a path, the correct notation is to end the directory name with a /. If the last item in the path is a directory, we can omit the /. Therefore, to indicate the absolute path to foxr's temp directory, we can either use /home/foxr/temp/ (where the trailing / indicates that temp is a directory rather than a file) or /home/foxr/temp. To change directory to temp, we can use either cd /home/foxr/temp/ or cd /home/foxr/temp. It is never incorrect to include the trailing / for a directory but it is seldom needed.

A relative path starts at the current working directory. The path itself specifies how to reach the destination directory from this location. That is, the path is the movements relative to the current directory. If the destination location is in the subtree beneath the current location, then the path starts with the subdirectory of the current directory that will lead us to the destination. Moving further down requires listing all of the subdirectories in order, each separated by /. The relative path does not start with a / as doing so would imply that we are specifying an absolute path. For instance, if the user is currently at /home/foxr and wishes to move downward into temp2, the command using a relative path is cd temp/temp2. Notice that the trailing / was omitted but could have been there if desired.

If the new location is up the hierarchy, the path starts with one or more .. where .. indicates the parent directory. If in temp2, moving up to home is accomplished using cd ../../.. where again a trailing / can be omitted.

What if the destination directory is in another branch of the hierarchy? That is, it is both up the hierarchy and then down the hierarchy? The relative path must start by moving up as far as necessary and then down. For instance, if at foxr and we want to move to zappaf, we would use cd ../zappaf. The .. indicates moving from foxr to home followed by zappaf to move down from home to zappaf. If we are currently at foxr and want to move to etc, we would use cd ../../etc where the first .. moves us to home, the second .. moves us to / and then etc moves us down into etc. All trailing / were omitted from the examples in this paragraph (and will be throughout this chapter).

Beginning Linux users often confuse using absolute and relative paths. As an example, a user currently in `foxr` who wants to move down to `temp2` might use `cd /temp/temp2`. This will likely yield an error because the path starts with / indicating an absolute path. `cd /temp` would take the user to the top-level directory called `temp` and as there likely is no such directory this will cause an error.

Another point of confusion is moving to another branch of the hierarchy. We cannot move down and then up; we must move up first. For instance, moving from `temp2` to `zappaf` requires moving up until we reach `home` and then down. The correct relative path is `../../zappaf`. While the path `../../../home/zappaf` is not incorrect, it is not as efficient as it goes up one too many levels before descending.

Should we use an absolute or relative path? This depends on where we are and where we are going. It is less typing using a relative path if we are moving strictly downward from our current position. For instance, if we are at `foxr` and want to go to `temp2`, `cd temp/temp2` is shorter than `cd /home/foxr/temp/temp`. If we are moving up, the relative path may be shorter but it depends on how deep we are in the hierarchy. If we are at `temp2` and want to move to `foxr`, the relative path is `../..` while the absolute path is `/home/foxr`. If we are at `etc` and want to move up to / then the choices are `cd ..` and `cd /`, which are similar in terms of number of keystrokes. Introductory Linux users may lack confidence or familiarity with the Linux directory structure and so the relative path may be more challenging.

3.3.2 Filename Arguments with Paths

Many Linux file commands can operate on multiple files at a time. Such commands will accept a list of files as its parameters. For instance, the file concatenation command, `cat`, displays the contents of all of the files specified to the terminal window.

Consider the command `cat file1.txt file2.txt file3.txt`, which concatenates the three files specified. Each file listed in the command refers to a file stored in the current working directory. What if some of the files are stored elsewhere? Then, the command must include the path to each of those files. Let's assume as an example that `file1.txt` is in `/home/foxr`, `file2.txt` is in `/home/foxr/temp`, and `file3.txt` is in `/home/zappaf`, with the current working directory being `/home/foxr`.

There are multiple ways to issue the above `cat` command. Figure 3.4 shows three possibilities. The `./` notation, as used in the third example, indicates "in the current directory" and can be omitted in this case (we require this notation when executing programs in the current directory but not when referencing files as parameters). Remember that `/home/foxr` and `/home/zappaf` can be expressed as `~foxr` and `~zappaf` respectively to shorten the amount of typing. Note that the second example should be on a single line.

Given a filename or a full path to a file, the instructions `basename` and `dirname` return the file's filename and the path respectively (`dirname` returns just the path). Figure 3.5 provides an example of using the two commands and the responses received. This example comes from Red Hat Linux as Debian-based Linux does not have a `sysconfig` directory. For the commands `basename ~` and `dirname ~`, we receive *`username`* and `/home` respectively where *username* is the current user's username.

```
cat file1.txt temp/file2.txt ../zappaf/file3.txt

cat file1.txt /home/foxr/temp/file2.txt
        /home/zappaf/file3.txt

cat ./file1.txt ./temp/file2.txt ../zappaf/file3.txt
```

FIGURE 3.4 Examples of paths for parameters in the `cat` command.

```
basename /etc/sysconfig/network-scripts/ifcfg-ens33
Response: ifcfg-ens33

dirname /etc/sysconfig/network-scripts/ifcfg-ens33
Response: /etc/sysconfig/network-scripts
```

FIGURE 3.5 Examples demonstrating `basename` and `dirname` commands.

3.3.3 The PATH Variable

Recall that one of the environment variables set up for us is `PATH`. This variable contains paths to various directories that might contain executable programs. For instance, `ls`, `cat` and `cd` are all stored in `/usr/bin`. If an executable program is not stored in a directory noted in `PATH`, or if we did not have a `PATH` variable, we would have to indicate the command using a path (absolute or relative).

Assume that we have no `PATH` variable. Issuing the `cat` command from Figure 3.4 would be more complex as we would have to specify the command as `/usr/bin/cat`. Our full command might become `/usr/bin/cat file1.txt temp/file2.txt ../zappaf/file3.txt`. Certainly, having the `PATH` variable set up for us is more than convenient.

Does the `PATH` variable save us from having to specify paths of filenames that are parameters to a command? Unfortunately, no. Even if `/home/foxr`, `/home/foxr/temp` and `/home/zappaf` were in our `PATH` variable, we would still have to include the paths to the files `file1.txt`, `file2.txt` and `file3.txt` in our previous `cat` command. But remember tab completion can save some typing!

We can modify the `PATH` variable using an assignment statement. Recall from Chapter 2 that the form of an assignment statement is *VAR=VALUE*. *VALUE* can include the variable's value itself, which is obtained using *$VALUE*, or `$PATH` in this case. The contents of `PATH` are a list of directories, each separated by a colon (:) with the next. To add to the `PATH` variable, we use `PATH=$PATH:`*new_directory* (or `PATH=`*new_directory*`:$PATH`). This notation says set `PATH` to be the current contents of `PATH` (`$PATH`), followed by `:` followed by the new directory, or set `PATH` to be the new directory, followed by `:` followed by the current contents of `PATH`.

Is there a preference between the ordering of the directories in `PATH`? Yes. Bash searches the `PATH` variable to locate an executable instruction. It searches for the instruction in each directory, in the order they are listed in `PATH`. It makes more sense to have the most commonly used directories listed early and the less used directories listed later. Assume we are not zappaf and we want to add his directory `/home/zappaf/bin` to our `PATH` variable. We would likely use `PATH=$PATH:/home/zappaf/bin` so that this new directory is placed at the end of our variable so that it is the last directory searched since most commands we execute will not be stored there.

3.3.4 Specifying Filenames with Wildcards

A wildcard is a way to express a number of items (files, directories, links) without having to enumerate all of them. The most common wildcard is *, which represents "all items". By itself, the * will match any and every item in the current (or specified) directory. For instance, `ls *` lists all items in the current directory. Note that `ls` and `ls *` respond with the same list.

We can combine the * with a partial description of item names. When appended with `.txt`, the notation matches all items whose name ends with `.txt`. When placed after `a`, as in `a*`, it matches all items that start with an `a`. The * can go anywhere in the filename specifier such as `f*.txt` to indicate all items that start with `f` and end with `.txt`. We can use multiple * in any filename specifier as in `f*.*`, which will match any item that starts with an `f` and contains a period.

The Bash interpreter performs an operation called *filename expansion*, or *globbing*, prior to executing an instruction. The interpreter replaces the filename(s) listed that contain wildcards with all matching items. After that replacement, or expansion, the command is executed on the listed

TABLE 3.1
Wildcards in Linux

Wildcard	Explanation
*	Match everything.
?	Match any one character.
[*chars*]	Match any one of the characters in the list (e.g., [aeiou]).
[*char1-char2*]	Match any one of the characters in the range from *char1* to *char2* (e.g., [0-9], [a-e], [A-Z]).
{*word1*, *word2*, *word3*}	Match any one of the words
[!*chars*]	Match any one character not in the list.

files. In the command `ls *.txt`, the interpreter first expands `*.txt` into the list of filenames ending with `.txt`, while `ls a*` has Bash first expand `a*` to the list of all filenames that start with an `a`. Table 3.1 lists the base wildcards in Bash. There are additional wildcards that can be used in Bash but require specific installation.

Table 3.2 provides a number of examples. At the top of the table are the files of the current working directory. Beneath this are several `ls` commands and the files that match.

Let's comment on a few of the examples from Table 3.2. The fourth, fifth and sixth examples use ? as part of the filename. The ? will match any single character. In the case of `file?.*`, the word `file` must be followed by exactly one character followed by `.` and any extension. In the case of `file??.txt`, the word `file` must be followed by exactly two of any character followed by `.txt`. For `file?.{dat,txt}`. The word `file` must be followed by any one character and then followed by a period and either `dat` or `txt`. The last entry, `[!a]*`, will match any name as long as its first character is not `a`.

Wildcards can be confusing. The best way to learn them is through practice. Later in the text, we look at regular expressions which also use the characters *, ?, and []. Unfortunately, the interpretation of * and ? differ between the wildcard usage and their usage in regular expressions, which may further complicate your ability to use them correctly.

TABLE 3.2
Demonstrating Wildcards

Files:

abc	bcd	aba	bb
file.txt	file1.txt	file2.txt	file2a.txt
file21.txt	file_a.txt	file_21	file.dat
file1.dat	file2.dat		

ls command	Files Listed		
ls *.txt	file.txt file2a.txt	file1.txt file21.txt	file2.txt file_a.txt
ls *.*	file.txt file2a.txt file.dat	file1.txt file21.txt file1.dat	file2.txt file_a.txt file2.dat
ls *	All files		
ls file?.*	file1.txt file2.dat	file2.txt	file1.dat
ls file??.txt	file2a.txt	file21.txt	file_a.txt
ls file?.{dat,txt}	file1.txt file2.dat	file2.txt	file1.dat
ls [abc][abc][abc]	aba	abc	
ls [!a]*	All files except for aba and abc		

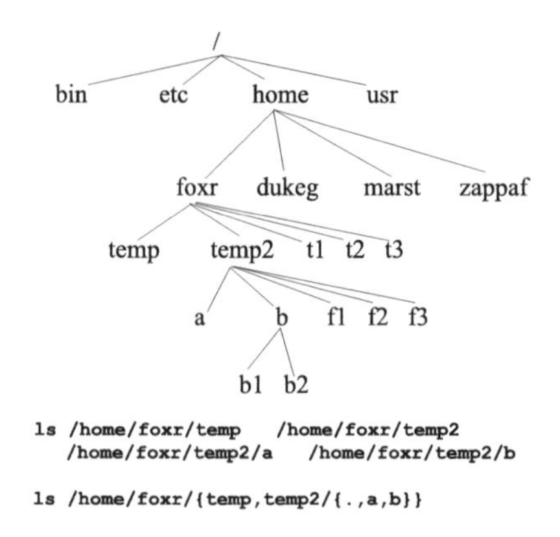

FIGURE 3.6 Example directory structure demonstrating brace expansion.

One of the wildcards shown in Tables 3.1 and 3.2 is the braces, {}. In Table 3.2, we see that we can specify multiple extensions, as in `.{dat,txt}`. Another usage is for files that have different names but the same extensions, as with `{aaa,file1,mystuff,target}.txt`. We use the curly brackets to enumerate a list of items more conveniently than expressing each item in full. We introduced this notation in Chapter 2 as brace expansion (refer back to Chapter 2.4.6 if needed).

Consider the directory structure shown in Figure 3.6. Assume we want to perform some operation, say `ls`, on all of the subdirectories of `foxr` (which consists of directories `temp`, `temp2`, `a` and `b`). We might list all of the directories individually as with the first `ls` instruction shown near the bottom of the figure. With brace expansion, however, we can reduce the instruction substantially, as shown with the second `ls` instruction at the bottom of the figure.

The Bash interpreter unfolds brace expansion so that the second `ls` instruction in Figure 3.6 is equivalent to the first `ls` instruction. With brace expansion, we do not specify full paths to each directory because they are implied with the previous parts of the instruction. For instance, `temp` is implied to be `/home/foxr/temp`. Obviously brace expansion requires less typing so can be more convenient but may also be challenging to use correctly.

Through the use of `..`, `.`, ~, *, ?, `[]` and {}, the user is provided a number of very useful shortcuts to simplify paths and filenames. With respect to these symbols, Bash first expands items found in braces. Bash then replaces ~ with the appropriate home directory name. Next, filename expansion takes place replacing wildcards with lists of matching files and directories. Finally, the instruction is executed on the resulting list(s) generated.

SECTION ACTIVITIES

1. If you are a Windows user, bring up the Control Panel (you can reach this by typing Control Panel in the Cortana search box). Select System and then from the menu list in the left pane, select Advanced system settings. From the System Properties pop-up window select Environment Variables… One user variable is Path, there is a similar variable for System. Look at the values in both of these Path variables. What directories are familiar to you? Look at the other environment variables for the user and see if any are of directories that you might use. If so, which ones?
2. Windows uses \ to separate directories, such as `C:\Users\foxr\My Documents`. Do a web search to find out why. You'll discover that / was already used for another purpose. What purpose?
3. Have you ever used a wildcard when searching, whether from a command-line instruction like `ls` or when doing an Internet-based search? If so, how did you use it and did you find it easy or challenging?

3.4 FILE COMMANDS

In this section, we examine many common Linux file commands. This section does not include commands reserved for system administrators. Keep in mind that we need proper permission to perform operations on files and directories that are owned by other users. We explore permissions in Section 3.5 of this chapter. We also visit some more special purpose file system commands later in this chapter.

One convenient aspect of Linux is that many of the file commands can operate on different types of items because Linux treats them all as files. The full list of file types is regular files, directories, symbolic links, named pipes, domain sockets, character and block devices. We discuss the difference between these types in Chapter 8 although we also look at symbolic links in this chapter.

The `file` command describes the type of entity passed to the command. For instance, `file /etc/passwd` will tell us that this is `ASCII text` (a text file), `file /` will tell us that `/` is a `directory` and `file /dev/sda1` will tell us that `sda1` is a `block special` (a special file for a block device). Other types that might be output from the `file` command include `empty` (an empty text file), a particular type of text file (e.g., C program, Bash shell script, Java program), `symbolic link` and `character special` (character-type device). Using the `-i` option causes `file` to output the MIME type of the file as in `text/plain`, `application/x-directory` and `text/x-shellscript`, among others.

Table 3.3 displays many of the commands we will cover in subsections 3.4.1–3.4.8 of this section, along with a description of each command and the more common options. Even though we have already looked at `pwd` and `cd` in Chapter 2, we also cover them in the first subsection. To gain a full understanding of each command, it is best to study the command's man page.

TABLE 3.3
Common Linux File Commands

Command	Description	Common Options
`pwd`	Display current directory.	
`cd`	Change directory.	
`ls`	List contents of directory.	`-a` (all), `-l` (long listing), `-r` (reverse), `-R` (recursive), `-S` (sort by file size), `-1` (display in one column)
`mv`	Move or rename file(s)/directory(ies).	`-f` (force), `-i` (interactive), `-n` (do not overwrite existing file)
`cp`	Copy file(s).	same as `mv`, also `-r` (recursive)
`rm`	Delete file(s).	`-f` (force), `-i` (interactive), `-r` (recursive)
`mkdir`	Create a directory.	
`rmdir`	Delete a directory.	
`cat`	Concatenate files (display to window).	`-n` (add line numbers), `-T` (show tabs)
`less`	Display file screen by screen.	`-c` (clear screen first), `-f` (open non-regular files)
`more`	Display file screen by screen.	–num # (specify screen size in # rows), +# (start at row number #)
`head`	Display first part of a file.	`-n #` (number of lines), `-c #` (number of bytes)
`tail`	Display end of a file.	`-n #` (number of lines), `-c #` (number of bytes)
`find`	Locate file(s).	Covered in Tables 3.11 and 3.12
`cmp`	Compare files.	`-i #.#` (start at the specified bytes), `-n #` (compare # bytes)
`diff`	Compare multiple files.	`-i`, `-E`, `-Z`, `-b`, `-B` (ignore case, tabs, trailing space, white space, blank lines)
`cut`	Remove portions of each line of a file.	`-b`, `-c`, `-d`, `-f` (control which portion of a line to cut)
`wc`	Word counter.	`-c`, `-w`, `-l` (output bytes, words, lines of the file)
`touch`	Update file's last access/modification date but also create an empty text file.	`-a`, `-m` (Update access time, modification time)

3.4.1 Directory Commands

We should always know our current working directory. Without this knowledge, we will not be able to use relative paths. The command pwd is used to print the current working directory. It is one of the simplest commands in Linux as it has only two options, neither of which are commonly used. These options are -L to display the contents of the PWD environment variable and -P to avoid reference to symbolic links. The command responds with the current working directory as output. Figure 3.7 shows the interaction with pwd.

We can also identify, at least partially, the current working directory by examining our user prompt. The default prompt only shows the name of the directory, not its path. In Figure 3.7, we see we are in ~, which is equivalent to /home/foxr. Imagine that we have changed directory to /etc/sysconfig. The prompt would change to [foxr@localhost sysconfig]$. We might recognize where we are because sysconfig is an uncommon name. But consider someone has created a subdirectory called TEMP. We cd into it. Now our prompt is [foxr@localhost TEMP]$. We may not know whose TEMP directory we are in if we do not remember. So even though the prompt is useful in identifying our location, pwd is simple enough to run whenever we are unsure.

In order to navigate around the file system, we need to change directory. The command to do that is cd. The command expects the path to the destination directory as its parameter. This path can be relative or absolute. The destination will always be a directory, not a filename. We provide a number of cd commands as examples in Table 3.4. Refer back to Figure 3.3 for the sample file space that these commands apply to. Notice in the table that we do not end any of the commands or directory names with / but we could.

If we have a lot of different directories that we wish to visit often, we might place them onto the *directory stack*. The dirs command displays all directories on the stack. To add a directory, use pushd *dirname*. The push operation places the new directory on the top of the stack, so we are building our list backward. To remove the top directory from the stack, use popd.

```
[foxr@localhost ~]$ pwd
/home/foxr
```

FIGURE 3.7 Using pwd.

TABLE 3.4
Example cd Commands with Absolute and Relative Paths

Current Location (top) and Destination (bottom)	Command with Absolute Path	Command with Relative Path
/home/zappaf /bin	cd /bin	cd ../../bin
/home/zappaf /home/marst	cd /home/marst	cd ../marst
/home/zappaf /home/foxr/temp	cd /home/foxr/temp	cd ../foxr/temp
/home/zappaf /home/foxr/temp2/a	cd /home/foxr/temp2/a	cd ../foxr/temp2/a
/home/foxr/temp2/b /home/foxr/temp2/a	cd /home/foxr/temp2/a	cd ../a
/home/foxr/temp2/b /home/foxr/temp	cd /home/foxr/temp	cd ../../temp
/home/foxr/temp2/b /home/dukeg	cd /home/dukeg	cd ../../../dukeg

Once we have moved to a new location, we may want to view the contents of the current location. This is accomplished through the `ls` command. As we already explored the `ls` command in some detail in Chapter 2, we will skip it here.

3.4.2 File Movement and Copy Commands

Now that we can navigate and view the directories of Linux, we want to be able to manipulate the files and directories. We do this by moving, renaming, copying creating, and deleting files and directories. The commands `mv`, `cp` and `rm` (move/rename, copy, delete) can work on both files and directories (`rm` can be used to delete directories only in a specific case, which we examine in the next subsection). For each of these, however, the user issuing the command must have proper permissions for the file(s)/directory(ies) listed. We will assume that the commands are being issued on items whose permissions will not raise errors.

The `mv` (move) command is used to both move and rename files/directories. The standard format for `mv` is `mv [options] source destination`. To rename a file, *source* and *destination* are the old name and the new name respectively. Otherwise, *source* and *destination* must be in different directories. If *destination* is a directory without a filename, then the file's name is not changed. For instance, `mv foo1.txt ~/temp` moves `foo1.txt` to the current user's `temp` subdirectory without changing the name. The instruction `mv foo1.txt ~/temp/foo2.txt` moves the file and renames it from `foo1.txt` to `foo2.txt`.

To indicate that a file from a directory other than the current directory should be moved to the current directory, we specify the destination location using period (.). For instance, `mv ~foxr/temp/foo2.txt .` moves foxr's file `foo2.txt`, which is in his subdirectory `temp`, to the current working directory. Notice the use of ~ for the specification of his home directory.

To move a file within the same directory, the command acts as a rename command. Otherwise, the file is physically moved from one directory to another, thus the file in the source directory is deleted from that directory. This physical "movement" is only of a link. The file's contents, stored in disk blocks, are usually not altered (unless the file is moved from one partition to another).

The `mv` command can operate on multiple files if we are moving them all to a new directory. For this to work, we must list all of the files and follow it with a destination directory. For instance, to move three files named `file1.txt`, `file2.txt` and `file3.txt` in the current directory to the `/home/zappaf` directory, we can use the command `mv file1.txt file2.txt file3.txt /home/zappaf`. We can reduce this instruction using wildcards to just `mv file*.txt /home/zappaf`.

We are not allowed to rename multiple files using a single `mv` command because when we rename a file, we are required to specify the destination as a filename, and `mv` only has a single destination parameter. Consider, for instance, `mv *.txt *.dat`. This instruction yields an error because the destination must be either a directory or a single file name. `*.dat` implies more than one destination.

Among the options for `mv` are `-i` and `-f` indicating *interactive* and *force* mode, respectively. The difference only arises if the `mv` operation is to move a file to a destination where a file of the same name already exists. In interactive mode, the user is prompted to see if the move should take place and thus overwrite the existing file in the destination directory. The user responds with a `y` or `n` answer. In force mode, which is the default, the file is moved and the existing file in the destination is overwritten (deleted). If the destination directory does not contain a file of the same name as the file being moved, the two modes operate identically.

The copy command is `cp`. Its format is much the same as `mv` where we specify a source and destination. Specifically, the syntax is `cp [options] source destination`. There are three different combinations of *source* and *destination* that can be used. First, the destination specifier can be another directory in which case the file is copied into the new directory, and the new file is given the same name as the original file. Second, if destination is both a directory and filename, then

```
cp foo.txt ~zappaf/foo1.txt
cp *.txt ~
cp ~zappaf/foo1.txt .
```

FIGURE 3.8 cp Examples.

the file is copied into the new directory and given the new filename. Third, the destination is a filename in which case the file is copied into the current directory under the new filename. Examples are shown in Figure 3.8 with explanations in the following paragraph.

The first example in Figure 3.8 copies the file `foo.txt` from the current working directory to `zappaf`'s home directory, changing the name to `foo1.txt`. The second example copies potentially multiple files. Filename expansion is used to expand `*.txt` to all `.txt` files in the current directory. Each of these files is copied to the user's home directory with names being unchanged. The third example copies the file `foo1.txt` from `zappaf`'s home directory to the current directory, leaving the name unchanged.

Table 3.5 lists some of the most commonly used options for `cp`. The use of `-i` and `-f` is the same as with `mv`. Some of the others should be self-explanatory like `-b`, `-I`/`-s` and `-u`. Let's explore the others. With `-L`, if the source is actually a symbolic link, the link is followed to the actual file and it is the file that is copied, not the link. Normally, `cp` does not report on what has been copied. Adding `-v` causes `cp` to output each file movement as it takes place.

Unlike `mv`, the file is physically copied so that two versions now exist. By default, when `cp` copies someone else's file, its ownership and permissions are updated to that of the user who issued the `cp` command. For instance, we copy a file owned by `zappaf` with `zappaf`'s default permissions. The result is that the new file is given our ownership and our default permissions. The `-p` option preserves the original ownership and permissions.

Let's consider a specific example. Assume `foo1.txt`, `foo2.txt` and `foo3.txt` are stored in `/home/zappaf` and are owned by `zappaf` with the group owner being `cit371`. We issue the instruction `cp ~zappaf/foo*.txt .` which copies the three files into the current directory (presumably ours). The copies have our username as the owner and our private group rather than `zappaf` and `cit371` respectively. These files are newly created so their creation dates and time are the current time. If we used `cp -p ~zappaf/foo*.txt .` then the resulting files have an owner of `zappaf`, group of `cit371`, permissions of the original files and a creation/modification date of the original files.

The remaining option of `cp` to explore is `-r`, which performs a *recursive* copy. A recursive operation is one that operates on the files and subdirectories of the current directory and also applies to the contents of any subdirectories of those subdirectories, recursively meaning that the same thing takes place on anything in the subdirectories of the subdirectories. A recursive copy then is used to

TABLE 3.5
Common Options for the cp Command

Option	Meaning
`-b`	Create a backup of every destination file.
`-f`	Force copy, as with `mv`'s `-f`.
`-i`	Prompt the user, as with `mv`'s `-i`.
`-I, -s`	Create hard/symbolic link rather than physical copy.
`-L`	Follow symbolic links.
`-p`	Preserve ownership, permissions, time stamp, etc.
`-r`	Recursive copy (recursion is discussed in the text).
`-u`	Copy only if source is newer than destination or destination missing.
`-v`	Verbose (output each step as it happens).

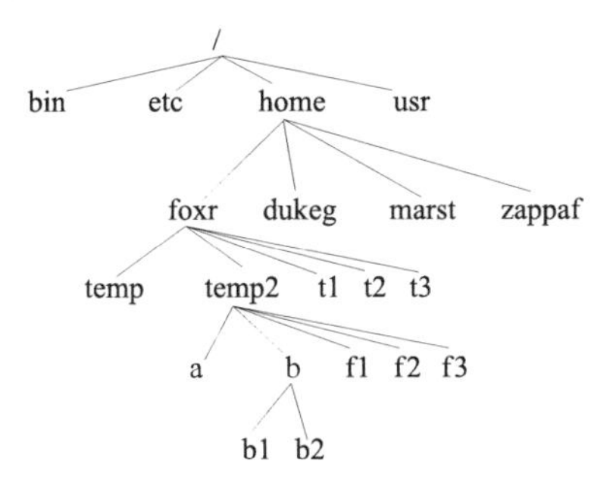

FIGURE 3.9 Example file space.

copy not just the contents of the specified directory but everything beneath it in the hierarchical file space. Consider the file space illustrated in Figure 3.9 (which is the same as that shown earlier in Figure 3.6). Assume `t1`, `t2`, `t3`, `f1`, `f2`, `f3`, `b1` and `b2` are files and all other items are directories.

The current working directory is `~dukeg`. The command `cp -r ~foxr/temp2 .` copies directory `temp2` with the files `f1`, `f2` and `f3`, subdirectories `a` and `b`, and the files `b1` and `b2` from directory `b`. These would be copied into `~dukeg` so that `dukeg` would now have a subdirectory of `temp2`, files `f1`, `f2`, `f3`, and `temp2` would have subdirectories `a` and `b`, where subdirectory `b` would have files `b1` and `b2`.

3.4.3 File Deletion Commands

`rm` is the remove, or delete, command. Its syntax is `rm [options] file(s)`. As with `mv` and `cp`, the `rm` command can work on multiple files either by listing each item (separated by a space) or using wildcards, or both.

Assume we have a directory with the files `file1.txt`, `file2.txt` and `file3.txt` (and no other files ending in `.txt`). To delete all three, we can use `rm file*.txt` or `rm file1.txt file2.txt file3.txt`. Obviously, the first instruction is simpler. Now assume the directory contains other files ending with `.txt` such as `file4.txt` and `file10.txt`. The former instruction would delete too much. We could use different wildcards to better control the `rm` instruction as in `rm file[1-3].txt` but we will have to be careful not to delete the wrong files.

As `rm` permanently deletes files, it is safest to use `rm` with the option `-i`. Unlike the interactive mode for `mv` and `cp` which only asks user permission if the command will overwrite an existing file, the interactive mode for `rm` will pause before each deletion to ask the user for permission. Users may not think to use `rm -i`, so it is common for a system administrator to create an alias of `rm` to `rm -i` so that, by default, `rm` is always run with the `-i` option.

If we are applying `rm` to a lot of files, for instance by using `rm *.txt`, we may not want to be bothered with having to respond to all of the prompts (one per item being deleted) and so we might use `rm -f *.txt` to override the prompting messages. This can be dangerous though, so only use `-f` when you are sure that you want to delete the file(s) specified. Force mode is the default.

The `-r` option is another recursive operation, like `cp -r`. Here, `rm` deletes the contents specified but if any are subdirectories, then `rm` applies to the contents of the subdirectory before deleting the subdirectory itself. Having probably aliased `rm` to be `rm -i`, performing `rm -r` results in being asked for permission to delete potentially a lot of items. We may want to delete an entire subhierarchy of our file space by using `rm -fr *` (or `rm -fr directory`). This will delete everything in the current or named directory including all subdirectories and their contents. Of course, we should explore the current directory before using such a command to ensure that we are not deleting needed content mistakenly.

The `rm` instruction is intended to delete files and not directories. If we specify `rm directory`, we receive an error. But if we use `rm -r directory` then the directory (and any subdirectories) will be deleted. Without rm -r, the only way to delete a directory is with rmdir (explained in the next subsection). Referring back to Figure 3.9, rm -fr /home/foxr/temp2 will delete the files and directories f1, f2, f3, b1, b2, a, b, temp2 in the order listed.

3.4.4 Creating and Deleting Directories

To create a directory, we use mkdir followed by the directory name as in mkdir TEMP. The directory is created in the current working directory unless the named directory contains a path. The directory is initially empty. The permissions for the directory default to our default directory permissions.

The mkdir command only has a few options. With -m or --mode=*MODE*, we can specify the initial permissions of the directory to override the default permissions. With -Z or --context=*CTX* we can specify the SELinux (security enhanced) context. The context, *CTX*, provides for rules that describe access rights. SELinux is discussed in Chapter 7.

We remove a directory with rmdir, but the directory must be empty. If we want to remove a directory that is not empty, we would have to delete its contents first using rm. Then we can apply rmdir as in rmdir *directory*. Attempting to delete a non-empty directory yields the error message rmdir: failed to remove `*dirname*': Directory not empty. where *dirname* is the name of the directory.

As seen in the previous subsection, we can also delete a directory using rm -r. To remove TEMP, rm TEMP yields an error even if TEMP is empty, but rm -r TEMP does not. Figure 3.10 provides a sequence of commands that we might use to delete the directory TEMP by first deleting its contents. Notice the cd command to move from TEMP to its parent directory, which in this case is our home directory, for deletion. Obviously, it is much easier to delete a directory using rm -r *directory*. Remember that we would be asked permission for each deletion if rm had been aliased to rm -i, so we might prefer to use rm -rf *directory*.

rmdir has an option, -p, to recursively delete parent directories as well as the current directory. In such a case, we specify the path of the directories to delete from top to bottom. Referring back to Figure 3.5, if all of the directories foxr, temp2 and b were empty, and we wanted to delete them all, then from /home we could issue the command rmdir -p foxr/temp2/b.

3.4.5 Textfile Viewing Commands

There are a number of commands that allow us to view the contents of files. We have already used cat in a number of examples. The problem with cat is that the contents are displayed in the window, and if the file is large enough, we only see the last screen's worth. The commands less and more let us step through the file, screen by screen. This can often be far more convenient.

The more instruction is the older of the pair. It displays the contents of one screen's worth and pauses. As it pauses, waiting for the user to press either <space> or <enter>, more sometimes displays the amount of the file we have already viewed as a percentage. <space> moves us forward by one screen and <enter> by one line. Upon reaching the end of the file, or typing q, more exits and returns us to our command line prompt. more has a number of options to help more precisely control the program's behavior, for instance by using -n *number* to specify the number of lines that appear in each screen or +*linenumber* to start the display at the given line number.

```
[foxr@localhost  TEMP]$ rm *
[foxr@localhost  TEMP]$ cd ..
[foxr@localhost  ~]$ rmdir TEMP
```

FIGURE 3.10 Deleting a non-empty directory.

The less command is more useful as it gives the user more control to step through the file being viewed. Specifically, less lets the user scroll both forward and backward through the file, for instance by using the arrow keys. less loads the file into a vi-like browser whereby we can move up and down using vi commands. Specific commands are <space>, <page down> and f to move forward by one screen, <page up> to move back a page, <up arrow> and <down arrow> to move up or down one line, d and u to move forward or backward one-half screen and *n*G to move to line *n* of the file. Typing q exits less and returns us to our command line prompt. Unlike more, reaching the end of the file does not exit from less as we are still in the vi-like browser. There are options that we can specify when launching less as well. For more movement commands and to see the available options, see less' man page.

What if we only want to view a few lines of a file? There are two commands that show only the first or last set of lines of the file: head and tail. By default, the two commands display the first and last ten lines of the given file. We can control the amount displayed with options -c and -n. Both options expect an integer number to follow the option where -c specifies the number of bytes (characters) to output and -n specifies the number of lines to output.

For head, we can precede the integer with a minus sign to indicate that the program should skip that number of bytes or lines. Similarly, for tail, precede the integer with a plus sign to indicate the starting point within the file. For instance, head -c -20 *filename* would stop at the 21st byte of the file, and tail -n +12 *filename* will display the file starting at line 12. Some additional examples are shown in Table 3.6. Assume that file.txt is a text file that consists of 14 lines and 168 bytes.

Another file operation is sort, to sort a given file line by line in increasing order. Use -r to sort in decreasing order. sort can work on multiple files in which case the lines are mixed, as if the files were first combined using cat. Another useful option for sort is -f which causes sort to ignore case so that 'a' and 'A' are treated equally (otherwise 'A' comes before 'a').

3.4.6 File Comparison Commands

The instructions cmp, comm and diff are all available to compare the contents of text files. Both cmp and comm compare two files where comm further expects the files' contents to be in sorted order. diff can compare two files, two directories or a file to the contents of a directory.

cmp receives two files as parameters and compares them byte by byte, line by line. It stops as soon as it finds a mismatch, returning the byte and line of the difference. We can force cmp to skip over a specified number of bytes for each file or stop after reaching a specified limit. For instance, cmp file1 file2 -i 100:150 -n 1024 compares file1 and file2, starting at byte 101 of file1 and 151 of file2, comparing 1024 bytes.

The default counting used with cmp -i (or --ignore-initial=*value*) is a value in bytes (characters), but we can also use values like MB (2^{20} or 1048576) and kB (10^3 or 1000). Notice that for the lowercase letter (kB), the value is rounded off while the uppercase letter specifies a value as

TABLE 3.6
Examples Using Head and Tail

Command	Result
head file.txt	Output first ten lines of file.txt.
tail file.txt	Output last ten lines of file.txt.
head -c 100 file.txt	Output first 100 bytes of file.txt.
head -n -3 file.txt	Output all but the last three lines of file.txt.
tail -n 5 file.txt	Output last five lines of file.txt.
tail -c +100 file.txt	Output all of file.txt starting at the 100th byte.

a power of 2. Output from `cmp` indicates where the mismatch occurred as in `file1.txt file2.txt differ: byte 912, line 36`. If there is no mismatch, `cmp` provides no output. Aside from `-i` and `-n`, a couple of other useful options are `-b` to output the actual mismatched characters and `-l` to force `cmp` to continue comparing the files, displaying all mismatches found.

`comm` receives two files which are expected to be sorted. If the contents of the files are not sorted, while `comm` will still work, we may not get the results expected. The `comm` instruction works through the two files, comparing them line by line and returning for each line whether content appeared in the first file, the second file or both files. This output is organized into columns where column 1 features lines only appearing in the first file, column 2 features lines common to both files, and column 3 features lines only appearing in the second file. Options are limited in `comm` and we might only use `-1`, `-2` or `-3`, which are used to output only lines unique to the first file, lines unique to the second file and lines that appear in both files, respectively.

As noted, `diff` can operate on two files, a file and a directory, or two directories. Depending on the types of the two parameters, `diff` operates differently. With two files, `diff` performs a line-by-line comparison, reporting every mismatch.

The notation of `diff`'s output is somewhat cryptic. For each line that is in the first file but not the second, the output is < and the line, and for every line that is in the second file but not in the first, it is output as > followed by the line. This output is preceded with the location (line number) and the *type* of mismatch in one of three forms, #a#, #c# or #d# where # is the line number(s). The `a` indicates that the given line needs to be added to the file to make the two files the same, the `d` means that the given line needs to be deleted from the file to make the two files the same, and the `c` means that the given line needs to be changed in the given file. In the case of `c`, multiple lines might be reported using the notation `#,#` where the first # is the starting line number and the second # is the ending line number.

Let's take a look at some sample output to better understand `diff`. We created a text file, `fruit1`, containing the name of a piece of fruit on each line. The specific words are `apple`, `banana`, `cherry`, `date`, `fig`, `grape`. We copied `fruit1` into `fruit2`. Running `diff fruit1 fruit2` generates no output indicating the files are identical. We modified `fruit2` by replacing `banana` with `Banana` and now `diff` outputs the report shown on the left side of Figure 3.11 telling us that both files contain a line (line 2 in both files) with a difference where the first file has `banana` and the second has `Banana`.

Resetting `fruit2` to its original form, we then deleted `date` from the file. Upon rerunning `diff`, our output changed to that shown in the center of Figure 3.11. This output informs us that we need to delete the fourth line of the first file to match the second file. The first file contains a line with `date` which is causing the mismatch.

Restoring `fruit2` one more time, we deleted `fig` from `fruit1` and reran our `diff` command. The result is shown on the right of Figure 3.11. The output we receive with this run is that we have to add after the fourth line in `fruit1` before the fifth line of `fruit2`, which consists of `fig`.

By default, `diff` compares character-by-character literally. Using the option `-i` we tell `diff` to ignore case when comparing letters. We can also have `diff` ignore spacing using `-E` (ignore tabs), `-Z` (ignore trailing spaces), `-b` (ignore differences in white space), `-w` (ignore all white space) and `-B` (ignore blank lines). We can alter the output with `-y` so that the two files are displayed in columns instead of line after line. In such a case, the `#letter#` report is omitted but missing lines are shown as blanks with < or > to indicate where the line should be inserted and | to indicate lines that have mismatched characters.

```
2c2           4d3          4a5
< banana      < date       > fig
---
> Banana
```

FIGURE 3.11 Output from running `diff` three times.

When comparing two directories, diff first compares file names to see if a file exists in one directory and not in the other. If this is the case, diff outputs a report like Only in dir1: file1.txt. This indicates that dir1 has file1.txt and the other directory does not. If the same named files exist in both directories, then diff compares the two files as described above. If we run diff on a file and a directory, then diff searches the directory for a file of the same name and either compares the two files or reports that the file was not found.

diff can also compare one file to several other files. There are two ways to specify this by either using --from-file= to specify the file to compare against others, or --to-file= to specify a list of files to compare one to. Assume we want to compare file1 to each of file2, file3, file4 and file5. We can use either diff --from-file=file1 file2 file3 file4 file5 or diff file1 --to-file=file2 file3 file4 file5. In either case, diff compares file1 to the other four files, outputting all results. When doing such a comparison, we are not told which of the four files (file2, file3, file4 or file5) contains the mismatch. Because of this, it might be better to run diff four times, once for each of file2, file3, file4 and file5.

Yet another comparison program is uniq. Unlike cmp, comm and diff, uniq operates on a single file, searching for consecutive duplicate lines. Based on the parameters supplied, we can use uniq to remove any such duplicates although uniq does not overwrite the file but instead outputs the file without the duplicates. We might then redirect the output to a new file, as in uniq file.txt > filewithoutduplicates.txt. As uniq only compares adjacent lines, it would not find duplicate lines that do not appear consecutively. Options allow us to output the found duplicates or count the number of duplicates rather than outputting the file without the duplicates, and we can specify that case be ignored or specify characters to skip over.

3.4.7 File Manipulation Commands

The join command is used to join two sorted files together when the files contain a common value (field) per line. When the two files each have a row that contains that same value, then those two lines are joined together. Lines that do not contain a matching first field are not joined. As an example, consider two files, first.txt and last.txt, each of which contains a line number followed by a person's first or last name. Joining the two files would append the value in last.txt to the value found in first.txt for the given line number. Figure 3.12 shows the contents of the two files and the result. Notice that the two files are in sorted order by line number because the first item on each line is 1., 2. or 3.

The options -1 *NUM* and -2 *NUM* allow us to specify which fields should be used on the first and second files respectively where *NUM* is the field number desired. The default is 1 (the first field). The option -i causes join to ignore case if the common values include letters. The option -e *STRING* uses *STRING* in place of an empty field, and -a 1 or -a 2 output lines from the first or second file which did not contain a match to the other file.

Another means of merging files together is through the paste command. Here, the contents of the files are combined, line by line. For instance, paste file1 file2 would append file2's first line to file1's first line, file2's second line to file1's second line and so forth. The paste

```
File first.txt:          File last.txt:
1.  Frank                1. Zappa
2.  Tommy                2. Mars
3.  George               3. Duke

Result of join first.txt last.txt:
1.  Frank Zappa
2.  Tommy Mars
3.  George Duke
```

FIGURE 3.12 Using the join command.

command is like the join command except that we do not need to have common field values. Referring back to Figure 3.12, if first.txt and last.txt had no line numbers, the instruction paste first.txt last.txt would yield the same result as join did except that the result has no line numbers, just the combined names. The option -s will serialize the paste so that all of file2 would appear after all of file1. However, unlike cat, paste -s would place the contents of file1 onto one line and the contents of file2 onto a second line.

The split instruction allows us to split a file into numerous smaller files. We specify the file to split and a *prefix*. The prefix is the name used for the new files where each new file's name is appended by a pattern like aa, ab, ac or 00, 01, 02. By default, split will place 1000 bytes in each new file.

Imagine that our original file consists of 3500 bytes. Using split on it would create four files where the first three files would have 1000 bytes each and the last would have 500. To override this, use -b *value* where *value* is the number of bytes per file. The option -d causes split to name the new files using digits (00, 01, 02, etc.) instead of letters as the extension to the prefix.

The cut command is used to remove portions of each line of a file. This is a useful way of obtaining just a part of a file. The cut command is supplied options that indicate which parts of a line to retain, based on a number of bytes (-b *first-last*), specific characters (-c *charlist*), a delimiter other than tab (-d '*delimiter*') or field numbers (-f *list*) if the line contains individual fields. Adding --complement will reverse the option so that only the cut portion is returned.

Let's consider a file of user information which contains, for each row, the user's first name, last name, username, shell, home directory and user ID (UID), all delineated by tabs. We want to output for each user their username, shell and UID (columns 3, 4 and 6). The command cut -f 3,4,6 *filename* will output just those fields. Had each row's contents been separated by spaces, we would add -d ' ' to indicate that the delimiter is a space.

We can also use cut to provide reduced output from other commands. Imagine that we want to view the permissions of the files of a directory but we do not want to see other information like ownership or size. The ls -l command gives us more information than desired. We can pipe the result to cut to retain only the permissions (characters 2–10 remembering that the first character of the long listing is the file's type) of each line. The command is ls -l | cut -c 2-10.

Now consider that we want the permissions *and* the filenames. We cannot specify both -c to indicate characters 2–10 and -f to indicate the ninth field. So, we will output the entire first field (including the file type) and the name using ls -l | cut -f 1,9 -d ' '. Unfortunately, our solution may not work correctly as the size of the listed items may vary so that some lines contain additional spaces. If one file is 100 bytes and another is 1000 bytes, the file of 100 bytes has an added space between the group owner and the file size in its long listing. This would cause field 9 to become the date instead of the filename. In Chapter 5, we will look at the awk instruction which offers better control than cut for selecting fields.

One last instruction to mention in this subsection is called strings. This instruction works on files that are not necessarily text files, outputting printable characters found in the file. The instruction outputs any sequence of four or more printable characters found between unprintable characters. The option -n *number* overrides four as the default sequence length, printing sequences of at least *number* printable characters instead.

3.4.8 Miscellaneous but Useful File Commands

The word counter, wc, outputs a count of the characters, words and lines in a text file. The number of characters is the same as the file's byte count (characters stored in ASCII are stored in one byte apiece). The number of lines is the number of new line (\n) characters found in the file. Word counts are produced by looking for whitespace between groups of characters.

We can control wc's output using options. The options -c or -m limit the output to just the character count (-m counts characters while -c counts bytes), -l (lowercase L) to output the line count

```
5     16    file1.txt
2     39    file2.txt
16    20    file3.txt
```

FIGURE 3.13 Output of `wc -l -L *.txt`.

and `-w` for the word count. We can combine these options to output multiple values such as `-c -w` to get the number of characters and the number of words. The option `-L` prints the length of the longest line. Output is a list of one or more numeric values (the count based on which parameters you used) and the file's name.

Let's assume we have a directory of three text files, `file1.txt`, `file2.txt` and `file3.txt`. We issue the command `wc -l -L *.txt` and receive the output shown in Figure 3.13. The first number is the number of lines in the file, and the second number is the length of the longest line in the file.

Similar to how we used `cut` with ls, we can pipe the result of a command to `wc` to count the results. If we want to know how many entries are in a directory, we can use the command `ls | wc -l`. This instruction takes the list as returned by `ls` and asks `wc` to count and output the number of lines. We could similarly find out how many users are currently logged in by using `who | wc -l`.

There are numerous options to create files. We have already seen that we can create a file using `vi` or combining the `cat` instruction with redirection as in `cat << string > filename`. We can also create a file by redirecting output of one or more commands to the file as in `ls -l > listing`. And of course many types of applications software can create files.

We can create an empty text file with the `touch` command. The command's true purpose is to modify the last access/modification time stamps of a file. However, if the file does not currently exist, an empty text file is created. This can be useful if we wish to ensure that a file exists which will eventually contain output from some other program such as a shell script. Or, `touch` can be used to start a text file that we will later edit using `vi`.

`touch` has options of `-a` and `-m` to modify the last access time and date and last modification time and date to the current time, respectively. The option `-t` allows us to specify a new access and/or modification date and time. We follow `-t` with the date and time specification using the format `[[CC]YY]MMDDhhmm[.ss]`. This means year, month and date as a 6- or 8-digit number (if CC is omitted, the year is implied to be from the current century, and if YY is omitted, the year is implied to be this year). The `.ss` notation permits seconds, if desired. For instance, `-t 2212051415.16` would be 2:15.16 pm on December 5, 2022. `touch` also has the option `-r filename` so that the given filename's time/date is used in place of the current time/date.

SECTION ACTIVITIES

1. Make a list of all of the instructions we covered in this section. Next to each, check mark those you have already used. Place an asterisk next to those that offer functionality similar to something you have used in other operating systems, whether a command-line instruction or a GUI-based instruction. Finally, rank the instructions in terms of which ones you would be most likely to use and those you are least likely to use. Use a scale from 1 to 5 where 5 is extremely likely and 1 is extremely unlikely.
2. Consult the man pages for `cat`, `cd`, `cp`, `head`, `less`, `more`, `mv`, `rm`, `tail` and `wc`. Rank them in order of easiest to hardest to use.
3. Create two similar text files and experiment with `cmp`, `comm` and `diff` to see how they work and how their outputs differ. Next, use the `cut` command to experiment with how you specify what should be retained using `-c` and using `-f`.

3.5 PERMISSIONS

If you have already been experimenting with some of the instructions described in Sections 3.3 and 3.4, you may have found that you did not have proper permissions to perform operations on some of the files. What are permissions, how do you know what permissions a file or directory has and how can you modify permissions? In this section, we explore the answers to these questions.

3.5.1 WHAT ARE PERMISSIONS?

Permissions are a mechanism to support operating system protection. *Protection* ensures that users do not misuse system resources. A system's resources include the CPU, memory and network. But as users primarily interact with files, the resources that we protect with permissions are directories and files. Permissions proscribe what access rights a user has to a file or directory. By establishing permissions, we can share files and directories or make sure that files and directories are inaccessible to other users.

Many operating systems implement file permissions through access control lists (ACLs). An *ACL* is attached to a specific resource (e.g., a file) and consists of the list of users and groups that can access the given resource. For each user/group, the list contains that user's/group's access rights to the resource. For instance, the owner of `file1.txt` may have read/write/execute access while four specific users and one group may have read/execute access and all other users have no access.

ACLs can be lengthy depending on the number of users and groups in the system. Linux provides a shorter version of permissions, establishing access rights for three categories of users: the owner, members of the group that owns the item and all other users of this Linux computer. We refer to these three as user (`u`), group (`g`) and other (`o`). `u` really means the item's owner but do not confuse this with `o` (other), which we also refer to as "the world".

All files (including directories and symbolic links) are given default permissions based on a setting placed on the user. This setting is established with a command called `umask` (which we look at in Chapter 7). The owner of a file, as well as root, can change the permissions at any time. Root can also change file ownership and group ownership. With adequate access rights, the file owner can also change the group ownership.

Forms of access are one or more out of read (`r`), write (`w`) and execute (`x`). These three forms of access differ depending on whether the item is a regular file or a directory. Table 3.7 describes these access rights. Other operating systems, such as Windows, may provide more access rights such as modify and full control.

Let's explore the role of the group by imaging first that we have no groups. A user who creates a file would then be able to set up two levels of permissions: their own access and everyone else's access. Consider that the file owner has read and write access to a text file and other has no access. But now imagine that there is a set of users that the file owner wants to give read access. Without a group, the file owner can either open read access to all Linux users or no Linux users, but not this specific list of users.

TABLE 3.7
File and Directory Access Rights

Access Right	Access for Files	Access for Directories
`r` (read)	File can be viewed, copied or opened as read-only.	Contents can be viewed by `ls`.
`w` (write)	File can be overwritten (modified).	Files can be written here; a file can be deleted from here if user has `w` access to the file.
`x` (execute)	File can be executed (needed for both programs and shell scripts).	User can `cd` into this directory.

We provide a third level of access by having groups. First, root has to create a group and populate it with the set of users that the owner wants to provide access to. The file owner (or root) modifies the file to give group ownership to this group of users. Finally, the file owner modifies the permissions on the file so that group has read access.

3.5.2 Altering Permissions from the Command Line

To change a file's permission, the command is `chmod`. The command's syntax is `chmod` *`permissions file(s)`* where *permissions* can be specified using one of three different approaches described below. The parameter *file(s)* is one or more files/directories. As with the previous file commands, multiple files are specified with spaces between their names. Wildcards may also be used.

The first approach for specifying permissions is to describe the changes to be applied as a combination of `u`, `g`, `o` along with `r`, `w`, `x`. To add a permission, use + and to remove a permission, use -. Let's assume `file1.txt` is currently readable and writable by `u` and `g` and readable by `o`. We want to remove writable by group and remove readable by other. The command is `chmod g-w,o-r file1.txt`. Notice there are no spaces within the permissions. We see that `g` is losing `w` (group is losing writability) and `o` is losing `r` (other is losing readability). If instead we want to add execute to owner and group, it would be `chmod u+x,g+x file1.txt`.

To change a permission for all three of `u`, `g`, `o`, we can use `a` instead of listing each individual change. For instance, given the original permissions of `file1.txt` above, we want to add execute for everyone. This could be done with `chmod a+x file1.txt`.

The second approach uses = instead of +/–. We assign new permissions rather than changing existing ones. We can specify any of `u`, `g` or `o` or some combination. Let's go back to the original permissions for `file1.txt` and instead make the file readable, writable and executable for user, readable and executable for group, and nothing for world. This instruction is `chmod u=rwx,g=r,o= file1.txt`. Notice that to specify no access rights, after the equal sign we leave it blank.

The approach using the equal sign can be limited to just the category(ies) we want to change. That is, we do not need to list all three of `u=`, `g=` and `o=`. For instance, to remove all access from group and other, we use `chmod g=,o= file1.txt`, omitting `u=` as we are not changing the owner's permissions.

The two approaches can be combined. We may use +/– to add/remove permissions from one of the three user categories while resetting another one using =. Again, returning to `file1.txt` having read/write access for user and group and read for world, we want to change user to have read/write/execute access, remove write access to group and reset world to nothing. Any of the instructions in Figure 3.14 will accomplish this. There are other possibilities as well.

The third approach requires specifying the full permissions for each of `u`, `g` and `o`, expressing the permissions as a 3-digit number. This is less typing although slightly more complicated in that we must first figure out the appropriate 3-digit number.

Each digit of the 3-digit number represents the permissions for one of `u`, `g` and `o`. The digit is computed by adding 4 for `r`, 2 for `w` and 1 for `x` access. For instance, read and write access is 4+2=6 while read and execute access is 4+1=5. If a file is to have read, write and execute for user, read and execute for group and nothing for world, the 3-digit number is 750 (7=4+2+1 or read/write/execute for user, 5=4+1 or read/execute for group, 0 means no access for world). The command is `chmod` *`number file(s)`*.

```
chmod u=rwx,g-w,o-r file1.txt
chmod u=rwx,g-w,o= file1.txt
chmod u+x,g=r,o-r file1.txt
chmod u+x,g-w,o= file1.txt
```

FIGURE 3.14 Example `chmod` instructions to accomplish the same changes to `file1.txt`.

Another way to derive the number is to think of the permissions as listed in the `ls -l` instruction. If an access right is granted, replace the letter (r, w, x) with 1, and if no right is granted, replace the – with 0. This gives us a 3-bit number for each of u, g and o. Now, replace each 3-bit number with its equivalent decimal (or octal) number. For instance, if user has read, write and execute access, this would be 111_2 or 7. If group has read and execute access, this would be 101_2, or 5. If other has execute access, this would be 001_2 or 1. Thus, this set of permissions is 751.

Let's understand better how we got 751. A file has permissions of `rwxr-x--x` (the user has read, write and execute access, the group has read and execute access and world has only execute access). Rewriting these nine permissions in binary gives us 111101001_2. Separating this binary number into three numbers gives us 111_2, 101_2 and 001_2. Now, converting each binary number to decimal (or octal) gives us 7, 5 and 1, thus these permissions are equivalent to the 3-digit number 751. Figure 3.15 illustrates the grouping of these first ten characters where the leading hyphen is the file's type (hyphen indicates a regular file).

Table 3.8 provides some examples of permissions and their 3-digit equivalents. Table 3.9 discusses several different 3-digit combinations and why they might be useful or not useful.

By default, any file or directory created by a user is owned by that user and assigned to that user's private group (if a user has no private group then the `users` group is used). Root can change any item's owner and/or group. As a non-root user, we can only change a file's group that we own as long as we are a member of the group we are assigning the file to. The two commands to change owner and group are `chown` and `chgrp`. The syntax is shown in Figure 3.16 along with some examples. Notice that `chown` can be used to change both file owner and group (see the last instruction in the figure).

In Figure 3.16, we see that all of the `.txt` files in `foxr`'s directory are being modified to be owned by `foxr`. As this is already likely to be the case, the instruction may not be worth executing. Only root can issue this instruction. In the second example, the file `project-data.txt` is being modified to be owned by the group `cit371`. If `foxr` owns the file (which is likely since the file is in his directory) and he is a member of `cit371`, he could issue this instruction instead of root. In the third example, all files in the apache2 `htdocs` subdirectory are being modified to be owned by user `www` and in `www`'s group. Again, this instruction can only be issued by root.

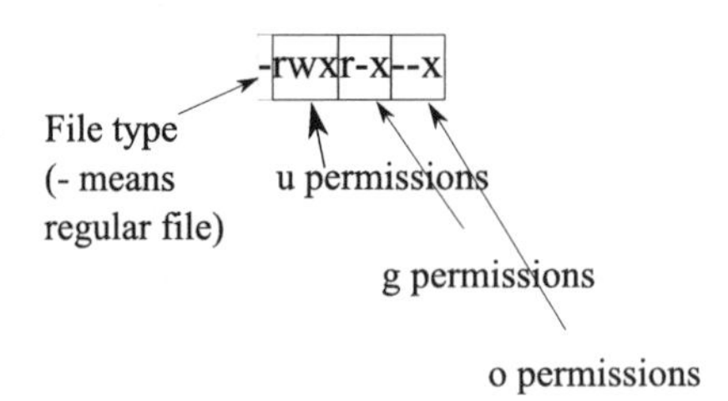

FIGURE 3.15 Understanding 3-digit permissions.

TABLE 3.8
3-Digit Permission Examples

`---------`	000	`--x--x--x`	111	`r--------`	400	`r--r--r--`	444
`rw------`	600	`rw-r-----`	640	`rw-r--r--`	644	`rw-rw----`	660
`rw-rw-r--`	664	`rw-rw-rw-`	666	`rwx------`	700	`rwx--x---`	710
`rwx--x--x`	711	`rwxr--r--`	744	`rwxr--r-x`	745	`rwxr-x--x`	751
`rwxr-xr-x`	755	`rwxrwxr--`	774	`rwxrwxr-x`	775	`rwxrwxrwx`	777

TABLE 3.9
Example 3-Digit Permissions

Permission	Explanation	Usefulness
000	No one has any type of access to the item (root always has access).	Not useful other than system files accessible only by root.
200	Write only.	Not useful; need to be able to read the file before writing back to it.
400	Read only for owner.	Useful if file contains confidential read-only information; not common.
444	Read only file.	Useful if a file should be readable by everyone; 644 or 664 is more common.
644	Read and write access to owner, read-only for all others.	Very common.
646	World can write but group members cannot.	Not useful; we would expect group to have equal or greater access than world.
660	Read and write access to owner and group, no access for world.	Common if file should not be accessed outside of group.
664	Read and write access to owner and group, read-only for world.	Common for file sharing with group members.
666	Read and write access to everyone.	Dangerous.
711	Full access for owner, group members and world can execute the file only.	Sometimes used for directories or some binary executable programs.
745	Full access to owner, group can read while others can read and execute.	Sometimes used for script files instead of 755.
755	Adds execute access to 644.	Very common for executable files and scripts as well as directories.
775	Adds write access to group on top of 755.	Less common than 755 but useful if group needs write access to directory or script file, or members of group might compile the program.
777	Full access to everyone.	Dangerous although found in the top-level directory `/tmp`.

```
chown newowner file(s)
chown newowner:newgroup file(s)
chgrp newgroup file(s)

chown foxr /home/foxr/*.txt
chgrp cit371 /home/foxr/cit371/project-data.txt
chown www:www /usr/local/apache2/htdocs/*
```

FIGURE 3.16 Syntax and examples for `chown` and `chgrp`.

3.5.3 Altering Permissions from the GUI

We can also modify a file's permissions through the GUI. The File Browser is a GUI program that displays the contents of a directory and allows us to navigate around the file space. It is similar to Windows' File Explorer. We can use this to view individual file information, including permissions. From the File Browser, locate the file or directory of interest. Right click on the item in the main pane and select Properties from the pop-up window. The Properties window appears. It has three tabs, the middle one being Permissions. Figure 3.17 illustrates a file's Permissions tab of its Properties window.

Notice that "Me" is the file's owner (this is displayed instead of the owner's name). We can see the current u/g/o permissions are set to read/write, read/write and read-only respectively. Each of

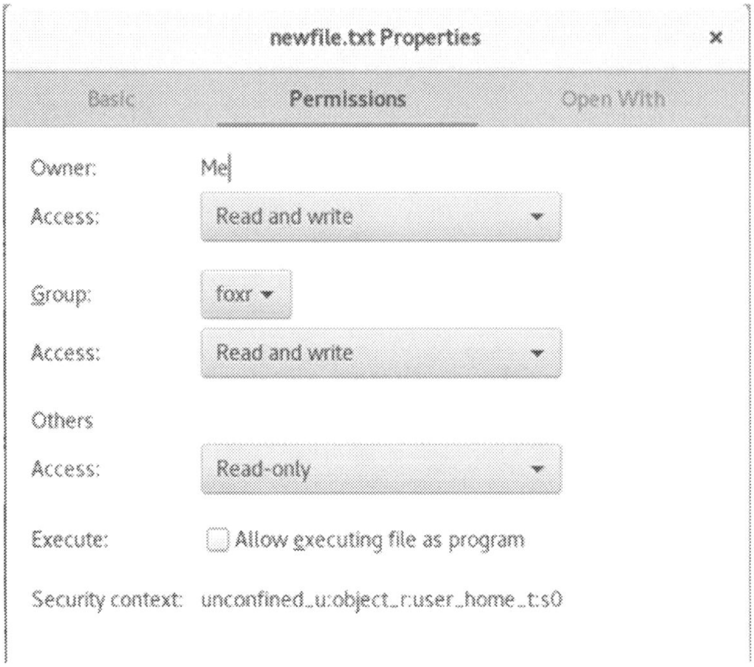

FIGURE 3.17 Viewing and changing permissions through the GUI.

these drop-down menus allows us to change the permissions. Choices for files are "None", "Read and write" or "Read-only". Execute access is not available through the drop-down menu but there is a checkbox to permit execute access. For directories, choices are "None", "List files only", "Access files" and "Create and delete files". Owner does not have a "None" selection for either files or directories. Additionally, for directories is the ability to change permissions of all items within the directory. Selecting this brings up a separate window whereby we can change owner, group and world access for files and for directories. Choices are the same as above.

The GUI has a drop-down box for group ownership so that the user can change group owner. If the user is root then owner also has a drop-down box to change the owner. Figure 3.17 shows that this GUI also displays the item's "Security context". This is part of SELinux which we will explore in Chapter 7.

3.5.4 Advanced Permissions

There are three other types of permissions to explore. One of these, SELinux, utilizes a far more complex scheme than the `ugo/rwx` mechanism described here and we will hold off on coverage of SELinux until later in the book. Another form of permission deals with the user ID bit and the group ID bit. This influences how a process will execute. This is reserved for executable files only and will be discussed in Chapter 4.

The last remaining form of permission is known as the *sticky bit*. Historically, the sticky bit was used to indicate whether a process, which was no longer executing, should remain in swap space (in case it was executed again in the near future). The use of the sticky bit was only provided in Unix and was largely abandoned. Linux implementations however have used the sticky bit for a completely different purpose. For Linux (and modern versions of Unix), the sticky bit applies only to directories and indicates that a writable directory has restrictions on the files therein.

Let's consider a scenario to motivate the use of the sticky bit. We want a directory to serve as a repository for group work of the users of the Linux system. We make the directory writable so that users can store files there. Thus, the directory's permissions are either 775 or 777 depending on whether only members of the directory's group or all Linux users can write to the directory. Let's assume this directory is for multiple groups so 775 won't be sufficient and thus the directory has permissions of 777. This can be a dangerous setting as it would allow the world to not only write new contents to the directory but also rename or delete the directory and any of its contents.

By setting the sticky bit, the files within the directory can be controlled by the file's owner or the directory's owner, but not the rest of the world. Specifically, users can write content to the directory. But users would not be able to delete or modify any files found there except files that they own. Now, the directory serves as a writable repository for all users but access to files is still dictated by the individual file permissions.

A file whose access right is 640 would be writable only by the owner of the file while being readable by members of the group. A file whose access right is 644 would be readable by everyone but could only be deleted or modified by the owner. A file whose access right is 660 would be readable, writable, modifiable and deletable by the owner and members of the group but not other Linux users.

To establish that a directory should have its sticky bit set, use `chmod` and set the permissions to `1777`. The initial `1` digit sets the world's executable status to `t`. Alternatively, set the sticky bit through the instruction `chmod o+t` *`directoryname`*. The value `t` is for the sticky bit. To remove the sticky bit, reset the permissions to 777 (or a more reasonable 755) or use `o-t`.

Let's consider an example, shown in Figure 3.18. The top portion of the figure displays a directory using `ls -al` so that we can see the current directory's (.) and parent directory's (..) permissions. This current directory is called `pub` (for public), which `foxr` wants to use as a repository for other users. Initially, the directory has `755` permissions. Into this directory, `foxr` places a file, `foo.txt`. This file is world readable. Both this directory and `foxr`'s home directory (the parent directory, or `..`) are world readable. These permissions make `foo.txt` accessible to every Linux user.

Now, foxr modifies `pub`'s permissions to 1777. This does two things. First, it makes the directory writable by group and world (access rights that they did not previously have) and also sets the sticky bit. We see this in the lower portion of the figure where the current directory's (.) permissions are updated (w's are now set for group and other, and the other has execute access of `t`). At some later point, zappaf copies or moves a file into this directory, `stuff.txt`. He would not have been permitted to place one of his files into this directory without these changes (although foxr could have copied it here).

Later, dukeg accesses the directory ~`foxr/pub` and performs `rm foo.txt` and `mv stuff.txt stuff2.txt`. Both commands fail because the directory is world writable but the sticky bit is set. Specifically, dukeg receives the error message `rm: cannot remove 'foo.txt': Operation not permitted` in response to trying to delete `foo.txt` and `mv: cannot move 'stuff.txt' to 'stuff2.txt': Operation not permitted` in attempting to rename `stuff.txt`. Had foxr only made the directory `777`, then both operations would have been permitted!

```
drwxr-xr-x.  2  foxr    foxr    4096  Jul 29  08:24  .
drwxr-xr-x.  2  foxr    foxr    4096  Jul 29  08:24  ..
-rw-r--r--.  1  foxr    foxr    1851  Jul 29  08:26  foo.txt
-----------------------------------------------------------
drwxrwxrwt.  2  foxr    foxr    4096  Jul 29  08:26  .
drwxr-xr-x.  2  foxr    foxr    4096  Jul 29  08:24  ..
-rw-r--r--.  1  foxr    foxr    1851  Jul 29  08:26  foo.txt
-rw-rw-r--.  1  zappaf  zappaf  1851  Jul 30  09:52  stuff.txt
```

FIGURE 3.18 Creating a world-writable directory using the sticky bit.

SECTION ACTIVITIES

1. Have you ever changed file permissions before? If so why? If not, come up with a reason why you might want to or need to do so. Note: If you are the only user of the computer, this question will not make sense so assume other people in your household share your computer.
2. Of the various ways to change permissions using chmod, which of the three approaches do you favor and why?

3.6 HARD AND SYMBOLIC LINKS

A directory stores items such as files and subdirectories. In reality, the directory stores *links* to these items. The links are formally known as *hard links*. Each hard link consists of the item's name and the item's dedicated inode number. The inode number is then used to index into or reach the inode, which is stored on disk. It is through the inode that the item's location in storage is reached. Thus, the name in the directory references an inode. The inode contains pointers to the file's physical location(s).

We generally have a single hard link per file and two hard links per directory. But we are not restricted to this pattern. We can create additional links and store them in other locations or by other names.

The reason for adding links is to make access easier on users. Let's assume /home/zappaf/mystuff has a file called guitar_solos.txt. We want to frequently access this file but do not want to have to remember its location or have to type in the full path. To avoid having to reference the full path to the file, we create a link, stored in our home directory, which points to /home/zappaf/mystuff/guitar_solos.txt. If the link is called zappastuff, then we can access zappaf's mystuff just using zappastuff in our home directory.

There are two types of links. As noted, entries in a directory are hard links which reference the inode of a file. In the previous paragraph, guitar_solos.txt is actually a hard link to its inode which itself has pointers to point to the file's disk blocks. If we create a hard link to guitar_solos.txt, then we are duplicating the original link. The hard link stored in our directory will be identical to the original hard link. We can also create a soft, or *symbolic*, link to guitar_solos.txt. If we do so, then this link only stores a pointer to the original hard link.

Figure 3.19 illustrates this concept where the original file, file1, is in foxr's home directory. zappaf creates both a symbolic link, file1a, and a hard link, file1b, both of which point to file1. In the figure, we see that both file1a and file1b are entries placed into zappaf's home

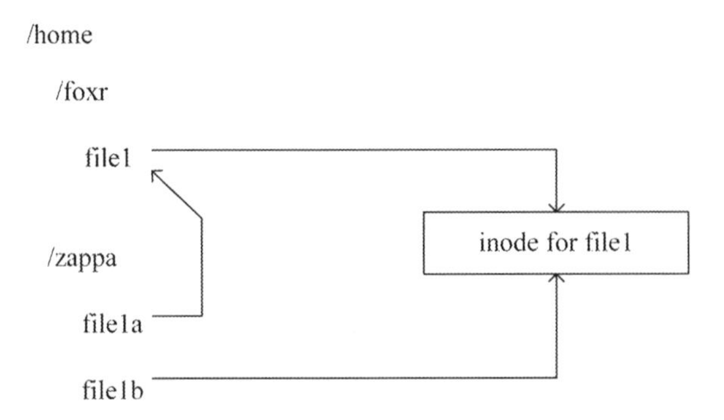

FIGURE 3.19 Hard vs symbolic link. (From Richard Fox, 2021, *Information Technology. An Introduction for Today's Digital World*, 2nd edition.)

directory. The difference is that file1b points directly at the file's inode whereas file1a points at the hard link file1 in foxr's directory. If this is confusing, we explore hard and symbolic links and inodes in detail in Chapter 8.

We saw earlier in the chapter that hard and symbolic links appear differently when viewed using ls -l. With the hard link, the only indicator we have that this item is a duplicate hard link for an entry elsewhere is the link count. With symbolic links, we clearly see that the item is a symbolic link (the l file type) and we see what is linked to as indicated by the name.

You might think that the symbolic links are preferred, but the hard link is more efficient. This is because we are merely duplicating the content of a directory entry. Both entries point at the same inode. Accessing an item through one hard link takes no more time than accessing the item through the other hard link. The symbolic link, as shown in Figure 3.19, also creates a new directory entry, like the hard link. The symbolic link adds another access slowing down the process as to reach the file through a symbolic link, we first must follow the symbolic link (pointer) to the hard link stored elsewhere to reach the inode to reach the contents of the file.

Creating a link is handled through the ln instruction. The syntax is ln [-s] *existingfile newfile* where the *existingfile* must permit links (there are permissions that can be established that forbid links to a file) and *newfile* is the newly created link. The option -s is used to create a symbolic link; otherwise, the link by default is a hard link.

There are two other points to raise to differentiate between the two types of links. Hard links can only point to contents within their current partition. If, for instance, we want to point to a file /etc/foo from /home/foxr, we cannot use a hard link because /etc and /home are situated in different partitions. On the other hand, from /home/foxr, we can create a hard link to point to /home/zappaf/mystuff because both the file and the hard link are in the /home partition.

The other issue is what happens when an item being linked to is deleted. Let's consider first that file foo has two hard links to it, one is the original hard link called foo and the other is called foo2. When we do a long listing on either, we see a link count of 2. If we were to delete either of these, we are removing a hard link and reducing the link count from 2 to 1. The link count is stored in the inode. The file itself is not yet deleted because there is at least one hard link pointing to it. All we have done is remove a hard link from a directory. Deleting the other hard link causes the link count to become 0 and now the file is deleted.

What happens when foo is a hard link to a file and foo2 is a symbolic link? Deleting foo2 has no affect at all on foo. Deleting foo though is disastrous for foo2 because foo2 was pointing at foo the hard link, not foo the file. Now foo no longer exists and so foo2 is pointing at nothing. When we view foo2 using ls the name has changed color to indicate a dead link. Should we try to follow foo2, for instance with the instruction cat foo2, we receive an error because the item being pointed at no longer exists.

We demonstrated creating a link to the file guitar_solos.txt so that our access to the file was simplified. System administrators will often create links for convenient access. But another role for a link is when content has been moved. In earlier versions of Linux, user programs were stored in one of two directories, /bin and /usr/bin, while system administrator programs were divided among /sbin and /usr/sbin. With systemd-based Linux (e.g., Red Hat starting with Red Hat 7 and Debian starting with Debian 8), all content from /bin has been moved to /usr/bin and all content from /sbin has been moved to /usr/sbin. /bin and /sbin remain merely as symbolic links so that a user who forgets and thinks something is in /bin or /sbin will follow the link to /usr/bin or /usr/sbin. We find many instances of links used for this purpose.

When a user creates a link, what permissions are used for the link? If the item being linked to is not accessible by the user, the instruction to create the link will fail. When creating a hard link, the new hard link takes on the same permissions, owner and group as the original. If zappaf tries to create the hard link in Figure 3.19 to point to foxr's file1, the ln instruction would likely fail because zappaf cannot create an item owned by foxr. Root could do this for zappaf if the system administrator found a reason to do so. On the other hand, if file1 is accessible to zappaf, then zappaf could create a symbolic link. The symbolic link would have an owner and group of zappaf and permissions of `rwxrwxrwx` as all symbolic links are given this level of permission. Notice that even though the permissions seem to indicate world writable, this is not actually the case.

The most common use of a hard link is to have multiple access points to directories, which have a link count of 2 at a minimum, but some have additional hard links. The most common use of symbolic links is to point to files that have been moved or provide easy shortcuts to those files. A user might set up a symbolic link to a commonly used resource. As an example, Linux contains a textfile dictionary in `/usr/share/dict` called `words`. We might use the instruction `ln -s /usr/share/dict/words ~/words` to set up our own symbolic link, `words`, to the file.

SECTION ACTIVITIES

1. Windows has its own version of a symbolic link called a soft link, and it is implemented as shortcut (desktop) icon. Have you created your own shortcut icons before? Did you have any difficulty creating them? What happens to the original file if you delete the shortcut icon? Give it a try.
2. If you use Windows, from your desktop right click on a shortcut icon and select Properties. From the pop-up window, explore the shortcut's properties which among other things tells you the command used to launch the process and where it is located. Had you ever looked at this before?
3. In Linux, perform `ls -l | less` on several directories, notably `/` and `/etc` and look for symbolic links. How many did you locate?

3.7 LOCATING FILES

In this section, we examine different mechanisms to search through the Linux file space for particular files or directories that match some criteria. We use several tools starting with the simplest, the File Browser GUI. We then move on to more challenging forms.

3.7.1 Search Using the File Browser

We referenced the File Browser GUI earlier in this chapter. This tool has a useful search feature. The search bar is indicated as a magnifying glass icon. Before using the search bar, we move to the directory where we want our search to begin from. Then, in the search bar, we type a string of what we are searching for. The string does not have to be a full filename.

As an example, we want to find the `gdm` (Gnome Desktop Manager) program. We open a File Browser window. By default, it opens in our user's home directory. We select Other Locations and from there, My Computer, which is equivalent to the topmost level, /. Now, in the search bar, we enter `gdm`. It searches for all items (directories and files) that contain the specified string as part of its name. Icons with an `X` in them indicate that we do not have permission to examine the item whereas items with a padlock on them indicate that the item is readable but not writable. See Figure 3.20 which shows a portion of the result when searching for `gdm`. For each item located, we are given its full name, the path for each and the type of item (file, directory).

The search box has a drop-down menu. Selecting this gives us options to specify a search date and a type of item. Search dates allow us to specify a time unit that the item was either last modified or last used. Selectable times are any time, 1, 3 or 5 days ago, 1, 2, 3, 4 weeks ago, 1, 4, 7 or 10 months ago, 1, 2, 3, 4 or 5 years ago. We can also insert our own date by using a calendar. The item's type can be files, folders (directories) or specific types of files such as a Document, Illustration, Music, PDF/Postrscript, Picture and Text File, among others. Finally, we can search by Full Text or File

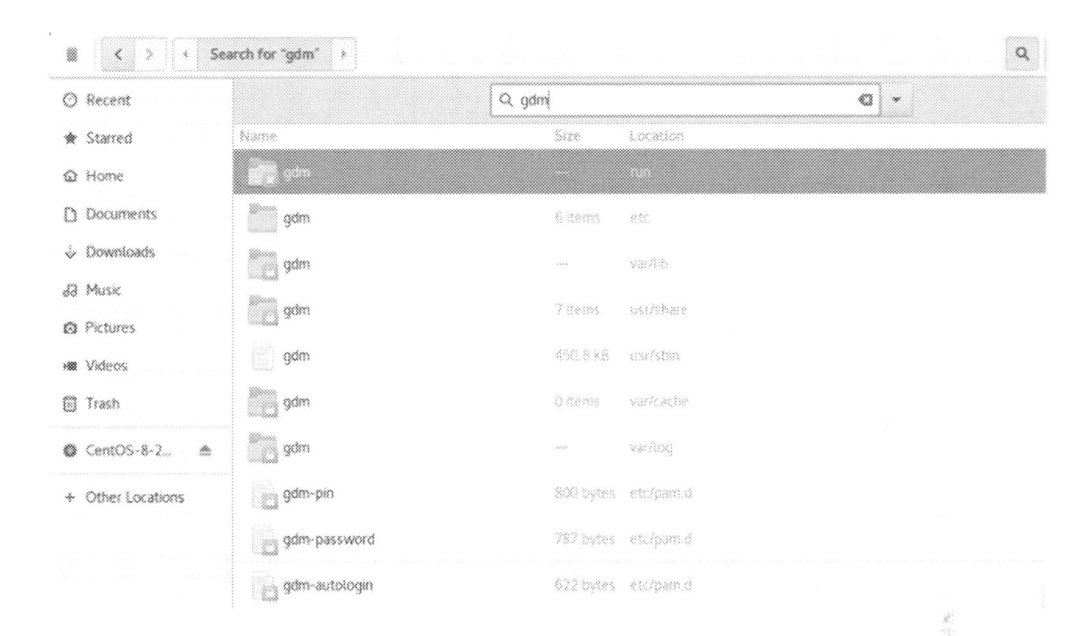

FIGURE 3.20 Searching through the file browser GUI.

Name. The search in Figure 3.20 is by filename. If we use Full Text, the search examines the search string against both filenames and contents found within the file. This only works for text files.

Although searching with the File Browser is convenient, it is not as powerful as using the `find` command. It is also only available if we have logged in through the GUI. So, let's turn to...

3.7.2 The `find` Command

The `find` program is a very powerful tool. Unfortunately, to master its usage requires a good deal of practice as it can be complicated. The `find` command's format is `find` *`directory options`* where *directory* is the starting position in the file space for the search to begin. There are several different forms of *options* that we discuss below but at least one form of the options (expression) is required.

The simplest form of `find` is to locate files whose name matches a given string. The string can include wildcard characters. For this, we use the option `-name` followed by `"`*`string`*`"`. For example, we would issue `find /etc -name "*.conf"` to locate all files ending in `.conf` in the `/etc` directory. A variation of `-name` is the option `-iname` which is a case insensitive version of the same search.

The above `find` command is similar to using `ls /etc/*.conf` except that the `ls` command only searches the `/etc` directory, not any of its subdirectories. We could add `-R` to the `ls` command to perform a recursive search so that it becomes equivalent to the above `find` command.

The `find` command provides numerous options which increase its usefulness over `ls`. Options are divided into three categories. The first set of options are the form of *search criteria* to perform, of which `-name` and `-iname` are two. The second set of options are like those of other Linux instructions that alter `find`'s behavior. The third type of option controls what `find` will do when it has located an item that matches the expression(s). We will refer to these three types of options as search criteria, options and actions.

Let's look at the search criteria first. Table 3.10 illustrates some of the more useful expressions available. The search options are expressed using `-`*`criteria`* such as `-anewer` or `-empty`. Most, but not all, of the search expressions require a parameter. In the table, *n* indicates an integer (time, size or UID/GID), *file* indicates a filename, *test* indicates a set of permissions, *type* represents either a file type or a file system type, *name* is a user or group name, and *pattern* is a regular expression.

TABLE 3.10
Search Options in Find

Expression	Meaning	Example
`-amin [+-]n`	Find all files accessed >,<or=*n* minutes ago.	`-amin +10`
`-anewer file`	Find all files accessed more recently than *file*.	`-anewer file1.txt`
`-atime [+-]n`	Find all files accessed >,<or=*n* days ago.	`-atime -1`
`-cmin [+-]n`	Find all files with a status change >,<or=*n* minutes ago.	`-cmin 60`
`-cnewer file`	Find all files with a status change more recently than *file*.	`-cnewer abc.dat`
`-ctime [+-]n`	Find all files with a status change >,<or=*n* days ago.	`-ctime +3`
`-mmin [+-]n`	Find all files last modified >, <,=*n* minutes ago.	`-mmin -100`
`-mtime [+-]n`	Find all files last modified >, <,=*n* days ago.	`-mmin 1`
`-newer file`	Find all files modified more recently than *file*	`-newer foo.txt`
`-empty`	Find all empty files and directories.	
`-executable`, `-readable`, `-writable`	Find all files that are executable, readable, or writeable.	
`-perm test`	Find all files whose permissions match *test*.	`-perm 755`
`-fstype type`	Find all files on a file system of type *type*.	`-fstype nfs`
`-uid n` `-gid n`	Find all files owned by user ID *n* or group ID *n*.	`-uid 1001`
`-user name` `-group name`	Find all files owned by user *name* or group *name*.	`-user zappaf`
`-regex pattern`	Find all files whose name matches the regular expression in *pattern*.	`-regex [abc]+\.txt`
`-size [+-]n`	Find all files whose size is >, <,=*n*; *n* can be followed by b (512-byte blocks), c (bytes), w (2-word bytes), k (kilobytes), M (megabytes) and G (Gigabytes).	`size +1024c` `size +1k`
`-type type`	Find all files of *type* where *type* is one of `b` (block), `c` (character), `d` (directory), `p` (pipe), `f` (regular file), `l` (symbolic link), `s` (socket).	

The notation `[+-]n` from Table 3.10 needs a bit of explanation. The time and size conditions use an integer number to compare against each file's property. If the integer is by itself, then `find` looks for an exact match. If the integer is preceded by +, then `find` looks for matches where the file's corresponding property is greater than the integer. With `-`, `find` looks for properties whose values are less than the integer. For instance, `-size 1000c` means "exactly 1000 bytes in size" while `-size -1000c` means "less than 1000 bytes in size".

Search criteria can be combined using `-and` or `-a` to represent ANDed criteria while `-o` or `-or` represent ORed conditions. find assumes multiple conditions are ANDed together if the `-and/-a` is omitted.

Let's take a look at several examples, shown in Figure 3.21. The first three are all equivalent; each seeking to find files in the user's home directory whose sizes are between 100 and 200 bytes. The

```
find ~ -size +100c -size -200c
find ~ -size +100c -a -size -200c
find ~ -size +100c -and -size -200c
-----------------------------------
find ~ -uid 1000 -o -gid 1000
-----------------------------------
find /dev ! -type c
find /dev -not -type c
```

FIGURE 3.21 Several `find` examples.

fourth example in the figure searches for files whose owner's ID (UID) is 1000 or whose group owner's ID (GID) is 1000. It might be the case that an account's UID and GID are not the same number.

The last two examples in Figure 3.21 show how to negate a condition when we want to find all files where the condition is *not* true. We see we can either use ! `expression` or `-not expression`. Both instructions search for the same thing: all files in `/dev` that are not of character type. If our expression contains numerous parts and we need to specify operator precedence, place them in parentheses. Without parentheses, not is applied first, followed by and, followed by or.

In addition to the search criteria are further expressions that control how the search criteria are applied. Adding `-daystart` to one of the time-based criteria (e.g., `-amin`, `-atime`) causes the time comparison to be not the current time but the beginning of the current day. To control the depth of the search, use `-mindepth` or `-maxdepth` where `-mindepth` is used to start the search at the number of levels below the current level (e.g., `-mindepth 1` would start the search not in the specified directory but in all subdirectories of the specified directory) and `-maxdepth` limits the search to no more than the specified levels including the specified directory (for instance, `-maxdepth 0` will only search the specified directory and no subdirectories). `-mount` indicates that any directories that are part of other file systems should not be searched.

There are few options available for `find`. The first three deal with the treatment of symbolic links. Option `-P` specifies that symbolic links should never be followed. This is the default behavior and so can be omitted. Option `-L` forces symbolic links to be followed. If an item in a directory is a symbolic link, then the expression is applied not to the link but to the item being pointed at. As an example, if a symbolic link points to a character-type file and the search criteria is `-type c`, then the file will match when using `-L` but not without `-L`.

The option `-H` is a little trickier to understand. Like `-P`, it prevents `find` from following symbolic links when searching for files unless, as we will see below, the item should have an action performed on it. That is, `-H` only follows symbolic links when the linked item will be enacted upon by the `find` command's action. Note that if we use `-L`, `-H` and `-P` in some combination, find only applies the last one that appears in the instruction.

There are a couple of forms of debugging/informational options. For debugging, the option is `-D` followed by one or more debugging options (`all`, `exec`, `help`, `opt`, `rates`, `search`, `stat`, `tree`). See `find`'s man page for more detail. We can also add `-warn` (which is applied by default) or `-nowarn` to turn on or off warning messages. Warning messages include noting directories that we do not have permission to access.

Ordering of search criteria can have an impact on the performance of `find`. For instance, if we have `-size +0 -a -atime -1` then it is probably more likely that the first criteria will match (non-empty files) than the second (accessed within the last hour). It might be more efficient to write this criterion as `-atime -1 -a -size +0` so that if a file does not match `-atime -1` then find does not bother to test the file's size.

Another option is to specify optimization of search criteria. This is done through `-On` where *n* is one of `0`, `1`, `2` or `3`. The default is level `1` which causes `find` to reorder search criteria so that names and regular expressions are applied first. Level `0` is equivalent to level `1`. Level `3` provides the greatest amount of optimization so that when using `-o` criteria that are likely to be true are tested first and when using `-a` criteria that are likely to be false are tested first.

The default action for `find` is to list all matches found to the terminal window. `find` can also perform other operations, as indicated in Table 3.11.

As the `find` program is far more complicated than the GUI-based File Browser, you might question its use. However, keep in mind that the File Browser takes more resources to run and would be unavailable if you are accessing a Linux machine through a text-only interface. Additionally, the `find` command offers a greater degree of control. Finally, `find` can be used in a shell script or as part of a command that involves a pipe. If you will serve as a system administrator on a Linux system, `find` becomes an essential tool. Just have patience when learning it as there are a lot of parts to learn.

TABLE 3.11
`find` Actions

Option	Meaning	Example Usage
`-delete`	Delete all files that match; if a deletion fails (for instance because the user does not have adequate permissions) output an error message.	`find ~ -empty -delete` Deletes all empty files.
`-exec` *`command`* `\;`	Execute *command* on all files found; as most commands would use the files found as parameters we indicate this using { }.	`find ~ -type f -exec wc -l {} \;` Executes wc –l on each file of type f (regular file).
`-ok` *`command`* `\;`	Same as `exec` except that `find` pauses before executing the command on each matching file to ask the user for permission.	`find / -perm 777 -ok chmod 755 {} \;` Ask permission and then change permission of found items to 755.
`-ls, -print`	Output found item(s) using the command `ls -dils`, or by full file name respectively.	`find ~ -name *core* -print`
`-printf` *`format`*	Same as `-print` except that we provide formatting specifiers to output such information as the found files' last access times, modification dates, sizes, depths in the directory tree, etc.	`find ~ -size +10000c -printf %b` Output size in disk blocks of all files greater than 10,000 bytes.
`-prune`	If a found item is a directory, do not descend into it.	`find ~ -perm 755 -prune`
`-quit`	If a match is found, exit immediately.	`find ~ -name *core* -quit`

3.7.3 Other Means of Locating Files

There are several other options to locate items in the Linux file space. To locate an executable program, use `which` *`name`* where *name* is the program's name. This works as long as the program is located in a directory that is part of the user's `PATH` variable. The program `whereis` is similar in that it will output a program's location. But `whereis` does not rely on the `PATH` variable. Additionally, if available, `whereis` also reports on all locations relevant to the program: the executable program (binary), the source code (usually stored in a `/lib` subdirectory) and the manual page (usually stored under `/usr/share/man`). The instructions `which cd` and `whereis cd` provided output of `/usr/bin/cd` (`which`) and `/usr/bin/cd /usr/share/man/man1/cd.1.gz /usr/share/man/man1p/cd.1p.gz` (`whereis`).

Another search program is `locate` which accesses a database storing file locations. Before using `locate`, the database must be updated or created with the `updatedb` command (which can only be run by root because it accesses a protected file). Keep in mind that updating the database can be time consuming and so should only be done when needed. To reduce the updating time, the administrator can use option `-U` to specify only a portion of the file space to update. For instance, `updatedb -U /home/foxr` would only update the portion of the database corresponding to user foxr's home directory.

Once the database is established, the `locate` instruction is simply `locate` *`[option] string`*, which responds with the location of all files that match the given *string* but within the directory(ies) specified when issuing the `updatedb` command. `locate` might output a good deal more items than `which/whereis`. `which` and `whereis` are used to find executable programs while `locate` finds all files of the given name. The common options for `locate` are `-c` to count the number of matches found and output the count rather than the matched files, `-d file` to use *file* as a substitute database over the default database, `-i` to ignore case, `-L` to follow symbolic links, `-n #` to exit after # matches have been found, and `-r regex` to use the regular expression (*regex*) instead of the string.

SECTION ACTIVITIES

1. Have you searched for files before using a GUI? How easy was the task? How long did it take?
2. The `find` program can be challenging to use well. Look at the various criteria you can use to find programs like "all programs older than this one" or "programs that are greater than X bytes". Try to come up with reasons for using the various criteria available for searching.

3.8 SECONDARY STORAGE DEVICES

Let's focus on secondary storage devices to get a better idea of what they are and how they operate. Storage devices differ from main memory in three ways. First, we use storage devices for *permanent* storage whereas main memory is used to store the program code and data that the processor is using, has used recently or will use in the near future. Second, storage devices are usually non-volatile forms of storage meaning that they can retain their contents even after the power has been shut off. This is not the case with DRAM (main memory) or SRAM (cache memory and registers), which are forms of volatile storage. Third, storage devices have a far greater storage capacity than main memory. The cost per byte of storage for hard disk is extremely low when compared to that of DRAM, and flash memory used for secondary storage continues to drop in cost. On the other hand, DRAM access time is far shorter than hard disk access time and somewhat shorter than flash memory access time.

There are several different technologies used for storage devices. The primary form is the hard disk drive. There are also optical disk drives, magnetic tape and flash memory. Although magnetic tape is seldom used by individual users today, large organizations may still use it for backups. Another form of storage, which is obsolete, is the floppy disk. Each of these technologies stores information on a physical medium except for flash memory. In the case of floppy disk, optical disk and magnetic tape, the media is removable from the storage device. In the case of the hard disk drive, the disk platters are permanently sealed inside the disk drive. Flash memory may be removable (USB drives which are removable from USB ports) or fixed (solid-state drives). Table 3.12 illustrates differences between these devices.

3.8.1 The Hard Disk Drive

The hard disk drive is the most commonly used form of secondary storage. Most laptop and desktop computers come with an internal hard disk drive. In most cases, the operating system will be stored here. Additionally, the user(s) of the computer will most likely install all software and store all data files onto this drive. Alternatively, some data files may be saved on other drives such as an external hard disk, optical disk or USB (flash) drive.

The hard disk drive is a self-contained unit that combines the media storing the files with the technology to access the media. The media are several *platters* of "hard" (rigid) disks. The disk drive unit also contains a spindle to spin the platters, two read/write heads for each platter (one above, one below) and the circuitry to move the read/write heads and perform the disk access. The surface of the hard disk is broken into logical segments called *tracks* and *sectors*. This segmentation is performed during disk formatting. See Figure 3.22 for an illustration of the hard disk drive.

TABLE 3.12
Characteristics of Storage Devices

Device	Maximum Transfer Rate	Common Size	Common Cost	Common Usage
Hard disk	200 MB/s	1–4 TB	$50-100	Store the operating system, application software and most user data.
Optical disk (CD, DVD, Blu-ray)	150 KB/s (CD) 20 MB/s (DVD) 54 MB/s (Blu-Ray)	650 MB (CD) 5 GB (DVD) 25 GB (Blu-ray)	$.50 (CD) $2 (DVD) $5 (Blu-ray) (for single discs, bundled packages are less)	Store multimedia data and software; offers portability.
Magnetic tape	300 MB/s	Up to 580 TB but commonly 15 TB	Varies based on size but commonly $50–100	Store backup and archival data.
Solid-State Drive	3.5 GB/s	Up to 4 TB but sizes are generally no more than 512 GB	Varies based on size, up to $700 for 2 TB, $200 for 256 GB	Used in place of or in addition to hard disk; often supports storage area networks (SAN).
USB (3.0)	500 MB/s	Up to 512 GB	$20–40	Store data files for portability; not impacted by magnetic fields.
Floppy disk	1 Mbit/s	1.44 MB	$0.50	Obsolete.

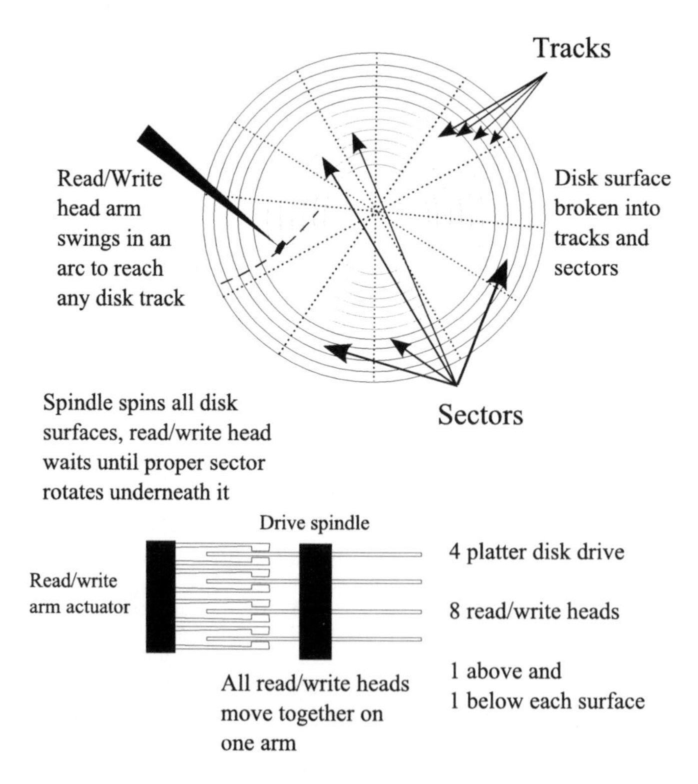

FIGURE 3.22 Hard disk technology. Access, read/write heads, spindle. (From Richard Fox, 2021, *Information Technology. An Introduction for Today's Digital World.*)

The hard disk drive uses a form of magnetic storage. Information is placed onto the surface of the disk as a sequence of magnetized spots, called *magnetic domains* or *magnetic moments*. These magnetized domains correspond to the magnetization of the atoms of the region. One bit is stored as a magnetized spot where the atoms in the region are aligned in one of two directions to represent the 1s and 0s of binary.

Hard disk drives spin their disks very rapidly (anywhere from 5400 to 15,000 revolutions per minute) and have a high transfer rate. The hard disk drive is sealed so that the environment (dust, particles of food, hair, etc.) cannot interact with the disk surface. Hard disk technology continues to improve in terms of both transfer rate and storage capacity.

Information stored on disk is broken into *blocks*. Each block is of a fixed size. This allows a file to be distributed in blocks across the surface of the disk and even across multiple surfaces. The advantages of this are twofold. First, because all blocks are the same size, deleting a file is handled by indicating that the file's blocks can be reused. Thus, the file is not physically deleted which allows us to possibly recover a deleted file if its blocks have yet to be reallocated. Second, as the disk drive rotates the disk rapidly, placing disk file content in contiguous locations would not be efficient because the transfer of data takes some time and so the disk will have rotated beyond the next disk block location when the next block is ready for reading/writing. The drive would have to wait for another revolution before access could resume.

Disk access time consists of three distinct parts: *seek time*, *rotational latency* (or rotational delay) and *transfer time*. Seek time is the time it takes for the read/write head to move across the surface of the disk from its current location to the intended disk track. Rotational latency is the time it takes for the desired sector to rotate underneath the waiting read/write head. Transfer time is the time it takes to read the magnetization of the disk's surface and move that information as binary data to memory, or the time it takes to receive the binary information from memory and write each bit as a new magnetic moment onto the disk.

Transfer time for most hard drives has been timed at between 100 and 200 Gigabytes per second. Seek time is dependent on where the read/write head is now and where it has to move to and the speed of the read/write head's arm. Rotational latency is dependent on which sector is currently under/over the read/write head, which sector is being sought, and how rapidly the disk is spinning. Since sectors take up more space as the read/write head moves outward from the center of the disk, rotational latency is also impacted by the specific track that the block in question is located on. Disk access for a block is measured in milliseconds (thousands of a second).

There are three drawbacks to the hard disk: expense, lack of portability and fragility. As of 2021, the cost per terabyte of storage is between $20 and $25 where a 4TB drive might cost around $100. Although this seems cheap (and it is), tape storage gives us a greater amount of storage capacity for less cost and is portable. However, tape, as we will see in the next subsection, is not a very efficient form of storage. The storage capacity per penny of an optical disk and a USB drive are both greater than the hard disk making both forms impractical for large-scale storage (e.g., hard disk backups). We comment on the third drawback shortly.

The hard disk and the floppy disk are similar in that they both use read/write heads, store magnetic information and rotate to provide access. The floppy disk is removable from its disk drive. This causes the floppy disk's surface to be exposed to the environment. Because of this, the floppy disk has a much lower density of magnetic regions such that floppy disk storage capacity is significantly less than any other form of storage media. Floppy disk technology has been obsolete for decades.

Hard disk drives are the primary form of storage today. We rely on them to store our operating systems, application software and data files. But hard disks are also more susceptible to failure than other components of our computer because they contain moving parts. Because of these two factors, it's important to back up our significant files onto some other form of storage. This might be an external hard disk or a hard disk available through cloud storage. If the amount of storage is not large, one or a few USB flash drives might be used. For large organizations, magnetic tape is often used.

3.8.2 Magnetic Tape

Magnetic tape is often thought of as obsolete, just like the floppy disk. Like the hard and floppy disk, magnetic tape stores information as magnetized spots, here on the surface of a specially coated tape. And like the floppy disk, magnetic tape is removable from the tape drive. But magnetic tape is more effective than floppy disk in that magnetic tape can store a great deal of information on one unit (tape). Today, storage capacities of tape cartridges exceed 15 Terabytes and can be as much as 150 TB. The advantage of tape is its ability to back up several hard disks' contents on a single tape cartridge. Although tape also offers a form of portability, tape is never used in this way because few users use tape and the tape must be compatible with the user's tape drive.

The drawbacks of tape are that to use it we need to also purchase a magnetic tape drive, and tape offers a much poorer access time than other forms of storage. The tape drive is a one-time purchase but can cost more than several hard disk drives. Performance is really the more significant issue.

A file is stored on tape in one contiguous block, unlike a disk file which is broken into blocks and distributed across one or more disk surfaces. While this sounds advantageous in terms of accessing the full file, it leads to the possibility that a deleted file's space could not be reused. Consider that we have three files stored consecutively on tape. We delete the middle file. Can that space be reused? If you have used either audio cassettes or videotapes to record material, you probably know the issue here. We do not want to record something in between two or more items for fear of running over the latter item. Recording over a part of a data file corrupts that file. Thus, reusing freed-up space is often not possible on tape.

A second performance issue is that the tape must be repositioned so that the file of interest is under the tape drive's read/write head. On disk, repositioning requires moving the read/write head's arm to the proper track and waiting for the sector to come underneath the read/write head, which can take milliseconds. On tape, repositioning requires fast-forwarding or rewinding the tape. Tapes can be as long as half a mile in length so that fast-forwarding or rewinding can take literally minutes of time.

Because of these drawbacks, tape tends to be used for only two purposes. The first is to *back up* the contents of a hard disk. As our hard disk usually stores all of our software and data files, losing any part of or the entire hard disk's contents can be devastating. Performing timely backups ensures that we will not lose much if our hard disk fails or is damaged. Large organizations often use magnetic tape for backups because it is more cost effective to buy a tape drive and a few to a few dozen magnetic tapes than it is to buy dozens or hundreds of external hard disks or one or a few file servers.

The second use of magnetic tape is related to the first, storing information as an *archive*. The archive is essentially data from a previous time period (e.g., last year) that the organization hopes to not need but stores anyway just in case it is needed. Alternatively, an archive might be used for research into previous customer trends, to perform an audit, or to answer some inquiry into transactions. For either the backup or archive, the items stored will be used infrequently enough that the poor access time is a worthwhile tradeoff to reduce cost.

3.8.3 Optical Discs

Optical disc was first introduced to the personal computer market in the 1980s. The technology centers around the use of a laser to shine onto an optical disc. The surface of the disk was normally reflective but pits would be burned onto the surface. A pit would swallow up the light whereas without a pit, light would shine back. In this way, 1s and 0s could be written onto a disk.

Optical disc was most commonly used as a form of read-only memory because most of the discs of that day had information permanently burned onto the surface of the disc. Burn marks cannot be removed. Thus, this technology was called a CD-ROM (compact disc read-only memory). The CDs would either store music, application software or multimedia data files. Once the technology became more commonplace, optical drives were being sold that could burn onto a blank disc. This form of disc was called a WORM (write-once, read-many).

A different form of optical drive was introduced in 1985 when the NeXT workstation offered a form of magneto-optical drive (CD-MO). Here, a laser would be used to read the magnetic state on the surface of the disc to pick up the corresponding bit. To write to the disc, first the given disk location would have to be erased and then written requiring two separate accesses. In addition to the laser, an electromagnet would be applied where the laser would heat up the disc and the electromagnet would adjust the bit stored in the given location. Although this technology was superior to the CD-ROM/WORM approach, it was too cost prohibitive to find a large audience.

In the late 1990s, a different form of readable/writable optical technology was released. The disc's surface comprised a crystalline structure which in its natural state is reflective. Instead of burning information onto the surface and permanently altering its contents, a heated laser is used to alter the crystalline structure so that it is no longer reflective. This allows us to erase the entire disc and reuse it, which has led to the more common CD-R and CD-RW formats. No matter which format of disc was used, it could only be erased wholly or not at all, unlike tape and hard disk which could be erased at the file/block level.

The optical disc led to both portable storage and a less costly storage device over the hard disk drives of the era. However, optical discs have a vastly limited storage capacity compared to hard disks, and this differential has grown over time as hard disk technology improves. Today, DVD and Blu-Ray formats allow greater storage capacity over CDs but they are still not very useful forms of storage as they are too slow in access time to replace the hard disk and too costly to be used for backups (it would take dozens to hundreds of disks to back up a single hard disk drive).

In today's market, optical discs are primarily used to store multimedia (music, movies/tv shows) and not normal computer data files. The exception is that operating system installation often comes on an optical disc because of the large size of the installation programs. Although optical disc drives were common in most personal computers and laptops, the use of broadband Internet and streaming audio/video has reduced the need for optical drives and so newer computers may not come with one.

3.8.4 Flash Memory Drives

Flash memory is a form of writable ROM (read-only memory) called EEPROM (electrically erasable programmable ROM). We primarily use flash memory to build two forms of storage: USB drives and solid-state drives (SSD). Although the technology is similar, their usage differs. USB drives are portable and plug into a computer via a USB port. SSDs are used either in place of a computer's hard disk drive or in addition to the computer's hard disk drive.

The main advantage of flash memory over hard disk storage is speed. The typical SSD is as much as 20 times faster than the typical hard disk drive. And the cost of SSD has been dropping significantly over the past decade so that an SSD is perhaps 2–3 times more expensive than a hard disk. Another advantage is that the SSD does not have moving parts. The hard disk drive is more likely to fail because of moving parts, especially if you move the computer while it is in use.

There are two drawbacks to the SSD. First, most SSDs are limited to no more than 512 GB of storage (thus having 4–16 times less storage space than a hard disk drive). The reason that the SSD is of such limited storage capacity is to keep the cost down, but we are also finding manufacturers are not willing to produce larger capacity SSDs (for now).

The other drawback is that EEPROMs have a limited number of accesses before the blocks stored in the EEPROM become unusable. There are both limitations in the number of reads and the number of writes. Estimates have shown that a specific block of flash memory can become unusable after perhaps 100,000 erasures. The actual number of erasures before a failure occurs will vary by specific device but it does indicate a limited lifetime for the device.

As a historical note, magnetic tape is the oldest of these storage media, dating back to the earliest computers. In those days, the media was reel-to-reel tape. Later came audio cassette and video cassette tape. The tape cartridges available today are far more cost effective because they can store as much as several to dozens of hard disk drives' worth of content. In the 1960s, magnetic floppy disk was introduced, followed by magnetic hard disk. Optical discs were not introduced until the 1980s. Flash memory was available in a rudimentary form in the 1980s, but the USB drive was not introduced until around 2000. SSDs became available in the 2010s.

The reason for the previous paragraph is to note that it's not only flash memory that has a limited number of accesses or duration. The fact is that that most magnetic tape, whether reel-to-reel or cassette, became unreadable after some decades of storage. Tapes stored in the 1950s and 1960s are no longer accessible and even those of 20–30 years ago may not be accessible. Optical disc also suffers from degradation over time (particularly CDs). In many cases, CDs have a lifetime of perhaps 10–20 years before data is no longer accessible. It is also thought that a hard disk will only last about 10 years.

3.8.5 Device Drivers

In order for the CPU to communicate with an attached device (whether storage, input, output or input/output), we need to install the proper *device driver.* A device driver is software whose role is to act as an intermediary between the CPU and the device itself. This is a requisite because the CPU will not know (nor should it) how to communicate with each type of device. Every device has its own functions, features and peculiarities. Therefore, the device driver program is required to translate from generic CPU commands to the specific actions needed to control the given device.

It used to be the case that aside from some hardware device drivers installed in ROM BIOS, we would have to install a device driver for every new piece of hardware. Today, many popular drivers are pre-installed in the operating system. Other popular device drivers are easily installed over the Internet. For Linux, most device drivers are pre-installed and available through kernel modules. We can find these modules under `/lib/modules/`version`/kernel/drivers` where *version* is the Linux version installed.

SECTION ACTIVITIES

1. Research and compare cost per byte of storage for hard disk, solid-state disk, USB drive and DVD.
2. One reason we still use tape is to perform backups of our hard disk. Do you back up your hard disk? If so, to where? An external hard disk? The cloud (which is also hard disk storage)? Optical disc? USB?

3.9 FILE COMPRESSION

We wrap up this chapter with a brief examination of file compression techniques and the file compression tools available in Linux. The reason to use file compression is to reduce the impact that a file's size has on disk storage and to reduce the time it takes to transfer large files over a network.

3.9.1 Types of File Compression

There are two forms of compression that we use to reduce file size: lossy and lossless compression. In *lossy compression*, information is discarded from the file. This loss cannot be regained and so if the information is important, the result is a degraded file. We typically use lossy compression for multimedia files where the loss will either be unnoticeable or deemed acceptable.

Audio lossy compression, for instance, will often eliminate data of sounds that are outside or at the very range of human hearing as such audio frequencies are not usually missed. Image and video lossy compression discard some of the image data resulting in perhaps a lesser quality image. With JPEG image compression, the image is slightly blurred as neighboring pixels are grouped together. The JPEG format can reduce a file's size to as much as 1/5th of the original file at the expense of having an image that does not match the original's quality. GIF image compression (which combines both lossy and lossless compression) uses a scaled-down palette of 256 colors so that, should the original image requires only colors of that palette, there is no loss in quality.

There are specialized formats for image, video and human speech recordings. In the case of both video and audio, there are compression algorithms to promote the streaming of content over the Internet. One idea for streaming video is to encode the starting frame's image and then describe, in a frame-by-frame manner, the changes that take place from the previous frame. We can describe the change between frame i and frame i+1 much more succinctly than we can describe frame i+1 in its entirety. Certain frames are stored as true images because, starting with the given frame, a new image/view/perspective is being presented in the video. Lossy compression of video and audio files can reduce file sizes as much as ten-fold for audio files while video files might be reduced by as much as 100-fold!

Lossless compression is a file compression technique whereby the contents of the file are being changed from one form to another such that the new form requires less storage space. The compressed file is not itself accessed like the original; instead, the compressed file must be uncompressed first before being accessed.

Most lossless compression algorithms revolve around the idea of searching for and exploiting *redundant* or *repetitive* data found within the original file. While this works fine for text files, it can also be applied to binary files although in many cases with less success. As many of our Linux data files are text files, we might look to compress these files to save disk space.

3.9.2 The Lempel–Ziv Algorithms for Lossless Compression

There are numerous algorithms that can perform lossless compression. Most are based on the idea of searching for repeated strings in the original file and then replacing those strings with shorter strings. Strings found in the file are placed into a *dictionary*. Their placement in the dictionary is then used as the replacement. For instance, a string like "The" might be replaced with 1 because "The" is the first string in the dictionary. Although the approach is more complex than this, this brief description will serve us adequately for this section.

The first example of a dictionary-based compression algorithm is Lempel–Ziv (LZ), named after the two mathematicians who derived the algorithm. Two popular versions of LZ are LZ77 and LZ78, numbered by the years the two algorithms were announced.

In LZ77, no explicit dictionary is built but instead when a string is found that matches an earlier string, the later string is replaced by a reference (pointer) to the earlier string. When searching for a matching string, LZ77 uses a "sliding window". Upon finding a string match, the window size is increased to see if the match can be made of a larger string. For instance, the algorithm has matched "The" but the current string is actually "There" so the window is enlarged by one character to see if the string being matched is "Ther" and if so, the window is increased again to see if the string being matched is "There".

Once a string has matched, it is replaced with a pointer to the original string. As the pointers will be small in size, removing a larger string for a shorter pointer always causes the document's size to decrease. The amount of reduction is dependent on repetitiveness found in the document and the maximum size of the sliding window. The maximum size of the sliding window is a parameter supplied to the algorithm.

With LZ78, an explicit dictionary is constructed. Replacement of strings matching entries in the dictionary is then made. With LZ78, the dictionary's size is restricted by a parameter supplied to the algorithm. So LZ78 has an upper limit to its usefulness directly dependent on the amount of repetition in the original document.

There are three types of comparisons we can make between the two algorithms: speed of compression, speed of decompression and amount of compression. LZ78 outperforms LZ77 at compression speed because LZ77's sliding window takes more time to utilize (when its maximum size is reasonably large) than the time it takes to produce the dictionary. LZ77 outperforms LZ78 at decompression speed because replacing the pointer with the matching string is fast while in LZ78, the dictionary must be recreated. Both algorithms will decompress faster than they will compress but LZ77's decompression is significantly faster than its compression.

In terms of the amount of compression, this is dependent on a number of factors including the size of the sliding window (LZ77), the maximum size of the dictionary (LZ78) and the amount of repetitiveness found in the document. With that said, the more effective compression algorithm is based on the type of data being compressed. In a standard text file, LZ78 will reduce the file's size more than LZ77. When tested on more repetitive patterns like DNA or bmp (image) files, LZ77 reduced the file's size to a greater extent.

In general, both algorithms trade compression time for compressed file size. The more time the algorithm can take to compress a file, the more likely the resulting file will be smaller. A user compressing a file may not want to wait a lengthy period of time. And therefore, the algorithm will have less time to search and so finds fewer combinations resulting in a less compressed file.

Numerous improvements have been implemented to both LZ77 and LZ78. These include Lempel–Ziv–Storer–Szymanski (LZSS) which offers an improvement over LZ77 that can, in some cases, produce more compressed files, and Lempel–Ziv–Welch (LZW) which is a modified version of LZ78 by having a pre-initialized dictionary so that all dictionary construction begins from the same starting point.

3.9.3 Other Lossless Compression Algorithms

Aside from the LZ algorithms, another popular algorithm is Burrows–Wheeler. This algorithm first sorts the text in strings to reorder the characters so that repeated characters appear multiple times. This allows those repetitions to be encoded into shorter strings. In order to obtain the greatest amount of compression, the algorithm looks for the longest repeated substrings within a group of characters (one word, multiple words, phrase, sentence). Once rearranged, compression can take place. Compression uses a substitution code like Huffman coding.

In *Huffman coding*, strings are replaced such that the more commonly occurring strings have smaller codes. For instance, if the string "The" appears more times than anything else, it might be given the code 110 while less commonly occurring strings might have four, five, six, seven, eight or even nine bits like 1000, 11100, 101011 and so forth. In Huffman coding, we have to make sure that codes can be uniquely identified when scanned from left to right. For instance, we could not use 00, 01, 10, 11, 000, 001, 011, 101 in our code because the sequence 000101101 would not have a unique interpretation. We might think that 0001011 is 00 followed by 01 followed by 011 followed by 01, but it could also be interpreted as 000 followed by 101 followed by 1. A code must be interpretable without such a conflict. We might therefore pick a better code using sequences like those shown in Figure 3.23.

In this way, 000001011001010101000010101010111101010101 could clearly be identified as starting with 000 001 011 001 010 10100001010101011101010101. As we have not enumerated the rest of the code, after the first five encoded strings, it is unclear what we have, but the first five are clearly identifiable.

Most common string: 000
Second most common string: 001
Third most common string: 010
Fourth most common string: 011
All other strings start with a 1 bit and are at least 4 bits long

FIGURE 3.23 Possible encoding scheme.

3.9.4 Compression and Decompression Programs in Linux

In Linux, there are several programs available for compression and decompression on text files. The most popular two are `gzip` and `bzip2`. `gzip` is based on the LZ77 and Huffman codes. The use of LZ77 performs string matching through a sliding window, and the use of Huffman codes permits the most commonly occurring strings to be encoded in as few bits as possible. To compress a file using `gzip`, the command is `gzip` *`filename`* which compresses the file, renaming it by adding `.gz` to the file's name. Notice that this changes the original file by compressing it and renaming it. The compressed file cannot be viewed until it is decompressed. The decompression command is either `gunzip` or `gzip -d`, and it will remove the `.gz` extension.

`bzip2` uses the Burrows–Wheeler. sorting algorithm combined with Huffman codes for compression. The `bzip2` program operates much the same as `gzip`/`gunzip` with the instruction being `bzip2` *`filename`* which renames the file to *`filename`*`.bz2`. Decompression is achieved using either `bunzip2` or `bzip2 -d`. As with `gunzip`, decompression removes the `.bz2` extension. There is also an older `bzip` program which is based on the LZ77 and LZ78 algorithms.

Another compression program is called `zip`, which uses the `DEFLATE` compression algorithm which combines both LZ77 and Huffman codes. Whereas `gzip` and `bzip2` compress/uncompress files, `zip` is used to both bundle and compress files. `zip`'s syntax is `zip` *`destination`*`.zip` *`sourcefile(s)`*. This differs from `gzip`/`bzip2` because it creates a new file which is the collection of zipped files bundled together. To unzip, use `unzip` *`filename`*`.zip`.

The Linux kernel is itself a partially compressed file (we examine this in Chapter 9). The compression algorithm used on it is a combination of LZ77, Huffman encoding and finite-state entropy. Some open-source compression programs use different algorithms. WinRAR uses LZSS. 7zip uses a variation of LZ77 called Lempel–Ziv Markov chain (LZMA). GIF images are compressed using LZW while PNG images use DEFLATE (LZ77 and Huffman codes).

SECTION ACTIVITIES

1. The operating system of your main computer (not your Linux computer) should have some form of built-in compression program for compressing/uncompressing files. What is it? Have you used it before? Compress a text file. By how much was it reduced in size?
2. In your Linux computer, compress the textfile `goose`, which you created in Chapter 2 when learning `vi` (if you did not create the file, select some other textfile) using `gzip` and `bzip2` to see which compresses the file more completely. Now try to compress a binary file (an image or music file, or an executable program like `/usr/bin/ls`) for a similar comparison.
3. Research streaming audio and video algorithms. Which forms of compression algorithms were you already familiar with from downloading or creating your own files and which are new to you?

3.10 CHAPTER REVIEW

Concepts and Terms Introduced in This Chapter

- Absolute path – path from root of the file space to the specified directory/file.
- Access control list (ACL) – form of resource protection by enumerating the users/groups access rights on the resource; see Linux permissions.
- Archive – setting aside data files for long-term/permanent storage.
- Backup – saving the significant files from the hard disk onto some other medium in case of data loss.

- DEFLATE – algorithm built into various Linux programs to perform lossless file compression/decompression; based on LZ77 and Huffman coding.
- Device driver – program that permits the CPU to communicate with a hardware device; drivers are installed as part of the operating system.
- Directory – a unit of organization which can contain files, subdirectories and links.
- File – the basic unit of storage in the logical file system; see also Linux file.
- File compression – process of reducing a file's size by either discarding data (lossy compression) or replacing strings with shorter replacement strings (lossless compression).
- File decompression (or decompression, uncompression) – restoring a compressed file to its original format; only available in lossless compression.
- File permissions – access rights placed on a file to control which users can access the file and the types of access.
- File space – the collection of all devices storing files; or logically the collection of all accessible files/directories/links.
- Filename – identifier given to a file by which we can reference it; operating systems have different rules and restrictions on names.
- Filename extension – a portion of a filename, typically following a period, and used to indicate the file's type such as .txt, .sh, .c, .gif or .html.
- Group – collection of specific users that share access rights to a file or directory; see also private group.
- Hard link – an entry in a directory which points at a file's inode; an inode can be pointed to by any number of hard links; when an inode is not pointed to by any hard link, the file is considered deleted and the inode is handed back to the file system.
- Home directory – a given user's dedicated storage space.
- inode – data structure storing file metadata and pointers to disk file blocks; covered in more detail in Chapter 8.
- Link – a pointer to an object in the file system; see hard link and symbolic link.
- Linux file – in Linux, "files" include regular files, directories and symbolic links, among other entities (which we explore in Chapter 8).
- Linux permissions – read, write and execute access assigned to the item's owner, group owner and everyone else (world); denoted as the second through tenth characters in a long listing.
- Logical volume manager (LVM) – a means of partitioning the file space logically rather than physical; covered in more detail in Chapter 8.
- Lossless compression – a form of file compression whereby the file's size is reduced while compressed but the file cannot be used until it is uncompressed; commonly applied to text files.
- Lossy compression – a form of file compression whereby content of the file is removed to reduce the file's size; the content cannot be reproduced but is typically not missed; commonly used to reduce the size of multimedia files.
- Mounting – the act of making a file system accessible.
- Parent directory – the directory in which the current directory is contained.
- Partition – a physical separation of the hard disk, each of which contains select directories and files; partitions can be operated upon independently.
- Path – the list of directories from the current location to the intended location; see absolute paths and relative paths.
- PATH variable – an environment variable which stores the directories that the Bash interpreter searches for an executable program when a command is entered.
- Permissions – access control to determine who can read, write or execute a file or directory.
- Private group – by default, every user is given their own private group; when they create files or directories, the new item is assigned to their ownership as user and their private group; we cover groups in more detail in Chapter 7.

- Recursive copy/delete – an operation to copy or delete not just the contents of a directory, but all subdirectories and their contents.
- Relative path – path specified from the current working directory.
- Root directory – denoted by /, this is the topmost directory; the top-level directories are those found within /.
- Sticky bit – used to place restrictions on file access in a directory that is world writable.
- Top-level Linux directories – a standard collection of directories that help Linux users and administrators know where to look.
- Wildcards – characters used to express a group of files with similar names; * matches all characters of a filename,? matches a single character of a filename, `[]` is used to match any character in brackets and {} is used to enumerate possible matching strings.
- Working directory – the directory that the user is currently interacting in.

Important Characters Covered in This Chapter

- * – wildcard meaning "match all"
- ? – wildcard meaning "match any single character"
- [...] – wildcard meaning "match any single character in the brackets"
- [!...] – wildcard meaning "match any single character that is not in the brackets"
- {...} – wildcard meaning "match any string in the brackets"
- ~ – user's home directory
- . – current directory
- .. – parent directory
- / – separate directories in a path

Linux Commands Covered in This Chapter

- basename – return just the name of an item (omitting the directory); see also `pathname`.
- bzip2/bunzip2 – compress/decompress files using Burrows–Wheeler and Huffman codes.
- cat – concatenate files; often used to display files' contents to the terminal window.
- cd – change the working directory.
- chgrp – change the group owner of a file; only available to root or if the user owns the file and is a member of the group being changed to.
- chmod – change the permissions of a file.
- chown – change the owner of a file; only available to root.
- cmp – compare two files byte by byte, displaying the point where the two files differ.
- comm – compare two sorted files, displaying all differences.
- cp – copy one or more files or entire directories to another location.
- cut – remove portions of every line of a file.
- diff – compare two (or more) files and display their differences.
- dirs – display all directories pushed to the directory stack (see also `popd`, `pushd`).
- file – output the type of entity, e.g., text file, a directory, a block device.
- find – locate files in the specified directory (and subdirectories) based on some criteria like name, permissions, owner, last access time.
- gzip/gunzip – compress/uncompress files using LZ77 and Huffman codes.
- head – display the first part of a file.
- join – join lines of files together which have a common value for one of its fields.
- less – display a file screen by screen; allows forward and backward movement.
- locate – locate files in the file system based on name; uses a pre-established database of file names generated through `updatedb`.
- ln – create a hard link; creates a symbolic link with the option `-s`.
- ls – list the contents of the specified or current directory.
- mkdir – create a new directory.

- more – display a file screen by screen; allows only forward movement.
- mv – move one or more files to a new location or rename a file.
- paste – unite files together by lines.
- popd – remove and return the top directory off of the directory stack.
- pushd – put the current directory onto the top of the directory stack for later recall.
- pwd – print the current working directory.
- rm – delete one or more files; can delete directories if the `-r` recursive option is used.
- rmdir – delete an empty directory.
- sort – sort the contents of a file.
- split – distribute a file's contents into multiple files.
- strings – output sequences of printable characters found in non-text-based files.
- tail – display the last part of a file.
- touch – modify access dates of a file; can also be used to create an empty text file.
- uniq – examine a file for duplicated consecutive lines.
- updatedb – create or update the file database as used with `locate/slocate`.
- wc – perform a character, word and line count on the given file(s).
- whereis – return the location of a program's executable file, source code file and man page.
- which – return the location of the specified executable program as long as the program is in a directory found in the user's PATH variable.
- zip – lossless file compression that uses DEFLATE.

REVIEW QUESTIONS

1. Name two things you would see when viewing the file space logically. Physically.
2. True/false: The file is the smallest physical unit stored within the file space.
3. In Linux, filenames are limited to how many characters in length?
4. Why might you not want to have blank spaces in filenames in Linux?
5. In order to create a hierarchical file space, which type of file construct do you use?
6. Which of the following is not a top-level directory in typical Linux systems?

```
/boot   /dev   /etc   /home   /local   /usr   /var
```

7. Which of the following Linux file system commands can you issue on a file? On a directory? Some of these operate on both so list all that qualify for both questions.

```
ls   cd   mv   rm   cp   sort   touch   less   cat   wc   chmod
```

8. You are user zappaf. You were in the directory `/usr` and then entered the command `cd /home/foxr/temp`. What values would you find in these environment variables? `PWD`, `OLDPWD`, `HOME`
9. What is the difference between `home/foxr` and `/home/foxr`?
10. What does ~ represent? What does `~zappaf` represent?
11. Given the directory layout in Figure 3.3, write the path from `etc` to `temp2` using a relative path. Using an absolute path.
12. Given the directory layout in Figure 3.3, write the path from `temp` to `zappaf` using a relative path. Using an absolute path.
13. Given the directory layout in Figure 3.3, if you are currently in `temp`, what does `../..` refer to?
14. Provide one additional example of how you can reference the three files referenced in Figure 3.4 that are different from the three `cat` commands in the figure.

15. Given the directory layout in Figure 3.6, write the path from a to zappaf using a relative path. Using an absolute path.
16. Given the directory layout in Figure 3.6, write the path from zappaf to b using a relative path. Using an absolute path.
17. Given the directory layout in Figure 3.6, write the relative path needed to move from temp to /.
18. What is the difference between . and .. ?
19. Assume the current directory contains the following items (all of which are files):

```
abc.txt     abc1.txt    abb.txt    bbb.txt
bbb.dat     bbb         bbbb       c123.txt    ccc.txt
```

What files from the list are listed for each of the following commands?

a. ls *
b. ls *.txt
c. ls a*.*
d. ls ab?.txt
e. ls bbb*
f. ls bbb.*
g. ls c??.txt
h. ls [ab][ab][ab].*
i. ls [a-c]??.???
j. ls [!c]*

20. Given the list of files in the top portion of Table 3.2, which files will appear with each of the following ls commands?
    a. ls f*
    b. ls *b*
    c. ls *[a-c]*
    d. ls {file,file1,file2}.txt

21. Refer to Figure 3.6 for this problem. Assume you are currently in the directory foxr and that you have proper permission for each operation.
    a. Write a command to copy the files f1, f2, f3 from their current location to directory b1.
    b. Write a command to copy the files t1, t2, t3 from their current location to directory zappaf.
    c. Write a command to copy all of the contents in directory temp (not the directory, just its contents) to temp2.
    d. Redo a to move the files.
    e. Redo b to move the files.

22. What is wrong with this mv command? mv *.txt *.dat
23. Is the command mv foo1.txt foo2.txt a movement or a renaming operation? How do you know?
24. Referring to Figure 3.6, and assuming you are currently in the directory temp2, write a command to copy the directory a, along with all of its contents, to zappaf. Assume you have proper permission.
25. How does the -i option differ between cp and rm?

26. Referring to Figure 3.6, you are currently in `temp2` and enter the instruction `rm -rf *`. In what order will the contents be deleted?
27. Why is it important to alias `rm` to `rm -i`?
28. Referring to Figure 3.6, assume `temp` is empty. What happens when you issue the instruction `rm temp`? If this does not delete the directory, how should you delete it?
29. Referring to Figure 3.6, you type `rmdir temp2`. What happens?
30. Which of `more` and `less` gives you more control over moving through a file?
31. Provide a command to view from line 10 to the end of the file `foo.txt`.
32. Provide a command to view the last 50 characters of the file `foo.txt`.
33. What is the difference between `tail -n 10` and `tail -n +10` when issued on a file of 100 lines?
34. What is the difference between `tail -c 10` and `tail -n 10` when issued on a file of 100 lines and 1000 characters?
35. We have run the command `diff a.txt b.txt` and receive the following outputs. Explain what the outputs are telling you.

```
0a1
> something
4c4
< aaaaa
---
> bbbbb
```

36. Match the Linux command with its usage.

a. `cmp`	i. compares the contents of two sorted files
b. `comm`	ii. determines if any lines of a file are repeated
c. `diff`	iii. compares the contents of two files for the first mismatch
d. `uniq`	iv. compares the contents of two or more files for all mismatches

37. What restriction must be true for the `join` command to work successfully?
38. We have two files as shown below where the left column is `file1`, the right column is `file2`. What does `paste file1 file2` return?

```
aaaa    bbbb
cccc    dddd
eeee    ffff
```

39. The file `foo.txt` consists of 2000 bytes. We issue the command `split foo.txt new -d -b 500`. What are the files created?
40. What are the three values returned by `wc` by default?
41. Assume `file1.txt` exists and `file2.txt` does not. What happens when we issue the following instruction? `touch file1.txt file2.txt`

42. Assume we have the following long listing result. The ... are other details omitted for brevity. Answer the questions that follow.

```
-rwxrw-r-- zappaf cool      ... file1.txt
-rw-rw-rw- zappaf zappaf    ... file2.txt
-rw-r----- zappaf cool      ... file3.txt
-rwxrwxr-- zappaf cool      ... file4.txt
-rwxr-x--x zappaf zappaf    ... file5.txt
-rwxr-x--- zappaf cool      ... file6.txt
```

   a. You are not zappaf and not a member of cool. Which files do you have read access to? Write access to?
   b. You are not zappaf but are a member of cool. Which files do you have write access to? Execute access to?
   c. For each of the six files, convert its permissions to the appropriate 3-digit number.
   d. You are zappaf. Issue a `chmod` command using +/– to change `file1.txt` so that it is group and world executable.
   e. You are zappaf. Issue a `chmod` command using the 3-digit approach so that `file5.txt` is not accessible by group or world.
   f. You are zappaf. Issue a `chmod` command using whatever approach you like so that `file4.txt` loses it's execute access for user and group but gains write access for world.
   g. You are zappaf. Issue a `chmod` command using +/– so that `file2.txt`'s permissions become `444`.
   h. You are zappaf. Issue a `chmod` command using = to change `file5.txt` so that group gains write access and all three areas lose execute access.
   i. You are zappaf. You want to change `file1.txt` to be in your private group. What command do you issue?
   j. True/false: zappaf cannot change the owner of any of the six files.
43. Under what circumstance might you expect to see a file with 000 permissions?
44. Why does a directory usually have execute access?
45. What does the permission 1777 mean? Under what circumstance might you want this form of permission?
46. The directory mystuff currently has permission 755. You want to modify it to 1777. Using the +/– method of `chmod`, how would you do this?
47. A file has a link count of 3. You delete the file. What is the result? Is the file deleted? If not, what is deleted?
48. A symbolic link is pointing at a file. The file gets deleted. What happens to the symbolic link?
49. Write `find` commands for each of the following. In each case, start searching from your home directory.
   a. Find all files newer than the file `foo.txt`.
   b. Find all files whose size is more than 1000 bytes.
   c. Find all files that are empty and executable.
   d. Find all files whose user is yourself and whose group is cool.
   e. Find all files that were modified less than 10 minutes ago.
   f. Find all files whose size is less than 1K or were accessed more recently than `foo.txt`.

50. You want to locate all files under your home directory that have been accessed less than 10 minutes ago and are writable and get their word count using `wc`. Write this `find` instruction.
51. You want to change all files under your home directory that are writable and change their permissions to 644, pausing before each change to ask permission. Write this `find` instruction.
52. Which is more likely if you bought a computer today, it would have a hard disk of at least 2TB or it would have an SSD of at least 512GB?
53. True/false: SSDs have a limited number of reads and writes before blocks become inaccessible.
54. True/false: Disk and flash memory store files in blocks, magnetic tape does not.
55. What is the most common use of optical disc today? What is the most common use of magnetic tape today?
56. Explain why a user may wish to invest in a magnetic tape drive and magnetic tape cartridges.
57. What is the difference between lossy and lossless compression? When compressing a text-file using a program like `zip` or `gzip`, which form of compression is used?
58. Research the following compression formats and specify whether they are lossy or lossless: flac, gif, jpeg, mp3, mpeg-1, mpeg-2, mpeg-4, png, tiff, wma.

4 Managing Processes

This chapter's learning objectives are to be able to:

- Differentiate between the forms of process management
- Utilize Linux commands to launch and control processes from the command line including moving processes to the foreground and background and changing process priorities, specifically commands `bg`, `fg`, `jobs`, `nice`, `renice`, and keystrokes `&` and `control+z`
- Obtain and use process PIDs and explain the role of EUIDs and GUIDs
- Use Linux commands to monitor running processes and process resource utilization, specifically `free`, `iostat`, `mpstat`, `ps`, `top`, `vmstat` and the `System Monitor`
- Terminate Linux processes using `kill` and `killall`

4.1 INTRODUCTION

A *process* is a running program. We use the term process rather than program because a program is a static entity while a process has a *state* which changes over time. The state of a process is described in many ways. Is it currently being run by the CPU or waiting? If waiting, where? What are the values of its variables? Where in memory is it currently stored (if in memory)? What resources does it have assigned to it? The answers to these questions are just some of the ways we describe a process.

In Linux, processes can be started from the GUI or the command line or by other running processes. Some processes interact with the user; some run to completion and report their results in the terminal window in which they were launched but otherwise have no user interaction; and some report their results to a file. Those processes that run with user interaction are known as *foreground* processes while those that run with no interaction are *background* processes.

Whenever a process runs, the Linux kernel keeps track of it through a process ID (*PID*). After the Linux kernel is loaded and running, the first process it launches is called `systemd`. `systemd` is responsible for starting the run-time environment and then monitoring the environment while in use. `systemd` is given a PID of 1. From that point on, each new process gets the next available PID. If your Linux system has been running for a while, you might find PIDs in the tens or hundreds of thousands.

In Linux, a process can only be created by another process (with the exception of `systemd`). We refer to the creating process as the *parent* and the created process as the *child* whereby the parent process *spawns* the child process. Spawning of a process utilizes a system call of the parent process to the Linux kernel. There are several forms of child creation system calls. We look at these in Table 4.1.

TABLE 4.1
Linux System Calls to Create Processes

System Call	Result of the Call
`clone()`	Like `fork` (see below) except there is more control over the child process produced with respect to what is duplicated and what is shared between parent and child.
`exec()`	Take an existing process and replace its image (executable code) with a new image.
`fork()`	Create a duplicate process of the parent but with its own PID, its own memory and its own resources; parent and child can run concurrently.
`vfork()`	Same as `fork` except parent is temporarily suspended and child might be permitted to use the parent's memory space.
`wait()`	Suspend parent process to wait for an event of a child process.

DOI: 10.1201/9781003203322-4

When a process is to be created in Linux, the current process (the parent) invokes one of the fork() or clone() system calls (note that the parentheses indicate that system calls are functions). There are several different *clone* system calls named clone(), clone2() and clone3(). The difference between clone(), fork() and vfork() is just how much will be shared between the parent and the child as the child is a duplicate of the parent. We would use one of the clone() system calls to spawn a thread (we cover threads in Section 4.2).

vfork() suspends the parent process while the child process runs. Another option is to create the child with fork() or clone(), and then have the parent invoke wait(). There are several *wait* system calls: wait(), waitpid(), waitid(). With wait(), the parent suspends until one child terminates. With waitpid() and waitid(), the parent suspends until the child with the specified PID terminates.

The wait()/waitpid()/waitid() system calls can be modified so that the parent resumes under other circumstances than the termination of a child (or the specified child). This makes wait() more flexible than vfork(). For instance, the parent might need a result from the child and so suspends itself until the result is available. This is known as a rendezvous, a term we will examine in more detail later in the chapter. The wait() command is also used to implement threads.

fork()/vfork() and clone() create a child that has the same executable code as the parent. For instance, if we run wc from a bash shell, it is bash that is the parent and when it calls fork()/vfork(), it creates a child that is also bash. This is where exec() comes in. Given a child process which is a duplicate of the parent process, we want to overwrite the child's image with the image of the program that the child should be. In the bash/wc example, exec() would cause the code of the child to be replaced by the code of wc. There are several versions of exec() known as the *exec family* of system calls. These are execl(), execle(), execlp(), execv(), execve() and execvp(). The differences between these system calls are the parameters passed to them, a detail we will not concern ourselves with.

It is common to pair up fork() and exec(). First, the parent process issues a fork() (or vfork() or clone()) system call to create a new process with a new PID. The child now invokes one of the exec() system calls to change the process to one of a different program.

Let's consider a more concrete example of parents and children. The first process to run once the Linux kernel has initialized itself is systemd. After systemd initializes the operating system including the user interface, we have a child process, the GUI environment itself; let's assume the GUI is Gnome and so the program is called gdm. To run some application, it is gdm that spawns a child process through fork() and exec(). If we are operating in a terminal window, then the terminal window represents one process spawned by gdm. The terminal window then spawns a shell process (for instance bash). Commands that we enter from the command-line prompt are spawned by bash. Usually, a process run from the command line executes and terminates before the command line becomes available again to the user. Thus, bash will spawn one process at a time.

Entering a command which contains a pipe actually spawns multiple processes. The instruction ls -al | grep zappaf | wc -l causes bash to spawn an ls process. The output from ls is saved and bash spawns a grep process. The saved output is redirected as input to grep. That output is saved and bash spawns wc, redirecting the saved output to it. That output is displayed in the terminal window. All three of the processes that bash spawns, ls, grep and wc, will have now terminated permitting bash to resume.

Let's assume we have logged into Linux using the GUI and opened a terminal window. Our default shell is Bash, so the terminal window, upon starting, runs bash. From the Bash shell, we launch a background process such as a script. Then we launch a foreground process, perhaps a find operation. We have created a chain of processes. The systemd process spawns the GUI which spawns a terminal window which spawns bash which spawns both a background and a foreground process. This example chain is shown in Figure 4.1. Notice in this case that bash has two child processes, both of which might be active depending on how long each runs.

There are different types of processes that run in the system. As a user, we are mostly interested in our own processes. From a system's point of view, there are also services. We will ignore services

```
systemd
     `-- gnome
              `-- gnome_terminal
                             `--bash
                                   `--./somescript
                                   `--find
```

FIGURE 4.1 Example of parent/child relationships in Linux.

in this chapter but cover them in detail in Chapter 9. In this chapter, we look at forms of process management to understand the different ways an operating system can run processes. Some of this is more of historic interest than of value in understanding Linux. We also look at a related topic, resource management. We then turn to Linux to see how we, as users, manage processes. Although there are GUI tools we will consider, we primarily use command-line instructions.

SECTION ACTIVITIES

1. Aside from family relationships and Linux processes, come up with one other type of relationships that we reference as parent/child.
2. Research the exec system calls. See if you can understand the differences between each specific call.

4.2 FORMS OF PROCESS MANAGEMENT

Before we specifically address processes in Linux, let's first consider what the process is and the different ways processes run. As defined in the previous section, a process is a running program. Consider that a program is a piece of software. It contains code and has memory space reserved for data storage. Most of this storage is dedicated to variables that are declared and used in the program. The program exists in one of two states: the original source code and the executable code. In either case, the program is a *static* entity.

The process, on the other hand, is an *active* entity. As the process executes, the values it stores in memory (variables) change. As the process executes, it may request resources from the operating system which it then holds, uses and returns to the operating system. Depending on the size of the process and the number of other running processes, the process may be wholly contained in memory or partially stored in swap space; as the process executes, parts of it may get swapped in and out of memory.

The CPU stores information about the running process via its registers. These register values change as the process executes. The program counter (PC) stores the address of the next instruction to fetch from memory while the instruction register (IR) stores the current instruction. Data registers store data currently in use. The results of the most recent operation are indicated using status flags in the status flag (SF) register (such as whether the last result was positive, zero or negative;had even or odd parity; or caused a carry or overflow to be produced). A run-time stack is maintained for this process, indicating subroutine invocation and parameter passing. The top of the stack is pointed to by the stack pointer (SP).

We refer to the running program as a process to differentiate it from the static program. But we also need to differentiate between them because we can have multiple copies of the same program running at a time. The programs are identical while the processes all differ because each will have its own unique state.

TABLE 4.2
Process Control Block Information

Item	Use
Process state	The run-time state of the process; usually one of `new`, `ready`, `running`, `suspended`, `terminating`, `waiting`.
Process ID	Usually a numeric designator to differentiate the process from others.
Other process data	Parent process (if there is one), process owner, priority and/or scheduling information, accounting data such as amount of CPU time elapsed.
Process location	Which queue the process currently resides in (ready queue, wait queue, job queue).
Process privilege/state	What mode the process runs in (user, privileged, some other).
Hardware-stored values	PC, IR, SF, SP (and possibly data register values) and interrupt masks.
Resource allocation	I/O devices currently allocated to the process; memory in use; page table.

The operating system must keep track of the process's state and does so through a data structure often called the *process control block* (PCB). The PCB is a collection of the most important pieces of information about the running process. These are described in Table 4.2.

Aside from keeping track of a process' status, the operating system is in charge of scheduling when processes run, and of changing the process' status as needed. For instance, a process which requires time-consuming input will be moved from "ready for CPU" to "waiting for input". Handling the startup and termination of, scheduling and movement of, and interprocess communication of processes is called *process management*. There are many different forms of process management which we look at in the next two subsections.

4.2.1 Single-Process Execution

In *single tasking*, the operating system starts a process and then the CPU's attention is entirely held by that one running process. The CPU executes this one process until either the process terminates or requests some operation by the operating system. In the latter case, the CPU switches context from the running process to the operating system. Upon completion of the request, the context switches back to the process. This means that the computer is limited to running only one process at a time.

Single tasking is the easiest form of process management to implement but has many drawbacks. Among them, the user is limited to running one task at a time and so there is no way to switch back and forth between multiple tasks. Additionally, if the process has to perform some time-consuming input or output, the CPU remains idle while the input/output (I/O) takes place. Early mainframe computers ran single tasking although this began to change by the end of the 1950s. Nearly all personal computers from the 1970s through the 1980s were only capable of single tasking.

Batch processing is much like single tasking with three differences. First, in batch processing, the computer is shared among multiple users such that users can submit processes to run at any time. Since the system can only execute one process at a time, like single tasking, a separate scheduling operation is required to order the waiting process requests, called *jobs*. Scheduling may be performed by some offline system, which is the second difference. Third, since the job may run any time from immediately to hours or days later, the job cannot be expected to run interactively.

To support the lack of interactivity, a user must submit the input with the job as well as the destination for the output. Typically, the input was stored as one or more files on magnetic tape or was encoded on punch cards and input through a punch-card reader. The computer would run the specified program and load the data from tape or card.

One approach, pioneered by IBM mainframes, was to specify operating system operations such as loading a file, inputting data off of cards and saving results to file on tape, through a separate set of operating system instructions. For IBM mainframes, the instructions were written in Job Control

Language (JCL). JCL instructions were themselves encoded onto punch cards and interspersed with the program code and data. Later versions of IBM mainframes accepted JCL input from keyboard. Today, if batch processing is used, the input is stored on a disk file whose name is provided when the process runs and output is typically saved to disk file.

With these early computers, a human was typically tasked with mounting and unmounting tapes. One tape might store a compiler. Another tape might store the executable program. A third tape might store the input and output files. After execution, the tape containing the output would be unmounted from the mainframe computer and remounted onto a printer to be printed. The tape would be known as a *print spool*, which is where the term *spooler* came from.

The offline scheduling system receives user job requests and schedules them using some type of *scheduling algorithm*. During the era of batch processing, a number of different scheduling algorithms were developed. The simplest, and perhaps fairest, is *first come first serve* (FCFS, also known as FIFO for *first in first out*). From a statistical point of view, ordering jobs from shortest to longest (known as *shortest-job first*) provides the shortest average wait time for the waiting jobs. This may or may not be fair as it penalizes jobs that would take a long time to execute. The opposite approach, *longest-job first*, statistically would yield the longest wait time but was thought by some to be a fairer approach. A *priority* scheme may be used, perhaps coupled with FCFS. A series of queues (waiting lines) are used to order the processes by priority. A university, for instance, might have as priorities administration, faculty, graduate students and undergraduate students, in that order. An interesting legend comes from MIT's IBM 7094 mainframe where it was discovered that a low-priority job submitted in 1967 was not executed until 1973!

4.2.2 Concurrent Processing

One of the great inefficiencies of both single tasking and batch processing is that of I/O. Imagine, for instance, that input is to be read from magnetic tape. Even if the tape is mounted before execution of the process begins, it is possible that the tape needs to be rewound or fast forwarded from its current location to another location. This could take minutes. The physical reading of the tape and transferring data to the computer may take many additional seconds of time. While I/O takes place, the CPU idles. It would be nice to give the CPU something to do while waiting.

The idea of switching from a waiting process to a process ready for execution leads us to an improved form of process management called *multiprogramming*. In multiprogramming, the CPU executes a single process, but if that process needs to perform time-consuming I/O then that process is set aside and another process is executed.

We noted queues in the last subsection. Here, we have at least two queues. One is the *ready* queue which contains those processes that are ready for execution; the other is a *wait* queue where processes are moved when they are waiting for I/O. There will likely be multiple waiting queues, one per I/O device. We might refer to these collectively as I/O queues. It is the operating system's job (specifically the scheduler) to select the process that the CPU should switch to. By switching between processes in this way, it appears to the user that the computer is executing multiple processes at a time. We call this *concurrent processing*. Multiprogramming is one form of concurrent processing; we look at others below.

The changing of the CPU from one process to another is known as a *context switch*. To perform a context switch, the operating system gets involved. First, the currently executing process' status must be saved. The process' status includes the values stored in various registers: PC, IR, SF and SP (and possibly others). These must be saved to a new location, usually memory. Now the operating system starts or resumes the next process in the ready queue by loading its saved status (PC, IR, SF, SP) from memory into the corresponding registers. Once the registers are loaded, the CPU resumes executing but now it is running a different process in a different area of memory. To the CPU, it is immaterial that it has changed processes as it just continues to fetch and execute program instructions. To the users though, the CPU is now executing a different process.

The context switch is not instantaneous in that there is a short interval while the CPU idles so that register values can be moved from CPU to memory and from memory to CPU. This is not nearly as time consuming as pausing the CPU to wait for I/O. High-performance computers avoid even this short switching time by having multiple sets of registers so that switching is handled solely within the CPU.

The idea behind multiprogramming is that a process will voluntarily surrender its access to the CPU because it has to wait for some event. Above, we referred to that event being the completion of some input or output task. Another event arises when a process has to wait for another process. For instance, one process may need to use a datum that is to be generated by another process. If the first process reaches the point of needing the datum but the second process has yet to generate it, the first process must wait. The point where one process waits for the other is called a *rendezvous*. In such a situation, the current process voluntarily yields the CPU to another process while it waits.

An example of a rendezvous is when we have two sets of processes where one group is known as *producers* and the other group is of *consumers*. As their names imply, producer processes generate data and consumer processes consume (use) that data. Typically, consumers and producers communicate through some shared storage medium like a buffer stored in memory. A consumer cannot continue if there is no data to consume in the buffer, and so must wait for a producer to produce more data. Similarly, if the buffer is currently full, a producer must wait before it can produce and store new data until a consumer consumes some of the existing data. Thus, producers and consumers have reasons to yield the CPU.

We noted that voluntarily yielding the CPU is a form of process management called *multiprogramming*. It can also be referred to as *cooperative multitasking*. In essence, the computer is multitasking (doing several things at once) by switching off between the tasks whenever the current process needs to wait. The CPU switches to another task and works on it in the interim. In multiprogramming, once the waiting process has completed its wait (whether for I/O or a rendezvous), the CPU switches back to this process to resume it and the process that was executing in the interim moves back to a waiting state.

The next step in the evolution of process management is to *force* a context switch so that no single process can monopolize the CPU. We add a piece of hardware called a *timer*, which counts clock cycles. The timer is initialized to some value, say 10,000. After each clock cycle, the timer is decremented. When it reaches 0, the timer alerts the operating system to perform a context switch. The running process is moved to the end of the ready queue, and the CPU resumes the process next in line in the ready queue.

This form of process management goes beyond multiprogramming (cooperative multitasking) because context switches now arise either voluntarily or when the timer expires. We refer to this form of multitasking as *preemptive multitasking*. Collectively, we refer to both cooperative and preemptive multitasking generically as multitasking, although historically they were called *time sharing*.

Multitasking requires some form of scheduling to select the next process to run/resume. The selection of the next process to execute when the CPU becomes free will be made by the operating system. For multiprogramming, scheduling may be accomplished via first come first serve, priority or a round-robin scheduling algorithm. For preemptive multitasking, operating systems typically use a round-robin scheduling approach modified by priority as described in the next paragraph. *Round-robin* scheduling sees all processes ready to run placed in the ready queue and processed in the order they appear in the queue. The processor switches off between these processes, looping back to the beginning of the queue once the last process in the queue has received some processor time.

The amount of time awarded to a process when the CPU begins or resumes it is known as its *time slice* (or *quanta*). We can add a priority scheme to preemptive multitasking so that the number of clock cycles that the timer is set to varies based on that process' priority. Imagine, for instance, that processes have a priority between 1 and 10 (10 being the highest, 1 being the lowest). The number of clock cycles that the timer is set to will be 1000 * process priority. A process with a priority of 10 is given 10,000 clock cycles each time it gets access to the CPU, while a process with a priority of 1 is given 1000 clock cycles each time it gets access to the CPU. In this way, while all ready processes get time with the CPU, higher priority processes get more time and so may complete sooner. We can alter this simple strategy to *age* processes over time so that a high priority process' priority lessens.

Figure 4.2 illustrates the concepts of the context switch in multitasking. In the top portion of the figure, we see how multiprogramming (cooperative multitasking) operates. Process P3 is executing and the CPU's registers store information about P3 (its current instruction in the IR, the next instruction's location in the PC, etc.). At some point, P3 needs to perform I/O, so the operating system saves P3's status information to memory and moves P3 to an I/O queue (not shown in the figure). The CPU then starts or resumes process P7, the next one in the ready queue. In the bottom of the figure, a timer has been added so that we can also implement preemptive multitasking. P3 executes and after each clock cycle, the timer is decremented. Upon reaching 0, the timer interrupts the CPU and the operating system performs the same context switch except that P3 moves to the back of the ready queue.

Notice in Figure 4.2 that both P3 and P7 have code stored in memory. Aside from switching contexts (register values), the operating system must also ensure that the resumed process is in memory and available for execution. It is possible that since P7 was last executed portions of its executable code have been removed from memory. Having to restore P7's registers takes a short amount of time. Having to load some of its code (from swap space on hard disk) takes a much longer amount of time.

Therefore, a context switch between processes could be time consuming and during that time, the CPU idles. *Threads* are similar to processes, but they represent different instances of the same running program. Threads then share some of the information and resources that separate processes would not share. For instance, two threads of the Mozilla Firefox web browser would use the same executable

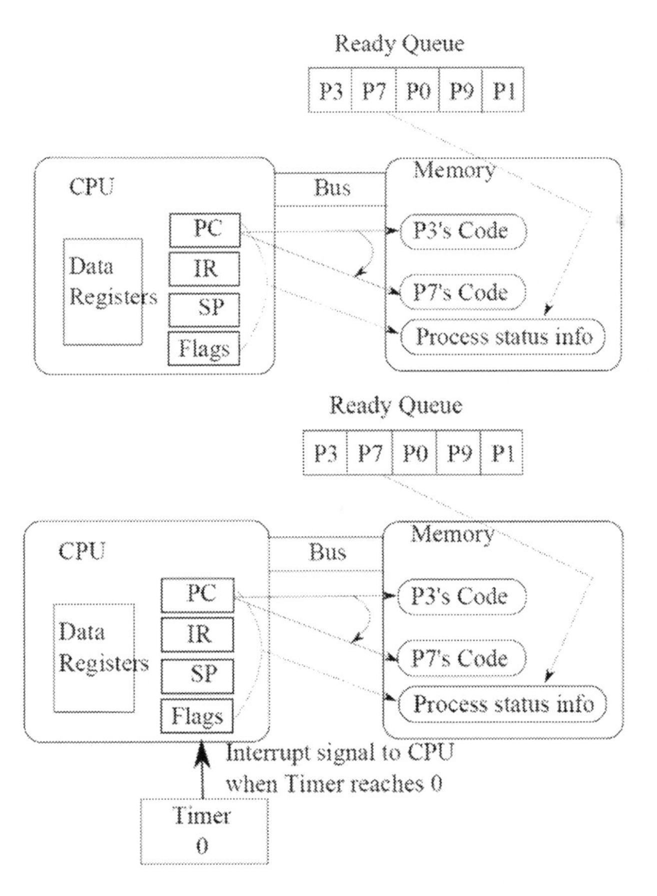

FIGURE 4.2 Context switch in multiprogramming (top) and multitasking (bottom). (From Richard Fox, 2021, *Information Technology. An Introduction for Today's Digital World*).

code and some of the same data. They would differ in that they would have different register values and their own unique pathways through the code. The advantage of using threads is that context switching between threads is often accomplished faster than context switching between processes.

Threads are a newer concept in programming but are supported by most operating systems. To take advantage of threads, the program must be written so that code within the program can run concurrently. That is, there must be a way of running multiple program threads at a time. A multitasking operating system that is switching off between threads is known as a *multithreaded* operating system. As this is the same basic mechanism that is used to support multitasking, we see that a multitasking system can also be multithreading. More specifically, all multithreaded operating systems perform both multithreading and multitasking. Older multitasking systems may not be multithreaded.

Another form of process management occurs when the computer has multiple processors (either multiple CPUs or multiple cores in a single CPU). This is known as *multiprocessing*. In multiprocessing, the operating system selects a processor to run a given process. Multiple processors can execute their own processes independently. Typically, once launched onto a processor, the process remains there. However, load balancing allows the operating system to move processes between processors as necessary. Some programs can be distributed across multiple processors in which case one process might execute in parallel on more than one processor, each processor in charge of some portion of the overall process.

Linux by default uses cooperative and preemptive multitasking and multithreading. Programmers wishing to take advantage of multiprocessing must include instructions in the program that specify how to distribute the process. Linux can also perform batch processing on request using the `batch` instruction.

4.2.3 Interrupt Handling

We mentioned in the previous subsection that the timer must alert the CPU when it has reached 0. Consider that left to itself the CPU repeatedly fetches and executes program instructions (what is known as the fetch-execute cycle) until it finishes executing the current process. The timer reaching 0 is one example of a situation where some hardware component of the computer needs the CPU's attention. We need a mechanism whereby the hardware component can interrupt the CPU's fetch-execute cycle and request that the CPU focus its attention on the hardware component to handle whatever situation arose. We naturally refer to the process of interrupting the CPU as an *interrupt* and the process of requesting an interrupt as an *interrupt request* (IRQ).

The IRQ may originate from hardware or software. For hardware, the IRQ is carried over a reserved line on the bus connecting the hardware device to the CPU (or to an interrupt controller device). For software, an IRQ is submitted as an interrupt signal.

Upon receiving an interrupt request, the CPU finishes the current fetch-execute cycle and then decides how to respond to the interrupt request. The CPU must determine which device (or software) raised the interrupt. The CPU acknowledges the IRQ to the interrupting device. To handle the interrupt, the CPU switches from the user process to the operating system (switching processor modes from user to privileged).

For every type of interrupt, the operating system contains an interrupt handler. Each *interrupt handler* is a piece of code written to handle a specific type of interrupting situation. The CPU performs a context switch from the current process to the proper interrupt handler, requiring that the CPU saves what it was doing with respect to the user process, as explained in the last subsection. The interrupt handler executes and upon completion, the operating system switches back to the process that had been running (or a new process if the interrupt were caused by the timer reaching 0), also changing modes back to user mode.

IRQs are prioritized. If the CPU is currently handling an IRQ and a higher priority IRQ arrives, the CPU will postpone handling the lower priority IRQ for the newer, higher priority IRQ. If the new IRQ is of a lower priority, the CPU will ignore it until the current IRQ is handled.

```
         CPU0       CPU1
0:       20342577   20342119   IO-APIC-edge     timer
12:      3139       381        IO-APIC-edge     i8042
59:      22159      6873       IO-APIC-edge     eth0
64:      3          0          IO-SAPIC-EDGE    ide0
80:      4          12         IO-SAPIC-edge    keyboard
```

FIGURE 4.3 Processor interrupts as stored in `/proc/interrupts`.

In Linux, interrupt handlers are part of the kernel. As the interrupt handler executes, information about the interrupt is recorded in `/proc/interrupts`. This file stores the number of interrupts per I/O device as well as the IRQ number for that device. A computer with multiple CPUs or multiple cores provides a separate listing for each of the CPUs/cores.

We see an example of the `/proc/interrupts` file in Figure 4.3. In this case, the computer has two CPUs (or two cores). The first column lists the IRQs by their number. These are currently occurring IRQs only. The next two columns are the number of interrupts that have arisen on each processor. Notice that IRQ 64 has only arisen on CPU0 while all of the other IRQs have arisen on both processors. The fourth column lists the type of interrupt, and the last column is the device's name that generated the IRQ. The timer is responsible for the most interrupts by far, which is reasonable as the timer interrupts the CPU hundreds to thousands of times per second.

The file `/proc/stat` also contains interrupt information. It stores for each IRQ the number of interrupts received. The first entry is the sum total of interrupts followed by each IRQ's number of interrupts. For instance, on the processor above, we would see an entry of `intr` where the second number would be 40684696 (the sum of the interrupts from the timer, IRQ 0). The entire `intr` listing will be a lengthy sequence of integer numbers. Most likely, many of these values will be zero indicating that no interrupt requests have been received by that particular device.

SECTION ACTIVITIES

1. How often do you run two or more applications at any one time? Would you find computers as useful if you could only run one at a time and only start another when the current application ended?
2. Research interrupts and enumerates a list of ten types of interrupts aside from the timer. How many of the interrupts that you listed are hardware interrupts?
3. Look at the interrupts listed in Figure 4.3. Aside from the first, which is listed as timer, can you identify what device or situation caused each type of interrupt?

4.3 STARTING, PAUSING AND RESUMING PROCESSES

Starting a Linux process from the GUI is much like starting a process in any other operating system. Depending on the version of Linux and the GUI being used, there are menu selections and icons. `systemd`-based Linux versions that use the Gnome GUI present a single menu called Activities. Selecting it brings up a list of "favorites", including for instance a web browser, a terminal window, a file browser, settings and applications. Selecting applications expands the available applications, presented as desktop icons. Selecting an icon, whether from the favorites list or from the desktop, will cause Gnome to submit the startup command for that application to the kernel.

As the user runs programs, some are spawned by the GUI. Some application processes will spawn additional processes. For instance, the user may run the Mozilla web browser. Opening a new tab requires spawning a child process (actually a thread). Sending a web page's content to the printer spawns another process (thread).

With a terminal window open (or when remotely logging in using a text-based login), a shell runs. The shell presents to the user a command-line prompt. The user enters commands at the prompt. Commands are programs which the shell interpreter must locate and execute.

4.3.1 Ownership of Running Processes

Whether text-based or GUI-based, the user submits requests to run programs. The parent process making the request might be the GUI, Bash, another running application or a part of the operating system (e.g., a service). The kernel's process manager creates the process and prepares it for execution.

Processes often require access to resources (notably files, but possibly other system resources). For instance, `vi` commonly has to open a file and/or save its contents to a file. The files opened and saved will likely be in the user's home directory. How does the `vi` process obtain access rights to that user's home directory?

When a process is launched by the user, the process runs under access rights based on its *effective user ID* (EUID) and *effective group ID* (EGID). By default, the EUID and EGID are the same as the user's ID (UID) and the user's private group ID (GID). These effective IDs make the process an extension of the user so that the process has the same access rights as the user. This is perfectly suitable in many cases.

Some applications require access to their own file space and their own files. For instance, the `lp` command (print files) copies the output to be printed to a printer spool file, located in `/var/spool`. This space is not accessible to the user. So how does `lp` write the data there?

When some software is installed, an account is created for that software. `lp`, for instance, has its own user account. Examine `/etc/passwd` (the file storing all accounts on the system) and you will find, aside from root and user accounts, numerous software accounts with users named `mail`, `halt`, `ftp` and `apache`, among others. For `lp`, it has a home directory of `/var/spool/lpd`. Other software accounts have home directories in the top-level directories of `/sbin`, `/proc` or `/var`. In running `lp`, the operating system switches from the user's access rights to `lp`'s (the software account). Now `lp` has access to its own file space.

Aside from a program running under the user's rights, another situation is where a process needs to access system resources. Consider the `passwd` command. This program is owned by root. When a user runs `passwd`, the program accesses the password file (which is stored in `/etc/shadow`, not `/etc/passwd`) first to ensure that the user has entered the correct current password. Then the program accesses the file again to modify the password. When a user runs `passwd`, it should run under the user's account. But the shadow file is only accessible to root.

The way Linux gets around this problem is to provide a special bit in the permissions that indicates that the process runs not under the user's access rights but under the rights of the owner of the executable file. In the case of `passwd`, this is root. Because this special permission bit is set, `passwd` runs under root's ownership. Now it can access the `/etc/shadow` file that is only accessible to root.

To establish that a program executes under the ownership of the file's owner, we alter the permissions (using `chmod` as described below). To determine if a process will run under the user's ownership or the file's ownership, inspect the file's permissions through a long listing. The owner's executable bit will either be `-`, `x` or `s`. These three forms of permissions represent that the file is non-executable, executable and run under the user's UID, and executable and run under the file owner's UID, respectively. Group executable access has the same three possibilities where `s` means that the program runs under the group owner's ID (GID). Setting the executable bit to `s` is known as *setting user permission ID* or *setting group permission ID* (SUID, SGID). Programs like `passwd`, `mount` and `su` are set up in this way.

Let's take `passwd` as a specific example. The program, stored in `/usr/bin`, has permissions of `-r-s--x--x`, and its owner and group are both `root`. When run, `passwd` runs under root's access rights to be able to read and write to files that we, as users, do not have permissions for. We, as users, have execute access to run the program (because it is executable by world), but the program itself runs under root's privileges.

Specifically, when a program runs, its EUID and EGID are set to either the UID and GID of the user, or if the program's execution bit for the owner is `s` then the EUID is set to the UID of the file's owner and if the program's execution bit for the group is `s` then the EGID is set to the GID of the file's group. You will not see an `s` for both owner and group although it is possible to set both of them.

In order to change the execution bit to `s`, use `chmod`. Recall from Chapter 3 that there are three ways to use `chmod`, `ugo+/-`, `ugo=` and the 3-digit permission. We can change the execution bit from `x` to `s` (or from nothing to `s`) and from `s` to `x` (or from `s` to nothing) using the +/- approach and the 3-digit approach. The `ugo=` approach cannot be used. With the `ugo+/-` approach, merely use `s` as in `u+s` or `g+s` and to reset the bit back to `x` use `u-s+x` or `g-s+x`. Remove execute access entirely with one of `u-s` or `g-s`.

The 3-digit approach is different in that we use a 4-digit number like we did to add the sticky bit to a directory. To establish `s` for the user, we prepend a 4 to the 3-digit number. To set group execute access to `s`, we prepend a 2 to the 3-digit number. As an example, an executable program's current permissions are `-rwxr-xr-x` and we want to change them to `-rwsr-xr-x` (we want to set the user permission ID). Normally we would use 755 but with `s` for the owner, we add a 4 giving us 4755. If instead we want to make the group's execute access `s`, we would specify 2755. Figure 4.4 provides several examples of either adding or removing an `s` from the permissions. The top six add the `s`, and the last two remove an `s`.

Among the programs with a SUID bit are those that access the password file (`/etc/shadow`) such as `chage`, `passwd`, `su` and `sudo`. Some other programs that have a set SUID are the scheduling programs `at` and `crontab` and the `mount` instruction. Fewer programs have their GUID bit set but include `slocate` and two email utilities (`postdrop`, `postqueue`).

4.3.2 Launching Processes from the Command Line

Launching processes from within a shell is accomplished by specifying the program name. We have already seen this in Chapters 2 and 3. If the executable file of the program is not in a directory stored in the PATH variable, then the command must include the directory path. This can be specified either with an absolute or relative path. Assume for instance that `/usr/sbin` is not in the user's PATH variable. To execute the `ip` program (to display information about network connectivity and IP addresses), we would enter `/usr/sbin/ip`.

Launching a program that exists in the current directory uses a slightly different syntax. The notation to run such a program is `./`*`filename`*. If the file is in another directory, we omit the `./` notation, and if the program is in the current directory but that directory is in our PATH variable, we may also omit the `./`. Keep in mind that programs include script files, so to execute a script in the current directory we would typically use `./`*`scriptname`*.

To this point, most of the programs we have examined run quickly. Because of this, the program starts, runs, outputs to the terminal window and exits, returning us to our command-line prompt. What happens if a program either takes some time to execute (such as the `find` instruction or a long-running script) or runs interactively (like `vi`)? In either case, we do not immediately regain our prompt. Let's take a closer look at how we can manage this situation.

```
4744 rwsr--r--          4754 rwsr-xr--
4755 rwsr-xr-x          6755 rwsr-sr-x
2750 rwxr-s---          2755 rwxr-sr-x
-------------------------------------------
0755 rwxr-xr-x          0710 rwx--x---
```

FIGURE 4.4 Setting/clearing the SUID/GUID bit.

Let's assume we have written a script called myscript which runs for several minutes. We enter ./myscript, and while it runs, we have no access to our prompt. If the script runs without user input, there is no real reason why we shouldn't have access to our prompt. But Bash prevents this from occurring because, should the program need user input, the command line cannot be given back to the user. Bash does not know which programs will require user input and which will not, but we do (or should).

Now let's consider a slight variation on our script. The script does indeed receive input but that input comes from a text file instead of keyboard. In order to run our script, we redirect input using ./myscript < mydata.txt. Now the script runs in batch mode as it has no user interaction. But even in this case, we do not get our command-line prompt back until the script terminates. Is there something we can do about this? Yes.

Let's differentiate between two types of process execution: interactive execution and batch mode. Two programs that demonstrate this difference are vi and find, the former is interactive and the latter displays its output and terminates. We could redirect find's output to a file (e.g., find / -name "*.txt" > all_text_files.txt) so that it has no interaction with the user at all.

For programs that run with no user interaction (either input or output), we can run them in the *foreground* or *background*. A foreground process is one that interacts with the user. Even if the process requires no interaction with the user the user does not regain their prompt until the process ends. A background process runs with no user interaction and so the prompt can be returned to the user to issue further commands.

From the command line prompt, all commands entered run in the foreground by default. To change this behavior, add an ampersand (&) after the instruction. This informs the Bash interpreter to run the process in the background. We would only choose to run processes that receive no interactive input as background processes. Output will be displayed when the process is able to display it, so this might interrupt what we are doing at the time the process ends, and so it makes the most sense to redirect output of a background process to a file for later viewing.

Let's revisit the find example from above. We do not need to view its output immediately and since we believe the process will take some time, we launch it into the background by appending an ampersand. We issue the instruction find / -name "*.txt" > all_text_files.txt &. Notice the placement of the & at the very end of the command rather than before the redirection operator.

Upon entering the above instruction, we receive a response that looks like [1] 313546. The [1] indicates that this is the first current job running and the large number is the process ID (PID). We now have access to our prompt and can enter other instructions. Upon the process terminating, the Bash interpreter informs us by displaying [1]+ Exit 1 find / -name "*.txt" > all_text_files.txt. This tells us that job [1] has exited with exit code of 1, and it reiterates the instruction (although notice the & is omitted). We explore what the + means shortly.

If we had executed the previous instruction as ourselves (rather than root), we would receive numerous Permission denied error messages because there are many directories in the file space that we do not have permission to access. The reason these messages are sent to our terminal window and not redirected to the output file is that it is the Bash interpreter generating these messages, not find.

Can we prevent these error messages? Yes, we can redirect error messages elsewhere. For instance, we might redirect all error messages to a file in our home directory using find / -name "*.txt" 2>~/errors > ~/all_text_files.txt &. The notation 2>~/errors specifies that all output to file descriptor 2, STDERR, should be redirected to the file ~/errors. The notation > ~/all_text_files.txt is the redirection of STDOUT, or the output of the find command. Thus, the output of find goes to one location and error messages to another. The & used in the instruction causes the process to run in the background.

We usually want to discard error messages rather than save them to a file. To do so, we can replace 2>*filename* with 2>/dev/null as /dev/null is a file that serves as the Linux trashcan (it is not in fact a file, just a location whereby items can be discarded). It is used precisely for purposes like this where we want to ignore messages, particularly those generated from a background process.

```
[foxr@localhost ~]$ ./long_process < data.txt > out.txt &
[3] 314429
[foxr@localhost ~]$ ls
...                (output from ls here)
[foxr@localhost ~]$ cd temp
[foxr@localhost ~]$ ls
...                (output from ls here)
[3] Done    ./long_process < data.txt > out.txt
```

FIGURE 4.5 Launching a process to the background.

Although the notation takes some getting used to, it is a convenient way of ensuring our background process does not interfere with our further use of the command line prompt.

So now we have another tool to use in our Bash sessions, launching background processes. In general, the interaction will look like what is shown in Figure 4.5. We enter a command into the background and are given a response of the job number and PID. We continue to interact and eventually when the process terminates, we are given an indication that the job has ended when the `Done` message appears.

4.3.3 Suspending and Resuming Processes from the Command Line

We may run as many processes as we wish in the background. A terminal window, however, can only have one process running in the foreground at a time. The foreground process is the one that provides interaction with the user, whether to receive user input or to output information to the window, or both. This does not mean that while a process is running in the foreground, we cannot launch other processes or move background processes to the foreground.

Before discussing how to move processes between foreground and background, we need a mechanism to identify the foreground and background processes in the terminal window. We obtain this information with the command `jobs`. The `jobs` command responds with all of the active jobs in the terminal window. These are processes that have been started from the command line but have not yet terminated.

A process launched from the command line will be in one of three states. It may be in the foreground which means that while it runs we do not have access to the command line. It may be in the background in which case we have no interaction with it and we receive a notification only when it terminates. Finally, a process may be in a *stopped* state. This means that the process has been running but is currently suspended. We can resume a suspended process. In doing so, we specify whether to resume the process in the foreground or in the background.

In order to stop a running, foreground process, type `control+z` in the terminal window. Upon doing so, we are presented a response like `[2]+ Stopped  find ~ -name *.txt`. Again, the number, [2] in this case, is the job number. We saw earlier that responses from the Bash interpreter include `Done` and `Exit`. The difference is that `Stopped` means the process has not terminated, `Done` means that it has terminated and `Exit` means it has terminated with errors.

Having stopped the foreground process, we can control this and other processes in terms of running them, stopping them and moving them between foreground and background. To see which processes are available in this bash shell, type `jobs`. We resume a stopped process using either `fg` (resume into foreground) or `bg` (resume into the background). The `jobs` command displays the job numbers. We use the job number with `fg` or `bg` to control which job we are resuming. We can also use `fg` or `bg` with no number to resume the most recently executing job. Let's step through an example, as shown in Figures 4.6 through 4.9.

In Figure 4.6, we launch three processes, `vi`, `man` and the script `somescript`. Each process is launched into the foreground and then suspended using `control+z`. We then type `jobs`, whose output is shown in Figure 4.7. We see the three jobs listed, all of which are stopped. Each listed job has a job number based on the order that the job was submitted. Notice the + and - in the listing. The + indicates the most recently active job, and the - indicates the job that was active before it.

```
[foxr@localhost ~]$ vi somefile
      (this launches vi, we type control+z)
[1]+ Stopped          vim somefile
[foxr@localhost ~]$ man vi
      (this launches man, we type control+z)
[2]+ Stopped          man vi
[foxr@localhost ~]$ ./somescript < input.txt > output.txt
      (while the script is running, we have no prompt, we type control+z)
[3]+ Stopped          ./somescript < input.txt > output.txt
```

FIGURE 4.6 Using `control+z` to stop running processes from the command line.

```
[foxr@localhost ~]$ jobs
[1]   Stopped     vim somefile
[2]-  Stopped     man vi
[3]+  Stopped     ./somescript < input.txt > output.txt
[foxr@localhost ~]$
```

FIGURE 4.7 Output from `jobs`.

At this point, we resume the `vi` job by typing `fg 1`. Had we entered `fg`, we would have resumed `somescript` instead (the most recently active job). We again type `control+z` to stop `vi` and type `jobs`. Now we see the slightly different output, as shown in Figure 4.8. If you don't look carefully enough, you won't spot the difference. What has changed? The placement of the + and the – have changed because `vi` (vim) is the most recently executing job and `somescript` is the second most recently executing job.

We now decide to move the script into the background. We type `bg 3`. The response from this is `[3]- ./somescript < input.txt > output.txt &`, and we obtain our command line prompt because we resumed a job to the background rather than the foreground. The meaning behind this response is that the second most recent job (-), job 3, has now been moved to the background (&). Typing `jobs` has only one slight change from Figure 4.8 which is that job 3 is no longer listed as `Stopped` but `Running`. Additionally, the `&` now appears at the end of the instruction. This interaction of using `bg` and `jobs` is shown in Figure 4.9.

If we want to now resume `vi`, we can type `fg` or `fg 1`. To resume `man`, we type `fg 2`. We can also move the script from the background into the foreground by typing `fg 3`. Doing any of these commands resumes a job to the foreground and thus we lose our prompt until we either exit out of

```
[foxr@localhost ~]$ jobs
[1]+  Stopped     vim somefile
[2]   Stopped     man vi
[3]-  Stopped     ./somescript < input.txt > output.txt
[foxr@localhost ~]$
```

FIGURE 4.8 Updated output from `jobs`.

```
[foxr@localhost ~]$ bg 3
[3]-  ./somescript < input.txt > output.txt &
[foxr@localhost ~]$ jobs
[1]+  Stopped     vim somefile
[2]   Stopped     man vi
[3]-  Running     ./somescript < input.txt > output.txt &
[foxr@localhost ~]$
```

FIGURE 4.9 Moving process to the background.

the process or type `control+z`. Assume we have moved job 3 back to the foreground and then typed `control+z` followed by `jobs`. The result is the same as that shown in Figure 4.9 except that the most recent (+) and second most recent (-) jobs have changed again. Now the script (job 3) has the + and `vi` (job 1) has the `-`.

Note that we can also regain our command-line prompt if we have a foreground process running by typing `control+c`. In this case, we are killing the current process, which is a more drastic means of obtaining our prompt. We would only kill a process in this way if we no longer wanted the process running. We explore killing processes later in the chapter.

By using `control+z`, `fg` and `bg`, we can run multiple processes in a single terminal window. For those of you who are used to opening windows whenever you have something new to do you might wonder why we need such mechanisms? Why not just open `vi` in one terminal window, `man` in a second and run the script in a third? There are three reasons for this. The first is that some Linux users operate in a text-only environment. This might be the case, for instance, if you have remotely logged in to a Linux computer using `ssh` (the secure shell program). The second is that opening up multiple windows uses additional resources. If you are working on an older computer with limited memory and a slower processor, you might want to reduce the load on the computer by limiting the amount of GUI interaction. Third is a far more subtle reason; the processes being run might need to share some attribute of a single shell, for instance an environment variable defined by one process and used by others. It may be the case that none of these restrictions qualify for your situation and so you are free to use multiple terminal windows and not have to resort to this, but you might find yourself in a situation where this approach is useful or even preferable.

Note that the job number is not in any way related to the PID. We explore methods of obtaining the PID shortly and use the PID to further control processes. The `fg` and `bg` commands only operate on job numbers. We left our Bash session from Figure 4.9 in a state whereby there are three stopped processes. Imagine we now type `exit` to leave this Bash shell and close the terminal window. We will be told `There are stopped jobs`. We will want to exit out of these processes before we exit out of Bash. We can exit out of each process by typing `fg` to resume the job and exit out of it (for instance `q` to exit out of `man`, `:q` to exit out of `vi` and `control+c` to kill the running script).

SECTION ACTIVITIES

1. Have you gotten comfortable with launching processes from the command line yet? If you had a choice to launch programs from the GUI or command line, would you always select the GUI? Come up with a list of reasons why launching a process from the command line might be better.
2. In the second subsection of this section, we explored how to *juggle* multiple processes in one terminal window. Does it seem like much ado about nothing? That is, would you prefer to just open multiple windows and have one process per window? Then you wouldn't have to use `control+z`, `fg`, `bg` and `jobs`. Do you think juggling of processes or using multiple windows is a more efficient use of your computer? Of your time?

4.4 MONITORING PROCESSES

We might find that that we seldom have to keep tabs on running processes. The operating system can run multiple processes efficiently with little or no user intervention. Yet there are times when we might want to see what a process is doing. For instance, a process may be using an inordinate amount of system resources or a process might have died on us.

4.4.1 GUI Monitoring Tools

There are several different tools available to monitor processes and system resources. The primary GUI tool is called the System Monitor. Launching System Monitor from the GUI has changed. In Red Hat 6 and Red Hat 7, we would start it from the System Tools selection under the Applications menu. In all of Red Hat 8, Debian 10, Fedora 33 and Ubuntu 20, we only have the Activities menu and from there we would select Applications. From the list of tiles that appear, we next select Utilities, and from those tiles, we can select System Monitor. As this tool might be useful, we could make it into a Favorite so that it appears directly with the Activities icons. Alternatively, we can launch the program from the command line as `gnome-system-monitor`.

There are two different versions of the System Monitor, one for Gnome and one for KDE (an alternative GUI desktop). Figure 4.10 shows the Gnome version. There are three tabs along the top of the GUI: Processes, Resources and File Systems. In the figure, the Processes tab is shown, displaying the active processes in the window beneath the tabs. They are ordered, in this case, by CPU utilization. The process currently requiring the most CPU attention is the Gnome terminal window (indicated as `gnome-shell`) followed by the System Monitor itself (`gnome-system-monitor`). Third on the list is `systemd`. Notice that these three jobs using the most CPU time are all GUI-based programs.

Other information in this window describes for each process the user who launched it, the process' ID and priority, and various resource utilization statistics. These are the CPU utilization, memory utilization and various disk access statistics. These values update on a preset interval of 3 seconds. We can modify this interval if desired.

Aside from viewing processes, the Processes tab allows us to select any single process and obtain its properties, memory maps and open files. We can also control the process by stopping the process, resuming the process, killing the process, ending the process and changing the process' priority. Ending a process terminates it as if we select exit from the process' GUI. Killing a process ends it immediately which may damage open files or cause processes spawned by this process to be abandoned. It is better to end a process rather than kill it unless the process has stopped responding. We explore priorities later in this chapter and omit further detail for the moment.

Processes Resources File Systems

Process Name	User	% CPU	ID	Memory	Disk read tota	Disk write tot	Disk read	Disk write	Priority
gnome-shell	foxr	4	2285	197.4 MiB	8.2 MiB	852.0 KiB	N/A	N/A	Normal
gnome-system-monitor	foxr	1	6279	14.4 MiB	792.0 KiB	N/A	N/A	N/A	Normal
systemd	foxr	0	2193	1.7 MiB	161.0 KiB	16.0 KiB	N/A	N/A	Normal
(sd-pam)	foxr	0	2199	5.2 MiB	N/A	N/A	N/A	N/A	Normal
pulseaudio	foxr	0	2212	1.5 MiB	616.0 KiB	8.0 KiB	N/A	N/A	Very High
gnome-keyring-daemon	foxr	0	2218	1.1 MiB	N/A	N/A	N/A	N/A	Normal
dbus-daemon	foxr	0	2225	1.5 MiB	N/A	N/A	N/A	N/A	Normal
gdm-wayland-session	foxr	0	2228	1.6 MiB	4.0 KiB	N/A	N/A	N/A	Normal
gnome-session-binary	foxr	0	2231	2.9 MiB	36.0 MiB	8.0 KiB	N/A	N/A	Normal
gvfsd	foxr	0	2308	1.1 MiB	316.0 KiB	N/A	N/A	N/A	Normal
gvfsd-fuse	foxr	0	2317	3.1 MiB	356.0 KiB	N/A	N/A	N/A	Normal
Xwayland	foxr	0	2327	10.5 MiB	N/A	24.0 KiB	N/A	N/A	Normal
at-spi-bus-launcher	foxr	0	2333	1.0 MiB	N/A	N/A	N/A	N/A	Normal
dbus-daemon	foxr	0	2338	544.0 KiB	N/A	N/A	N/A	N/A	Normal
at-spi2-registryd	foxr	0	2343	1012.0 KiB	N/A	N/A	N/A	N/A	Normal
ibus-dconf	foxr	0	2350	812.0 KiB	24.0 KiB	N/A	N/A	N/A	Normal

End Process

FIGURE 4.10 Process listing shown in system monitor.

The Resources tab provides a summary of CPU, memory/swap space and network utilization over time (the last minute). An example is shown in Figure 4.11. This tab is strictly output (we cannot control any aspect of the system from here, unlike in the Processes tab). Notice that there is almost no swapping indicated in the figure. This is because the system consists of a single user running few applications. Network utilization has increased, which was caused by the user starting the Mozilla web browser.

The File Systems tab displays statistics about the file system. This information is similar to that provided by the `df` command (which we examine in Chapter 8). Also available by clicking on the System Monitor icon (upper-left corner of the System Monitor window) is a menu that includes Preferences. The System Monitor Preferences window, like the System Monitor, contains three tabs, one for each of Processes, Resources and File Systems. We can modify the Processes preferences by specifying what information is displayed for each process, for Resources the type of graph displayed and for File Systems the type of information displayed about the file systems. All three allow us to modify how quickly the System Monitor refreshes itself.

Another GUI tool that provides some of the same information is Cockpit. Recall Cockpit is a new feature in Red Hat Linux to provide GUI control over the system through your web browser. Bring up a web browser and enter the URL *`ipaddress`*`:9090` where *ipaddress* is your IP address. You will be asked to log in. By default, the Overview window shows your computer's name/domain name, the version of Linux running, the health of the system (only if registered), the amount of CPU and memory usage, other system information and configuration details. Some of the menu selections for Cockpit include Storage to display a graph of the amount of disk operations (reading and writing) and file system information much like the File System tab of the System Monitor, and Networking which shows the amount of message traffic being sent and received, again in a graph. There are other controls in Cockpit that offer various functionality such as starting and stopping services, but we omit further detail because the remaining functionality doesn't provide us with detail of running processes.

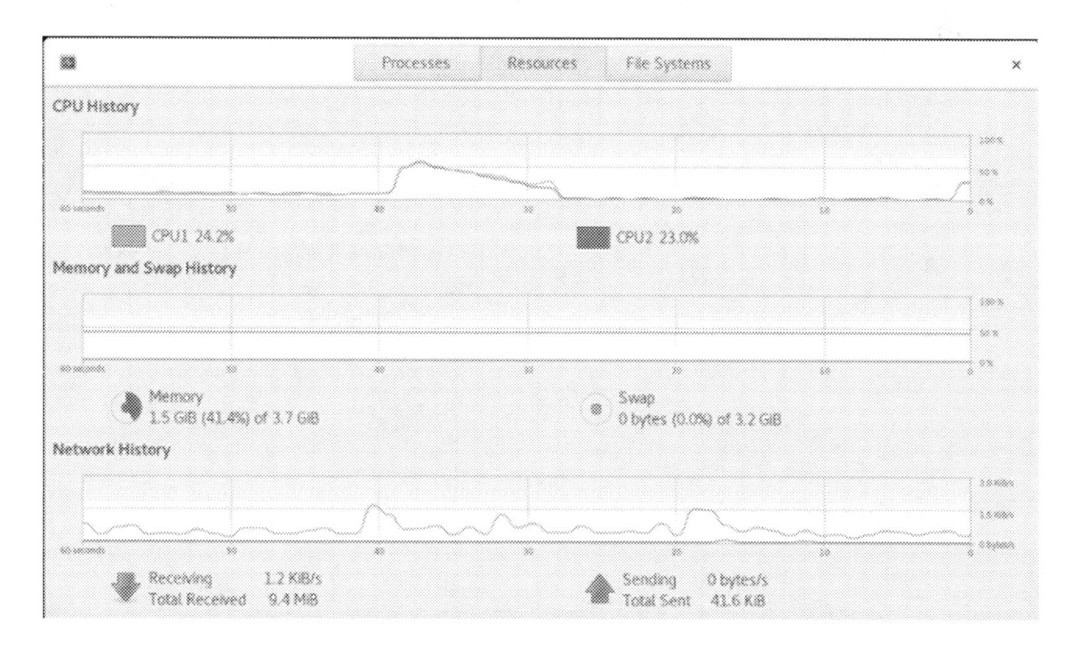

FIGURE 4.11 Resource utilization as shown through system monitor.

4.4.2 Command-Line Monitoring Tools

While the System Monitor provides a convenient way to view system usage and control running processes, we might be more interested in exploring system usage from the command line for two reasons. First, similar to the idea of opening multiple windows to run multiple processes that we explored in the last section, using the System Monitor takes more resources than command-line programs. Second, the information we can get from the System Monitor is not as comprehensive as what can be found from some of the command-line programs. So, we turn to two of the primarily monitoring tools: `top` and `ps`.

The `top` program is launched from the command line but unlike most of the commands we have viewed, this command fills the terminal window with its output and remains running. `top` is an interactive program that updates its output in a specified refresh rate and changes its appearance based on input keystrokes. The default refresh rate is 3 seconds. This rate can be altered by adding the option `-d` *`s.t`* when `top` is launched, where *s* is the number of seconds and *t* is the number of tenths of seconds. For instance, `-d 1` would change the delay from 3 seconds to 1 second while `-d 0.5` would change the delay to every half second. We can provide an even shorter interval if desired (for instance, `-d 0.001`) but that is probably unnecessary and also resource consuming. The value supplied must be positive or 0.

`top`'s output is illustrated in Figure 4.12. Along the top of `top`'s output are system statistics. The first row of output displays the current time, the amount of time the system has been running since its last boot, the number of logged in users, and the average system load which provides the average load over the last minute, five minutes and fifteen minutes. The next line displays the number of total tasks (processes/threads) and then a breakdown of those tasks by running, sleeping, stopped and zombie (we explore the term zombie later in the chapter).

The next three lines show system resource utilization: CPU, memory, swap space. For CPU, the fraction of CPU usage is divided between user usage, system usage, nice usage (the amount of CPU time being given away because of low priority), idle time, time the CPU waits while I/O takes place, time spent servicing hardware and software interrupts, and time stolen by the VM

```
top - 09:22:02 up 1 day, 20:34,  1 user,  load average: 0.06, 0.02, 0.00
Tasks: 290 total,   3 running, 283 sleeping,   4 stopped,   0 zombie
%Cpu(s):  0.2 us,  0.0 sy,  0.0 ni, 99.6 id,  0.0 wa,  0.0 hi,  0.3 si,  0.0 st
MiB Mem :   1789.5 total,     79.3 free,   1107.6 used,    602.6 buff/cache
MiB Swap:   2048.0 total,   1623.4 free,    424.6 used.    504.4 avail Mem

  PID USER      PR  NI    VIRT    RES    SHR S  %CPU  %MEM     TIME+ COMMAND
  944 root      20   0  575544   6772   5948 S   0.3   0.4  67:03.88 vmtoolsd
 2299 foxr      20   0 4361240 162476  40476 S   0.3   8.9   1:33.92 gnome-s+
27078 root      20   0       0      0      0 I   0.3   0.0   0:03.92 kworker+
27122 foxr      20   0  275272   5228   4368 R   0.3   0.3   0:00.15 top
    1 root      20   0  245520   6640   4464 S   0.0   0.4   0:05.05 systemd
    2 root      20   0       0      0      0 S   0.0   0.0   0:00.10 kthreadd
    3 root       0 -20       0      0      0 I   0.0   0.0   0:00.00 rcu_gp
    4 root       0 -20       0      0      0 I   0.0   0.0   0:00.00 rcu_par+
    6 root       0 -20       0      0      0 I   0.0   0.0   0:00.01 kworker+
    9 root       0 -20       0      0      0 I   0.0   0.0   0:00.00 mm_perc+
   10 root      20   0       0      0      0 S   0.0   0.0   0:01.07 ksoftir+
   11 root      20   0       0      0      0 I   0.0   0.0   0:06.45 rcu_sch+
   12 root      rt   0       0      0      0 S   0.0   0.0   0:00.07 migrati+
   13 root      rt   0       0      0      0 S   0.0   0.0   0:00.01 watchdo+
   14 root      20   0       0      0      0 S   0.0   0.0   0:00.00 cpuhp/0
   15 root      20   0       0      0      0 S   0.0   0.0   0:00.00 cpuhp/1
   16 root      rt   0       0      0      0 S   0.0   0.0   0:00.03 watchdo+
```

FIGURE 4.12 Example of `top` program output.

hypervisor (if any). Memory and swap space usage are broken into total, free, used and buffered/cached (for memory) and available memory (for swap space).

The remainder of the top's output is a list of some of the running processes. By default, processes are ordered by CPU usage so that the list of processes changes whenever top refreshes. Processes that are not using much CPU time would likely not appear as there are more processes in the system than can be displayed in the window. The displayed information for each process defaults to showing that process' ID, user, priority and niceness values (covered in the next section), various amounts of memory usage, process state, CPU utilization, memory utilization, the amount of time it has run, and the command that launched it.

Interaction with top, aside from refreshing itself based on an interval, is through single keystroke inputs. The more useful controls for top are given in Table 4.3. In addition to these keystrokes, pressing the enter key will automatically refresh the output.

Many of the commands whose keystrokes are shown in Table 4.3 can be activated from the command line as options when top is launched. These include -H to show threads, -i to show idle processes and -u *username* to show processes started by *username*. The user can specify -n # where # is an integer to indicate the number of refreshes that top should perform before terminating. For instance, top -n 10 will only update itself nine times after the first display before it terminates. If we do not specify -n, then we must stop top on our own. This can be done by either typing a q or pressing control+c.

The ps command provides a detailed examination of the running processes as a *snapshot* (an instance in time). Unlike top, it is not interactive. It displays the information and exits. The ps command by itself (with no options) displays only the running processes of the current user in the current terminal window. As we will enter ps from the command line, it will display ps and bash but the only other processes it will display are those that are suspended or running in the background.

Figure 4.13 provides an example of what we might find when running ps (with no options). In this case, we see only two processes, the ps command and bash. By the time the output is displayed, the ps command will have terminated and the only process is bash. The information displayed when using ps without options is also minimal: the process ID (PID), the terminal window (TTY),

TABLE 4.3
Keystroke Commands for top Instruction

Keystroke	Meaning
A	Use alternate display mode.
d	Alter interval time.
l (lower case L)	Turn on/off load statistics.
t	Turn on/off task statistics.
m	Turn on/off memory statistics.
f, o	Add/remove fields or alter display order.
H	Show threads.
U	Show specified user owned processes only.
n	Show a specific number of processes only.
q	Quit.
i	Turn on/off including idle processes.

```
  PID    TTY     TIME      CMD
16922    pts/0   00:00:00  bash
24042    pts/0   00:00:00  ps
```

FIGURE 4.13 Output from ps command.

the amount of CPU time (`TIME`) being used by the process and the command name (`CMD`). Note that `TTY` stands for *terminal type* and indicates the terminal window in which a process is running. In early days of Unix, users would access the Unix computer remotely and the TTY would literally be a network connection or a dumb terminal. In Linux, the notation `pts/#` indicates an open window where # is a number. As we launched `ps` with no options, we are only seeing the processes running in this terminal window and thus both processes are in `pts/0`.

By itself, `ps` did not show us much. With options though we can view more information about running processes and view different subsets of running processes, or all of the processes running in the system. `ps` has a great number of options making it somewhat more challenging than the typical Linux command. `ps` originated with early Unix and was modified for BSD Unix and again for the GNU's project. In Linux, all of these options are available. Options from the original Unix version of `ps` are preceded by hyphens while options from the BSD version use no hyphens and options from the GNUs version are preceded by double hyphens (--). It is important to remember which options require which notation. Not only does this complicate `ps`, but understanding the meaning behind these options and how they differ can be a challenge.

Table 4.4 describes some of the many options available. Options are combined in one row when they are synonymous, as with the option to show parent/child relationships using "ASCII art" (`f`, `-H`, `--forest`). Note the terminology for `ps -a` of *session leader*. Processes are grouped into sessions with the parent process of the group being the session leader. The session ID is equal to the session leader's PID.

Let's take a closer look at some of the options from Table 4.4. As you will have seen, there are multiple ways to accomplish the same or similar tasks. To show all running processes, use any of `ps -A`, `ps -e` or `ps ax`. To show all processes of a given user, use any of `ps -u` *`user,user,user`*, `ps U` *`user,user,user`* or `ps --user` *`user,user,user`* where each *user* is either the user's ID (EUID) or username. Notice that each user is separated by a comma but no space. The EUID is the user that the process is running under. We can also use `-U` or `--User` to list the processes of the user who launched the process for situations where the EUID differs from the real UID (RUID). For instance, recall that `passwd` runs not under our UID but root's.

We can request the status of specific processes by listing the process' PIDs or the commands by name. The former uses `p`/`-p`/`--pid` followed by a list of PIDs separated by commas but no spaces. To output the processes by command name(s), use `-C` *`commandlist`* where the *commandlist* is a list of command names separated by commas but no spaces.

Notice that `ps` (and `top`) show us PIDs. So, to use `p`/`-p`/`--pid` we already need to know the PIDs and may have to obtain them with a previous `ps` (or `top`) command. For this reason, `-C` is easier to use than one of `p`/`-p`/`--pid`. Consider as an example that we have several terminal windows open and are running processes like `bash`, `vi`, `man` and `find`. We want to view the process information of all of these processes. The simplest approach is to type `ps -C bash,vi,man,find`. Keep in mind that `-C` will find processes of all users, not just our own, unless we further restrict the list by adding one of the user options (`U`/`-u`/`--users`).

When we ran `ps` by itself, the information provided on the processes was limited, as shown in Figure 4.13. Whether we run `ps` on our processes of the current TTY, different TTYs or all users, we can vary the information provided in the output. These are known as *output formats* and *output modifiers*. If we do not want to enumerate our own output format, the simplest option is to add option `u` for user-oriented format. Variations are `s` and `v` (also `X` for Register format). Modifiers alter how the information is presented with, for instance, `c` showing true command names and `f` showing parent/child relationships.

If we want to specify our own format, we use one of `o`, `-o` or `--format` followed by formatting information. The formatting information can be specified by "normal" description or "code". When using normal specifiers, the list of formatting options is separated by commas but unlike when specifying users (`u`/`-u`/`--user`) can include blanks after the commas. When using the code specifiers, place all of the codes in quote marks separated by blank spaces but no commas. Table 4.5 shows the options available.

TABLE 4.4
Some of the Many ps Options

Option	Meaning	Comments
pid, *-pid* --pid *PID*	Show the process as specified by PID, format is 1234 or -1234.	Multiple PIDs are separated by commas with no spaces.
-A, -e	Show all processes.	The combination ax is equivalent.
-a, -d	Show all processes except session leaders and those without a TTY; show all processes except session leaders.	
a	Lift the "only yourself" restriction to show all processes in any TTY.	Show all processes as long as they are in TTYs (non-GUI).
-C *commands*	Show all processes of the given command(s) or pid(s).	With multiple commands or pids, separate them by commas but no spaces.
c	Display commands by true name.	
--context	Add SELinux context to output.	
-f, -F	Show full-format listing.	Also called extra format listing.
f, -H --forest	Output parent/child relationships using "ASCII art".	Uses / and \ to show parent/child relationships.
H, -L, m, -m, -T	Show threads.	These options all show threads, but the output differs by order and possible types of data output.
k *specifier*	Use *specifier* to indicate order that output is sorted.	Uses a key (e.g., pid, ppid, comm (command)) to indicate sort criteria and precede it with – to indicate descending instead of ascending order.
L	List all format specifiers.	
-l l (lower case L)	Long format.	
-M Z	Add column of security (SELinux) information.	
-N *condition*	Show all processes except those that fulfill the specified condition.	Also available as --deselect condition.
o *format* -o *format* --format *format*	Use format to indicate the types of information displayed.	Format described in the text; similar options are O/-O which combine format with defaults.
p *pids* -p *pids* --pid *pids*	Show all processes of the given pid(s).	Also available in "quick mode" using q pids, -q pids or --quick-mode pids.
--ppid *pids*	Same as --pid except lists the processes whose parents are specified.	
r	Only show running processes.	Do not show sleeping or suspended processes.
S	Add together values of dead child processes into the parent.	Only sums up some information like CPU usage.
s u v	Different preset formats: signal, user-oriented, virtual memory.	Do not confuse u with one of the format options (o/-o/--format) options.
t *ttys* -t *ttys* --tty *ttys*	Show all processes of given TTY(s).	Use – to indicate all processes not associated with a TTY.
U *users* -u *users* --user *users*	Show all processes of users as specified by users.	Users is specified by EUID or name; use -U or --User when specifying real user ID (RUID).
w, -w	Wide output.	
x	Lift the "must have a TTY" restriction.	When combined with a shows all processes (ax is equivalent to -A and -e).

TABLE 4.5
Specifiers for ps Formatting

Code	Normal	Meaning
%a	args	Show command with all arguments (options, parameters).
%C	pcpu	Process' CPU utilization (amount of CPU time given to this process divided by amount of time process has been running).
%c	comm	Same as %a/args.
%G	group	Show the process' EGID.
%g	rgroup	Show the process's group by name.
%n	nice	Show the process' nice value (explained in the next section).
%P	ppid	Show the parent process's PID (the PPID).
%p	pid	Show the process's PID.
%r	pgid	Show the process group ID (the session leader's PID).
%t	etime	Show the elapsed time since this process started.
%U	user	Show the process' effective user by name.
%u	ruser	Show the process' user by name.
%x	time	Show the process' cumulative CPU time.
%y	tty	Show the TTY to which this process is attached, ? if none.
%z	vsz	Show the process' amount of virtual memory being used (we explore virtual memory later in the chapter).

Now let's look at the types of information that ps provides as output. ps outputs row after row of process information where the columns are made up of the requested (or default) specifiers such as PID, PPID, CPU utilization, command, TTY and state. The top row is a header describing the output of each column where the header is abbreviated. Table 4.6 shows many of the headers we would see and what each stands for. Some entries, which directly coincide with the options as listed in Table 4.5, are omitted for brevity, but a few of the more significant ones are still listed in Table 4.6 so there is some overlap.

Before we look at more ps output, let's focus on STAT (process state) and its possible values. The process state is indicated using one or more characters describing the status of the process. The most common values are R (running), S (interruptible sleep) and T (stopped). Interruptible sleep means that the process is waiting for an event before it can resume. Less common states include D (uninterruptible sleep, process waiting for I/O), W (paging taking place), X (dead) and Z (zombie). In addition to these states, modifiers may be added to indicate priority (< for high, N for low), that the process is a session leader (s), that the process has pages locked in memory (L), that the process is threaded (l) and that the process is running in the foreground (+).

Figure 4.14 illustrates three different outputs from the ps command. The first portion demonstrates ps with no options. Notice that there are more processes listed aside from bash and ps because, in this case, several foreground processes have been suspended. The second portion of the figure shows three different excerpts from ps aux, here showing the header and the first process listed (systemd), other root-owned processes and then some processes run by user foxr. The final portion shows a part of the output from ps f to demonstrate "ASCII art", showing parent/child.

Another command, pstree, captures the parent-child relationship between processes similar to ps -f. The hierarchical relationship is easier to view in the pstree output, but pstree is far less expressive in terms of available options and output statistics when compared to ps.

TABLE 4.6
Headers for Various ps Options

Header	Explanation	Comments
%CPU	CPU utilization.	CPU time/total time that the process has been admitted to system.
%MEM	Memory utilization.	Memory usage of process/total memory size.
CMD, COMMAND	Command name.	May include arguments depending on format selected.
NI	Niceness value.	Niceness and priority are explained in the next section.
PID	Process PID.	
PPID	Process parent's PID.	
PRI	Process priority.	Niceness and priority are explained in the next section.
RSS, RSZ	Amount of non-swapped memory the process has used.	
SCH	Scheduling policy of this process.	Displayed as a number from 0 to 6 (see the ps man page for detail).
SESS, SESSION, SID	Session ID.	This is the PID of the process serving as the session leader.
SIZE	Approximate size of swap space needed if all writable pages are swapped out.	See section 4.7.1 on virtual memory.
START, STARTED	Time/date process started.	Format differs for processes started within the last 24 hours/last year.
STAT	State of the process.	Described in the text.
SZ	Size of the full process in pages.	
TIME	Amount of total CPU time so far spent on process.	Given in seconds; if less than one second then it appears as 0:00.
TTY	Terminal to which this process belongs.	? for processes not associated with a terminal.
VSZ	Size of full virtual memory for the process.	In KBs.

SECTION ACTIVITIES

1. If you have not yet gained experience from the programs covered in this section, run Linux, open a terminal window and compare the various forms of the ps command to top and to the System Monitor. Of the three, which gives you the most information? Which is the easiest to use? Which do you prefer?
2. If you are a Windows user, start the Windows Task Manager. Compare it to both top and the System Monitor. Make a list of what each of these three programs tells you and what types of operations each allows you to do on running processes. Rank the three in terms of usefulness and usability.

4.5 MANAGING PROCESS PRIORITY

By default, Linux runs processes and threads together by switching off between them. As noted earlier, this is known as multitasking/multithreading. Because of the speed and power of modern processors, the user is typically unaware of the time elapsing as the processor moves from one process to another and back. Literally, the processor moves through the processes in the ready queue and back to the first in under a millisecond (thousandths of a second).

```
   934 avahi      20   0   85284   2676   2544 S   0.3   0.1   0:00.23 avahi-d+
[foxr@localhost ~]$ ps
    PID TTY          TIME CMD
   2923 pts/0    00:00:00 bash
  27134 pts/0    00:00:00 vim
  27141 pts/0    00:00:00 man
  27152 pts/0    00:00:00 man
  27153 pts/0    00:00:00 less
  27603 pts/0    00:00:00 ps
[foxr@localhost ~]$ 
```

```
USER         PID %CPU %MEM    VSZ   RSS TTY      STAT START   TIME COMMAND
root           1  0.0  0.3 245520  6636 ?        Ss   May24   0:05 /usr/lib/syst
emd/systemd --switched-root --system --deserialize 17
root         904  0.0  0.0 143916  1756 ?        S<sl May24   0:00 /sbin/auditd
root         906  0.0  0.0  48516  1084 ?        S<   May24   0:00 /usr/sbin/sed
ispatch
root         928  0.0  0.0  17744  1680 ?        Ss   May24   0:00 /usr/sbin/mce
log --ignorenodev --daemon --foreground
root         929  0.0  0.1  50220  3476 ?        Ss   May24   0:00 /usr/sbin/sma
rtd -n -q never
polkitd      930  0.0  0.7 1772536 13160 ?       Ssl  May24   0:06 /usr/lib/polk
it-1/polkitd --no-debug
foxr        2923  0.0  0.2 235336  5116 pts/0    Ss   May24   0:00 bash
foxr       14727  0.0  0.4 436708  8604 ?        Ssl  May25   0:00 /usr/libexec/
gvfsd-metadata
root       16451  0.0  0.4  94128  8124 ?        Ss   May25   0:00 /usr/lib/syst
emd/systemd-journald
root       26144  0.0  0.0      0     0 ?        I    07:42   0:00 [kworker/3:0-
events]
root       26977  0.0  0.0      0     0 ?        I    09:08   0:00 [kworker/1:0-
events]
root       27133  0.0  0.0      0     0 ?        I    09:21   0:00 [kworker/3:2-
events]
foxr       27134  0.0  0.4 256776  8248 pts/0    T    09:21   0:00 vim foo
foxr       27141  0.0  0.1 228536  3568 pts/0    T    09:21   0:00 man top
foxr       27152  0.0  0.0 228536  1004 pts/0    T    09:21   0:00 man top
foxr       27153  0.0  0.1 219436  2500 pts/0    T    09:21   0:00 less
```

```
    PID TTY      STAT   TIME COMMAND
   2923 pts/0    Ss     0:00 bash
  27134 pts/0    T      0:00  \_ vim foo
  27141 pts/0    T      0:00  \_ man top
  27152 pts/0    T      0:00  |   \_ man top
  27153 pts/0    T      0:00  |   \_ less
  27727 pts/0    R+     0:00  \_ ps f
  27728 pts/0    S+     0:00  \_ less
   2224 tty2     Ssl+   0:00 /usr/libexec/gdm-wayland-session --register-session
 gnome-session
   2229 tty2     Sl+    0:00  \_ /usr/libexec/gnome-session-binary
   2299 tty2     Sl+    1:50      \_ /usr/bin/gnome-shell
   2351 tty2     Sl+    0:00      |   \_ /usr/bin/Xwayland :0 -rootless -termina
te -accessx -core -listen 4 -listen 5 -displayfd 6
   2381 tty2     Sl     0:03      |   \_ ibus-daemon --xim --panel disable
   2385 tty2     Sl     0:00      |       \_ /usr/libexec/ibus-dconf
   2386 tty2     Sl     0:01      |       \_ /usr/libexec/ibus-extension-gtk3
   2598 tty2     Sl     0:00      |       \_ /usr/libexec/ibus-engine-simple
   2485 tty2     Sl+    0:00      \_ /usr/libexec/gsd-power
   2486 tty2     Sl+    0:00      \_ /usr/libexec/gsd-print-notifications
   2489 tty2     Sl+    0:00      \_ /usr/libexec/gsd-rfkill
   2491 tty2     Sl+    0:00      \_ /usr/libexec/gsd-screensaver-proxy
   2495 tty2     Sl+    0:00      \_ /usr/libexec/gsd-sharing
```

FIGURE 4.14 Output from various combinations of ps options.

We mentioned the difference between processes and threads in Section 4.2, but let's explore this again. A process is a standalone entity; it has its own code, data and status information. A thread, sometimes called a *lightweight* process, shares its code and data (or at least some data) with other threads. Threads are, in essence, portions of a process. Threads will communicate with each other through shared data. For the processor to switch between threads, it does not have to load new program code into memory as it is already present. Switching between processes often requires additional overhead.

Let's also define an *application*. As users, we run programs like Mozilla Firefox or OpenOffice Writer. Large applications like these consist of not a single process but many different processes. Some processes are used to support the application, and others are parts of the application. Additionally, some of the application's processes might spawn child processes or threads. For instance, the Apache web server runs multiple child processes, each assigned to handle a different incoming HTTP request.

By default, Linux executes processes and threads using multitasking and multithreading. However, this does not mean that all processes run in this fashion. As a user, we can dictate how processes run. The Linux command `batch`, for instance, forces a process to run in batch mode. A batch process is one that does not interact with the user. Therefore, any batch process must be provided its input at the time the process is launched. Output from a process run with the `batch` command is either sent to a file or is emailed to the user. With `batch`, the launched process will execute only when the system load drops below a preset amount, usually 80%.

There are other forms of control the user has over running processes and threads. As described earlier in the chapter, we can suspend (stop) any running process from the command line and resume it later. We can also alter a process' priority. Priorities establish the amount of CPU time that the process is given each time the CPU switches to it. The higher the priority, the more attention the CPU provides the given process.

In Linux, a process' priority is established by setting its *niceness* value. Niceness refers to how nice a process is with respect to other processes. With a higher niceness value, a process will offer some of its CPU time to other processes. Thus, the higher the niceness value, the less CPU time it will take when it is that process' turn, and thus higher niceness means lower priority. This might seem counterintuitive in that we would expect a higher value to mean a higher priority, but in Linux, it is just the opposite. A lower niceness value means that the process takes CPU time away from other processes and thus has a higher priority.

There are several ways to set or change a process' niceness value. From the System Monitor GUI, under the Processes tab right click on a process and select Change Priority. This provides a submenu with choices of Very High, High, Normal, Low, Very Low and Custom. We see an example of changing a process' priority in the top portion of Figure 4.15. If we select Custom, a pop-up window appears like that shown in the bottom part of Figure 4.15. Slide the slider to the desired priority and select the Change Priority button. Notice that the slider sets the process' niceness rather than a priority. Moving the slider to the left lowers the process' niceness while moving the slider to the right raises the process' niceness.

As a normal user, we are not allowed to lower a niceness value as this would impinge on other process' CPU times. This is true whether we select a higher priority from the list (e.g., Very High) or through the slider. We can lower the niceness of any of our processes. As root, we can raise or lower any process priority.

Another way to establish or change a process' priority is to use one of two instructions: `nice` and `renice`. We use `nice` when launching a command and `renice` to change a priority of a running process. In both cases, we specify the priority as a niceness value. Niceness values range between +19 (most nice or lowest priority) and –20 (least nice or highest priority) with the default of 0. Normal users cannot launch a process with a niceness value lower than 0.

The `nice` command requires that we specify the command as one of its parameters along with the niceness value. The syntax of the instruction is `nice -n #` *`command`* where # is the niceness value. If we omit the niceness value, then `nice` launches the process with a niceness of +10 (instead

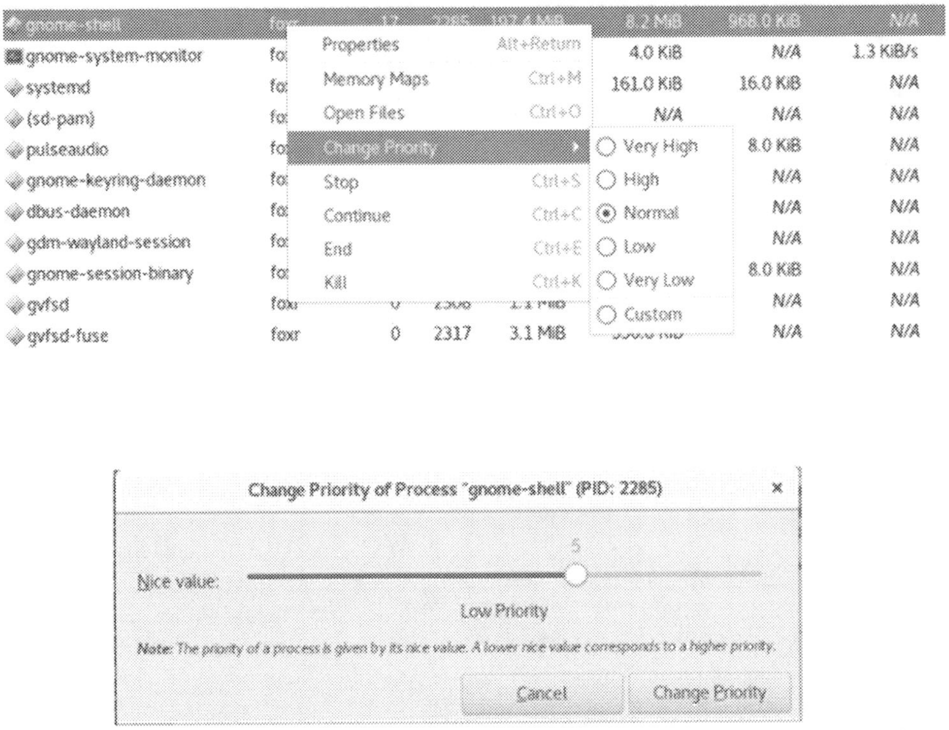

FIGURE 4.15 Changing process priority through the system monitor.

of 0). If a normal user selects a niceness value less than 0, an error message is returned and the niceness value of the process is set to 0. If the niceness value is above +19, the niceness is set to 19.

Figure 4.16 shows some examples of nice commands (also renice which we cover next). The first example launches a `find` command in the background. The second launches `myscript` into the background. As both are background processes, their outputs are being redirected. Notice that the `find` command has a negative niceness and so if not launched by root would be given a niceness of 0.

The `renice` command allows us to change the niceness value of one or more existing processes. The syntax is `renice [-n]` *`niceness`* `[-g|-p|-u]` *`identifier`*... The notation indicates that after the command is the new niceness value, which may but does not have to be preceded by `-n` followed by one or more identifiers to indicate the processes to modify. The identifiers are specified using any combination of `-g` (for GIDs), `-p` (for PIDs) and `-u` (for UIDs/usernames), each followed by one or more GIDs, PIDs or UIDs/usernames. Notice that `-g` only permits GIDs while `-u` permits either UIDs, usernames or some combination. Multiple identifiers are separated by spaces but no commas.

We can combine `-g`, `-u` and `-p` so that, for instance, we can `renice` some processes based on username, others based on groupname and still others based on PID. If specifying PIDs, the `-p` can be omitted, as shown in the third and fourth examples of Figure 4.16. In the fourth and fifth examples, multiple types of identifiers are used (the fourth combines PIDs, without `-p`, and usernames; the fifth combines GIDs and UIDs). Notice for the fourth and fifth examples, the new niceness is

```
nice -n -15  find / -name *.txt > found_files &
nice -n 19 ./myscript < inputfile > outputfile &
-------------------------------------------------
renice -n 19 18311
renice -n -5 53113 -u foxr zappaf
renice -n -10 -g 1001 1002 -u 1003 1004
```

FIGURE 4.16 Example `nice` and `renice` commands.

```
[foxr@localhost ~]$ ps aux | grep bash
root 1094      ...          ?    S   ...
/bin/bash/sbin/ksmtuned
foxr  202445  ...  pts/0   Ss ...  bash
foxr  328525  ...  pts/0   S+ ...  bash
foxr  328564  ...  pts/1   Ss ...  bash
foxr  328629  ...  pts/1   S+ ...  grep bash
[foxr@localhost ~]$ renice -n 5 202445 328525 328564
```

FIGURE 4.17 Using ps and grep to obtain PIDs for renice.

lower than 0 indicating that this command has been launched by root. Another indication that these two were commands launched by root is that a normal user is not able to adjust the niceness values of any process not owned by that user. Since the fourth and fifth commands are adjusting nicenesses of multiple users/groups, they must be root-issued instructions.

How can we obtain the PID of a process to use in `renice`? For this, we might use `top` or `ps`. Both commands list processes by PID. If the process is part of the given terminal window, `ps` by itself would suffice. If the process is in another terminal window, or if we are root and looking for another user's process, we might use `ps aux`. Unfortunately, `ps aux` will result in dozens or hundreds of processes being listed. To output just the processes that we are interested in, we might use the `grep` program, which searches its input for all lines that match some string or regular expression and outputs those that match. We cover regular expressions and `grep` in the next chapter, so for now we look at just a simple example.

We want to obtain the PIDs of all running `bash` sessions. We use the command `ps aux | grep bash`. We see the output of this command in Figure 4.17 (some of the detail is replaced with …). Notice that there are more matches than just `bash` as `grep` finds all lines from the `ps aux` command that contains "bash" anywhere in the line, including the `grep` command itself. The PIDs for the `bash` processes are 202445, 328525 and 328564, and we use these in a `renice` command to raise their niceness (assuming they started with a default of 0).

We can also obtain PIDs using the `pidof` command. `pidof` returns the PIDs of processes matching the command name given. For instance, `pidof bash` will return the PID of every `bash` process running. Unlike `ps` which returns one process per line, `pidof` returns all of the PIDs as a single list. The parameter `-s` causes `pidof` to exit after it outputs the first found PID. We can supply `pidof` with a list of as many programs as desired. However, `pidof` outputs all of the PIDs matching the list of program names without differentiating between the programs. That is, the output is just the list of PIDs without indicating which PID is of which program.

Another way to obtain PIDs based on process names is with `pgrep`. When run with the name of a process, like `bash`, it responds with the PIDs of all running processes of that name. We can add `-U` *`user`* to restrict the list to just those run under the given user's UID or `-u` *`user`* to restrict the list to just those run under the given user's EUID.

Aside from `renice`, there are many other reasons for acquiring a process' ID. One such example is the `lsof` command. This command provides a list of open files attributed to the process. The command's structure is `lsof -p` *`PID`*. We visit another use of the PID in the next section.

SECTION ACTIVITIES

1. Have you ever changed a process' priority? If so, for what purpose? If you raised a process' priority, did you notice an impact on how it performed? Did the change noticeably impact other running processes' performances?
2. Is the idea of niceness confusing? Would it be better to just use priority values? If so, should priority values be numbers (e.g., 1–100) or words?

4.6 PROCESS TERMINATION

Typically, a Linux process runs until it reaches a normal termination point. In a program like `ls`, termination occurs once the program has output all of the content that matches its parameters. In a program like `cd`, this occurs when the `$PWD` and `$OLDPWD` variables have been modified. Interactive programs run until the user chooses to exit out of the program. With `top`, the user enters `q`, and in `vi` the user enters `:q`. A GUI program exits when the user selects the Exit menu selection, typically found under the File menu.

Some processes do not terminate normally. A process that reaches a terminating error will simply stop running, possibly producing an error message but not always. When a user-run process terminates with an error, it leaves behind a *core dump*. This file, named `core`, is an image of the process' memory at the time the termination occurred. The core dump is produced to help a programmer debug the programmer. There are exceptions to a program error generating a core dump. If the file system does not have enough space to store the file (core files are typically large) or if the process does not have access to the current working directory, the file is not produced.

4.6.1 ORPHANS AND ZOMBIES

When a process does terminate, if it has spawned any children, this creates orphans. An *orphan* is a process whose parent has terminated. This creates a problem because processes report back to their parent at least when they terminate if not more frequently. Without a parent, the process cannot report. Upon a parent process terminating, any orphaned child processes must be re-parented or adopted. Linux handles this by using the process whose PID is 1, which is either `systemd` (in more recent versions of Linux) or `init` (in older Linux distributions).

The child process could be orphaned because of a number of different situations. First, the parent process may unexpectedly terminate because of some erroneous situation. Second, a child process may be launched to run in the background as a long-running job. In such a situation, it is not necessarily expected that the parent would run for the same duration. If the parent is a shorter-running process, it could very well terminate before the child. The third situation arises when the parent process creates the child process using one of the `exec` system calls. With `exec`, the parent terminates and is replaced by the child. The child in essence takes over from the parent and so the terminating parent's PPID (parent PID) becomes the child's PPID.

Another situation arises when a child process is ready to terminate but the parent is not currently awake. The parent may have executed a `wait` system call and so is sleeping for some situation. Before the child can completely leave the system, the parent must wake to "clean up" after the child. With the parent unavailable, the child must remain in existence yet we do not want the child to continue to use system resources. So, the child surrenders all resources provided to it including its place in the ready queue and any memory allocated to it. Now, the child exists in the system only in name, state and with a PCB. We refer to the child in such a state as a *zombie* process, denoted with a STAT of `Z`. We can view zombie processes using `ps` command (depending on which options are specified) because they continue to have a PCB in the system.

Users cannot get rid of zombie processes directly because they don't really exist as normal processes. But the operating system will remove each one once the zombie's parent runs and is able to receive the child process's terminating status (report). In the meantime, zombies can accumulate. As they have freed up their resources, the only practical issues of having zombies in the system are that each one takes up a small amount of memory to store their PCB and each is still assigned a PID (Linux has a finite number of PIDs to use). In order to attempt to remove a zombie, we can wake up the parent process by sending it the signal SIGCHLD. We discuss signals shortly.

4.6.2 KILLING PROCESSES

This leads us to the discussion of killing processes. Why would we want to kill a process and how does this differ from exiting a process? In some cases, killing a process is the same as exiting a process.

This is especially true if the process does not currently have any open resources. Consider for instance using vi. We would exit vi normally by saving any open file and typing :q. In saving the file, we are assured there is no data corruption, and by exiting, we ensure that the file is closed. If the program does not have an open file, then exiting the program and killing it leave the system in the same state.

We might kill a process if that process was not responding. This happens frequently in the Windows operating system but is less common in Linux, although it can still happen. In addition, a process launched into the background has no interface with the user. Aside from changing the process to the foreground to end the process, a user could kill the process instead as it might be faster to do so. Finally, if an emergency situation arises requiring that the system be shut down quickly, the system administrator may have no choice but to kill all running processes.

To kill a process, the command is naturally kill. kill requires that we specify *what* we are killing and *how* it should be killed. The what is one or more PIDs. The how is a signal. *Signals* are messages that we send to processes to interrupt them. Upon receiving a signal, the process operates on that signal. Different types of signals cause processes to respond in differing ways. For instance, we mentioned the signal SIGCHLD in the last subsection; this signal is used to inform a parent process that it needs to inspect its children.

The syntax for the command is kill [-*signal*|-s *signal*|-p] *pid(s)*. *Signal* can be specified by number or name. There are dozens of signals which we explore momentarily. We are allowed to specify any number of PIDs in one kill command; we separate multiple PIDs by commas but no spaces. In place of a PID, we can also specify 0 to indicate all processes of the current process group, -1 to indicate all processes with a PID larger than 1 (all processes other than systemd/init) or -*pid* to indicate all processes in the process group where *pid* is the session leader's PID.

kill does not necessarily kill a process. Instead, its role is to send the signal to the process(es). Some signals will kill processes depending on how the process is set up to handle the given signal. There are 64 signals. To view the signals by name and number, issue the instruction kill -l (or -L) with no PID(s) listed. We can also specify a number between 1 and 64 after the -l/-L to see the name of that specific signal. For instance, the command kill -l 17 returns CHLD, short for SIGCHLD.

When using kill, we can specify signals by their number, full name (e.g., SIGCHLD) or abbreviated name (e.g., CHLD). The signal can be indicated as -s *SIGNAL* or as -*SIGNAL*. For instance, kill -s 9 *PID*, kill -s SIGKILL *PID*, kill -s KILL *PID*, kill -9 *PID*, kill -SIGKILL *PID* and kill -KILL *PID* all do the same thing.

Let's now focus on some of the more significant signals and what they do when used in the kill command. Table 4.7 presents the ones we are interested in. The table provides both the signal number and its name, along with its meaning. The last column gives a brief indication of what the signal is used for or what can cause it.

Of the signals available, only SIGKILL (9) is guaranteed to kill a process. Other signals will cause the process to perform a different type of operation. For instance, a network program receiving SIGHUP will terminate the network connection but leave the process running. With SIGIO (29), a process that was waiting for I/O can resume. In some cases, a process will still terminate when receiving a signal other than SIGKILL, for instance if a process generates an illegal memory address and generates a SIGBUS (7) signal.

Most modern programming languages permit exception handling code. An *exception handler* is a piece of code in a program that, upon receiving an exception signal, executes that code to handle the specified exception. A Java program, for instance, might have an exception handler to deal with floating-point exceptions so that the SIGFPE signal would not terminate the process. If the program did not have such an exception handler then it would terminate upon receiving SIGFPE.

Many of the signals will likely cause the process to terminate and in some cases cause the kernel to generate a core dump. This would happen for instance with signals like SIGABRT, SIGBUS, SIGFPE and SIGILL. Other signals do not result in termination but instead the process responds to the signal such as with SIGCHLD or SIGALRM.

To kill all processes of a given program, we can use pkill. The syntax is the same as kill except that we list the process(es) by name instead of PID. For instance, pkill -9 bash will kill all running

TABLE 4.7
Notable Linux Signals (There Are 64 Total Signals)

Number	Name	Meaning	Sources/Causes of the signal
1	SIGHUP	Hang up.	Terminal connection lost.
2	SIGINT	User interrupt.	User sends interrupt to process.
3	SIGQUIT	User quit signal.	User sends quit to process.
4	SIGILL	Illegal instruction.	Program contained a machine instruction not understood by CPU.
5	SIGTRAP	Send trap.	Mainly invoked for debugging; causes interrupt to operating system.
6	SIGABRT	Abort.	Process sends abort signal to itself (emergency stop).
7	SIGBUS	Bus error.	Incorrect memory access attempted.
8	SIGFPE	Floating-point exception.	Illegal mathematical error.
9	SIGKILL	Terminate the process.	The only signal that will definitely result in process termination.
13	SIGPIPE	Pipe closed.	Generated when a process attempts to write to a closed pipe.
14	SIGALRM	Alarm clock.	Sets a "wake-up call" to the process.
15	SIGTERM	Software termination.	Generated from some software such as the kill command.
17	SIGCHLD	Signal parent about child.	Used to let the parent know that a child has died or exited.
29	SIGIO/ SIGPOLL	I/O is waiting.	Signal sent to a process to let it know I/O operation is completed (input may be waiting).
30	SIGPWR	Power management.	Used if power has switched to a short-term supply (e.g., uninterruptible power supply).

bash interpreters. Be careful with pkill because it may kill more processes than you expect. Issuing pkill on bash for instance would cause all open windows running bash to close on you!

Another variation of kill is killall. This instruction will also kill all instances of the named process(es). For instance, killall man vim will kill all man and vim processes. As with kill, killall accepts a signal. However, if none is given, the signal defaults to SIGTERM (signal 15). The killall command has an interactive mode, specified with option -i, which pauses to ask the user for permission before killing each process. Each process is presented to the user by name and PID. Aside from specifying programs by name, we can specify users in killall using the option -u *userlist*. This will kill all processes of the named user(s). Only root would be able to kill processes owned by other users.

We can also kill or exit processes using the System Manager tool. From the Processes tab, select a process, right click and select one of Stop, End or Kill. Stop does the same thing as control+z; that is, it suspends the process instead of exiting the process. End permits the process to exit gracefully (close files, etc.), while Kill kills the process which is the same as sending it the SIGKILL signal.

4.6.3 Shutting Down Linux

To wrap up this section, we look at the methods available to shut down the Linux operating system. The reason we want to use one of these methods to shut down the computer as opposed to just turning it off is that the shutdown mechanisms will ensure that the system shuts down properly. Among the steps that the system will go through is making sure all files are closed, all services are stopped and all processes other than systemd are ended. Should these steps not be taken files could be corrupted and data lost.

Under normal circumstances, only the system administrator should have privileges to shut down and/or reboot a Linux system. This is because Linux is intended to be a multiuser system and so a user shouldn't be able to shut down the system when other users are currently logged in. Even if users are not logged in, they may have left processes running or have processes scheduled to run. Giving a normal user the ability to shut down the system is a violation of the rights of other users. But Linux is commonly run on personal computers where there is rarely more than one person using the computer at a time. And so, for convenience, all users are able to shut down or reboot the system.

Earlier versions of Linux used several different commands to shut down the system. Specifically, the commands were `shutdown`, `poweroff`, `halt` and `reboot`. All of these commands are now symbolic links pointing to `/usr/bin/systemctl`. The `systemctl` utility is used for a number of different actions including starting and stopping services and changing run-time modes. We explore `systemctl` in detail in Chapter 9. For now, we will consider how to use it simply to execute these four operations.

One type of operation that `systemctl` runs is to change the *run-time mode*. The command is `systemctl` *mode* where *mode* is one of the modes listed in Table 4.8. Depending on the type of Linux, either normal users will be unable to issue this command without authenticating with the root password or will be limited to just issuing a few of the modes such as `halt`, `poweroff`, `reboot` and `suspend`.

When issuing the `systemctl` instruction on any of `halt`, `kexec`, `poweroff` and `reboot`, the option `--force` can be added. This specifies that services should not be shut down. This saves some time and might be more useful with `reboot`, which may permit other options based on the specific processor and firmware that your system uses. Notice that there is no run-time mode for `shutdown` so instead of using this word, we use one of `halt` or `poweroff`.

The more common way to shut down or reboot the system is through the GUI. In the Gnome desktop, this control is found in the upper right-hand corner of the desktop. Clicking the power button brings up a window of various settings, as shown in Figure 4.18. From here, we can change the speaker volume, change the network connection (indicated as wired in the figure), switch to another user or click on one of three icons at the bottom. These icons are respectively from left to right to system settings, lock the screen or shut down the system. This last selection brings up a window with three options: Cancel, Restart (reboot) and Power Off. Most users will find this approach easier than using `systemctl`.

TABLE 4.8
`systemctl` Run-Time Modes

Mode	Meaning
`default`	Switch to default mode (usually full mode with GUI).
`emergency`	Switch to single-user (root) text-based mode with no services or mounted file systems, running only `systemd` and a shell.
`exit`	Shut down the service manager (`systemd`) leaving the system in a state similar to `poweroff`.
`halt`	Shut down and halt the system.
`hibernate`	Saves the state of the machine (memory image) to swap space and shuts the machine down.
`hybrid-sleep`	Perform both suspend and hibernate.
`kexec`	Similar to `reboot` but the reboot is handled by the kernel so that ROM BIOS (including POST) is skipped (see chapter 8 for more information on system booting).
`poweroff`	Like `halt` but further powers off the system.
`reboot`	Shut down the system but instead of entering `halt` or `poweroff` mode, perform a reboot.
`rescue`	Same as emergency except that it starts some basic services and mounts all file systems.
`suspend`	Put the system into sleep mode (low power mode).
`switch-root`	Change to a different root directory (specified in the command).

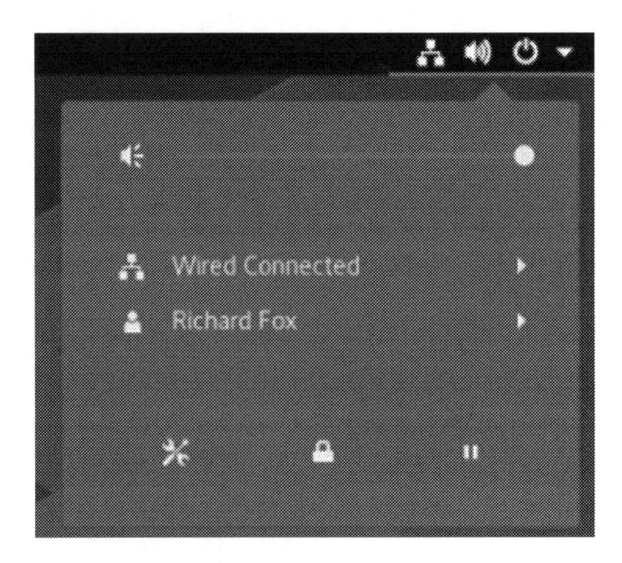

FIGURE 4.18 Shutting down the system through the GUI.

SECTION ACTIVITIES

1. Have you ever had to kill a running program (i.e., end it without selecting the quit option)? If so, for what reason and in what operating system?
2. Look at the list of signals in Table 4.5 and see how many you understand. Aside from SIGKILL, are there any you would use? If so, which ones and for what reasons?
3. Come up with a list of three reasons why a user should not be allowed to shut down, halt or reboot a Linux computer.

4.7 A LOOK AT SYSTEM RESOURCES

All processes use system resources. At a minimum, every process has a PCB stored in the operating system's portion of memory and some amount of memory space allocated to it, plus a place in a queue and access to the CPU. Other resources that a process may request are access to I/O, storage and communication devices, and interprocess communication with other processes through domain sockets, named pipes and/or remote procedure calls. Here, we explore some of the resources that you should know to better understand Linux.

4.7.1 Memory and Virtual Memory

As user processes are given their own memory space, the computer must ensure that one process does not try to access memory allocated to another process. The operating system maintains tables of process memory space in the form of *memory maps*. There are many different memory maps utilized by the Linux kernel. Some of these maps are used to map devices and files to memory locations. Others are used to specify process-allocated memory.

As main memory (DRAM) is limited in size, we often find that the processes we want to run will not fit wholly in main memory, especially because we tend to run a large number of processes. We handle the need for more memory than is available by extending main memory with a reserved area on the hard disk (or solid-state drive) called *swap space*. Portions of our running programs reside in swap space rather than memory so that we can accommodate both a number of processes and large processes at one time. We refer to this extension of memory as *virtual memory*.

The implementation details for virtual memory differ operating system to operating system. The typical strategy though is to divide programs into fixed-size units called *pages* and memory into fixed-sized units called *frames*. One page fits precisely into one frame. A program's pages are loaded when needed (on demand) by the operating system.

A data structure called a *page table* contains the information about which of a process' pages are in which frames of memory. The page table also indicates which pages are not currently stored in memory. Each process has its own page table, and all of the page tables are stored in the operating system's reserved portion of memory.

As a process runs, the CPU generates memory addresses of instructions and data. A CPU-generated address is the location of where the item is to be found within the program, which is not the same as where it will be located in memory. For instance, the CPU might generate address 0 for the very first instruction of the program. But this instruction is likely not located in memory address 0.

We refer to a CPU-generated address as a *virtual address*. Its physical location in memory is known as its *physical address*. We need a process to translate from a virtual address to a physical address. This mapping is handled by exchanging the address' page number for the frame number containing that page of the program. See Figure 4.19, for process 15 which currently has several pages stored in DRAM in various frames. Also located on the figure is the process' page table.

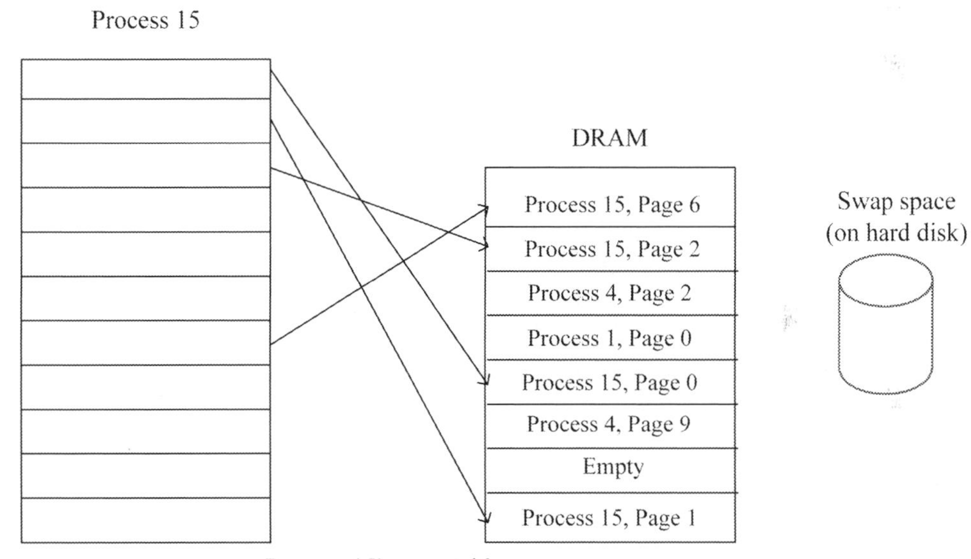

Process 15's page table

Page	Frame	Valid
0	4	1
1	7	1
2	1	1
3	-----	0
4	-----	0
5	-----	0
6	0	1
7	-----	0
8	-----	0
9	-----	0
10	-----	0

FIGURE 4.19 Virtual to physical address mapping.

In the figure, we see that process 15 currently has four pages in memory. Page 0 is located in frame 4, page 1 at frame 7, page 2 at frame 1 and page 6 at frame 0. The seven pages of process 15 that are not currently resident in memory are stored in swap space. Other processes have pages residing in memory and at this point in time, one frame is empty and so is available for whatever page may be loaded next.

Imagine that the CPU has generated address 500 in process 15's page 2. The CPU has to use the page table to map this virtual address to its physical address. It does so by retrieving from the page table the location of page 2 in memory, which is frame 1. The address within a page or in a frame is known as an *offset*, and the value is the same in both the page and the frame since they are of equal sizes. Thus, the virtual address of page 2, offset 500 is mapped to the physical address of frame 1, offset 500. The CPU now sends the address of frame 1, offset 500 to memory (encoded as a binary number).

The example shown in Figure 4.19 is unrealistic for several reasons. DRAM will likely consist of tens of thousands or more frames and processes could contain hundreds, thousands or millions of pages. We would tend to see far more processes in memory than just three. We would also expect to see more frames of each process in memory. The operating system kernel would have its own area of memory reserved and likely starting at frame 0, where process 15 has a page. Finally, as we use our computer, we fill up memory. It is unlikely that we would have any free frames, at least for long.

When an access is made to a page that is not in memory, a *page fault* arises. For instance, if the CPU generates for process 15 the address page 7, offset 0, this is currently an invalid reference since page 7 is not in memory. A page fault generates an interrupt which causes the processor to switch to the operating system to handle the page fault. The page fault requires obtaining the needed page from swap space on disk and moving it into a free frame in memory. In the case of Figure 4.19, page 7 can be loaded into frame 6. Once the frame has been loaded into memory, the page table is updated to reflect the change. We refer to the process of moving pages into and out of memory as *page swapping*.

Unfortunately, as memory is limited in size, it is likely that memory will be full and so there would be no free frames available. Before swapping in the requested page, the operating system will first use a *replacement strategy* to select a page to discard from memory. There are many types of replacement strategies, but we would like to use one that best predicts a page which won't be used again for the longest amount of time. This type of prediction is known as *least recently used* (or LRU for short). The idea is that any page that hasn't been used recently won't be used again any time soon. Unfortunately, selecting the least recently used page can be time consuming, so various approximation algorithms have been created to be used in place of least recently used.

Another strategy to employ is to select a page that has not been modified. Any modified page cannot simply be discarded from memory because we would lose whatever modifications were made. Instead, we have to write such a page back to swap space. Removing a modified page from memory requires two disk accesses: one to save the current page to disk and one to copy the requested page from disk into memory. Two disk access would further slow the swapping process.

When implementing virtual memory, operating system designers have many choices to make. Upon starting a process how many pages of the program are loaded into frames in memory? Do all processes get the same number of frames? Which of the program's pages should be loaded (usually it will be the first *n* pages if n pages are to be loaded)? When swapping must occur, does the operating system discard from memory a page of the current process or a different process? In the latter case, a process' size in memory can change over time as it might acquire more frames or lose frames to other processes.

How many pages and frames might there be? This depends on the page/frame size, the size of a program and the size of memory. Let's assume that we have a computer whose page/frame size is 4096 bytes (4 Kbytes). We want to run a program that is 32 MBytes in size on a computer with 8 GBytes of memory. The number of pages is 32 MBytes / 4KBytes = 8K pages or 8192 pages, and the number of frames is 8 GBytes/4KBytes = 2M or roughly 2 million frames. So, this computer has over 2 million frames, and this program requires 8192 of those frames to fit the entire program.

Virtual memory is a great idea in that it allows us to run multiple programs whose total size is larger than that of main memory. There is a price to pay with virtual memory and that is the time to perform swapping. Disk access is among the slowest activities of a computer. We want to minimize disk access as much as possible. Virtual memory provides a compromise between spending more money on memory and having a less efficient processor due to waiting on page swapping. Another potential compromise is to use a solid-state drive (SSD) to store virtual memory. While SSDs tend to have less storage capacity than hard disk drives, using one for virtual memory provides more than adequate storage space while also reducing the impact on swapping because SSD access is so much faster than hard disk access.

4.7.2 Linux Commands to Inspect System Resources

Earlier in this chapter, we looked at the `System Monitor`, `top` and `ps` to explore process usage. Here, we look at some of the many Linux commands available to both normal users and root to explore system resource usage. Two of these programs may not be pre-installed into your Linux distribution. If you try to experiment with the instructions below and find that `mpstat` and `iostat` are unavailable, you can install them using `dnf` (or `yum`) `install sysstat` (Red Hat distributions) and `apt-get install sysstat` (Debian/Ubuntu distributions). You must be root or use `sudo` to perform this operation.

`mpstat` is used to display multiprocessor statistics. Although the instruction's primary usage is to display the load across all processors, it can also be used whether our computer has multiple or a single processor. It gives a snapshot much like `ps` does but of processor load. Figure 4.20 provides an example.

Most of the terms should be familiar from our previous exploration of `top` and `ps`. Consult `mpstat`'s man page to learn about some of the other values. Omitted from the figure is the time that the command was run (which appears prior to CPU and all in both lines). Notice here that the output is combining (averaging) the results of both CPUs. We can also obtain the results of each CPU individually. Most of this information is available via `top` but here we can break it down by processor if desired.

The `iostat` command reports on both process and I/O statistics. Without options, it outputs most of the same information as `mpstat` with respect to processor usage, and like Figure 4.20, it provides averages of the processors. But `iostat` also outputs for various devices several key statistics. These are transactions (accesses) per second, amount of KBs read and written per second and the number of KBs read and written since the system was last rebooted. Options allow us to view other I/O statistics.

Two other instructions of use are to view memory and virtual memory usage. These commands are `free` and `vmstat`, respectively. With `free`, we receive a report on the total amount of memory and swap space, the amount each is currently using and the amount free. We also see the amount of memory shared between processes and used for buffers/cache. An example of this output is shown in the top portion of Figure 4.21. The bottom portion of the figure provides a sample output of `vmstat`. `vmstat` provides averages since the last system boot and as you can see in the figure provides a breakdown not only of virtual memory usage but also CPU usage and processes.

The output for `free` is given in KBs and should be self-explanatory. The output for `vmstat` is extremely cryptic. The values `r` and `b` are the number of runnable processes and the number of processes in uninterruptible sleep. The values under the `memory` heading indicate the amount of free memory, the amount of buffer and cache space, and the size of the swap space in use. The two

```
Linux 4.18.0-193.28.1.el8_2.x86_64 (locahost.localdomain) 06/14/2021
_x86_64_ (2 CPU)

CPU %user %nice %sys %iowait %irq %soft %steal %guest %gnice %idle
all 12.11  2.53 16.34  5.21  2.88 15.74   0.00   0.00  0.00  45.19
```

FIGURE 4.20 Output from `mpstat` showing processor usage statistics.

```
[foxr@localhost ~]$ free
              total        used        free      shared  buff/cache   available
Mem:        1832412     1216820      171304       30680      444288      420452
Swap:       1048572      284160      764412
[foxr@localhost ~]$ vmstat
procs -----------memory---------- ---swap-- -----io---- -system-- ------cpu-----
 r  b   swpd   free   buff  cache   si   so    bi    bo   in   cs us sy id wa st
 4  0 284160 171304     56 444256    0    1     8    17   35    8 98  2  0  0  0
[foxr@localhost ~]$ 
```

FIGURE 4.21 Output of `free` and `vmstat` showing memory and virtual memory statistics.

values under `swap` indicate the number of pages swapped in and out, respectively, while the `io` section indicates the number of blocks input and output from a block device, respectively. Under `system`, the two values are the number of interrupts per second and the number of context switches per second. Finally, the `cpu` section provides statistics on CPU utilization broken down by time spent running user code, kernel code, processor idle time, time spent waiting for I/O and time stolen by a virtual machine. You will notice that this is a lightly loaded and recently booted system (98% of the CPU time is spent idling). We will take a closer look at all four of these programs (`mpstat`, `iostat`, `vmstat`, `free`) in Chapter 12.

We wrap up this section and chapter with one additional Linux instruction, `pmap`. As noted earlier, the operating system maintains many memory maps. The `pmap` instruction allows us to explore the memory map of a specified process. When run, `pmap` outputs for the process the breakdown of the memory it uses showing us blocks of memory. As a process may run other helper processes, `pmap` also shows us those. We ran `pmap` on a running `bash` shell (whose PID was 4568) and received the output shown in Figure 4.22.

For each of the processes related to `bash` (including `bash` itself), we find four pieces of information. The first is the starting memory address, given in hexadecimal notation. The second is the amount of memory used. In this case, each page is a 4K block so all of these values are multiples of 4K. The third column displays the mode associated with the page. We explain the modes momentarily. The last item is the name of the process or helper utility.

You might wonder why some items appear multiple times. This is because the item has been allocated to several different locations in memory. The total at the bottom shows the total space used by this process. The mode indicates if a block of memory is readable (`r`) writable (`w`), contains executable code (`x`), is shared (`s`) and if mapping is private (`p`). Options are available to show additional information about the process' memory utilization.

SECTION ACTIVITIES

1. How much should the normal computer user know about computer memory, virtual memory and paging? If you think the computer user should understand these concepts, come up with a list of up to three situations where such information would be useful. Have you come across these concepts in your everyday computer usage?
2. Of the commands introduced in this last subsection, which are ones you feel you should use as a user to explore how your computer is running? If you haven't tried them out yet, do so to see which one(s) you find most useful.

```
4568:   bash
000055d7c7a00000   1056K r-x-- bash
000055d7c7d07000     16K r---- bash
000055d7c7d0b000     36K rw--- bash
000055d7c7d14000     40K rw---   [ anon ]
000055d7c95c9000   1944K rw---   [ anon ]
00007f8cb124e000   9040K r--s- passwd
00007f8cb1b22000     40K r-x-- libnss_sss.so.2
00007f8cb1b2c000   2044K ----- libnss_sss.so.2
00007f8cb1d2b000      4K r---- libnss_sss.so.2
00007f8cb1d2c000      4K rw--- libnss_sss.so.2
00007f8cb1d2d000   2528K r---- LC_COLLATE
00007f8cb1fa5000   1760K r-x-- libc-2.28.so
00007f8cb215d000   2048K ----- libc-2.28.so
00007f8cb235d000     16K r---- libc-2.28.so
00007f8cb2361000      8K rw--- libc-2.28.so
00007f8cb2363000     16K rw---   [ anon ]
00007f8cb2367000     12K r-x-- libdl-2.28.so
00007f8cb236a000   2044K ----- libdl-2.28.so
00007f8cb2569000      4K r---- libdl-2.28.so
00007f8cb256a000      4K rw--- libdl-2.28.so
00007f8cb256b000    164K r-x-- libtinfo.so.6.1
00007f8cb2594000   2044K ----- libtinfo.so.6.1
00007f8cb2793000     16K r---- libtinfo.so.6.1
00007f8cb2797000      4K rw--- libtinfo.so.6.1
00007f8cb2798000    160K r-x-- ld-2.28.so
00007f8cb2954000    332K r---- LC_CTYPE
00007f8cb29a7000     20K rw---   [ anon ]
00007f8cb29af000      4K r---- LC_NUMERIC
00007f8cb29b0000      4K r---- LC_TIME
00007f8cb29b1000      4K r---- LC_MONETARY
00007f8cb29b2000      4K r---- SYS_LC_MESSAGES
00007f8cb29b3000      4K r---- LC_PAPER
00007f8cb29b4000      4K r---- LC_NAME
00007f8cb29b5000      4K r---- LC_ADDRESS
00007f8cb29b6000      4K r---- LC_TELEPHONE
00007f8cb29b7000      4K r---- LC_MEASUREMENT
00007f8cb29b8000     28K r--s- gconv-modules.cache
00007f8cb29bf000      4K r---- LC_IDENTIFICATION
00007f8cb29c0000      4K r---- ld-2.28.so
00007f8cb29c1000      4K rw--- ld-2.28.so
00007f8cb29c2000      4K rw---   [ anon ]
00007ffc54fd4000    132K rw---   [ stack ]
00007ffc54ffa000     16K r----   [ anon ]
00007ffc54ffe000      8K r-x--   [ anon ]
ffffffffff600000      4K r-x--   [ anon ]
 total            25644K
```

FIGURE 4.22 Output of pmap showing memory map for bash process.

4.8 CHAPTER REVIEW

Concepts and Terms Introduced in This Chapter

- Background – processes which run with no user interactivity.
- Batch processing – a form of process management whereby the operating system assigns a single process to the CPU to execute; as the process may be scheduled at any time, there is no user interactivity and all input must be submitted with the job.
- Child – a process spawned (started) by another process; in Linux, all processes form a parent-child relationship with only one process, `systemd`, having no parent.
- Concurrent processing – any form of process management in which multiple processes are available to run and the CPU switches off between them.
- Consumer-producer processes – processes where one group produces data that are consumed by the other group; such processes need to be synchronized so that a consumer does not try to consume data that has not yet been produced; synchronization is typically handled through a *rendezvous*.
- Context switch – the CPU switching from one process to another by swapping process status information.
- Cooperative multitasking – a form of concurrent processing in which the current process voluntarily gives up the CPU for another process; also known as *multiprogramming*.
- Core dump – a file containing an abnormally terminating program's memory image; used for debugging purposes.
- Effective group ID (EGID) – group owner under which a process runs; the EGID is used to determine the resources accessible to the running process.
- Effective user ID (EUID) – the user owner under which a process runs; the EUID is used to determine the resources accessible to the running process.
- Exec – a family of system calls in Linux used to create a new process in place of the current process; the new process inherits the old process' PID and environment.
- First come first serve (FCFS, also known as first in first out or FIFO) – a simple scheduling strategy for ordering waiting processes (jobs) in a queue based on the order that the jobs arrived in the queue.
- Foreground – processes which run with user interactivity.
- Fork – a family of system calls in Linux used to start a new process which is a duplicate of the parent process.
- Frame – a fixed-size chunk of memory used to store a program page; pages and frames are used in operating systems with virtual memory.
- Interrupt – a signal received by the CPU to suspend the current process and perform a context switch to an interrupt handler in order to handle the interrupting situation.
- Interrupt handler – a part of the operating system kernel set up to handle a specific interrupting situation.
- Interrupt request (IRQ) – a request made by a hardware device or running software to gain the CPU's attention; each IRQ is assigned to an interrupt handler.
- Job – generic term for a submitted or running process.
- Least recently used (LRU) – replacement strategy to discard the page in memory that is least likely to be used in the future.
- Memory map – a mapping of a device or process to the locations in memory that are reserved for its use.
- Multiprocessing – a form of process management in which processes are divided up by the operating system to run on different processors (or cores).
- Multithreading – multitasking over threads as well as processes.
- Niceness – how much time a process is willing to give back to the system for other processes; a higher niceness is a lower priority; niceness values in Linux range from –20 (least nice, highest priority) to +19 (most nice, lowest priority).

- Orphan – an active process whose parent has terminated; as processes report to their parent before they terminate, any orphan must be readopted.
- Page – a fixed-size piece of a program; one page fits into one frame in memory.
- Page fault – in virtual memory, only a portion of a program is placed into memory; a page fault (interrupt) arises if an address is generated by the CPU of a page that is not currently in memory; the page fault requires that the operating system swap the needed page into memory from swap space.
- Page table – operating system data structure storing the location (frame in memory or not in memory) of all of the pages of a process.
- Parent – the process that spawned this process; all processes in Linux have a parent except for `systemd` (or `init` in older versions of Linux), the first process to run.
- Physical address – the CPU generates a virtual address which specifies the page of the process and an offset which must be converted into the physical address by replacing the page number for that page's frame number.
- Preemptive multitasking – in addition to cooperative multitasking, an operating system can preempt the running process forcing a context switch when the process' time limit has expired.
- Priority – a value used by a scheduler to determine when a process should execute and how much CPU time should be allocated to the process.
- Process – a running program; we differentiate the process from the program because the process is not just code but has a state including its values stored in memory, resources allocated to the process and a process state.
- Process control block (PCB) – data structure maintained by the operating system for every running process which stores process information such as its PID, state, resources allocated to it and page table location.
- Process ID (PID) – a unique identifier assigned to a process; in Linux, PIDs are assigned sequentially starting at 1 and increasing with each new process; many Linux instructions reference processes by PID.
- Process management – how the operating system manages running processes; most modern operating systems use concurrent processing in the form of cooperative and preemptive multitasking and multithreading.
- Process state – the state of a process; in Linux states including running, sleeping/waiting, uninterruptible sleep and dead among others.
- Queue – a waiting line; used in operating systems to store processes waiting for access to the CPU or some device like a printer.
- Rendezvous – situation where one process reaches a point in its execution that it must wait for another process to reach an equivalent point so that both are ready to communicate with each other.
- Round robin – scheduling algorithm where the CPU operates on each process of the ready queue one at a time and wrapping around to repeat after visiting the last process in the queue.
- Scheduling – the task of an operating system to organize waiting processes to decide the order that the processes will gain access to the CPU.
- Signal – a means of passing a message to a Linux process for action; some signals cause the process to terminate; signals are used in the `kill` command.
- Single tasking – an old form of process management whereby the operating system only permits one process to be active and in memory at a time (aside from the operating system).
- Sleeping – a process' status when it is either waiting for an event or waiting some amount of time before resuming.
- Spawn – term used to denote one process creating another process; in Linux, we refer to the processes as the parent (spawning process) and child (spawned process).
- Swap space – a dedicated area of hard disk used to back up main memory to support virtual memory.

- Swapping (or page swapping) – the process of moving a page between swap space and memory.
- Thread – a lightweight process which shares code and (at least some) data with other threads of the same process; context switching between threads is faster than between processes.
- Time slice – the amount of time a process gets with the CPU before the CPU is forced to perform a context switch to another process.
- Timer – hardware device used to count the number of clock cycles until a context switch occurs; used in preemptive multitasking.
- Virtual address – the address that the CPU generates specifying a location within a program; because of virtual memory, this address will not be the same as the physical address and so requires mapping into a physical address using the process' page table.
- Virtual memory – the extension of main memory to the swap space on hard disk so that the computer can run more and larger processes than could fit in main memory.
- Wait – a Linux system call used to put a process into a sleep mode.
- Zombie – a process which has terminated and freed up its resources but is still in the system because it must report back to a sleeping parent.

Linux Commands Covered in This Chapter

- & – used after a command to launch the process into the background.
- 2>/dev/null – when added to an instruction sends error messages to the Linux trash can.
- bg – resume a suspended process in the background.
- control+c – kill the running process.
- control+z – suspend (stop) a running process to provide the user access to the command-line prompt.
- fg – resume a suspended process to the foreground.
- free – display free and used memory and virtual memory.
- gnome-system-monitor – launch the System Monitor GUI from the command line.
- iostat – output CPU and I/O usage statistics.
- jobs – list the processes that are suspended or running in the background of this terminal window.
- kill – send a signal to one or more processes; commonly used to terminate processes.
- killall – terminate a collection of processes based on process name, user or group who launched the process.
- mpstat – display processor (or multiple processor) usage statistics.
- nice – launch a process with a non-default priority (niceness).
- pgrep – list PIDs of all processes of a given name.
- pkill – like kill but send signals to processes by name rather than PID.
- pmap – display the memory map of the specified process(es).
- ps – print the process snapshot of running processes as specified by options, or those in the current terminal window when no options are provided.
- pstree – output running processes indicating the parent/child relationships.
- renice – change the niceness of a running process.
- System monitor – GUI program displaying running processes, system resource utilization and file system usage.
- systemctl – utility to start and stop services and change run-time modes including `poweroff`, `halt`, `reboot` and `suspend`.
- top – interactive, text-based program to display process status and computer resource usage.
- vmstat – display virtual memory statistics.

  NOTE: `mpstat` and `iostat` may not be installed; if this is the case, you need to install the package `sysstat`.

REVIEW QUESTIONS

1. Compare a program and a process. Why do we need two different terms to describe them?
2. What is the first process to always run in Linux (assume Red Hat 8)?
3. From Bash, we run a program `foo`. This launches a new process for `foo`. Answer the following.
   a. What is foo's parent if any?
   b. To create the new process requires one or more system calls to exec and fork. Which of these is(are) needed and if both, what order do they run in?
   c. Assume we run foo in the background and eventually its parent ends. What happens to foo?
   d. Assume we run foo in the background and its parent goes to sleep. What happens when foo attempts to terminate?
4. In examining Figure 4.1, what is the relationship between `gnome` and `gnome_terminal`? Between `bash` and `find`? Between `somescript` and `find`?
5. Redo Figure 4.1 assuming that from your terminal window, you type `bash` to open an inner session and from there, run the program `./somescript` which itself launches the program `./somescript2`. You do not run `find`.
6. From the desktop, you start a terminal window. Your default shell is Bash. From the terminal window, you type a `mv` command but add `&` after and then type `ps`. Show the tree of processes.
7. List four pieces of information that you would find in a process control block.
8. Which form of process management would you find in early personal computers?
9. Batch processing comes closest to which of these forms of processing, single tasking, preemptive multitasking or multiprocessing?
10. Which of shortest-job first, longest-job first, first come first serve and priority would provide the minimal average wait time for the jobs waiting?
11. Name a situation in which a process would voluntarily surrender the use of the CPU.
12. What does the timer count? What happens if the timer reaches 0?
13. What does the CPU do while a context switch is taking place?
14. What does the CPU need to store and restore during a context switch?
15. In preemptive multitasking, if we alter a process' priority, what specifically changes?
16. What scheduling algorithm is used whereby processes are visited one at a time in a queue such that after the CPU executes some instructions of the last process it loops around and resumes from the beginning of the queue?
17. How does a thread differ from a process?
18. True/false: All multitasking systems are also multithreading systems.
19. True/false: IRQs always originate from running programs.
20. Where are interrupt handlers stored?
21. In which file can you find information about interrupts that have arisen in Linux?
22. A process is running, and its EUID and EGID are of your user account and your private group account. What does this mean?
23. How do you change an executable program from running under your user account to that of the file's owner.
24. With `chmod`, what would it mean if we use the value 4755 to change an executable file's permissions?

   For questions 25–28, assume we have the following partial long listing of three executable programs:

permissions	owner	group	filename
-rwsr-xr-x	user1	group2	foo1
-rwsr-sr-x	user1	group2	foo2
-rwxr-xr-x	user1	group2	foo3

25. If user foxr runs `foo1`, `foo2` and `foo3`, which will run under EUID `user1`?
26. If user foxr runs `foo1`, `foo2` and `foo3`, which will run under EGID `group2`?
27. Assume `foo1`, `foo2` and `foo3` all access files that are readable and writable only to `user1`. If foxr attempts to run all three programs, which can successfully read and write to this file?
28. If user `user1` wishes to change the permissions of `foo3` to match those of `foo1`, what 4-digit value would `user1` submit with the `chmod` command?
29. True/false: When launching a command from the command line, if the program is not stored in the current directory then you must specify the full path to the location of the program.
30. Under what circumstance might you run a program using the notation `./` as in `./myscript`?
31. You have issued the command `man ps` from the command line. While you do not want to stop the command, you want to regain access to your prompt. What do you do?
32. You type `jobs` and see the following. Assume each question below is independent of the others (i.e., each question asks about the output as shown below).

```
[1]  Stopped     vim foo
[2]- Stopped     vim bar
[3]  Running     ./myscript
[4]+ Stopped     man ps
```

   a. What command should you enter to resume `vim foo`?
   b. What command should you enter to resume `man ps`?
   c. For `./myscript`, why is it listing as Running instead of Stopped?
   d. How can you move `vim bar` to the background? Why would you not want to do that?
   e. After resuming `vim foo`, you stop it with `control+z` and type `jobs`. How does the output differ from what is listed above?

33. You are in a Bash shell and type `exit` to close it and are told `There are stopped jobs`. What does this mean? Does the Bash session end?
34. How does a job number, listed when you issue the `jobs` command, differ from a process' PID?
35. From the Resource Monitor GUI under the Processes tab, you right click on a process. List four things you can do from the pop-up menu.
36. Aside from the Processes tab, what are the other tabs in the Resource Monitor?
37. Which of these tools shows current resource utilization and updates itself frequently?
    `iostat`, `mpstat`, `ps`, `System Monitor`, `top`
38. True/false: The `top` program, by default, will update its output every 3 seconds but this behavior can be changed.
39. True/false: `ps u` shows all processes of the current user.
40. Which option or options in `ps` is(are) equivalent to `ps -e`?
41. Which option(s) in `ps` show(s) parent/child relationships?
42. You want to obtain the process snapshot for all running `bash` and `man` processes. What command should you enter?
43. How do the `NI` and `PRI` fields differ in the `ps` output?
44. A process has the state of `Ss+<`, what does this mean?
45. A process has the state of `RN`, what does this mean?
46. A process has the state of `Z`, what does this mean?
47. In order to lower a process' priority, do you increase or decrease its niceness?
48. True/false: A normal user cannot lower the niceness of one of their processes.
49. What is the difference between using `nice` and `renice`?
50. How do you modify all of marst's processes to have a niceness of 10 (assuming you are either marst or root)? Provide the instruction.

51. True/false: A zombie continues to use system resources even though it is no longer running.
52. Which signal (by name, not number) would you use to ensure that a process terminates when using `kill`?
53. What does the `SIGPIPE` signal indicate?
54. Why might a `SIGBUS` signal be sent?
55. Write a `kill` command to send to process `123456` the signal `SIGHUP`.
56. You are running a script call `myscript.sh` in your terminal window and so have no access to your command line prompt. You realize the script is erroneous and you want to terminate it. Explain how to do this in a step-by-step manner.
57. Write an instruction to shut down and power off your Linux system. Is this something that a normal user can do or just root?
58. What is the difference between `poweroff` and `halt`?
59. What is a memory map? What Linux instruction might you use to display the memory map of a process?
60. Using the words frame and page, fill in the blanks. A program is broken into fixed-size ______ and main memory is broken into fixed-size ______.
61. True/false: The size of a page is always equal to the size of a frame.
62. In order to convert a virtual address into a physical address, the CPU consults what?
63. A page fault causes what to happen?
64. Which command would you use to inspect the amount of used and available memory?
65. The frame size of a computer is 1KB. A program is 128KB in size. How many pages does the program consist of?

5 Regular Expressions

This chapter's learning objectives are to be able to

- Read and write regular expressions
- Use the `grep/egrep` program for pattern matching
- Use the `sed` program for simple substitution
- Use the `awk` program for basic processing

5.1 INTRODUCTION

A *regular expression* (regex) is a string that expresses a pattern. The regex consists of a combination of literal characters and metacharacters. The *metacharacters* modify the literal characters to provide variability. The result is that the regex can match more than just a literal string. As a simple example, "aciou" is a string of all literal characters and can only match the string "aeiou". The regex "[aeiou]{5}" will match any string of five lowercase vowels. The characters "[ ]" and "{5}" are metacharacters expressing how to apply the literal characters of "aeiou".

We use regular expressions to search strings for patterns of interest. Using the previous vowel example, we might ask if there are any words in the dictionary that contain five consecutive vowels (in any order) and use that regex to perform *pattern matching*. There are programs and programming languages that can apply regular expressions to scan strings for such matches. In this chapter, we concentrate on three Linux programs that apply regular expressions: `grep`, `sed` and `awk`. But before we learn these programs, we have to understand regular expressions. And to learn about regular expressions, we must understand the metacharacters.

The reason this topic is important is that regular expressions provide us with a powerful tool whether we are Linux users or Linux administrators. Let's motivate this statement with a couple of examples. Imagine that we have a file that contains various pieces of information about people but all we want from this file are the full names. The names are not in any particular location within the file, such as the first two columns of each row, so we have to search the file. But we can't search using a literal string because we don't know what the names are. Instead, we have to define a regular expression that expresses the pattern of characters we expect for full names.

To simplify this task, we'll assume that a full name is a first name, a space and a last name. Each of the first and last names starts with a capital letter followed by lower case letters. As this pattern expresses any capitalized word, we differentiate by supplying a regex that consists of two names with a single space in between. Specifically, the regex will consist of any uppercase letter followed by some number of lowercase letters followed by a space followed by another uppercase letter followed by some lowercase letters.

We can use the `grep` program for this task. We pass to `grep` the regular expression of the pattern of interest and the file(s) to search. `grep` searches the file, line by line, attempting to match the regex to any part of the line. If it matches, the entire line is output. Note that neither `grep` nor our regex ensures that the names found are actually people's names. `grep` could also match and return lines that contain two-word cities, states or countries like Green Bay, South Dakota and Great Britain.

Let's look at a more elaborate example. We want to build a *spam filter*. This is a program that will search for words and terms in an incoming email (text) that would lead us to believe that the email is spam. One such word that is often found in spam messages is "lottery". Searching for the literal strings "Lottery" and "lottery" is easy enough. Clever spammers, however, will try to disguise words that might trigger a spam filter by substituting, adding or removing characters.

DOI: 10.1201/9781003203322-5

Our emails are full of words and phrases that are disguised in an attempt to avoid being caught by our spam filters. We might find for instance that "lottery" appears using different cases as in "lOTTERY" or "LoTtErY" or that numbers have been substituted for some of the letters as in "l0ttery" or "lo44ery" or that characters have been placed between the letters of the word as in "l.o.t.t.e.r.y" or "l-o-t-t-e-r-y".

To define every possible string would be nearly impossible. Spam filters instead apply regular expressions to specify a pattern. We might use a regex that contains the letters in "lottery" but where each letter can appear in upper or lowercase and where the "L" might be replaced by a "1", the "o" with a "0" and the "t" with a "4". The regex would also allow one or more characters to appear between each letter.

Keep in mind that the spam filter is not tasked with just recognizing if the word "lottery", or a variant, appears in the email, but any of a number of words and phrases. The filter should also search for words and phrases like "apply now", "click here", "double your income", "no strings attached", "order now" and "you are a winner!", along with variations. We would have to come up with regular expressions for each of these phrases to match the phrases and variations such as "aPPly N0w" and "dooooouble your incoooooome".

Recall the wildcard characters we examined in Chapters 2 and 3, such as `*`, used for filename expansion. Through a wildcard, we have a convenient means to specify multiple file/directory names where they overlap in some way such as filenames starting with an `a` and ending in `.txt`. Wildcards present one way to express patterns, but regular expressions allow us to express more complex patterns. Unfortunately, this also makes regular expressions more challenging to use correctly. To complicate matters further, some of the wildcard characters are also used as metacharacters in regular expressions but their usage differs.

In this chapter, we first study metacharacters to understand how to read and write regular expressions. We then learn how to apply regular expressions in the programs `grep` (`egrep`), `sed` and `awk`.

SECTION ACTIVITIES

1. Does your email server use a spam filter? If so, how accurate do you find it to be? How often do you find an email in your spam filter that should not have been caught? How often do you receive an email that should have been caught by the spam filter? Are there specific words you find that the spam filter catches that it shouldn't?
2. Aside from the phrases listed in this section, come up with several other phrases or words that you tend to find in spam emails.
3. Come up with as many different ways as possible to spell "click here" and "apply now" that a spammer might use to defeat a spam filter.
4. There are many possible applications for regular expressions aside from spam filters. Two are in a password manager program determining if a new password is too similar to an older password and as a simple chatbot which looks for types of syntactic structures and words in a sentence. Research uses of regular expressions and identify three more applications.

5.2 METACHARACTERS

The *Portable Operating System Interface for Unix* (POSIX) was defined by the IEEE Computer Society as a series of standards that operating system developers should implement to ensure that Unix (or Unix-like) operating systems shared features. Among the definitions provided in this standard were signals, the use of pipes, and C library files. In terms of regular expressions and this chapter, POSIX defined the set of regular expression metacharacters. Unix adopted POSIX although most Linux distributions did not fully adopt POSIX, but they all have adopted the POSIX-specified regular expression sets.

The original metacharacters as proscribed in POSIX make up the *basic regular expression set*. Some of these metacharacters required being preceded by the backslash (\) to enforce that the character was to be treated as a metacharacter. Later, an *extended regular expression set* was defined in POSIX. This set includes the same metacharacters but without the backslash. After we introduce the metacharacters, we will use only those versions defined without the backslash for simplicity. Outside of POSIX, other metacharacters are defined that overlap those in POSIX. In this section, we will first look at the POSIX metacharacters and then present other versions.

Table 5.1 lists the POSIX metacharacters. The notation `B:` indicates that the metacharacter is defined in the basic set while `E:` indicates that this is how it is defined in the extended regular expression set. Where there is an `E:` but no `B:` entry, it means that the metacharacter is *only* defined in the POSIX extended regular expression set and where there is no `B:` or `E:`, the metacharacter is defined identically in both sets. Notice how several of the metacharacters overlap the wildcard characters that we introduced in Chapters 2 and 3.

The brackets are used to specify a list of possible matching characters. Examples include `[aeiou]` to match any lowercased vowel, `[a-e]` to match any of `a`, `b`, `c`, `d` or `e` and `[[:lower:]]` to match any lowercase letter. This last example uses one of the POSIX *character classes*. These are listed and described in Table 5.2.

We will introduce the metacharacters in an order that lets us build upon previously introduced metacharacters. For each metacharacter, we will examine a number of examples to more fully

TABLE 5.1
Regular Expression Metacharacters

Metacharacter	Explanation
*	Match the preceding character if it appears 0 or more times.
E: +	Match the preceding character if it appears 1 or more times.
E: ?	Match the preceding character if it appears 0 or 1 time.
.	Match any one character.
^	Match if regex is found at the beginning of the string.
$	Match if regex is found at the end of the string.
[*chars*]	Match if the next character in the string contains any of the characters listed in `[]`.
[ch_i-ch_j]	Match if the next character in the string contains any characters in the range from ch_i to ch_j (e.g., a–z, 0–9).
[[:*class*:]]	Match if the next character in the string is a character that is part of the `:class:` specified; see Table 5.2 for the available POSIX classes.
[^*chars*]	Match if the next character in the string is not one of the characters listed in `[]`; the `[]` can include a range or a class.
\	Escape the meaning of the next character; matches if the next character of the string is the specified character; e.g., `\$` means "match a $" instead of "match at the end of the string".
B: \{*m*\} E: {*m*}	Match if the string contains *m* consecutive occurrences of the preceding character; *m* is a number.
B: \{*m*,*n*\} E: {*m*,*n*}	Match if the string contains between *m* and *n* consecutive occurrences of the preceding character; *m* and *n* are numbers.
B: \{*m*,\} E: {*m*,}	Match if the string contains at least *m* consecutive occurrences of the preceding character; *m* is a number.
E: |	Match any of these sequences of characters; this differs from [...] which matches one character.
B: \(...\) E: (...)	The items in the parentheses are treated as a group; match the entire sequence; used when we need to apply a metacharacter to an entire sequence.

TABLE 5.2
POSIX Character Classes

Class Name	Full name	Meaning
[:alnum:]	Alphanumeric	Any letter (either case) or digit.
[:alpha:]	Alphabetic	Any letter (either case).
[:blank:]	Blank	Blank space or a tab; see [:space:].
[:cntrl:]	Control	Any control character.
[:digit:]	Digit	Any digit (0–9).
[:graph:]	Graphic characters	Any visible character (any non-control character).
[:lower:]	Lowercase	Any lowercase letter.
[:print:]	Printable	Any printable character (combines [:graph:] and [:space:]).
[:punct:]	Punctuation	Any punctuation mark.
[:space:]	White space	Any white space (blank space, tab, enter/new line).
[:upper:]	Uppercase	Any uppercase letter.
[:word:]	Word character	Any alphanumeric character or underscore (_) (this character class is only available in Bash).
[:xdigit:]	Hexadecimal	Any hexadecimal digit (0–9, A–F/a–f).

illustrate its usage. You might find this material to be challenging as this is a complicated topic, so working through the examples presented should help.

5.2.1 Controlling Repeated Characters through *, + and ?

The first three metacharacters from Table 5.1 are all used to express a variable number of times that the *preceding character* in the regex can appear in the string. With *, the preceding character can appear zero or more times. With +, the preceding character can appear one or more times. With ?, the preceding character can appear zero or one time exactly. We can think of *, + and ? as meaning that the preceding character can occur *any* number of times, *some* number of times and *once optionally*, respectively.

Consider the three regexes 1*0*, 1+0+ and 1?0?. The first of these will match any sequence of any 1s followed by any 0s, the second will match any sequence of some 1s followed by some 0s and the third will match any sequence of zero or one 1 followed by zero or one 0. The first regex, 1*0*, will match 11111100000, 111111110, 1111111, 10, 0, and the empty string among others. The *empty string* is a string of no characters.

The second regex, 1+0+, will match 111111100000, 111111110 and 10 but will not match 11111111, 0 or the empty string. With the +, the character 1 and the character 0 must appear at least one time so that a string that contains no 1s or no 0s will not match. The third regex, 1?0?, is interesting because it will match only a finite list of strings as the 1 and the 0 must appear either zero or one time. Specifically, 1?0? will only match one of these four strings: the empty string, 1, 0 and 10.

Let's consider these three metacharacters from a different perspective. Table 5.3 lists a number of strings containing the characters of a, b and c in different numbers and in one case in mixed order. Which of these regular expressions, a*b*c*, a+b+c+, and a?b?c?, will match the strings in the table? The table lists the regexes in the last three columns. A checkmark in a column indicates that the regex will match the string on that row.

From Table 5.3, first notice that all strings match a*b*c* except for abacab because the letters are not in the correct sequence. a*b*c* will match all of the other strings because each of these strings has the letters of a, b and c in that order in some number (including no occurrences of each letter). For a+b+c+, there are only three matches because each of the letters, a, b and c,

TABLE 5.3
Example Regular Expressions with *, + and ?

	a*b*c*	a+b+c+	a?b?c?
aaaabbbc	√	√	
aaaacccc	√		
abc	√	√	√
ab	√		√
abacab			
bbbb	√		
abbbccc	√	√	
Empty string	√		√

must appear at least once and some of the strings omit some of the characters such as ab and bbbb. The regex a?b?c? also only matches three of the strings, those that contain zero or one of each of the letters in order. Thus, abc and ab match and since each letter can occur zero times, and the empty string matches as each of the letters in the regex is optional.

A regex can contain literal characters without metacharacters modifying them. For instance, the regex jpe?g specifies that the j, p and g must appear and the e may appear zero or one time making the e optional. Thus, this regex matches either jpeg or jpg. The regex hello!? will match either hello or hello!. Each of the letters must appear and the exclamation mark (!) *may* appear.

5.2.2 Using and Modifying the . Metacharacter

Perhaps the easiest of the metacharacters to understand and utilize is the period (.). It is a true wildcard metacharacter meaning that it matches anything, but unlike the * wildcard used by Bash in globbing, the metacharacter . only matches a *single* character. The regex b.t will match any string that has b followed by any one character followed by a t, keeping in mind that the "any one character" can be a letter, digit, punctuation mark, white space or even a control character (non-printable). Thus, b.t will match each of bat, bbt, bit, b0t, b#t and b t. It would not match batt or bt because the regex is seeking to match exactly three characters.

Let's consider the regular expression ... which will match any three characters no matter what those characters are. This regex can match, for instance, abc, a b, 123, 3*5 or ^%#. Notice the second string contains a blank. In fact, ... could match three blank spaces.

Some metacharacters can modify other metacharacters. This is the case with the *, + and ? when coupled with the . metacharacter. This allows us to specify any sequence of any characters, some sequence of any characters and an optional character of any kind. These are specified using the .*, .+ and .? respectively.

The expression b.?t means b followed by zero or one of anything followed by t. This would match all of the earlier listed strings, bat, bbt, bit, b0t, b#t and b t, but will also match bt because the . can be applied zero times. What will the regex .+ match? This regex matches any sequence of some characters, that is, a sequence of one or more characters. The characters can be any character where the characters do not need to match each other, for instance abc, abcdefg, a b c d, and aaaaa. The difference between .+ and .* is that the latter will match the empty string while the former does not. The regex b.+t will match a b followed by any sequence of some characters followed by a t, including each of bat, bbbbbbbt, b t, b!#%&!$t, bait, and b00t, but not bt.

5.2.3 Controlling Where a Pattern Matches

In the previous two subsections, we have been purposefully misleading when we noted what can and cannot match a given regex. For instance, `a*b*c*` will match the string `abacab` just as `b.t` will match match `batt`. We were misleading for simplicity, but now we have to understand just how regular expressions work.

A regular expression attempts to match any substring of a string. A *substring* is a contiguous sequence of characters found within the larger string. Consider the string `bbbbbbbattttttt`. The regular expression `b.t` will match this string because it matches a substring of the string. Specifically, it will match the last `b`, the `a` and the first `t`. That sequence, `bat`, allows `b.t` to match `bbbbbbbattttttt`. Put another way, the regex does not have to match the *entire* string.

There are many regular expressions that will find a match within `bbbbbbbattttttt`, including `b?.t+`. Figure 5.1 shows just where `b?.t+` matches within the larger string. The last `b` in the sequence of `b`'s in the string starts the match. That last `b` is followed by a single character (the `a`) to match the `.` in the regex. Next, the regex attempts to match one or more `t`'s, which it finds in the string. So even though the string has seven `b`'s, the `b?` does not have to match all seven `b`'s (and in fact can match no `b`'s since it is modified by the `?`). Although the regex does not match the *entire* string, it does match several substrings including `at`, `att`, `atttt`, `bat`, `batt` and `battttttt`. Note also in the figure is `^b?.t+`. We describe this momentarily.

Fortunately, we have three ways to control where the regex will match. We can force the regex to match starting from the beginning of a string. We can force the regex to match up through the end of a string. And we can force the regex to match the entire string. Referring back to the example in Figure 5.1, if we wanted to try to match from the beginning of the string, the regex would not match because `b?.t` indicates that there can be no more than two `b`'s to start the string (where the period is serving as the second `b`). We see this in the bottom of the figure where `b?` matches the first `b`, the period matches the second `b`, but then `t+` does not match as the next character is not a `t`.

Before we describe how we control where a match takes place, let's be more specific about what we mean by string and substring. The *string* is the item we are matching against. In a program, a string will be stored in a variable or enclosed in quote marks such as "Hello world!". In a file, we might consider a string to be any sequence of characters that is separated from another string by a *delimiter* (separator) like a space or tab. Alternatively, as we will see in `grep` and `sed`, a string is taken to be an entire line of a file (the delimiter between strings in this case is the `\n` end-of-line character). The *substring* is any consecutive characters found within the string, including the empty string. With a regular expression, we are looking to see if the regex will match *any* substring of the string.

By this new definition, a regular expression that matches the empty string (no characters at all) will match any and every string no matter what appears in that string. For instance, the regex `a*` will match zero or more `a`'s. If there are no `a`'s in a string, `a*`still matches. Any regex of the form `char*` where *char* is any character would match every string.

Let's revisit the regex `a*b*c*` from Table 5.3. This string matches `abacab` (the only string in the table that it seemed to not match) because we find an `a` and a `b` followed by zero `c`'s consecutively. As the `*` metacharacter seems to match anything and everything, it seems somewhat useless. For this reason and others, it is important to be able to control *where* a regular expression is allowed to match within the string (or line).

And now we turn to our next two metacharacters: `^` and `$`. The `^` metacharacter indicates that the regular expression can only match from the start of the string, and the `$` metacharacter indicates

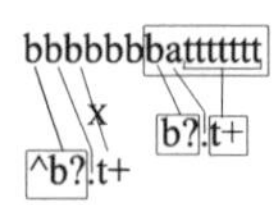

FIGURE 5.1 Controlling where an expression matches.

that the regular expression can only match at the end of the string. When using both metacharacters, the regex must match the full string.

Let's return to our previous regular expression (from Figure 5.1) but add the ^ to the beginning. ^b?.t+ will match a string that *starts with* zero or one b followed by any one character followed by one or more t's. This will not match the string of bbbbbbbattttttt because ^b forces the regular expression to begin matching this string from the beginning and b? insists that there be no more than one b. What causes ^b?.t+ to fail to match the string bbbbbbbattttttt is that after zero or one b we expect any character (which will be the next b) followed by at least one t, but the next character in the string is another b.

While the regex ^b?.t+ does not match bbbbbbbattttttt, the regex b?.t+$ will match this string. Scanning the string from left to right, we eventually find zero or one b followed by any single character, the a, and then we find one or more t's to end the string. Although there are additional b's earlier in the string, our regex has matched a substring of the string which is sufficient.

Using both ^ and $ allows us to restrict the regex to match the full string. For instance, ^b?.t+$ will match any string that starts with zero or one b, has any character and ends with one or more t's. This regex will match the strings battttttt, bbtttttt and attttttt (in this case there are zero b's) but will not match the string from Figure 5.1.

Let's move on to the regex ^0a*1+. This expresses a pattern for a string that starts with a 0 followed by any number of a's and at least one 1. This regex will match any of these strings (among others): 0a1, 0aaaa1111, 011111 and 01. It can also match strings where there are characters after the 1 because we are not restricting this regex to "end the string" using $. So, it will also match 0a1a and 011110. What it cannot match are strings like 0a (there must be at least one 1), 0000aaaa1111 (the start of the string must be a single 0), 0aaaab (there must be at least one 1 after the a's) or 0aba1 (there can be nothing but a's between the 0 and 1 if any characters appear at all).

The variation ^0a*1+$ adds a closing $. Now the regex can only match those strings that matched the previous regex but with no characters after the 1('s). Thus, 0aaaa1111 and 01 will match but 0a1a and 011110 will not.

The regex ^$ is interesting as there are no characters present between the ^ and $. Using both ^ and $ forces the regex to match the entire string, but the entire string, in this case, is nothing. Thus, this regex is literally expressing the pattern "any string that has nothing between the start and end". There is only one such string, the empty string.

5.2.4 Matching from a List of Options

So far, our expressions have allowed us to match against strings that have a variable number of characters, but only if the characters appear in the order specified. For instance, a+b+c+ can match some number of a's followed by b's followed by c's, but the letters must appear in that order. What if we want to match a string that contains any number of a's, b's and c's, but in no particular order? If we want to match any three-character sequence that consists of only a's, b's and c's, as in abc, acb, bca and so forth, we will need a way to indicate that the character must match "any of these characters".

The [] metacharacters, referred to as *brackets*, *straight brackets*, or *braces*, allow us to specify a list of options. The list indicates that the *next* character in the string must match *any single character* from the list. Inside the brackets we can specify lists using three notations: an enumerated list as in [abcd], a range like [a-d] or a *class*, as described in Table 5.2. An enumerated list is simply a list of characters. The characters can be listed in any order, and there is no separator between the characters (such as a comma). A range specifies the earliest and latest characters that can match, separated by a hyphen. We can only use a range when the characters have an explicit ordering, as is the case with letters and digits. We cannot express a range of punctuation marks.

Consider the regex [abc] [abc] [abc]. This expression will match any string that contains three consecutive characters that are a's, b's or c's in any combination. This expression will match abc, acb and bca. And because the regex does not restrict the number of times any of one of the three characters appear, it will also match aaa, bbb, aab and aca, among others. Additionally, since we did not use the ^ and/or $, this will match larger strings that contain a sequence of any three a's, b's and/or c's as with zyxaaa, @aaa@, aaaaaaaaa and 1cba2. As the letters have an explicit ordering, we could use a range to indicate a, b and c using the regex [a-c][a-c][a-c].

We can combine any characters in the brackets as in [abcxyz], [abcd1234] or [abcdABCD]. If we have a number of characters to enumerate, a range is more practical. We would certainly prefer to use a range like [a-z] rather than to list all of the lowercase letters.

We can combine ranges and enumerated lists. For instance, the three sequences above could also be written as [a-cx-z], [a-d1-4] and [a-dA-D], respectively. Now consider the list of all lowercase consonants. We could enumerate them all as [bcdfghjklmnpqrstvwxyz], or we could combine ranges and enumerated lists in the form of [b-df-hj-np-tv-z].

If a hyphen is one of the characters we want to include in an enumerated list, its placement within the list is important. Imagine that we want to match any of a, b, c or a hyphen. If we write this as [a-bc] then the hyphen is interpreted as part of a range and not as a hyphen, making [a-bc] equivalent to [abc]. We resolve this by placing the hyphen at one end of the list (either end) so that we can express this list as either [-abc] or [abc-]. Now consider [+-*/]. This is erroneous because the regex is interpreted as the range from + to * and / but punctuation marks have no ordering so we cannot express a range. The proper regex to match any of these four arithmetic operators requires that we move the hyphen to one end, as in [+*/-].

To specify a class, we place the class indictor, such as [:alpha:], in another set of brackets giving us [[:alpha:]] to match any letter, [[:digit:]] to match any digit and [[:upper:]] to match any uppercase letter. This last regex is equivalent to [A-Z] while [[:alpha:]] is equivalent to [a-zA-Z] or [A-Za-z]. As noted, there is no range available to express a group of all punctuation marks and so to recognize any punctuation mark, we either must enumerate them all in a list or use the [:punct:] class, which is obviously preferable.

We can combine [] with *, + and ? to control the number of times we expect the characters in the brackets to appear. For instance, [abc]+ will match any string that contains a sequence of one or more characters in the set a, b, c while [abc]* matches the same strings but also matches the empty string. And remember that without ^ or $, using * will allow the regex to match an unlimited number of string. Thus, [abc]* will match anything including for instance 12345 because the string contains zero occurrences of a's, b's and c's but ^[abc]*$ will only match strings that consist solely of 0 or more a's, b's and/or c's.

Now we have a means of expressing a regular expression where order of the characters is not important. The expression [abc]+ will match any order of any of the three characters in brackets, as long as there is at least one character. This regex will match any of aaaabbbbcccc, abcccc, accccc, aaaaaabbbbbb as well as abcabcabcabc, abacab, aaaaaccccc and cccccbbbbbbaaaa.

Let's now combine all of the metacharacters we have learned with some examples. We want to find a string that consists only of letters. We can use ^[a-zA-Z]+$ or ^[[:alpha:]]+$. If we had used only [a-zA-Z]+ then it could match any string that *contains* letters but is not restricted to match strings of *only* letters. For instance, [a-zA-Z]+ matches any of abc123, 123abc, abc!def as well as ^#!$a*%&. See Figure 5.2.

We could similarly match a string of only binary digits, 0 and 1. Instead of [a-zA-Z] or [[:alpha:]], we use [01]. The regex ^[01]+$ matches strings that consist solely of binary digits. Again, we use the ^ and $ to force the expression to match entire strings, and we use + instead of * to disallow the empty string. If we wanted to match strings that solely comprise digits, but any digits, we would use either ^[0-9]+$ or ^[[:digit:]]+$. If we want to match a string of only punctuation marks, we would use ^[[:punct:]]+$.

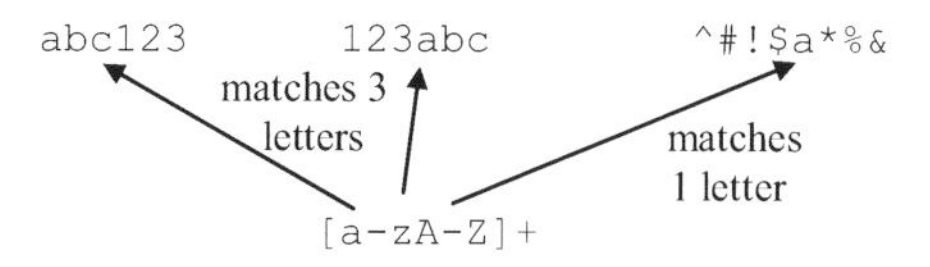

FIGURE 5.2 Demonstrating why [a-zA-Z]+ does not match "only letters".

Let's build a regex to match any string that consists only of letters but ends with one punctuation mark. Both ^[A-Za-z]+[[:punct:]]+$ and ^[[:alpha:]]+[[:punct:]]+$ capture this pattern. Now consider that we want to match a string that consists only of letters and punctuation marks in any order. This problem could be easily resolved if we were looking for any combinations of letters and digits because we have a class, [:alnum:], that expresses this. Our regex would be ^[[:alnum:]]+$. We could also use ranges as in ^[A-Za-z0-9]+$. However, there is no single class that combines letters and punctuation marks, and there is no range to express all punctuation.

One solution is to enumerate all of the characters within the two classes of alphabetic and punctuation marks. The regex would be something like ^[A-Za-z!@#$...=-]+$ (where the ... consists of the remaining punctuation marks, note the hyphen is placed at the end). This is not particularly attractive because there are 32 punctuation marks in all.

Fortunately, we can still resolve this problem by making a list in brackets where the list consists of the two classes we are interested in. This gives us the regex ^[[:alpha:][:punct:]]+$ (or alternatively ^[[:punct:][:alpha:]]+$). These two regexes express that the entire string (^...$) consists of one or more (+) of the characters in the brackets. The characters in the brackets are not an enumerated list but two different classes, [:alpha:] and [:punct:]. By placing both in the same outer brackets, we are listing alternatives, much like we would have with [ab] or [01].

To motivate the use of regular expressions, we talked about identifying people's names in a file. Let's expand the pattern slightly to a person's name of the form *first middle last* where *middle* is the person's middle initial (a single letter), which is optional. We won't bother with the period after the middle initial, at least for now.

The first and last names consist of letters where the first letter is uppercase and the remaining letters are lowercase. We express these using [A-Z][a-z]+. The middle initial will be a single uppercase letter if it appears. This can be expressed as [A-Z]? where the question mark permits the initial to be optional. Our full regex is [A-Z][a-z]+ [A-Z]? [A-Z][a-z]+. Should we want to match a string that consists only of a name, it becomes ^[A-Z][a-z]+ [A-Z]? [A-Z][a-z]+$. There is a problem with this regex in that if there is no middle initial we have a regex that looks for a first name, two blank spaces (one before the omitted middle initial and one after) and the last name. We will return to this example later in this section.

5.2.5 Matching Characters That Must Not Appear

In some cases, we need to express a pattern whereby the string should not contain a specified character. For instance, we might want to match a string that has no blank spaces in it. You might think to use ^[...]+$ where the ... is "all characters except the blank space". That would require enumerating quite a list as it would have to include every letter, every digit and every punctuation mark. There is a more convenient way to specify "must not include" which is to use [^]. The ^, when used inside of brackets means "match if the next character in the string does not match anything in brackets". The blank space after ^ indicates that the only character we do not want to match against is a space.

Unfortunately, our regex [^] has the same flaw as earlier expressions when we did not force a specific location. The regex says "match any string that contains a character other than a space". Thus, this regex matches every string that is not solely blank spaces.

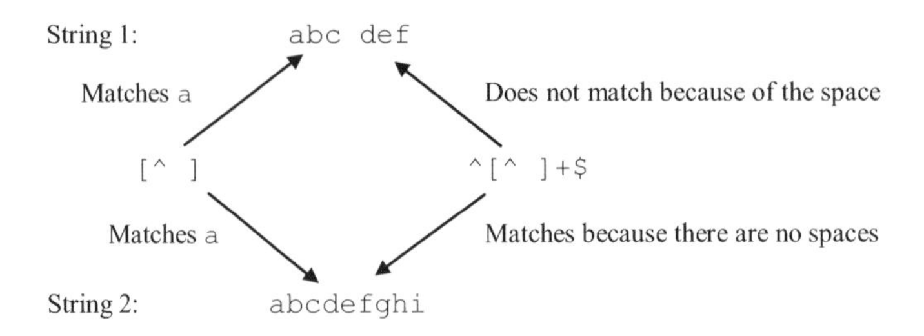

FIGURE 5.3 Demonstrating proper use of [^].

We can solve this problem by first indicating that the regex must match the entire string using ^ and $. Next, we have to specify that the string can contain multiple characters as long as none are spaces, which we do by using + to modify "no blank space". Our regex is ^[^]+$. We would replace + with * if we want to allow the empty string to also match. The use of the [^...] metacharacter is among the most challenging to apply correctly. We will find that the grep program has an easier way to express "not".

We compare the two regular expressions [^] and ^[^]+$ in Figure 5.3, showing why the first fails and the second succeeds. In the figure, [^] will match both strings because it matches any non-blank. The first string has several non-blank characters (e.g., a) while the second has no blanks. The regex ^[^]+$ will not match the first string because there is a blank somewhere in the string.

5.2.6 Matching Metacharacters Literally

All of our metacharacters are punctuation marks. What happens when the regular expression requires one of these punctuation marks as a literal character? For instance, what if we want to match a decimal point or dollar sign? Let's write a regex in the form of a dollars and cents amount where there is at least one digit for the dollar and exactly two digits for the cents. Our first inclination might be to express this as $[0-9]+.[0-9][0-9]. This regex uses two punctuation marks, neither of which is intended to serve as a metacharacter. We can do this but only if we indicate that the punctuation marks should be interpreted as literal characters and not metacharacters.

To use a punctuation mark that is a metacharacter but to be found literally, we need to *escape* the meaning of that punctuation mark. We use the *escape character*, \, for this. We change the above regex to read \$[0-9]+\.[0-9][0-9]. An alternative to using the backslash is to place the punctuation mark in brackets. Thus, the regex could also be written using the notation [$][0-9]+[.][0-9][0-9].

For another example, let's define an expression to match a string in the form of a simple addition problem. The pattern we seek is number + number = number where number is any integer number. We use the plus sign in two different ways here. First, a number is expressed as one or more digits using [0-9]+. Second, we want to match a plus sign.

If we are not careful, we might use the regex [0-9]+ + [0-9]+ = [0-9]+. The problem here is that the second plus sign is treated as a metacharacter, and therefore, it is modifying a blank space. Instead, we must enforce that the second plus sign should literally match a plus sign only. We can do this using either [0-9]+ \+ [0-9]+ = [0-9]+ or [0-9]+ [+] [0-9]+ = [0-9]+.

As the equal sign is not a metacharacter, it does not need to be escaped. Keep in mind that while these regexes will match addition problems, they do not guarantee that the addition problems are mathematically correct. These regexes will match both 13+27=40 and 13+27=4.

Let's expand the regex to match any simple arithmetic problem. That is, instead of a +, the operator can be any of +, -, *, /. Because the operator can be one of several choices, we place them in brackets. In doing so, we do not need to use the backslash. But recall that when a hyphen appears in brackets as one of the optional characters, it must be moved to one of the ends of the sequences of characters so that it is not confused to be part of a range. Thus, `[0-9]+ [+-*/] [0-9]+ = [0-9]+` is incorrect while `[0-9]+ [-+*/] [0-9]+ = [0-9]+` and `[0-9]+ [+*/-] [0-9]+ = [0-9]+` are correct. We can place the hyphen anywhere in brackets if we use the escape character, thus we can also express this regex using `[0-9]+ [+\-*/] [0-9]+ = [0-9]+`.

We can omit the \ when expressing some metacharacters under limited circumstances. Returning to our earlier example of dollars and cents, the `$` was placed at the beginning of the regex. The `$` as a metacharacter is used to indicate "ends the expression" and so we expect to see the metacharacter in place only at the end of the regex. If a `$` appears anywhere other than at the end of the regular expression, it is treated literally as a dollar sign and therefore does not require the backslash. This allows us to rewrite the earlier regex as `$[0-9]+\.[0-9][0-9]`.

In a similar vein, the ^ metacharacter is expected to appear in only two positions: at the beginning of an expression or as the first character inside the `[]`. If a ^ appears anywhere else in a regex, it is treated literally. On the other hand, the characters `[` and `]` must be preceded by the \ if we wish to treat them literally. We would not be able to avoid this like we did the plus sign by inserting the characters in brackets because we would wind up with `[[]]` and that is not valid.

5.2.7 More Precisely Controlling Repetition

Through the use of * and +, we indicate that a pattern of characters can be repeated. What we can't do using these metacharacters is control the *number* of repetitions. With ?, we have slight control in that the number of occurrences is limited to either zero or one. What if we wanted to indicate that a character could appear multiple times but a limited number of times?

To control the number of repetitions beyond one but less than infinity, we use {}, called the *curly brackets* or *curly braces*. Within the curly braces, we specify one or two integer numbers. These numbers indicate a minimum and maximum number of occurrences. There are three formats. The first is {m} to match exactly m occurrences of the preceding character. Second is {m,n} to indicate that the preceding characters must match between m and n occurrences (where $m<n$). Third, we can use {n,}to indicate at least n occurrences of the preceding character. There is no need for {,m} which indicates no more than m occurrences as we will describe shortly.

Let's consider `[0-9]{5}`. This will match exactly five digits which we might use for a US zip code. A US phone number (without area code) could be written as `[0-9]{3}-[0-9]{4}`. Notice that we did not need to use the backslash with the hyphen because the hyphen is only interpreted as a metacharacter when found in brackets. Both the zip code and phone number examples do not require {m} as we could enumerate the m occurrences by hand, as in `[0-9][0-9][0-9][0-9][0-9]`, but obviously `[0-9]{5}` is simpler.

Now let's consider a person's grade on an exam. The largest grade is 100, and the smallest is 0. Apparently, we want to match any one-, two- or three-digit number. We can express this as `[0-9]{1,3}`. Unfortunately, this can match numbers that would not be test scores like 999 and 000. We will return to this problem later.

Let's consider a pattern that will result in a similar problem. We want to define a regex to match an IPv4 address. IPv4 addresses consist of four octets (numbers between 0 and 255), each separated by a period. A novice at regular expressions might try to specify an IPv4 address-matching regex as `[0-255].[0-255].[0-255].[0-255]`. This regex is not correct.

The first problem with this regex is that `[0-255]` does not express "any number between 0 and 255". Recall what is placed in brackets is a list of characters that can match the *next* character of the string. `[0-255]` will only match a single character. The list of items consists of a range (`0-2`) followed by two enumerated characters, both of which are the digit `5`. Thus, `[0-255]` will match any one digit out of 0, 1, 2, 5 and 5. The second problem is that the period is a metacharacter that can match any single character. While the regex could match `5.2.0.1` (which is a legal IP address), it can also match 0a0a0a0 which is not an IPv4 address. It would also fail to match many legal IPv4 addresses like 10.11.12.13 and 123.279.8.16.

To fix our regex, we need to specify that an octet is any sequence of one digit, two digits or three digits. We can do this using [0-9]{1,3}. Unfortunately, like our grade example, this will match three-digit numbers greater than 255 while also permitting numbers like 000. We will fix these problems later in this section. To fix the issue with the period, we use the escape character. Our (flawed) regex is [0-9]{1,3}\.[0-9]{1,3}\.[0-9]{1,3}\.[0-9]{1,3}, which is better than our previous attempt.

We noted earlier that there is no need for {,m}. Why not? Let's consider what [0-9]{,4} means. It says match any number of digits but no more than four of them. As with the *, "any number of" includes zero. Thus, [0-9]{,4} will match no digits all the way up to four digits. We could accomplish the same thing using [0-9]{0,4}. Alternatively, if we want to match up to four digits, we likely would not want to match zero digits, so we might use [0-9]{1,4}. As we clearly have a way to express "up to m digits", we do not need {,m}.

To wrap up this subsection, let's consider one additional example. The regular expression ^[^0-9]*[0-9]{5}[^0-9]*$ says "match a string that consists entirely of zero or more non-digits followed by five digits followed by zero or more non-digits". This regular expression will match any of abc12345abc, 12345abc, abc12345 and 12345. On the other hand, the expression will not match 1234567 or abc123def because neither of these strings contains exactly five digits in sequence.

5.2.8 Selecting between Sequences

Now that we can control the exact number of repetitions that we expect to see, let's revisit our regular expression to match a zip code. There are two formats for US zip codes, a 5-digit zip code, which we previously matched using [0-9]{5}, and a 9-digit zip code which consists of five digits, a hyphen and four additional digits. To match a 9-digit zip code, we could use the regex [0-9]{5}-[0-9]{4}. Unfortunately, if we use the latter regex, it would not match a 5-digit zip code. We could use just the 5-digit zip code regex as any 9-digit zip code has as part of it a 5-digit zip code, but we could match things other than zip codes as well.

Let's assume a zip code will always end a string, so we want to add the $ at the end. Our regex would become [0-9]{5}$. With this change, we can still match 5-digit zip codes at the end of a string but it would no longer match a 9-digit zip code because such a zip code does not end the string with five digits. We need a metacharacter where we can express "match this pattern *or* that pattern".

The metacharacter to specify "OR" is the vertical bar (|). Unlike the "OR" of the list of characters in brackets, this version of "OR" is used between two or more regular expressions. To express "match a 5-digit zip code OR a 9-digit zip code to end the string", we use [0-9]{5}$|[0-9]{5}-[0-9]{4}$. This regex expresses "match a string that ends with either a 5-digit number or a 5-digit number followed by a hyphen followed by a 4-digit number".

Let's revisit the telephone number example we covered previously. Phone numbers in the US are either local numbers, consisting of a 3-digit number, a hyphen and a 4-digit number, or a long distance number in which this 7-digit phone number is preceded by a parenthesized 3-digit area code followed by a space. We can define two regular expressions for this and insert an "OR" between them.

One set of metacharacters that we haven't addressed yet are the parentheses. To match literal parentheses, we must use the escape characters. An open parenthesis would be recognized using \(and a close parenthesis by \). To express a phone number as either a 7- or 10-digit number is [0-9]{3}-[0-9]{4}|\([0-9]{3}\) [0-9]{3}-[0-9]{4}. The blank space after the area code is required.

Let's consider another example. The Cincinnati metropolitan region extends into three states, Ohio, Kentucky and Indiana. These states have two-letter abbreviations of OH, KY and IN respectively. What if we want to write a regular expression to match any one of these three? The first thought might be to use the brackets as in [IKO][NYH]. While this will match any of IN, KY, and OH, it will also match any of IY, IH, KN, KH, ON and OY. By using the | we can avoid this problem through IN|KY|OH.

We noted above that the parentheses are metacharacters. What are they used for? They indicate that the items in the parentheses should be treated as a group. In this way, we can modify an entire sequence of characters with another metacharacter. For instance, we might use (...)+ to indicate that the group may repeat, or (...)? to indicate that the group is optional.

Let's use the parentheses to redo our previous examples of the 5/9-digit zip code and the two forms of telephone numbers. In the 5/9-digit zip code, the hyphen and extension are optional. Thus, we require a 5-digit number, and then we group together the hyphen and 4-digit number in parentheses and modify it with the question mark. This gives us [0-9]{5}(-[0-9]{4})?$. We interpret this as "the end of the string must match a 5-digit number, optionally followed by both a hyphen and 4-digit number".

For the telephone number, because the area code appears in parentheses already, and because we want to indicate that the parentheses and 3-digit number are optional, we wind up having nested sets of parentheses, or (\([0-9]{3}\))?[0-9]{3}-[0-9]{4}. This notation is certainly more of a challenge to read because of all of the metacharacters involved. Notice that the blank space is inserted in the outer parentheses so that if there is no area code there would be no blank space preceding the 7-digit phone number.

Now we have the tools to fix our earlier problem of whether a person's full name has a middle initial or not. We can also add the period with the middle initial. We will consider a full name to consist of a first name, an optional middle initial and a last name where each part is separated by a blank space. The middle initial will be an uppercase letter and a period, and there would be a space both before and after it. As the middle initial is optional, if it is omitted, we would only have one blank space, between the first and last names.

Our regex must indicate that one blank space must appear while the other blank space should only appear if the middle initial does. We create a group consisting of the middle initial (with the period), and the added space. We make this group optional with the ?. Our regex is [A-Z][a-z]+ ([A-Z]\.)?[A-Z][a-z]+.

A sentence consists of some number of words separated by one blank space between each word. The sentence ends with a punctuation mark. For simplicity, assume the punctuation will be a period, question mark or exclamation mark. The words of the sentence will consist of letters. We will assume that the words do not contain or end in punctuation marks (for instance, we assume no apostrophes, commas, colons or semicolons). We do this not because our regex will be useful but to simplify the example we are presenting. One final assumption is that the only word with an uppercase letter will be the first word of the sentence.

Our sentence then consists of a first word, capitalized, additional words and a punctuation mark. Each word is separated by a blank space. We will define this pattern as a regex of three items. First is a capitalized word, expressed as [A-Z][a-z]*. This is followed by a list of one or more words, each of which start with a blank space and each word consists only of one or more lowercase letters. We express this using ([a-z]+)*. We use the * because a sentence could conceivably be a single word. The sentence ends with a punctuation mark, which we denote using [!?.]$. The full regex is ^[A-Z][a-z]*([a-z]+)*[!?.]$.

Let's assume that we want to recognize sentences that consist of more than one word. We would replace the * after the parentheses with a +. If we want to recognize sentences that consist of between two and six words, we replace the * with {1,5}. Why did we use {1,5} instead of {2,6}? Because the first word is already expressed separately in the regex.

Let's consider a variation of our sentence where the first word is not capitalized. Apparently, we can now recognize all words of the sentence using the same notation, [a-z]+, with this regex modified by {2,6}. Unfortunately, this is incorrect because it would omit the blank space. So we refine

```
Test score breakdown
        0-9                     [0-9]
        10-99                   [1-9][0-9]
        100                     100
Regex: [0-9]|[1-9][0-9]|100

IP address octet:
        0-9                     [0-9]
        10-99                   [1-9][0-9]
        100-199                 1[0-9][0-9]
        200-249                 2[0-4][0-9]
        250-255                 25[0-5]
Regex for an octet: [0-9]|[1-9][0-9]|1[0-9][0-9]|2[0-4][0-9]|25[0-5]

Full IP address regex:
        ([0-9]|[1-9][0-9]|1[0-9][0-9]|2[0-4][0-9]|25[0-5]\.){3}
            [0-9]|[1-9][0-9]|1[0-9][0-9]|2[0-4][0-9]|25[0-5]
```

FIGURE 5.4 Solving test score and IPv4 address regexes.

this so that after each word is a space. This gives us `^([a-z]+ ){2,6}[!?.]$`. Unfortunately, this regex is also flawed because the last word should not have a space before the punctuation mark. So, we have to revert back to our earlier regex where we had a first word indicated separately from the remainder. This gives us `^[a-z]+( [a-z]+){1,5}[!?.]$`, or a lowercased word followed by one to five additional lowercased words, each starting with a blank space, and the entire string ending with one of the three punctuation marks.

Earlier, we saw that our regular expressions for a test score and an IP address would match items that were not test scores or IP addresses. Now that we have learned all of the metacharacters, we have the tools available to fix our earlier regexes. Figure 5.4 steps through the solution to both problems. In the figure, we list the various possibilities for the ranges of numbers and the individual regex for each of the possibilities. We then group the individual regexes together using OR (|). For the IP address, we break down an octet and then show the full IP address regular expression (on two lines because of limited width).

We have covered all of the metacharacters as defined in POSIX. There are other metacharacters defined outside of the POSIX standard that may still be applicable depending on the programming language or program we are using. For instance, Java and Perl have different ways to express classes like letters. The `vi` text editor can also search by regular expression which uses a few additional metacharacters. Table 5.4 presents alternative metacharacters comparing them to equivalents in POSIX and where they are defined.

SECTION ACTIVITIES

1. Of the metacharacters you learned in this section, which do you think will be the easiest to use? The hardest? Which do you find easier to remember, the classes denoted as `[:class:]` or the abbreviations found in Perl/`vi`/Java like `\u` or `\S`? Should these abbreviations be made available in Linux?
2. Aside from Java and Perl, what other programming languages have some form of regular expression matching? Research this.
3. Do you think you will ever use regular expressions outside of Linux, for instance in Java or Perl? If so, for what purpose? What about in Windows?

TABLE 5.4
Non-POSIX Metacharacters Found in Perl, vi and Java

Metacharacter	POSIX Equivalent (If None, Then Description)	Used In
\b, \<, \>	Word boundaries (blanks and other delimiters)	Perl and Java use \b \< and \> in vi
\B	Non-word boundaries	Java
\d	[:digit:]	Perl, vi, Java (which also permits \p{Digit})
\D	[^0-9] (that is, non-digit)	Perl, vi, Java
\l (lower case L), \p{Lower}	[:lower:]	vi Java uses \p{Lower}
\p, \p{Print}	[:print:]	vi Java uses \p{Print}
\p{Alnum}	[:alnum:]	Java
\p{Blank}	[:blank:]	Java
\p{Cntrl}	[:cntrl:]	Java
\s	[:space:]	Perl, vi, Java (which also permits \p{Space})
\S	Non-white space	Perl, vi, Java
\u, \p{Upper}	[:upper:]	vi Java uses \p{Upper}
\w	[:alnum:] (but also includes the underscore)	Perl, vi, Java

5.3 EXAMPLES

To master regular expressions, it is not enough to read about the metacharacters but to write your own regular expressions and apply them. Later in this chapter, we will look at Linux programs whereby you can gain such experience. But before we do so, it might be useful to step through additional examples. In this section, we present a number of problems, describe how to break each problem down to understand the pattern, and then solve them through a regex.

Let's start with a look at some brief examples. Table 5.5 provides a number of descriptions of problems and then a regular expression that could be applied to match the given description. Each of these deals with numbers and non-numbers.

Let's focus on the third entry in Table 5.5. We want to match any number no matter if it has a decimal point or not. Preceding the decimal point, if there is one, must be at least one digit. After the decimal point, there may or may not be digits. Thus, the regex could match `1234`, `1.234`, `1234.` and `1`. What it would not recognize is `.1234`, `12.3.4` or the empty string.

TABLE 5.5
Sample Problems and Regular Expressions for Numbers

Problem	Regex	Comments
Match a number.	`[0-9]+`	Matches a string that contains at least one digit.
Match a string that consists solely of a number.	`^[0-9]+$`	We force the regex to match the entire string.
Match a string that consists of a number which may include a single decimal point.	`^[0-9]+\.?[0-9]*$`	We assume that if there is a decimal point, preceding it is at least a digit as in `0.123`, but not necessarily any digit after the decimal point (as in `123.`).
Match a string that contains no digits.	`^[^0-9]+$`	This regex matches the entire string whereby every character is a non-digit.

Could we revise this regex so that if there is a decimal point then there *must* be a digit afterward? For instance, instead of `1234.`, we would require `1234.0`. Yes, we could. Our first inclination might be to use `^[0-9]+\.?[0-9]+$`. Here, we have modified the previous regex by changing the `*` to a + for any digit(s) that follow the optional period. The problem with this revision is that it now requires that our number be at least two digits long because both of the `[0-9]` are being modified with a + whereas our original regex modified the second `[0-9]` with `*`.

To fix our revised regex, we need to provide a slightly more challenging version. We have to indicate that *if* there is a period then there must be at least one digit that follows. We group together the period and the second `[0-9]+` and modify the group with the `?` to indicate that it is optional. This gives us the regex `^[0-9]+(\.[0-9]+)?$`. Now, if there is a decimal point, it must be followed by at least one digit.

The last regex from Table 5.5 matches an entire string where all of the characters are non-digits. Without the ^ and `$`, our regex would match a string that contained at least one non-digit. As we've discussed, the two are not the same. Using `[^ ]` can be tricky and we'll see an easier solution when we look at `grep`.

Table 5.6 presents some additional examples; this time dealing with words. We present these examples in the same way as Table 5.5 (problem, regex, comments). Explanations follow.

Let's take a closer look at the final example in Table 5.6. We want a word to include uppercase and lowercase letters, apostrophes and hyphens. The apostrophe and hyphen are included among the letters so that there is the possibility that the word can consist of multiple occurrences, for instance as with the words `play-by-play` or `y'all's`. By adding the hyphen in brackets, we have to ensure that it appears at the beginning or ending of the list of characters. Alternatively, we could modify the hyphen with the escape character as in `[A-Za-z\-']`.

In most programming languages, an identifier consists of letters, digits and underscores where the first character cannot be a digit. In Java, the identifier could also start with a dollar sign. Some other languages permit hyphens (e.g., COBOL). In Perl, the characters `$`, `%` and `@` are used to start identifier names. Let's define a regex that can match legal identifier names. First, we create one that

TABLE 5.6
Sample Problems and Regular Expressions for Words

Problem	Regex	Comments
Match a string which consists of any sequence of uppercase and lowercase letters.	`^[A-Za-z]+$`	The string has at least one letter. This regex does not permit internal punctuation. Notice that this will recognize a list of letters but the letters may not qualify as a word in English (or another language).
Match a word that consists of some consonants, a single vowel and some consonants.	`^[b-df-hj-np-tv-z]+[aeiou][b-df-hj-np-tv-z]+$`	We assume only lowercase letters in this example for brevity.
Match several (at least two) words with blank spaces between them.	`^([A-Za-z]+ )+[A-Za-z]+$`	This regex insists that words be separated by a blank space aside from the last word. Words may be any combination of upper and lowercase letters.
Match several (at least two) words where a punctuation mark may follow any word.	`^([A-Za-z]+[[:punct:]]? )+[A-Za-z]+[[:punct:]]?$`	We end each word with an optional punctuation mark, a required space occurs after each word except the last.
Same except that words can include hyphens and apostrophes.	`^([A-Za-z'-]+[[:punct:]]? )+[A-Za-z'-]+[[:punct:]]?$`	Included in each word is the possibility that contains hyphens and apostrophes.

includes only the letters, underscores and digits. We then revise our regex to include those that can start with `$` (Java) and `$`, `%`, `@`, (Perl). Figure 5.5 provides two regular expressions, one for simple identifier naming rules and one that fits Java and Perl.

Some programming languages limit the number of characters permitted in an identifier name. We could revise our rules above by replacing the `*` with `{0,31}` for instance if an identifier name was limited to 32 characters (which is the case with some languages). We would also have to ensure that there were no additional characters before or after the identifier, perhaps by embedding the full regex in ^ and `$`.

In the last section, we provided a regex to match a person's name of the form *first name middle initial last name* where the middle initial was optional. Let's now consider how to match a US postal address. Assume a postal address consists of the street address, a city, state abbreviation and zip code. Let's further assume that the postal address ends the string. The street address itself will consist of an optional street number, a street name and an optional apartment indicator.

Figure 5.6 provides four examples demonstrating a range of allowable addresses. In the last example, there is no street number. The first and third examples have apartment specifiers, but the first includes a letter and the third uses # instead of "Apt" or "apartment". Notice that in the first example, the apartment specification follows a comma but in the third example there is no comma before the apartment number. The third example has a 9-digit zip code. There's a lot going on here. Let's break this down part by part.

The first part of our address is the street address. The street address may start with a number, which we specify as `([0-9]+)?`. The question mark makes the street number optional. Before the street name, we expect a blank space but only if there was a street number. Therefore, we modify the street number to be `([0-9]+ )?`.

The street name, as can be seen in the examples, consists of uppercase letters, lowercase letters and blank spaces, but can also have periods and numbers as shown in the third example. The easiest way to express this variability is with a single list of characters that can appear some number of times as in `[A-Za-z0-9. ]+`. This is likely not the best way to express a street name as we would not expect, for instance, a string name of `9.a.1.B. CD.4`, but the regex will match all of the examples in Figure 5.6.

The apartment specifier is optional, so however we encode it, the apartment portion will be modified by a ?. Making the regex more challenging, the apartment specifier might follow a comma

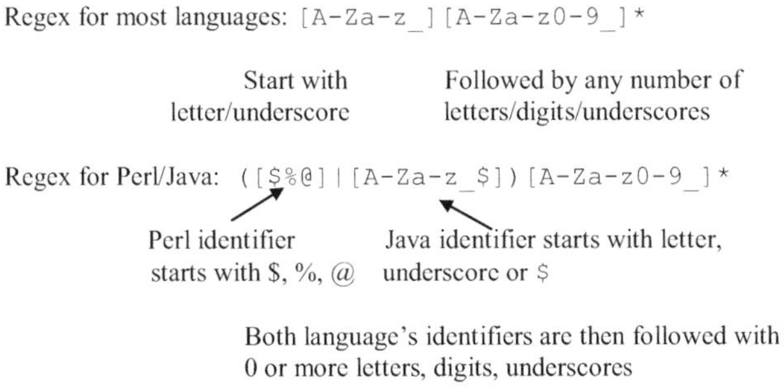

Regex for most languages: `[A-Za-z_][A-Za-z0-9_]*`

Start with letter/underscore — Followed by any number of letters/digits/underscores

Regex for Perl/Java: `([$%@]|[A-Za-z_$])[A-Za-z0-9_]*`

Perl identifier starts with $, %, @ — Java identifier starts with letter, underscore or $

Both language's identifiers are then followed with 0 or more letters, digits, underscores

FIGURE 5.5 Regular expressions for programming language identifiers.

911 Pine Street, Apt B7, Plano, TX 75025
1023 Inca Road, Los Angeles, CA 90017
616 E. 3rd St #99, Cincinnati, OH 45204-1533
Observatory Drive, Decatur, IL 30032

FIGURE 5.6 Example addresses to create an address-matching regex.

(as seen in the first example of Figure 5.6) but this is also optional. The indication that the item is an apartment will use one of Apt, apt, Apt., apt., Apartment, apartment or #. The apartment number itself will consist of digits but may also have letters. We can express all of these variations with the regex `(,? ([Aa]pt\.?|[Aa]partment|#) [A-Z0-9]+)?`.

There's a lot to unpack in this sample regex. First, this entire statement is modified by `?` meaning the apartment specifier is optional. If there is an apartment specifier, it may start with a comma and then must contain a blank space. It is then followed by one of `Apt`, `apt`, `Apartment`, `apartment` or # where `Apt/apt` may be followed by a period. Following this term is a blank space. Finally, the apartment number itself is some combination of digits and uppercase letters.

Typically, the word `apartment` (or `apt`) is separated from the apartment number with a space but the # and apartment number are not separated by a space (see the third example in Figure 5.6). We should make this portion of the regex more precise by enforcing a blank space after `apartment/apt` but not after #. We do so by defining two separate expressions, one with the words and one with the #, and place an OR (|) between them. For the version using the word, it is followed by a blank, and for the version using #, it is followed by an optional blank. Our regex becomes `(,? ([Aa]pt\.? |[Aa]partment |# ?) [A-Z0-9]+)?` where the `# ?` indicates that the space following the `#` is optional.

Whether there is an apartment specifier or not, we have a comma and space followed by the city, a comma, the state, a space and a zip code. The city can be multiple words, each starting with a capital letter. The city is indicated by `[A-Z][a-z]+( [A-Z][a-z]+)?`. The state is simply `[A-Z]{2}`. The zip code is `[0-9]{5}(-[0-9]{4})?`. To finish off this portion of the regex, we connect the three parts but add a comma and space prior to the city/state/zip and a comma and space between the city and state. Putting all of this together gives us the regex shown in Figure 5.7. We have broken this into multiple lines, offering brief explanations for each part. The full regex is shown at the bottom of the figure, although divided into multiple lines because of a lack of space. The regex would be a single, contiguous string.

The introduction to this chapter described a spam filter where we said that the various terms we would search for might be specified using regular expressions. Spam filters use other tools aside from or in additional to regular expressions, but we are only interested in regular expressions in this chapter. Table 5.7 lists some of the terms we might expect to find in a spam message and how we might express them using regular expressions. A comment field shows some of the variations of the words that the regex will find.

```
([0-9]+ )?                    Street address, optional, followed by a blank if it appears

[A-Za-z0-9. ]+                Street name which can include digits, spaces, periods
(,? ([Aa]pt\.? |[Aa]partment |# ?)  Apartment specifier, which may start with
                              a comma and then one of apt/apt./apartment/# and a space
                              (optional space with #), and the apartment number (which
                              might include letters), the entire apartment portion is optional
[A-Z0-9]+)?
, [A-Z][a-z]+( [A-Z][a-z]+)?,  A comma, space and the city name which can be
                              multiple words separated by spaces, ending with a comma
[A-Z]{2} [0-9]{5}             and a 2-letter state abbreviation, blank and a 5 or 9-digit
(-[0-9]{4})?$                 zip-code

The regex in full:
([0-9]+ )?[A-Za-z0-9. ]+(,? ([Aa]pt\.? |[Aa]partment |# ?)?
[A-Z0-9]+)?, [A-Z][a-z]+( [A-Z][a-z]+)?, [A-Z]{2} [0-9]{5}
(-[0-9]{4})?$
```

FIGURE 5.7 Postal address regex.

TABLE 5.7
Possible Regular Expressions for Spam Terms

<table>
<tr><th>Term</th><th>Regex</th><th>Possible string matches</th></tr>
<tr><td>Act now</td><td>[Aa]ct n[o0]w!?</td><td>Act now! act n0w</td></tr>
<tr><td>Click here</td><td>[Cc]lick h[e3]r[e3]</td><td>Click here click h3r3</td></tr>
<tr><td>Lose</td><td>[Ll].*[Oo0].*[Ss].*[Ee]</td><td>L*o*s*e l0se LLLLOSSSSE</td></tr>
<tr><td>Lottery</td><td>[Ll][o0][t4][t4][e3]ry</td><td>l044ery Lott3ry</td></tr>
<tr><td>Order now</td><td>[Oo]rder [Nn]ow!?</td><td>Order Now order now!</td></tr>
<tr><td>Viagra</td><td>[Vv][Ii1!][Aa@][Gg9] [Rr][Aa@]</td><td>Viagra v1@gr@ V!A9RA</td></tr>
<tr><td>You are a winner!</td><td>[Yy]ou (may already be|are) a winner!?</td><td>You are a winner!
you may already be a winner</td></tr>
</table>

There are many different variants we might create when trying to recognize spam words and phrases. In most of the examples in Table 5.7, we allow some letters to be uppercase or lowercase, particularly to start the phrase or to start each word. We also allow variations of some letters that can also look like other characters such as 'o' with '0' and 'i' with '1' or '!'. If we expect that there might be an exclamation mark at the end of the phrase, we used !? to indicate that the exclamation mark is optional.

Look at the entry for "Lose" in the table. One strategy by spammers is to put extra characters in the word. For instance, "Lose" may be written as "L*o*s*e" or "LLLLOSSSSE" or some other variation. The inclusion of `.*` between letters in our regex allows any number of extra characters to appear. We could do this for just about all spam words but limited it to just the one entry for brevity's sake.

SECTION ACTIVITIES

1. Come up with three or four types of patterns that you might want to search for. These might be dates, names of schools, band names or chemical compounds, as just a few examples, and then try to write a regex that will recognize each. How challenging did you find this exercise?
2. Believe it or not, there is regex humor. See if you can make sense of these jokes. The first two might be more challenging.
   a. What did one regex say to the other? .+
   b. What regex are you most likely to see at Christmas? [^L]
   c. If you put a million monkeys at a million keyboards, one of them will eventually write a Java program, the others will write Perl programs.
   d. Some people when confronted with a problem think "I know, I'll use a regular expression." Now they have two problems.
   e. Popular regular expression t-shirt: [--~]. What does this represent?

5.4 `grep`

The name `grep` comes from *global regular expression print*. The `grep` program searches one or more files line by line to match each line against a specified regular expression. Each line is treated as the string and `grep` searches for any substring of the line that matches the regex. If a match is found, by default `grep` outputs the line. `grep` will search the entire file or files, outputting all matching lines.

grep only applies the POSIX basic regular expression set. To apply the extended regular expression set, we use egrep. Note that although deprecated, grep permits the option -E which is equivalent to running egrep. There is also a variant of grep called fgrep (which can be run in grep using the deprecated option -F). With fgrep, we apply a literal string instead of a regular expression. As egrep can do everything that grep and fgrep do, there is no need to use anything other than egrep. Therefore, for the remainder of this section we only consider egrep.

5.4.1 Using egrep

The syntax of the egrep instruction is egrep *[options] regex file(s)*. The *regex* is a string that can but does not have to include metacharacters. We typically place the regex in single quote marks to avoid confusing the Bash interpreter if any of the metacharacters could be interpreted as wildcards. We explore this later. As you might notice in the syntax, we can specify a single file or multiple files. Files can be specified by listing them (separated by spaces) or by using wildcards, or both.

Let's start with a few simple examples, as provided in Table 5.8. In the table, we have egrep instructions and explanations for what each will search for. In the first two examples, we did not quote the regex because the expression contains no metacharacters that are also wildcards. All of the examples operate on all .txt files in the current directory.

Why do regular expressions that include wildcard characters need to be quoted? To answer this question, we need to recall how the Bash interpreter operates on a command. Among the things that the Bash interpreter will do is replace aliases with the instructions they alias, replace variables with their values, handle redirection if applicable and apply filename expansion (globbing) on wildcards.

A variable might be listed in an instruction using the notation *$VAR* while redirection operators are <, <<, >, >> and |. Globbing wildcards include *, ? and [ ]. The first example in Table 5.8 involves no Bash-specific characters, so there is no interpreting for Bash to perform. Thus, when executed, egrep literally tries to match the string, 2022. In the second example, the only metacharacter is ^, which is not a character that the Bash interpreter will operate upon (at least in isolation, it will apply the ^ when used in [ ]), and so again the regex does not require quote marks.

Now consider the fourth example that contains | which is also recognized by Bash as the pipe symbol. Without quote marks around the regex, the Bash interpreter will see the | as a pipe and set up a pipe between two instructions before it even executes the egrep instruction. The [] are used as wildcard characters. The instruction egrep [Ss]mith *.txt (without quote marks around

TABLE 5.8
Example egrep Statements

Instruction	Explanation
egrep 2022 *.txt	Find all lines that contain 2022.
egrep ^a *.txt	Find all lines that start with an a.
egrep '[Ss]mith' *.txt	Find all lines that contain either Smith or smith.
egrep 'Duke|Zappa' *.txt	Find all lines that contain either Duke or Zappa.
egrep '[0-9]+' *.txt	Find all lines that contain at least one digit.
egrep '[A-Za-z]+[[:punct:]]' *.txt	Find all lines that contain letters followed by any punctuation mark anywhere in the line.
egrep '^[0-9]*[^0-9]+$' *.txt	Find all lines that if they have digits are found at the beginning of the line.
egrep '^[A-Z][a-z]+ [A-Z][a-z]+$' *.txt	Find all lines that contain exactly two words where both words are capitalized (perhaps first and last names).
egrep '[a-z]+ [a-z]+ [a-z]+' *.txt	Find all lines that contain at least three words (lowercase letters separated by spaces).

the regex) will cause the Bash interpreter to first perform filename expansion on `[Ss]mith`. If there was a file named either `Smith` or `smith` then the regex would be replaced by the filename leaving us with an instruction like `egrep  smith  *.txt` whereby `egrep` only looks for the string `smith` in the `*.txt` files. If there were no matching file, the instruction becomes `egrep  *.txt`, which is erroneous because there is no regex provided.

Now let's consider the instruction `egrep  fa*[0-9]  *`. The instruction asks `egrep` to find all lines that contain an `f` followed by zero or more `a` followed by a digit among all files in the current directory. The Bash interpreter, however, attempts to perform filename expansion on both `fa*[0-9]` and `*` before executing the `egrep` instruction. Assume the current directory has files `fa1`, `fa2`, `fab1`, `fab99`, `fb34`, `file1` and `file2.txt`.

Filename expansion on `fa*[0-9]` causes Bash to replace this with `fa1  fa2  fab1 fab99` since all start with `fa` following by anything followed by a digit. It does not match `fb23`, `file1`, `file2.txt`, because the filename must start with `fa` and it cannot end with a `.txt` extension. The full instruction appears as `egrep  fa1  fa2  fab1  fab99  fa1  fa2  fab1  fab99  fb34  file1  file2.txt`. Now, `egrep` executes attempting to locate the literal string `fa1` in each of the files `fa2`, `fab1`, `fab99`, `fa1`, `fa2`, `fab1`, `fab99`, `fb34`, `file1` and `file2.txt`. Not only did Bash change our regex into the name of the first matching file but it created a list of files that include duplicates (`fa2`, `fab1`, `fab99`).

In order to prevent Bash from overstepping its authority, we need to quote our regex. Should we use single or double quotes? Remember that when the Bash interpreter confronts anything in single quotes, it treats the items in quote marks literally while items found in double quotes are interpreted. For instance, imagine we have a variable, `FIRST`, storing the name `Frank`. Then `'$FIRST'` is treated literally as `$FIRST` but `"$FIRST"` is interpreted to be `Frank`. To ensure that any Bash characters found in a regex are treated as regex metacharacters and not interpreted first by Bash, we use single quote marks.

We do not always need to use single quotes around our regex as we saw in Table 5.8. The rule is that if the regex has no characters that Bash will interpret, we can omit the quotes. But it is a good habit to get into so that we do not make mistakes or forget. So, for this reason, always use single quote marks around the regex. The one exception to this rule is if the regex is strictly a literal string of letters and/or digits with no spaces in it.

Now let's take a look at `egrep` output. Figure 5.8 provides two examples. In the first, we search for all users who are assigned the Bash shell by using the command `egrep  bash  /etc/passwd`. The `passwd` file stores all user accounts and information like their shells. In the second example, we search all files in the current directory whose names start with an `f` for lines that consist solely of two capitalized words (first and last names).

In both examples from Figure 5.8, `egrep` responds by listing the lines of matching substrings. The output shows the matching substring highlighted (they would appear in red but because the

```
[foxr@localhost ~]$ egrep bash /etc/passwd
root:x:0:0:root:/root:/bin/bash
foxr:x:1000:1000:Richard Fox:/home/foxr:/bin/bash
underwoodr:x:1002:1002:Ruth Underwood:/home/underwoodr:/bin/bash
boothc:x:1003:1003:Christina Booth:/home/boothc:/bin/bash
zappag:x:1005:100:Gail Zappa:/home/zappag:/bin/bash
[foxr@localhost ~]$ egrep '^[A-Z][a-z]+ [A-Z][a-z]+$' f*
f1:Richard Fox
f1:Tommy Mars
f1:Mike Keneally
f2:George Duke
f2:Christina Booth
f2:Richard Fox
```

FIGURE 5.8 Two example `egrep` outputs.

```
/etc/dnsmasq.conf:#alias=192.168.0.10-192.168.0.40,
/etc/hosts:127.0.0.1    localhost localhost.localdom
main4
Binary file /etc/ld.so.cache matches
/etc/networks:default 0.0.0.0
/etc/networks:loopback 127.0.0.0
/etc/networks:link-local 169.254.0.0
/etc/resolv.conf:nameserver 172.28.102.11
/etc/resolv.conf:nameserver 172.28.102.13
```

FIGURE 5.9 Using `egrep` to search for IPv4 addresses in files in `/etc`.

image is black and white they simply appear in a lighter shaded font). The first command's output shows the matching lines while the second also includes the file that contained the matching lines (`f1` or `f2` in this case). When `egrep` searches multiple files, its default behavior is to include the filename as part of the output.

Let's take a look at another example. We issue an `egrep` command using our regex for IPv4 addresses as defined in Figure 5.4 on the files in the directory `/etc`. Unfortunately, we receive numerous error messages because many of the items in `/etc` are subdirectories and `egrep` cannot work on the subdirectory unless we request a recursive search. Additionally, there are some directories that we, as normal users, do not have access to. We append our instruction with the notation `2>/dev/null` to send any error messages generated from the instruction to the Linux trash can. Our instruction, with the regex replaced by *ip_regex* to save space, is `egrep '`*ip_regex*`' /etc/* 2>/dev/null`. There are dozens of matches. We show only a small portion in Figure 5.9.

The figure does not show the complete width of the output. The reason we covered this example is to spotlight a particular output. Look at the fourth line in the figure (which is really the third line of output), `Binary file /etc/ld.so.cache matches`. This message informs us that the regex matched a binary file. Although we define our regular expressions in ASCII text, ASCII characters are stored in binary. Thus, the regex could also match the contents of a binary file. However, outputting part of a binary file would be meaningless to most of us since we won't know what the binary values represent. So, in this case, `egrep` informs us of a match but does not show us the matched content. We return to this in the next subsection.

5.4.2 Useful egrep Options

Let's take a deeper look at `egrep` (and its `grep` and `fgrep` variants). There are numerous options available that can help improve our interaction with `egrep`. One example comes from our example at the end of the last subsection (and Figure 5.9) where `egrep` reported on a match to a binary file. We can force `egrep` to process binary files using `-a`.

Another option when searching all files of a directory is to apply the option `-r` to cause `egrep` to recursively descend all subdirectories. In Figure 5.9, we searched all files in `/etc` (that were not binary files and were accessible to our user account). Had we added `-r`, `egrep` would have also descended accessible subdirectories. Although we do not cover all of the options, many are explained in Table 5.9.

Let's explore some more of these options. One of the most useful is `-c` so that, rather than being given a list of all of the matches, we get a count of the number of matches. If there are matches among multiple files, we get a count for each file (those that have no matches output a 0). Figure 5.10 provides two examples, one showing the number of matches in a single file, and one showing matches among multiple files. The first example counts the number of lines in `/etc/resolv.conf` that match our regex for an IP address (note: we are using the simpler IP address regex here). The second, run by root, searches files in `/var/log` for any lines that contain `foxr`. The full command is `egrep -c 'foxr' /var/log/* 2>/dev/null`. In the first example,

TABLE 5.9
grep/egrep Options

Option	Meaning	Comments
`-A` *N* `-B` *N* `-C` *N*	Add context.	Output *N* lines after, before or both before and after the matching lines to add context.
`-a`	Process binary files.	Force any match in a binary file to be output.
`-c`	Count matches.	Return the number of matches, not the lines that matched.
`-e` *regex*	Specify the *regex* explicitly.	Allows us to avoid using quote marks around the regex.
`-f` *file*	Use *file* instead of regex.	The regex is stored in a file so that we can use it repeatedly in many `egrep` instructions.
`-G`	Use only the basic regex set, not the extended regex set.	Whereas `grep -E` is the same as `egrep`, `egrep -G` is the same as `grep`.
`-H`	Output filenames.	Output filenames with the matching lines; this is the default when `egrep` searches multiple files as shown in the bottom part of Figure 5.8 and all of Figure 5.9.
`-h`	Do not output filenames.	Do not output filenames; this is the default if only one file is searched as shown in the top part of Figure 5.8.
`-i`	Ignore case.	Allows us to avoid using `[A-Za-z]` or `[Aa]`, for instance.
`-l`, `-L`	Output file names only.	Do not output lines that match/do not match, only output filenames (`-L` is like `-v` and `-l` combined).
`-m` *n*	Only output first *n* matches.	Useful if we only want to see a few matches.
`-n`	Add line numbers.	Output the line number of each matching line along with the other output.
`-o`	Output only the matched substring.	We can view just the substring that matched the regex.
`-P`	Apply Perl metacharacters.	Permit the use of metacharacters defined in Perl (see Table 5.4)
`-r`	Recursive search.	If any files are directories, process them recursively.
`-v`	Invert the match.	Output all the items that do not match; discussed further in the text.
`-x`	Match the entire line.	Let's us omit `^` and `$` in the regex.

```
[foxr@localhost ~]$ egrep -c '[0-9]{1,3}\.[0-9]{1,3}\.[0-9]{1,3}\.[0-9]{1,3}' /etc/resolv.conf
4

maillog-20210110:0
messages:2
messages-20201220:42
messages-20201227:26
messages-20210103:17
messages-20210110:21
private:0
qemu-ga:0
rhsm:0
sa:0
samba:0
secure:0
secure-20201220:22
secure-20201227:17
secure-20210103:60
secure-20210110:26
speech-dispatcher:0
```

FIGURE 5.10 `egrep` examples using option `-c`.

we only receive the number, not the filename, because we are only searching one file. In the second example, files with no matches are output with the number 0.

Had we issued the `egrep` instruction on `/etc/resolv.conf` with `-H`, the output would have included the filename, as in `/etc/resolv.conf:4`. Had we issued the `/var/log/*` instruction

```
foxr:x:1000:1000:Richard Fox:/home/foxr:/bin/bash
----------------------------------------------------
tcpdump:x:72:72::/:/sbin/nologin
foxr:x:1000:1000:Richard Fox:/home/foxr:/bin/bash
zappaf:x:1001:1001:Frank Zappa:/home/zappaf:/bin/bash
```

FIGURE 5.11 Output from `egrep` without and with `-C 1`.

with `-h`, we would have received a list of numbers without file names. Seeing numbers like 0, 2, 42, 26, 17, 21, 0, 0 on separate lines would not inform us of much.

Let's take a look at using `-A`, `-B` and `-C`. We want to view the line in the `/etc/passwd` file storing the record for user foxr. The output from `egrep 'foxr' /etc/passwd` is shown in the top portion of Figure 5.11. If we want to see context of this line (lines above and/or below it), we use `-A`, `-B` or `-C`. In the bottom portion of Figure 5.11, we use `-C 1` so that we can also see the line immediately before and after the matching line.

Although the context may not be particularly useful in this case, it could be valuable if our search requires that we understand not just the matching line but information about that line. For instance, imagine that we are attempting to debug some issue in a program. We might want to see the line(s) immediately before and after the line with an error. Another useful option is `-n` to display line numbers to help us identify the location of a match within the file.

The option `-i` can be among the most valuable, or biggest time savers. This option tells `egrep` to *ignore case*. This means that `egrep` will accept a match of a letter no matter the case. This allows us to specify our regex with only upper or lowercase letters and not both. If, previously, we used `[A-Za-z]`, then with `-i` we can simplify this to just `[A-Z]` or `[a-z]`. Let's reconsider a word like `Lose` as discussed in Table 5.7. To express all possible combinations of uppercase and lowercase letters for `Lose`, we would have to specify `[Ll][Oo][Ss][Ee]`. With `-i`, we need only specify `LOSE`, `lose` or `Lose`.

Recall the challenge of properly using `[^...]` to denote *not*. A far easier solution is to add `–v` to our `egrep` command. Literally, the meaning of this option is to *invert the match*. When we specify a regex using `-v`, `egrep` returns all lines that did not match. For instance, if we want to search for all lines of a file that do not contain a period, the regex would be `'^[^.]+$'`. Instead, we issue `egrep –v '\.' somefile`, which is far simpler.

One last option of note is `-m`, to specify the number of matches that should be output. Why might we want to limit the output? This option could be useful for one of three reasons. First, we want to test a regex out to see if it works and only care to see the first match, so we add `-m 1`. Second, we might be interested not in which lines match but in which files contain a match. Again, using `-m 1` will show us the first match but of each file. Now we know which files contain the match (alternatively we could use `-c` to only output the count of matches per each file). Finally, if we suspect there is a problem early in a file but don't want to look at a lot of output, we might use `-m` *n* where *n* is a small number like 5 to make the output more manageable.

We conclude this section with a notable comment about `egrep`. Recall that POSIX did not define {,*n*}. We explained earlier that we did not need this notation because it is the same as {0,*n*}, and in practice, we would likely want to find one to *n* matches and so would use {1,*n*}. However, the GNU implementation of `egrep` permits the use of {,*n*} if we have a need for it. This notation may not be available in some Unix dialects since it was included as part of the GNU project, but it should be available in `egrep` in all or just about all versions of Linux.

5.4.3 Examples: Searching the Linux Dictionary

Let's reinforce everything we've learned to this point of the chapter by using `egrep` to search the Linux dictionary, stored in `/usr/share/dict/linux.words`. In the same directory is a symbolic link called words which points to linux.words. We will assume for simplicity that our current directory is /usr/share/dict so that our egrep commands only require the filename words.

We start by looking for all 30-letter words with egrep '[[:alpha:]]{30}' words. The output of this command is two words, dichlorodiphenyltrichloroethane and pneumonoultramicroscopicsilicovolcanoconiosis. These words are of different lengths, and neither is 30 letters long. To obtain exactly 30-letter words, we have to match a string from the start of the line to the end. We enhance our command by adding -x to force the regex to match the entire line (we could also add ^ and $ to the regex). We receive no output indicating there are no 30-letter words in the dictionary.

Next, let's look for 20-letter words where the words start with a capital letter and the rest of the word is lowercased. This command is egrep -x '[A-Z][[:alpha:]]{19}' words. We use 19 in curly brackets because A-Z matches the first letter. By including -x here, we receive exactly 20-letter words as output. Our output is shown in Figure 5.12. Had we not used the -x option, we would receive eight matches.

The dictionary contains many words that include non-letters (digits and punctuation marks). Words found in this file contain punctuation marks like hyphens and apostrophes with some words also containing commas, ampersands, periods and slashes, among others. Let's look for all words that contain an apostrophe. The regex is simply ' but unfortunately as we place our regular expressions in ' ' we can't just say egrep ''' words. We also cannot use \' to escape the meaning of the apostrophe because the apostrophe is not a metacharacter but instead syntax used to prevent Bash from interpreting characters like $.

In order to search for the apostrophe, we can do one of three things. First, we can place the regex in a file and use -f to reference the file. This is a little silly since our regex is a single character. Second, we can use double quote marks instead of single quote marks around the regex as in egrep "'" words. Lastly, we can place the quote mark in brackets, as in egrep '[']' words.

Some words in the dictionary contain digits. Let's look for words with multiple digits. For instance, catch-22, 1080 and 30-30 are all found in the dictionary. In this case, let's look for words where the digits are not consecutive. You might think that [0-9].+[0-9] would work as we are asking for two digits with something (anything) between them. Unfortunately, the . can match a digit, leading to for instance a match like 1080. Instead, we use the regex [0-9][^0-9]+[0-9] to match any string with two digits as long as there is at least one non-digit between them. The full instruction is egrep '[0-9][^0-9]+[0-9]' words. We receive just three responses: 2,4,5-t, 2,4-d and 30-30. In each case, the digits are separated by a comma or hyphen or both.

Over 48,000 words appear in the dictionary that include some form of punctuation. We can see this using egrep -c '[[:punct:]]' words. How many of these words contain a punctuation mark that is not found at the end of the word? For this, we need to elaborate upon our regex that we have at least one punctuation mark but that the last character is not one. To indicate that a character is not a punctuation, we use [^[:punct:]]. To indicate that this character is at the end of the line, we add $ at the end. This gives us egrep -c '[[:punct:]][^[:punct:]]$' words.

Unfortunately, our regex from the last paragraph is not sufficient because it requires that the punctuation mark occur *immediately before* the last character. As we want the punctuation mark to appear anywhere except the end, we indicate add .* to indicate that anything can occur before the last character. This gives us egrep -c '[[:punct:]].*[^[:punct:]]$' words. Our regex

```
Archaeopterygiformes
Biblicopsychological
Chlamydobacteriaceae
Llanfairpwllgwyngyll
Mediterraneanization
```

FIGURE 5.12 20-letter words found in the Linux dictionary.

now says "find a punctuation mark followed by anything ending with at least one non-punctuation mark". As we used -c, we find that there are nearly 45,000 words in the dictionary that match our regex.

Let's look at one more example. We want to find all "words" listed in the dictionary that have no letters. To search for non-letters, we might try to use [^...]. In fact, the regex [^A-Za-z] will not work because it will match any single non-letter meaning rather than full words of no letters. The proper regex is ^[^A-Za-z]+$. To simplify our regex, we can add -i (ignore case) and reduce it to ^[^a-z]+$. However, we can also use the -v option to simplify things. We want to search for all lines that *contain* a letter and then return all lines that do not match. Our command is egrep -iv '[a-z]' words. We find there are only three entries.

5.4.4 Using egrep to Control the Output of Other Linux Commands

In Chapter 4, we briefly introduced the grep command to reduce the output from ps aux. Specifically, we piped the result from ps to grep to only output lines that contained a particular string. We can employ grep (or egrep) to reduce the output of any Linux instruction that might output a lot of content by similarly piping the result of the operation. In this subsection, we look at a few examples. These examples are provided to illustrate that egrep can further our command-line capabilities rather than view an output that lists items we are not interested in.

Let's start with the ls command. We can find all files of a directory that contain certain characters through wildcards such as a* to find all files that start with an a or *.txt to find all files ending in .txt. What if we have a more involved pattern to search for? We saw in Chapter 3 that we could use {} and [] to specify either a list of options or alternative characters as in *.{dat,txt} to search for all files ending with .dat or .txt, and file[0-9].txt to find all files starting with the word file followed by a digit followed by .txt.

To motivate our use of egrep though, let's perform ls on files whose names have a 3- or 4-letter extensions (we don't care what the extensions are). A 3-letter extension would match the wildcard .??? (recall that as a wildcard, ? matches any one character as opposed to its role in a regular expression). To match a 4-letter extension, we could use the wildcard .????. How do we combine them? We could issue the instruction ls *.??? *.???? so that Bash expands all filenames that match either set of wildcards.

As an alternative, we can use egrep and provide a slightly more concise instruction. The regex we search for is '\....?$' where the first period is treated as the period of the extension, the middle three periods match any character of the extension, and the last period, used to match a fourth character, is optional. The $ ends the regex enforcing that the regex matches only at the end of the filename. We could also use '\..{3,4}$'. The full instruction is ls | egrep '\..{3,4}$'. Notice that we do not specify any files for the egrep command itself because its input is the list of files from ls. When we type ls, it often displays the contents of a directory in columns, however, when piping the output to another command, each item of the ls output is provided on its own line.

Let's consider two simpler uses of egrep. We do not care to see the output of a command like ls or ps aux but merely the number of items that match some pattern. We can use egrep -c to count for us. What if we want to count all of the output, not just the items that match some regex? egrep still requires a regex so we use a regex to indicate everything, or .*. We might issue the command ls | egrep -c '.*' or ps aux | egrep -c '.*' to output the number of items in the current directory or the number of running processes.

To wrap up this subsection, we present several additional examples in Table 5.10. The examples use both ls -l and ps aux. The ls -l examples all list contents from the /etc directory. Brief explanations are provided in the table. You might try some of these on your own.

Let's take a quick look at the last entry in the table, ps aux | egrep -v '0:00'. You might wonder why we didn't try a regular expression like [0-9]:[0-9][1-9]. This regex would match 0:01 or 5:31, but it would not match 5:30. It takes a bit of work to express "not 0:00" as, like

TABLE 5.10
Additional `egrep` Examples

Command	Explanation
`ls -l /etc | egrep '^-rwx'`	Output all regular files whose owner permissions are `rwx` (notice the ^ to start the regex).
`ls -l /etc | egrep '.{13}[^1]'`	The first 13 characters of the `ls -l` output are the filetype, permissions, a period and two spaces; the next entry is the number of hard links; this command outputs all items that have more than one hard link.
`ls -l /etc | egrep -v 'root root'`	The third and fourth fields of the `ls -l` command are the file owner and group owner; this instruction searches for all entries where both are root and inverts the match (return all files that are not both owned by root and in root's group)
`ps aux | egrep '\?'`	Find all processes not tied to a TTY (listed as `?`); the `?` should only appear in the TTY column but since `?` is a metacharacter, we have to escape its meaning.
`ps aux | egrep ' Ss '`	`Ss` is a state of some processes and this command locates just those processes; notice we use spaces around `Ss` because without the spaces we might find processes whose states have other characters in them like `Ss+` or `Ssl`; additionally, `Ss` could potentially appear in a user's name or process name so by adding the spaces we ensure that `Ss` is found by itself, which should only occur in the `STAT` column.
`ps aux | egrep -v '0:00'`	Output all processes whose CPU time is more than 0 seconds (in fact all processes are more than 0 seconds but few are as much as 1 second), again we use `-v` to invert the match.

we did earlier with legal test scores and IP addresses, we would have to break this into all possible occurrences that we want to match. If we want to match anything other than 0:00, we have to be able to match 0:01–0:09, which is the regex provided above, but also 0:10–0:99 and 10:00 through 99:99. Using `-v` is much easier!

Let's also look more thoroughly at the second example, `ls -l /etc | egrep '.{13}[^1]'`. We want to match any entry in `/etc` whose 13th character is not 1. The 13th character is either a 1-digit hard link count if the file has fewer than ten hard links, or the second digit of a hard link count that is at least two digits long. If the file has less than ten hard links, then this regex works as expected. If, however, the file has say 11 or 21 hard links, then the regex will not match the file. Fortunately, we don't really have to worry about this error in practice because no files in `/etc` have that many hard links.

Let's try one other example of piping to `egrep`. Recalling the Linux dictionary, we want to find all words that contain a punctuation mark other than a period, hyphen or slash. While this seems simple enough in that we could write `egrep '[...]' words` where … are all of the punctuation marks aside from the three we are not interested in, this is a lot of work. Further, `egrep -v '[./-]' words` does not work because this will locate all words that do not have one of these three characters but not necessarily words that have other punctuation marks.

The solution is to pipe the output of one `egrep` instruction to another `egrep` instruction. The first `egrep` might be used to find all words with punctuation marks. The second `egrep` might then be used to remove all entries that contain one of the three punctuation marks we don't want to see by using the `-v` option. Our instruction is `egrep '[[:punct:]]' words | egrep -v '[./-]'`. We could have also accomplished this by having the two `egrep` instructions in opposite order, as in `egrep -v '[./-]' words | egrep '[[:punct:]]'`.

SECTION ACTIVITIES

1. The grep program is available in Windows (it is part of the Mac OS already in that it was implemented in the Mach Operating System which the MacOS is built upon). If you are a Windows user, do you have any interest in installing it? Why or why not? Keep in mind that in Linux, most of our data files are text files. This may not be the case in Windows.
2. In this section, we looked primarily at egrep but also mentioned fgrep. Another variation is called pgrep. Compare grep, egrep, fgrep and pgrep. You can read their man pages or research them online. Which of these do you feel you would use most of the time and why? Which do you think you might never use and why?
3. If you did not try out some of the egrep commands in this section, in your Linux computer cd to the Linux dictionary directory and try some of the examples presented. Then, come up with some of your own. Make a note of those that did not work as expected and try to fix them.

5.5 sed

sed is a *stream editor* which is a program that takes a stream of text and modifies it. A *stream* is short for an I/O stream meaning the stream of text characters that are being input from one source and output to another. The role of sed is to manipulate the text in the stream en route from input to output. This process can be useful if we want to perform mass substitutions on a text file without having to manually edit the file. In this section, we take an introductory look at how to use sed. sed can do many things, but we concentrate on using regular expressions to find strings to modify.

5.5.1 Basic sed Syntax

The syntax for sed is sed [*options*] *script file(s)*. There are numerous options, some of which we will explore in this subsection and others we will look at in later subsections. The *script* can be one of several types of commands but the one we are most interested in is the search and replace script of the form 's/*pattern*/*replacement*/'. *Pattern* is a search string, which can be a literal string or a regular expression, and *replacement* is the replacement string, which may include special designators to recall parts of the matching portions of the string. sed operates much like egrep; it explores the file(s) line by line looking for a match. Upon finding one, the matched item is replaced with the replacement string. All lines, whether modified or not, are output.

By default, sed outputs to the terminal window so that the original file is unaffected by the sed command. We can redirect the output to a file but not to the same file because that file is currently being used as an input file. If we want to modify the input file, we must add the option -i. In doing so, sed does not actually write over the existing file while running. Instead, it executes, saving to a temporary file which is then moved in place over the original input file once sed terminates. Alternatively, we can redirect the output of sed to a new file.

Let's start with a very basic sed instruction. We have a file, mydata.txt, that contains some type of data, including the current year. We want to modify the year, updating every instance of 2022 to 2023. We use the instruction sed 's/2022/2023/' mydata.txt. Each line of mydata.txt is processed by sed and if it finds 2022 on a line, it is replaced by 2023. The output is sent to the terminal window.

The default behavior of sed is that once it finds a match on a line, the replacement is made, the line is output, and the next line is processed. That is, sed only makes one replacement per line.

To modify *all* instances of the search pattern, we have to indicate a *global* replacement. We accomplish this by adding g at the end of the script, as in 's/*pattern*/*replacement*/g'. Redoing our previous command, it now reads sed 's/2022/2023/g' mydata.txt. With this change, all instances of 2022 are replaced by 2023.

Let's try a variation of the same problem. Again, assume mydata.txt stores information including years. We want to find all years of the form 20xx and replace them with just xx. That is, we want to eliminate the 20. This would change 2022 to 22 while leaving years like 1995 intact. As we want to replace all occurrences, we make the command global by adding the g. We want to save the result to a new file, mydata2.txt. We come up with the command sed 's/20//g' mydata.txt > mydata2.txt.

The above instruction causes sed to search for all instances of 20 and replace each with nothing. Thus, 10/05/2021 becomes 10/05/21 while 10/05/1995 remains unchanged. Unfortunately, this is not a particularly wise instruction as it removes all instances of 20 no matter where the 20 appears. For instance, the 20 might be part of a phone number or address. Even worse, consider the date 10/20/1995. sed would replace this date with 10//1995.

To solve our problem, we need a more sophisticated regex to capture 20 as we are only interested in the 20s that appear as the first two digits of a year. We might think that 20[0-9][0-9] would suffice, but again, this could match a phone number or address.

Let's assume all of the dates in the file are stored using the format xx/xx/xxxx. Now we know that the 20 in the year will always follow a /. We can modify our regex to search for /20[0-9][0-9]. Unfortunately, this has its own problem because the forward slash is part of the sed syntax and must be handled differently if we want to use it as part of our regex. This means that sed 's//20[0-9][0-9]//g' mydata.txt > mydata2.txt does not work as expected.

Even if our solution worked with the forward slash, it has another problem which is perhaps worse. Our replacement string is empty; we are not just removing the 20 but the *entire year*. The date 10/05/2022 does not become 10/05/22 but instead 10/05. We resolve this problem in the next subsection.

Let's take a look at another example. We want to spell out the word apartment in place of situations where an address contains an abbreviation such as Apt. or apt. We will assume that the abbreviation can appear as any of apt., Apt., apt, Apt and #. Our regex can be expressed as [Aa]pt.?|#. Our instruction is sed 's/[Aa]pt.?|#/apartment/g' file.

There are a couple of issues with this sed statement. One is that the | metacharacter is part of the extended regular expression set. This is not available in sed by default. We can resolve this in one of two ways. First, we can use the escape metacharacter (\) prior to any extended regular expression set metacharacter (e.g., \| or \+). Alternatively, we can use the option -r (or -E). Our sed instruction becomes one of sed 's/[Aa]pt.?\|#/apartment/g' file or sed -r 's/[Aa]pt.?|#/apartment/g' file.

The other issue we have with this instruction is that we would expect to see a space appear after an abbreviation like Apt. but probably not after a #. In the former case, sed replaces a string like Apt. B7 with apartment B7, but #B7 becomes apartmentB7. We would like to insert a space after the word apartment but when we match # we do not want to insert a space.

Fortunately, we can tackle this by specifying *multiple* scripts. We will use one script to search for [Aa]pt.? and replace this with apartment with no space as we expect the space to already exist before the apartment number (for instance, Apt B7 becomes apartment B7). A second script will search for # and replace it with apartment with a space after the word.

To use multiple scripts, sed requires that we place the option -e before each script. Our revised instruction contains two scripts, each preceded by the -e option. Our instruction becomes sed -e 's/[Aa]pt.?/apartment/g' -e 's/#/apartment /g' file > file2. The replacement in the second script is the word apartment followed by a blank space. Notice that that we no longer need -r in this version of our instruction because we are no longer using the | metacharacter.

5.5.2 Placeholders

Let's return to our example of changing 20xx to xx. We want to remove 20 but only in one context: when the 20 appears in the year. The date has the form ##/##/#### but let's assume that months January through September (1–9) may but do not have to have a leading zero and that 1-digit dates (like January 5) may but do not have to have a leading zero. This gives us dates in four forms: #/#/####, #/##/####, ##/#/#### and ##/##/####. We will use the ? to indicate that the leading digit for the month and date are optional. Our new regex is [01]?[0-9]/[0-3]?[0-9]/20[0-9][0-9]. Notice that if the year were prior to 20xx, the regex does not match and so sed will not alter such a date. One issue with our regex, as we noted earlier, is that the slash is a character used by sed so we have to escape its meaning using the backslash. Our regex becomes [01]?[0-9]\/[0-3]?[0-9]\/20[0-9][0-9].

We have tackled one issue, but the other issue still remains. Our previous replacement string was the empty string to simply remove 20. What we need is to locate /20xx among the entire regex and remove just the 20 while leaving the remainder of the date. How?

One way to tackle this problem is to enumerate multiple scripts, each of which searches for a specific year. The replacement string can be the year without the 20. For instance, if we only expected three years, 2022, 2023 and 2024, then we provide three scripts. Each script locates the specified year and replaces it with the last two digits. For instance, one script would be 's/2022/22/g'. This is not a practical solution though because we might need 100 scripts (one each for the years in the range of 2000–2099). It also can mistakenly change phone numbers and addresses.

Instead, what we need to do is indicate that a portion of our regex should be recalled as part of the replacement string. We indicate that portions of the pattern are to be remembered through *placeholders*. We divide our regex into regions, indicating each region as a placeholder. Our replacement string can then reference some of these placeholders so that we output those regions that we want to retain.

Placeholders are indicated as part of our regex itself. We embed the regex inside \(...\) notation. We recall a substring stored in the placeholder using the notation *n* where *n* is a number indicating which placeholder we want to recall. *n* can range from 1 to 9 giving us up to nine placeholders in any script. More concretely, imagine that we want to search for strings of two patterns and output them so that the two matching substrings are in reverse order. Our regex would look like \(*first pattern*\)\(*second pattern*\) and our output string would be \2\1.

Returning to our example, our replacement string should include the entire date aside from the 20 that starts the year. We remember the date as three parts: the portion before the year (\1), the 20 portion of the year (\2) and the last two digits of the year (\3). Our replacement string will consist of the first and third placeholders and so is expressed as \1\3. Our command is shown in Figure 5.13. It is hard to read thanks to a proliferation of forward and backward slashes so the figure also shows the command broken into component parts.

Let's piece apart the instruction in Figure 5.13. The search pattern is broken into three parts. The first part consists of the month which is either a 1- or 2-digit number and the date, which again can be a 1- or 2-digit number. Note that because the leading digit is optional but can include 0, the month and date may appear as 1, 01 and 10, for example. The forward slash between the month and date and the forward slash after the date must be escaped and thus appear as \/. This entire portion of the search string is placed into \(...\) to be referred to by the placeholder \1.

The second part of the pattern is literally 20, referred to by the placeholder \2. We will not be recalling \2 in the replacement string as this is the portion of the date that we want dropped. The third part of the pattern is the last two digits of the year after the 20. This portion is being retained via the placeholder \3. Our replacement string is \1\3 indicating the portion of the string that matched the first pattern followed immediately by the portion of the string that matched the third pattern. As the / that separates the date from year is part of the first placeholder, we do not need to specify any of the slashes in the replacement string.

```
Command: sed 's/\([01]?[0-9]\/[0-3]?[0-9]\/\)\(20\)
                \([0-9][0-9]\)/\1\3/' file

Breakdown of command:
            sed 's/\([01]?[0-9]\/[0-3]?[0-9]\/\)
                   └──────────────┬──────────────┘
      First placeholder (the \/ indicates a / in the regex)

                         \(20\)
                         └─┬─┘
            Second placeholder (to be discarded)

                     \([0-9][0-9]\)
                     └─────┬──────┘
                    Third placeholder

                     /\1\3/g' file
                     └────┬────┘
          Replacement pattern and end of instruction
```

FIGURE 5.13 "20" removal sed command explained.

All three parts of the pattern must match or else sed ignores this string and continues searching the line for another match. The script includes g to indicate that it should replace all 20xx years found. The command ends with the file's name. sed may have difficulty when there are placeholders and we have specified the option -r (or -E). If we need to use both extended regular expression metacharacters and placeholders, it is best to escape those metacharacters using \ rather than using -r. We see an example shortly.

Now let's consider a file containing names where some names contain first, middle and last names, some names are first and last names with a middle initial, and some names are just first and last names. We want to take every name, no matter its appearance, and rearrange it into the form lastname, firstname, eliminating the middle name/initial if there is one. To accomplish this, we need three scripts, one for each of the expected patterns of names. We make the assumption that every line of the file starts with the name so that we don't have to worry about multiple matches per line or lines that contain two- or three-word capitalized names as in San Diego.

The search pattern for all three names is [A-Za-z]+ [A-Za-z]+ [A-Za-z]+. The search pattern for a first and last name with middle initial is [A-Za-z]+ [A-Z]\. [A-Za-z]+. The search pattern for just a first and last name is [A-Za-z]+ [A-Za-z]+. Each of these patterns needs to include multiple \(...\) so that we can recall the proper portions using placeholders.

We insert a blank space between the last and first names in our replacement string. Because of this, the pattern for the first and last names should not have a blank space, so we will place blank spaces in portions of the regex dealing with the middle name/middle initial. As the third pattern does not have a middle name/initial, we still need to capture the blank space by itself, so it will have its own placeholder as in \(\). Thus, all three patterns have three placeholders, and our replacement strings will always be the content of the third, a comma and blank space, and the content of the first. Thus, all three scripts will have the same replacement string of \3, \1. We conjoin our three scripts using the -e option. The full sed command is given in Figure 5.14.

```
sed -e 's/\([A-Za-z]\+\)\( [A-Za-z]\+ \)\([A-Za-z]\+\)/\3, \1/'
    -e 's/\([A-Za-z]\+\)\( [A-Z]\. \)\([A-Za-z]\+\)/\3, \1/'
    -e 's/\([A-Za-z]\+\)\( \)\([A-Za-z]\+\)/\3, \1/' file
```

FIGURE 5.14 sed command to rearrange names.

Examining the command in Figure 5.14, we find that the + metacharacter is being escaped because it is a part of the extended regular expression set and so not available directly in `sed`. We do not use the `-r` option because when it is combined in a `sed` instruction with placeholders it leads to an error message like `sed: -e expression #1, char 54: invalid reference \3 on 's' command's RHS`. This error appears when we use placeholders like `\1` with the `-r` option. Fortunately, `sed` allows the escape character when referencing extended regular expression set metacharacters to avoid this problem.

`sed` has other forms of placeholders. The simplest is `&` which returns the matching string. We might use this if we do not want to rearrange or remove parts of the matching substring but instead want to enhance it.

For instance, imagine we have a file that contains dollars of the form `$###` where the number of digits may be between 1 and 5. We want to add `.00` to every dollar amount. Any time we have a match, we can recall the original string as `&` and we add to it `.00` as in `&.00`. We add `g` to make sure that every occurrence of the dollar amount is replaced. Our instruction is `sed -r 's/\$[0-9]{1,5}/&.00/g' file`. Here, we use the escape character to escape the meaning of `$` as it is a metacharacter, and we include the `-r` option because `{1,5}` uses the extended regular expression set.

There are several modifiers that can be applied to the `&` reference to alter how the matching substring will appear in the replacement string. There are five specifiers, `\U`, `\u`, `\L`, `\l` and `\E`. The first four are used to take the letters in the matched string and entirely uppercase them, capitalize the first letter, entirely lowercase them and lowercase the first letter respectively. The `\E` specifier is used to stop applying any previous specifier if `\U` or `\L` were used.

Let's look at an example. We have a file in which each line starts with a date and all of the dates are specified in the form `March 15, 2022`. The months are specified inconsistently. Some are capitalized, some are all lowercase letters, some appear in all uppercase letters and some are a mixture of uppercase and lowercase letters. We want all of the months to output as capitalized (that is, only the first letter is uppercase). We might use `sed -r 's/^[A-Za-z]+/\u&/'` *`filename`* to solve this problem where the matching month has only the first letter uppercased. The regex includes +, an extended regular expression set metacharacter, so we use the `-r` option, although we could also use \+. As the date starts each line, we use ^ to ensure that the regex matches at the beginning of the line. We do not use the global (`g`) modifier because we do not want any other words to match and be capitalized.

To round out this subsection, we take a brief look at several additional examples, as shown and described in Table 5.11. Assume the file used in each instruction, `names.txt`, is a file of lists of names and only names. Some names will have middle initials, others will not.

TABLE 5.11
Example `sed` Commands

`sed` Command	Explanation
`sed 's/[aeiou]/\u&/g' names.txt`	Uppercase all vowels (notice the `/g` to uppercase every occurrence of every vowel).
`sed 's/[A-Z][a-z]*/\U&/' names.txt`	Fully uppercase all names.
`sed 's/[A-Z]\.//g' names.txt`	Remove all middle initials (this will not remove middle names, just middle initials).
`sed 's/ /\t/g' names.txt`	Replace all spaces with tabs.
`sed 's/[A-Za-z. ]+/ &\n&/' names.txt`	Each line consists of letters, spaces and possibly a period; output the line followed by `\n` (new line) and the line again (`&\n&`); this `sed` command duplicates all lines.

5.5.3 Other sed Capabilities

In this subsection, we look at two other sed options and then other forms of sed scripts aside from the s (search/replace) script. The option -f is followed by the name of a file. This file is the script file. With this option, we do not place the script in the command but instead in the separate file. This is a convenient option if we are developing a complex script that we need to experiment with to get it right. By editing the script in an editor like vi, it is going to take less effort than editing the script from the command line.

When we run sed on multiple input files, the entire set of files is treated as one combined stream. The option -s runs sed individually on each file. We might use this option when specifying line numbers to operate upon.

Now let's turn to other types of scripts in place of the 's/search/replace/' scripts we've already seen. Other types of scripts use different letters in place of s. Some of these are covered in Table 5.12 (more complex ones are omitted lest this subsection be too lengthy; review any online sed manual for more detail).

While the s command searches for a *string* as a pattern to be replaced, the y command searches for a *character* to replace. The *source* and *destination* are lists of characters. For instance, if the first character of *source* is found, it is replaced by the first character of *destination*. Consider the string Hello World! and the command sed 'y/lo!/10L/'. The result is that each l (lower case L) is replaced by 1, each o by 0 and the ! by L. We get He110 W0r1dL as output. Notice that we do not use g to indicate a global replacement as sed will search all characters, one at a time, even when there is a match on a line.

The commands a and i can be used to insert text. They insert text on a separate line immediately after and before the specified line(s) respectively. The command sed '2,4a ****' file will cause a line with **** to be output after the second and fourth lines. To add text using both a and i, we add the -e option and specify two separate scripts, as in sed -e '2,4a ****' -e '1,3i @@@@' file. When using c, we either specify a single line or a range of lines. The text specified replaces either the single line or the entire range as a single line. For instance, sed '2,4c foo' file replaces lines 2–4 with the single line foo.

There are many other uses for sed. We do not have the space to examine sed in further detail. To explore more about sed, check out the sed online user's manual at https://www.gnu.org/software/sed/manual/sed.html.

SECTION ACTIVITIES

1. How useful do you find sed? While it is a convenient way to manipulate text, would you prefer to use sed or edit a textfile by hand? Another possibility is to use a word processor/text editor's search and replace feature. Explain your answer.
2. Research the Liddiard Stream Editor and compare its usage to sed.

TABLE 5.12
Other Types of sed Scripts by Command Letter

Command (by letter)	Meaning	Syntax
a	Append *text* to each line.	a *text*
c	Replace lines with *text*.	c *text*
d	Delete line *n* or lines *m* through *n*.	*n*d or *m*,*n*d
e	Execute *command* whose output becomes part of the stream.	e *command*
i	Insert *text* before the line (opposite of a).	i *text*
y	Transliterate matching characters (see discussion in the text).	y/*source*/*destination*/

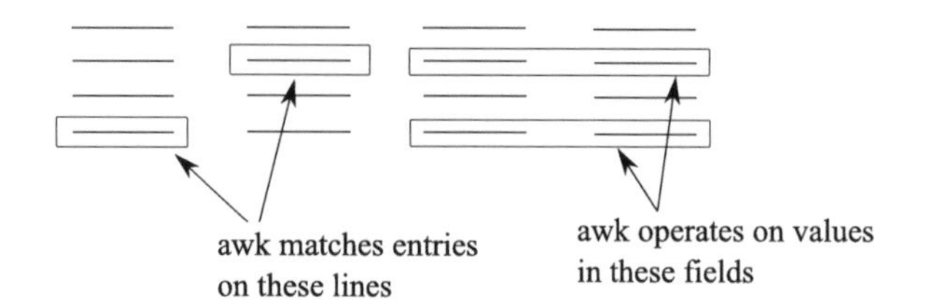

FIGURE 5.15 awk operates on fields within rows of a file.

5.6 awk

The awk program, like sed and egrep, will match a regular expression against a file of strings. Whereas egrep returns the matching lines and sed replaces matching strings, the awk program allows us to specify actions to perform on matching lines. Actions can include outputting information from the matching line, performing calculations on numeric values of the line, performing calculations on variables or calling built-in functions. Additionally, awk can locate strings or lines of interest with conditions aside from regular expression matching.

Overall, awk (named after its authors Aho, Kernighan and Weinberger) is a more powerful tool than either egrep or sed, giving us many of the same types of operations as a programming language. We are able to search files for literal strings, regular expressions and conditions such as if a particular value is less than or greater than another. Based on a matching line, we can specify instructions to operate on the field(s) that matched, on other fields of the same line, or perform other operations entirely. awk commands can reference variables to store values, perform calculations on those variables and output results of calculations.

Unlike egrep and sed, awk expects that the text file is not just a sequence of lines but that the lines are organized into fields (columns). The intention is that awk operates on tabular information rather than ordinary text files. While we could use awk on any textfile, say a text document or an email message, its real power comes from processing spreadsheet or database like data. Figure 5.15 illustrates this idea where we see two matching lines with awk operating on the values of other fields from those lines.

5.6.1 awk Condition-Action Pairs

The awk command typically involves one or more condition-action pairs somewhat reminiscent of an s/pattern/replacement/ script in sed. When executing, awk acts like a nested if-then-else statement found in most programming languages. awk works line by line testing each of the conditions to that line. The first condition that matches causes the corresponding action to execute. The remainder of the conditions are skipped and awk moves on to the next line, testing the conditions again in order until one matches. Should no condition match, no action takes place and awk moves on to the next line.

There are two different forms of conditions. One form is referred to as a pattern and consists of one or more literal strings or regular expressions, each embedded in /.../ notation. We will refer to these as *pattern-action pairs*. If a pattern matches, the action executes. Each pattern is tested on the current line, one at a time, until a pattern matches or there are no more patterns. Figure 5.16 illustrates the syntax of an awk instruction using pattern-action pairs. The BEGIN and END sections are optional, and we will address them in the next subsection. There is no limit to the number of patterns and there can be any number of files that awk works upon.

A pattern might be /2022/ or /20[0-9][0-9]/; the first will only match 2022 and the second will match any four-digit number starting with 20. Multiple patterns can be combined using && (and) and || (or), and a condition can be preceded by ! to express NOT. Each pattern is in its own /.../ notation. Table 5.13 provides several example patterns and explanations.

```
awk 'BEGIN {action0}
     /pattern1/ {action1}
     /pattern2/ {action2}
     /pattern3/ {action3}
     ...
     /patternn/ {actionn}
     END {actionn+1}' filename(s)
```

FIGURE 5.16 Syntax of an awk statement.

TABLE 5.13
Example awk Patterns Using /.../

Pattern	Explanation
`/^[A-Z]/`	String starts with an uppercase letter.
`!/[0-9]+/`	No digits are found in the string.
`/^[A-Z][a-z]*$/ \|\| /^[a-z]+$/`	String consists solely of letters where the first letter may be capitalized (string must contain at least one letter).
`/^[A-Za-z][a-z]*$/`	Same as above but simplified to omit the ||
`/^[0-9]/&&/[0-9]$/`	String begins and ends with a digit (has at least two digits) but can contain other characters.
`/[0-9]{5}-[0-9]{4}/`	String contains nine digits with a hyphen between the fifth and sixth (perhaps a 9-digit zip code).
`/[A-Z]\./`	String is a capitalized letter and period (perhaps an initial).

Fields of the files that awk operates on are separated by *delimiters* which by default are either spaces or tabs. The field can be referenced by number, denoted as `$n` where the first field starts at 1. The field `$0` constitutes the entire line. The other form of awk condition references a field and compares it to another field, a literal value or a value stored in a variable. This type of condition is not placed in /.../ notation but instead specified like a relational expression in a programming language such as with `$1 != $2`, `$3 > 0` or `$1 == x` (x is a variable). This form of condition can also contain multiple parts, connected using `&&` and `||`. We use this form to specify *condition-action pairs*.

Consider, for instance, that a file consists of payroll information where the first two columns are the employee's first and last names, and the next two columns are the hours worked and hourly wages. To determine if an employee has worked overtime, we use the condition `$3 > 40`.

Actions are placed in curly braces ({}), and multiple actions in braces are separated by semicolons. The primary forms of actions are assignment statements and output statements. Assignment statements follow C syntax and can include arithmetic operations, string concatenation, prefix/postfix increment/decrement operators and function calls.

Output statements use the `print` command and are followed by a list of items. The items, when output, are concatenated together as one string. The items can include literal values placed in double quote marks, values stored in variables and reference to a field using `$n`. For instance, in our payroll file, we might output for each employee a message like `print $1,$2,"'s earnings are $pay"` where `$1` is the first name and `$pay` is a variable used to store the employee's pay. We explore the role of the commas shortly.

Let's build an awk command to compute and output each employee's pay. We need to differentiate whether an employee worked overtime or not. Normal pay means that the hours worked were less than or equal to 40, so we use the condition `$3 <= 40`. Normal pay is computed as hours * wages, or `$3 * $4` in this case. The full condition-action pair might read `$3 <= 40 { pay = $3 * $4 }`. This means that if hours worked are less than or equal to 40 hours, compute pay as hours * wages. Note that the spaces are not needed and this condition-action pair could be written as `$3<=40 {pay=$3*$4}`.

We need a second condition-action pair for employee's who worked overtime hours. This can be written as `$3 > 40 { pay = 40 * $4 + ($3 - 40) * $4 * 1.5}` (again, the spaces can be omitted). This instruction first tests to see if hours worked is more than 40 hours and if so, stores in the variable pay the portion of the employee's salary that is the normal pay (40 * $4) and adds to it the overtime pay ($3 – 40 is the overtime hours, multiply this by wages * 1.5).

We would probably want to output the pay so we would follow each assignment statement with `print pay` at a minimum. We might prefer to have a more elaborate output like `print $1,$2 " pay is $" pay`. Notice that the comma between `$1` and `$2` adds a space whereas the space before the variable pay causes the print statement to *not* output a space. By having a second instruction in our action portion of both condition-action pairs (the `print` statement which follows the assignment statement), we need to separate the two instructions with a semicolon.

Our full awk command will include both condition-action pairs. This might be written using the awk statement found in the top portion of Figure 5.17. Unfortunately, the output that this instruction produces is not particularly informative as it just outputs the value stored in pay. We should revise the output statement so that we get the employee's first and last names and even a dollar sign.

When there are multiple items to output in a `print` statement, separating them by blank spaces causes them to output with no space between them ironically. Instead we separate items with commas. However, to ensure no space between the dollar sign and the amount, we omit the comma. The space immediately before the variable pay is optional. We see this revised version of our awk command at the bottom of Figure 5.17, where we have also output the employees' full names along with the text "pay is $".

Let's consider another example. We have sales information for a collection of salespeople stored in the file `sales.txt`. Each row of the file contains one employee's sales information for a month, consisting specifically of the month, the employee's last name, the amount in sales earned for the month, the employee's commission rate and the states that the employee covered in that month. Employees might appear on multiple lines, one per month. Figure 5.18 demonstrates what the file might look like. Notice the file contains a row of headers that we do not want to process.

We want to compute each employee's commission for the month. This will simply be the sales * commission rate. We can specify an awk statement with no pattern or condition and just an action as `awk '{print "$" $3*$4}' sales.txt` but this will attempt to compute the commission for the header (first row) and we want to avoid that. All of the other lines of the file contain at least one digit, so we can modify the instruction to use the pattern `/[0-9]/` giving us the instruction `awk '/[0-9]/ {print "$" $3*$4}' sales.txt`.

```
awk '$3>40 {pay=$3*$4; print pay}
     $3<=40 {pay=40*$4+($3-40)*$4*1.5; print pay}'
     file
-----------------------------------------------------------
awk '$3>40 {pay=$3*$4; print $1,$2," pay is $"}
     $3<=40 {pay=40*$4+($3-40)*$4*1.5;
          print $1,$2," pay is $" pay}' file
```

FIGURE 5.17 Sample awk instructions for payroll computation.

```
Month      Salesman   Sales  Commission rate     Region
Jan        Zappa      3856   .15                 CA, OR, AZ
Feb        Zappa      6158   .20                 CA, OR, WA
Jan        Mars       2994   .12                 IN, KY, OH
Jan        Duke       5752   .15                 OH, WV
Feb        Duke       4215   .12                 MD, WV, NY, NJ
Mar        Duke       5822   .15                 OH, PA
```

FIGURE 5.18 `sales.txt` file for awk examples.

Notice that we are not storing the computed value in a variable. There is no need to do so since we only want to print it out. Also notice that, like our first instruction in Figure 5.17, we are just outputting the commission amount, not the employee's name or month. We should enhance our `print` statement to also output `$2` and `$1` (probably in that order as we would prefer to see the employee's name prior to the month). We might use `{print $2, $1, "$" $3*$4}`.

Now let's consider that we want to output the employee name, month and commission amount for all employees who earned a 15% commission rate. We write 15% as `.15`. This seems to imply that the pattern should be `/.15/`. Unfortunately, the period is treated as the metacharacter meaning "match anything". This causes `awk` to process not just the three rows where commission rate is 15% but also Zappa's February and Duke's February entries as well.

Why? Both have a string of the form `x15` where `x` is any character (a 6 for Zappa, a 2 for Duke). We need to escape the period and so redo our pattern as `/\.15/`. Alternatively, we could test for the value `.15` using the condition `$4==.15`. With this latter condition, we do not worry about the decimal point being a metacharacter.

We noted earlier that `awk` has built-in functions. One is `length`. This returns the length of the item specified, which can include a field. As a trivial example, we could output the length of each employee's name using the instruction `awk '{print $2, length($2)}' sales.txt`. Unfortunately, as we do not have a pattern or condition, this action will also execute on the file's header and so print `Salesman 8`. We should instead use a pattern to prevent the header line from being included. We change our command to `awk '{/[0-9]/ print $2, length($2)}' sales.txt`.

A built-in variable called `NF` is the number of fields of a line. Re-examine Figure 5.18. Although it looks like there are just five fields per line, `awk` considers a field as being separated by a tab or space. Thus, the number of fields varies based on the number of states listed. Let's write an `awk` statement to output for each line the employee's name, the month and the number of states covered. The number of states can be computed as `NF - 4` because NF is the total number of fields per line and four of the fields consist of the month, name, sales amount and commission rate. Our command is `awk '/[0-9]/ {print $2,$1,NF-4}' sales.txt`.

Earlier, we computed commission amounts for every employee but did so month-by-month. Imagine that we want to compute this amount for one employee across all months. We will do so for Zappa. You might think to use the command `awk '/Zappa/ {print $1, "$" $3*$4}' sales.txt`. While this gives us Zappa's amount for each row where his name is listed, it does not give us a total.

Having computed the commission for a month ($3*$4), we need to add this to some *running total*, which we will store in a variable called `total`. Our action must become `{total = total + $3*$4}`. We can then output the total using `print $1 " earned $" total`. Our full command is `awk '/Zappa/ {total=total+$3*$4; print $2 " earned $" total}' sales.txt`.

Unfortunately, as Zappa appears twice in the file, this `awk` command outputs two lines, once with the total as computed for just `Jan` and once with the total for both `Jan` and `Feb`. Had Zappa appeared 12 times, we would have had 12 outputs, each with a larger total. We only want one total. Should we use two pattern-action pairs, one to compute total and the other to output the result? Consider the variation of the previous instruction shown in the top part of Figure 5.19.

```
awk '/Zappa/ {total=total+$3*$4}
     /Zappa/&&/Dec/  {print  $2  "  earned  $"  total}'
     sales.txt
------------------------------------------------------------
awk '/Zappa/&&/Dec/ { total=total+$3*$4;
               print $2 " earned $" total}
     /Zappa/ {total=total+$3*$4}' sales.txt
```

FIGURE 5.19 Solving the Zappa total pay problem.

There are two problems with our solution in the top portion of Figure 5.19. First, we are assuming Zappa will have a December entry. Second, recall that awk continues looking at pattern-action pairs until one matches. Had Zappa worked in December, awk would never have reached the second pattern because the first pattern will match Zappa's December entry in the file.

In the bottom half of Figure 5.19, we resolve the second issue by reversing the two pattern-action pairs and combine assignment statement and output as part of this action. Unfortunately, it does not solve the first problem. If Zappa didn't work in December, we get no output. We need another mechanism to solve this problem, which leads us to BEGIN and END sections.

5.6.2 BEGIN AND END SECTIONS

The BEGIN and END sections are optional. If provided, a BEGIN section always executes before any lines of the file are examined. If provided, an END section always executes after the file's lines have been examined. Thus, we can use these sections as start-up and wrap-up code to be performed before and after the condition-action/pattern-action pairs are tested, even if no lines of the file match our patterns/conditions.

We primarily use a BEGIN section to initialize any variables if necessary (variables are automatically initialized to 0, so we would only need this if we wanted to initialize a variable to a non-zero value) and perhaps output a header to describe what the instruction is doing. An END section will often be used to finalize any computation and output results. Actions in these sections appear in {} just as with any pattern-action or condition-action pair.

Let's return to the example from the end of the last subsection. We want to compute Zappa's commission for each month he appears and output just the final total. We simplify the awk statement in one way. We only have one pattern-action pair because we no longer need a special case for December. We add an END statement to take care of the output. The first instruction shown in Figure 5.20 will accomplish what we need.

Let's enhance this awk command to not only print out Zappa's total pay but his average pay. The average is computed by the total/count where count is the number of months that Zappa worked. We have to make three enhancements. The first is that the pattern-action's action must be modified to add one to count. We noted earlier that actions can include incrementing operations. awk, being based on C syntax, can add 1 to a variable using the ++ operator. Our choices are count++ and ++count (they will do the same thing in this case), or we can write the instruction as count=count+1.

The second modification is that we have to compute the average. We only want to compute the average once, at the end of processing, so we add the statement average=total/count to the END statement. Finally, we have to modify our print statement to output both total and average. We can use two separate print statements or a single print statement with a \n (new line). We use the latter to demonstrate the new line character. This revised version of our command is shown in the lower half of Figure 5.20.

The print statement should appear on one line but is placed on two lines due to space restriction. Notice that we never initialized count. If we needed to, we could use a BEGIN statement for that. But awk initializes count to 0, which is exactly what we want.

```
awk '/Zappa/ {total=total+$3*$4}
     END {print "Pay for Zappa is $" total}' sales.txt
----------------------------------------------------------
awk '/Zappa/ {total=total+$3*$4; count++}
     END {average=total/count;
          print "Pay for Zappa is $" total
          "\nAverage pay for Zappa is $" average}'
     sales.txt
```

FIGURE 5.20 awk commands to compute Zappa's total pay and average.

```
awk '/Zappa/ {total1=total1+$3*$4; count1++}
     /Duke/ {total2=total2+$3*$4; count2++}
     END {average1=total1/count1;
          average2=total2/count2;
          print "Pay for Zappa is $" total1
          "\nAverage pay for Zappa is $" average1
          "\nPay for Duke is $" total2
          "\nAverage pay for Duke is $" average2}'
     sales.txt
-----------------------------------------------------
awk '/Zappa/||/Duke/ {total=total+$3*$4; count++}
     END {average=total/count;
          print "Pay for Zappa & Duke is $" total
          "\nAverage  pay  for  Zappa  &  Duke  is  $"
          average}' sales.txt
-----------------------------------------------------
awk 'BEGIN {print "Employees/months who worked OH:"}
     /OH/ {print $2,$1; count++}
     END {print "Total who worked OH was " count}'
     sales.txt
```

FIGURE 5.21 Three additional awk commands.

There is one problem with this revised awk statement. What if Zappa appeared in no lines of the file? The pattern-action pair would never execute but the END statement would. The END statement would try to compute average but with count being 0, we get a division-by-zero error. We will solve this issue in the next subsection.

Let's consider a variation of the previous problem. Instead of outputting Zappa's total and average, we want to output the total and average of two employees, Zappa and Duke. To accomplish this, we need two pattern-action pairs, one for Zappa and one for Duke. We also need extra variables. We rename total, count and average to total1, count1 and average1 and use these for Zappa, and add total2, count2 and average2 for Duke. The END statement will compute both employees' averages and have two print statements, one for each. This solution is shown in the top portion of Figure 5.21.

Another variation of this problem is to compute and output the total and average of both Zappa and Duke combined. This is a simpler problem. We return to our earlier solution as shown in the bottom half of Figure 5.20. We only need one variable for total, one for count and one for average. We make two modifications to the instruction from Figure 5.20. First, our pattern becomes /Zappa/||/Duke/ so that we compute the pay and add to total and count for either Zappa or Duke. Second, our output lists both Zappa and Duke by name. This solution is shown in the middle portion of Figure 5.21.

Now let's output all employees who worked in the state of OH. We will output the names and months along with a total number of employees/months who worked in OH. Our awk instruction will include both a BEGIN and an END section. The BEGIN will be used to output a header describing the output. Our pattern is simply /OH/. The corresponding action will output this employee's name and the month, and also add one to a count variable. The END statement will be used to output count. The command is given in the bottom portion of Figure 5.21.

5.6.3 Other Forms of Control

The awk instruction provides a number of other operations making it like a full-fledged programming language. These operations include input statements, selection statements and loops. We postpone our look at input for a couple of subsections.

The selection statements are if and if-else statements which are similar in syntax to those found in C. The if statement is written as `if(condition) statement;` and the if-else statement is written as `if(condition) statement; else statement;`. An if/if-else statement must appear as an action inside {}. If the if-clause or else-clause consists of more than one statement, the statements must be enclosed in {}.

The role of the if statement is the same as the condition-action pair. The condition is tested line by line and if it is true for a given line, the action executes on the data of that line. If the statement is an if-else and the condition is false for a given line, then the else action executes. With the if-else structure, we do not need any condition-action pairs. Figure 5.22 illustrates a revised version of the second awk command from Figure 5.17 to compute payroll. We flesh out the command to include both BEGIN and END sections and an average pay, using the file payroll.dat. Note that we do not need to initialize count in the BEGIN clause.

We can include an if statement (or if-else) in our END statement. With this, we can prevent a division-by-zero error in our previous example that computed Zappa's average pay (as shown in the bottom half of Figure 5.20). This solution is given in Figure 5.23. Notice that the if-clause is placed in {} because there are two operations to perform in the clause, but the else-clause, which only has one statement, does not require them. The close } on the second to last line closes the END statement, not the else-clause.

awk also can utilize loops. There are three types of loops available: while, do-while and for. All three have syntax like that of C. The while loop might look like `awk '{i=1; while(i<=NF) { print $i; i++ }}' file`.

This awk instruction has no condition-action pairs. Instead, it operates on every line of the file. First, i is set to 1. The while loop executes while i is less than or equal to the number of fields on the line. It prints out $i (the field whose number is equal to i) and increments i. In effect, this command outputs the file not as a table but with each field of each line on a separate line.

The do-while loop is similar but the syntax is `do{…}while(condition);`. The difference between the while and do-while loops is that the while loop's body will not execute if the condition is initially false but the do-while loop's body always executes at least one time.

```
awk 'BEGIN {total_pay=0.0;count=0}
     {if ($3>40) {current_pay = ($3-40)*$4*1.5+40*$4;
                  total_pay+=current_pay; count++;
                  print $1,$2 "\t $" current_pay}
           else {current_pay = $3*$4;
                 total_pay += current_pay;
                 count++;  print $1,$2 "\t $" current_pay}
          }
      END {print "Average pay is $" total_pay/count}'
         payroll.dat
```

FIGURE 5.22 awk instruction using if-else instead of condition-action pairs.

```
awk '/Zappa/ {total=total+$3*$4;count++}
     END {if(count>0) {
          average=total/count;
          print "Pay for Zappa is $" total
             "\nAverage pay for Zappa is $" average}
          else print "Zappa did not work this year"}'
     sales.txt
```

FIGURE 5.23 Using an if-else in the END statement.

Referring back to the `sales.txt` file from Figure 5.18, we might want to output each employee's name and the states they worked in. The name is `$2` but the states' fields vary, starting with `$5` and going through the remainder of the line. We can employ the for-loop to iterate through the states and output them. Let's further assume that we only want to output those employees who worked in 3 or more states for the month. In this case, `NF` needs to be at least 7. A command to accomplish this is `awk '{NF>=7 {print $2; for(i=5;i<=NF;i++) print $i}}' sales.txt`.

Loops tend to be of less value in awk than the other types of operations because awk already has an implicit loop, iterating once per line of file. Thus, a loop when used in awk is in reality a nested loop. The above example, printing each state, has awk searching each line (an outer loop) and then outputting the states within the lines that match (the inner loop).

awk also features a `case` statement (like a C or Java switch statement), `break` and `continue` statements, which we omit. Another instruction available in awk is `next`. This instruction ends any processing of the current line of the file and moves on to the next line. This can be used, for instance, if while processing the current line a condition is met that indicates the line should not be considered any further.

Returning to our number of states example, imagine that we want to print all states whose name comes before NY. We can modify the for-loop's body to be `if($i<"NY") print $i; else next`. The logic here assumes that the states are listed in alphabetical order.

Related to `next` is `nextfile`. awk can operate on any number of files. The `nextfile` command causes awk to discontinue processing the current file and resume processing with the first line of the next file. There is also an `exit` statement which causes awk to immediately move to the END statement and then terminate.

5.6.4 awk Command Line Options and Arguments

As with most Linux commands, awk contains several useful options that can be employed from the command line. Perhaps the most useful is `-f` which is followed by a filename. As awk commands can quickly become large and complicated, it is worthwhile to enter the command in an editor (e.g., vi) and save this to a textfile. The `-f filename` option uses that file for the command. The format becomes `awk -f filename file(s)` where *filename* is the file containing the portion of the awk command that had previously been in single quote marks and *file(s)* is(are) the file(s) to execute upon. Related to `-f` is `-e` where we can combine part of the instruction from the command line with a portion of the command that is in a file. Another option is `-E` in which any other options on the command line are ignored.

As we've seen, awk can use variables that are assigned the first time they are referenced. We can also initialize variables using a `BEGIN` statement. Yet another option is to initialize variables from the command line. In this way, any variables we might reference in the command are already pre-initialized. We do this through the option `-v` and follow it with the initialization in the form `var=value`. Each variable must be defined and initialized in its own `-v` clause, as in `-v x=1 -v y=2 -v z=3`.

The delimiter used by awk defaults to a space or tab. We can override this using `-F delimiter`. The *delimiter* can be a string or a single character and can but does not need to be placed inside of single or double quote marks. The delimiter is stored in a special variable called `FS`. There are several other options that we will not cover. Consult awk's man page for details.

Aside from options, we can also specify that a variable should be reset with each file being processed. To assign a variable a value, we saw above that we can use the `-v` option. But with this option, any variable(s) assigned is only assigned before awk commences. With each successive file applied, the variable(s) continue from the value(s) at the end of the previous file. What `-v` does not do is *reinitialize* a variable with each new file. If we want to reinitialize the variables, we can do use by simply listing the assignment of the variable between the files.

Let's explore this in detail. First, we start with the following awk command, which simply outputs each line of a file but with each line preceded by a line number. awk -v count=0 '{count++; print count, $0}' file. If we had multiple files instead of one, count continues to increase file-by-file without starting over. We can change this performance by reinitializing count between each file as with awk -v count=0 '{count++; print count, $0}' file1 count=0 file2 count=0 file3. Notice that we do not need the -v option when resetting count.

5.6.5 Non-File Input to awk

We wrap up this brief but complicated look at awk by discussing two other forms of input aside from files. All of our examples so far have expected at least one file from which awk will process. If no filenames are provided, then awk provides us with a prompt to enter input. The format of this prompt might vary and often is just an empty line. We type in a line of input as if it were a line of a file. Upon typing control+d, awk processes the line.

As a simple example, we enter awk '$1 > 0 {print $2 * $3}'. With no file at the end of the command, awk puts us into a buffer. In looking at the command, we are expected to enter at least three numbers. If the first number is greater than 0, then awk will compute and output the product of the second and third numbers. Numbers should be separated by the default delimiter (space or tab). If, for instance, we type 1 2 3 <control+d>, we receive output of 6. If we type 0 1 2 <control+d>, we receive no output. If we type 5 10 15 20 25 30 <control+d>, we receive 150. The excess fields are not used.

Another alternative is to pipe the output of another instruction to awk. Likely candidates are ls -l and ps aux. Let's write a Linux command that takes the long listing of a directory and outputs those files whose owner is not the same as the group. The condition we want to test is that the third field (owner) does not equal the fourth field (group), or $3 != $4. The action is {print $0} (the full line). Our awk command is ls -l | awk '$3 != $4 {print $0}'. We could also use {print $9} to just print the filename.

Another task we can execute quite easily with awk is to sum up the size of files in a directory. We could do this with another Linux command, du, so let's make this a little more interesting. We want to sum up all of the regular files which are writable by either group or world. We use a regular expression to search for files whose long listings respond with permissions that are either -....w.... or -.......w. where the periods will match any character. We want to obtain just the file size of matching files, which is the fifth field ($5) of ls -l. Our awk instruction is awk '/^-....w..../||/^-.......w./ {total=total+$5} END {print total}'. We could reduce this slightly to awk '/^-....w/||/^-.......w/ {total=total+$5} END {print total}' as we are only interested in finding the w in either position.

Let's do one final example, this time using the output of ps aux. Here, we want to determine the smallest and largest PIDs of all processes whose CPU run time is greater than 0:00. With ps aux, the PID is the second field and the CPU time is the tenth field. Our condition is $10!="0:00". We are treating this as a string, thus the quote marks. In order to find the smallest and largest PIDs, we keep track of a minimum and a maximum. We need to retain both the minimum and maximum PIDs and the process names of those PIDs. We will use four variables, min, name1, max and name2, to store these four values, respectively.

We will initialize min to a large value and max to 0 using a BEGIN statement. With each process whose CPU time is not 0:00, we compare its PID ($2) to min and max to see if this new PID should replace the current min or max. We not only replace the min or max value but also remember the process' name in name1 or name2. We use an END statement to output the result. We must make one further enhancement to this instruction. ps aux has a header line that we do not want to process. Another useful variable is NR, which indicates the number of records so far processed. If NR is 1, then this is the first record and we want to ignore it. The full instruction is provided in Figure 5.24.

```
ps aux | awk 'BEGIN {min=1000000;max=0}
   NR>1 && $10!="0:00" {if($2<min) {min=$2;name1=$11}
                        if($2>max) {max=$2;name2=$11}}
           END {print "minimum PID is ", min,
             " for process ", name1,"\nmaximum PID is ",
             max, " for process ", name2}'
```

FIGURE 5.24 Combining `ps aux` and `awk`.

SECTION ACTIVITIES

1. Between `sed` and `awk`, which do you find more challenging? Which do you think could be more useful?
2. Which would be a more convenient way to process data in a file, writing a program or writing an `awk` instruction? Why do you feel that way?
3. As we discussed in the last subsection, we will often use `awk` as part of a larger instruction when issuing Linux instructions. Aside from `ls -l` and `ps aux`, list three other instructions you have learned to this point of the textbook where you might want to "peel off" just one part of the response by piping the result of the instruction to `awk`. We will find several examples as we move forward in the textbook.

5.7 CHAPTER REVIEW

Concepts and terms introduced in this chapter

- Basic regular expression set – a set of metacharacters originally defined by POSIX.
- Character class – an abbreviated way to specify a list of all characters within a classification such as alphabetic, `[:alpha:]`, uppercase letter, `[:upper:]`, digit, `[:digit:]` and punctuation mark, `[:punct:]`.
- Character range – an abbreviated way to specify a list of characters that are expressed as the smallest and largest separated by a hyphen; ranges are permissible for letters and digits but not other types of characters.
- Empty string – a string that contains no characters.
- Escape character – used to indicate that the given character should be treated literally and not as a metacharacter.
- Enumerated list – specifying a list of characters to match from; the lists are placed in `[]`.
- Extended regular expression set – metacharacters added to the basic regular expression set to extend the capabilities of Unix/Linux regular expressions.
- Field – used in `awk` to indicate the column that a value is found in; fields can be used in conditions as in `$4 > 0` or in actions like `pay=$3*$4`.
- Literal character – a character in a regular expression that should be matched as is.
- Metacharacter – special character that is not interpreted literally but used to express how other character(s) are to be applied when matching against a string.
- Placeholder – used in `sed` to denote a part of the matched string that can be recalled through one of several special indictors such as `&` or `\1`.
- POSIX – Portable Operating System Interface for Unix, a set of standards used to implement Unix and Linux operating systems; among the various standards are the metacharacters defined in the basic and extended regular expression sets.
- Regex – abbreviation for regular expression.

- Regular expression – a string comprising literal characters and metacharacters; used in pattern matching.
- Spam filter – common program added to an email server or client used to identify if an email is spam (unwanted) or legitimate; many spam filters use regular expressions.
- Stream – input or output characterized as a sequence of characters that can be intercepted and operated upon.
- Stream editor – a program that searches for and replaces strings in a stream (usually a file).
- String – any set of characters (including the empty string).
- Substring – any subset of consecutive characters of a string (including the empty string).

Linux commands and files covered in this chapter:

- awk – program that searches a file line by line for matching patterns or conditions and applies operations to the items in that line; operations include output and assignment statements containing arithmetic operations.
- grep/egrep/fgrep – program that searches for matches of a regular expression to every line in one or multiple files and report on the matches found.
- sed – a stream editor to search each line of a file for one or more matching strings of a regular expression and replace all matched items with a replacement string.
- /usr/share/dict/words – the Linux dictionary.

REVIEW QUESTIONS

1. What is the difference between `[0-9]+` and `[0-9]*`?
2. What is the difference between `[0-9]?` and `[0-9]+`?
3. What is the difference between `[0-9]` and `[^0-9]`?
4. How would `.*` be interpreted? How would `.+` be interpreted?
5. Is there any difference between `[0-9]` and `[[:digit:]]`?
6. How does `[[:digit:]]` differ from `[[:xdigit:]]`?
7. Rewrite the regex `[A-Za-z0-9]` using a character class.
8. Write a regular expression to match a word that has two vowels in a row. Write a regular expression to match a word that has two of the same vowel in a row.
9. We want to match against any sequence of exactly five digits. Why does `[0-9]{5}` not work correctly?
10. Is there any difference between `[0-9]{1,}` and `[0-9]+`?
11. Which of these metacharacters let's you precisely control the number of times an item should match? `*` `+` `?` `{m,n}` `{m,}` `{m}`
12. How does `.?.?.?.?.?` differ from `.{1,5}`?
13. Of the following regular expressions, which have a finite number of strings it could match against?
    a. `[A-Z][a-z][a-z]*`
    b. `[A-Z][a-z][a-z]+`
    c. `[A-Z][a-z][a-z]?`
    d. `[A-Z][a-z]{2,5}`
14. The following regular expression should match a person's name in the form first name middle initial last name. What is wrong with it? Assume names start with an uppercase letter and have no other uppercase letters. `[A-Z][a-z]+ [A-Z]. [A-Z][a-z]+`
15. Write five strings that will match the regex `^[0-9].+[^0-9]$`.
16. Write three different regular expressions that will match a string that consists solely of letters (upper and/or lowercase) and spaces as long as the string does not start or end with a space. The string must contain at least two characters which are letters.

17. A last name might have multiple capital letters like McCartney or MacArthur. Write a regular expression that will match a last name that will start with a capital letter and might have another capital letter in the third or fourth position, but no others, and no more than two capital letters in total. Assume the full name will be at least four letters long.
18. Interpret `[0-999]`. If we truly wanted to match any number from 0 to 999, how would we express it correctly?
19. Imagine that we want to match the fractional value of `.10` (that is, 10%). What is wrong with using this expression , `.10`?
20. We want to match against any arithmetic expression of the form `X op Y = Z` where `X`, `Y`, and `Z` are any numbers and `op` is any of +, -, *, or /. Write the proper regular expression.
21. To find four words in a sentence, we might use `([[:alpha:]]+ ){4}`. Why is this incorrect? How would you fix it?
22. Write a regular expression that will match a sentence that has at least two words where words are letters, the only uppercase letter in the sentence starts the first word, words can have hyphens and apostrophes in them but not before the first letter, words can end with a comma, semicolon or colon, all words are separated by one blank space and the last word ends with a period, exclamation mark or question mark.
23. Write a regex for each of the following patterns.
    a. Find a string that contains a 2-digit number.
    b. Find a string that contains a 2-digit number but has other characters before and after it.
    c. Find a string that contains a 2-digit number but has non-digit characters before and after it.
    d. Find a string that consists solely of a 2-digit number.
24. Write a regex for each of the following patterns.
    a. Find a string that contains a punctuation mark but not as its first or last character.
    b. Find a string that contains only punctuation marks.
    c. Find a string that contains no punctuation marks.
    d. Find a string that contains two punctuation marks but not consecutively.
25. Write a regex to match a string that consists solely of a person's name of the form Title First Middle Last, Extension. The title is one of Dr., Miss, Mrs., Ms., Mr., but may be omitted. The middle name, if given, is a full name, but may be omitted. The extension will be one of Sr., Jr., or I, II, III or IV and if there is no extension, then the last name does not have a comma after it.
26. We want a regular expression that will match any string that does not start with a capital letter. We use `[^A-Z]`. Why is this wrong and how would you fix it?
27. We want a regular expression that will match any string that contains only digits. We use `^[0-9]*$`. What is wrong with this and how would you fix it?
28. What is wrong with the following regular expression? `[A-Z][+-*/][0-9]` How would you fix it?

For questions 29–32, imagine that we have a file that lists student information, one row per student. Among the information for each student is every state that the student has lived in. For instance, we might have one entry with OH and another with OH, MO, NY (states are arranged by the order the student lived in them, not alphabetically). Further assume that the only use of commas in the entire line will be to separate the states.

29. Why would `[OK][HY]` be an incorrect way to specify people who have lived only in OH or KY?
30. Write a regex to match students who have lived in either OH or KY correctly.
31. Write a regex to match students who have lived in neither OH nor KY.

32. We want to find all students who have lived in at least three states. Write such a regular expression. Hint: consider the use of the commas.

For questions 33–36, assume a file contains a list of information about people, row by row, where each row starts with a person's first name, a comma, and the person's last name followed by a colon. Names always start with an uppercase letter and are followed by lowercase letters. After the colon are other pieces of information about the person so do not assume that the only letters appear in the two names.

33. Write a regular expression to find anyone whose last name starts with either D, E, F or G.
34. Write a regular expression to find anyone whose first and last names are both exactly six letters long.
35. Write a regular expression to find anyone whose first and last names do not contain any `A/a`'s.
36. Is it possible to write a regular expression to find anyone who has the same letter to start their first and last names? Why or why not?
37. Explain why in `egrep` it is better to use `-v` than `[^...]` when trying to match strings that do not contain some character or set of characters.
38. Why should you place your regex in `' '` when using `egrep`? Under what circumstance(s) can you avoid using the single quote marks?
39. Write an `egrep` command to output the number of lines in all `*.txt` files that match the following patterns.
    a. Contains a string of only letters.
    b. Contains a string of only non-letters.
    c. Contains at least three distinct strings separated by a space each.
    d. Contains a string that starts with letters (at least two), ends with digits (at least two) and can contain anything in between.
    e. Contains any sequence of letters or words as long as there is an `a`, `b`, `c` and `d` (uppercase or lowercase) where they appear in that order (not necessarily consecutively). For instance, this should match `A bat can die` and `abacad` but not `Becaused`.
    f. Contains exactly four digits but can contain any other characters before, after or in between the digits.
    g. Contains both a period and a question mark in any order and characters (including additional periods and question marks) can appear between them.
40. Explain the role of each of these `egrep` options: `-f`, `-H`, `-h`, `-l`, `-L`
41. The `egrep` option `-x` allows you to avoid using which metacharacter(s)?
42. Provide a reason for using any of `-A`, `-B` or `-C` in `egrep`.

For questions 43–51, use the Linux dictionary found in `/usr/share/dict/words` and `egrep`; question 52 asks about the dictionary.

43. Write an `egrep` instruction to find words that have two consecutive punctuation marks.
44. Write an `egrep` instruction to find any words whose entries contain a digit with letters on either side of it.
45. Write an `egrep` instruction to find all words that begin with an `a/A` and end with a `z/Z`.
46. Write an `egrep` instruction to find all five letter words that end with a `c` (but not a `C`).
47. Write an `egrep` instruction to find all five letter words that start with an `a/A` and end with a `z/Z`.
48. Write an `egrep` instruction to find any entries that contain a `q/Q` which is not followed by a `u/U`, as in `Iraqi`.

49. Write an egrep instruction to find all entries that contain two x's somewhere in the word (uppercase or lowercase).
50. Write an egrep instruction to find all entries that contain two o's as long as they are not consecutive and not at the beginning or ending of the word.
51. Write an egrep instruction to find all entries that are no more than three letters long (note: these words may only contain letters, not digits or punctuation marks).
52. Is it possible to use a regex to find the longest word in the dictionary? Explain.
53. What is wrong with the following egrep instruction? `egrep abc* foo.txt`
54. Assume we have the statement `egrep 'someregex' *.txt`. Should we use the option `-H`? Explain.
55. When using the `-c` option, will egrep output a 0 for a file that contains no matches?
56. Explain the difference between `egrep [^abc] somefile` and `egrep -v [abc] somefile`.
57. What regular expression metacharacters are not available in grep but are available in egrep?
58. For the following, write a Linux instruction which pipes `ls -l` to egrep.
    a. Find all files in /dev which are block or character devices (these start with b or c as the first character of the permissions).
    b. Find all files in /etc which have at least two hard links (this is the number which follows the permissions). Note that '2' is not an appropriate regex for this task.
    c. Find all files in the current directory whose sizes are not 0.

Questions 59–68 pertain to sed.

59. What is erroneous about the following sed command?
    `sed 's/aaa/bbb/' file1.txt > file1.txt`
60. Write a sed command to remove every line break (new line, \n) in the file somefile. These should be replaced by blank spaces. Output should be sent to the terminal window.
61. What is the difference between \U and \u when used in a sed command?
62. Explain what each of the following sed commands does.
    a. `sed 's/ //' somefile`
    b. `sed 's/ //g' somefile`
    c. `sed 's/[[:digit:]]\+[[:digit:]]/0/g' somefile`
    d. `sed 's/^[A-Z]+/\L&/' somefile`
    e. `sed 's/\([0-9]\+\)\([A-Za-z]\+\)/\2\1/' somefile`
63. Write a sed command to reverse any capitalized word to start with a lowercase letter and whose remaining characters are all upper case. For instance, Dog becomes dOG while cat remains cat (because cat is not capitalized). Use the file somefile. Output should be sent to the terminal window.
64. Write a sed command to replace every occurrence of 1 with one, 2 with two and 3 with three in the file somefile. Output should be sent to the terminal window.
65. Look at the sample sales.txt file in Figure 5.18. Write a sed command which will replace each percentage from the form .xx to xx% (for instance, .10 becomes 10%). Save the result to the file sales2.txt.
66. Write a sed command to add the word done after the first, fifth and eighth lines of the file somefile. Output should be sent to the terminal window.
67. Write a sed command to delete the second through sixth lines of the file somefile. Output should be sent to the terminal window.
68. Write a sed command to place the text of the file somefile into a code as follows. Every a becomes a z, every z becomes a q, every q becomes an x, every x becomes a p, every p becomes an a, every b becomes a 1, every c becomes a b and every 1 becomes a 0. Use transliteration to accomplish this. Output should be sent to the terminal window.

Questions 69–92 pertain to awk.

69. What is the difference between a pattern-action pair and a condition-action pair in awk? Provide an example awk statement to demonstrate the difference.
70. What is a delimiter? What are the default delimiters in awk? How can you change this to specify your own delimiter?
71. What does the following awk command do?

```
awk '/^[A-Z]/ {count1++}
/^[a-z]/ {count2++}
 /^[0-9]/ {count3++}
END {print count1, count2, count3}' somefile
```

72. Use the sales.txt file from Figure 5.18 and write an awk command for each of the following. The header line should not match in any of these questions.
    a. Output every employee's name and month and the number of states served for every line.
    b. Compute the total number of sales entries for the month of February and output this with a useful message. This will be number of employees who worked in the month.
    c. Repeat b but compute and output the total sales amount (the third field) for the month.
    d. Count the number of employees who worked both OH and KY and output the count.
    e. Output the employee's name and month for every employee who worked either OH or KY.
    f. Output the employee's name, month and salary for every employee whose sales totaled more than 5000.
    g. Output the employee's name and month for every employee who did not work in both NY and NJ for the month.
    h. Repeat g but for employees who did not work in either NY or NJ for the month.
73. True/false: Input to awk can only originate from a file.
74. One use of the BEGIN clause in an awk command is to initialize variables.
    a. Why do we not need to use this to initialize numeric values to 0?
    b. What other way(s) can you initialize variables when using awk?
75. We do not generally use loops in awk statements. Why not?

For questions 76–82, assume we have a file payroll.dat which contains employee wage information where each row contains

```
first_name    last_name    hours    wages    week
```

where week is a number from 1 to 52 indicating the week of the year. Employees can occur in the file multiple times each occurrence of an employee will be of a different week and employees may not occur for all weeks.

76. Write an awk command to output the first and last names of all employees who worked during week 5. NOTE: make sure the 5 is not part of the hours or wages or a week like 15 or 35.
77. Write an awk command to compute and output the total pay for employee Frank Zappa assuming overtime receives the same pay rate as normal hours (that is, do not compute overtime). Assume no other employee has the last name of Zappa.
78. Revise the awk command from question 77 so that the computation includes overtime at 1.5 times the wages specified.
79. Write an awk command to compute the average number of hours worked for each week that Frank Zappa appears. Assume in this case that there may be other employees with the last name of Zappa.

80. Write an awk command to compute the number of times any employee worked overtime.
81. Write an awk command to compute the average wage of all records in the file.
82. Write an awk command to output the employee who worked the most total hours during the year.

Questions 83–92 all pertain to piping an operation to awk.

83. The du command outputs both the size of the contents of a given directory as well as the directory's name. We may want to obtain just the size, which is the first of the two outputs, and discard the name. Show how we can do this with the current directory by piping du to awk.
84. The wc (word count) program outputs the number of lines, words and characters of one or more files. We can use -l, -w, -c to only output the lines, words or characters. But in every case, the output includes the filename. Write a command which takes the output of wc -w and pipes it to awk, counts and outputs the total number of words in all of the files specified. For instance, wc -w *.txt | awk '...'. Note that wc returns a "total" at the bottom. Make sure you do not include this line in your output.
85. Similar to question 84, use wc on a number *.txt files and pipe the output to awk to determine which file has the greatest number of words. Output that file's name and the number of words of that file. Again, make sure this does not include the total at the end of the wc output.
86. Write a Linux instruction that pipes ls -l to awk and outputs the name of any file that is readable, writable and executable by owner and has no other permissions.
87. Write a Linux instruction that pipes ls -l to awk and counts the number of files found whose type is not regular (-).
88. Write a Linux instruction that pipes ls -l to awk and finds and outputs the largest file found.
89. Write a Linux instruction that pipes ps aux to awk to find all processes whose state is R (running) and output those process' names. The state should not contain any other characters aside from R. This will require that you determine which field stores the process name.
90. Revise your answer to question 89 so that the state includes at least one other character aside from R such as RN or R<.
91. Write a Linux instruction that pipes ps aux to awk and computes and outputs the average PID. This requires obtaining every process' PID, adding it to a sum, adding one to a counter and computing and outputting the average. Remember that ps outputs a header line that you will want to ignore.
92. Write a Linux instruction that pipes ps aux to awk and determines which process has been given the most current CPU utilization and outputs that process' name, PID and CPU time.

6 Shell Scripting

This chapter's learning objectives (all pertaining to Bash scripting) are to be able to:

- Use variables and parameters
- Write input and output statements
- Explain and write conditional statements
- Explain and write selection statements
- Explain and write loop statements
- Use arrays and string operations
- Invoke and write functions

6.1 INTRODUCTION

Most computer programs are written in a high-level language (e.g., C++, Java) and compiled into an executable program. Users run those executable programs. A script is a program that is interpreted instead of compiled. An interpreter takes each instruction in the program, translates it to an executable statement and executes it. The interpreter runs in an environment so that the result of each instruction is saved in that environment. With the interpreter and this environment, we can develop a program one instruction at a time, experimenting with instructions as we work to write your program.

The downside of running a program using an interpreter is that it is far less efficient. The end-user must wait while each instruction is translated. Compiled programs have already been translated so the end-user can run the program without waiting for translation.

Many Linux users write and run programs that are compiled, using languages like C, C++ or Java. If the user is a developer, then the compiled version of the program can be distributed to end-users as a product. But for many Linux users, there may be a need to solve some problem with a program where that program would only run locally on the user's computer. In such a case, the program can be interpreted instead of compiled and so the user can write the program in an interpreted language like the Bash scripting language or Perl (which we look at it in supplementary reading material).

Small, interpreted programs are usually referred to as scripts. A *script*, as the name implies, is a set of operations that the computer will step through to accomplish some needed task. A script, for instance, might be written to search the file system for files that have poor permissions and report on those found, or determine if and when to backup stored content. Although scripting is not essential for the Linux user, it becomes a valuable tool. For the system administrator, writing scripts is a common way to solve problems.

A Bash session has its own environment whereby definitions such as aliases, variables, functions and the history list persist. As we enter instructions from the command line, we modify this environment. But Bash is not just an interpreter and environment. The Bash interpreter is able to execute instructions beyond those of Linux commands. It has its own high-level programming language. We can enter instructions of the Bash language from either the command line or place them in a file (script) to be executed.

We see in this chapter that the Bash shell scripting language is a programming language similar to other high-level languages like C and Java. It contains many of the facilities that we expect in a programming language such as variables, input and output statements, selection statements (if-then, if-then-else), loops, arrays, functions and function calls.

DOI: 10.1201/9781003203322-6

It is assumed that the reader is already at least somewhat familiar with programming and understands such concepts as variables, assignment statements, conditions, selection statements and loops. While we explore these and other aspects of the Bash scripting language, we do not spend much time introducing such concepts.

SECTION ACTIVITIES

1. Early scripting languages include Multics' active functions, IBM's Job Control Language (JCL), the Thompson shell from Unix and COMMAND (later named EXEC), written for IBM's CP/CMS operating systems. An early PC scripting language was MS-DOS. Research MS-DOS and JCL and compare them to what you know of high-level programming languages.
2. The word script usually references the words and actions for actors/actresses in a play, tv show or movie. Another application of the term was coined in artificial intelligence to represent the stereotypical sequence of actions accompanying some event like eating in a restaurant. Do a web search on the "restaurant script schank abelson" (Schank and Abelson were the two AI researchers who came up with the script concept). How does this usage of the term script differ from programming?

6.2 SIMPLE SCRIPTING

6.2.1 Scripts of Linux Instructions

In Linux, every shell script must start with a comment that specifies the interpreter which will run the script. To use the Bash interpreter, that first line is `#!/bin/bash`. The remainder of the script can consist of Linux instructions, Bash instructions and comments (comments are anything that follows #). To write `csh` and `tcsh` scripts, the first line would begin with one of `#!/bin/csh` or `#!/bin/tcsh`. Script files must be executable. We generally use a permission of 755 or 745 for our scripts.

Among the simplest scripts, we can write are those that perform sequences of Linux operations. For instance, a user might want to start off each login with a status report consisting of the time/day, disk utilization output and a listing of any user files that are empty. Such a script is presented in the top half of Figure 6.1. A portion of the output is shown in the bottom half of the figure. Notice that the output is not particularly readable.

6.2.2 Running Scripts

Assume that we save the script in Figure 6.1 under the name `start`. After changing the permissions of the file `start` to 745 (or 755) using `chmod`, we can execute this script with the command

```
#!/bin/bash
date
du -s ~
find ~ -empty
-----------------------------------------------------
Mon Feb  1 08:22:26 EST 2021
189604    /home/foxr
/home/foxr/.mozilla/extensions
/home/foxr/.mozilla/plugins
/home/foxr/.mozilla/firefox/fk60hcwz.default
default/.parentlock
```

FIGURE 6.1 Simple script comprising Linux instructions.

./start (assuming our current working directory stores the file start). The start script outputs information to our terminal window.

Figure 6.1 shows only the first six lines of output from the script. There could be far more empty files than is listed (in fact there were 68 empty files when we ran this script). As we noted, the output is not particularly readable because the script just dumps the output without telling us what the output means. With 68 empty files listed, the output scrolls down the screen so that we do not see the whole thing. We can resolve this latter problem by piping the script's results to less as in ./start | less.

Let's enhance the script itself to improve the output. We can add echo statements that output literal messages that explain what the output is such as "your disk utilization is" and "your empty files are". Inserting empty echo statements (i.e., echo statements with no parameters) provide blank lines in the output. We might add such echo statements between each pair of Linux instructions.

We can also redirect some or all of the output to one or more files. This would allow the user to view the output later or collect output for long-term storage. To redirect the output of a Linux instruction or echo statement, we use > *filename* (or >> *filename* to append to an existing file). We've enhanced the script from Figure 6.1 by adding echo statements both those that indicate what the output is and empty ones to insert blank lines, adding comments, and a redirection operator for the find instruction's output to send the empty files list to a file. This is shown in Figure 6.2.

With the revised script in Figure 6.2, we will only see some of the output as we are sending the list of empty files to the file empty_report.txt. We would have to view its contents separately. An alternative to redirecting output from within the script is to redirect the entire script's output. Let's assume we have removed >> empty_report.txt from the last instruction in the script. Now, when we run it, we redirect the script's output to a file as in ./start > login_report.txt. As we might run this report every time we login, we might use >> to append to the file so that we accumulate reports over a matter of days or weeks.

How do we run this script every time we login? While we could schedule its execution using one of the scheduling programs (e.g., at or crontab, programs we look at later in the textbook), recall that .bashrc executes every time we log in and start a Bash shell. We could place the above instruction (./start >> login_report.txt) there (or alternatively in .bashrc_profile).

6.2.3 Scripting Errors

One benefit that arises with using a compiler is that it can locate syntax errors for us. Interpreters do not have this compilation step and so instead run the program even if there are syntax errors present in the code. All errors are reported along with any output. It becomes harder to debug an interpreted program when we receive a list of *all* errors rather than just syntax errors. Additionally, the Bash interpreter does not necessarily provide us with meaningful error messages, only the location of the errors.

Let's consider a couple of errors that we might come across. The script in the top part of Figure 6.3 has an assignment statement to assign the variable FULL_NAME a value and an echo

```
#!/bin/bash
echo Login report for $USERNAME
echo The date and time are:
date
echo      # output a blank line
echo Disk utilization report:
du -s ~
echo      # output a blank line
echo Sending report on empty files to ~/empty_report.txt
find ~ -empty >> ~/empty_report.txt  # redirect output
```

FIGURE 6.2 Enhanced script.

```
#!/bin/bash
FIRST=Frank
LAST=Zappa
FULL_NAME=$FIRST $LAST
echo $FULL_NAME
----------------------------------------------
./error: line 4: Zappa: command not found
```

FIGURE 6.3 Example Bash script with error.

statement to output the result. We already saw this specific example in Chapter 2 so hopefully you can remember the issue. Assume this script is stored in the file `error`. Upon running the script, we receive the error shown in the bottom part of Figure 6.3.

This error message indicates that when the interpreter attempted to execute the instruction in line 4 of the file, which is `FULL_NAME=$FIRST $LAST`, something went wrong. The error message looks cryptic because there is no instruction `Zappa` in the script. Why did it try to run something called `Zappa`? We explored this problem in Chapter 2 where we saw that an assignment statement's right-hand side needs to be enclosed in quote marks if it has a blank space. In this script, the variable `FULL_NAME` is assigned the value `Frank`. The interpreter next tries to execute the next item as if it were a separate instruction. Therefore, it tries to execute `$LAST` which stores the value `Zappa`, which is not a valid program name.

As another example, the script in Figure 6.4 executes three Linux `wc -l` commands, one on each of three files. Unfortunately, as a normal user, we do not have access to `/etc/shadow` and so this instruction generates a different type of error. We present the error message at the bottom of the figure. In this case, the error message is not as cryptic: we are told that the `wc` command does not have permission to access `/etc/shadow`. We will explore other errors that may arise as we introduce Bash programming instructions in later sections of this chapter.

```
#!/bin/bash
wc -l /etc/passwd
wc -l /etc/group
wc -l /etc/shadow
----------------------------------------
wc: /etc/shadow: Permission denied
```

FIGURE 6.4 Another example Bash script with error.

SECTION ACTIVITIES

1. Is there any value to the simple script shown in Figure 6.1? Consider some sequence of actions that you take whenever you start up or log into your computer. Could those be captured in a script and if so, would that be preferable to you performing the actions by hand?
2. If you have experience programming, you most likely work in an IDE (integrated development environment) where compiler-generated error messages help you debug your code. Bash does not have an IDE and instead it is the interpreter that generates error messages. Will this make learning and using the Bash scripting language harder for you?

6.3 VARIABLES, ASSIGNMENTS AND PARAMETERS

6.3.1 Bash Variables

A shell script can store information in and recall information from variables. There are two types of variables: environment variables that, because they have been exported, are accessible within your script, from the command line and in other scripts, and variables defined within your script that are only available in the script.

All variables have names. Names are commonly just letters although a variable's name can include digits and underscores (_) as long as the name starts with a letter or underscore. For instance, we might name variables `x`, `y`, `z`, `first_name`, `last_name`, `file1`, `file_2` and so forth. We would not be able to use any of the following names in a Bash script: `1_file`, `file 2` (spaces are not allowed) or `file$3` (the `$` is not allowed). Variable names are *case sensitive*. If we assign the variable `x` to have a value and then reference `X`, we are referencing a different variable. Note that all reserved words (e.g., `if`, `then`, `while`) are only recognizable if specified in lowercase. Unlike most programming languages, variable names can be the same as reserved words such as `if`, `then`, `while`, although there is no reason to do this as it would make for confusing code.

Variables in Bash *only* store strings. Numbers are treated as strings unless we specify that we want to interpret such a value as a number. If we do so, we are limited to integer numbers only as Bash cannot perform arithmetic operations on values with decimal points (floating-point numbers). Bash also permits arrays, which we examine later in this chapter.

Variables in Bash do not have to be declared; we can just assign a variable a value when we first need it. This is unlike languages like C or Java, but similar to more recent languages like Python. We can, however, declare variables if we choose to. The syntax for declaring a variable is `declare [options] name[=value]`. Both *options* and an initial *value* are optional.

Options are denoted using + to turn an attribute of the variable *off* and – to turn it *on*. This is counterintuitive and somewhat of a challenge to get used to. The attributes are `a` (array), `f` (function), `i` (integer), `r` (read-only) and `x` (exported). Read-only means that the variable, once assigned a value, cannot be changed. In other languages, a read-only attribute is handled by declaring the item to be a *constant*.

Consider `declare -rx ARCH=386`. This creates a read-only variable, `ARCH`, with the value `386` and exports it beyond the current shell. While we can make a variable read-only with `-r`, we are not allowed to change a read-only variable into a writable variable; that is, `+r` does not work.

6.3.2 Assignment Statements

In order to assign a variable a value, we use an assignment statement. The basic form of the assignment statement is *`VARIABLE=VALUE`*. No spaces are allowed around the equal sign. As noted earlier, if there are any spaces in *VALUE* then all of *VALUE* must be placed inside quote marks. We can use double quote marks (`""`) or single quote marks (`''`) although using the single quotes is less common because the content of the string is treated differently (we explore this shortly). The value on the right-hand side (RHS) can be one of five things as described in Table 6.1

Let's explore Table 6.1 in more detail. Literal values should be self-explanatory. If the value is a string with no spaces, then quote marks are optional. If the value is a string with spaces, quote marks must be included. There are two types of quote marks: single quote marks (`''`) and double quote marks (`""`). The difference between them only arises if the string contains items that the Bash interpreter would interpret. This will include, for instance, Linux instructions embedded in `$( )` or `` ` ` `` notation and values in variables referenced by `$`. When placed in double quotes, the Bash interpreter does interpret those items as expected.

For instance, `X="$FIRST $LAST"` causes `X` to store the value stored in `FIRST`, a blank space, and the value stored in `LAST`. When embedded in single quote marks, the Bash interpreter takes the items literally so that `X` would store `$FIRST $LAST`.

TABLE 6.1
The Types Permitted on the RHS of a Bash Assignment Statement

RHS Type	Explanation	Examples

Literal string or integer	If the string has one or more spaces, it must be placed in quote marks; numbers with a decimal point are treated as strings.	`X=Hello`
		`NAME="Frank Zappa"`
		`Y=5.1` (a string)
		`AGE=25` (stored as a string but can be treated as a number)

Value stored in another variable	To obtain the value stored in the variable, precede it with `$`.	`Y=$X` `Y=$FIRST_NAME`
Result of an arithmetic or string operation	For arithmetic operations, place the expression in `$(( ))` notation.	`A=$((B+5))` `hi="Hello $FIRST_NAME"`
Result of a Linux command	We can invoke a Linux instruction and store its return value (we explore the notation shortly).	`X=$(date)`

Any combination of the above	If we are dealing with numeric values, we combine the above through arithmetic operations; if they are strings, we concatenate them.	`X=$((Y+Z))`
		``X=$((Y+`function1 5`))``
		`NAME="The date and time are $(date)"`

Table 6.2 provides several assignment statement examples. The table shows the result and explains the instruction when necessary. Assume `FIRST` stores `Frank` and `LAST` stores `Zappa` for the last three entries.

Once a variable has been given a value, there are several ways to alter it. To assign it a different value, use a new assignment statement. For instance, if `X` was set to `0`, `X=1` will change it to `1`. To remove a value from a variable, we can either assign it the NULL value which is accomplished by having nothing on the right-hand side as with `X=` or we can use the `unset` command as in `unset X`. Either way, `X` will no longer have a value.

6.3.3 Executing Linux Commands from within Assignment Statements

The *`VALUE`* portion of the assignment statement can include executable statements. Consider the assignment statement `DATE=date`. This literally sets the variable `DATE` to store the string `date`. This would not differ if we use either `"date"` or `'date'`. However, if `date` is placed within either `` ` ` `` marks or `$( )` marks, then the Bash interpreter *executes* the string as if it were a Linux command (which it is). We revise our previous assignment statement to be ``DATE=`date` `` or `DATE=$(date)`. This instruction causes the Linux `date` command to execute with the response stored, as a string, in the variable `DATE`.

If the RHS of our assignment statement includes multiple parts such as literal text, values in variables and the results of Linux operations then we would embed the entire RHS in quote marks. For instance, the instruction `DATE="Hello $FIRST_NAME, today's date and time is $(date)"` stores into the variable `DATE` a literal string, the value stored in `FIRST_NAME` and the result from the `date` command.

Had we used single quotes, we would have a problem because the RHS's literal text also includes a single quote mark (`today's`). The single quote mark in `today's` is permissible from within the double quote marks but not from within single quote marks. Assuming we remove that single quote, using single quotes instead of double quotes would still yield an incorrect result. `DATE` would wind up storing `Hello $FIRST_NAME, todays date and time is $(date)` because we ask the Bash interpreter to not interpret `$FIRST_NAME` or `$(date)`.

TABLE 6.2
Example Assignment Statements

Example Assignment	Explanation	Value Stored in X
X=5	Store a literal value.	5 (as a string, but we can treat it as a number)
X=Frank	Store a String literal value.	Frank (a string)
X="Frank Zappa"	Store a String literal value with a blank space.	Frank Zappa (with the space)
X=Frank Zappa	An error arises because Zappa is not a legal instruction.	Frank
X=FrankZappa	Store a String literal value.	FrankZappa
X=`ls *`	Store the result returned by executing ls *.	The contents of the current directory and all subdirectories.
X=ls *	An error arises (bash: Desktop: command not found...).	Need quote marks for this instruction to succeed.
X="ls *" X='ls *'	Store the string ls *.	This stores a string but based on how we view the string, we can obtain the ls command's output; see the text in Section 6.3.3.
X=1.2345	Store a string literal value.	1.2345 (as a string)
X="$FIRST $LAST"	Use the values stored in two other variables along with a blank space.	Frank Zappa
X=$FIRST $LAST	Error arises (bash: Zappa: command not found).	Frank
X='$FIRST $LAST'	Do not interpret $FIRST, $LAST.	$FIRST $LAST

Let's consider another example of executing a command on the RHS of an assignment statement. We saw in Table 6.2 that we can issue the instruction X=`ls *` or X=$(ls *). This allows us to store the listing of directory items in X as a string. But, the contents of the string may not appear as we expect them to. Assume the current directory stores four files, file1.txt, file2a.txt, file3ab.txt and file4.txt. The command X=$(ls *) will result in these four filenames stored as one string in X. The command echo $X will produce the output file1.txt file2a.txt file3ab.txt file4.txt, that is, the output all appears on a single line. The command ls * would instead produce the same output but with the files listed on separate lines.

Also shown in Table 6.2 are the assignment statements X="ls *" and X='ls *'. These two instructions store the string ls * in X. Interestingly though is what happens when we either specify $X or echo $X. With $X, we are asking Bash to execute $X which is ls *, so $X on the command line by itself causes ls * to execute. We get the listing as if we had typed ls * on the command line. With echo $X we do not see ls * but instead what ls * would return. But in this case, because we executed this in an echo statement, like what we saw in the previous paragraph, the listing appears on a single line.

What if X had stored something other than a Linux command, for instance X=Frank? If we place $X on the command line by itself, we get an error because Bash tries to execute $X, or Frank, which is not an instruction. However, echo $X outputs Frank.

Let's consider the instruction LIST=*.txt. LIST will store all filenames in the current directory with a .txt extension. Given the list from a few paragraphs previous, LIST would literally store the list of file1.txt, file2a.txt, file3ab.txt and file4.txt. This is because the Bash interpreter performs filename expansion on *.txt before executing the assignment statement.

Other than cute tricks with wildcards, the types of operations we place on the RHS of our assignment statement generally break into two categories: string operations and arithmetic operations. We explore arithmetic operations in the next subsection and for now concentrate only on one string operation.

The most commonly used string operation is string *concatenation* which is the conjoining of multiple strings to create one larger string. Concatenation is accomplished by listing each string on the RHS of the assignment statement. If the strings are to be separated by blank spaces or they contain blanks themselves, then the entire sequence must be enclosed within quote marks. The items concatenated can be any combination of variables, literal strings and Linux operations embedded in ` ` or `$( )`. We saw an example earlier of such a message being stored in the variable `DATE`. Notice that if we have values in variables `X`, `Y` and `Z`, we can concatenate them as `A=$X$Y$Z` with no quote marks since there are no blank spaces on the RHS (assuming that `X`, `Y` and `Z` store values without blank spaces).

We need to make sure that any command placed inside ` ` or `$( )` notation is a legal command which the user, who is running the script, has access to. For instance, `Greeting="Hello $NAME, today is $(dat)"` yields an error because `dat` is not a legal instruction.

6.3.4 Arithmetic Operations in Assignment Statements

For arithmetic operations, we have to denote that the values stored in variables are to be interpreted as numbers and not strings. We indicate such operations using one of two notations. First, we can precede the assignment statement with the word `let`, as in `let a=n+1`. Second, we embed the arithmetic operation inside the notation `$(( ))`, as in `a=$((n+1))`. In either case, `n`, a variable which appears on the RHS of the assignment statement, is not preceded by a `$`. We are allowed to include the `$` before `n` with either notation but it is not necessary.

The Bash interpreter can perform many arithmetic operations on integer values. It cannot however perform arithmetic operations on values with a decimal point (floating-point values). The operators available are shown in Table 6.3.

Several examples are shown in Table 6.4. In these examples, assume `A=2`, `B=5` and `N` has no value currently. Each example starts with these initial values (i.e., do not assume that values carry from example to example). We do not cover all of the operators in Table 6.3, and we omit the

TABLE 6.3
Bash Arithmetic Operators

<table>
<tr><th>Arithmetic Operators</th><th>Meaning</th></tr>
<tr><td>+, -, *, /</td><td>Addition, subtraction, multiplication, division</td></tr>
<tr><td>**</td><td>Exponent (e.g., x**2 is x^2)</td></tr>
<tr><td>%</td><td>Modulo (division retaining the remainder only)</td></tr>
<tr><td>~</td><td>Negation of 1's complement value</td></tr>
<tr><td><, >, <=, >=, ==, <>, !=</td><td>Relational operators, the last two are available for not equal; see the comments about these operators in the text.</td></tr>
<tr><td><<, >>, &, ^, |</td><td>Bit operations (left/right shift, AND, XOR, OR)</td></tr>
<tr><td>&&, ||, !</td><td>Logical AND, OR, NOT</td></tr>
<tr><td>=</td><td>Assignment</td></tr>
<tr><td>+=, -=, *=, /=, %=</td><td>Reassignment (e.g., x=x+y can be written as x+=y)</td></tr>
<tr><td>&=, ^=, |=, , >>=</td><td>Reassignment using &, ^, |, <<, >></td></tr>
</table>

TABLE 6.4
Arithmetic Examples

Example	Result of Operation	Explanation
`N=1`	N stores 1	Simple assignment.
`N=$((B+1))`	N stores 6	Computes 5+1, stores result.
`let B+=5`	B stores 10	Same as `B=$((B+5))`.
`Y=$B+1`	Y stores 5+1	Without the `(())` or `let`, the Bash interpreter treats this as a String concatenation operation so `Y` gets the value `$B`, the + and a `1`.
`let Y=B%A`	Y stores 1	`B%A` is the remainder of 5 / 2 which is 1.
`N=$((B**A))`	N gets 25	`B**A` is 5**2 or 5^2 or 25.
`A=$((A+1))`	A stores 3	We add `1` to `A`, or `2+1`, and store the result back into `A`; note that `A+=$((1))` does not work as expected.
`N=$((A<<2))`	N stores 8	`A<<2` is `A` (which is 2, or 0010 in binary) shifted two bits to the left (which is 1000, or 8).

relational and logical operators for now. Of particular note are the reassignment operators which are similar to those found in C and related languages.

Let's focus on the relational operators from Table 6.3. You are no doubt used to these symbols when you have written conditions in if-then or while statements in other languages. For instance, in C, we might have code like `if(x<y) …;` In Bash, these operators are not used in conditions. We will see later in this chapter that in place of these operators, we use abbreviations like `-lt` for less than. So, what are these operators used for?

Like in C, Bash can evaluate a relational operator and return a numeric value. In C, false is 0 and true is any non-0 value. In Bash, these operators will return either `0` (false) or `1` (true). For instance, if `X` is `5` and `Y` is `4`, then `$((X>Y))` returns `1`. If `X` had been `4`, then `$((X>Y))` returns return `0`. We might use such an expression on the right-hand side of an assignment statement such as `N=$((X>Y))` to store a `0` or `1` in `N` that might be used later in another condition or an assignment statement.

We must specify the arithmetic operations correctly. For instance, a statement like `X=$((Y+Z*))` will yield an error because the operation is lacking an operand. The specific error will read `Y+Z*: syntax error: operand expected (error token is "*")`.

We can specify multiple assignment statements on the same line. This is accomplished by separating each assignment with a space. Figure 6.5 illustrates assigning variables separately (on the left of the figure) and in one instruction (on the right). Notice that we can even use variables assigned values earlier in the same instruction.

Also available in Bash are the prefix or postfix increment/decrement operators as found in C-like languages. These appear slightly different than in C or Java though. To increment or decrement the variable `X`, we use one of `((X++))`, `((++X))`, `((X--))` or `((--X))`. The `$` *must* be omitted in this case. We can also increment `X` using `X=$((X+1))`, `let X=X+1`, or `let X+=1`. Notice that without `let`, `X+=1` does not work as expected as it performs string concatenation. If `X` was `3` then `X+=1` sets `X` to `31` (concatenating string 1 onto string 3).

```
X=1                          X=1 Y=2 Z=$((X+Y))
Y=2
Z=$((X+Y))
```

FIGURE 6.5 Multiple assignments in one instruction.

When using any form of arithmetic operation, any variable referenced must currently store an integer to work correctly. If not, we do not necessarily receive an error but the result may be unexpected. Assume X stores the string Hello. The operation Y=$((X+1)) will result in Y storing the number 1 because X is considered to have no numeric value, or 0. The operation ((X++)) will leave X unaffected. We can use the prefix/postfix increment/decrement from within the RHS of an assignment statement, in which case the syntax differs. The operation Y=((X++)) yields an error while Y=$((X++)) does not.

If you are unfamiliar with C syntax, you might wonder what the difference is between having ++ before the variable (*prefix*) and after (*postfix*). There is no difference when using the increment/decrement operator outside of an assignment statement. That is, ((X++)) and ((++X)) do the same thing. But when used in an assignment statement, the two have different results. The prefix increment/decrement performs the increment/decrement first, and then performs the remainder of the assignment statement. The postfix increment/decrement has the assignment statement execute on the current value of the variable being incremented/decremented, and then the increment/decrement takes place.

Let's consider an example where X currently stores 1. Y=$((X++)) results in Y equal to 1 and X incremented to 2. With Y=$((++X)), the increment takes place first, so X becomes 2. Then, the assignment takes place and Y is set to 2. We can combine the increment/decrement operator and other operators in an assignment statement. For instance, Y=$((X++*4)) will set Y to X*4 and then increment X. If X were originally 1, then Y would store 4 and then X would be incremented to 2. With Y=$((++X*Z++)), first X will be incremented, Y will then be set to X*Z using the incremented value of X, and then Z would be incremented.

Another way to invoke arithmetic expressions is with the expr command. This instruction performs the operation specified and returns the resulting value. We use expr in both arithmetic and string operations (as shown in the next subsection). The use of expr is different from the earlier examples because we do not necessarily use it in an assignment statement. Instead, the value computed is returned to be used in some other way.

Let's try this out. Assume N stores 3. The instruction expr $((N+1)) returns 4 but does not change N. When run from the command line, expr returns the new value. To store the result, we place the expr statement in either ` ` or $( ). See Figure 6.6 which first shows the result of using expr without storing the value and then using it in an assignment statement.

6.3.5 String Operations Using expr

The expr command is commonly used with string operations. Three string operations are substr, index and length. The format of the expr statement for string operations is expr *command string params* where *command* is one of the commands (substr, index, length), *string* is either a literal string or a variable storing a string (or a Linux operation that returns a string) and *params* depend on the type of command as we will see below.

The substr command performs a *substring* operation. We use the syntax expr substr *string start length*. *string* is the string to operate on, *start* is an integer indicating the position in the string to start the substring and *length* is an integer indicating the number of characters that should make up the substring. The expr statement returns the selected substring, or the

```
[foxr@localhost ~]$ N=3
[foxr@localhost ~]$ expr $((N+1))
4
[foxr@localhost ~]$ X=`expr $((N+1))`
[foxr@localhost ~]$ echo $X
4
[foxr@localhost ~]$ echo `expr $((N+1))`
4
```

FIGURE 6.6 Arithmetic examples using expr.

length characters starting at *start*. For instance, `expr substr abcdefg 3 2` would return the string `cd`. C-like languages start indices at 0 for strings. In Bash, there are two different techniques for retrieving a substring. The approach with `expr` starts counting at 1 (later, we will visit a second approach where the first character is at index 0).

Any of *string*, *index* and *length* can be stored in variables, in which case we would reference the variable using `$VAR`. If `NAME` stores `"Frank Zappa"`, `start` is 7 and `length` is 3, then `expr substr "$NAME" $start $length` would return `Zap`. Notice that `$NAME` is placed inside of double quote marks because the value stored in `NAME` contains a blank. If `NAME` had stored `FrankZappa`, the quotes would not be necessary (although could still be used).

The `index` command is given two strings and it returns the location within the first string of the first character found in the second string. We can consider the first string to be the master string and the second string to be a search string. The index returned is the location in the master string of the first matching character of the search string. For instance, `expr index abcdefg gcd` returns `3` because the first character to match from `gcd` is `c`, at index 3. Notice again that the index starts counting at 1 instead of 0.

The `length` command returns the length of a string. Unlike `substr` and `index`, there are no additional parameters other than the string itself. For instance, we might use `expr length abcdefg` to obtain the length of the string `abcdefg (7)`. If `NAME` stored `"Frank Zappa"` then the instruction `expr length "$NAME"` would return `11`.

Each of these string operations will only work if executed in an `expr` command. As with using `expr` with arithmetic expressions, these can be used as the RHS of an assignment statement or an `echo` instruction by placing the entire `expr` statement inside `` ` ` `` or `$()`. Figure 6.7 illustrates a few examples where `PHRASE` stores `"hello world"`. In the first example, we output the result of the `expr` instruction, and in the latter two, we store the result in `X` and then output `X`.

6.3.6 Command-Line Parameters

Parameters are specified by the user at the time that the script is invoked. Parameters are listed on the command line after the name of the script. For instance, a script named `a_script` which expects numeric parameters might be invoked as `./a_script 32 1 25`. In this case, the values `32`, `1` and `25` are made available to the script as parameters.

The parameters themselves are not variables. They cannot change value during the script's execution. But parameters can be referenced in the script to access these values. To access a specific parameter, use the notation `$`*n* where *n* is the index of the parameter. For instance, `$1` is the first parameter and `$2` is the second parameter.

As an example, a script named `sub` is shown in the top portion of Figure 6.8. This script returns substrings of a string. The user specifies the starting index and length of the substring as parameters. The script then uses `expr` to obtain the substrings of three strings defined in the script (`STR1`, `STR2`, `STR3`). Notice that we are not using `echo` statements in the script because the return value from the `expr` statements is automatically output. Beneath the script in the figure, we see the command to execute the script run on parameters 2 and 3 along with the script's output.

```
[foxr@localhost ~]$ echo `expr substr "$PHRASE" 4 6`
lo wor
[foxr@localhost ~]$ X=`expr index "$PHRASE" nopq`
[foxr@localhost ~]$ echo $X
5
[foxr@localhost ~]$ X=`expr length "$PHRASE"`
[foxr@localhost ~]$ echo $X
11
```

FIGURE 6.7 String examples using `expr`.

```
#!/bin/bash
STR1="Hello World"
STR2="Frank Zappa Rocks!"
STR3="abcdefg"
expr substr "$STR1" $1 $2
expr substr "$STR2" $1 $2
expr substr $STR3 $1 $2
------------------------------
[foxr@localhost ~]$ ./sub 2 3
ell
ran
bcd
```

FIGURE 6.8 Accessing parameters in a script.

A script does not have to use all of the parameters passed. If we had invoked the script in Figure 6.8 as `./sub 2 3 4 5 6`, then `4 5 6` are ignored. On the other hand, if we invoke `sub` with no parameters, we receive error messages. The errors arise because `$1` and `$2` have no values and yet we are trying to use them as parameters in an `expr substr` instruction, which expects exactly two integer numbers.

Once a parameter has been used in a script, we can discard it (if we no longer need it). We do so with the `shift` instruction. The result of `shift` is that all parameters are rotated down one position so that `$2` becomes `$1` and `$3` becomes `$2`, and so forth. This can simplify a script that will use each parameter one at a time. In this way, after each `shift`, we only need to reference the current parameter using `$1`. The proper usage of `shift` is in the context of a loop. We will examine this later in this chapter.

There are four additional parameter references of note that we can access in any script even if no parameters are passed to the script. The notation `$0` returns the name of the script itself and `$$` returns the PID of the script's running process. The notation `$#` returns the number of parameters that the user supplied. Finally, `$@` returns the entire group of parameters as a list. Both `$#` and `$@` are discussed in more detail in Sections 6.5 and 6.6.

SECTION ACTIVITIES

1. It is assumed that you know another programming language. How does that language differ from what you've learned so far of Bash with respect to variables and assignment statements? Is Bash's use of strings for all storage going to be a problem in writing scripts?
2. Bash originated with an ALGOL-like syntax but has adopted some C-like syntax as well. Will the two forms of syntax present a detriment to using Bash correctly or will the shortcuts available with the C-like syntax be helpful? You should keep this in mind as you learn more about the Bash scripting language because at times the syntax may look familiar and at others it will not.
3. Which languages can receive command line parameters? You might be surprised that it is more languages than you might think of. Research this.

6.4 INPUT AND OUTPUT

6.4.1 OUTPUT WITH `echo`

We have already seen the output statement, `echo`. This instruction is supplied with a list of items to output. These items are a combination of literal values, variables (with `$` preceding each variable's names) and Linux commands placed in ` `` ` or `$()` marks. Any variable without `$` and any Linux command not placed in ` `` ` or `$()` will be output literally. Figure 6.9 provides a few examples showing the result of echo statements where $NAME is Frank and the user is zappaf, currently in his home directory.

The instructions shown in Figure 6.9 are issued from the command line but we would see the same output if these were in a script that we executed. Note that $PWD and $HOME are environment variables available from either the command line or from within a script. $NAME is not an environment variable and so if accessed in a script would either have to be exported from wherever it had been defined (presumably the command line) or it would have to be assigned a value in the script.

The echo statement permits the use of *escape* characters. These are characters preceded by a backslash (\). We must force echo to apply the backslash correctly if present. This is done through the -e option. Table 6.5 displays some of the more common escape characters available.

To use the escape characters (with the exception of \\), the string, including any literal characters, must be placed within double quote marks. Table 6.6 provides a few examples of using the -e

```
[foxr@localhost ~]$ echo Hello $NAME, how are you?
Hello Frank, how are you?
[foxr@localhost ~]$ echo Your current directory is $PWD
Your current directory is /home/zappaf
[foxr@localhost ~]$ echo Home is ~ or $HOME
Home is /home/zappaf or /home/zappaf
[foxr@localhost ~]$ echo the date is $(date)
the date is Tue Feb 1 14:51:50 EST 2022
[foxr@localhost ~]$ echo Seconds per year = $((365*24*60*60))
Seconds per year = 31536000
```

FIGURE 6.9 Example echo statements.

TABLE 6.5
Some Useful Escape Characters in Bash

Escape Character	Meaning
`\\`	Backslash
`\a`	Bell (alert)
`\b`	Backspace
`\n`	New line
`\t`	Horizontal tab
`\v`	Vertical tab
`\0###`	ASCII character matching the given octal value ###.
`\xHH`	ASCII character matching the given hexadecimal value HH.

TABLE 6.6
Example echo Statements Using -e and Escape Characters

Example echo Statement	Output
`echo -e "Hello\nWorld!"`	Hello World!
`echo -e "Jan\tFeb\tMar"`	Jan Feb Mar
`echo -e "\x68\x65\x6C\x6C\x6F"`	hello
`echo -e Hello\\World`	Hello\World

option with escape characters. Notice that we do not need to place the last example within quote marks while the others require them.

Aside from using escape characters, we seldom need quote marks in our output statements. That is, unlike the assignment statement where spaces cause problems on the RHS if a string has spaces, we are free to list as many items, including spaces, in an `echo` statement without the use of quote marks. With that said, when we use quote marks we should use double quote marks to ensure that the Bash interpreter interprets `$VAR` and `` ` ` `` or `$()` correctly. Table 6.7 presents several example `echo` instructions without quote marks, with single quote marks and with double quote marks. Assume `$FIRST` stores `Frank` and `$LAST` stores `Zappa`.

Of particular note in Table 6.7 are the last four examples. `\n` is the escape character for a new line. Without using `-e`, the Bash interpreter ignores the special purpose of `\n` and outputs it literally. In the third to last example, we have added `-e` and our output is on two lines, as we would expect. In the second to last example, we use single quote marks and so while `\n` is applied, the `$` before the variables are not and we get `$FIRST` and `$LAST` on two lines. The last example shows the use of `\b` (backspace). We ask Bash to output `$FIRST`, back up two characters and then output `$LAST`, and so the result omits the last two characters stored in `$FIRST` because they are overwritten by the first two characters of `$LAST`.

Let's return to the idea of storing the result of a Linux operation in a variable. We can, for instance, store the result of `ls *.txt` in a variable. We do so using ``LIST=`ls *.txt` ``. Printing `$LIST` differs depending on whether we use quote marks or not. We show this in the top portion of Table 6.8. Now let's consider what would happen if we store in `LIST` the value `*.txt` as in `LIST=*.txt`. In this instruction, much like executing `ls` on `*.txt`, we are asking the Bash interpreter to perform *globbing* on `*.txt` and store the result in `LIST`. We show in the bottom portion of the table the results of `echo` statements outputting this version of `LIST`. Assume in all cases that there are three files, `file1.txt`, `file2.txt` and `file3.txt`, in the current directory.

As we might expect, using single quotes we literally output `$LIST` rather than the value stored in `$LIST`. Without quote marks, the output shows either the result of the `ls` operation (top portion) or the globbed filenames (bottom portion) but in both cases the output appears on one line rather than on separate lines. When using the double quote marks, the result of the `ls` operation outputs the results of executing the `ls` operation as if we launched the `ls` operation from the command line (that is, the list is on separate lines). The interesting result is the output for `"$LIST"` when `LIST` stores `*.txt`. Here, we do not receive the list of files that `*.txt` expanded into but instead we literally receive `*.txt`. Globbing does not take place because we are outputting the value stored in the variable, not what that value does.

TABLE 6.7
Using `echo` with Quote Marks

Example echo Statement	Output
`echo $FIRST $LAST`	`Frank Zappa`
`echo "$FIRST $LAST"`	`Frank Zappa`
`echo '$FIRST $LAST'`	`$FIRST $LAST`
`echo "$FIRST\n$LAST"`	`Frank\nZappa`
`echo -e "$FIRST\n$LAST"`	`Frank` `Zappa`
`echo -e '$FIRST\n$LAST'`	`$FIRST` `$LAST`
`echo -e "$FIRST\b\b$LAST"`	`FraZappa`

TABLE 6.8
Additional echo Examples

Operation	Output
`` LIST=`ls *.txt` ``	
echo $LIST	list1.txt list2.txt list3.txt
echo "$LIST"	list1.txt list2.txt list3.txt
echo '$LIST'	$LIST
LIST=*.txt	
echo $LIST	list1.txt list2.txt list3.txt
echo "$LIST"	*.txt
echo '$LIST'	$LIST

By default, the echo statement will cause a line break at the end of the line's output. We can prevent echo from outputting the line break by using the option -n. We might use this if we have several echo statements in our script, and we want a later echo's output to appear on the same line as the current echo statement. We may also find this to be useful when we want to prompt the user for input, as we explore in the next subsection.

6.4.2 Input with read

The input statement is read. The read statement is then followed by one or more variables. If multiple variables are listed, they are separated by spaces. The variable names are listed without $ preceding them. read can be used to input any number of variable values. As with assignment statements that store literal values in variables, all values input into variables are strings. Once stored in a variable, we can ask Bash to interpret the value as a number by using $((…)) as we saw in the last section. Table 6.9 presents a few examples of the read instruction. Each is shown with the presumed input and the result.

Let's focus on the last two examples in the table. If there are more inputs than variables in the read instruction, then the last variable takes on the collection of all remaining inputs, stored as a list. This is why C stores the list 15 20 25 in the second to last example. Both A

TABLE 6.9
Example read Instructions

read Instruction	Input	Result
read A	5	A stores the string 5.
read A	Frank	A stores the string Frank.
read A	Frank Zappa	A stores the list Frank Zappa.
read A	5 10 15	A stores the list 5 10 15.
read A B	Frank Zappa	A stores the string Frank and B stores the string Zappa.
read A B C	5 10 15 20 25	A stores the string 5, B stores the string 10 and C stores the list 15 20 25.
read A B C	Frank Zappa	A stores the string Frank, B stores the string Zappa and C has the value NULL.

Script code	Terminal window I/O
echo Enter your name read NAME	Enter your name Frank
echo -n Enter your name read NAME	Enter your nameFrank
echo -n "Enter your name " read Name	Enter your name Frank

FIGURE 6.10 Adding a prompt before the `read` instruction.

and `B` store one value each and the remainder is stored in `C`. If there are fewer inputs than variables, then the values input are placed in each of the first variables and the remaining variables are given the value `NULL`. Outputting a variable's value which is `NULL` results in a blank output (nothing appears). So, for instance, if we execute `echo $C` after the last read instruction in Table 6.9, we get a blank line.

Executing a `read` statement results in the script pausing while the cursor blinks in the terminal window. This does not inform the user of the expectations behind the input statement. Therefore, it is wise to always output a prompting statement that explains what the user is expected to do.

There are two ways to prompt the user before an input. The first is through an `echo` statement preceding the input statement. We see three versions of this in Figure 6.10. In the first version, the prompting message appears on one line and the input typed by the user on a separate line. By using `echo -n`, the prompting message and input are on the same line, as shown in the second example of Figure 6.10. There is no space between the prompting message and input in this case, which may look somewhat sloppy. We fix this in the third example by adding some blanks. For Bash to output these blanks as part of the `echo` statement, we put the entire prompt in quote marks.

The other approach to prompting a user before input is to add `-p "`*`prompt`*`"` to our `read` instruction, prior to the list of variables. The `-p` option outputs whatever message follows in quotes before the input takes place. We can add blank spaces at the end of the prompt so that the blanks appear between the prompt and the user's input. We can replace the three versions of our output and input from Figure 6.10 with the instruction: `read -p "Enter your name" NAME`.

`read` has a few other noteworthy options. Option `-N` allows us to specify an integer number so that Bash only accepts that many characters for the input string. For instance, `read -N 10 NAME` will only read up to the first ten characters to be stored in `NAME`. The `-N` option will only store the input into one variable, so we would not use it if our input was for multiple variables.

The `-t` option allows us to specify a number of seconds whereby if the input is not received from keyboard within that number of seconds the instruction times out. A timeout of 0 causes `read` to return immediately and thus not input any value. The `-s` option indicates silent mode in which any characters typed by the user are not echoed to the terminal window. This can be useful when inputting a password or other sensitive information. Finally, if we perform `read` with no variables, then the input is stored in the environment variable `REPLY`.

```
#!/bin/bash
read X
read Y
read Z
echo $((X+Y-Z))
```

```
#!/bin/bash
read X Y Z < datafile.txt
echo $((X+Y-Z))
```

FIGURE 6.11 Sample script with input from keyboard (left) versus file (right).

By default, the `read` statement accepts input from standard input, the keyboard. Recall from Chapter 2 that one of the forms of redirection is <. Through this form of redirection, the specified Linux command accepts input from a file rather than keyboard. Since most Linux commands expect their input to come from disk file, < is seldom used. The one instance where we might clearly want to use it is to change input for the `read` instruction from keyboard to disk file. We can do this in a couple of ways, the simplest is by redirecting the input location when the entire script is invoked from the command line. Another approach is to use redirection in each `read` statement from within the script.

Consider the two versions of the script `input.sh` shown in Figure 6.11. The version on the left expects input from keyboard. We could run it as `./input.sh`, and the `read` instructions pause the script to receive keyboard input. If we run this script as `./input.sh < datafile.txt`, then input for `X`, `Y` and `Z` are each obtained from the file `datafile.txt`. As `input.sh` expects the three variables to be input separately, the file `datafile.txt` must have a value on each of the first three lines. And as we are using these values in an arithmetic expression, the three values need to be integers. If the values are not on three lines or are not integers, the script will generate an error. If `datafile.txt` has more than three lines of values, the extra lines are ignored by this script.

The revised version of the script on the right-hand side of Figure 6.11 has the single `read` instruction inputting from `datafile.txt` instead of the keyboard. We have to make two changes to accomplish this. First, in the script we see all three inputs are handled by one `read` instruction. Second, the `datafile.txt` file now has to have the three numbers on the first line. The reason for these changes is that if we had used three separate `read` statements like we did in the script on the left, then each `read` would re-open the same file and input the first datum each time. By inputting all three variables with one `read` statement, we can read three consecutive entries in `datafile.txt`, but only if they are on a single line. Another disadvantage of the approach taken by the script on the right side of the figure is that it lacks flexibility; we are always reading from the same file, `datafile.txt`.

SECTION ACTIVITIES

1. There are many cryptic facets to the Bash scripting language like whether to use single and double quote marks or neither, and when to place a `$` before a variable name. Write down all of the syntax rules you have learned to this point of the chapter and rank them from easiest to hardest to remember. Share these with other students to see how their lists compare to yours.
2. We saw in this section that we can prompt the user with an `echo` statement, with `echo -n` and with `read -p`. Do you have a preference? If so, what and why?

6.5 SELECTION STATEMENTS

Selection statements are instructions used to make decisions. Based on the evaluation of a condition, the selection statement selects which instruction(s) to execute. Languages typically provide several forms of selection statements called if-then (test one condition and execute the then clause or not), if-then-else (test one condition and execute the then or the else clause), nested if-then-else (test a condition, if true execute the then clause; otherwise, the else clause is itself an if-then or if-then-else statement) and a multiway selection often called a case or switch. In Bash, we have each of these although the nested-if-then-else uses the reserved word `elif` in place of the words "else if". Before we focus on these forms of selection statements, let's examine conditions in Bash.

6.5.1 Conditions for Strings and Integers

A *condition* is a test that compares values, evaluating to either true or false. The comparison can be of a variable to a value, a variable to another variable, a variable to the value of an expression or the values of two expressions. For instance, does x equal y? is x greater than 5? does y+z not equal a−b? In Bash, we compare strings, compare integers, test variables to see if they have values and test file attributes.

The Bash scripting language was based on the Bourne shell and its syntax originally drew from the language ALGOL. But Bash also incorporated syntactic features similar to C. As a result, there are two different ways to express conditions. The first approach places conditions in square brackets, `[]`, and the second places conditions in nested sets of parens, `(())`. Both forms have different syntactic rules.

Let's start our look at Bash comparisons with the use of the `[]`. Inside the brackets, we list the items to compare. To compare strings, use == (equal to) and != (not equal to). We cannot compare strings using greater than or less than. To compare integers, we use two-letter abbreviations in the form of `-eq`, `-ne`, `-lt`, `-gt`, `-le`, `-ge` (equal to, not equal to, less than, greater than, less than or equal to, greater than or equal to, respectively). We explore file comparisons later. Within the brackets, the items being compared and the comparison operator *must* all be separated by spaces. Table 6.10 provides several examples with correct and incorrect syntax. For those that are incorrect, the last column explains the error.

Let's focus on the fourth example, `[ $X != "Frank Zappa" ]`. This will not always cause an error. When attempting to compare `$X` to a literal string, if either `$X` or the literal string have a blank space, then the item must appear in quote marks to avoid receiving an error listed as "too many arguments". Obviously, `Frank Zappa` has a space and thus we put the string in quote marks. We could use single or double quote marks for the literal string.

What about `$X`? It may or may not have a space. If it does not have a space then no error arises. However, if it is equal to `Frank Zappa` then it will have a blank space, so we expect `$X` to have a blank space at least some of the time. To avoid the error from arising, we use `"$X"`. Notice here we would not want to use single quotes because `'$X'` causes Bash to literally compare the string `$X` and not the value stored in `$X`.

TABLE 6.10
Condition Examples

Condition	Meaning	Error Explanation (If Any)
`[ $X == Frank ]`	Does variable X store the value `Frank`?	
`[$X == Frank]`	Same.	No spaces after `[` and before `]`.
`[ $X -eq Frank ]`	Same.	Use `-eq` for numbers only
`[ $X != "Frank Zappa" ]`	Does variable X store anything other than `Frank Zappa`.	Since `$X` is expected to store a string with a space, we need to embed `$X` in double quotes, as in `"$X"`.
`[ $a -gt $b ]`	Is the value in a greater than the value in b?	
`[ $a > $b ]`	Same.	Can't use > in this notation.
`[ $a -gt $((b*3)) ]`	Is a>b*3?	
`[ $a -ne "$b" ]`	Does the value in a not equal the value in b?	b is treated as a string because of the quote marks and so can't be compared using `-ne`.

As you can see, there are many syntactic restrictions when using `[]`. To avoid these restrictions, we can use `(())` to enclose a condition, but only when comparing numeric values, not strings. The conditions do not have the archaic spacing requirements of `[]`, we use the more familiar relational operators of >, <, ==, !=, >= and <= and we can even omit the `$` preceding variable names. Some example conditions are `(($x==5))`, `((x!=y+z))`, `(( $x + 1 < y * 2 ))` and `(($x >= $y))`.

String comparisons must still use `[]` and require proper spacing. If we compare strings using the `(())` notation, the condition *always* evaluates to true. Thus, while performing a string comparison in `(())` does not result in a syntax error, there is little point in using this notation.

Compound conditions are those that perform multiple comparisons. In order to obtain a single true/false value, the comparisons are joined together using one of two Boolean operators: AND and OR. In Bash, AND is denoted as `&&` and OR as `||`. For instance, `$X -ne $Y && $Y -eq $Z` is true if the value in `X` does not equal the value in `Y` and the value in `Y` is equal to the value in `Z`. To further complicate the syntax, a compound conditional cannot be placed in a single set of brackets, `[]`, but instead must be nested in two sets, `[[]]`, or each condition must be placed in its own set of brackets. When using `(())`, we do not have such restrictions. See Table 6.11 for examples.

A variation of evaluating a condition is through the command `test`. This command is followed by a condition to be evaluated but without either `[]` or `(())` symbols. If the condition is a compound condition, conjoin the conditions with `-a` (AND) or `-or` (OR). For instance, `test $AGE -gt 12 -a $AGE -lt 30` tests to see if `AGE` is greater than `12` and less than `30`. The result of `test` is not used within another instruction (like an if-then statement) but instead sets the special parameter `$?`. If the condition is true `$?` is set to `0` and if false then `$?` is set to `1`. We explore `$?` later in the chapter but omit any further discussion of the `test` command.

6.5.2 File Conditions

One feature that is somewhat unique to Bash's scripting language is the ability to test files for properties. As scripts generally perform administrative tasks, perhaps on files and directories, the ability to query a file for its property is quite useful. Consider a script that is set up to run other scripts, as shown in Figure 6.12. Should `script1`, `script2` or `script3` not exist or not be a regular file or not an executable file, the script would yield some errors. In order to safeguard our script from generating such errors, we can test each file before trying to execute it. We examine how to do this in the next subsection.

TABLE 6.11
Example Compound Conditions

Condition	Explanation
`[[ $X -ne 0 || $Y -ne 0 ]]`	True if `X` or `Y` is not 0.
`(($X!=0\|\|$Y!=0))`	Same with alternate syntax.
`(( $X != 0 || $Y != 0 ))`	Same where spacing is inserted to make it easier to read.
`[[ $X -gt $Y && $Y -gt $Z ]]`	True if `X>Y` and `Y>Z`.
`(( $X>$Y && $Y>$Z ))`	Same with alternate syntax.
`[[ $FIRST == Frank && $LAST == Zappa ]]`	True if the values of `FIRST` and `LAST` are `Frank` and `Zappa` respectively.
`[ $FIRST == Frank ] && [ $LAST == Zappa ]`	Alternative syntax where each condition is in its own set of brackets.
`(($FIRST == Frank && $LAST == Zappa))`	Always evaluates to true.

```
#!/bin/bash
./script1 > output.txt
./script2 >> output.txt
./script3 >> output.txt
```

FIGURE 6.12 Script which executes other scripts.

The file comparisons use two formats. We use `[ comparison filename ]` to test if *filename* has the stated property, and we use `[ ! comparison filename ]` to test if *filename* does not have the stated property. The entry *filename* can either be the literal name of a file, such as `script1`, or a variable that is storing a filename. The name does not have to be a regular file but could be a directory or link, for example.

The list of operators that we use to test properties is shown in Table 6.12. Each of the options tests if the file exists as well as the specific property (e.g., `-d` tests to see if the item both exists and is a directory). Therefore, `-e` is only useful when testing if the item exists but irrespective of any other property.

In addition to the comparisons in Table 6.12, three other operators are available that compare two files. These comparisons determine if the first file is older than the second (`-ot`), the first file is newer than the second (`-nt`) and the two files are the same (`-ef`). We would use the condition `[ f1.txt -ot f2.txt ]` to see if `f1.txt` is older than `f2.txt`. Table 6.13 provides several examples with explanations.

Focusing on the last example in Table 6.13, if we wanted to test to see if `$FILE1` equals `$FILE2`, we could simply compare the two variables' values using ==, as in `[ $FILE1 == $FILE2 ]`. This would be true if the two variables are storing the same string. What the condition in the table is testing is if the two files are the same either because they are the same file or because the two file names are hard links of the same file.

Two last conditions can be used to test a variable. The first tests to see if a variable has a value using `[ -n var ]`. This is true if *var* is set. The second tests to see if the variable currently stores `NULL` (has no value) using `[ -z var ]`. We can also test for non-`NULL` using the condition `[ $?var ]`. We might use one of these tests to avoid a run-time error when we attempt to perform an operation on a variable that may not yet have a value.

TABLE 6.12
File Operators

Comparison Operator	Meaning
`-e`	File exists.
`-f`	File is a regular file (not a directory, device type file, link, etc.).
`-d`	File is a directory.
`-b`	File is a block device.
`-c`	File is a character device.
`-p`	File is a named pipe.
`-h, -L`	File is a symbolic link.
`-S`	File is a domain socket.
`-r, -w, -x`	File is readable, writable, executable.
`-u, -g`	User ID/Group ID is set (executable permission for user or group is `s` instead of `x` or `-`).
`-O, -G`	The file is owned by this user/this user's group.
`-N`	File has been modified since last read.

TABLE 6.13
File Condition Examples

File Condition	Meaning
`[ -f file1.txt ]`	Is `file1.txt` an existing regular file?
`[ -h file1.txt ]`	Is `file1.txt` a symbolic link to another file?
`[[ -r file1.txt && -w file1.txt ]]`	Is `file1.txt` both readable and writable?
`[ ! -e $FILENAME ]`	Does the file whose name is stored in `FILENAME` not exist?
`[ -O file1.txt ]`	Is `file1.txt` owned by the current user?
`[ ! file1.txt -nt file2.txt ]`	Is `file1.txt` not newer than `file2.txt`?
`[ $FILE1 -ef $FILE2 ]`	Is the file whose name is stored in `FILE1` the same file as the one whose name is stored in `FILE2`?

6.5.3 The if-then and if-then-else Statements

Now that we have viewed the types of comparisons available, we can apply them in selection statements. The first type of selection statement is the if-then statement, or a *one-way* selection. If the condition is true, the *then clause* is executed, otherwise the *then clause* is skipped and the if-then statement ends. The syntax for the if-then statement is `if [ `*`condition`*` ]; then `*`action(s)`*`; fi` where the action(s) is one or more instructions that make up the then clause.

Notice the use of the semicolons to separate the condition from the word `then` and the action(s) from the `fi` statement. In fact, these semicolons can be removed if each of the items is on a separate line. We see two ways to write the same if-then statement in Figure 6.13. The condition tests to see if `file1.sh` is an executable file and if so, executes it. The first example shows the instruction on one line and thus requires the semicolons. The second example shows the instruction on multiple lines and so the semicolons can be and are omitted. The indentation is strictly to help us read the instruction and not required by the interpreter.

Re-examine Figure 6.12 where we executed three scripts. As noted, if one of those scripts either did not exist or had improper permissions, the script in Figure 6.12 would generate an error. We can avoid such errors using a simple if-then statement as shown in Figure 6.13 to first test that the script is executable (and exists) and then execute it if the condition is true.

The then clause can consist of one or more instructions. If there are multiple instructions and they are on the same line, they must be separated by semicolons. If they are on individual lines, the semicolons can be omitted. Figure 6.14 demonstrates an if-then statement with multiple instructions in the then clause. Again, the indentation is strictly for our readability. The `then` can line up anywhere on the second line, and similarly, the `fi` can be intended or not.

In both examples examined so far (in the two figures), the then clause action(s) is performed or not. These are one-way selection statements executing the then clause only if the condition evaluates to true. In many situations, there are two possible operations to execute: one if the condition is true and one if the condition is false. These are known as *two-way* selection statements, and we write them using if-then-else instructions.

```
if [ -x file1.sh ]; then ./file1.sh; fi
-----------------------------------------
if [ -x file1.sh ]
     then ./file1.sh
fi
```

FIGURE 6.13 Two ways to write an if-then statement.

```
if [ "$NAME" == "Frank Zappa" ]
      then
           PAY=50000
           AGE=53
           echo Processing Zappa
           ./pay_scale $PAY
fi
```

FIGURE 6.14 If-then statement with multiple instructions.

The syntax of the if-then-else statement differs only slightly from the if-then statement. Prior to the `fi` statement, we add the word `else` followed by an *else clause*. As with the if-then statement's then clause, the else clause can be one or more instructions. If the clause consists of multiple instructions and they are on the same line, then they must be separated by semicolons. Figure 6.15 provides three examples. Hopefully, you can understand these three scripts with no additional explanation.

6.5.4 Nested Statements

The need for an *n-way* selection statement arises when there are more than two possible outcomes. This does not occur from a situation where we have a single condition, which evaluates to true or false, but when we have multiple different conditions to test.

A commonly cited example for using an n-way selection statement is to determine a student's letter grade based on the student's numeric score. Assume we use a 90/80/70/60 grading scale. Given the student's score, which range does it fall into? If 90 or above then A, else if less than 90 but greater than or equal to 80 then B, else if less than 80 but greater than or equal to 70 then C, else if less than 70 but greater than or equal to 60 then D, else F. The way we expressed this in text is nearly identical to how we can write this in code where we use if-then-else structures.

We refer to such logic as a *nested* if-then-else statement because the then clause and/or the else clause can contain its own if-then or if-then-else statement. Nested logic can be a challenge to learn but is an invaluable programming tool. One notable difference between our English description and how we implement this in Bash is that the expression "else if" is written as `elif`.

Four versions of the code to determine the letter grade given a numeric score are shown in Figure 6.16. Not all of these are correct and only one is preferred. Let's take a look at each. In the first set of code, we use five separate if-then statements. In each, we test to see if `SCORE` falls in

```
if [ $SCORE -ge 60 ]; then GRADE=pass; else GRADE=fail; fi
------------------------------------------------------------
if [ $Y -ne 0 ]
    then
        Z=$((X/Y)); echo Quotient is $Z
    else
        echo Cannot divide by 0
fi
------------------------------------------------------------
if [ $FILE == script1.sh ]
    then
        echo input two parameters
        read x y
        ./$FILE $x $y
    else
        echo input a parameter
        read x
        ./$FILE $x
fi
```

FIGURE 6.15 Example if-then-else statements.

```
if [ $SCORE -ge 90 ]; then GRADE=A; fi
if [[ $SCORE -ge 80 && $SCORE -lt 90 ]; then GRADE=B; fi
if [[ $SCORE -ge 70 && $SCORE -lt 80 ]; then GRADE=C; fi
if [[ $SCORE -ge 60 && $SCORE -lt 70 ]; then GRADE=D; fi
if [ $SCORE -lt 60 ]; then GRADE=F; fi
-------------------------------------------------------
if [ $SCORE -ge 90 ]; then GRADE=A; fi
if [[ $SCORE -ge 80 && $SCORE -lt 90 ]; then GRADE=B; fi
if [[ $SCORE -ge 70 && $SCORE -lt 80 ]; then GRADE=C; fi
if [[ $SCORE -ge 60 && $SCORE -lt 70 ]; then GRADE=D;
else then GRADE=F; fi
-------------------------------------------------------
if [ $SCORE -ge 90 ]
     then GRADE=A;
     elif [[ $SCORE -ge 80 && $SCORE -lt 90 ]]
          then GRADE=B;
     elif [[ $SCORE -ge 70 && $SCORE -lt 80 ]]
          then GRADE=C;
     elif [[ $SCORE -ge 60 && $SCORE -lt 70 ]]
          then GRADE=D;
     else GRADE=F;
fi
-------------------------------------------------------
if [ $SCORE -ge 90 ]
     then GRADE=A;
     elif [ $SCORE -ge 80 ]
          then GRADE=B;
     elif [ $SCORE -ge 70 ]
          then GRADE=C;
     elif [ $SCORE -ge 60 ]
          then GRADE=D;
     else GRADE=F;
fi
```

FIGURE 6.16 Four possible solutions to determining letter grade.

the expected range for the letter grade and if so, assign the proper letter grade to GRADE. This first example works correctly but is not preferred because it requires more effort to write. It also tests every condition of the five if-then statements which is not necessary should, for instance, the first condition evaluate to true.

The second set of code is nearly identical except that the last statement has been changed from an if-then statement to the else clause of the previous if-then statement, making the second to last statement into an if-then-else statement. This code, although similar to the first, uses incorrect logic. Let's see what happens if SCORE is 92. The first if-then condition is true, and GRADE is assigned an A. Because this was an if-then statement, when it ends, we move on to each of the successive if-then and if-then-else statements. The next two if-then statements have false conditions and so do not execute their then clauses. But the last statement is an if-then-else and with the condition being false, the else clause executes. This changes GRADE from A to F. Even though SCORE is 92 and the correct if-then statement executed its then clause, the improper logic caused the if-then-else statement's else clause to execute. A student with a SCORE of 92 gets an F.

It is the last two examples that we will concentrate on because these both use nested-if-then-else logic that we want to explore. In the third example of the figure, we have used a similar approach as to the second example except that rather than having three if-then and one if-then-else statement, the entire set of code is a single nested if-then-else statement. In this case, the first else is an elif which has its own if-then-else statement as its then clause. The second else is another elif with its own if-then-else statement as its then clause. The third else is yet another elif with an if-then-else as its then clause. The last else is for the else clause of the last elif.

Although the code in this third example is correct, it is not the preferred solution because it is both too much work for the programmer and for the computer. Let's consider logically what happens if SCORE is less than 90. The first condition is false, and the computer executes the first elif clause. To reach this elif clause, we already know SCORE is not greater than or equal to 90. There is no need to compare SCORE to 90 since we already know that it must be less than 90 to reach this elif clause. Therefore, the first elif need only test $SCORE -ge 80. This is true of the remaining conditions as well. By requiring that each condition test SCORE to two values, we make the computer do more work and it is more work for the programming to enter the code.

The fourth example in Figure 6.16 takes advantage of the nested logic. If the first condition is false ($SCORE -ge 90 is not true) then for our second condition, we only need to test if $SCORE -ge 80. Should this also be untrue, then our next test (the second elif clause) needs to only compare $SCORE -ge 70 because we already know SCORE is not greater than or equal to 80. The first and third examples are logically correct, but it is the fourth example that uses the if-then-elif-else structure properly and thus is the most efficient and logical set of code.

Let's consider a more complicated example. We input three numeric values and store them in X, Y and Z. We want to determine and output the largest of the three. To determine which of the three is largest, we have to compare each variable against the other two. We can do this using three separate compound conditionals. The first might test to see if X is the greatest of the three by using [[$X -gt $Y && $X -gt $Z]]. We could then use three if-then statements, each with a condition similar to the above except that each one is testing a different variable. Alternatively, we could use two conditions and have an if-then-elif-else statement where the if tests to see if X is greatest and the elif tests to see if Y is the greatest. If we reach the else clause, we know that both conditions failed and so Z is greatest.

While either approach described in the previous paragraph is simple enough, neither is as efficient nor as elegant as handling the problem by using nested logic. We show the first simple approach and the fully nested approach in Figure 6.17. The top script, although more concisely written, does not take full advantage of the capabilities of the instruction. The bottom script is the preferred version. Notice that the first script in the figure contains three if-then-fi statements. The second script instead has a single if-then-elif statement where the then and elif clauses contain if-then-else statements. This is why there are two fi statements, one for the inner if-then-else and one for the outer if-then-elif.

There are three issues with the first script in Figure 6.17. The first is that it isn't as efficient as the second. If the first if-then statement executes, there is no need for the second or third to even be considered. We can resolve this by changing the code to an if-then-elif-else. The second issue is that it tests up to six conditions, whereas in the second script, we will only test at most three. Thus, the first script is less efficient.

```
#!/bin/bash
read -p "Enter three numbers:  "   X Y Z
if [[ $X -gt $Y && $X -gt $Z ]]; then Largest=X$; fi
if [[ $Y -gt $X && $Y -gt $Z ]]; then Largest=Y$; fi
if [[ $Z -gt $X && $Z -gt $Y ]]; then Largest=Z$; fi
echo Largest is $Largest
----------------------------------------------------
#!/bin/bash
read -p "Enter three numbers:  "   X Y Z
if [ $X -gt $Y ]
     then
          if [ $X -gt $Z ]
               then Largest=$X
               else Largest=$Z
          fi
     elif [ $Y -gt $Z ]
          then Largest=$Y
     else Largest=$Z
fi
echo Largest is $Largest
```

FIGURE 6.17 Two ways to determine the largest of three values.

The third issue is a logical problem. Should the two largest values (or all three) be the same, the logic behind the first set of code fails. `LARGEST` would not store a value and so the output would tell us the largest value was and then output no value! We could resolve this easily enough by changing all of the comparisons from `-gt` to `-ge`.

Imagine instead of testing three variables we are testing six or eight. The first approach becomes overly burdensome while the second approach, while becoming increasingly more complex and nested, only requires comparing two variables at a time so is ultimately a better approach. Thus, we might say that the first approach is not *scalable*.

Let's put all of these ideas together and write a script. We want to write a simple calculator script. The script expects three parameters: a number, an arithmetic operator and a second number. The script first tests to make sure we received three parameters (recall `$#` is the number of parameters supplied to the script). The script then tests the arithmetic operator and performs the operation on the first and third parameters. We reference the three parameters as `$1`, `$2` and `$3`.

We use a nested if-then-elif-else statement to determine whether the number of parameters is erroneous and if not, what operation to perform based on the value stored in `$2`. The script outputs an error message if the script received the wrong number of parameters, the response of the operation if we received three parameters and the operator was a legal operator, or an error message if operator was not one of those expected. The script can accept one of five arithmetic operators: +, –, *, /, % (mod). The script is shown in Figure 6.18 along with a couple of example outputs. Assume the script is called `calculator`.

The logic behind this script should be self-explanatory. There is one logical flaw with the script that has nothing to do with the code itself. Recall that * is a wildcard character in Bash. When used from the command line, the Bash interpreter performs filename expansion on it, thus replacing the * with all files in the current directory.

Imagine that the directory has three files, the calculator script named `calculator`, and files `file1.txt` and `file2.txt`. Bash will insert those three filenames into the command in place of *. Thus, `./calculator 3 * 5` becomes `./calculator 3 calculator file1.txt file2.txt 5`. Obviously, the script will find five parameters instead of three and output the illegal input message. In order to avoid this problem, we have the user use `m` in place of * and change the third `elif` statement to be `elif [ $2 == "m" ]; then echo $(($1*$3))`.

6.5.5 Case Statement

The `case` statement is another form of n-way selection which is similar in many ways to the if-then-elif-else statement. The instruction compares a value stored in a variable (or the result of an expression) against a series of enumerated lists. Upon finding a match, the corresponding action(s)

```
#!/bin/bash
     if [ $# -ne 3 ]
          then echo Illegal input, 3 parameters expected
          elif [ $2 == "+" ]; then echo $(($1+$3))
          elif [ $2 == "-" ]; then echo $(($1-$3))
          elif [ $2 == "*" ]; then echo $(($1*$3))
          elif [ $2 == "/" ]; then echo $(($1/$3))
          elif [ $2 == "%" ]; then echo $(($1%$3))
          else echo Illegal arithmetic operator
     fi
-----------------------------------------------------------
[foxr@localhost ~]$ ./calculator 315 / 12
26
[foxr@localhost ~]$ ./calculator 41 + 815
856
```

FIGURE 6.18 Example calculator script.

is(are) executed. The case statement (known as a *switch* in C/C++/Java) is more restricted in other languages than it is in Bash. For instance, in C, the expression can only be an *ordinal* type (e.g., an integer or character). Most languages require that an enumerated list be just that a list of specific values. If we wanted to implement the previous grading script (as described in Figure 6.16) using a case statement, most languages would require listing all of the integer values between 90 and 100, or between 80 and 89, etc. The Bash version though gives us a bit of a break in that we can employ wildcard characters to reduce the size of any enumerated list.

The Bash form of the `case` instruction is shown in Figure 6.19. The expression can be the value stored in a variable (in which case we have to precede the variable name with a `$`) or an arithmetic or string expression as in `$((X*5))` or `` `expr substr $string 5 10` ``. Each list can be a single item, a number of items separated by the OR operator (`|`), and/or items that use wildcard characters such as `*` , `[]` and `[!]`. Following the list and a close parenthesis is the action(s) to take place. There can be multiple actions but if on the same line the actions are separated by semicolons (if actions are on separate lines the semicolons can be omitted). The last instruction in the action list may end with the notation `;;` as described shortly. The entire `case` statement ends with `esac`.

Revising our previous calculator program using the `case` statement yields the script in Figure 6.20. Notice that we have replaced * with `m` not only for the reason discussed in the end of the last subsection but because * serves as a wildcard that would cause the Bash interpreter to select it whenever it was reached and thus the code would never reach the final three choices. The * can and is often used in `case` statements, as a default, or else, clause. We see this as the last choice in Figure 6.20 so that the `case` statement can output an error message. The `case` statement in this example is embedded in an if-then-else so that we can ensure three parameters were passed to the script.

Let's look at some additional examples as shown in Figure 6.21. The script in the top portion of the figure asks the user to input their favorite color and, based on the response, outputs a different message. The scripts in the middle and bottom portions of the figure demonstrate the use of wildcards (other than *). In the second script, we ask the user a yes/no question and compare the result against various forms of answers (`Y`/`N`/`yes`/`no`). The third example expects a directory as a parameter and iterates through the directory, executing a different compiler on each file based on the file's extension. We explore the for-loop in the next section so for now you can ignore that aspect of the script.

```
case expression in
          list1) action(s);;
          list2) action(s);;
          ...
          listn) action(s);;
esac
```

FIGURE 6.19 Structure of `case` statement.

```
#!/bin/bash
if [ $# -ne 3 ]
     then echo Illegal input, 3 parameters expected
     else case $2 in
          +) echo $(($1+$3));;
          -) echo $(($1-$3));;
          m) echo $(($1*$3));;
          /) echo $(($1/$3));;
          %) echo $(($1%$3));;
          *) echo Illegal arithmetic operator;;
     esac
fi
```

FIGURE 6.20 Calculator script using `case` statement.

```
#!/bin/bash
echo What is your favorite color?
read color
case $color in
     red ) echo Like red wine ;;
     blue ) echo Do you know there are no blue foods? ;;
     green ) echo The color of life ;;
     purple ) echo Mine too! ;;
     * ) echo That's nice, my favorite color is purple ;;
esac
-------------------------------------------------------
#!/bin/bash
echo -n Do you agree to the licensing terms of the program?
read answer
case $answer in
     [yY] | [yY][eE][sS] )  ./program ;;
     [nN] | [nN][oO]  ) echo I cannot run the program ;;
     * )  echo  Invalid response ;;
esac
-------------------------------------------------------
#!/bin/bash
if [ $# -eq 0 ]; then echo Error, no directory provided
   else
      for file in $1; do
          case $file in
               *.c ) gcc $file ;;
               *.java ) javac $file ;;
               *.sh) ./$file ;;
               *.* ) echo $file is of an unknown type;;
               * ) echo $file has no extension to match;;
          esac
      done
fi
```

FIGURE 6.21 Example scripts with `case` statements.

A last example of using `case` is shown in Figure 6.22. This script is used to control the starting and stopping of a service. This type of script was prevalent in older versions of Linux but has largely been replaced in `systemd`-versions of Linux (something we cover in later chapters). The script expects the user to supply one parameter of either `start`, `stop`, `restart` or `status`. The parameter is referenced as `$1`. Depending on the word supplied, the script invokes `someservice` passing it parameter `-s` (start it), `-k` (stop it), both (restart it) or `-d` (obtain the service's status). For `restart`, the `case` statement performs two operations. As the two instructions are on separate lines, we do not need to insert a semicolon between the two. Also, the `echo` statement should be on one line but due to space restriction is shown on two lines.

```
#!/bin/bash
if [ $# -eq 0 ]; then echo Error, no command supplied
  else
    case $1 in
      start ) /usr/sbin/someservice -s ;;
      stop ) /usr/sbin/someservice -k ;;
      restart ) /usr/sbin/someservice -k
                /usr/sbin/someservice -s ;;
      status ) /usr/sbin/someservice -d ;;
      * ) echo Usage: ./script [start | stop | restart |
          status ] ;;
    esac
fi
```

FIGURE 6.22 Script to control a service.

You will have noticed that every pattern/action in our `case` statements ends with two semicolons. The idea behind a `case` statement is that the first pattern that matches will cause the action to execute and the `case` statement ends. However, with C's switch statement, the implementers decided that after an action executes, the switch statement should continue testing additional patterns. In Bash, we can specify this same behavior by using `;&` instead of `;;`. That is, if an action ends with `;&`, then the Bash interpreter continues executing the `case` statement by examining the next patterns, but when an action ends with `;;`, then the Bash interpreter ends the `case` statement.

6.5.6 Conditions Outside of Selection Statements

We end this section with one additional form of selection statement. This is a shorthand notation for an if-then statement by using a concept called *short-circuiting*. Most languages employ short-circuiting to reduce computation that is unnecessary. Consider the statement `$A -gt $B && $A -gt $C`. If `A` is not greater than `B` then there is no point in comparing `A` to `C` because the entire statement is already determined to be false. By skipping the second comparison, we are short-circuiting its evaluation. When using `&&`, if the first condition is false, the second condition remains untested and similarly for `||`, if the first condition is true then the second condition is untested.

In a Bash script, a conditional operation has the form `[ some-condition ] && some-action` or `[ some-condition ] || some-action`. With the first notation, if the condition is false the action is not performed. Using the second notation, if the condition is true the action is not performed.

How might we put this to use? The most common situation is to combine a condition that tests a file to see if it exists or is accessible with an instruction that accesses the file. We saw previously that we used `-x` to see if a script existed and if so, then we executed it. Using this shorthand notation, we can specify `[ -x somescript.sh ] && ./somescript.sh` to execute the script should it exist rather than using the lengthier if-then statement.

We can similarly safeguard accessing a variable without risking an error should the variable not have a value. We do so by testing if a variable is not yet set and if so, short-circuit the remainder of the statement. This might appear as `[ -z $b ] || a=$((b+1))`. Here, if `b` has no value, the assignment statement does not execute, preventing an error from arising.

SECTION ACTIVITIES

1. Once again Bash has multiple ways to express something, conditions in this case. You can use `[]` and you can use `(())`, but the types of relational operators you use differ between them. Which approach do you prefer? Why?
2. Most programming languages do not have file condition tests. Why do you suppose these were included in Bash? Research other scripting languages to see if any have file conditions like "is a directory", "is readable", etc. If you find any, what were those languages?
3. Between if-then, if-then-else, nested if-then-else and case statements, which of these have you used in programming before? Compare the version you used before to the form available in Bash. Do you feel that your previous experience will help you use the related form in Bash?

6.6 LOOPS

Loops, also called *repetition* and *iteration* statements, are used to execute a set of statements numerous times. The Bash scripting language has three types of loops: *conditional* loops, *counter-controlled* loops and *iterator* loops.

A conditional loop is a loop controlled by a condition where conditions are as we described them in Section 6.5. There are two types of conditional loops: *while* loops and *until* loops. Both of these loops are known as *pre-test* loops because the condition controlling whether the loop body executes or not is tested *before* the loop body executes. The difference between the while and until loops is in the semantics of the condition. The while loop iterates *while* the condition is true. The until loop iterates *until* the condition becomes true, or iterates while the condition is false.

Counter-controlled loops were added to Bash. They use a variable to count the number of iterations. The Bash counter-controlled loop is similar to that of C in syntax. The iterator loop iterates through all of the items in a list. In this section, we explore all of these loops.

6.6.1 Conditional Loops

The syntax of the two conditional loops is shown in Figure 6.23. The `while` loop is shown in the top half, and the `until` loop is shown in the bottom half. Both loops use a condition like those of if-then statements. Both loops end with the word `done`. The semicolon following the condition is only needed if the word `do` appears on the same line. Similarly, the semicolon(s) after the action(s) is(are) only needed if multiple actions appear on the same line or the word `done` appears on the same line as an action.

Let's examine an example illustrating both loops. There are two scripts in Figure 6.24 which demonstrate how to sum up a list of numbers input from the user. We use a `read` statement to prompt the user for a number and input it. We use a loop to repeat while the user enters "legal" values. Any negative input value indicates that the loop should end.

Notice that we will require two input statements, one before the loop to obtain the user's first number, and one at the bottom of the loop to obtain the next number. Without the first input, there is no initial input value to test, and without the second input, the input value never changes resulting in an infinite loop (a topic we visit shortly). Compare the two loops' conditions and you will see

```
while [ condition ]; do
     action(s);
done
--------------------------
until [ condition ]; do
     action(s);
done
```

FIGURE 6.23 Syntax for `while` and `until` loops.

```
#!/bin/bash
SUM=0
read -p "Enter first number, negative to exit " VALUE
while [ $VALUE -ge 0 ]; do
     SUM=$((SUM+VALUE))
     read -p "Next number, negative to exit " VALUE
done
echo The sum is $SUM
------------------------------------------------------
#!/bin/bash
SUM=0
read -p "Enter first number, negative to exit " VALUE
until [ $VALUE -lt 0 ]; do
     SUM=$((SUM+VALUE))
     read -p "Next number, negative to exit " VALUE
done
echo The sum is $SUM
```

FIGURE 6.24 Two loop-based scripts to sum numbers.

that they are opposites. The `while` loop uses `[ $VALUE -ge 0 ]` and the `until` loop uses `[ $VALUE -lt 0 ]`.

The conditional loops from the scripts in Figure 6.24 demonstrate a type of control called a *sentinel* loop. This type of loop uses an input value both for some computation performed in the loop and to determine whether to continue executing the loop. An alternative to inputting a negative value to exit the loop is to explicitly ask the user if they have another input to add. This version of the script is shown in Figure 6.25. We only consider a version with a `while` loop. While this version of the loop might use simpler logic, it is more work for the user who now has to enter both a yes/no and a number.

Another style of loop is one controlled by a computed value. Figure 6.26 provides an example script which computes all of the powers of 2 less than 1000. We control the loop by comparing the computed value, `VALUE`, to 1000, exiting once `VALUE` is greater than or equal to 1000. We again use a `while` loop in which we output `VALUE` first and then multiply it by 2. By doing the two operations in this order, we ensure that we only print powers of 2 less than 1000. Notice that we initialized `VALUE` to `1`.

We mentioned an *infinite* loop earlier. This is a type of loop where, once executing the loop body, the loop never terminates. This occurs because the body of the loop does not alter whatever variable(s) is(are) being tested in the loop's condition. In Bash, `while` and `until` loops can create infinite loops.

For instance, if the loop body of Figure 6.26 only contains the `echo` statement, then `VALUE` never changes from `1` and so the condition is always true. A more subtle cause of an infinite loop arises in this script if we initialize `VALUE` to `0`. If this is the case, then `VALUE=$((VALUE*2))` always results in `VALUE` being `0` and so it will always be less than `1000`.

Re-examine both scripts from Figure 6.24. If we omitted the second `read` instruction from either then `VALUE` remains the original input number and if this was not a negative, the loop executes forever. In fact, infinite loops may not actually execute forever. In this case, unless `VALUE` was `0`, it would continually add itself to `SUM` and eventually overflow `SUM` causing the script to terminate with an error.

```
#!/bin/bash
SUM=0
read -p "Do you have numbers to sum? " ANSWER
while [ $ANSWER == Yes ]; do
     read -p "Enter the next number " VALUE
     SUM=$((SUM+VALUE))
     read  -p  "Do  you  have  additional  numbers? " ANSWER

done
echo The sum is $SUM
```

FIGURE 6.25 Another script to sum numbers.

```
#!/bin/bash
VALUE=1
while [ $VALUE -lt 1000 ]; do
     echo $VALUE
     VALUE=$((VALUE*2))
done
```

FIGURE 6.26 Script to compute and output powers of 2.

6.6.2 Counter-Controlled Loops

The counter-controlled loop in Bash is a variation of the C/Java for-loop, using similar syntax. In this loop, we specify three pieces of information: a variable *initialization*, a loop *continuation condition* and a variable *increment*. The format of this instruction is shown in Figure 6.27 along with several examples.

The counter-controlled for-loop starts by performing the initialization step. This places into the specified variable, called the *loop variable*, the starting value. In Figure 6.27, the loop variable in the three examples is always `i` and the starting value is `0`, `1` and `100`, respectively, for the three loops. The initialization component ends with a semicolon.

The next part of the for-loop is the condition to be tested before each loop iteration. If the condition is true, the loop body executes. The first example tests `i<100`, while the second tests `i<=100`. The third, a downward counting loop, executes while `i>j`. The condition ends with another semicolon. Notice that the conditions are already nested inside of `(())` and so need no further notation. As a counter-controlled loop expects the loop variable to be an integer, we are free to use operators like < and <= and forego the `$` before the variable names.

The third component in the loop describes how the loop variable will change. Here, we can provide a full assignment statement such as `i=i+2` or a shorthand notation like the postfix increment or decrement. Again, as the increment is within `(())` and we are operating on integers, we are free to express these without special notation like `$((i+1))`.

All three components of the for-loop are placed in double parentheses and followed by the word `do`. Following `do` is the *loop body*, or the instructions that make up the loop. These are the instructions executed if the condition tested is true. Ending the loop is `done`.

In examining the first two examples in Figure 6.27, notice that semicolons are placed before the word `do` and after the instruction that makes up the loop body (e.g., `echo $i`). If `do` was on a separate line from the condition and `done` on a separate line from an instruction in the loop body, the semicolons could be omitted, as we see in the third example. The three loops iterate from 0 through 99 printing each number, iterate from 1 up to but not including `j` by 2s, adding each value to `sum` and iterate from 100 down to 1 printing each number, respectively.

6.6.3 Iterator Loops

The final loop available in Bash is another for-loop and is in fact the original for-loop. This version of the for-loop is an iterator loop. An iterator loop is one that executes the loop body one time per item in the specified list. This for-loop is similar to the *foreach* loop as found in C# and Perl. The syntax is `for` *`VAR`* `in` *`LIST`*`;` `do` *`action(s)`*; `done`. The variable, *VAR*, is our *loop variable*, which is assigned to each element of *`LIST`* one at a time. As with the other loops, the semicolons

```
for (( initialization; condition; increment ))
     do
          loop body
     done
-----------------------------------------------------
for (( i=0; i<100; i++ )); do echo $i; done
-----------------------------------------------------
for((i=1;i<=j;i=i+2));  do  sum=$((sum+i)); done
-----------------------------------------------------
for ((i=100;i>0;i--))
     do
          echo $i
     done
echo $sum
```

FIGURE 6.27 Bash counter-controlled for-loop syntax and examples.

can be omitted if do is on a separate line from the `for ... LIST` and if the loop body instructions are on separate lines.

The list can be an enumerated list such as (1 2 3 4 5), the result of a Linux command which generates a list (e.g., `ls`, `cat`), a list generated through filename expansion (for instance `*.txt` would expand into all file names of `.txt` files in the current directory), the list stored in a variable (assuming the variable stores a list) or the list of parameters supplied to the script using `$@`. There is also a sequence command, `seq`, that we will examine later, which can generate lists of numbers

Two example scripts in Figure 6.28 demonstrate the use of both the iterator for-loop and `$@`. The first script expects some numeric parameters, and the script adds each of these parameters to `SUM`, outputting the result. The if-then statement is used to ensure that the user supplied some parameters.

The second script in Figure 6.28 iterates through all of the `*.txt` files in the current directory, runs the `wc` program on them and outputs the total number of words and lines found in these files. Notice that we could run `wc` on the `*.txt` without writing a script but doing so would not provide the summary. By default, `wc` outputs the number of lines, words and characters as well as the file's name. We use `awk` to pull out just the first and second numbers.

Earlier, we introduced the instruction `shift`. Now that we have seen the use of `$@`, we can reconsider `shift`. The easiest way to iterate through the parameters in a list is through the for-loop iterating over `$@` as in `for var in $@; do ...` We can similarly use `shift` and iterate through a list using a counter-controlled for-loop. We see an example of this in Figure 6.29 whereby, because of `shift`, the next parameter will always be located in `$1`. This script is similar to that in the top half of Figure 6.28, but instead of using the iterator for-loop, we use a counter-controlled for-loop, shifting each parameter down one position after we use the current `$1`. Also, unlike the script in the first part of Figure 6.28, we do not use an if-then-else statement to ensure that some parameters were passed to the script.

Let's put together everything we have learned to this point of the chapter to build a script that will count the number of times a particular word occurs in a list of files. For this script, we will input from the user the word we are searching for. The list of files will be supplied as parameters that we will iterate through via the iterator for-loop and `$@`. For each file, we will use the `cat` command to obtain all of the words in the file. We will use another iterator for-loop to iterate through that list and count the number of times the user's input word appears. We situate the two for-loops so that they are nested. The outer loop iterates over each file, and the inner loop iterates over the strings found

```
#!/bin/bash
SUM=0
if [ $# -eq 0 ]; then echo ERROR, no parameters provided
else
   for VALUE in $@; do
      SUM=$((SUM+VALUE))
   done
   echo Total of parameters is $SUM
fi
------------------------------------------------------
#!/bin/bash
LINES=0
WORDS=0
for FILE in *.txt
   do
        LINES=$((LINES+`wc $FILE | awk '{print $1}'`))
        WORDS=$((WORDS+`wc $FILE | awk '{print $2}'`))
   done
echo Number of lines found: $LINES
echo Number of words found: $WORDS
```

FIGURE 6.28 Sample scripts using iterator for-loops.

```
#!/bin/bash
NUMBER=$#
SUM=0
for (( i=0; i<NUMBER; count++ )); do
     SUM=$((SUM+$1))
     shift
done
echo $SUM
```

FIGURE 6.29 A summation script using `shift`.

within the current file. Between the outer and inner loops, we reset our counter so that each file's count starts at 0. The script is shown in Figure 6.30.

Let's step through the script in Figure 6.30. First, it obtains the user's word to search for, stored in the variable `input`. Next, it iterates through all of the parameters. An if statement ensures that each of these parameters is a readable file. If so, the variable `count` is set to `0` to start a new count.

The inner for-loop obtains all of the words in this file using the `cat` instruction and iterates over these words, storing each word in the loop variable `word`. If `word` is equal to `input`, `count` is incremented to reflect another instance of `input` was found. When done iterating through the words of this file, the script outputs the filename and the value of `count`. If the file was not readable, the script instead outputs an error message. By using "word", we mean a string delimited by white space (spaces, tabs, new lines).

We could alter the script so that instead of inputting a word from the user, the search word is the first parameter passed to the script. We would make two slight changes to the script, replacing the `read` instruction with `input=$1` and then executing `shift` to remove that item from the parameter list.

We could also accumulate all of the counts so that we could output the total number of matches of all of the files after executing the outer loop. We would accomplish this by adding a variable, `total` (initialized to 0 before the outer for-loop) and adding the instruction `total=$((total+count))` after the inner for-loop (between the `done` of the inner loop and the first `echo` instruction). After the outer for-loop ends, we would add an `echo` statement to output `total`.

While the script from Figure 6.30 illustrates several of the ideas we have covered, it is not a necessary script as we could use `grep -c` to obtain the same information (although not the total). Alternatively, we could write a script with a single for-loop to iterate through all of the parameters (files). The loop's body would use `grep -c` piped to `awk` to retrieve the number of times the user's word appears in that file and add it to a running total.

Iterating over a list of files is a common application for Bash scripts. Iterating over the parameters passed to the script, as we did in Figure 6.30, is just one way we could perform such an operation. To iterate through files in the current directory, we can use `for file in *` or `for file in `ls`

```
#!/bin/bash
read -p "Enter the word you are searching for " input
for file in $@; do
    if [ -r $file ]; then
       count=0
       for word in `cat $file`; do
           if [ $word == $input ]
               then count=$((count+1))
         fi
       done
       echo $input occurs $count times in $file
    else
       echo $file is not a readable file
    fi
done
```

FIGURE 6.30 Example script to count a specific word in a group of files.

*`. Yet another option is to obtain a directory name from the user via a read instruction and then iterate through the files found in that directory using a loop like `for file in `ls $input``.

Figure 6.31 receives a list of directories as parameters. It iterates through the parameters and, assuming the current item is a directory and accessible, an inner loop iterates through the contents of the directory and sums up the size of the items, counting the number. It then outputs each directory's name and number of and size of its contents.

6.6.4 Using the seq Command to Generate a List

We noted that the iterator loop will iterate over any list. We can supply the list ourselves as in (1 2 3 4 5) but we can also generate a list. The seq command returns a list of numbers in a sequence. We control the sequence by specifying the first number in the sequence, the last number in the sequence and a *step size* (the amount that the sequence increases by with each new number). The last number is required. If the first number is not included, the sequence starts at 1. If the step size is not included, the sequence increases by 1.

Figure 6.32 shows the three forms of syntax for the seq command followed by a number of examples issued from the command line. Each of *first*, *step* (the step size) and *last* can be literal numbers, numbers stored in variables or numbers returned from arithmetic operations. Assume in the figure that A=2 and B=9.

```
#!/bin/bash
for dir in $@; do
    if [[ -d $dir && -r $dir ]]; then
       count=0
       total=0
       for item in `ls $dir`; do
           size=`ls -l $item | awk '{print $5}'`
           total=$((total+size))
           count=$((count+1))
       done
       echo Directory $dir has $count items of size $total
    fi
done
```

FIGURE 6.31 Script to iterate through directories and output size.

```
seq last
seq first last
seq first step last
----------------------------------------------------
[foxr@localhost ~]$ seq 10
1 2 3 4 5 6 7 8 9 10
[foxr@localhost ~]$ seq 3 9
3 4 5 6 7 8 9
[foxr@localhost ~]$ seq 0 5 25
0 5 10 15 20 25
[foxr@localhost ~]$ seq $A $B
2 3 4 5 6 7 8 9
[foxr@localhost ~]$ seq 1 $A $B
1 3 5 7 9
[foxr@localhost ~]$ seq $((A+B))
1 2 3 4 5 6 7 8 9 10 11
[foxr@localhost ~]$ seq $A $((A+1)) $((A*B))
2 5 8 11 14 17
```

FIGURE 6.32 The seq command.

We can use seq to control a for-loop in one of two ways. First, we can store the result of seq in a variable, such as LIST=`seq 1 10` and then reference the variable in the iterator for-loop. Second, we can insert the `seq` instruction in the iterator for-loop itself. Figure 6.33 provides an example of this, first showing a counter-controlled for-loop and then the equivalent iterator for-loop. Which version you might use is based on your own preference of which you find easier.

We need to be careful with the types of operations that we place inside `` marks in a for-loop. For instance, if we use `ls -l *`, the for-loop iterates through each item in the long listing of each item in the directory. Thus, the for-loop would step through not only the files but their permissions, hard link counts, owner and group, and so forth.

6.6.5 The while read Statement

We've seen both the while loop and the read statement. We can combine the two instructions when we want to repeatedly input values using read. We most commonly use this form of input to iterate through the contents of a file, although we could also use this pair of instructions to iterate over values input from keyboard.

The top portion of Figure 6.34 demonstrates the syntax of this instruction and an example script called foo. If we invoke the script as ./foo (i.e., without redirecting an input file to it) then the while read inputs from keyboard until the user enters control+d. If we invoke the script as ./foo < somefile then the while read iterates once per line of input. The script in Figure 6.34 inputs pairs of values and adds the larger to sum, outputting sum once all inputs have concluded.

The while read consumes all values of one line of the input from the keyboard ending with <enter> or one line of input from a file per loop iteration. If the number of inputs does not match the number of variables in the read statement, we may not necessarily receive an error. Instead, if there are more inputs than variables then all variables get one value except for the last variable which receives the remaining inputs as a list of values. If there are fewer inputs than variables, then each input is placed into one variable until we run out of variables and the remaining variables receive NULL. Unfortunately, this *could* cause errors in other instructions.

```
#!/bin/bash
for (( number=1; number<=10; number=number+2)); do
     echo $number
done
for number in `seq 1 2 10`; do
     echo $number
done
```

FIGURE 6.33 Comparing counter-controlled for-loop to iterator for-loop.

```
while read varlist; do
     loop body
done
-----------------------------------------
#!/bin/bash
sum=0
while read x y; do
     if [ $x -gt $y ]; then
          sum=$((sum+x))
     else
          sum=$((sum+y))
     fi
done
echo Sum of largest of each pair is $sum
```

FIGURE 6.34 Syntax and example using while read.

Consider in the example script in Figure 6.34 if a line of input has three numbers. The variable x receives the first number and y receives a list of the second and third numbers. This is not an error. But when testing [$x -gt $y] we do receive an error because $y is expected to be a single value but is a list.

If we want to input from disk file, we could also indicate the redirection of the read instruction from the file directly in the script itself. This is done not by redirecting the file from the command line, as in ./script < inputfile.txt, but instead using done < *filename* in the done statement that closes the while read instruction. We revise the script from Figures 6.34 to 6.35 showing this variation. The input file, in this case, is named pairs.txt. The disadvantage with this approach is that the script will only run on this one file.

SECTION ACTIVITIES

1. Assuming you know another programming language, which of the types of loops does that language have? Iterator for-loop? Counter-controlled for-loop? Pretest conditional loop (while, until)? Posttest conditional loop (e.g., C's do-while)? Which of these do you feel a language should have?
2. With the iterator for-loop and the seq command, the counter-controlled for-loop is not necessary. Yet it was added to Bash. Why do you suppose it was added? Come up with an example demonstrating why the counter-controlled loop would be useful instead of the iterator loop with the seq command.
3. Have you ever written an infinite loop (unintentionally)? If so, at what point did you realize it was an infinite loop and how long did it take you to debug the reason?
4. Change the while loop in Figure 6.24 into an infinite loop by removing the second read instruction. When done, run the script and enter a positive number when asked. Does the script run forever or does it terminate with an overflow error?

6.7 ARRAYS

Arrays are data structures that store multiple values. In most languages, arrays are *homogenous* meaning that all of the elements of the array must be of the same type of value. In Bash, all variables store strings (including numbers) so the idea of a homogenous structure is unimportant.

As arrays store multiple values, we need to indicate the specific value we want to access. This is accomplished through an array index. The *index* is the numeric position in the array of the element. Like C/C++ and Java, array index numbering starts at 0 in Bash making the first element at position 0, the second element at position 1 and the nth element is at position n−1.

```
#!/bin/bash
sum=0
while read x y; do
    if [ $x -gt $y ]; then
        sum=$((sum+x))
    else
        sum=$((sum+y))
    fi
done < pairs.txt
echo Sum of largest of each pair is $sum
```

FIGURE 6.35 Revised script to input from specific file.

6.7.1 Declaring and Initializing Arrays

Unlike most programming languages, array variables do not need to be declared in Bash. Instead, we can just start assigning values to the array. To place values into an array, we either place individual values into array locations or we initialize the entire array. However, if we want to create an array whose size is larger than the items we are currently placing into the array, then we need to declare it. Figure 6.36 demonstrates six possibilities of creating an array named `ARRAY`. In four of these six cases, we also initialize some of the array locations.

The first example in the figure creates the array by assigning it a list of values. In the second example, we create an empty array. For the third example, we create an array piecemeal by assigning some elements values but not all. In each of these cases, Bash understands that `ARRAY` is an array because the RHS of the assignment statement denotes the item being stored with parentheses. In the third example, since we did not specify elements for indices 1 and 3, those two array locations are empty or are initialized to the value `NULL`.

In the fourth example, we assign each of five elements of `ARRAY` individually. This allows us to build an array over several instructions. Notice that we could have skipped elements 1 and 3 so that the assignment statements would be equivalent to the third example.

In the last two examples, we explicitly declare `ARRAY` using the `declare` instruction, using option `-a` to indicate that the variable will be an array. The fifth example only declares the variable as an array but does not initialize any element. This is equivalent to the second example. The last example declares and initializes the array.

6.7.2 Accessing Array Elements and Entire Arrays

How do we get values out of the array? For any other variable, the approach is to reference the variable using `$variable`. With arrays though, the notation differs. To retrieve an item from an array, we must reference the array location in brackets. However, the notation is slightly more cumbersome. We wrap `ARRAY[]` inside `${}`. This gives us a notation like `${ARRAY[0]}` and `${ARRAY[1]}`. If the index is stored in a variable, we place the variable's name in `[]` but without a `$`, such as with `${ARRAY[i]}`.

There are special notations that allow us to obtain the number of elements in an array or the entire array's elements as a list. These are explored in Table 6.14 where the array is stored in the variable `ARRAY`. Assume `ARRAY` is defined as the third example shown in Figure 6.37 in which the elements are `apple`, `cherry` and `fig` at indices 0, 2 and 4, respectively, and indices 1 and 3 are `NULL`.

In the last section, we used loops to iterate through lists of values. We can similarly use loops to iterate through the values in an array. There are several ways we can accomplish this. The most

```
ARRAY=(apple banana cherry date fig)
---------------------------------------------
ARRAY=()
---------------------------------------------
ARRAY=([0]=apple [2]=cherry [4]=fig)
---------------------------------------------
ARRAY[0]=apple
ARRAY[1]=banana
ARRAY[2]=cherry
ARRAY[3]=date
ARRAY[4]=fig
---------------------------------------------
declare -a ARRAY
---------------------------------------------
declare -a ARRAY=(apple banana cherry date fig)
```

FIGURE 6.36 Different ways to declare and initialize an array.

TABLE 6.14
Obtaining the Entire Array's Values or Its Size

Notation	Meaning	Result
${#ARRAY[@]} ${#ARRAY[*]}	Return the number of elements in the array (NULL elements are not counted).	3
${ARRAY[@]} ${ARRAY[*]}	Return all elements of the array as a list.	apple cherry fig
"${ARRAY[@]}"	Return all elements of the array as a list where each element is individually quoted.	"apple" "cherry" "fig"
"${ARRAY[*]}"	Return all elements as a single quoted string.	"apple cherry fig"

common two are to use the iterator for-loop where the list is denoted as `${ARRAY[@]}` and a counter-controlled for-loop where we iterate from `0` up to `${#ARRAY[@]}`, accessing each element using `${ARRAY[i]}` assuming `i` is our loop index.

Figure 6.37 demonstrates both approaches where the loop body outputs each element. For the counter-controlled loop, `i` is our loop index and our array index. It iterates from 0 up to but not including `${#ARRAY[@]}` which is the number of elements in the array. Notice that this will not work correctly if the array elements are not stored consecutively. For instance, if the array stores items at indices 0, 2 and 4, `${#ARRAY[@]}` is 3, but the loop iterates over elements 0, 1 and 2 instead of 0, 2 and 4. The iterator for-loop, as shown in the top portion of Figure 6.37, is both easier to use and logically more correct should the array not store values consecutively.

6.7.3 Example Scripts Using Arrays

To wrap up arrays, let's look at a few example scripts. In each script, we iterate through the array and perform some operations on the given array elements. We start with a script that will download a list of files from webservers using the `wget` command. `wget` is a non-interactive way to download files from a webserver. The script is shown in Figure 6.38. We might use this script as part of a *web crawler*, which is a component of a search engine. Notice that we hardcode the file names (including their full URLs) in the script. We could also have placed the URLs in a file and input each URL one at a time (in which case we wouldn't need an array).

We could implement the script in Figure 6.38 using a counter-controlled for-loop instead of the iterator for-loop, like we saw in the bottom half of Figure 6.37. If we did so, we would change two things. First, the loop itself would be written as `for (( i=0; i<${#list[@]}; i++));` and second, we would need to retrieve the ith element of the array in our `wget` statement as in `wget ${list[i]}`. We would probably prefer to place `${list[i]}` in a variable for easier reference both

```
for item in ${ARRAY[@]}
     do
          echo $item
     done
----------------------------------
for (( i=0; i<${#ARRAY[@]}; i++ ))
     do
          echo ${ARRAY[i]}
     done
```

FIGURE 6.37 Iterating through the elements of an array.

```
#!/bin/bash
     list=(www.nku.edu/~foxr/CIT371/file1.txt
          www.uky.edu/cse/welcome.html
          www.uc.edu/cs/csce310/hw1.txt
          www.xu.edu/welcome.txt
          www.cs.ul.edu/~wul/music/lyrics.txt)
     for i in ${list[@]}; do
          wget $i
          if [ ! -e $i ]; then
               echo Warning, $i not found!
          fi
     done
```

FIGURE 6.38 Using an array to download files.

in the `wget` statement and the `if` statement that follows, so we might use `file=${list[i]}` followed by `wget $file` and `if [ ! -e $file ]` instead.

For another example, let's use an array to control which users are permitted to execute another script called `foo`. We hardcode a list of usernames into an array. We obtain the current user's username using the environment variable `USERNAME`. If `USERNAME` is one of the users in the array, then the script will run the script `foo`. This script is shown in the top half of Figure 6.39.

We implement this script by using a Boolean flag. The idea behind a *flag* is that it will store the result of some condition that we can test later. Here, our flag, `isLegal`, is set to 0 (false). We iterate through the array looking for a match of the array element to `USERNAME`. If found, we change the flag to true. The flag is then tested to determine whether this user can run the script. This is another form of logic commonly applied in programming.

One drawback to both of the scripts we've examined in this subsection is that the array is hardcoded into the program. Another approach is to input values into the array, either from the user or from file. We might do so using a `while read` statement.

The script in the bottom of Figure 6.39 provides an enhancement to the script from the top half of the figure. The enhancement is that we input the values into the array before searching the array. To insert into an existing array, we add an element using the notation *`array+=(newvalue)`*. In this

```
#!/bin/bash
legalUsers=(foxr zappaf underwoodi dukeg marst)
isLegal=0
for user in ${legalUsers[@]}; do
     if [ $user == $USERNAME ]; then isLegal=1; fi
done
if [ $isLegal -eq 1 ]; then
     ./foo
     else echo $USERNAME is not authorized to run foo
fi
------------------------------------------------------
#!/bin/bash
legalUsers=()
while read user; do
     legalUsers+=($user)
done
isLegal=0
for user in ${legalUsers[@]}; do
     if [ $user == $USERNAME ]; then isLegal=1; fi
done
if [ $isLegal -eq 1 ]; then ./foo
     else echo $USERNAME is not authorized to run foo
fi
```

FIGURE 6.39 Using an array to test for legal user.

script, we add `$user` to `legalUsers` where `user` is input in the `while read` statement (either from keyboard or a file depending on whether this script had a file redirected to it from the command line). As we noted with the first example, we do not need an array in this example either as we could just input each name and compare it. We omit further examples but a couple of exercises at the end of the chapter provide examples for why you might choose to use an array in a script.

SECTION ACTIVITIES

1. One use of an array is to store a list of values which the program then sorts. Do you know any sorting algorithms? If not, research sorting, you might be surprised just how many different algorithms there are. If you know only one, research others. Why do we want to sort items?
2. Arrays are a form of data structure. Another form is the record. Research the record type in a language like C (where it is called a struct) and compare its syntax to that of an array. Also compare its usage. We use arrays to store lists. What do we use records/structs for?
3. Whereas the array stores homogenous elements, the record stores heterogeneous elements (members can be of different types). Python has the tuple, which is like an array but it is heterogeneous. Can you think of a reason where the items you want to store will be of different types? If not, research the Python tuple to find examples.

6.8 STRING MANIPULATION

6.8.1 Substrings Revisited

The notation we used to access arrays can also be used to access strings more conveniently than using `expr`. Specifically, we can use `${item}` on strings. If `item` is a string, we can obtain a substring using `${`*`string`*`:`*`start`*`:`*`length`*`}` where *start* and *length* are the starting index and length of the substring. As with `expr`, these values can be literal integers or integers stored in variables.

A difference between `expr substr` and the `${`*`string`*`:`*`start`*`:`*`length`*`}` notation is that the starting index is counted differently. Recall when using `expr` that the first character starts at index 1 but with `${`*`string`*`:`*`start`*`:`*`length`*`}`, the first character is counted as location 0. For instance, if `string` stores abcdefgh then `${string:2:4}` returns the substring cdef while `` `expr substr $string 2 4` `` returns bcde instead. We now have two ways of obtaining a substring. Let's put this into practice.

Figure 6.40 demonstrates two ways to obtain the initials of a person's name by generating the substring that is the first character of the three strings: `First`, `Middle` and `Last`. Four examples are shown, two using `expr` and two using `${…}`, where the first two provide the initials without periods and the latter two generate initials with periods. The first and third versions, both using

```
Initials="`expr substr $First 1 1``expr substr $Middle 1 1`
`expr substr $Last 1 1`"
------------------------------------------------------------
Initials=${First:0:1}${Middle:0:1}${Last:0:1}
------------------------------------------------------------
Initials="`expr substr $First 1 1`.`expr substr $Middle 1 1`.
`expr substr $Last 1 1`."
------------------------------------------------------------
Initials=${First:0:1}.${Middle:0:1}.${Last:0:1}.
```

FIGURE 6.40 Comparing substring approaches.

expr substr, should appear on one line. For the version with expr substr, we have to enclose the entire RHS of the assignment statement in double quotes.

When using the ${*string*:*start*:*length*} notation, if the value for *start* is greater than the length of the string, nothing is returned but it does not yield an error. If the value for *length* is larger than the remaining length of the string, it returns the rest of the string. For instance, if name="Frank Zappa" then ${name:6:15} returns Zappa. If we omit *length*, the substring operation defaults to returning the remainder of the string. For instance, ${name:0} returns Frank Zappa and ${name:3} returns nk Zappa.

A negative value for *start* indicates that counting will start from the end of the string rather than the beginning. For this to work, we must either place the negative value in parentheses or add a space after the colon, as in either ${name:(-3)} or ${name: -3}, which returns the substring ppa. The notation ${name:(-7)} returns the substring k Zappa. If we fail to include the space or parenthesis, the entire string is returned.

We can combine the negative number with a length which operates as you would expect; it returns *length* characters starting from the right side as indicated by the negative start value. For instance, ${name: -7:3} returns k Z and ${name:(-4:3)} returns app.

The notation to obtain a string's length is ${#*string*}. Using name from the previous examples, echo ${#name} will output 11 (don't forget the space in the name) and X=${#name} sets X to 11.

6.8.2 String Regular Expression Matching

When we covered expr earlier in the chapter, we omitted one of the available commands, match. The match command compares a string to a regular expression, returning the portion of the string that matches. The syntax is `expr match *string* '\(*regex*\)'`. Here, *string* is either a quoted string, as in "abcdefg" or a string stored in a variable.

The *regex* is placed not only in single quote marks but also the notation \(\) making it look highly confusing. match can apply any of the extended regular expression metacharacters but only by escaping those metacharacters with \. Unlike grep, the default for match is to match from the beginning (leftmost part) of the string. Table 6.15 provides several examples including the output. Each example is placed in an echo statement.

The first example attempts to match any characters of a, b, c, d and g. As the string starts with abcd, this is the portion that matches. The second example is the same except that the regex ends with a $. This forces the match to end the line. You might think that the output would be g, but because match starts from the beginning of the string, this match is like using ^ and $ in which the entire string must match, and since the regex will not match the e or f, no part of the string matches.

The third example matches all lowercase letters; it matches all letters from the string. The fourth example matches any group of letters followed by a single digit and so ends once the first digit is reached in the string. The fifth example matches one letter followed by one digit but as it will only

TABLE 6.15
Examples Using the match Command

Command	Output
echo `expr match "abcdefg" '\([a-dg]*\)'`	abcd
echo `expr match "abcdefg" '\([a-dg]*$\)'`	No output
echo `expr match "abcdef123" '\([a-z]*\)'`	abcdef
echo `expr match "abcdef123" '\([a-z]*[0-9]\)'`	abcdef1
echo `expr match "abcdef123" '\([a-z][0-9]\)'`	No output
echo `expr match "abcdefg123" '.*\([0-9]\)'`	3
echo `expr match "abcdefg123" '.*\([0-9]\{3\}\)'`	123

match from the beginning of the string, it matches nothing because the second character of the string is not a digit.

For the last two examples, we need to understand the notation .* added to our regex. By including this notation, we override `match`'s default behavior of matching from the beginning of the string to instead force it to match from the end of the string. The notation .* is added immediately after the opening single quote around the regex. We see this with the sixth example in Table 6.14 which returns `3` because it is looking to match a single digit from the right. The seventh example revises this statement by seeking three digits and so matches all three of the digits at the end of the string. Notice that {3} appears as \{3\} because the {n} notation is part of the extended regular expression set.

If we omit the \(\) around the regex, then the operation returns the location ending the substring that matched. For instance, `expr match "abcdefg" '[a-dg]*'` returns 4 because the regex matched up through the fourth character. With `expr match "abcdefg123" '.*[0-9]\{3\}'` we get `10` because the regex matched the characters at indices 10–12. Notice that the index starts counting at 1, not 0, as we saw with `expr substr`.

We can replace the word `match` and place a colon between the string and the regex as a shorthand notation as in ``echo `expr "abcdef" : '[a-z]*'` ``. The shortened notation we used for substring and length provides other useful variants. These are described in Table 6.16 with a few examples where the variable `str` stores `abcdef123`.

SECTION ACTIVITIES

1. We can obtain the length and substring of a string more easily using `${ }` than use `expr`. We can perform regular expression matching using both forms of syntax. Would you ever need to use `expr` given the `${ }` shorthand notation?
2. Which is more convenient to apply a regular expression, using a program like `egrep` or using the `expr` or `${...#/%regex}` approach?

TABLE 6.16
Removing a Matching Portion of a String

Notation	Meaning	Example
`${string#regex}`	Return *string* having removed the shortest matching substring from the beginning of string.	`${str#[a-z]*[a-z]}` returns cdef123
`${string##regex}`	Return *string* having removed the longest matching substring from the beginning of string.	`${str##[a-z]*[a-z]}` returns 123
`${string%regex}`	Return *string* having removed the shortest matching substring from the end of string.	`${str%[a-z0-9]}` returns abcdef12
`${string%%regex}`	Return *string* having removed the longest matching substring from the end of string.	`${str%%[a-z0-9]}` returns the empty string

6.9 FUNCTIONS

A *function* is a standalone piece of code that is executed upon being invoked from some different program unit. The function is a type of *subroutine*. All high-level programming languages have some subroutine mechanism whether the subroutines of the language are known as *functions*, *procedures* or *methods*.

The subroutine is a means of breaking programs into smaller chunks of code. The subroutine supports a concept called *modularity* (writing code in modules). Through modularity, it is often easier to design, implement and debug your program code.

In Bash, subroutines are called functions although they do not necessarily return a value. Functions are not mandatory in our Bash scripts, but they do provide a convenient mechanism to support *reusable* code. We might, for instance, define a number of functions in a file and then load that file to call those functions from the command line or from scripts.

As an example, consider a script that searches a directory for files with bad permissions, reporting those found. Such a function could be defined once and then called by the user from the command line or from within their `.bashrc` file. An example of this approach is taken with the function `pathmunge`, defined in `/etc/profile` and used to construct user `PATH` variables.

6.9.1 Defining Bash Functions

In Bash, the function is defined somewhat differently from the subroutines of most other languages. As Bash is interpreted, the placement of the function definition can be anywhere, whether in a file by itself, amid other instructions in a script, defined at the command line, or in a file of other functions. Once the function has been defined in the interpreter's environment, it can be called. If exported, it can then be called from any other script.

Functions receive parameters much like scripts receive parameters. Unlike the notation used in most programming languages, parameters are not listed in parentheses when the function is called but instead enumerated after the function's name in the function call like we do when we invoke a script. In C, a function might be invoked as `someFunction(a, b, 10);` but in Bash we would use `someFunction $a $b 10`.

Similarly, when we define the function itself, we do not list parameters in parentheses. So, `someFunction` might be defined as `void someFunction(int a, int b, int c)` in C but would be defined in Bash as `someFuncion()`. As the function header does not define the parameters by name, within the function the parameters are referenced using `$1`, `$2`, etc.

There are two ways to define a function in Bash. The first, as noted in the previous paragraph, is to list the function name followed by `()`. The function's body (its code) is then placed inside {}. A second format is to place the word `function` prior to the function's name and omit the `()`. The code is still placed in {}. Figure 6.41 demonstrates the syntax using both approaches. The figure then provides three example functions: `foo`, `bar` and `adder`.

The three sample functions in Figure 6.41 do not provide much indication for why we will want to use functions. The function `foo` merely outputs two messages. The second function, `bar`, outputs a message based on the first parameter passed to it. The third function, `adder`, adds up all of the parameters passed to it and outputs the sum. We explore more useful functions as we move through this section.

6.9.2 Using Functions

The placement of a function's definition is not necessarily important just as long as the function is defined *before* it is called in our code. If we place a function in a file of functions, say `functions.sh`, then any script that will use those functions requires having the functions defined. We can cause the functions to be defined with the instruction `source functions.sh`.

```
someFunction() {
          ...
}
-------------------------------------
function someFunction {
          ...
}
-------------------------------------
foo( ) {
     echo Function foo
     echo Illustrates syntax
}
-------------------------------------
function bar {
     echo Hello $1
}
-------------------------------------
adder() {
    Sum=0
    for Value in $@; do
        Sum=$((Sum+Value))
    done
    echo Sum is $Sum
}
```

FIGURE 6.41 Syntax for defining functions and some sample functions.

If our function is in the same file as the code which calls the function, then the function's definition *must* be defined in the file earlier than the function call. However, the script can place that function after other script code. Figure 6.42 illustrates the location of a function's definition among other code. There is no code in this figure, just comments explaining code location.

Functions can but do not have to return a value. Most Bash functions receive parameters and operate on those parameters, outputting results (if any) but not returning any value. Functions that return something can either return a computed value or return an error code. Functions for the most part then are similar to ordinary scripts. They can contain input, output, assignment, selection and iteration instructions. Mostly, functions will be short.

Figure 6.43 demonstrates a script with a function. The main part of the script, located at the bottom, iterates through the parameters passed to the script. For each parameter, it is passed to the function `runIt`. The function tests to see if the parameter is both a regular file and executable and if so, runs it. If the file cannot be run, a warning message is provided. The function returns no value.

Obviously, there is no need to have the function `runIt`. We could have just as easily (or more easily) placed the `if` statement inside of the for-loop and replaced `$1` with `$file`. The function `runIt` might be useful though if we feel that other shell scripts might utilize it.

```
#!/bin/bash
# some script code here

# define function foo here

# source somefile.sh which defines function bar

# code that calls foo and bar here

# other code may also go here
```

FIGURE 6.42 Placement of function definitions in a file.

```
#!/bin/bash
runIt( ) {
   if [[ -f $1 && -x $1 ]]
      then ./$1
      else echo Error when running $1
   fi
}

for file in $@; do
     runIt $file
done
```

FIGURE 6.43 Example script with function.

6.9.3 Functions and Variables

Local variables are variables that are only applicable in the function they were declared. Unlike any other variable used in a script, these variables must be explicitly declared using the `local` statement as in `local a b c`. The reason to define local variables is to ensure that the variables no longer exist outside of the function once the function ends.

The reason to use local variables is to avoid side effects. A *side effect* arises when we invoke a function which inadvertently changes a value in a variable used outside of the function. This can happen if a variable used in a function happens to be the same name as a variable used outside of the function.

Let's see how a side effect might occur. The script shown on the left side of Figure 6.44 initializes a script variable, `X`, to `0`. After this assignment statement, the function `foo` is defined. Execution continues with the first `echo` instruction following `foo`'s definition. Next, the code invokes `foo`, passing it `5`. In the function, `X` receives the value of the parameter (`5`), increments and outputs it. After `foo` terminates, the script resumes with the last `echo` instruction.

The problem with the code in Figure 6.44 is that `X`, in `foo`, is the same variable as `X` outside of `foo`. Thus, while we would expect the last `echo` instruction to output `0` because `X` in the script was set to `0`, it actually outputs `6`, the value `X` was set to in the function `foo`. On the right side of the figure we see the output we would expect to receive and the output that is actually generated.

To avoid this side effect, we must make `foo`'s `X` a local variable. We change the first instruction in `foo` to `local X=$1`. Now there are two `X`'s involved in this script, the `X` outside of `foo` and the inside `X` (the local `X`). The outside `X` starts at and remains `0`. The inner `X` starts at `5`, is incremented and output as `6`, and when `foo` terminates the inner `X` is removed from memory. Thus, no side effect arises because outside `X` is not modified. Side effects are unpredictable and so we prefer to avoid them.

Let's consider a *swap* function, shown as part of the script in Figure 6.45. The role of a swap function is to exchange the values stored in two variables. We commonly use a swap routine as part of a sorting algorithm although it could have many other uses. Here, we have `X` and `Y` defined, storing `5` and `10`, respectively. We call `swap` and upon exiting, we output `X` and `Y`. In this case, the `X`

```
#!/bin/bash
X=0

foo()  {
     X=$1
     X=$((X+1))
     echo $X
}

echo $X
foo 5
echo $X
```

Expected output:
0
6
0

Actual output:
0
6
6

FIGURE 6.44 Example script demonstrating side effect.

```
#!/bin/bash
X=5
Y=10

swap() {
      TEMP=$X
      X=$Y
      Y=$TEMP
}

swap
echo $X $Y
```

FIGURE 6.45 Function with intended side effect.

and Y outside of swap are intended to be used by swap. That is, swap is supposed to modify the X and Y of the outer code. We rely on the side effect to accomplish our task. In fact, without the side effect, swap would do nothing useful for the code outside of the function in the script.

One interesting aspect of the swap function is that it introduces a new variable, TEMP. Referring back to the script in Figure 6.44, the variable X used in function foo is the same as the variable X defined before foo. But in the script in 6.45, the variable TEMP did not previously exist. As TEMP was not declared locally, what is the result of TEMP after swap terminates? The answer is that TEMP still exists and has retained the value it had while swap executed.

Specifically, since X is 5, TEMP is set to 5. Had we modified the echo statement at the bottom of the script to read `echo $X $Y $TEMP` then the output would be `10 5 5`. If TEMP had been declared a local variable in swap then TEMP outside of swap would have no value (NULL) and the output statement would be `10 5`.

6.9.4 exit AND return STATEMENTS

All functions return an exit status. If the function executes without error, the exit status by default is 0. We can force a function to terminate and return an exit status with the exit instruction. We might use this, for instance, as the result of some if-then statements testing a condition to ensure that the function can run correctly. If a condition fails, we exit with a number representing the "error".

For instance, if a function expects exactly two parameters and receives none, we might use `exit 1`. If it receives exactly one parameter, we might use `exit 2`. If it receives three or more parameters, we might use `exit 3`. These exit codes then represent "error, no parameters", "error, too few parameters" and "error, too many parameters", respectively. Outside of the script, we might want to see why the function did not work correctly by viewing the error code.

It is up to us to define exit numbers. Most Linux commands respond with an exit number, used as an error code. In the command's man page, you can see how these codes are defined. See for instance the man page for useradd which has nine codes defined (exit code of 0 means success). When used in a script, the exit statement not only terminates the function but terminates the running script. We can examine the exit code from the command line using `echo $?`.

Similar to exit is the return statement, which causes the function to terminate and return a value. But unlike exit, the remainder of the script still executes. It is through the return statement that lets us *return* a value from a function. We might choose to use return rather than relying on non-local variables to store results. The instruction is `return` *n* where *n* is a numeric value.

The script in Figure 6.46 illustrates both an exit and return statement in a function. The function's role is to return the largest value among the parameters. If no parameters are received, the script immediately exits with an error code of 1. Otherwise, the script stores the first parameter ($1) in local variable max and shifts all parameters down so that the first parameter is discarded. Now, an iterator for-loop iterates through the remainder of the parameters, and if any value is greater than max, we replace max with this value. At the end of the function, max is

```
#!/bin/bash

maximum() {
     if [ $# -eq 0 ]; then exit 1
     else
          max=$1
          shift
          for num in $@; do
               if [ $num -gt $max ]; then max=$num; fi
          done
     fi
     return $max
}

maximum 5 1 2 6 4 0 3
echo The maximum value is $?
```

FIGURE 6.46 Example function with both `exit` and `return`.

returned. In the script, after calling the function, we print out $?. Notice that if the `exit` instruction executes, the entire script terminates and we receive no output at all. This will inform us that the exit code should be examined.

Although we have covered a lot of material in this chapter, to truly understand Bash scripting you will need experience. Much of the syntax in Bash will be foreign to you no matter what languages you might have already learned. If you are interested in learning another scripting language, the C-shell language is the default language when using `csh`. The supplemental readings found on the textbook's website include a small unit about C-shell, comparing it to Bash.

SECTION ACTIVITIES

1. Subroutines are used to help us design and implement large programs. Research the term *top-down design*. Have you used this approach to programming? What are the advantages of doing so?
2. Bash' syntax for parameters and for returning a value differs from that of most other languages. Will this be an issue for you in learning to write and use Bash functions?

6.10 CHAPTER REVIEW

Concepts and terms introduced in this chapter:

- Array – a homogeneous data structure that stores multiple values; reference to a specific value requires specifying its location (an index).
- Array index – a value to specify the specific array element to be accessed.
- Assignment statement – instruction that assigns a variable a value.
- Case sensitive variable names – some languages, including Bash, differentiate between uppercase and lowercase letters in variable names.
- Case statement – a type of n-way (or multiway) selection statement in which a variable is compared to lists of values to select the action(s) to execute.
- Compiling – the process of translating an entire high-level language program into executable code before it can be run.
- Compound condition – a condition which has multiple parts that are combined through Boolean operators AND and/or OR.

- Condition – a comparison that evaluates to true or false, often used in an instruction to decide what to do.
- Conditional loop – an iteration statement which repeats while a condition remains true (or false in instructions like `until`).
- Constant – a variable that once assigned a value, cannot change; in Bash, these are variables with the read-only attribute.
- Counter-controlled loop – an iteration statement which repeats by counting in a sequence from a start to stop value with an optional step size.
- Else clause – the set of code corresponding with the else statement in a two-way selection statement; this clause executes when the condition evaluates to false.
- Exit code – an integer returned from a function which terminates early; the exit code is used to express that the function terminated successfully or had an error.
- File condition – a test of a file to determine if it matches some attribute (e.g., is a directory, is executable).
- Function – a set of code to be invoked from different areas of a script or program; more generally a function is a *subroutine*.
- Function call – an instruction that invokes a function.
- If-then/If-then-else statement – a selection statement that tests a condition and if true executes the instructions in the then clause; if there is an else clause, it executes if the condition is false.
- Infinite loop – a type of loop that, once it starts to iterate, never stops because the condition never changes.
- Interpreting – translating one instruction into an executable operation and executing it; the Bash interpreter does this for both instructions entered at the command line and scripts.
- Iterator loop – an iteration statement which repeats once per item in a specified list.
- Local variable – a variable that is accessible only from within a function.
- Loop body – the instructions that execute each time the loop repeats.
- Loop variable – a variable assigned to a counter in a counter-controlled loop or a list item in an iterator loop; the variable can be referenced in the loop body.
- Loop statement – a programming language instruction which repeats a body of code based on either a condition, the length of a list or the sequence of counting values supplied.
- N-way selection – a selection statement which has n conditions to select between one of n groups of actions.
- Nested statement – embedding a loop inside a loop body or an if-then/if-then-else statement in a then clause or else clause; nested statements permit more complex logic.
- One-way selection – the simplest form of selection statement which tests a condition and if true, executes the action(s) listed; if-then statements are one-way selection statements.
- Parameter – a value passed to a script or a function that the script/function might utilize during execution.
- Postfix/prefix increment/decrement – a shortcut operation to add or subtract 1 to/from a variable.
- Repetition statement – another name for an iteration statement or loop.
- Script – a small program executed by an interpreter.
- Selection statement – a programming language instruction which evaluates one or more conditions to select which corresponding action(s) to execute.
- Short-circuiting – when evaluating a compound condition, if the condition can be determined without evaluating the entire condition, it ends; used to reduce the computation needed when the first part of an AND is false or the first part of an OR is true.
- Side effect – a situation in which a function alters a variable which exists outside of the function; side effects are generally unpredictable and can lead to erroneous code.

- Subroutine – a piece of standalone code that can be invoked from other code; used to support modularity.
- Substring – a contiguous group of characters from a larger string.
- Syntax error – an error that arises because code is not used with grammatical correctness.
- Then clause – the set of code corresponding to the then statement in a one-way or two-way selection statement; this clause executes when the condition evaluates to true.
- Two-way selection – a selection statement with one condition that executes one of two sets of code based on whether the condition is true or false; if-then-else statements are a form of two-way selection.
- Variable – an abstraction of a memory location storing a value; the variable's name is used to provide access to the value.
- Variable condition – a condition to test if a variable is storing a value.

Linux (Bash) commands covered in this chapter:

- #!/bin/bash – the instruction that starts every Bash script to indicate the interpreter to use.
- $1, $2, $3, ... – notation for referencing parameters in a script or function.
- $@ – the entire list of parameters.
- $# – the number of parameters.
- ${#*str*} – shorthand notation to obtain the length of string *str*.
- ${*str*:*value1*:*value2*} – shorthand notation to obtain the substring of string *str* where value1 is the index and *value2* is the length of the substring.
- ${*arr*[@]}, ${*arr*[*]} – shorthand notation to obtain the list of values in the array *arr*.
- [ ], [[ ]], (( )) – symbols used to embed a conditional.
- $(( )) – symbols used to embed an integer computation.
- ==, != – notation used to compare two strings for equality/inequality.
- -eq, -ge, -gt, -le, -lt, -ne – numeric comparison operators (equal, greater than or equal, etc.).
- &&, ||, ! – Boolean operators for AND, OR and NOT.
- case – n-way selection statement.
- declare – used to declare a variable or change a variable's attributes.
- echo – output statement.
- exit – used in a function to indicate that the function should terminate with a specified error code; 0 is the default for no error, a positive integer is used to denote an error.
- expr – perform an arithmetic or string operation.
- for…do…done – syntax of a counter-controlled and iterator for-loop.
- if…then…fi – syntax of an if-then statement.
- if…then…else…fi – syntax of an if-then-else statement.
- if…then…elif…else…fi – syntax of a nested if-then-else statement.
- length – an `expr` operation to return the number of characters stored in a string.
- let – used to assign a variable the result of an arithmetic computation.
- local – used to declare a variable to be local in a function.
- read – input statement.
- return – used to exit a function and return an integer value.
- seq – generate a list of the numeric sequence given the first and last values in the sequence and a step size; first and step size are optional and default to 1.
- shift – used to move all parameters down one position ($2 becomes $1, $3 to $2, etc.).
- unset – used to remove a value (uninitialized) from a variable.
- until…do…done – syntax of a conditional loop that iterates while condition is false.
- while…do…done – syntax of a conditional loop that iterates while condition is true.
- while read…do…done – syntax of a conditional loop which inputs values and executes as long as there are inputs.

REVIEW QUESTIONS

1. Write a Bash script to execute the `date` command, `ls` on the user's home directory and the who operation, sending all output to the terminal window.
2. For the script in #1, how would you redirect the output of the script when running from the command line so that the output goes to the file `info.txt`? How would you modify the script itself to redirect all output to the file `info.txt`?
3. You have written the script `foo.sh` in your current directory. You go to run it. What command do you enter to run it? Upon running it you get a permission denied error. Why might this arise and what should you do?
4. You have entered the following instructions:
```
FIRST=Frank
LAST=Zappa
NAME=$FIRST $LAST
```
The third instruction causes an error. Why and how would you fix it?
5. Which of the following are legal variable names in Bash? VARIABLE A_VARIABLE VARIABLE_1 1_VARIABLE variable A-VARIABLE while WHILE

For questions 6–13, assume `X=5`, `Y=10`, `Z=15` and `N=Frank` for each of the problems. Do not carry a value computed in one problem onto another problem.

6. What will `echo $X $Y $Z` output?
7. What will `X` store after executing `X=$Y+$Z`?
8. What is wrong with each of the following statements either syntactically or logically?
   a. `Q=$X`
   b. `Q=X`
   c. `$Q=$X`
9. Does the following result in an error? If not, what value does `X` obtain? `X=$((Y+N))`
10. Does the following result in an error? If not, what value does `X` obtain? `let X=$N`
11. What is the value stored in `X` after the instruction `((--X))`?
12. What is the value stored in `N` after the instruction `((N++))`?
13. What is the result of the instruction `echo $((Y++))`? Would `echo $((++Y))` do something different?

For questions 14–18, assume `A=one` and `B=two`.

14. What is output with the statement `echo $A $B`?
15. What is output with the statement `echo "$A $B"`?
16. What is output with the statement `echo '$A $B'`?
17. What is output with the statement `echo $A$B`?
18. What is output with the statement `echo "A B"`?
19. What is the difference in the output of the following two statements?
```
ls
echo `ls`
```
20. You have entered the following instruction: `./foo.sh 5 10 15 20`. What are the values of each of the following? `$0`, `$1`, `$2`, `$3`, `$4`, `$5`, `$#`, `$@`.

For questions 21–26, assume `X="Hello World"`, `Y=3`, `Z=5` and `S="acegiklnor"`.

21. What is output from the statement `` echo `expr substr "$X" $Y $Z` ``?

22. What is output from the statement `echo `expr substr "$X" $Z $Y``?
23. What is output from the statement `echo `expr substr $S $Y $Z``?
24. What is output from the statement `echo `expr index "$X" $S``?
25. What is output from the statement `echo `expr length "$X"``?
26. What is output from the statement `echo `expr index $S "$X"``?
27. Write an `echo` statement to output on one line the current user's username, home directory and current working directory.
28. Write an `echo` statement to output "the current date and time are " followed by the result of the `date` command.
29. Assume we have variables `location` and `work` which store the city name and company name of the current user. Write an `echo` statement to output a greeting message including the user's username, the city name and the company name.
30. Locate a listing of all ASCII characters. Using the notation \x*HH*, write an `echo` statement to output the string `Fun!` using the ASCII values for each of the characters.
31. What does the following instruction do assuming `NAME` stores your name?
    ```
    echo Hello $NAME, how are you today? >> greeting.txt
    ```
32. The script below will input two values from the user and sum them. What is wrong with the script? (there is nothing syntactically wrong). How would you correct it?
    ```
    #!/bin/bash
    read X Y
    SUM=$((X+Y))
    echo The sum of $X and $Y is $SUM
    ```
33. What is the `-p` option used for in `read`?
34. Write a script to input five values from the user via keyboard and output the average of the five values. The average will be an integer.
35. What would you need to do to the script from #34 so that it would input the data from the disk file `numbers.dat`?
36. Write a conditional to test to see if `HOURS` is greater than or equal to `40`.
37. Write a conditional to test to see if the user's name is not `zappaf`.
38. What is wrong with the following conditional? There are several errors.
    ```
    [ $X > $Y && $Y > $Z ]
    ```
39. Write a conditional statement to test if the variable `NUM` is between `90` and `100`.
40. Write a conditional statement to test to see if the file `foo.txt` does not exist.
41. Write a condition to determine if the file `foo.txt` is neither readable nor executable.
42. Write a condition to determine if the file `foo.txt` is owned by both you the user and your private group.
43. If a person is 21 years or older, they are considered an adult, otherwise they are considered a child. Write a statement that tests the variable `AGE` and assigns the variable `STATUS` to one of the strings `adult` or `child`.
44. Write a script which inputs a user's name and state (location) as a two-letter abbreviation. If the user lives in either `AZ` or `CA`, they are charged a fee of `$25`. If the user lives in either `OH`, `IN` or `KY`, they are charged no fee. Otherwise, they are charged a fee of `$10`. If their name is `Zappa`, `Duke` or `Keneally`, their fee is doubled. Output their name and fee.

45. Write a script to determine the amount of disk space remaining and output this percentage and the proper response based on the following table:

Amount Available	Response
100%–61%	No need for any action
60%–41%	Keep an eye on the system
40%–26%	Delete unnecessary software
25%–16%	Implement disk quotas on all users
15%–6%	Add a new disk drive
5%–0%	Panic

    Assume the program `diskcheck -r` returns the number of bytes remaining free in the current file system and `diskcheck -a` returns the total capacity of the file system. You will have to use both of these to obtain the percentage of disk space remaining to determine what the user should do. NOTE: do not just divide the amount remaining by the total because this will be a fraction less than zero and as an integer division will be 0. Instead, multiple the amount remaining by 100 first, and then divide by the total to get a percentage remaining. For instance, if `diskcheck -r` returns `127713651` and `diskcheck -a` returns `1000000000`, then the program should output 12%, Add a new disk drive.
46. Revise the calculator program from the top portion of Figure 6.18 to add the operators ** (exponent), << (left shift) and >> (right shift). Since **, << and >> can all be interpreted by the interpreter as Bash wildcards or redirection, you need to use some other symbol(s) for these operators (e.g., E for exponent, L for left shift and R for right shift).
47. Explain why you would not want to rewrite the SCORE/GRADE if-then-elif-else statements from Figure 6.16 (the fourth version of the code) using a `case` statement.
48. Rewrite the "favorite color" case statement from Figure 6.21 using an if-then-elif-else statement.
49. If the variable `FOO` has a value, we want to output it. Write such a statement using the `[`*`condition`*`] &&` (or `||`) *action* syntax.
50. Like with question #49, write a statement to set `FOO` to 1 if it currently has no value.
51. Write a script that contains a while loop and inputs numbers from the user until the user enters `0`. The loop should take each input and test to see if it is the number `100` and if so, count it (add one to a counter). After the loop, output the count. That is, this script outputs the number of times the user entered `100`.
52. Redo #51 using an until loop instead of a while loop.
53. Redo #51 where the user supplies a parameter to the script which will be the number to compare each input to rather than `100`. For instance, if this script is called as `./count.sh 50` then the script will count the number of times the input is `50`.

    For questions 54–57, assume `X=5` and `Y=10`. How many times will each loop iterate?
54. `while [ $X -lt $Y ]; do ((X++)); done`
55. `while [ $X -gt $Y ]; do ((X++)); done`
56. `while [ $X -lt $Y ]; do ((X++)); ((Y++)); done`
57. `while [ $X -lt $Y ]; do ((X++)); ((Y--)); done`
58. Rewrite the following iterator for-loop as a counter for-loop.
    `for num in (1 2 3 4 5); do … done`
59. Rewrite the following iterator for-loop as a counter for-loop.
    ``for num in (`seq 5 3 15`); do … done``
60. Rewrite the following counter-controlled for-loop to count downward instead of upward.
    `for ((i=0;i<n;i++)); do … done`

61. Write a script which receives a list of parameters and iterates through them, counting the number of times the parameter is greater than 0. The script will output the number greater than 0, or an error message if no parameters are supplied.
62. An infinite loop is a loop which never exits because the condition always causes it to continue to execute. We can write infinite loops using either while, until or counter for-loops (see the example below), but not iterator for-loops. Why could an iterator for-loop never be an infinite loop? For both cases below, assume x is initially 0.
    Infinite while loop: `while [ $x -lt 10 ]; do echo $x; done`
    Infinite for-loop: `for ((i=0;x<10;i++)); do echo $i; done`
63. A palindrome is a string which reads the same forward as backward, as in `radar` or `madamimadam` (Madam, I'm Adam) but not `abacab`. Write a script to determine if a parameter is a palindrome and output the parameter and whether it is or not.
64. Repeat #63 so that the script receives a list of strings and outputs all of those that are palindromes. Output an error message if the script receives no parameters.
65. Write a script which receives a list of parameters, computes and outputs the average of the list, or an error if no parameters are supplied.
66. Write a script which receives a list of parameters and outputs the number that is the smallest. If no parameters are supplied, output an error message.
67. Assume the array `users` stores a list of usernames. Write a loop to iterate through the usernames and count the number who are currently logged in. You can determine if a user, x, is logged in by using `who | grep $x` and see if the response is `NULL` or a value.
68. Assuming `array` is a variable storing an array of values, what is the difference (if any) between `${array[@]}` and `${array[*]}`? Between `"${array[@]}"` and `"${array[*]}"`?
69. Write a loop to output each letter in the string `str` using the notation `${str:i:j}`.
70. Write a script which receives a list of values input from a file, storing each value in an array. Next, sort the array in increasing order. You may use any bubble sort, insertion sort or selection sort. If you do not know any of these sorting algorithms, do a web search for one of them. Wikipedia, for instance, will list pseudocode versions of each.
71. Write a script which receives a list of values input from either file or keyboard, placing them in an array. Next, use an iterator for-loop to iterate through each value in the array. Inside this loop is a nested iterator for-loop (or a counter-controlled loop if you prefer) to iterate through the values in the array so that you can count how many equal the current value. Retain whichever value occurs most frequently. The result of the script is to output the value which appeared most frequently and the number of times it occurred.
72. Write a function which receives a list of values as parameters and computes and outputs their sum.
73. Write a function which receives a list of values as parameters and computes and returns their average.
74. Revise your function from #73 to return an error code of `9999` if the function receives no parameters and so cannot compute an average, otherwise return the average.
75. Rewrite your palindrome script from #63 so that the palindrome checking code is in a function which returns a `1` if the parameter passed to it is a palindrome and a `0` if it is not a palindrome.
76. Write a script which has a while loop and inputs from the user strings until the user enters `done`. For each string, pass it to the palindrome function you wrote in #75 and output the string and whether it is a palindrome or not.
77. Redo #76 so that the strings input are first stored in an array. Next, iterate through the array, passing each array value to the palindrome function.

7 User Accounts

The learning outcomes of this chapter are to be able to:

- Explain the role of user and group accounts
- Change user using `su`
- Create, modify and delete user and group accounts
- Use password management tools
- Describe the roles of the `passwd`, `group` and `shadow` files
- Use the various Linux facilities to establish user resources
- Set up access to and use `sudo`
- Explain the role of and modify SELinux contexts
- Create user account policies

7.1 INTRODUCTION

The user account is the mechanism by which the Linux operating system is able to handle the task of *protection*. Protection is needed to ensure that users do not maliciously or accidentally destroy (delete), manipulate or inspect resources that they should not have access to. In Linux, there are three forms of user accounts: root, user (human) accounts and software accounts.

The root account has access to all system resources. The root account is automatically created during Linux installation. Most software and configuration files are owned by root. No matter what permissions a file has, root can access it.

As root lies at one end of the spectrum of access rights, software accounts typically are at the other end. Most software does not require its own account. However, if the software has its own files and directory space then the software is often given its own account. A software account, unlike a user account, usually has no login shell. Thus, if a hacker attempts to log into a Linux system under a software account, the hacker would be unable to issue commands from that account.

You might recall from Chapter 4 that we can assign the execution permission to be `s` rather than `x` so that the software runs under the executable file's owner (or group) permissions rather than the user's permissions. The software account allows the software to run under more restricted access rights than those of root should the software use `s` for its execute permission.

The user account lies in between these extremes. User accounts are meant for human users. Most user accounts are created with the same set of initial resources and attributes, although as we will explore in this chapter, the defaults can be overridden. Table 7.1 presents the typical user attributes and resources in Linux.

In this chapter, we look at the programs available to create, modify and delete users and groups. We look at the mechanisms available to the system administrator to automatically establish users with initial files. We look at controlling passwords (i.e., using Linux tools to require that users update their passwords in a timely fashion). We provide an overview of SELinux. Finally, we discuss user account policies.

DOI: 10.1201/9781003203322-7

TABLE 7.1
User Account Attributes

Attribute	Meaning/Usage	Comments
Username	Symbolic name assigned to this user's account.	Used for login and in commands like `su`, `passwd`; files created by this user are given the username for ownership.
User ID (UID)	Numeric value assigned to the user's account.	Can be used in lieu of the username in some instructions (e.g., `killall`); used in some shell scripts.
Password	Assigned by the user; may have restrictions depending on the strength requirement established by root.	Encrypted and stored in `/etc/shadow` and used as part of the authentication process (e.g., logging in, `su`, `passwd`).
Private group	For group ownership of files owned by this user.	All files and subdirectories created by this user are, by default, assigned to this group account.
Group ID (GID)	Numeric value assigned to the private group.	Note that UID and GID are not necessarily the same value.
Home directory	Automatically generated when the user's account is created.	Populated by the files and subdirectories in the `/etc/skel` directory.
Default shell	Login shell that the user is automatically placed into upon opening a shell.	Defaults to `/usr/bin/bash` but others are available.
Accounting information	Entries in `/etc/passwd`, `/etc/group`, `/etc/shadow`.	Automatically generated when the user account and private group are created; `shadow` stores the user's encrypted password and password control information.

SECTION ACTIVITIES

1. Do you share a computer with other family members? If so, do you each have your own account or do you share one account? If you share one account, can you think of any advantages of setting up the computer this way? If you have your own account, are there any disadvantages to this?
2. In your Linux account, what is your UID and GID? If you installed Linux and created an account for yourself during installation, it is likely that you are the first user and assigned UID/GID of 1000. If not, look at the file `/etc/passwd` for your account information. Do your UID and GID share the same number?

7.2 CREATING ACCOUNTS AND GROUPS

There are two approaches to creating user accounts and groups: through the GUI and from the command line. In Red Hat Linux, the GUI tool has transitioned from the User Manager tool to being part of the Cockpit settings and administration tool. Cockpit is also available for Debian/Ubuntu distributions. Both GUI account creation tools are simple to use but not necessarily preferred.

There are three reasons why the GUI-based approach to creating users is a drawback. First, it is less flexible than the command-line programs in terms of creating accounts/groups that do not take on default values. Second, it can only create users with their private groups; we are not able to create other groups. Third, to create accounts for many users requires a good deal of interaction. Instead, Linux offers several programs that can not only be run from the command line program but also invoked from shell scripts.

In this section, we begin with a brief look at both the Users and Groups GUI and Cockpit. We then turn to the command line programs of `useradd`, `groupadd` and `newusers`. Finally, we build a script to automatically generate user accounts.

7.2.1 Creating User and Group Accounts through the GUI

In Red Hat 6, the GUI User Manager program could be launched from the menus along the top of the desktop by selecting System>Administration>Users and Groups. In both Red Hat 6 and Red Hat 7, the GUI can be launched from the command line as `/usr/bin/system-config-users`. Either way brings up the tool shown in Figure 7.1. To use this GUI, we must provide the root password when prompted.

In Figure 7.1, there are three (human) users already created: Student, foxr and zappaf. Shown in the GUI for each user is the user's username, UID, primary (private) group, full name (absent in this case), login shell and home directory location. The system accounts (e.g., root, adm, bin) and software accounts are not shown. From the User Manager GUI, we can add a user, delete a user, modify a user, add a group, delete a group or modify a group. Here, we will concentrate on users. Later we will look at groups.

Creating a new user is accomplished by clicking on the Add User button. This causes the Add User pop-up window to appear, shown in Figure 7.2. In this window, we enter the new username, user's full name and an initial password. The user's full name is placed into a field called the gecos field (GECOS dates back to early Unix days when Unix computers used general comprehensive operating system, GECOS, hardware devices). Both the user's full name and password are optional but recommended. The item to be placed in the gecos field is referred to as a "comment" as it can also be assigned some string other than a name. For instance, software accounts sometimes use the field to specify the name of the software.

Through the GUI, we can alter the login shell from the default (`bash`) to any of the other available shells. Typically, Linux comes with `sh` (the original bourne shell), `csh` and `tcsh`. If we add other shells (e.g., `korn`, `ash`, `zoidberg`), they should also appear. One last choice is `/sbin/nologin`. This is the choice for any software account as we do not want anyone to be able to log in as software and issue commands via a shell. `nologin` is a program which displays that the account is unavailable. The reason to restrict software accounts is that they commonly have no password and so could be a form of security breach. By assigning the account no login shell (or more specifically, the shell of `nologin`), a hacker is not able to gain entrance to our computer by trying to log in under a software account.

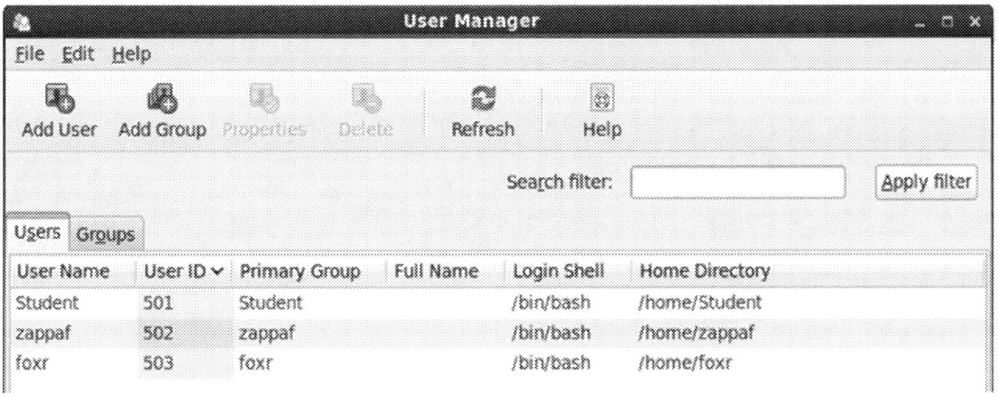

FIGURE 7.1 User Manager Tool.

FIGURE 7.2 Adding users.

As we enter the username in the Add User window, the home directory is automatically filled out as `/home/`*`username`*. We can override this. For instance, we might segment user home directories based on type of user such as `/home/faculty`, `/home/staff` and `/home/student`. Alternatively, software is often either given no home directory or a home directory under `/var`. Unchecking the box for `Create a home directory` will create the user account without a home directory.

The last three checkboxes default to the selections as shown in the figure. `Create a primary group` is selected, and `Specify user ID manually` and `Specify group ID manually` are unselected. The primary group is the user's private group, which is, by default, given the same name as the username. There are few occasions where we would not want to give a user account its own private group.

The default UID and GID are one greater than the last UID and GID issued. Unless we are creating a software account or have a specific policy regarding numbering, we will likely leave these as the default. Note that the default UID and GID may not be the same number depending on whether we altered previous defaults, created users without private groups or created non-private groups previously.

In Red Hat 6, normal users account IDs start at 500. Starting with Red Hat 7, the numbering scheme has changed so that user account IDs start at 1000. It is recommended that software accounts be given UIDs in the range of 200–999, with values less than 200 reserved for administrator accounts. As a system administrator, we might come up with a different numbering scheme. For instance, Table 7.2 provides an example showing how numbers might be distributed to Linux users at a small university.

Group creation through the GUI is even simpler than user creation. Clicking on the Add Group button brings up the add group pop-up window. Here, we specify the group name and optionally alter the GID. The Group tab in the User Manager GUI provides a listing of all of the groups, their

TABLE 7.2
Example Distribution of UIDs

Range of UIDs	User Classification	Description
1000–1199	Administration	Deans, chairs, etc.
1200–2000	Faculty	All instructors
2001–2999	Staff	All other employees
3000–4999	Graduate students	Reserved for students in a graduate program
5000–9999	Undergraduate students	All other students

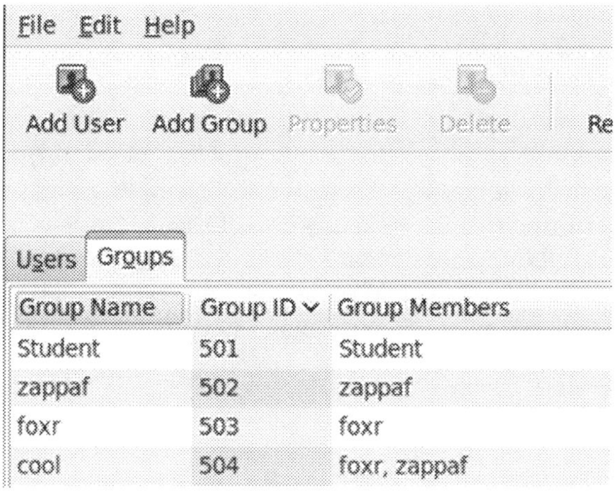

FIGURE 7.3 Viewing group membership.

GIDs and the users in each group. An example is shown in Figure 7.3. In this example, we see three private groups: Student, zappaf and foxr. Additionally, a non-private group has been created so that users can alter file group ownership to have a finer-grained control over file and directory permissions. Here, we see the group cool contains members zappaf and foxr but not Student.

Just as UIDs are generated automatically, so are GIDs with the new group given the GID that is one greater than the previously used GID. We noted earlier that UIDs and GIDs can diverge. In Figure 7.3, we see that cool was given the next sequential GID, 504. The next user account to be created would, by default, be given the UID of 504 and the GID of 505. Note that the examples shown in Figures 7.1–7.3 are from Red Hat 6. As noted, the default starting in Red Hat 7 is to start human user UIDs/GIDs at 1000.

Starting with Red Hat 8, the User Management GUI is no longer available, and instead the GUI tool for user management is Cockpit. If Cockpit is not already installed, as root you can install it using `dnf -y install cockpit`. Cockpit is also available for Debian and Ubuntu distributions. To install, again as root, use apt as in `apt-get install cockpit`. If you do not have access to the root password, use `sudo apt-get install cockpit`. Once installed, the cockpit service may not be running so you will need to start it using `systemctl start cockpit` (or `sudo systemctl start cockpit`).

Cockpit is accessed through a web browser using the URL `ipaddress:9090` where *ipaddress* is our computer's IP address. To view your computer's IP address, use `ip addr` and look at the entry for `inet` (`ip addr` is an instruction we examine in Chapter 10). As Cockpit uses HTTPS and our computer likely has no digital certificate, we receive a security warning. We must accept the risk in order to proceed.

We may also need to adjust the firewall. If, upon connecting to the above URL you receive an error in your browser, then make sure the Cockpit service is permitted through your firewall. In Red Hat, we use `firewalld` to manipulate the firewall with the following two commands.

```
firewall-cmd --add-service=cockpit --permanent
firewall-cmd --reload
```

The first of these adds the Cockpit service to the permanent firewall, and the second then reloads the permanent firewall into the runtime firewall. For Debian/Ubuntu versions of Linux using `ufw`, use the instruction `sudo ufw allow 9090`. We examine `firewalld`, `firewall-cmd` and `ufw` in Chapter 10. If you had to modify the firewall, refresh your browser at this point.

Upon reaching Cockpit, we are asked to log in. If we login using our user account, we do not have sufficient privileges to create or manipulate existing accounts. So, we log in as root. Alternatively, if our user account has system administrator privileges (as is the case in Ubuntu and some Debian distributions where we do not have access to the root password), we can select `Reuse my password for privileged tasks`. Once logged in, we select Accounts from the menu on the left. If we log in as ourselves without privileged access, Create New Account is greyed out. Similarly, selecting an existing user displays information about that user but does not let us modify the user. Logging in as root gives us full access.

Figure 7.4 shows the interface for modifying an existing user (top) and creating a new user (bottom). Notice selections include the option to delete the existing user's account and set or force a

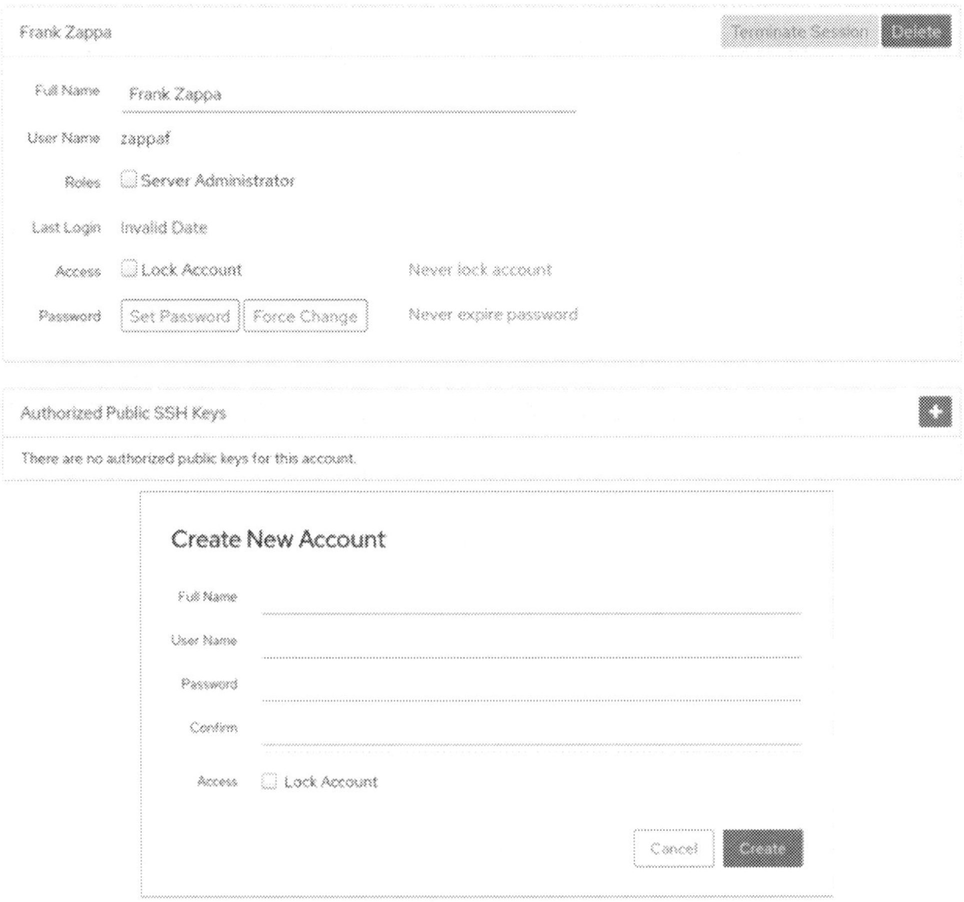

FIGURE 7.4 Using Cockpit to manage user accounts.

password change. These options are unavailable if we have logged in as a normal user. Compare the new user display with the User Manager's Add User window from Figure 7.2, and you will find that Cockpit has fewer choices available. In fact, we cannot alter any of the default values like directory location or default shell.

Ubuntu also has a facility to create user accounts through Settings (which are available from the Applications Software desktop icons). We could use this approach rather than installing and using Cockpit.

7.2.2 Creating User and Group Accounts from the Command Line

The command line-based approach to creating users (and groups) is often preferred because the GUI is less easy to use when creating numerous new accounts or when the new accounts are to be created with non-default values. Consider that we have to create 50 new user accounts. We could issue the first command from the command line and then use command line editing to quickly alter the command 49 additional times. More effectively would be the creation of a simple script that reads usernames from a textfile and calls upon the appropriate command or commands to generate the accounts.

In order to create accounts from the command line, we must be logged in as root. It is unsafe to log into the GUI as root because we might forget that we are serving as root and issue commands that cause damage. Instead, the best approach is to log in under our normal user account and then from the command line, switch to root when needed. We do so using the command `su` (substitute user).

The `su` command with no parameter tries to switch us from our current user account to root. We are asked to supply the root password. We can also change from our current account to any other account by using `su` *`username`*. We are asked to supply the password for *username*. One exception to this is if we are already root; when we use `su` to change from root to another user's account, we are not asked to supply a password. To return to our previous account, we enter `exit`.

Let's explore `su` with an example. Assume we are currently logged in as zappaf. We type `su foxr`, and if we successfully enter foxr's password, we are now logged in as foxr. If we type `who` and `whoami`, we get two different responses. `who` returns `zappaf` but `whoami` returns `foxr`. From foxr's account, we type `su` and enter the root password. We are now root. Typing `who` and `whoami` return `zappaf` and `root`, respectively. We now type `su dukeg`. We are not asked to provide dukeg's password because we issued the `su` command as root. `who` and `whoami` now return `zappaf` and `dukeg`, respectively. `exit` returns us to root. Another `exit` returns us to foxr. One final `exit` returns us to zappaf. Typing `exit` one more time would close the shell.

Remember to use `su` whenever needed to switch to root. Upon completing the task at hand, `exit` from root so that you don't forget and mistakenly enter a command that could do harm. For instance, entering `rm -fr * /home` as root would delete all user directories but in our own account would do nothing because we don't have access to perform that operation.

Another command that we can use that will let us issue commands as another user is through `sudo`. As noted in the previous subsection, Ubuntu and some Debian users may not know the root password so cannot `su` to root. In such cases, the first user account created in these distributions is given full administrator access via the `sudo` command. In order to use `sudo`, we issue the command as if we were root but add `sudo` before the command. The first time we use `sudo` in a session, we are asked to provide our account's password. This is remembered when executing `sudo` in the future.

Through `su` (and `sudo`), we have the means to enter system administrator commands. Now, let's learn some of these commands. We start with two commands to create user accounts. The first is `useradd`. The other command might be used to create bulk accounts with non-default values, `newusers`. We start with and concentrate on `useradd`.

At a minimum, `useradd` requires the new user's username. Likely we will want to create a home directory for this new user's account. The command for this is `useradd -m` *`username`*. If we are planning to create multiple accounts, we could issue the first command, type `control+p`, `escape+b` (which places us at the beginning of the user's name), `control+k` (to delete the first user's name), *username* <enter> to create a second account. We could then repeat this for each additional account.

The useradd command has a number of options. Most of these alter default attributes although one, -c, fills in the gecos (comment) field. Only a few of these options are available through the Cockpit GUI. See Table 7.3 which describes most of the more useful options and provides examples.

When a user is created, an entry is added to the /etc/passwd file storing for this user their username, comment field (name), UID, login shell, home directory and an x in place of their password. Early versions of Unix used the passwd file to also store user passwords in an encrypted form. However, the passwd file has permissions of -rw-r--r-- making it readable by the world. This was felt to be a security flaw in spite of the passwords being in an encrypted format as one might experiment with their own password to try to crack the encryption algorithm.

Passwords are now kept in /etc/shadow, which is not accessible by anyone except root. Whenever a new user is created, an entry is also added to /etc/shadow and in place of the encrypted password in the /etc/passwd file is an x. All of the values of a user are on a single line of the file with each field separated by a colon (:). We explore the shadow file later in this chapter.

TABLE 7.3
Common useradd Options

Option	Meaning	Example
-c *comment*	Fills gecos field with *comment*; used to specify user's full name.	"Richard Fox" (quote marks are necessary if the value has a space)
-d *directory*	Alter the user's home directory location from /home/*username* to *directory*.	-d /home/faculty/foxr
-D	Print or alter default values for common attributes including home directory location, expiration date, default shell, default skeleton directory and whether to create an email storage location.	
-e *date*	Set account expiration date to *date*.	-e 2023-05-31
-g *GID*	Alter private group ID to this value; otherwise, GID defaults to one greater than the last issued GID.	-g 1101
-G *groups*	Add user to the listed groups; groups are listed by name or GID and separated by commas with no spaces in between.	-G faculty,staff,admin
-k *directory*	Change default skeleton directory (explained in Section 7.6.1).	-k /etc/students/skel
-l	Do not add this user to the lastlog or faillog log files; this permits an account to go "unnoticed" by authentication logging mechanisms, which constitutes a breach in security.	
-m	Create a home directory for this user.	
-M	Do not create a home directory for this user (this is the default so can be omitted).	
-N	Do not create a private group for this user (user is placed in the users group instead).	
-o	Used in conjunction with -u so that the UID does not have to be unique; see -u.	-u 999 -o
-p *passwd*	Set the user's initial password; *passwd* must be encrypted.	
-r	Create a system account for this user.	
-s *shell*	Use specified *shell* rather than the default shell; for software, use /usr/sbin/nologin.	-s /usr/bin/csh
-u *UID*	Give user the UID of *UID*; the default UID is one greater than the last UID; can be used with -o so that two users share a UID.	-u 1045

TABLE 7.4
Example `useradd` Commands

Command	Comments
`useradd foo1`	With no `-m`, this user account is not given a home directory.
`useradd -m foo2`	This account is given a home directory.
`useradd -m -d /home/students/foo3`	This user's home directory is not in the default location.
`useradd -m -s /usr/bin/csh foo4`	This user's login shell is `csh` instead of `bash`.
`useradd -m -u 1500 foo5`	We alter the default UID to `1500`.
`useradd -m -o -u 1500 foo5jr`	foo5jr will share the `1500` UID with foo5.
`useradd -m -N -e 2025-12-31 foo6`	foo6 has no private group and has an expiration date.
`useradd -l -m -N -d /var/somesoftware -s /usr/sbin/nologin softwaretitle`	Presumably, a software account which has a home directory under `/var`, no private group and no login shell.
`useradd -m -l -r -N backdoor`	This account is given administrator privileges (`-r`) and is not being recorded in various log files.

Table 7.4 provides several examples of `useradd` instructions along with comments to describe the instruction. The first example is of a user who is not given a home directory. We might want to provide no home directory for guest accounts or for some software accounts. On the other hand, the second to last example shows a software account being created with a non-default home directory. Can you figure out what the last account might be used for? Read the comment and look at the account's username. Note that even though the creation of this account will not be recorded in a log file, the `/etc/passwd` and `/etc/shadow` files will contain an entry for this account, so `backdoor` is probably not a very good name!

The command `useradd -D` is used to either display or modify the `useradd` defaults. With no additional parameter, `useradd` lists the default values. Figure 7.5 shows us what the output might look like. Adding one of the other `useradd` options lets us change the default for the value that the option specifies. For instance, `useradd -D -s /usr/bin/csh` changes the default shell from `bash` to `csh`. The `useradd` default values are recorded in the file `/etc/default/useradd`.

We might modify a default to create a number of new accounts with a non-default value. This allows us to forego using an option to override the default when issuing the `useradd` instructions for the new accounts. We can later reset the default value to its original after we create the new accounts. For instance, we are preparing to create one hundred new accounts, all of which will have a home directory in `/home/sales/`*`username`* instead of `/home/`*`username`*. We issue

```
GROUP=100
HOME=/home
INACTIVE=-1
EXPIRE=
SHELL=/bin/bash
SKEL=/etc/skel
CREATE_MAIL_SPOOL=yes
```

FIGURE 7.5 `useradd -D` defaults.

useradd -D -d /home/sales and then create the 100 new accounts. When done, we issue useradd -D -d /home to reset the default.

Let's explore some of the values in Figure 7.5 other than the obvious ones of the home directory (HOME) and login shell (SHELL). SKEL indicates the location of the skeleton directory. This directory is used to generate the new user's home directory contents. We explore the contents of this directory in Section 7.6.1. As a system administrator, we might set up multiple skeleton directories depending on the role of the user and therefore may need to either change the default or use the -k option with useradd. The mail spool file is the user's email file. Specifying yes for CREATE_MAIL_SPOOL causes useradd to create the file for the user.

Some of the other defaults pertain to account activity. INACTIVE specifies the number of days that must elapse after a user's password has expired when that account becomes inactive. An INACTIVE value of -1 means that there is no inactivity date. EXPIRE indicates the default date by which new user accounts will expire. This might be set to say 12-31-2025 if we know that all of the accounts we are about to create should expire on that date.

One last default is specified as GROUP=100. The value for GROUP indicates which group to assign any new user account should that account not be given its own private group. We specify no private group with the useradd option -N. Group 100 is named users. Therefore, unless we alter the default, any user account not given its own private group is placed into the users group. We might choose to give guest accounts and some software accounts no private group.

The groupadd instruction is simpler than useradd. The syntax is groupadd *groupname*. The options available for groupadd are limited. Table 7.5 describes the most common ones. The most significant option is -g to override the default GID. As noted earlier, creating groups outside of private groups and generating accounts without private groups will throw off future numbering for GIDs so that they do not match UIDs. For this reason, if nothing else, we might create non-private groups with GIDs in a different range.

Let's consider an example scenario. We want to place any user who is not issued a private group account into a group other than users (GID 100). We will call this group others, which we will assign a GID of 505. First, we create the new group using groupadd -g 505 others. Next, we reset the default GROUP value 100 to 505 with useradd -D -g 505. Note that for the -D -g combination to work, the GID must be an already existing group ID. Now, when we use useradd -N *username*, *username* is automatically placed in group 505 instead of 100. Later, we might reset GROUP to 100 with useradd -D -g 100.

Whenever a new group is created, an entry is added to /etc/group. Since most user accounts generate a private group, a new user account will usually also add an entry to /etc/group. And since groups can have passwords, their passwords are placed in the file /etc/gshadow. Thus, new users will likely cause a new entry in each of /etc/passwd, /etc/shadow, /etc/group and possibly /etc/gshadow.

TABLE 7.5
Common groupadd Options

Option	Meaning
-f	Force groupadd to exit without error if the specified *groupname* is already in use, in which case groupadd does not create a new group.
-g *GID*	Use the specified *GID* in place of the default; if used with -f and the GID already exists, it will cause groupadd to generate a unique GID in place of the specified GID.
-o	Used with -g so that two groups can share a *GID*.
-p *passwd*	Assign the group to have the specified *passwd*.
-r	Create a system group.

7.2.3 Creating a Large Number of User Accounts

Consider that we want to add three new accounts: Mike Keneally, George Duke and Ruth Underwood. We want to use the defaults for all three. We might enter the first instruction as `useradd -c "Mike Keneally" -m keneallym <enter>`. Now, we use command line editing to alter this instruction and modify it for the next user. The sequence of steps is given in Table 7.6. We would repeat Table 7.6 but with `underwoodr` as the username and `Ruth Underwood` as the comment field.

Alternatively, a nice little shell script would allow us to generate these user accounts without the command-line editing effort. We place the users' first and last names in a file. We then run the script, shown in Figure 7.6, redirecting input to come from the file of user names. We make the assumption that we want usernames to be the user's last name followed by their first initial, as in `zappaf`. We also assume that the names in the file are fully lowercased so that the generated usernames are also fully lowercased.

This script should be easy to understand. First, the script iterates over each row of the file, inputting the two values of the row (first and last name of a user) and storing them in the variables `first` and `last`, respectively. Next, the value for the variable `name` is composed of the values in `first` and `last` with a space between them. The `name` variable is used to fill the comment field.

The script then forms the `username` as last name and first character of the first name, as in `underwoodr`. Now `$name` and `$username` are used in the `useradd` instruction to create the account. The reason why `$name` appears in quote marks in `useradd` is that `-c` expects a string that has no spaces, but `name` has a space. To force the parameter to be treated as a single string,

TABLE 7.6
Modifying the First `useradd` Instruction to Create Next User

Keystrokes	Command Line (Cursor Position Indicated with an Underline When Not at End of Line)	Explanation
`control+p`	`useradd -c "Mike Keneally" -m keneallym`	Recall last instruction.
`escape+b`	`useradd -c "Mike Keneally" -m keneallym`	Move to beginning of the username.
`control+k`	`useradd -c "Mike Keneally" -m`	Delete the username.
`dukeg`	`useradd -c "Mike Keneally" -m dukeg`	Add the new user's username.
`control+a`	`useradd -c "Mike Keneally" -m dukeg`	Move to beginning of line.
`escape+f` twice, `control+f` twice	`useradd -c "Mike Keneally" -m dukeg`	Move cursor to beginning of the comment field.
`escape+d` twice	`useradd -c "" -m dukeg`	Delete comment field.
`George Duke`	`useradd -c "George Duke" -m dukeg`	Insert new name as a comment.
`<enter>`		Submit command, returns us to our prompt.

```
#!/bin/bash
while read first last; do
   name="$first $last"
   username="$last${first:0:1}"
   useradd -c "$name" -m $username
done
```

FIGURE 7.6 Simple user account creation script.

```
n=`egrep -c $username /etc/passwd`
n=$((n+1))
username=$username$n
```

FIGURE 7.7 Added code to derive unique username.

we must use quote marks. Using '$name' would result in the comment field containing $first $last rather than the user's first and last name.

Assuming this script is called create_users.sh and our list of user first and last names is stored in new_users.txt, we run the script as ./create_users.sh < new_users.txt. Notice that this simple script does not attempt to alter any defaults. We consider how to do this shortly.

What happens if there is already a user in the system with the same username? Let's assume we already have an existing user whose name is Tom Fowler and with a username of fowlert. We now try to add Tim Fowler. We will receive an error generated by useradd because fowlert already exists. We modify the script from Figure 7.6 with three additional instructions, as shown in Figure 7.7. These instructions are inserted immediately before the useradd instruction.

These new instructions slightly alter our naming scheme for usernames by appending an integer at the end, as in fowlert1 or fowlert2. The number makes every username unique. The first new instruction is an egrep -c instruction to search the /etc/passwd file on the username which, at this point in the code, is last name first initial (e.g., fowlwert). If there already exists one or more users whose usernames include this string, then egrep returns a value greater than 0, otherwise it returns a 0. We add 1 to whatever value egrep returned and tack this number onto the already formed username.

Let's be more concrete about this. We already have two user accounts whose names start with fowlert (fowlert1, fowlert2). The egrep instruction will return 2, and the script stores this value in the variable n. n is then incremented (to 3). Now, $n is appended to username to give us fowlert3, a unique name. If there had been no previous fowlert's in /etc/passwd, then n will be 0 and then incremented to 1. Thus, the new user would be fowlert1.

What if we want the script to generate user accounts with non-default values? Our script is too simplistic as is our input data file. We make the following assumptions. Users fall into one of three categories: faculty, staff and students. Faculty will have home directories under /home/faculty, staff under /home/staff and students under /home/students. All accounts will use bash unless tcsh is explicitly listed in the data file. We modify the input data file to include the user's category after the last name, and tcsh as a fourth item on any line if the user has requested that tcsh serve as their default shell.

We modify the while read instruction in the script by adding two new variables: category and tcsh. category will always have a value (faculty, staff, student) whereas the variable tcsh will either be the string tcsh or NULL (empty). We add two if-then statements prior to the useradd instruction in the script. These are shown in Figure 7.8. The first tests the value of category to determine which directory to use, and the second tests the value of tcsh to decide whether to use bash or tcsh. We then modify the useradd instruction, as shown in the figure. Note that none of the quote marks are needed in the script aside from those around $name in the useradd instruction but are added here to clearly denote that we are specifying strings.

The script would become more complex still if we wanted to modify the UID/GID defaults. If our numbering scheme was based on the user's category, we would not only have to test the category to determine which range the new UID/GID should be within but also find the largest UID and GID within that range and add 1 to it. For instance, if faculty have UIDs/GIDs in the range 1200–1999, it's not enough just to determine that the new user is a faculty. We would then have

```
if [ $category == faculty ]; then dir="/home/faculty"
  elif [ $category == staff ]; then dir= "/home/staff"
  else dir="/home/students"
fi

if [ $tcsh == tcsh ]; then shell="/usr/bin/tcsh"
  else shell="/usr/bin/bash"
fi

useradd -c "$name" -d "$dir" -s "$shell" $username
```

FIGURE 7.8 Added code to the account creation script to change default settings.

to search the /etc/passwd file to find the largest UID between 1200 and 1999, add 1 to it, using that new value for this new account's UID. We would have to similarly search /etc/group for the largest GID between 1200 and 1999 and add 1 to it. This enhancement would involve a good deal more logic than what we see in Figure 7.8. We leave it to the reader to think about how to implement such code.

Creating the initial script from Figure 7.6 and adding the revised code found in Figures 7.7 and 7.8 are simple enough. We might use such a script to generate user accounts where all users get default values or allow a few modest forms of non-defaults. If we want to create users with other non-defaults, the script is still writable but more complex. As we permit more and more non-default specifiers in the useradd instruction, we have to add more content in the input data file. If we want to go to such an extreme, the better choice is to use the program newusers.

newusers performs mass account creation. We provide the command with a data file that contains details for each new user account. Unlike the script we developed in Figures 7.6–7.8 which only required first and last names and possibly a category and an optional tcsh, newusers requires that the data file contain seven values describing settings for each new user's account. The values are username:passwd:uid:gid:comment:dir:shell. Notice that each value is separated by a colon (:). All but the username may be omitted, in which case default values are used. Omitting values will have the appearance of two or more colons as in ::: if two consecutive values are omitted.

The password is in plain text, and newusers will encrypt it for us. The UID and GID can be specified by number or name. If the UID/GID already exists, then the UID/GID is shared between the old and new accounts. When there is no entry, the default value UID/GID is used (one greater than the last one used). The comment field should be the user's full name.

The newusers command has some modest error checking, for instance by allowing duplicate UIDs and GIDs. If the path specified in the directory is erroneous, newusers does not terminate but continues without generating the directory. Thus, the new user will not have a home directory and newusers sends an error message to STDERR so that the system administrator can resolve the problem and create a home directory by hand. If the path is valid but the directory does not exist, then it is created. No checking is performed on the shell specified. If the shell is erroneously listed, then the entry in /etc/passwd will also be erroneous.

As newusers will automatically take the text-based password specified for a user and encrypt it, we are able to specify which encryption algorithm to use. This is accomplished using the -c *method* option, where *method* is one of DES, MD5, NONE, SHA256 or SHA512. The latter two are only available if the proper library is available. An additional option is -r to create a system account although this would apply to all accounts created with this submission.

Although this program's executable file is set to world executable status, only root can actually run it because only root has write access to the /etc/passwd and /etc/shadow files. Further, the input file should be protected with proper permissions because it stores unencrypted initial user passwords. As `/root` is typically not accessible to any user other than root, it is best to store the `newusers`' input text files underneath `/root`.

SECTION ACTIVITIES

1. Provide a list of reasons why you should use `useradd` (or `newusers`) to create user accounts instead of the GUI.
2. Assume you are administrator of a Linux computer that you share with family and friends. How often do you think you would need to use non-default values when creating new user accounts? Would you find it easier to create user accounts by using `useradd` or would you prefer to write a shell script like that of Figures 7.6 and 7.7?
3. If you are learning Linux for school or work, find a system administrator responsible for creating user accouånts on school/work computers (whether its Windows or Linux or other) and ask if he/she creates accounts by hand or uses some form of script. If a script, did this person write the script himself/herself? See if he/she will show you the script so that you can compare it against the script developed in this section.

7.3 MANAGING USERS AND GROUPS

With users and groups created, we must manage them. Management includes modifying user accounts (e.g., changing shells, home directories) and groups (e.g., adding users to groups). A useful program to inspect a user's information is `id`. This program returns the given user's UID, GID of the given user's private group and other groups that the user is a member of. For instance, `id foxr` might result in output of `uid=1003(foxr) gid=1003(foxr) groups=1003(foxr),1004(cool)`. If we do not supply the username, `id` returns the current user's information, including the SELinux security context.

Both the older User Manager tool and the newer Cockpit interface permit modest modifications to user accounts. While the system administrator can use Cockpit, command-line instructions are far more flexible, and so we concentrate on these instructions. The commands we will examine in this section are `usermod`, `userdel`, `groupmod`, `groupdel` and `newusers`.

The `usermod` operation has similar options to `useradd` but rather than overriding defaults, `usermod` changes the existing value(s). With `-d` *`directory`*, we change the user's home directory to the newly specified directory. This does not move any of the user's old directory's contents but instead just updates the entry in `/etc/passwd`. Add `-m` to physically move the contents. We can also specify `-l` *`username2`* to change the user's username to the *username2*. Other options include `-L` and `-U` to lock and later unlock this user's account. The `groupmod` instruction has the same options as `groupadd` except that it also has a `-n` *`newgroup`* option to change the group name to *newgroup*, similar to `useradd`'s `-l` option.

The `newusers` command can be used to modify existing user accounts. It is used identically to how we create initial accounts. We specify a data file but in this case the file will contain usernames of existing users. We include any user whose account we wish to modify by placing any or all of a new password, comment, UID, GID, home directory and default shell in the file for the select user(s). We can combine the creation of new user accounts with the modification of existing user accounts in one file.

`usermod` is probably the better instruction to use when modifying users as `usermod` allows us to make a greater variety of changes. With that said, if we find that we have to create some new accounts and modify some existing accounts, we might combine the tasks into one file and one `newusers` command.

The `userdel` command is used to delete user accounts. The command is `userdel` *`username`* where *username* must be an existing user. If the user is logged in, we will be given an error and the user will not be deleted, unless we *force* the deletion with the option `-f`. User files (home directory and its contents, email file) are not automatically deleted when we delete the user. To delete these types of user files, add the option `-r`. Notice that this will not delete files owned by this user that are outside of the user's home directory.

Deleting a user removes the user's entry in `/etc/passwd`, `/etc/shadow`, and if the user has a private group both `/etc/group` and `/etc/gshadow`. The user's name is also removed from all non-private groups.

The `groupdel` instruction is perhaps one of the easiest in Linux. There are no options and the only parameter is the group to delete. The command is simply `groupdel` *`groupname`*. The group is deleted from the `/etc/group` and `/etc/gshadow` files. There may be files owned by this group in which case we would have to manually search for those files and change their group ownership (or delete them).

Both `userdel` and `groupdel` return a success code. Although we expect them both to be successful (0), there are other codes worth knowing. These are shown in Table 7.7. Any value other than 0 means that the instruction failed to perform the deletion.

Care should be taken when deleting a user or a group. Your organization should have policies established that determine when a user or group can be deleted. Section 7.9 discusses user account policies. Before deletion, ensure that the user is not logged in or that the group is not a private group. Finally, after performing the deletion, check the system to make sure that no remnant files owned by the user or group exist.

Just as we wrote a script to create user accounts, we might similarly create a script to delete user accounts. The top portion of Figure 7.9 presents an example of such a script. Assume that the user account names are stored in a text file and we redirect this file to the script to delete each of the listed accounts. Some questions to answer before writing this script are whether we should force deletion

TABLE 7.7
`userdel` and `groupdel` Exit Codes

`userdel` Exit Code	Meaning	`groupdel` Exit Code	Meaning
0	Success.	0	Success.
1	Cannot update passwd file.	2	Invalid command.
2	Invalid syntax.	6	Group does not exist.
6	User does not exist.	8	Group is an existing user's private group.
8	User currently logged in.	10	Cannot update group file.
10	Cannot update group file.		
12	Cannot remove home directory.		

```
#!/bin/bash
while read username deletion; do
     if [ -z $deletion ]; then userdel $username
     else userdel -r $username
     fi
done
-----------------------------------------------
Added instruction for error handling
if [ $? -ne 0 ]; then echo "$username" >>
          /root/not_yet_deleted.txt; fi
```

FIGURE 7.9 Account deletion script.

in case any of these users are currently logged in and whether we want to remove the users' home directories and email space.

The script presented in the figure assumes that we do not want to force deletion under any situation and we will delete the user's directory/email only if there is an entry of `yes` in the input file after the username. The script requires that we input up to two values from each row of the file, the user's username and an optional value which will either be `yes` or empty to indicate whether user files are to be deleted. We will store these in variables named `username` and `deletion`. The script will test to see if `deletion` has a value using `[ -z $deletion ]`. The else clause performs a delete with the option `-r`.

As we did not force deletions, this script fails to delete users who are logged in and/or have processes currently running. If the system administrator were to run this script, it may not complete the task fully. We could enhance the script by keeping a record of any user who, when we attempted to delete them, we received an error code of 8 (user logged in).

We revise the script by adding an if-then statement immediately before the `done` that ends the while loop. The condition tests `$?` which is the result of the last instruction executed, `userdel`. `$?` will store 0 if the `userdel` was successful. If it is not 0, then the `userdel` failed and we should make a note of this. The code is shown in the bottom part of Figure 7.9 where the then clause appends the username to a file in root's home directory that lists all users who have not yet been deleted. We could also output the value of `$?` so that the administrator has more information about why the `userdel` failed.

SECTION ACTIVITIES

1. What is a real-world reason why you might want to user `usermod` to modify an existing account?
2. Provide a reason for why you would want to delete user files when deleting a user's account and a reason for retaining user files when deleting a user's account. Is there a situation whereby you would want to delete a user's account while the user is logged in? If so, why?

7.4 PASSWORD MANAGEMENT

In this section, we look at password management. *Password management* involves three tasks: initial generation of passwords, enforcing a policy whereby passwords are modified in a timely fashion and enforcing a policy requiring strong passwords.

7.4.1 Automatically Generating Passwords

The `apg` program is an easy way to generate random passwords. It provides six passwords of eight characters each, using randomly generated characters from one of the two programs `/dev/random` or `/dev/urandom`. The `apg` program can check the quality of a password by comparing it to a dictionary file to ensure that no random passwords come close to matching a dictionary entry. The program has many options, the most useful of which are listed in Table 7.8.

As `apg` may not be part of the initial Linux installation, we might have to install it. As an alternative, we can use one of the random number generator programs to generate our own passwords. `/dev/urandom` is a random number generator program serving as a device. It generates any ASCII characters, including non-printable characters. To use `urandom` to create passwords, we need to remove characters that are not alphabetic or alphanumeric.

TABLE 7.8
Useful apg Options

Option	Meaning
`-a 0` or `-a 1`	Select between `/dev/random` (0) and `/dev/urandom` (1); defaults to 0.
`-n` *num*	Change the number of passwords produced from the default of 6 to *num*.
`-m` *min*	Change the minimum length of the passwords produced from 8 to a minimum of *min* characters.
`-x` *max*	Change the maximum length of the passwords produced from 8 to a maximum of *max* characters.
`-M` *mode*	Change the *mode* which specifies the types of characters generated; modes are `S` (include at least one non-alphanumeric character), `N` (include at least one digit), `C` (include at least one capital letter).
`-E` *string*	Do not include any characters specified in the given *string*; not available for algorithm 0.
`-y`	Generate passwords and then encrypt them for output.

We can remove characters using the `tr` program (translate or delete characters). For `tr`, we add the `-cd` option to specify that we want to delete all but the characters provided in the included set. The set should be one of `[:alpha:]` or `[:alnum:]`, or a range of characters as in `a-z`, `a-zA-Z` or perhaps `bcdfghjklmnpqrstvwxyz` if for instance we wanted to generate a string with no vowels. We issue the `tr` instruction and redirect input to come from `/dev/urandom`. We pipe this result to `head` (or `tail`) to leave us with just eight characters. The full instruction is `tr -cd '[:alpha:]' < /dev/urandom | head -c8`, which generates a password of eight random letters. Alter the option of `-c8` to some other number for longer (or shorter) passwords and use `[:alnum:]` to generate passwords of letters and digits.

Another approach to generating a password is to obtain the current time and date using the `date` command and pass this to a hashing function. A *hashing function* takes a string and returns a numeric value by combining the characters of the string using some formulaic approach. We use hashing functions as part of modern encryption.

The `sha256sum` program takes a string and produces a hash value consisting of hexadecimal digits. The generated password would then consist of digits 0–9 and letters a–f. We could pipe the result to `head` (or `tail`) to reduce the size from 64 hexadecimal digits (256 bits) to 8 digits. Our instruction is `date +%s | sha256sum | head -c8`. Another option is to use an encryption program like `openssl` to generate random characters. One such instruction is `openssl rand -base64 8`, where the `8` indicates the number of characters to produce.

One other approach to generate a password is to write our own script that draws random entries from the Linux dictionary and combines random excerpts from the selected entries. For example, we might generate four random numbers to select four different words from the dictionary and then generate four additional random numbers to select pairs of characters from those strings, conjoining them together for a password of eight characters. There is a built-in Bash function called `$RANDOM` that can generate random numbers between 0 and 32,767 that we could employ to assist us with the endeavor (although the file contains nearly 500,000 words so we could only draw a fraction of the words from the dictionary). As the script would be far more complicated than the approaches described above, we omit further detail.

We can use an assignment statement to store the generated password having used any of the previously described approaches. The format of the instruction is `password=`...`` where ... is the instruction used to generate the password. For instance, we might use ``password=`apg -n 1 -m 8` `` to obtain a single password of eight characters. We then use the `passwd` command to give the new user a password.

Figure 7.10 provides the code we might add to our script from Figure 7.6 to generate and assign a random password to the new account. The first instruction generates the password. We chose to use `apg` simply because it is a brief command. In order to assign the password to the current account (`$username`), we cannot use the `passwd` command as is because `passwd` expects the password

```
password=`apg -n 1 -m 8`
echo $password | passwd --stdin $username
echo "$username $password" >> /root/tempPasswords
```

FIGURE 7.10 Password generating code for our account generating script.

to be entered interactively. Instead, we add the option `--stdin` to indicate that the password will be piped to the instruction, which we do using `echo $password`. Finally, the last instruction stores both `username` and `password` to a file in `/root` so that the administrator can later inform the user of his or her initial password. Without this, the administrator would have no idea what the initial password was.

Given that any of our approaches that generate passwords will create passwords that are combinations of random letters or letters and digits, the user will have a challenge memorizing it. We will want the user to change their password the first time they log in. Fortunately, we have a means to force the user to do just that. We also want to make sure that the password the user selects is deemed a strong password. So, let's move on to explore how we control password management.

7.4.2 Managing Passwords

Password management requires that passwords are modified in a timely fashion. To enforce such a policy, we turn to two Linux programs: `chage` and `passwd`. The `chage` program allows the system administrator to change user password expiration dates for a specific user. Through `chage`, the system administrator can force a user to change their password by a specific date or have their account locked. The format of this instruction is `chage [`*`options`*`]` *`username`*.

Many of the available options require a day or date. The day is a number representing the number of days that have elapsed since January 1, 1970 (known as the *epoch*). The date is specified using the format YYYY-MM-DD as in 2023-05-31 for May 31, 2023.

The `chage` program modifies entries in `/etc/shadow` which stores not only encrypted passwords but also password expiration information. The format of the `shadow` file for each entry is shown in Table 7.9. The option column indicates the `chage` command's option that allows us to change that particular value. More detail on `chage` is presented shortly.

Two entries from the `/etc/shadow` file are shown in Figure 7.11, with the encrypted passwords replaced by … for brevity. For zappaf and foxr, they are allowed to change passwords daily (minimum days between passwords is 1) and are required to change their passwords within 35 and 28 days, respectively. A warning is issued to the user in advance of the password expiring. In this case, the days until the warning issued are 25 and 26, respectively. Notice that the value in the

TABLE 7.9
`/etc/shadow` Fields and `chage` Options

Field	Option
Username	
Encrypted password	
Days since January 1, 1970, that the password was last changed	`-d`
Days before the password may change again	`-m`
Days before the password must be changed	`-M`
Days before a password expires that a warning is issued	`-W`
Days after password expires that account is disabled	`-I`
Days since January 1, 1970, that the account will become disabled	`-E`

```
zappaf:...:15558:1:35:10:20:365:
foxr:...:15558:1:28:2:10::
```

FIGURE 7.11 Two entries for /etc/shadow.

TABLE 7.10
Common chage options

Option	Meaning
`-d day`	Set the number of days of when the password was last changed; this is automatically set when a user changes password but can be altered; if never changed, this date is the number of days since the epoch.
`-E day`	Set the day on which the user's account will become inactive (will expire), specified as a date (YYYY-MM-DD) or the number of days since the epoch; this option can be used to provide all accounts with a lifetime; to remove a previously established expiration date, use `-1` for the day.
`-I day`	Set the number of days of inactivity after a password has expired before the account becomes locked; to remove a previously established expiration date, use `-1` for the day.
`-l` (lower case L)	Show this user's password date information.
`-M days`	Set the number of days remaining before the user *must* change their password; should be used with `-W`.
`-m days`	Set the minimum number of days between which a user is allowed to change passwords; a value of 0 means that the user is free to change password at any time; if the value is greater than 0, this limits the user in terms of how often the password can be changed.
`-W days`	Set the number of days prior to when a password must be changed that a warning is issued to the user to remind the user to change password; for example, `-W 7 -M 28` contacts the user 7 days prior to the password's expiration, or in 21 days.

/etc/shadow file is not the number of days until a warning is issued but the number of days prior to the password expiring that warning is issued.

Continuing with our examination of the /etc/shadow file, we see numbers 20 and 10, respectively, for zappaf and foxr. These values indicate the number of days after the password expiration date that their accounts will still be accessible. This gives users a grace period. Upon logging in during this grace period, the user must change their password. After this time period has elapsed, the user account becomes inaccessible and the user needs to contact the system administrator to resolve the issue. If this entry were missing, it means that the grace period extends forever.

The last number for zappaf is 365, the number of days until his account expires. foxr has no such entry, as denoted by `::`, and so foxr's account is not set to expire. There is space for one more entry (after the last colon) but it is reserved for future use.

The chage command has eight options, one being -h to obtain help. The other seven are described in Table 7.10. If no options are provided to chage, then chage operates in an interactive mode, prompting the user for each field.

Many of the same options in chage are available in passwd, although they are specified differently. Table 7.11 lists available passwd options, with reference to equivalent options in chage (refer back to Table 7.10 as needed). passwd has several options not available in chage such as -d, -l and -u. passwd also has option -e to expire the current password immediately. Referring back to the script excerpt from Figure 7.10, we might add this option to the passwd command causing the new password to expire which would force the new user to change password the first time he or she logs in.

TABLE 7.11
Options for passwd

Option	Meaning
-d	Disable the password (make the account password-less).
-i *day*	Same as chage -I *day*.
-k	Only modify the password if it has expired.
-l	Lock the account (user cannot login until unlocked).
-n *day*	Same as chage -m *day*.
-S	Output status of password for the given account, similar to chage -l.
-u	Unlock the locked account.
-x *day*	Same as chage -M *day*.
-w *day*	Same as chage -W *day*.

SECTION ACTIVITIES

1. How many different user accounts do you have (between your own computer(s), school and work computer(s) and website accounts)? How do you remember all of your passwords?
2. Hackers often try to break into accounts by using the most popular passwords or variations thereof. Each year, the most popular passwords are published. See for instance https://nordpass.com/most-common-passwords-list/. Look at this list of 200 passwords. Have you ever used any of these?
3. Have you ever had a computer account disabled? If so, for what reason? If not, seek out your school's system administrator and ask him or her if they have a policy for disabling accounts for some reason and what reason(s) they might have.

7.5 PAM AND ENFORCING STRONG PASSWORDS

PAM (or pam) is short for Pluggable Authentication Module. Programs that deal with authentication in any form (chage, passwd, su, sudo, login, ssh, etc.) need to handle such activities as obtaining the password from the user, authenticating the username and password with a password file, maintaining a session during which the user does not need to reauthenticate and logging events. PAM is an *extensible* piece of software in that the system administrator can configure how a given authentication program will perform its authentication duties. This is handled through a configuration file that lists for specific applications' duties, which PAM modules to call upon. PAM is also extendable by adding our own modules or third-party modules.

Let's consider as an example the passwd program. This program requires that the user authenticate before changing their password. Authentication is a two-step process of first obtaining the password from the user and then testing the password against the /etc/shadow file. passwd then inputs a new password from the user and if it passes whatever strength tests are required, passwd's final task is to modify the /etc/shadow file. We are able to specify how each of these steps can be handled through PAM modules.

Another program that requires user authentication is login, which is called whenever a user attempts to log in from either the GUI or a text-based program like ssh. The login program differs from passwd because it adds the task of maintaining a *session* while the user is logged in. The session, in this case, permits the accessibility of services without requiring that the user reauthenticate. For instance, upon logging in, the user's session interacts with auditd and the user's /home directory. In addition, the session maintains the user's namespace within the user's context (i.e., the session is run under the PID of the user with the user's access rights).

The strength of PAM is in the modules which implement the functions responsible for authentication operations. One function is called pam_open_session() which can be invoked through the module pam_unix_session.so. The collection of modules provides the proper support for most programs that require authentication. It is up to the system administrator to tie together those modules that will be used for the given program.

Before we look at PAM in more detail, keep in mind that PAM is already configured for the authentication programs available in Linux. We would only add our own configuration specification if we install some new piece of software that has its own authentication process. We might modify an existing configuration if we find the default configuration lacking in some way. For instance, imagine some type of operation is not logged and we want to add logging to the authentication process. Another change might be if a program has an authentication session which is maintained until the user logs out, and we want to change this to have some upper time limit.

All of the existing configurations are placed in separate files and stored in the directory /etc/pam.d. Each of these configuration files is named after the authentication program that the configuration supports, such as atd, cockpit, passwd, su and sudo. We would add our own configuration file named after a new piece of software as needed.

These configuration files consist of directives that specify the steps that the authentication process requires for the program to execute. These directives consist of three parts: the module's type, a control flag and the module's filename (these are all .so files).

There are four module *types*. The auth module type invokes the appropriate module to obtain the user's password and match it against the appropriate password file. Most modules are capable of using the /etc/shadow file while other modules might call upon password files specific to software (such as a password file used specifically for an Oracle database). The account module type handles account restrictions such as whether access is allowed based on a time of day or the current status of the system. The session type maintains a user session, handling such tasks as logging the start and end of session, confirming that the session is still open (so that a repeat login is not required) and maintaining sessions with remotely mounted file system. The password type deals with changing passwords, as with the passwd command.

The configuration file might include several directives with the same module type. A collection of directives with the same module type are known as a *stack*. Within a stack, we have the possibility that all of the modules must run and succeed or that only some of them have to succeed. The *control flag* is used to indicate what pattern of modules needs to succeed.

There are four types of control flags. With requisite, the given module must succeed for access to be permitted. The next control flag type is required. If multiple directives within a stack are listed as required, then all of the corresponding modules must succeed or it is considered a failure. If there is only one directive in a stack, we can use requisite or required. The third control flag type is sufficient. Here, if any single module labeled as sufficient succeeds within the stack then it is considered a success. A stack of required directives acts as a logical AND operation while a stack of sufficient directives acts as a logical OR operation.

If the module(s) succeeds, this is passed on to the application so that the application can proceed onto its next step which will be the next stack of modules. If there is a failure of a module leading to the current stack failing, then control is returned to the application to deal with the failure and decide how to proceed.

A fourth control flag type is optional. This flag does not impact the success of the module but can be applied to perform other tasks. For instance, an optional directive whose module succeeds might indicate that logging should take place. Should that module fail, it does not return a failure to the application but logging would not take place.

As configurations are available for a number of applications, we can "piggy-back" on prior configurations. This is accomplished using either include or substack in place of a control flag.

The include statement causes the specified configuration file to be included as if the directives of that file were located in place of the include statement. The substack statement is the same but places restrictions on that configuration's usage.

Among the configuration files that are included or substacked are system-auth and postlogin. system-auth is utilized in a number of different configuration files including chfn, chsh, gdm-autologin, gdm-launch-environment, login, passwd, polkit-1, su, sudo, systemd-user and vlock. postlogin is used in cockpit, gdm-autologin, gdm-fingerprint, gdm-launch-environment, gdm-password, gdm-pin, gdm-smartcard, login, passwd, remote, sshd and su. The use of the include directive prevents a system administrator from having to recreate the same strategy over and over again.

The last *required* entry in a directive is the module itself. All of the modules are stored in /usr/lib64/security, and their names are of the form pam_xxx.so where *xxx* is the remainder of the module's name. When using third-party modules or modules that exist in other libraries, we must specify the full path name. Otherwise, the path (/usr/lib64/security) may be omitted. The module name is replaced by a configuration file name if the control flag is include.

Some modules utilize arguments as further specifications in the configuration statements. After the module name, it is permissible to include parameters such as tests passed to the module so that the module has additional values to work with. For instance, we might have a condition such as uid < 1000 or user != root. Parameters can themselves be followed by an option. Options include such terms as quiet, revoke, force, open, close and auto_start.

Figure 7.12 presents the default PAM configuration for the su command, stored in the file /etc/pam.d/su. We see in this configuration file each of the four module types in use: auth (authentication), account, password and session. Notice that the sixth line, containing use_uid quiet, is part of the fifth entry but wrapped around due to limited space.

Let's examine the su configuration in detail. There are four auth directives. First, the pam_env.so module is invoked and must succeed to continue. Next, if pam_rootok.so succeeds, su moves on to the account section. If it fails, then the directives in both system-auth and postlogin must succeed.

Under account, if pam_succeed_if.so succeeds, su continues with the password module. Three arguments are passed to pam_succeed_if.so, uid=0, use_uid and quiet. The argument quiet informs PAM not to log a failure login attempt in the system log. If this attempt is not successful, then PAM attempts authentication through system-auth. The password section again calls upon system-auth, as does session, which also utilizes postlogin. Finally, although not required, pam_xauth.so runs.

For more information on PAM, view the man pages for pam, pam.conf and any of the modules in question. All modules have corresponding man pages. This allows us to see the possible parameters and their meanings.

```
auth        required        pam_env.so
auth        sufficient      pam_rootok.so
auth        substack        system-auth
auth        include         postlogin
account     sufficient      pam_succeed_if.so uid=0
                                      use_uid quiet
account     include         system-auth
password    include         system-auth
session     include         system-auth
session     include         postlogin
session     optional        pam_xauth.so
```

FIGURE 7.12 PAM configuration for su.

We have mentioned strong passwords before but have not yet described what the term means. In today's climate of hacking attacks, we want to ensure that users' passwords are hard to guess. Hackers try many types of approaches to crack passwords including using all the words in a dictionary. Another approach is to generate all possible collection of characters of varying lengths. The time it takes to do so increases dramatically as the length of the password and type of characters used in the password increase. For instance, generating all possible six-letter passwords consisting entirely of lowercase letters requires a little over 300 million different combinations. If a password includes at least one uppercase letter, this number grows to nearly 19 billion!

Strong passwords are those that will be hard to crack because they include different types of characters and are sufficiently long. We might require passwords that are at least 12 characters that contain at a minimum one non-lowercase letter. We might strengthen the password by requiring at least one uppercase letter, one digit and one punctuation mark. Each additional requirement increases, usually exponentially, the number of possible combinations which then makes it that much harder to crack a password.

`PAM` does not itself enforce strong passwords. This is left up to a `PAM` module called `pam_pwquality`. It is through the file `/etc/security/pwquality.conf` that we can specify the directives to control the strength of passwords. Table 7.12 lists the directives available for this configuration file. The directives specify one of two things. Some specify absolute settings such as `minlen` while others are used to add up "credit" that can offset the need for a minimum length. Two of the options refer to `cracklib`. This is a Linux library (set of functions) that tests passwords to make sure they do not contain dictionary words and are not easy to crack because they are, for instance, repetitive.

TABLE 7.12
Directives to Control Strong Passwords

Directive	Meaning
`badwords`	A list of strings that qualify as illegal passwords; each must be at least four characters long and each is separated by a space.
`dictcheck`	If not `0`, use the `cracklib` dictionary to look for the password.
`dictpath`	Location of `cracklib` dictionary.
`dcredit`	Maximum number of digits to generate credit.
`difok`	Minimum number of characters that must differ from the old password.
`enforce_for_root`	Enforce the enclosed password policy for normal users and root alike; upon failing the policy the password is rejected.
`enforcing`	Enforce the enclosed password policy for normal users; upon failing the policy the password is rejected.
`gecoscheck`	If not `0`, compare the user's GECOS string to the password.
`lcredit`	Maximum number of lowercase letters to generate credit.
`maxclassrepeat`	Maximum number of characters in a row that can be of the same class (e.g., if set to 2 then no more than 2 digits, punctuation marks, uppercase letters or lowercase letters may occur in a row).
`maxrepeat`	Maximum number of times a character can be repeated.
`minclass`	Minimum number of distinct classes that must appear in the password.
`minlength`	Minimum length for a valid password.
`ocredit`	Maximum number of non-digits/letters to generate credit.
`remember`	Specify number of past passwords to remember that cannot be reused.
`ucredit`	Maximum number of uppercase letters to generate credit.
`usercheck`	If not `0`, check if the password includes the user's username.

```
minlength=12
remember=6
difok=4
usercheck=1
minclass=2
maxrepeat=4
ocredit=1
dcredit=1
ucredit=-1
enforcing=1
```

FIGURE 7.13 Sample `/etc/security/pwquality.conf` directives.

Characters for the password are categorized as lowercase letters, uppercase letters, digits and other (punctuation marks). Specifying a `minclass` greater than `0` means that the user must use at least two of these classes. The actual classes used are not specified so that, for instance, if two are required a user might use all uppercase letters and digits or all digits and punctuation marks.

Credit is applied to reduce the `minlength` value. For instance, if given credit for two digits, then the password can be two characters less than the minimum length. If `minlength` is `10` and both `ocredit` and `dcredit` are set to `1`, then a password whose length is eight characters is acceptable as long as there is at least one digit and one other character. Negative numbers are allowed for any of these credit directives which means that at least one character of this class is required. No matter how much credit is applied, a password cannot be less than six characters to pass the `pwquality` check.

Figure 7.13 illustrates a set of directives that we might apply to maintain strong passwords. It is currently recommended that passwords be at least `12` characters long. However, with a mixture of other characters, this size can be reduced.

The configuration specified in Figure 7.13 can be explained as follows. The minimum length of a password is 12 characters but the minimum can be reduced by one for a digit or other (punctuation) character. The password must contain at least two classes including an uppercase letter (`ucredit=-1`). Six previous passwords are remembered, and the new password cannot repeat any of these. The new password cannot have more than four repeated characters from the last password. The user's username cannot be part of the password. The password cannot have a character occur more than four times. Finally, with `enforcing=1`, a new password fails should any of these tests fail.

SECTION ACTIVITIES

1. The last section had an activity asking you about your different accounts and how you remember all of those passwords. Of the accounts you listed, how many require the use of a strong password? Of those, how often are you required to change those passwords and are you able to repeat passwords or parts of passwords?
2. Go to https://passwordsgenerator.net/ and use the website to generate some strong passwords. Play around with the various selections like including digits and not including similar characters. Generate a few dozen passwords. Did it generate any that you feel would be easy to remember? Would you use any of these?
3. Look at some of the `PAM` configuration files. See how much you can understand and what you don't understand, such as options passed to a module, look up in the `PAM` or the module's man page. Do this for two or three configuration files in an attempt to better understand `PAM`.

7.6 ESTABLISHING COMMON USER RESOURCES

For a system administrator, the creation of accounts can be a tedious task. For instance, in a large organization, the system administrator may find that he or she is creating dozens or hundreds of accounts at a time and perhaps has to do this weekly or monthly. A shell script to automate the process is only one feature that the system administrator will want to employ. There are other mechanisms to help establish user accounts, including the automatic generation of user files and the establishment of environment variables and aliases. These items are all controlled by the system administrator.

7.6.1 Populating User Home Directories with Initial Files

The /etc/skel directory is, by default, the directory that the system administrator will manipulate to provide initial files for the user. Anything placed in this directory is copied into the new user's home directory when the directory is generated. The only change made to these items is the owner and group which are altered from root to the user's username and private group name.

Typically, the files in /etc/skel will be limited to user shell startup files and common software configuration files and subdirectories. The CentOS Stream Linux installation has three files (.bashrc, .bash_profile and.bash_logout) and one subdirectory (.mozilla/, which contains subdirectories for extensions and plugins). The system administrator is encouraged to keep these three files as is and instead place system-oriented startup instructions in such files as /etc/bashrc and /etc/profile. This makes it far easier for the administrator to update startup values like environment variables.

Consider defining an environment variable for the default text editor. This is called EDITOR, which we might set to /bin/vi. As a system administrator, we might define EDITOR in either /etc/bashrc (or /etc/profile) or in /etc/skel/.bashrc. The difference between these three files is that /etc/skel/.bashrc is placed into every new user's home directory as their initial .bashrc file. By placing this environment variable's definition in /etc/skel/.bashrc, we might later have a problem should we decide to modify this environment variable, say to vim. Updating the entry in /etc/skel/.bashrc is easy enough but any previously created user accounts will have EDITOR=/bin/vi in their .bashrc files. How would we modify all of those files? Instead, placing this assignment statement in /etc/bashrc allows us to make just one modification and have it affect all users.

The .bash_profile file will usually consist of just an if-then statement to test if the user's .bashrc file exists and if so executes it. The .bashrc file uses an if-then statement to test if /etc/bashrc exists and if so executes it. It then modifies the PATH variable by adding $HOME/.local/bin and $HOME/bin. The added directories, if they exist, are used by users to store their own binary (executable) files. The file's last instruction is to export the updated PATH variable. There are comments to explain to the user what other modifications they might want to make including adding aliases and functions.

Each user is free to modify and add to either or both of their .bash_profile and .bashrc files. It is the system administrator who would modify /etc/bashrc (as well as /etc/profile). The user might, for instance, add further environment variables and aliases to either .bash_profile or .bashrc. The user could also unalias system-wide aliases (although this is not a particularly good idea) and place commands to execute their own scripts.

The /etc startup script files are far more involved than the .bash_profile and .bashrc files. These script files include /etc/bashrc, /etc/profile, /etc/csh.cshrc and /etc/csh.login. The /etc/profile script is executed whenever a user logs in, irrelevant of whether the log in is via the GUI or command line, and irrelevant of any shell utilized. This script first defines the PATH variable. This variable is established based on the type of user (system administrator or other). Included in the path will be /sbin, /usr/sbin, /usr/local/sbin for root, and /bin, /usr/bin, /usr/local/bin for normal users, among other directories.

The variables USER, LOGNAME, MAIL, HISTCONTROL, HISTSIZE and HOSTNAME are established and exported. Next, umask is set (discussed in the next subsection). Finally, the .sh scripts in /etc/profile.d are executed. Most of these files establish further environment variables related to the type of terminal (or GUI) being used, the language of the user and specific software (e.g., vi). The system administrator can add further variables and aliases to the /etc/profile startup script.

7.6.2 Initial User Settings and Defaults

One of the actions that /etc/profile takes is to establish the user's umask value. The umask instruction sets the default file permissions for any newly created file or directory. The umask instruction is set for each user. In /etc/profile, we want to establish a umask value for the given user. The value though depends on whether the user is root or software account, or human account.

The format for umask is umask *value* where *value* is a 3-digit number. This is not the actual permissions (e.g., 755) but instead a number that is first NOTted and then ANDed to either 777 (for directories) or 666 (for files) to establish the initial permissions.

We might want all users to have permissions of 664 for all files and 775 for all directories that they create. We would set the umask value to 002 (000 000 010) which would be NOTted to 111 111 101. When ANDed with 777 (111 111 111) and 666 (110 110 110) it results in permissions of 775 and 664. The value for root's umask is typically set to 022 so that the default values for files is 644 and for directories is 755. An examination of the /etc/profile file shows that the umask value is set based on the if-then-else statement shown in Figure 7.14.

The if-then-else statement in Figure 7.14 tests the user's UID and user and group names. The command id -gn provides the user's group name, and the command id -un provides the user's username. The condition tests to see if the UID is greater than 199 (making the account a non-administrator account) and if the user's name and group name match (which will be the case for most human user accounts). Software accounts are generally given UIDs in the range of 200–999 but may have different username and private group name (if they have a group), and all administrator accounts will have a UID of less than 200. Thus, root and at least some software accounts will have a umask value of 022, while all other accounts will have a umask value of 002.

Earlier in this chapter, we explored the /etc/default/useradd file to establish default values for useradd. Another file storing default values is called /etc/login.defs. This file defines a number of default items that are utilized by various Linux commands. Table 7.13 lists many of the directives available that the system administrator can use to specify desired system defaults.

Directives like CREATE_HOME, UID_MIN, UID_MAX, GID_MIN and GID_MAX will impact the useradd instruction. Other entries impact operations like usermod, userdel, groupadd, groupmod, groupdel, login, passwd and su. The defaults can also impact the contents of the files /etc/passwd, /etc/group, /etc/shadow and the mail spool files.

Another default that the system administrator is able to apply to users is restrictions that they have on their shell usage. This can be done through ulimit which does not impact *currently* running shells or anything launched via the GUI but will impact future shells. Through ulimit we can, for instance, limit the size of a file created or the amount of memory usage permissible. To view the current limitations, use ulimit -a. To alter a limit, use ulimit *option value* where *option* is the proper option for the limit. Table 7.14 illustrates some of the more useful options.

```
if [ $UID -gt 199 ] && [ "`id -gn`" = "`id -un`" ];
     then
          umask 002
     else
          umask 022
fi
```

FIGURE 7.14 Setting umask in /etc/profile.

TABLE 7.13
Directives Available for /etc/login.defs File

Directive	Usage	Possible Values
CREATE_HOME	When creating a user account, does useradd default to automatically creating a home directory or not creating a home directory?	yes, no
DEFAULT_HOME	If user's home directory is not available (e.g., not mounted), is login still allowed?	yes, no
ENCRYPT_METHOD	Encryption algorithm used to store encrypted passwords.	SHA512, DES (default), MD5
ENV_PATH, ENV_SUPATH	To establish the initial PATH variable for logged in users, for root.	PATH=/bin:/usr/bin
FAIL_DELAY	Number of seconds after a failed login that the user must wait before retrying.	0, 5, etc
MAIL_DIR, MAIL_FILE	Default directory, filename of user mail spool files.	/var/spool/mail .mail or username
MAX_MEMBERS_PER_GROUP	Maximum number of users allowable in a group; once reached a new group of the same name is created adding a new line in /etc/group that shares the same name and GID (and password).	0 (unlimited)
PASS_ALWAYS_WARN	Whether a warning is made when a weak password is provided (if weak passwords are allowed).	yes, no
PASS_CHANGE_TRIES	Maximum number of attempts to change a password if password is rejected (too weak).	0 (unlimited), 3
PASS_MAX_DAYS, PASS_MIN_DAYS, PASS_MIN_LEN, PASS_WARN_AGE	Maximum, minimum number of days a password may be used, minimum password length, default warning date as with chage -W.	Numeric value, 99999 for max, 0 for min are common defaults
UID_MIN, UID_MAX, GID_MIN, GID_MAX	Range of UID, GID available for useradd, groupadd.	1000, 60000
UMASK	Default umask value if none is specified (as in the /etc/profile).	022

TABLE 7.14
Useful ulimit Options

Option	Meaning
-c	Maximum core file size, in blocks.
-e	Scheduling priority for new processes.
-f	Maximum size of any newly created file, in blocks.
-m	Maximum memory size useable by a new process, in kbytes.
-p	Maximum depth of a pipe (pipe size).
-r	Real-time priority for new processes.
-T	Maximum number of threads that can be run at one time.
-v	Maximum amount of virtual memory usage, in kbytes.
-x	Maximum number of file locks (number of open files allowable for the shell).

SECTION ACTIVITIES

1. Come up with two items you, as a user, might want to place in your `.bashrc` file.
2. The `/etc/skel` directory has only a few items. What might you, as a system administrator, want to add to this directory to share with new users? If you can't come up with anything, try to do a web search for ideas.

7.7 THE sudo COMMAND

Earlier in the chapter, we learned about `su` to change to another user account. Normal users will not have this ability because they will not know root's password and should not know other users' passwords. But there is still a mechanism by which users can execute instructions as other users, `sudo`, which you might think of as standing for `su`-*do* (`su` and *do*) or *superuser do*.

With `sudo`, a user can execute a command as if they were another user. Although we will primarily set up `sudo` so that a user can access some root-level commands, we can use it in another way. Let's assume that zappaf has created an executable program which reads data files from zappaf's home directory. zappaf does not want to give read access to other users for these files so he has set their permissions so that only he has access. The program, however, is one that he wants dukeg to be able to use. Instead of creating a group account of which zappaf and dukeg are members, zappaf decides to provide dukeg access via the `sudo` command. Now, dukeg can run this program through `sudo`.

The format of sudo is `sudo [-u` *`username|uid`*`] [-g` *`groupname|gid`*`]` *`command`*. The *username/UID* or *groupname/GID* is that of owner of the program to be executed, not of the user wishing to run it. For instance, dukeg could issue `sudo -u zappaf` *`program`* where *program* is the program that zappaf wants to share with dukeg. Notice that the user and group are optional. If not provided, then sudo runs *command* under root.

With `sudo` we have a means of allowing ordinary users to run programs that are normally restricted to root. This seems dangerous. Why would we want to give access to root-level programs to ordinary users? The reason for this is that we, as system administrators, can control just who gets access and to only the commands we specify. That is, we are able to restrict `sudo` usage. We place these restrictions in the file `/etc/sudoers`.

As an example, let's consider the `groupadd` instruction with permissions of `rwxr-x---`. As `groupadd` has an owner and group of root, this makes `groupadd` inaccessible by any non-root user. However, `groupadd` is a relatively harmless instruction. Why shouldn't we let other users create groups? So, we decide to allow select, trustworthy users access to `groupadd`. How exactly do we set this up?

To provide administrator access to a command first requires that the system administrator edit the `/etc/sudoers` file to add an entry with the format *`username(s) host=command`*. *Username(s)* is the username(s) of the user(s) who we will allow access to the specified command through `sudo`. If a user is not listed under the username, any attempt to use *command* via `sudo` will result in an error message with the event logged. To specify multiple users, list their usernames separated by commas but with no spaces.

We can define aliases for users. User aliases are defined as `User_Alias` *`alias=users`* where *alias* is the aliased name and *users* are separated by commas. We do this so that we can provide numerous commands for a select group of users. That is, once we define the alias, we reuse it in several `sudoers` listings.

As an alternative to using a list of users, we can specify `%group` to indicate all users in the listed *group* are able to use the given command. Another option is to use `ALL` if the command should be accessible by all users. We might use this option to let any and all users mount the optical drive.

The value for *host* indicates for which machine this `sudo` command is permitted. `ALL` is used to indicate that this `sudo` command is valid on all computers for a networked system; otherwise, we would use `localhost` for this computer (or this computer's name if given a hostname). If `/etc` is not shared among multiple computers, there would be no reason to use `ALL` although doing so would not impact who could use `sudo`.

The command(s) listed in the `sudoers` file must include full paths to the instruction as `sudo` executes without access to the user's `PATH` variable. Additionally, the command(s) can include options and/or parameters as needed. Multiple commands can be specified if separated by spaces.

Let's allow users foxr and zappaf access to `groupadd`. The directive in the `sudoers` file would be listed as `foxr,zappaf localhost=/usr/sbin/groupadd`. Once edited, we save the `sudoers` file.

With the `sudoers` file updated, now those specified users can use the instruction. To do so, the user precedes the instruction with `sudo`, as in `sudo groupadd my_new_group`. The `sudo` program requires that the user submit his or her password to ensure that this user is authorized to execute the command. This password is timestamped so that future usage of `sudo` during the current user session will not require the password. We see later that the timestamp can be reset.

Now that users can create their own groups, can they start populating those groups with users? The `groupadd` command does not allow us to insert users into a group. For that, we would need to provide access to `usermod`. We do not want users to have full access to `usermod` as they could use it to alter existing accounts. However, we could restrict users to access one part of `usermod`, `usermod -G`. We add to the `/etc/sudoers` file a new entry of `foxr,zappaf localhost=/usr/sbin/usermod -G`. By including the option `-G` in the entry, foxr and zappaf can issue `usermod` commands but only if they include the `-G` option.

Unfortunately, we cannot restrict foxr or zappaf from adding other options after `-G`. By giving users access to `groupadd` and `usermod`, we have created a security problem. Consider the pair of instructions in Figure 7.15 where zappaf first creates a new group and then assigns himself to that group. This is exactly the type of situation we want to permit via `sudo`, but a closer look at the new group, `hackers`, shows why this is an issue. zappaf has used options in `groupadd` to create a new group which shares the same GID as the root user group and then added himself to that group.

There are legitimate reasons to use `sudo`. Table 7.15 provides a few such examples of `/etc/sudoer` entries that we might want to create that would be both more valuable and potentially safer than the example shown in Figure 7.15.

We have noted that in some Debian distributions, the initial user account is given full system administration privileges. This is done because the root password is generated randomly and not given to any user. In order to provide this account root privileges, the installation process sets up an entry in the `/etc/sudoers` file of *`username`* `localhost=ALL` where *username* is the name of the initial user account. The thought is that by having to specify `sudo` before any root-level command, it helps remind the user that he or she is serving as root for that command and this is deemed safer than having a root password that might be shared among multiple users.

```
sudo /usr/sbin/groupadd -g 0 -o hackers
sudo /usr/sbin/usermod -G hackers zappaf
```

FIGURE 7.15 Using sudo to create and add to a group may not be safe.

TABLE 7.15
Example /etc/sudoer Entries

Entry	Explanation
`ALL localhost=/usr/bin/cat /etc/sudoers`	Allow all users to view the contents of the sudoers file.
`ALL localhost=/usr/sbin/shutdown -r +15 "shutting down in 15 minutes"`	Allow all users to reboot the system with a 15-minute grace period.
`ALL localhost=/usr/sbin/mount /cdrom`	Allow all users to mount the optical drive.
`ALL localhost=/usr/bin/updated`	Allow all users to run updatedb before locate.

The sudo command comes with a number of options, a few of which might be useful. The -b option runs the command in the background. This could also be accomplished by appending & to the instruction. With -H, sudo uses the current user's HOME environment variable rather than that of the program's user. From our earlier example, if dukeg ran zappaf's executable program without -H then the current directory for the program is /home/zappaf. If run with -H, then the current directory for the program would be /home/dukeg. This could be useful if the program were to utilize or store files for the given user rather than the owner.

The option -K is used without a command (i.e., sudo -K). The result of -K is that the user's timestamp reverts to the epoch. In essence, this wipes out the fact that the user had successfully run sudo in recent the past and will now be required to submit their password the next time they use sudo. The option -k resets the user's timestamp. The -k option, unlike -K, can be used as part of a sudo command.

Although root can directly edit /etc/sudoers with vi, it is best to only edit this file using the program visudo. This program loads /etc/sudoers into vi but additionally checks /etc/sudoers for syntax errors prior to closing it. In this way, we can be informed of potential errors that we can fix at the time we are editing the file, rather than discovering errors when users attempt to use sudo at a later point in time.

sudo has a configuration file of /etc/sudo.conf, and /etc/sudoers has its own set of configuration files under /etc/sudoers.d (which by default will be empty). For more information, consult the man pages for sudo, sudoers and sudo.conf.

SECTION ACTIVITIES

1. As root, read through the /etc/sudoers file to see if there are any initial entries. How much of the file do you understand? Review the sudo and /etc/sudoers man pages if you are unsure of any of the entries or directives.
2. As noted in this section, Ubuntu and some other Debian do not permit the initial user to know the root password but instead give that account full root access through sudo. Research the rationale behind this decision. List one advantage and one disadvantage to this approach.
3. Table 7.15 illustrated a few examples of sudoers entries. Do a web search on example sudoers entries and try to find other examples. For each that you find, would you implement this in your own Linux computer?

7.8 SELinux

SELinux is *security-enhanced* Linux. This addition to Linux allows one to implement security options that go far beyond the simple rwx permissions available for data files, executable files and directories. SELinux provides mandatory access control through a series of *policies*. Each policy defines access rules for the various users, their roles and types onto the operating system objects. We will define these entities as they are used in SELinux below, but first let's look at SELinux more generally.

SELinux became a standard part of Linux with kernel version 2.6 and has since been added to most Linux distributions. SELinux implements a number of policies with names such as `targeted` (the default), `mls` (multi-level security protection) and `minimum`. A fourth policy, `strict`, has been merged into `targeted`. Upon system initialization, SELinux should be enabled with the policy of `targeted`.

The SELinux configuration file (`/etc/selinux/config`) defines two directives: `SELINUX=enforcing`, `SELINUXTYPE=targeted` The `SELINUX` directive can also take on values of `permissive` and `disabled`. The `SELINUXTYPE` directive can take on the value of any of the available policies. We can change the mode of SELinux with the `setenforce` command followed by `Enforcing` (or `1`) or `Permissive` (or `0`). To see the current mode of SELinux, use `getenforce`.

7.8.1 SELinux Components

With these basics stated, let's now consider SELinux in some detail. First, we define the types of entities that we deal with in SELinux: users, roles, types, objects. We throw in another entity called a context.

A *user* is someone who uses the system. The user is not equivalent to any specific user. That is, we do not define a context for user foxr and another for zappaf. Instead, there are *types* of users. We might have users of type user (general human user), guest and root. SELinux also defines a type of user called *unconfined* to provide a user with broader access rights. Usernames are typically specified using the form *`name`*`_u` or *`name`*`_x` where *name* is one of types listed above. For instance, already established will be usernames of `guest_u`, `root`, `staff_u`, `unconfined_u`, `user_u` and `xguest_x`. We can view all of the usernames with the instruction `seinfo -u`.

Upon logging in, the normal setting is to establish the user as an `unconfined_u`. A user's type can change during a Linux session by running various programs. For instance, the type will change if the user successfully issues a `su` or `sudo` command.

The *role* allows SELinux to provide access rights (or place access restrictions) on users based on the role they are currently playing. The idea behind a role comes from role-based access control as used in database applications. In large organizations, access to data is controlled by the role of the user. In such an organization, we might differentiate roles between manager, marketing, production, clerical and research. Some files would be accessible to multiple roles (manager for instance might have access to all files) while some users might only be assigned to one role such as people in production who would only have access to production-oriented data. The type of access would vary based on the legitimate use of a particular data for that role. Some data may be made accessible as read-only to one role and read/write/modify to another role while a third role may have no access whatsoever.

In SELinux, roles are generally assigned to the user categories such as `unconfined_r`, `guest_r`, `user_r` and `system_r`. However, it is possible for a user to take on several roles at different times. The user `unconfined_u` may operate using the role `unconfined_r` at one time and `system_r` at another time.

The *type* specifies the level of enforcement. The type is tailored to the type of object being referenced, whether it is a process, file, directory or other. For instance, types available for a file include read and write. When a type is placed on a process, it is sometimes referred to as a domain. The *domain* dictates which processes the user is able to access. As with users and roles, SELinux contains predefined types of `auditadm_t`, `sysadm_t`, `guest_t`, `staff_t`, `unconfined_t` and `user_t` among others.

We can now apply the users, roles and types. We specify a *context* as a set of three or four values. These are, at a minimum, a user, a role and a type, separated by colons. One

TABLE 7.16
Example Contexts

Context	Usage
unconfined_u:object_r:user_home_dir_t:s0	User home directory
unconfined_u:object_r:user_home_t:s0	User file
unconfined_u:unconfined_r:unconfined_t:s0-s0:c0.c1023	Running user process
system_u:object_r:bin_t:s0	/usr/bin directory
system_u:object_r:shell_exec_t:s0	/usr/bin/bash file
system_u:object_r:shadow_t:s0	/etc/shadow file
system_u:object_r:fixed_disk_device_t:s0	Hard disk (device file)
unconfined_u:unconfined_r:unconfined_t:s0-s0:c0.c1023	Running root process

context is unconfined_u:object_r:user_home_t which defines the context for the user unconfined_u on the object object_r with the type user_home_t. This context is the one defined for a users' home directory.

The previous context is incomplete. The fourth entry in a context is optional but specifies the security level. The *security level* is specified as a sensitivity or a range of sensitivities, optionally followed by a category (or categories). For instance, we might find security levels of s0 to indicate sensitive level of s0, or s0-s3 if the sensitive level includes all of s0, s1, s2 and s3. The category can be a single category value such as c0, a list of categories such as c0,c1,c2 or a range of categories denoted as c0.c3. The full security level might look like s0-s3:c0.c2.

We have to define what each sensitivity level and category represent. As an example, we might define c0 as meaning general data, c1 as being confidential data, c2 as being sensitive data and c3 as being top secret data. Now we have a mechanism for ranking data so that a context on a particular file can be defined.

Table 7.16 provides examples of pre-established contexts in Red Hat 8. For each, we see the defined context and an example of where that context is used. We see contexts for two users, unconfined_u for normal users and system_u for root. Within these, objects are either object_r (files/directories) or unconfined_r (processes). Types vary by specific type of entity (file/directory, process, system directory, shell, shadow file, device, root process). Based on all of these, we see two different security levels of s0, the lowest of the security levels, and s0-s0:c0.c1023 which is a root-level security level. Contexts are already defined for most situations that we will face whether they are user processes, root processes, software-owned processes, user files and directories or system files and directories.

7.8.2 A Closer Look at Contexts

We can obtain the context of a directory or file using ls -Z and the context of a running process through ps --context. The directory /etc/selinux/targeted/contexts contains predefined contexts stored in files for various types of processes and in files in subdirectories for objects and user objects. Table 7.17 shows the already defined contexts found in the file user_u which is located beneath contexts in a subdirectory called users. We show this merely to give you an idea of some of the predefined contexts. If you want to explore SELinx further, take a look at some of the other predefined contexts (e.g., default_contexts) and types (default_type), as well as file contexts under files/file_contexts, all of which are in the contexts subdirectory.

Contexts under the targeted policy fall into one of four categories. First are *confined processes*, which run under their own domain. A confined process is one that is confined to just its domain. This limits the damage a process might do if it is compromised. Many network services

TABLE 7.17
Predefined Contexts in /etc/selinux/targeted/contexts/users/user_d

system_r:init_t:s0	system_r:local_login_t:s0	system_r:remote_login_t:s0
system_r:sshd_t:s0	system_r:cockpit_session_t:s0	system_r:crond_t:s0
system_r:xdm_t:s0	user_r:user_su_t:s0	user_r:user_sudo_t:s0
user_r:user_t:s0		

that listen to the network fall into this category, for instance, sshd and httpd. Tasks run by a user which run under root privileges also fall into this category, such as passwd and sudo.

The next defined context is the *unconfined process*. These processes run in unconfined domains. If the unconfined domain has rules, these are applied. Otherwise, the unconfined process falls back to the permissions tied to the user/effective user. For instance, if the user has read access to a file then an unconfined process of that user only has read access to that same file.

The last two contexts are *confined user* and *unconfined user*. All users are mapped to a user type which will be in one of these two contexts. An unconfined user can execute processes in unconfined domains. An unconfined user can also execute a process that can transition from an unconfined to a confined domain. Within the confined domain, the process has restrictions.

We can modify the context of an object in one of two ways. First, many file commands allow us to add -Z *context* to alter the object's context. A user's file may have the context unconfined_u:object_r:user_home_t:s0. Let's assume we want to copy a_file to /tmp and alter its context to unconfined_u:object_r:user_tmp_t:s0. We issue the command cp -Z unconfined_u:object_r:user_tmp_t:s0 a_file /tmp.

The other means of modifying a context is with the chcon instruction. This instruction allows us to change the context of a file or directory. We can specify the new context, as we saw in the above cp example, or we can specify any of the user portion through -u *user*, the role portion through -r *role*, the type through -t *type* or the security portion through -l *security*. The -R option for chcon operates on a directory recursively so that the change is applied to all files and subdirectories of the directory specified.

We can restrict chcon to not follow symbolic links (-P, the default) or to impact symbolic links but not the files they link to (-h). If we change the context of an object and later either discover that we made a mistake or want to change the context back, we can use the instruction restorecon item as in restorecon /tmp/a_file.

7.8.3 Rules

With the contexts defined, this leads us to rules. In SELinux, a *rule* is a mapping of a context to the allowable (or disallowable) actions of the user onto the object. We might have the rule allow user_t user_home_t:file { create read write unlink }; which permits users under user_t to create, read (open), save or write to and remove a link from a file.

Aside from *allow* rules, there are also four *type enforcement* rules types: type transition rules, type change rules, type member rules and type bounds rules. Other types of rules include *role allow* rules which define whether a change in a role is allowed, and *access vector* rules (AV rules) which specify the access controls allowable for a process.

Let's consider a type transition rule, which we use to specify that an object can be moved from a source type to a target type. The syntax of a type transition rule is type_transition *source_type target_type*:*class default_type [object_name]*;. The *object_name* is optional and might be included if we are dealing with a specific object (file).

Figure 7.16 provides an example of the transition and allow rules necessary for a user to create a file in their home directory. The user's process is denoted as user_t, the directory as

```
type_transition user_t user_home_t:file user_home_t;

allow user_t user_home_t:dir { read getattr lock search
     ioctl add_name remove_name write };

allow user_t user_home_t:file { create open getattr
     setattr read write append rename link unlink ioctl
     lock };
```

FIGURE 7.16 Sample transition and rules in SELinux.

user_home_t and the new file as user_home_t. The transition statement is used to identify the type of entity that the rules apply to. This is followed by two allow rules. The first allow rule applies to the directory, and the second allow rule applies to the file.

The first allow rule in Figure 7.16 shows that the directory needs such permissions as the ability to be read, have its attributes read, locked, searched, added to (have a named entity saved into the directory), an item removed from it and written to. The second allow rule applies to files, permitting files to be created, opened, read from, written to, appended, renamed, linked to, unlinked from and locked. The names of the access rights listed in the access rules of this example are the names of system calls (refer back to Chapter 4.1 for a discussion on system calls).

Aside from allow rules, as seen in Figure 7.16, we can also set up rules using other forms of permissions. These include auditallow, dontaudit and neverallow. The first two of these permit and deny logging of events of the specified item. With neverallow, we specify permissions that are not allowed to appear in an allow rule.

Rules are placed together to make up a *policy*. Policies already exist for a number of scenarios. Policies are placed into policy packages and are found in the files under /etc/selinux/targeted/modules/active/modules. These are binary files and not editable directly. To modify a policy package, we use the program audit2allow. We can create our own policies using checkpolicy. The checkpolicy program will examine a given policy configuration and, if there are no errors, compile it into a binary file that can be loaded into the kernel for execution. We omit the details of creating or modifying policies here as they are both complex and dangerous. A warning from Red Hat's Customer Portal says that policy compilation could render your system inoperable!

SECTION ACTIVITIES

1. Research a reason to turn SELinux off (turn off enforcing). If you were to turn off enforcing, would you do this at home? At work? In a school environment?
2. It is likely that you could use your Linux system and never even notice SELinux. Have you come across a situation where SELinux blocked your attempt to do something? (this is different from the normal Linux permissions blocking your attempt). If so, what was it and if not, try to research a situation where this could happen.

7.9 ESTABLISHING USER AND GROUP POLICIES

To wrap up this chapter, we briefly consider user account policies. These policies might be generated by the system administrator, or by management, or by management with the help of the system administrator. Once developed, it will be up to the system administrator to implement the policies. Policies will impact who gets accounts, what access rights they have, whether they can access their accounts from off-site, if they can use on-site computers to access off-site material (e.g., personal email) and how long the accounts remain active, to name a few.

In order to establish user and group policies, we should ask a number of questions. Do we need to assign different levels (types) to users? Will users have different types of software that they need to access and different files that go along with them? Will different levels of users require different resources, for instance different amounts of hard disk space? Will different users have different processing needs?

The answers to these questions help inform us how to handle the users. If we have different levels of users, then we might have different types of accounts. A university might, for instance, provide different disk quotas and resource access to the different categories of users (administrators, staff, faculty and students). In an organization that clearly delineates duties to different software, we might establish accounts to access specific software so that only certain users have access. If users are assigned different categories, do we also establish priority levels for user processes based on category?

We have to look at our resources and ask whether we have sufficient resources so that restrictions are not necessary. Will we need to enforce disk quotas? If so, do we place quotas on all partitions or on specific partitions? Can different users have different quota limits? See Chapter 8 for a discussion on disk quotas.

We next need to determine how users will access computers. Will users have sole access to their own workstation? Will workstations be networked so that users could potentially log into other workstations? Will resources be sharable? If users have sole access to their workstation, are users allowed to install software on their own workstations?

We also need to have a policy that dictates the duration of user accounts. Will user accounts exist for a limited amount of time or be unlimited? What do we do with the account once the user is no longer with the organization? In a university setting, we might allow students to retain their accounts for some time after they graduate and faculty to retain their accounts for some time after they retire or should they leave for another position.

What password policy will we enact? No organization should ever use anything other than strong passwords. But just how long should the password be? How often should passwords be changed? Can passwords be repeated and if so with what frequency? Can changed passwords be similar to previous passwords?

The policies that we establish will be in part based on the type and size of the organization. Larger organizations will have more resources but also have greater demand for those resources. Smaller organizations may not be able to afford a more costly file server and so may have greater restrictions on file space usage.

Policies will also be driven by the results of some form of risk assessment and management. *Risk assessment and management* is the process of identifying organizational assets and those assets' vulnerabilities and threats. Assets are not limited to the physical hardware such as computers, file servers and printers but include personnel and other non-hardware items. But here, it is the hardware that are the assets under the purview of the system administrator.

Even more critical than hardware assets are the data gathered by the organization. A company with clients would not want the client data to be accessible from outside as at least some of the collected data is likely to be of a confidential nature (e.g., credit card numbers). Any such access would constitute a breach in the privacy of the client's data which can then ruin the organization's reputation.

Once assets, vulnerabilities and threats are identified, management requires the creation of policies to safeguard those assets from the identified threats. This leads to IT policies. The system administrators then enact the IT policies and perhaps once a year, management and the system administrators evaluate their policies to improve them.

Let's be more concrete about user and group policies and couple them with other means to protect and safeguard an organization's assets in a Linux system. We break policy issues into three categories: user accounts, passwords and file space. For each of these, we discuss possible options.

We need to identify the type of user we will have in our organization to proscribe the duration of user accounts. Its considered bad practice for users to share accounts. We would want to enforce this by never using `-o` with `useradd`. We enforce account expiration dates using `chage`. Our policy will need to tell us what we should do with user files before deleting accounts. `userdel` allows us to either delete or retain files in the user's home directory and email.

We have other decisions to make when creating user accounts. We will use the `/etc/skel` directory to establish initial user resources. We might have multiple skeleton directories if we have different categories of user. We would likely assign all users the default shell of Bash but will we allow users to request other default shells? We also need a policy that determines under what circumstances we might establish `sudo` access for specific users, if ever.

Next up are password policies. We should generate an initial password for each new user account at the time the account was created. This is better than having initial accounts with no passwords. We saw a number of possible approaches such as using the `apg` program. Once accounts are created, we use `chage` and/or `passwd` to enforce the frequency by which users must update their passwords. We use a `PAM` module to enforce strong passwords. Among our choices are the length of a strong password, the amount of non-lowercase letters that must occur and whether passwords or parts of passwords can be repeated.

The last category is the user's file space. Every user will likely be given their own home directory. Early on, we need to decide whether users' directories will be maintained on their own workstations or whether the workstations will mount a common `/home` partition over the network. By giving users their own `/home` directories on their own workstations, we do not need to worry about disk quotas. But the downside to this is that backing up the `/home` partitions across multiple workstations is more of an effort than backing up a single, shared, `/home` partition. The disadvantages of remote mounting a single `/home` partition are that we would likely have to place disk quotas on the users and should the remote partition be unavailable, users would likely not be able to get much work accomplished.

Our policy may enforce limitations of what types of files users are allowed to store. For instance, users may not be allowed to store personal files or documents downloaded from unauthorized websites. Should the system administrator be tasked with examining user files to make sure they are legitimate (we would hope not!)? Using firewalls and proxy servers can help enforce that some content is not downloadable. We explore the Linux firewall in Chapter 10. The Squid proxy server is covered in supplemental readings.

Earlier in the text, we examined permissions on files and directories through `chmod` (as well as `chown` and `chgrp`). We also looked at SELinux to establish more specific types of access controls on files and directories. Do we, as system administrators, need to enforce proper file permissions? For organizations where employees share data, we need to identify the location where the shared data will be stored. We might want to set up special permissions for such data either through SELinux or by creating software accounts and access control lists specific to the data (e.g., accounts for an Oracle database management system to ensure only Oracle users can access the shared data). We also talked about various methods to scan for poor permissions (whether through `find` or our own script, for instance). We should use this tool from time to time to make sure users are not using ill-informed permissions.

On a related topic, how do we protect the organization's data? Encryption can be placed on any partition, such as each workstation's `/home` partition or the remotely mounted `/home` partition. This option is established at the time we create the partition, which will likely be when we install Linux. Do we want to enforce encryption? If we do not set up `/home` to be encrypted, then it is up to users to encrypt their own data files. We need to provide tools to accomplish this.

We need backup policies and procedures for the `/home` directories. We introduce how to perform backups in Chapter 8 and cover it in more detail, along with backup policies, in Chapter 12. We also look at disk quotas in Chapter 8. We look at encryption in Chapter 12.

SECTION ACTIVITIES

1. If you have a school or work account, are you aware of the policies? Do you have a disk quota attached to your account? Are you allowed to use the account for personal use such as email and web surfing? If so, do you know if the system administrator has the right to look at your email or web surfing behavior?
2. Speak to a system administrator at school or work and ask for a copy of the IT policies used, if one is available. Read through it and see if you agree with the policies. Are there policies that you would not have thought to create?
3. How many accounts do you have that you no longer use? If you have an account that you haven't used in at least 5 years, does that account still exist? Should it?

7.10 CHAPTER REVIEW

Concepts introduced in this chapter:

- Context – SELinux description of the user's type, user's role and type of object.
- Default shell – the shell assigned to a user when a terminal window is opened.
- Epoch – the date January 1, 1970, used in some Linux commands to count the number of days until an action should take place (e.g., modifying a password).
- GID – the ID number assigned to a group; used for bookkeeping.
- Group – a collection of users that permits a third level of access rights beyond owner and everyone else; Linux users can be assigned to various groups; see also private group.
- PAM – pluggable authentication module allows a system administrator to tailor how a program or service will achieve authentication by calling upon any number of modules.
- Password – a means of implementing access control by pairing a username with a password known only to that user; passwords, for security purposes, are stored in an encrypted manner; see also strong password.
- Password management – ensuring users modify their passwords from time to time; in Linux, we use `chage` and `passwd` to control password management options.
- Private (primary) group – a group account generated for a specific user whose name matches that of the user; the private group should contain only a single user that of the owner; files and directories created by a user default to being owned by this private group.
- Protection – the role of an operating system to ensure that a user is only able to access resources that the user has specific access rights to; without this, users could potentially change or erase resources of the system or other users.
- Risk assessment and management – the process whereby an organization evaluates its assets for vulnerabilities and threats to generate policies; system administrators are tasked with implementing IT-related policies.
- Root user – the system administrator account; root has access to all resources.
- SELinux – Security-Enhanced Linux was developed for Red Hat in 2005 as a more sophisticated form of protection beyond file permissions; it has since been added to most Linux distributions.
- Skeleton directory – a directory containing files and subdirectories used to populate the home directory of new users.
- Strong password – a set of restrictions placed on passwords to make them hard to crack typically consisting of at least eight characters (although many prefer at least 12) and a combination of upper and lowercase letters and at least one non-alphabetic character.
- UID – the ID number assigned to a user; used for bookkeeping.

- User – a person who uses the computer or runs processes.
- User account – the account generated so that a user can log into the computer and have access to their own and some system resources; three types of user accounts are normal users (humans), system administrator (root) and software.
- User manager – the Linux GUI application used to create, modify and delete user and group accounts which has since been replaced by Cockpit; see also the Linux commands `groupadd`, `groupdel`, `groupmod`, `newusers`, `useradd`, `userdel`, and `usermod`.
- User policies – usage policies established by management in conjunction with the system administrator(s) and implemented by the system administrator to dictate such aspects of user capabilities such as software access, download capabilities, file space quotas, website and email usage and password management.

Linux commands covered in this chapter:

- apg – third-party software to automatically generate random passwords.
- chage – control password expiration information.
- Cockpit – web-based interface to control user and system settings; can be used by root to create user accounts.
- exit – resume the previous user account after having used `su`.
- getenforce – output whether SELinux is being enforced or not.
- groupadd – create new group.
- groupdel – delete existing group.
- groupmod – modify existing group (use `usermod` to add users to a group).
- id – output UID/GID information on the current user.
- newusers – generate new and/or modify user accounts given a text file of user account data.
- passwd – modify a user's password and also control password expiration information similar to `chage`.
- setenforce – change whether SELinux is being enforced or not.
- su – allow a user to change to another account, temporarily.
- sudo – allow a user to execute a program as another user; most commonly used so that the system administrator can give access to some root programs to other users.
- system-config-users – launch the User Manager GUI.
- tr – translate characters from one format to another; couple with `/dev/urandom` to randomly generate characters for use as random passwords.
- ulimit – establish limits on resources in the shell session.
- umask – set default permissions when new files and directories are created.
- useradd – create a new user.
- userdel – delete an existing user.
- usermod – modify attributes of an existing user.
- visudo – open the `/etc/sudoers` file in `vi` for editing and syntax checking.

Linux files covered in this chapter:

- .bash_profile – script placed in user home directories and executed whenever a user opens a new Bash session; users can modify this file to add environment variables, aliases and script code.
- .bashrc – script placed in user home directories and executed by `.bash_profile`; users can modify this file to add environment variables, aliases, and script code.
- /dev/urandom – software serving as a device which provides random number generation.
- /etc/bashrc – script executed whenever a user starts a new Bash session; controlled by the system administrator.

- /etc/default/useradd – stores the default values used by useradd; use useradd -D to view or modify the default values.
- /etc/group – file storing all of the groups defined in the system and the groups' members; readable by the world.
- /etc/login.defs – default values used by a number of different programs such as useradd.
- /etc/pam.d/ – directory of configuration files used by PAM.
- /etc/passwd – file storing all user account information e.g., user's username, home directory, login shell, UID, full name) but not passwords; readable by the world.
- /etc/profile – script executed whenever a user logs into the system; controlled by the system administrator.
- /etc/security/pwquality.conf – configuration file to implement strong password policy.
- /etc/selinux/targeted/modules/active/modules/ – directory containing pre-existing modules for SELinux policies.
- /etc/shadow – file storing password information for all users; passwords are encrypted; also contains password expirations information; accessible only by root.
- /etc/skel/ – directory controlled by the system administrator containing initial files and directories to duplicate when a new user is added to the system; anything stored here is copied into the new user's home directory upon user account creation.
- /etc/sudoers – file storing sudo access rights; should only be opened using visudo.

REVIEW QUESTIONS

NOTE: in questions dealing with usernames, assume usernames will be of the form lastname followed by first initial like foxr, zappaf or dukeg.

1. Of the types of attributes of a user's account, which is(are) optional?
   username, UID, password, private group, GID, home directory, default shell
2. Using the User Manager GUI, which of the following is not possible when creating a new user account? If all are possible, use "none of these".
   a. Changing UID from default
   b. Changing GID from default
   c. Not generating a private group for the user
   d. Not generating a home directory
   e. Changing home directory location
   f. Changing the login shell
   g. Assigning an initial password
   h. Assigning a value for the comment field
3. Repeat question #2 with the Cockpit interface for creating a new user account.
4. True/false: By default, the next UID used will be one greater than the largest existing UID and the next GID used will be one greater than the largest existing GID.
5. Provide a reason for not using the default UID and/or GID value.
6. Assume we have always used the default UID and GID values for user accounts. Why might we find that UID and GID values for a given user's account are not the same number?
7. How has the UID numbering scheme changed between Red Hat 6 and Red Hat 7?
8. True/false: When using su you always have to enter the new user account's password.
9. How does su foxr differ from su?
10. Having used su to login as root, you type whoami and who. Do you get the same responses? If not, how do they differ?
11. With useradd, which option do you use to specify the gecos field?
12. The useradd option -D, when used with no other options or parameters, does what?
13. Which of the useradd options -g and -G is used to add this user to existing groups?

14. In the instruction `useradd -k /etc/somedirectory -m someuser`, what does `-k /etc/somedirectory` do?
15. Assuming that `useradd`'s `CREATE_HOME` environment variable is set to `yes`, which of the following is the default for `useradd`, `-m` or `-M`?
16. True/false: `useradd` has `-N` to not create a private group for this user but no `-n` to signify that a private group should be created for this user.
17. Which option would you use in `useradd` to create a system account?
18. Provide `useradd` instructions to create a new account for each of the following users. Each user is given a home directory and should have their gecos field filled with their name.
    a. Chad Wackerman using all of the defaults.
    b. Chester Thompson using all of the defaults except giving his account a login shell of `/bin/csh`.
    c. Terry Bozzio using all of the defaults except giving his account a home directory of `/home/musicians/`*`username`* and adding him to the group `musicians`.
    d. Adrian Belew whose starting home directory will obtain its contents from `/etc/skel2`.
19. What is wrong with the following instruction? `useradd -c Frank Zappa -m zappaf`
20. What is wrong with the following instruction? `useradd -m -o underwoodi`
21. Provide the `useradd` instruction needed to create a new account for the software `audacity` which will have its own home directory of `/var/media/audacity`, a login shell of `/sbin/nologin` and no private group account.
22. By default, the account expiration date is set to `-1` indicating that there should not be an initial expiration date. You want to change this so that the expiration date is automatically set for new users to be December 31, 2025. How can you do this?
23. You have just issued the following instruction: `useradd -c "Ed Mann" -s /usr/bin/csh -m manne`. You want to use this same command to create an account for Tommy Mars but with the login shell of `/usr/bin/tcsh`. Explain in a step-by-step manner how you will do this solely with command line editing.
24. What is the group with GID 100 used for?
25. Provide the `groupadd` instruction to create the new group `mystudents`.
26. Provide the `groupadd` instruction to create the new group `citMajors` with a GID of `1999`.
27. Put together the account creation script from Figures 7.6 and 7.7 into a single script. Create a file that consists of a few first and last names and then run your script using this file as input. If you wind up with errors, fix them and try again.
28. Revise the account creation script from question #27 so that an optional third value is available in the input file that lists the student's major. If the student is a CSC major, use the default shell of `/usr/bin/tcsh`, and if the major is a ECE major, use the default shell of `/usr/bin/csh`, otherwise the shell should be `/usr/bin/bash`.
29. Revise the account creation script from question #27 so that the input file has a third value, one of faculty, staff or students. Faculty are given UIDs in the range 1000–1999, staff 2000–2999 and students 3000 or higher. Given the category, your script will have to identify the largest UID within the given range and then assign the user account a UID one greater. For instance, if the largest number between 1000 and 1999 is 1205 and this user is of type faculty, then this user will be given the UID 1206. However, if this user is staff and the largest number between 2000 and 2999 is 2221, then this user's UID will be 2222.
30. Revise the account creation script from question #27 so that the input file has a third value, the user's role, which will be one of `Administrator`, `Database`, or `Network`. If `Administrator`, add this user to the existing group called `management`. If `Database` or `Network`, add the user to the existing group called `technical`. If `Database`, also add the user to the existing group called `oracle`.

31. When we enhanced the account creation script in Figure 7.7, we used the following instructions. Explain what each does individually and then why we used them.

```
n=`egrep -c $username /etc/passwd`
n=$((n+1))
```

32. Provide an entry for the `newusers` input file to create the new user account zappaf with no initial password, the default UID and GID, a comment field of Frank Zappa and the default directory and shell.
33. Repeat number #32 but in this case specify an initial password of `osfa93`!! and a login shell of `/usr/bin/tcsh`.
34. Under what circumstance might you issue the instruction `userdel -f someuser`?
35. Under what circumstance might you issue the instruction `userdel -r someuser`?
36. How does `usermod -m` differ from `useradd -m`?
37. What happens if you issue the instruction `usermod -d /home/faculty/foxr foxr`? How does `usermod -d /home/faculty/foxr -m foxr` differ?
38. You issue `userdel foxr` and receive an exit code of 6. What does this mean? Did the instruction execute?
39. You issue `userdel foxr` and receive an exit code of 8. You want to delete this account. What should you do?
40. Examine `/etc/passwd`. What are the home directories established for these users?
    a. root
    b. bin
    c. lp
    d. mail
    e. halt
41. Examine `/etc/passwd`. What login shell is specified for the following users?
    a. root
    b. bin
    c. sync
    d. shutdown
    e. mail
42. Examine `/etc/group`. What users are members of each of these groups?
    a. root
    b. bin
    c. adm
    d. tty
    e. lp
43. Type `which apg` to see if `apg` is installed. If not, as root type `dnf -y install apg` (or `yum -y install apg` on Red Hat 7 or earlier, or `sudo apt-get install apg` if running Ubuntu or Debian Linux). Run each of the following commands and list the responses.
    a. `apg`
    b. `apg -n 2`
    c. `apg -m 12`
    d. `apg -n 2 -M "NSC"`
    e. `apg -n 1 -y`
44. Issue the instruction `tr -cd '[:alpha:]' < /dev/urandom | head -c8`. Now reissue the instruction changing `alpha` to `alnum`. How does the output differ?
45. Issue the instruction `tr -cd '[:digit:]' < /dev/urandom | head -c10`. What is output? Explain the instruction.

46. In Figure 7.10, we defined code to add initial password generation to our account creation script. Explain the role of the two echo statements.
47. You forget to specify an expiration date for a new user, martinb. You want his account to expire on June 1, 2025. How can you do this using usermod?
48. Repeat question #47 using chage.
49. What is the epoch? How is it used with reference to information stored in /etc/shadow?
50. As root examine /etc/shadow. What does the second field (which will be the longest entry) of each entry signify?
51. As root examine /etc/shadow. You will likely find many of the entries after the third to be empty. Why?
52. Imagine that an /etc/shadow entry has the following (the second field is omitted for brevity). What do the numbers of 2, 30, 5 and 60 represent?

```
foxr:…:99999:2:30:5:60::
```

53. Provide a chage command for each of the following users for the purposes described, or state that it cannot be done with chage.
    a. Set keneallym's password to expire in 30 days, sending a warning 5 days in advance.
    b. Set underwoodr's password to expire in 90 days, sending a warning 10 days in advance.
    c. Lock the account for dukeg.
    d. Change the minimum time between password changes to 7 days for user marst.
54. Redo question #53a-c using the passwd command.
55. What does it mean to lock an account using the passwd command?
56. Which of the following can you not do using chage?
    a. Lock and unlock an account
    b. Establish maximum days before a password must be changed
    c. Warn the user in advance of when a password must be changed
    d. Set the number of days when the password was last changed
57. Repeat question #56 for passwd.
58. Using man, determine what each of the following modules do.
    a. pam_env.so
    b. pam_rootok.so
    c. pam_succeed_if.so
    d. pam_xauth.so
59. In a PAM configuration file, you find three entries with the module auth listed. These are collectively referred to as a(n) _____.
60. A PAM configuration file for some program, foo, has modules for auth only. What would you think foo does?
61. What is the difference between the module types auth and account in a PAM configuration file?
62. A PAM configuration file for some program, bar, contains a module of type password. What would you think bar does?
63. What is the difference between a PAM directive listed as required versus sufficient?
64. Assume there are four session directives for a given PAM configuration file. Three are listed as required and the fourth as optional. What happens if the third directive's call to the given module fails? What happens if the fourth directive's call to the given module fails?
65. What is the role of the directive include when used in a PAM configuration file?

66. In a PAM configuration file is a reference to a module listed as follows: `pam_unix.so nullok`. What does `nullok` mean? (hint: you might need to examine the man page(s) for `pam.d` and/or `pam_unix.so`).
67. Specify the directives for `pwquality.conf` to enforce strong passwords whose minimum length is ten, must include at least two classes including an uppercase letter, cannot repeat more than four characters and cannot match any of the last eight passwords.
68. How do the directives `difok` and `remember` differ, as used in `pwquality.conf`?
69. With your Linux account, look at the directory `/etc/skel`. What do you find there?
70. Which script first creates the user's `PATH` variable?
71. As a user, you want to add the directories `/opt/bin` and `~/myscripts` to your `PATH` variable. You want to do this in one of the available script files. Which file and how would you do this?
72. What permissions would a file have if you created the file after doing `umask 026`?
73. Your `umask` is set to `002` and you create a file. What are its permissions? You create a directory. What are its permissions?
74. User `umask`s default to `002`. Consider instead `007` and `077`. Provide an argument to use `007` instead of `002` and an argument for `077` instead of `002`.
75. In your own words, explain the if-then-else statement from Figure 7.14.
76. One of the directives in `/etc/login.defs` is `DEFAULT_HOME=no`. What does this mean?
77. How would you, as a system administrator, control the number of seconds that would elapse before a user can attempt to login again after a failed attempt?
78. Notice that `/etc/shadow` is not readable by anyone but root. As root, you want to give access to read this file to user zappaf on the local computer via `sudo`. What entry would you add to the `/etc/sudoers` file to establish this?
79. Why should you use `visudo` to modify the `/etc/sudoers` file rather than directly editing it in `vi`?
80. You have just used `sudo`. You use it again 10 minutes later. Will you have to provide your password?
81. After having used `sudo` twice recently, you want to remove the timestamp. How can you do this?
82. Aside from the examples covered in Section 7.7, name three additional examples of why you might set up `sudo` privileges for one or more users.
83. What command can you use to determine if SELinux is currently being used (enforced)? What command can you use to shut SELinux off (cause it to not be enforced)?
84. True/false: In SELinux, the user as specified in a context differs from the user's username and UID.
85. List four already defined users in SELinux.
86. An SELinux context has both `user_u` and `user_r`. What is each of these?
87. Provide the SELinux type that would be used to identify a normal user's home directory and a file owned by a normal user (note: normal means not root).
88. Which of the following security levels is more restrictive? `s0` or `s0-s3:c0.c1023`
89. What option will you use in `ls` to obtain the SELinux context of a file or directory?
90. What option will you use in `ps` to obtain the SELinux context of a process?
91. Which is more restrictive: a confined user or an unconfined user?
92. What do each of the following SELiux rules mean?
    a. `allow user_t bin _ t:file { read execute };`
    b. `allow system_u system_t:dir { read getattr search };`
    c. `allow kernel_t filesystem_type:filesystem mount;`

93. Assume you are the sole system administrator for a small company of 10–20 users. Your organization has a single file server which primarily stores software and group documents. User documents can either be stored on the file server or, more often, user workstations. The company currently has no restrictions placed on users in terms of personal use of company computers. Answer the following regarding user account policies for this organization. Offer a brief explanation to each answer.
    a. Should there be a policy prohibiting users from downloading software on their workstations?
    b. Should users have quotas placed on their file server space should they decide to store files there?
    c. Should the company install a proxy server to store commonly accessed web pages and also use it to restrict access to websites deemed unrelated to company business? (you might need to research what a proxy server does)
    d. Should users be able to access the file server from off-site computers?
94. You have been asked to propose user account policies for a large company of 250 employees. The company utilizes file servers to store all user files. In addition, the company has its own webserver and proxy server. Employees are permitted to have their own websites if they desire them. The company has a policy restricting computer usage to "professional business" only. Answer the following. Offer a brief explanation to each answer.
    a. Should users have disk quotas placed on them for either their file server space or for web space?
    b. Should you establish rules that prohibit certain types of content from employee websites?
    c. Should the proxy server be set up to restrict access to websites such as Facebook and personal email servers (e.g., Gmail, yahoo mail)? (you might need to research what a proxy server does)
    d. Should the system administrator be allowed to examine individual employee storage to see if they are storing information that is not related to the business?
95. As a system administrator, describe a policy for enforcing strong passwords (e.g., minimum length, restriction on types of characters used, duration until passwords need to be changed, etc.).
96. List ten Linux commands that will help you enforce IT policies pertaining to creating/modification/deletion of user and group accounts, password management, determining if users are using too much disk space and if files are poorly protected (i.e. is, using unwise permissions).

8 Administering Linux File Systems

This chapter's learning objectives are to be able to:

- Differentiate between the types of objects that Linux treats as files
- Describe the role of and the contents of the inode
- Utilize Linux file system commands of `badblocks`, `cpio`, `df`, `du`, `dump`, `mkfifo`, `mount`, `stat`, `umount` and `tar`
- Explain what partitions are and the steps in partitioning and repartitioning, and how partitions differ from using a logical volume manager
- Establish disk quotas
- Describe the roles of the various top-level Linux directories

8.1 INTRODUCTION

In Chapter 3, we viewed Linux files from a user's perspective. Here, we examine Linux files from a system administration point of view. Administrators should know the concepts of partitions and file systems, inodes and the types of Linux files, top-level directories and mounting. There is a suite of Linux commands available to handle these and other tasks. We examine many of them in this chapter.

The chapter first re-examines the file in Linux. Recall that Linux calls several different entities files; those we learned about in Chapter 3 were regular files, directories and symbolic links. We add to this list several other "file" types.

We then move on to partitions and file systems. In Chapter 1, when we explored how to perform a Linux installation, we omitted manual partitioning relying instead on the default partitions. In this chapter, we look at manual partitioning, repartitioning and the use of a logical volume manager to avoid partitioning, and how to add a new disk drive to our file space. In our look at partitioning, we also look at how to mount and unmount partitions, set up disk quotas and select file system types.

This chapter also provides a detailed look at the top-level directory structure. Although we introduced some of the top-level directories in Chapter 3, we now know enough about Linux to explore these directories in detail. Finally, we look at other administrative duties related to our files, partitions, and disks.

SECTION ACTIVITIES

1. If you are a Windows user, which of these tools have you used?
   - Disk Management to add or delete partitions or change partition sizes
   - A backup utility to automatically back up your hard disk
   - Storage Settings to view storage devices and their current capacities, change where content should be stored or optimize your drives
   - A disk utility to perform disk defragmentation and/or search for bad blocks

DOI: 10.1201/9781003203322-8

8.2 STORAGE ACCESS

A collection of storage devices presents a *file space*. This file space exists at two levels: a logical level defined by directories and files, and a physical level defined by partitions, file systems, disk blocks and pointers. Users and system administrators primarily view the file space at the logical level. The physical level is one that the operating system handles for us.

The devices that make up the file space include hard disk drives, optical discs placed into optical drives, solid-state drives and removable USB drives, and in some cases magnetic tapes placed into magnetic tape drives. For faster access, some items from storage might be moved into memory in the form of a ramdisk or other mechanisms whereby memory mimics file storage. *Ramdisks* were introduced around 1980 to provide a faster access to files by preloading some file content into memory. Ramdisk usage in Linux has largely been discontinued with a few notable exceptions that we will discuss briefly in Section 8.7.4 of this chapter and in the next chapter.

Collectively, these storage devices provide us with our file space. In this space, we store executable programs (including the operating system) and data files. There are a limited number of operations that we perform on storage devices. We might format a storage device. We might mount or unmount a storage device. We might open a file to read from or write to it and then close it. In writing to a file, we may be creating a new file or we may be appending to/altering an existing file.

8.2.1 Disk Storage and Blocks

As hard disk storage is the most common form of storage among our storage devices, we will concentrate on it, although some of the concepts presented here also apply to other forms of storage. To store a file on disk, the file is decomposed into fixed size units called *blocks*. Figure 8.1 illustrates a small file (six blocks) and the physical locations of those blocks. Notice that the last block of the file may not fill up an entire disk block so it leaves behind a small *fragment*.

The operating system must be able to manage the distribution of a file into its blocks in three ways. First, given the block number within a file (for instance, the file's first block), the operating system must map that block number into a physical location on some disk surface and do so quickly. Second, the operating system must be able to direct the disk drive to access the particular block requested through a movement of both the disk and the drive's read/write head. Third, the operating system must be able to maintain a list of available blocks so that blocks can be easily allocated for the next disk write. The available blocks should include not only the unused blocks of the disk drive but all file blocks returned from files that have been deleted. All of these operations are hidden from the user and system administrator.

Let's consider how a disk file might be broken into blocks. Let's assume that our hard disk already has hundreds or thousands of files stored on it. The files are distributed across all of the disk's surfaces (recall from Chapter 3 that a hard disk drive will contain multiple disk platters and each platter has two surfaces, a top and bottom). Given a new file to store, the file is broken into blocks. The first block is placed at the first available free block on disk.

Where should the next disk block be placed? If the next block after the first is available, we could place the block there, giving us two blocks of *contiguous storage*. This may or may not be desirable. The disk drive spins the disk platters very rapidly. If we want to read two blocks, we read the first

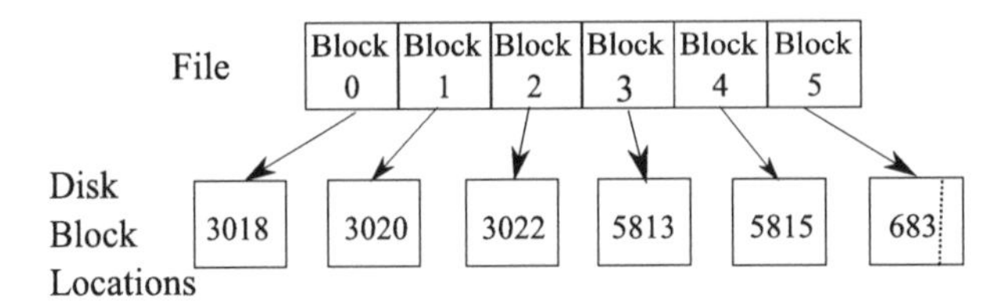

FIGURE 8.1 File decomposed into blocks.

block and transfer the contents to memory. During that transfer, the disk continues to spin. When we are ready to read the second disk block, it is likely that this block has spun past the read/write head and now the disk drive must wait to finish a full disk revolution before reading again.

One way to support contiguous allocation of blocks is to allocate all blocks for a file at the time the file is first stored. If the file uses fewer blocks than allocated, the result is internal fragmentation. We might choose to therefore limit the number of blocks given to the file initially to prevent this but this leads to its own problem. If we allocate space for a file and then another file is stored immediately after it, there is no room for this first file to grow.

To resolve this latter issue, we could allocate a new, larger space and move the file from its original location to the newly allocated space. Now we have a gap among our files because the originally allocated space is no longer being used. A new file could be stored there but only if the new file was smaller than the space available. This could leave behind its own fragment of unusable space.

The alternative to contiguous allocation is *linked storage.* When a file needs to grow, its next disk block is allocated from the list of free blocks. This new block may reside anywhere within the physical disk drive. The blocks are linked together by pointers. At the end of block i's storage is a pointer to the disk block storing block i+1.

In Figure 8.1, we see that the first three disk blocks are located near each other but not contiguously. We might assume that the operating system allocated three initial blocks but placed them slightly apart to resolve the first issue of being able to access the blocks without having to wait for the platter to make a full revolution. Additional blocks were allocated at a later point in time as the file grew.

Using the linked approach will result in file blocks becoming more widely scattered across the disk surfaces over time. This will lead to its own inefficiency in that access time grows because the distance between block i and block i+1 might require moving the read/write head to a new track while waiting additional time for the disk to spin. Thus, both seek time and rotational latency are lengthened.

Again, referring back to Figure 8.1, we might assume that the next available block after 3022 is at 5813 and so as this example file continues to grow its next block lies at 5813 followed by 5815. As the next block of this file, according to the figure, lies at 683, we might surmise that 683 had been part of another file that was deleted and so that block was returned as a free block to be reallocated.

8.2.2 Block Indexing Using a File Allocation Table

The operating system maintains a table of the starting position of every file. If the file were allocated with contiguous blocks, block i can be located by adding i to the file's starting position. But if a linked allocation approach were used, how do we locate block i?

One approach is to follow the pointer at the end of the file's first block to its second block. Now, we obtain that second block's pointer to the third block. We continue to do so i–1 times to reach block i. Unfortunately, each block requires a separate disk access and so to reach block i we must perform i disk accesses. Instead, we should use some mechanism to perform block *indexing.*

MS-DOS and early Windows systems used a *file allocation table* (FAT) for indexing. For every disk block in the file system, the next block's location is stored in the table under the current block number. That is, block i's successor location is stored at index i of the FAT. If that block were at location j, then the FAT would store j under index i.

The FAT is loaded from disk into memory at the time the file system is mounted (e.g., at system initialization time). Now, a search of memory is used to track down a particular disk block's location. For instance, if the file started at block 1500, we would look at FAT location 1500 to find the file's second block location. If it were 1505, we would look at FAT location 1505 to find the file's third location and so forth.

In Figure 8.2, a partial listing of a FAT is provided. Here, assume a file starts at block 151. Its second block is at 153 followed by 156, which is the end of the file (denoted by `EOF`). To find the file's third block, the operating system will examine the FAT starting at location 151 to find 153 and

File Allocation Table (portion)

Block	150	151	152	153	154	155	156
Next location	381	153	Bad	156	155	732	EOF

FIGURE 8.2 Excerpt of a file allocation table.

then look at location 153 to find 156, the file's third block. Another file might start at block 154. That file's second block is at 155 and its third block is at 732. The entry `Bad` indicates a bad block on the disk which should not be used.

More recent versions of Windows operating systems use NTFS (new technology file system) which utilizes a data structure called a B+ tree for indexing. In Linux, indexing is handled through the *inode* data structure. We explore this in Section 8.3 where we will see that every inode contains pointers to disk blocks that either store data or additional pointers.

8.2.3 Other Disk Storage Details

To this point in the text, we have referred to the files in storage as our file space rather than calling it a file system. The term *file system* means the implementation of how files will be stored and retrieved on the physical media of the storage device. There are numerous types of file systems. Some devices can only use one (or a few) types of file systems. Others are more flexible. Operating systems use a default form of file system but often can utilize other types. Some of the more common file systems are listed in Table 8.1. The table shows, for the given file system, the type of storage medium it applies to, the operating systems that use it and some of its more noteworthy features.

There are some terms introduced in Table 8.1 that need some explanation. The B-tree stores pointers mapping file block numbers to disk blocks similar to the FAT and inodes. However, the pointers are arranged in such a way as to optimize the search. That is, searching for the pointer will take a minimal amount of time. The FAT stores pointers to a file's disk blocks as a linked list so that searching for the last file block requires first traversing the entire list. The B-tree has a hierarchical arrangement of pointers instead. As you can see throughout Table 8.1, B-trees, or the B+ tree variant, are a popular means of indexing.

As a file is saved, it tends to grow. As it grows, its file blocks may become more and more distributed across the surface of the disks. This can lead to poorer performance. *Defragmentation* is an operation whereby distributed blocks of a file are brought closer together. During defragmentation, the drive is unavailable.

The term *journaling* means that disk operations that have been requested but not yet performed are logged. Should the system crash, journaling will have saved the intended operations so that they can still take place once the system is brought back up. With journaling, we are reasonably assured that an operation will take place eventually.

Snapshots are memory images stored on disk. The role of the snapshot is to capture the state of a computer system at the current point in time. If the system should crash or the user make unintended changes, we can roll back to the state as stored by the snapshot. Snapshots tend to be large and so this type of operation is less common than others found in the table. Operating systems may have their own ability to save snapshots and roll back to a previously saved one.

The *extent* is an allocated group of contiguous disk blocks. Recall that under most situations, disk blocks are not contiguous. However, distributing a file's disk blocks across surfaces of the disk drive can lead to less efficient access when an entire file is being loaded into memory. With an extent, a file is placed in contiguous blocks up to a preset amount. As additional blocks are needed, they can be provided separately (or with additional extents). Using extents limits the amount of defragmentation needed. Extents will cause some internal fragmentation.

TABLE 8.1
Common File Systems

File System Name	Type of Medium	Operating Systems That Use It	Notable Features
APFS (Apple File System)	Hard disk, solid-state drive, USB drive	MacOS, iOS	inodes and B-trees; supports snapshots and encryption; Linux permissions
btfrs (b-tree file system)	Hard disk	Linux	B-trees; extents; automated defragmentation; snapshots; RAID (levels 0, 1 and 10)
ext, ext2, ext3, ext4	Hard disk	Linux; ext2, ext3, ext4 available with restrictions in other operating systems including FreeBSD, MacOS, Windows	inodes (hashed B-trees are used for large directories); journaling introduced in ext3; extents (ext4); Linux permissions
FAT, FAT16, FAT32	Floppy and hard disk, USB drive	MS-DOS/early Windows, most OS can use FAT and many devices can use it	Indexing by FAT (linked lists of file blocks); easy way to exchange data between different types of devices
ISO 9660	Optical discs	Cross-platform	Hierarchical directory tree (similar to B-tree); extents
JFS (Journaled File System)	Hard disk	IBM operating systems; Linux	B+ tree; inodes with extents available; journaling
NFS (Network File System)	n/a	Just about any operating system that uses TCP/IP	Used for network access to a remotely stored file system
NTFS	Hard disk, solid-state drive	Windows 3.1 and later versions; later made available in Linux, some Unix dialects and MacOS (read-only)	B+ tree; performs journaling; access control list implementation for permissions; inclusion of hard links; strong encryption and disk quotas available
Transactional NTFS	Hard disk	Windows (starting with Vista)	Same as NTFS but adds atomic (uninterruptible) transactions
XFS	Hard disk	Silicon Graphics OS (Unix), Linux	B+ trees; journaling; extents; defragmentation; snapshots

The file system also dictates details like the naming schemes for files. It is common today for names to permit just about any character including blank spaces. Older file systems had limitations such as eight-character names and names consisting only of letters and digits (and perhaps a few types of punctuation marks like the hyphen, underscore and period). Some file systems do not differentiate between uppercase and lowercase characters while others do. Most file systems permit but do not require file name extensions.

File systems also maintain information about the entries (files, directories) like creation date/time, last modification date/time and last access date/time, owner (and group in many cases) and permissions or access control list. This collection of information is often referred to as *metadata*.

8.2.4 File Storage and Object Storage

We noted that files are broken into blocks for storage. We refer to this form of storage as *block storage*. Most storage devices offer block storage, and in Linux, we use block storage. Another form of storage is *file storage* (not to be confused with file systems), which offers storage at the level of files. That is, when loading or saving content, the storage device operates on full files at a time.

Whereas in block storage we rely on the operating system to treat a collection of blocks as a file and where the operating system maintains directories to point at the files' blocks, file storage stores all files wholly and places them into directories. In order to access a file, we need only provide the path to its location. This provides a more effective means of accessing files as they can be located without using some complex indexing scheme.

File storage is not without its own costs. We noted that we use block storage because all blocks are of a fixed size so that, upon deleting a file, we can reuse the blocks for other files. But what about in file storage? A deleted file will reallocate its space but can we reuse it? Only for a file whose size is smaller, and in doing so, it leaves behind a fragment. File storage is also not scalable. Rather than adding storage space, we have to add completely new storage units.

Let's consider a specific case whereby a user is modifying a file by adding content to the end. Upon saving the file in block storage, a new block is allocated and attached to the last block of the file by pointer. Saving one block is fast. In file storage, the file must be copied anew into a new location in storage, marking the old version as obsolete.

File storage tends to be used for network-attached storage, particularly when used for centralized file sharing and data repositories. File storage is implemented with its own operating system so that it does not require creating a file system before using it. Because of this, file storage is usually simpler to set up and start using.

A newer form of storage has been developed that is primarily used for cloud storage. *Object storage* is useful for storing collections of unstructured data. By unstructured, we mean that the files store items of differing types without a prespecified structure. For instance, we frequently see object storage used to store collections of multimedia files. Object storage is also commonly used to store large data sets for big data processing. Objects are stored as whole files, like that of file storage. What differentiates object storage is two added components: a globally unique identifier and metadata.

The globally unique identifier is how objects are located as object storage is flat storage (no directories). Object files are accessible using HTTP/HTTPS by using the globally unique identifier. This makes object files available over the Internet without a need to specially mount a storage device.

Metadata is what makes object storage a powerful tool in computing. All three forms of storage provide metadata but to different degrees. In block storage, the block itself stores no metadata. It is up to the file system to maintain metadata on the file by storing it separately. In Linux, the metadata describing a file are stored with the inode which is separate from the data that make up the file. File storage stores similar types of metadata as block storage. With object storage, the metadata goes far beyond the data describing the file by type, size, permissions and date.

Let's consider a file that stores an image of a cat. The metadata will likely include the file's format (e.g., "jpg"), creation date and owner. But the metadata could also include the file's source ("photograph"), a description of its content ("cat"), date the photograph was taken (as opposed to the creation date of the file itself) and even the person's name who took the photograph. With such a complex set of metadata, software can search for files not based solely on name, type and permissions (like the `find` comment) but based on some context.

SECTION ACTIVITIES

1. Research ramdisks. What computers popularized their usage in the 1980s? What value did ramdisks provide? Were there any drawbacks? We have outgrown ramdisks because we no longer rely on slow floppy disks.
2. If you have a Windows computer, what type is its file system? Likely NTFS. Research NTFS and compare it to xfs of Linux.
3. We mentioned B-trees and B+ trees in this section. What are they? They are variations of binary search trees. They have a few interesting properties like being height-balanced and where a node in the tree can have more children than two (which is the limit for a binary tree). Do a web search for the difference between binary trees and b-trees and take a look at any images you find.

8.3 LINUX FILES

In the Linux operating system, everything is treated as a file except for users and processes. What does this mean? Among other things, Linux file commands and redirection operators can be applied to entities that are not traditional files. These non-file file types include directories, physical devices (of two types), named pipes, domain sockets and symbolic links. Aside from physical devices, there are also some special purpose programs that are called devices and so treated like files (for instance, the random number generator). Let's consider each of the types of files in turn.

8.3.1 Files versus Directories

The directory should be familiar to the reader by now. It is a named entity that contains files (or devices, links, etc.) and subdirectories. The directory offers the user the ability to organize their files in some reasonable manner, giving the file space a hierarchical structure. Directories can be created just about anywhere in the file system and can contain just about anything from nothing to files to directories that themselves contain files and directories.

The directory differs from the file in a few significant ways. First, we expect directories to be executable. Without that permission, no one (including the owner) can `cd` into the directory. Second, the directory does not store content like a file; instead a directory stores a list of names of its contents each with a pointer (hard or symbolic link) to either an inode or a hard link. Third, there are some commands that operate on directories and not files (e.g., `cd`, `pwd`, `mkdir`) and some commands that operate on files but not directories (e.g., `wc`, `diff`, `less`, `more`). We do find that most Linux file commands will operate on directories themselves including for instance `cp`, `mv`, `rm` (using the recursive version), and wildcards apply to both files and directories.

8.3.2 Non-File File Types

Many devices are treated as files in Linux. These devices are listed under the `/dev` directory. We categorize these devices into character devices and block devices. *Character devices* are those that input or output streams of characters. These will include the keyboard, the mouse, a terminal (as in terminal window) and serial devices such as older MODEMs and printers. The random number generators (`/dev/random` and `/dev/urandom`) are also character devices even though they are not physical devices.

Block devices communicate via blocks of data. The term block is traditionally applied to disk drives where the files are broken into fixed sized blocks. However, here, block is applied to any device that communicates by transmitting chunks of data at a time (as opposed to the previously mentioned character-type). Aside from hard disks, block devices include optical disc and flash memory.

Aside from the quantity of data movement, another differentiating characteristic between character and block devices is how input and output are handled. For a character device, a program executing a file command must wait until the character is transferred before resuming. For a block device, blocks are buffered in memory so that the program can continue once the instruction has been issued. Further, as blocks are only portions of entire files, it is typically the case that a file command can request one portion of a file. This is often known as *random access*. The idea is that we do not have to request block 1 before obtaining block 2. Having to read blocks in order is known as *sequential access*. In random access, we can obtain any block desired, and it should take no longer to access block j than block i.

Another type of file is the *domain socket*, also referred to as a local socket. This is not to be confused with a network socket. The domain socket is used to open a connection between two local processes in order to facilitate interprocess communication (IPC). IPC is how two processes can share data. For instance, one process might produce data that another process is to utilize. This would be the case when some application software is going to print a file. The application software produces the data to be printed, and the printer's device driver uses the data. We refer to these processes as producers and consumers. IPC is also used to create a rendezvous between two processes where a process must wait for some event from another process.

There are several distinctions between a network and domain socket. The network socket is not treated as a file while the domain socket is (although the network interface is a device that can interact via file system commands). The network socket is created by the operating system to maintain communication with a remote computer while domain sockets are created by application software. Network sockets provide communication lines between computers rather than between processes.

Yet another type of file entity is the *named pipe*. We have already explored the pipe but the named pipe differs in that it exists beyond the single usage of a pipe between two Linux commands. To create a named pipe, use the `mkfifo` operation. The expression FIFO is short for "first-in-first-out". FIFO is often used to describe a queue (waiting line) as queues are generally serviced in a first-in, first-out manner. Once the pipe exists, we can assign it to be used between any two processes. Unlike an ordinary pipe that must be used between two Linux processes in a single command, the named pipe is typically used in separate instructions.

Let's consider an example. First, we define our pipe using the instruction `mkfifo a_pipe`. As with any file or directory, `a_pipe` has permissions, user and group owner, creation/modification date and a size (of 0). With the pipe available, we use it to pipe between the `ps aux` instruction and a `cat` instruction.

In a terminal window, we enter `ps aux > a_pipe`. Unlike performing `ps aux | less`, redirecting the output to the named pipe does not seem to do *anything* when executed. Our terminal window hangs as there is no output, but neither does the command-line prompt return to us. What we have done is opened one end of the pipe. Until the other end of the pipe is opened, there is nowhere for the `ps aux` instruction's output to "flow". To open the other end of the pipe, we issue a command that can retrieve from it. In this case, our command is `cat a_pipe` (we could also have issued commands like `less a_pipe`, `more a_pipe`, etc.). This instruction causes the contents of `a_pipe` to "flow" out of the pipe, emptying it.

In effect, we have opened the pipe with `ps aux` and emptied the pipe with `cat`. But notice that redirecting the output of `ps aux` into the pipe causes that terminal window to not respond. We cannot issue the `cat` command in that same window because we did not get our prompt back. Instead, we open the pipe from a second window where we enter the `cat` command. The output of the `ps aux` command appears in that second window. Upon emptying the pipe, we get our prompt back in the first window, and once the `cat` command completes, we receive a prompt in our second window.

You might ask why we should use a named pipe. In fact, the named pipe is being used much like an ordinary pipe. One difference is that the named pipe persists so that we can use it many times. Another difference is that we can control the usage of a pipe whereas processes control their communication when using a named pipe. Further, the named pipe reaches across terminal windows and shell sessions. Another interesting difference is that the `mkfifo` instruction allows us to fine tune the named pipe's performance. Specifically, we can assign permissions to the result of the pipe using the option `-M mode`. *Mode* is a set of permissions as in `-M 600` or `-M u=rwx,g=r,o=r`.

The named pipe does roughly the same thing as a domain socket; it is a go between for IPC. There are several differences between the named pipe and the domain socket. Perhaps the most visible difference from a user's point of view is that the named pipe is created by the user while the domain socket is generated through software through various system calls. For instance, `socket()` can create a domain socket, assigning it to a variable (of type file descriptor). The socket needs to be bound to an address using `bind()`. Once created, processes can now communicate to each other through the socket via the specified address.

Processes might listen to the socket for incoming messages using `listen()` and accept calls from another process via `accept()`. Alternatively, processes can communicate using some combination of `send()`, `recv()`, `read()` and `write()`. When a process is through using the socket, it will use `close()` to disconnect from the socket.

The named pipe always transfers one byte (character) at a time. The domain socket is not limited to byte transfers. Domain sockets can be shared between more than two processes at a time. The named pipe is merely a conduit for data to travel from one process to another. The domain socket permits sharing of parameters passed between processes including UID/GID values and file descriptors. Finally, the named pipe is strictly a one-way communication (one process generates data used by another process). With a domain socket, although one process may strictly listen it does not have to be that way. Domain sockets provide a facility for two-way communication.

8.3.3 Links as File Types

The final file type is the link. There are two forms of links: hard links and symbolic (or soft) links. A hard link is stored in a directory and points to an inode which itself uses pointers to point to the file's blocks. The hard link stores the file's name and the inode number. When creating a new hard link, it duplicates the original hard link, storing the new link in a different directory. The symbolic link is also stored in a directory but merely stores a pointer to a hard link.

The difference between the two types of links is subtle but important. Using a symbolic link to access a file requires an extra level of indirect access. The operating system must first access the symbolic link, which is a pointer. The pointer provides access to the hard link. This link then provides access to the file's inode, which then provides access to the file's disk blocks.

It may sound like the symbolic link's drawback makes it less useful than the hard link. However, hard links cannot link files that exist on separate partitions. Another key difference is that the hard link is always up to date. If we move the original object, all of the hard links are modified at the same time. If we delete or move a file that is linked by a symbolic link, any hard links are modified but not any symbolic link, thus we may have an out-of-date symbolic link. This can lead to errors at a later time.

With either type of link, it is used so that we can refer to a file that is stored in some other location. Links can be useful when we do not want to add the file's location to our PATH variable. For instance, imagine that zappaf has created a program called `my_program`, which is stored in `~zappaf`. We want to run the program (and zappaf was nice enough to set its permissions to 755). Rather than adding `/home/zappaf` to our PATH, we create a symbolic link from our home directory to `~zappaf/my_program`. Now we can issue the `my_program` command from our home directory.

A more tangible use of symbolic links arises when changes are made to Linux. Imagine, for instance, that a file in `/etc` has been moved to `/usr/lib` (which is the case with several files when Red Hat moved from Red Hat 6 to Red Hat 7). Already written programs and scripts that expect the file to be located in `/etc` would no longer function correctly. In place of the moved files are symbolic links pointing to the new files' locations. In this way, these programs/scripts can still function without having to be updated to specify the files' new locations. To a program that references a file, it doesn't matter if the specified location is that of the file or that of a link.

Table 8.2 lists some of the entities moved between Red Hat 6 and Red Hat 8. In each case, it is a symbolic link left behind. This is not an exhaustive list of moved content.

When we view a directory's contents through the `ls` command, we are either seeing a hard link to the inode or a symbolic link to a hard link to the inode. In the former case, `ls -l` shows us the file's type, for instance `-` for a regular file and `d` for a directory. We can determine that an item shares a hard link to the inode by looking at the hard link count. This number will never be less than one because if it is zero it means there are no hard links and so the file would not exist. However, the number could be larger than one. Deleting any of the hard links will reduce this number. If upon deleting a hard link this number becomes zero, then the file's inode is returned to the file system for reuse, and thus access to the file is lost. Additionally, all disk space blocks are reallocated to the file system for reuse.

TABLE 8.2
Items Moved between Red Hat 6 and Red Hat 8 and Referenced by Symbolic Link

Old Location/Name	New Location/Name	Reason for Movement
`/bin`	`/usr/bin`	There's no longer a reason to separate programs between `/bin` and `/usr/bin`.
`/bin/yum`	`/usr/bin/dnf`	Replacing yum with `dnf`.
`/etc/grub2.cfg`	`/boot/grub2/grub.cfg`	Better location for GRUB2's configuration file.
`/lib`, `/lib64`	`/usr/lib`, `/usr/lib64`	Same as `/bin`.
`/etc/mtab`	`/proc/self/mounts`	Keep mount table in the `/proc` directory.
`/sbin`	`/usr/sbin`	Same as `/bin`.
`/sbin/init`	`/usr/sbin/systemd`	Transitioned from `init` to `systemd`.
`/sbin/lv*`, `/sbin/vg*`	`/usr/sbin/lvm`	Various `lv` and `vg` programs have been replaced by `lvm` including `lvchange`, `vgchange`, `lvcreate`, etc.
`/var/run`	`/run`	Creation of new directory `/run` makes `/var/run` superfluous.

If the item is a symbolic link, its type and name will differ from other types of files. The type is indicated by an `l` (for link), and the name will contain the symbolic link's name, an arrow (->) and the location of the file being linked. Symbolic links are not detectable by looking at the file being linked to as they do not impact the hard link count.

8.3.4 Reviewing the File Types

Collectively, all of these special types of entities are treated like files in the following ways. First, each item is stored in a directory and thus listed by the `ls` command. Second, each item can be operated upon by various file commands including `mv`, `cp`, `rm` and redirection operators work on them. Third, each item is represented by an inode (in the case of symbolic links, they are not stored as inodes but they point at other items that are represented by inodes). A file's type can be determined when using `ls -l`. The first character of the ten-character permissions is the file's type. In Linux, the seven types are denoted by the characters in Table 8.3 (remember that a hard link is not a type by itself).

To illustrate the types of files, see Figure 8.3. We see excerpts from three directories. At the top is a subdirectory of `/run` where we see subdirectories (`d`) and domain sockets (`s`). In the middle is a short excerpt from the `/dev` directory where we find block-type devices (`b`) and character-type devices (`c`). There are other file types found in `/dev` including subdirectories (`d`), regular files

TABLE 8.3
Characters for Each Type of File

Character	Type
-	Regular file
d	Directory
b	Block device
c	Character device
l	Symbolic link
p	Named pipe
s	Domain socket

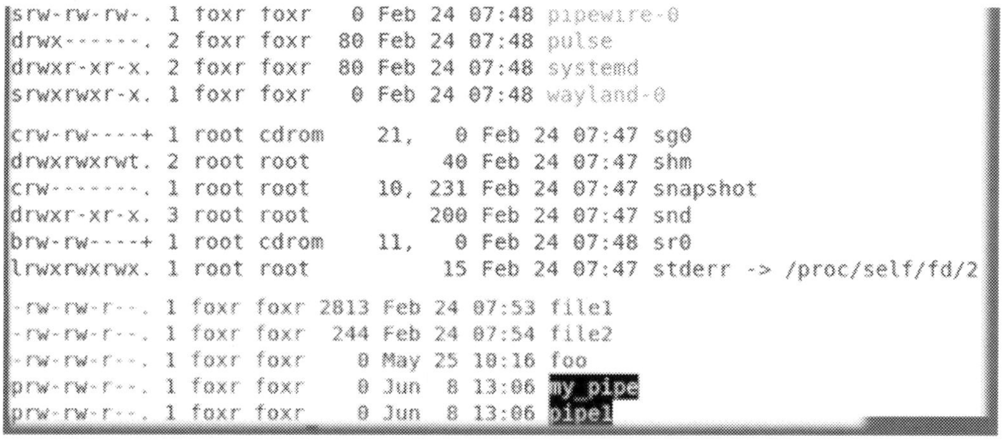

```
srw-rw-rw-. 1 foxr foxr    0 Feb 24 07:48 pipewire-0
drwx------. 2 foxr foxr   80 Feb 24 07:48 pulse
drwxr-xr-x. 2 foxr foxr   80 Feb 24 07:48 systemd
srwxrwxr-x. 1 foxr foxr    0 Feb 24 07:48 wayland-0

crw-rw----+ 1 root cdrom    21,   0 Feb 24 07:47 sg0
drwxrwxrwt. 2 root root          40 Feb 24 07:47 shm
crw-------. 1 root root     10, 231 Feb 24 07:47 snapshot
drwxr-xr-x. 3 root root         200 Feb 24 07:47 snd
brw-rw----+ 1 root cdrom    11,   0 Feb 24 07:48 sr0
lrwxrwxrwx. 1 root root          15 Feb 24 07:47 stderr -> /proc/self/fd/2

-rw-rw-r--. 1 foxr foxr 2813 Feb 24 07:53 file1
-rw-rw-r--. 1 foxr foxr  244 Feb 24 07:54 file2
-rw-rw-r--. 1 foxr foxr    0 May 25 10:16 foo
prw-rw-r--. 1 foxr foxr    0 Jun  8 13:06 my_pipe
prw-rw-r--. 1 foxr foxr    0 Jun  8 13:06 pipe1
```

FIGURE 8.3 Long listings illustrating file types.

(-) and symbolic links (`l`). The bottom portion of the figure comes from a user's home directory where there are regular files and two named pipes (`p`). Notice that the fonts are of differing colors to emphasize types or other properties. The named pipes appear with a black background.

Every file (no matter the type) is stored in a directory. The directory maintains the entities stored in it through a list. The listing is a collection of hard and symbolic links. A hard link of a file stores the file's name and the inode number dedicated to that file. The symbolic link is a pointer to a hard link stored elsewhere. As the user modifies the contents of the directory, this list is modified.

New files require new hard links pointing to newly allocated inodes. The deletion of a file causes the hard link to be removed and the numeric entry of hard links to a file to be decremented. The inode itself remains allocated to the given file unless the hard link count becomes 0.

SECTION ACTIVITIES

1. Pipes, named pipes and domain sockets all facilitate interprocess communication. Look up the term interprocess communication. What software do you regularly use that communicates with other software?
2. We explored several uses of symbolic links but hardly any uses of hard links. How frequently are hard links used in Linux? Use `ls -l` to view the hard link counts of various files in your system. You'll find in nearly every case that the items with multiple hard links are either directories or devices. If you can, find some regular files whose link count is greater than one. What did you find?

8.4 THE INODE

When the file system is first established, it comes with a set number of inodes. It is the inode that we use to access files, and it is the inode that is manipulated when files are moved or deleted. When files are appended to or copied, physical disk blocks must be allocated and again the inode is modified. Let's explore the inode in detail.

8.4.1 INODE METADATA

The *inode* is a data structure used to store file information. There are two types of file information. The first is *metadata* describing the file. The metadata is explained in Table 8.4. The other type of information that inodes store is a collection of pointers that point to physical disk blocks.

TABLE 8.4
Contents of an inode

Item	Meaning/Usage
Device	Device number where the inode resides; if the inode points at a device then also recorded is the device number consisting of a major ID and minor ID.
File type	Type as described in Table 8.3.
File permissions	User, group, world access rights as in `rwxr-xr-x`, also any special indicator such as sticky bit or set UID bit.
File owner	UID of file owner.
File group	GID of group owner.
File size	Entity's size in bytes; directories are typically 4096 bytes, symbolic links' sizes are the lengths in characters of the path/name of the item pointed to in bytes.
inode number	Numeric value attributed to the inode itself.
Number of blocks allocated	Current number of blocks allocated to the file; blocks by default are 512 bytes.
Preferred block size	Preferred size of a block for efficient access.
Timestamp • last modification date/time • last access date/time • creation date/time • last status change	Creation date/time cannot be altered; last modification is updated upon any file write, and last access is modified on each file read; last status change indicates when metadata in the inode has been modified such as owner, group, permission.
Link count	Number of hard links that point at the inode, must be at least 1.
Pointers to file blocks	Described in Section 8.4.2 and in Figure 8.4.

8.4.2 INODE POINTERS

The inode gives physical access to the file through pointers. The number of pointers directly available in the inode depends on the file system type. Typically, there will be 15 pointers broken into four types. *Direct pointers* point directly at disk blocks storing portions of the file. *Indirect pointers* point to disk blocks that do not store content but instead each stores additional direct pointers. *Doubly indirect pointers* point to disk blocks of indirect pointers. *Triply indirect pointer* point to disk blocks of doubly indirect pointers.

The typical breakdown of pointers in an inode are 12 direct pointers, one indirect pointer, one doubly indirect pointer and one triply indirect pointer. See Figure 8.4 which illustrates the inode and pointers. Mode stores the file's permissions, and flags represent different features of the file such as whether it is immutable (cannot be altered), whether the file should be compressed, whether file updates should be applied synchronously and whether the file uses extents, to name just a few.

Let's take a look at how to access a Linux file through the inode. We will make a few assumptions. First, our Linux inode will store 12 direct pointers, one indirect pointer, one doubly indirect pointer and one triply indirect pointer. Next, we assume that a block is 512 bytes in size and pointers are 32-bits. Blocks of pointers then can store up to 128 pointers no matter whether they are blocks of indirect, doubly indirect or triply indirect pointers. Finally, we assume that we have a file which consists of 1500 blocks, numbered 0–1499. Our example file then can store up to 1500 * 512 bytes = 768,000 bytes (approximately 750KB).

Table 8.5 shows how to access a given block. The first column lists the range of blocks accessible through the pointer, the second column describes access and the third column shows how many disk accesses are required to reach the requested disk block. Remember that the inode is stored on disk so any disk access requires a minimum of two disk accesses, one to access the inode via the hard link's pointer and one to access the block. When indirect/double indirect/triply indirect blocks are used, the access count increases. As our file in question only contains 1500 blocks, we will not use the triply indirect pointer at all.

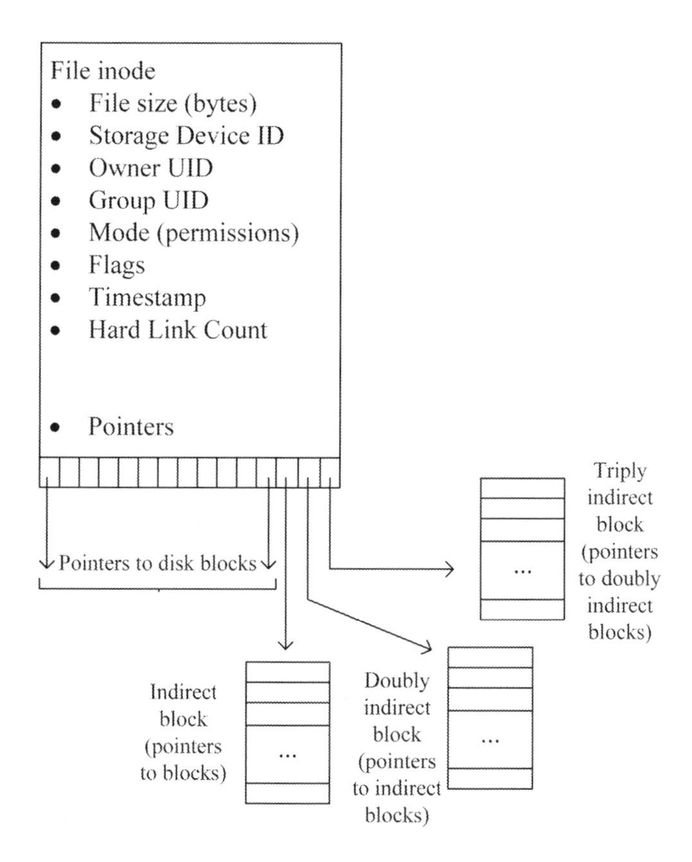

FIGURE 8.4 Inode structure with pointers to disk blocks. (From Richard Fox, 2021, *Information Technology. An Introduction for Today's Digital World*, 2nd edition.)

TABLE 8.5
Accessing a File through inode Pointers

Block Range	Block Accessed as Follows	Number of Disk Accesses
0–11	Follow direct pointer 0–11 to disk block.	2
12–139	Follow indirect pointer to indirect block; follow direct pointer to disk block.	3
140–16,523	Follow doubly indirect pointer to doubly indirect block; follow corresponding pointer to indirect block; follow pointer to disk block.	4
16,524–2,113,675	Follow triply indirect pointer to triply indirect block; follow corresponding pointer to doubly indirect block; follow corresponding pointer to indirect block; follow pointer to disk block.	5

When a file system is created, it comes with a set number of inodes. The actual number depends on the size of the file system, type of file system and the size of a disk block. Typically, there is an inode for every 2–8KB of file system space. If we have a 1TB file system, we might have as many as 128K (approximately 128,000) inodes. The remainder of the file system is made up of disk blocks dedicated to file storage and pointers. Unless nearly all of the files in the file system are very small, the number of inodes should be more than sufficient for any file system usage.

Let's take a look at how file operations work with respect to disk blocks and inode pointers. We look first at file creation. A new inode must be allocated from the available inodes that the file system maintains. The next available inode in the list is used. The inode's information is filled in consisting of the file type, initial permissions (defaulting from the user's `umask` value), owner and group of the user, the file system's device number and a timestamp for file creation. If the file is to store some initial contents, disk blocks are allocated from the file system and the direct pointers are modified in the inode to point at these blocks. Finally, the content can be saved to those blocks. The directory is modified to store a new hard link to the inode.

What happens to a file which is modified? The typical modification is made at the block level, by changing an existing block or adding new blocks. When modifying a block, the block on disk is accessed by pointer(s) and rewritten to contain the new content. If a new block is required, it is allocated by the file system and a pointer is used to point at it. The pointer selected depends on the current size of the file. For instance, referring back to Table 8.5, if the new block is the file's eighth, it will be pointed to by a direct pointer. If the block will be the file's 120th, then an access must be made to the indirect block storing pointers for blocks 12–119 and a new pointer is stored in this block to point at the new block.

What happens to an inode when a file is copied, moved or deleted? Copying a file means that the file is duplicated in storage. Each of the file's blocks is copied from one location to another. At the destination, disk blocks equal in number to the file need to be allocated and each block of the original file needs to be copied into the newly allocated blocks. An inode is allocated from the file system of the destination directory, and its pointers are set to point at the newly allocated blocks. Finally, the destination directory stores a hard link to the new inode.

Moving a file differs depending on where we are moving the file to. If we move a file within the same directory, then this is merely a renaming operation and the only change made is to the entry in the directory. If we move a file within the same partition, then all that happens is the hard link is moved from the source directory to the destination directory. The inode and file are not altered. If we move a file to a different partition, then the hard link cannot simply be moved (recall that hard links cannot reference items on different partitions). Instead, the file is copied to its new partition and a new inode is used. The original file is then deleted as described below.

Deleting a file requires removing the hard link from the directory listing and decrementing the file's inode's link count. Only if the link count becomes zero is the file physically deleted. File deletion does not physically delete the file from disk but instead returns the allocated disk blocks back to the file system. The inode itself is also returned to the file system to be reused. By not removing the content from the disk blocks, we might be able to scavenge the disk at a later time to reclaim a deleted file's contents. This is possible if those disk blocks have not been reallocated.

The system administrator is in charge of administering the file system. Rarely, if ever, will the system administrator have to worry about inodes or pointers. The Linux file system commands instead operate at a higher level of abstraction, allowing administrators and users alike to operate on the file entities (files, directories, links, etc.). It is the device drivers, implemented by system programmers, which must deal with inodes and the location of disk blocks.

8.4.3 Linux Commands to Inspect inodes and Files

There are several tools available to inspect inodes. We examine several commands that provide inode information. The `ls` command has an option, `-i`, to display the file's inode number along with the file's name. This might be useful if we want to compare file inode numbers but otherwise knowing the file's inode number is not particularly useful.

The `df` and `stat` commands provide more useful information. `df` displays file system usage statistics. Figure 8.5 demonstrates the output of the `df` command with the `-k` and `-i` options. With `-k`, shown at the top of Figure 8.5, we view the number of blocks used and available in 1K chunks. With `-i`, shown at the bottom of Figure 8.5, we see the number of used and available inodes.

The `stat` command is used to display file metadata or file system statistics. The `stat` command when used without options responds with the name of the file, the size of the file, the number

```
[rfox@localhost ~]$ df -k
Filesystem     1K-blocks    Used Available Use% Mounted on
/dev/sda5        5849088 4088400   1760688  70% /
devtmpfs         1926472       0   1926472   0% /dev
tmpfs            1941156     156   1941000   1% /dev/shm
tmpfs            1941156   50056   1891100   3% /run
tmpfs            1941156       0   1941156   0% /sys/fs/cgroup
/dev/sda1        1038336  455324    583012  44% /var
/dev/sda3        1038336  308640    729696  30% /home
tmpfs             388232      12    388220   1% /run/user/1000
[rfox@localhost ~]$ df -i
Filesystem      Inodes  IUsed   IFree IUse% Mounted on
/dev/sda5      2929664 135359 2794305    5% /
devtmpfs        481618    373  481245    1% /dev
tmpfs           485289     10  485279    1% /dev/shm
tmpfs           485289    523  484766    1% /run
tmpfs           485289     16  485273    1% /sys/fs/cgroup
/dev/sda1       524288   5223  519065    1% /var
/dev/sda3       524288   7942  516346    2% /home
tmpfs           485289     30  485259    1% /run/user/1000
```

FIGURE 8.5 Results of `df -k` (top) and `df -i` (bottom).

```
[foxr@localhost temp]$ stat file1
  File: file1
  Size: 2813            Blocks: 8          IO Block: 4096   regular file
Device: 10306h/66310d   Inode: 1578        Links: 1
Access: (0664/-rw-rw-r--)  Uid: ( 1000/    foxr)   Gid: ( 1000/    foxr)
Context: unconfined_u:object_r:user_home_t:s0
Access: 2021-02-24 07:53:46.371216881 -0500
Modify: 2021-02-24 07:53:46.373216916 -0500
Change: 2021-02-24 07:54:18.044768995 -0500
 Birth: -
```

FIGURE 8.6 Output of the `stat` command.

of blocks (and their size) allocated to the file, the file's type, the device storing the file (specified as a device number), the inode number of the file, the number of hard links to the file, the file's permissions, UID, GID in both name and number, the SELinux context and the timestamps (last access, modification, status change, and creation). Figure 8.6 demonstrates the output using `stat` on a text file. Notice `Birth` is not recorded because it is *unknown*. This is because Linux does not support this field.

The `stat` command has a number of useful options. With `-L`, `stat` will follow a symbolic link and output information about the inode being pointed to otherwise `stat` outputs information about the symbolic link itself. The option `-f` displays information about the entire file system containing the specified file. The option `-c` allows us to specify the metadata we want to view. We use `-c` with a string of formatting characters in quote marks. These formatting characters are single letters following a percent sign (`%`) as in `%b` to indicate the number of blocks allocated to the given file. Table 8.6 lists many of the formatting characters. The top portion of the table displays the descriptors for files, and the bottom portion shows descriptors for the file system.

Let's take a closer look at `stat` with two examples. First, we view information of several regular files. We use `stat` to provide for us the size of each file in blocks and bytes, the file name, the inode

TABLE 8.6
Formatting Characters for -c, Bottom Half for -c -f

Formatting Character	Meaning
%b, %B	Number of blocks (file size in blocks), size of blocks
%d, %D	Device number storing file in decimal, in hexadecimal
%f	File type
%g, %G, %u, %U	GID, group name, UID, username
%h	Number of hard links
%i	inode number
%n	File name
%s	Size in bytes
%x, %y, %z	Time of last access, modification, status change
%a	Free blocks available to ordinary users
%b	Total blocks in file system
%c	Total file inodes in use
%d	Free inodes
%f	Total free blocks
%l	Maximum allowable file name length
%s	Block size
%T	Type of file system

```
8 361 courses.txt 530970 2013-03-06 07:59:03.458188951 -0500
8 117 mykey.pub 530991 2013-03-06 07:59:03.436319789 -0500
8 413 names.txt 530974 2013-03-06 07:59:03.426421446 -0500
8 80 s1 531019 2013-03-06 07:59:03.440076584 -0500
8 48 s2 531021 2013-03-06 07:59:03.426421446 -0500
```

FIGURE 8.7 Sample output from stat -c on some files.

number of the file and the time of last access. We use the command stat -c "%b %s %n %i %x". A partial output is shown in Figure 8.7. Each file consists of only 8 blocks with sizes that vary from 48 bytes to 413 bytes. The inodes are all in the 53x,xxx range. Finally, the last access time and date are given.

Next, we use stat to inspect devices from /dev. The command is stat -c "%d %u %h %i %n %F". The command outputs the device number, UID of the owner, number of hard links, inode number, file name and file type, written using an English description (%F). The output for this command is shown in Figure 8.8.

Let's focus briefly on Figure 8.8 where we see devices of autofs, dm-0, dvdrw, input, log, pts and vcs. All except pts are located on device number 5. All except vcs are owned by user 0 (root); vcs is owned by vcsa, the virtual console memory owner account. Most of the items have

```
5 0 1 12556 autofs character special file
5 0 1 6106 dm-0 block special file
5 0 1 10109 dvdrw symbolic link
5 0 3 5341 input directory
5 0 1 2455364 log socket
11 0 2 1 pts directory
5 69 1 5185 vcs character special file
```

FIGURE 8.8 Sample output from stat -c /dev.

only one hard link, found in /dev. Both input and pts are directories and have more than one hard link. The inode numbers vary and are all small except for log which was created after Linux installation and has an inode number of 2,455,364. The file type demonstrates that Linux files are made up of a wide variety of entities from block or character files (devices) to symbolic links to directories to domain sockets. This last field varies in length from one word (directory) to three (character special file, block special file).

We might want to explore the inode numbers in our file system to see how diverse they are. Each new file is given the next inode available. As our file system is used, we will find newer files have higher inode numbers unless a file has been given an inode of a previously deleted file. A script is shown in Figure 8.9 to output the largest and smallest inode numbers of a list of files passed to it as parameters. Notice in this script the use of the shift command so that all of the parameters after the first are shifted down. Since we had already processed the first parameter ($1), we no longer need it.

Whenever any file is used in Linux, it must first be opened. The opening of a file requires a special designator known as the file descriptor. The *file descriptor* is an integer assigned to the file while it is open and referenced by the process when communicating requests to the Linux kernel. Three file descriptors are assigned by default as 0 for stdin (the default input device), 1 for stdout (the default output device) and 2 for stderr (the default device to report errors from a program). Any remaining files that are utilized during the program's execution need to have file descriptors assigned before the files can be accessed.

When a file is to be opened, the operating system kernel gets involved. First, it determines if the user has adequate access rights to the file. If so, it then generates a file descriptor. The kernel creates an entry in the system's file table, a data structure that stores file pointers for every open file. The location of this pointer in the file table is equal to the file descriptor generated. For instance, if the file is given the descriptor 15, then the file's pointer will be the 16th entry in the file table (remember that the first item, stdin, has a file descriptor of 0). The pointer itself will point to an inode for the given file. As devices are treated as files, file descriptors also exist for every device; such entities include the keyboard, terminal windows, the monitor, the network interface(s) and the disk drives.

To view the file descriptors of a given process, look at the fd subdirectory of the process' entry in the /proc directory (e.g., /proc/16531/fd). There will always be entries labeled 0, 1 and 2 for STDIN, STDOUT and STDERR, respectively. Any other file descriptor assigned to the process will also be listed.

```
#!/bin/bash
largest=`stat -c "%i" $1`
largestFile=$1
smallest=`stat -c "%i" $1`
smallestFile=$1
shift
for item in $@; do
     number=`stat -c "%i" $item`
     if [ $number -gt $largest ]; then
          largest=$number; largestFile=$item;
     fi
     if [ $number -lt $smallest ]; then
          smallest=$number; smallestFile=$item;
     fi
done
echo The largest inode from the files provided is
echo $largest of file $largestFile.  The smallest
echo inode from the files provided is $smallest
echo of file $smallestFile
```

FIGURE 8.9 Sample script using stat command.

To illustrate, we opened `vi` on a new file, `foo.txt`. Five file descriptors have been assigned to this process, three of which are `0`, `1` and `2` as noted above. Two additional file descriptors are `3`, which is of the file `/var/lib/sss/mc/passwd`, a cache entry indicating that the user has authenticated, and `4`, which is a backup file for `foo` called `.foo.txt.swp`. All of these file descriptors are actually symbolic links. We can also view all open files using the `lsof` command.

SECTION ACTIVITIES

1. Which provides more efficient disk access, a B+ tree or an inode? We saw in this section the potential number of indirect blocks that must be accessed to access a particular disk block by following inode pointers. See if you can find a similar analysis of B+ tree accesses to compare the most number of disk accesses that might be needed to access some random block of a file.
2. Does it make sense to treat the devices (in `/dev`) as files? Research why this is done in Unix/Linux operating systems.

8.5 PARTITIONS AND FILE SYSTEMS

The disk drive(s) making up our Linux storage space is(are) divided into partitions. Multiple partitions may be placed on one physical device or a single partition can be distributed across multiple devices. The distribution of partitions to physical device(s) should be transparent to the user, and unless a device fails, it should be transparent to the system administrator.

Each partition contains an independent file system. The file system implements how files are stored and accessed within the partition. We might view this breakdown as devices which are divided into partitions each of which is implemented as a file system and within each partition are directories that contain files and subdirectories.

The terms partition and file system are not interchangeable, yet when we refer to an existing partition it will have associated with it a type of file system, and when we refer to a specific file system we are also referring to the partition that it resides upon. As we describe access to and mounting of a partition, we will sometimes also refer to the object as a file system. For more information on types of available file systems, refer back to Section 8.2.3 and Table 8.1.

8.5.1 Why Partition?

The Windows operating system by default divides physical devices to their own partitions. The internal hard disk is usually labeled as the C: partition, the optical drive as the D: partition and each externally mounted device assigned to the next available letter such as F: and G:. This naming convention dates back to the earliest days of the IBM PC when there were two floppy disk drives labeled as the A: drive and B: drive.

While we are able to repartition the C: drive into multiple partitions, people rarely do. In comparison, the Linux operating system's file space is divided into several partitions. The reason to partition at a more finely grained level is that partitions all act independently of each other. We can perform an operation on one partition without impacting other partitions. As an example, we might encrypt one partition but not another or unmount a partition in order to back it up while leaving other partitions accessible. Having the file space partitioned should also reduce damage that might be done maliciously.

Each partition is assigned a type of file system where the partitions can use different types if desired. In this way, we can select a file system type that is most advantageous for the type of files that will be stored there.

Partitioning does not come without a risk. We set up partitions at the time we install Linux. At the time we partition the storage space, we have to specify the size of each partition. Once a partition is created, it is limited to the physical size we allocated to it. Imagine that we have a 1TB hard drive and we have allocated 200 GB for the /home partition. After adding several dozen user accounts, we find that the partition is nearly full. What can we do? One solution is repartitioning but doing so requires taking disk space from another partition, and that could damage files stored in that other partition. Another solution is to use a logical volume manager (LVM). We explore repartitioning and logical volume managers in later subsections.

The segmentation of the storage devices into partitions is a physical division of space. When doing so in Linux, the partition creates a new *drive*. This is not a new physical device but is treated as an independent device and given its own name. Let's assume our internal hard disk is labeled as /dev/sda. Partitioning divides this into devices labeled /dev/sda1, /dev/sda2 and so forth.

While Linux treats each of the partitions as a separate drive logically, they may be stored on one physical disk. SD, by the way, stands for *S*CSI *d*evice. In older Linux and Unix systems, HD stood for IDE drives. Linux has largely abandoned the use of HD in favor of SD. As SATA became popular and USB drives were introduced, they continued to be referenced using SD. The letter at the end indicates the drive. For instance, sda would be considered the first drive while sdb would be the second. The number that follows the name indicates a partition within the drive so that /dev/sda1 is the first partition of sda.

8.5.2 Viewing the Available Partitions

How can we determine which device maps to which partition? One way is to examine the contents of the /etc/fstab file. This file is the *file system table*, which specifies which partitions should be mounted at system initialization time. An example of an /etc/fstab file is shown in Figure 8.10. The table stores for each partition to be mounted the device's name, the device's mount point, the file system type and the mount options.

In Figure 8.10, the file system's device name is given in the first column. In more recent versions of Linux, we tend to see the device's universally unique ID (UUID) specifier instead of the device name. The UUID is a string of 32 hexadecimal digits like 7AE480BA-C911-1287-15B0-8ED67C3A021B. The second column shows the *mount point*. This is the location within our logical view of the file space that we use to access the partition. Stated another way, this is the directory where the partition is mounted. Notice that the swap partition has no mount point. Swap space is not directly accessible by users.

```
/dev/sda1                 /                 xfs     defaults                  1 1
/dev/sda3                 /home             xfs     defaults                  0 0
/dev/sda2                 swap              swap    pri=2000                  0 0
/dev/sda5                 /var              xfs     defaults                  0 0
/dev/sda6                 /boot             ext4    defaults                  1 2
proc                      /proc             proc    defaults                  0 0
/dev/cdrom                /media/cdrom      auto    ro,noauto,user,exec       0 0
tmpfs                     /dev/shm          tmpfs   defaults                  0 0
www.someserver.com:       /home/coolstuff   nfs     rw,sync                   0 0
  /home/stuff
```

FIGURE 8.10 Example /etc/fstab entries.

The third column is the file system type. ext4 had been the default type in Linux for a number of years. Most Linux distributions have moved from ext4 to xfs. Figure 8.10 comes from a Red Hat 8 implementation. Notice that the /boot partition still uses ext4. The fourth column contains mounting options. Most of the entries in the figure list defaults but the optical drive is read-only, not automatically mounted at system initialize time, can be mounted by the user and content within that file system can be executed. The final column contains two integer numbers used to specify the order by which partitions will be examined by the fsck program and will be archived by the kdump program. The most common entry is 0 0.

The last entry in the table (the bottom row) is a remotely mounted partition. It has been mounted with options of rw (can be written to) and sync (meaning that access to the partition must be synchronized). We explore these options in more detail below. Also notice for this remotely mounted partition that we are using NFS, the network file system. This is not the file system of the remote partition but the means by which we connect to it.

Let's take a closer look at the options, also called the *mount options*. The defaults option is common and the obvious choice if we do not want to make changes. If we want to specify something other than the default, we list each option separately. To indicate whether a file system is mounted at boot time or not, we use auto or noauto. The option user/nouser specifies whether ordinary users can mount the given partition (user) or whether only root can mount it (nouser). Typically, we find the user option attached to devices that a user should be able to mount and unmount after system initialization, such as with an optical disc (cdrom) or a USB flash drive.

The exec/noexec option indicates whether binary programs can be executed from the partition or not. We would add the option noexec if we do not want users to be able to run programs stored in a particular partition. As an example, we may want to place noexec on the /home partition. This would prevent a user from writing a program, storing it in their home directory and executing it from there. Instead, any such program would have to be moved to another partition, for instance /usr (which is likely part of the root partition, /). This restriction may be overly cautious if we expect our users to write and execute their own programs. One reason for this precaution though is that it would prevent users from downloading executable programs into their home directories and executing them.

Two other options are ro/rw (read-only, read/write) and sync/async. In the former case, the ro option means that data files on this partition can only be read. We might use this option if the files are all executable programs, as found in /usr/bin and /usr/sbin. Partitions with data files such as /var and /home would be rw. The latter option indicates whether files have to be accessed in a synchronized way or not. With sync, the operation is performed at the time that the command is issued. Other operations to the same device must wait. With asynchronous access, operations are not necessarily performed in the order submitted but instead are based on the most efficient access to the device.

Notice that swap has an option of pri=2000. This option is known as a *configuration priority*. When multiple partitions have a priority, mounting is handled in the order of priority where the larger the number the higher the priority. As swap is the only partition with a priority, the number we provide is immaterial, but if other partitions had priorities it would make sense for swap to have a higher priority than some of the others.

We saw the df command in the last section. It outputs for each mounted partition that partition's used and available capacity. Comparing /etc/fstab to the output of df we can see that df outputs more partitions. Many of these partitions are not physical partitions but *virtual partitions* stored in memory. These virtual partitions include, for instance, /dev and /run.

Another way to view partitions is through the file /etc/mtab. Where /etc/fstab is the file system mount table used during system initialization, /etc/mtab is the table of currently mounted partitions. We find a great many virtual partitions in mtab that we do not find in fstab or df. In most recent distributions, /etc/mtab has been moved to /proc/mounts leaving /etc/mtab as a symbolic link.

8.5.3 Creating Partitions

Partitions are created when we first install the operating system. We skipped over most of the details of partitioning in Chapter 1 because Linux is set up to install its preferred or default partitioning for us. Should we want to specify our own set of partitions, we do so during installation. We can modify partitions later using repartitioning. We cover both of these activities in this subsection and the next subsection respectively.

How do we decide upon the partitions for our system? What size should we reserve for each partition? What type of file system should each partition be assigned? Where should we mount each partition?

Answers to the above questions are in part driven by the type of and usage of our computer. A standalone workstation for a single user might not need as fine-grained a partitioning as a workstation that will host multiple users or a server which will mount potentially dozens of drives. Consider a single-user system which can have a small `/home` partition versus a computer that hosts hundreds of users and so will require a very large `/home` partition. A server will likely make extensive use of the `/var` directory and so will require a large `/var` partition; a workstation will not. Additionally, there is a question of how much application software will be installed in `/usr` versus perhaps having some application software mounted remotely.

Fortunately, these partitioning decisions are fairly uniform in that we do not have to consider too many other questions. This simplifies the partitioning process and lets us have confidence in our decisions. Table 8.7 shows a possible breakdown of partitions. Again, based on answers to the earlier questions you may or may not have separate partitions for `/var`, `/usr`, `/usr/bin` or `/opt`. The default Red Hat 8 installation provides just three partitions: `/boot`, `/` and `swap` (where `/home`, `/var` and `/usr` are all placed with `/`).

Let's step through how to create our own partitioning, specifically with Red Hat 8. Refer back to Chapter 1 as needed. When installing Red Hat, we are presented with an Installation Summary window (see Figure 1.7). From this window, we can specify the location of the Linux installation source, the type of installation and software to install, connection to the network and installation destination. The installation destination includes a selection for partitioning.

By default, the installation process will create the partitions for us. This default typically divides the file space into the three partitions already mentioned (root partition, boot partition and swap space). Selecting `Custom` for Storage Configuration takes us to the Manual Partitioning window.

TABLE 8.7
Possible Breakdown of Partitions

Partition Name	Mount Point	Usage
Root partition	`/`	Contains a majority of the OS including `/etc`, `/root`, and possibly `/usr` (which itself contains `/usr/bin` and `/usr/sbin`).
Boot partition	`/boot`	Contains `GRUB`, the Linux kernel (`vmlinuz`) and any `initramfs` images needed.
User partition	`/home`	Contains all user home directories excluding `/root`.
Var partition	`/var`	Although part of the operating system, this directory stores variable data files and so should be separate as it can grow in size unlike the directories under `/`.
Swap	N/A	System swap space.
Software partition	`/usr` (or possibly just `/usr/bin`)	A separate partition for software may be preferred to keep that partition separate from the root partition.
Third-party software	`/opt`	If we plan to install third-party software, we might want to place it in a separate partition so that it cannot impact the operating system.

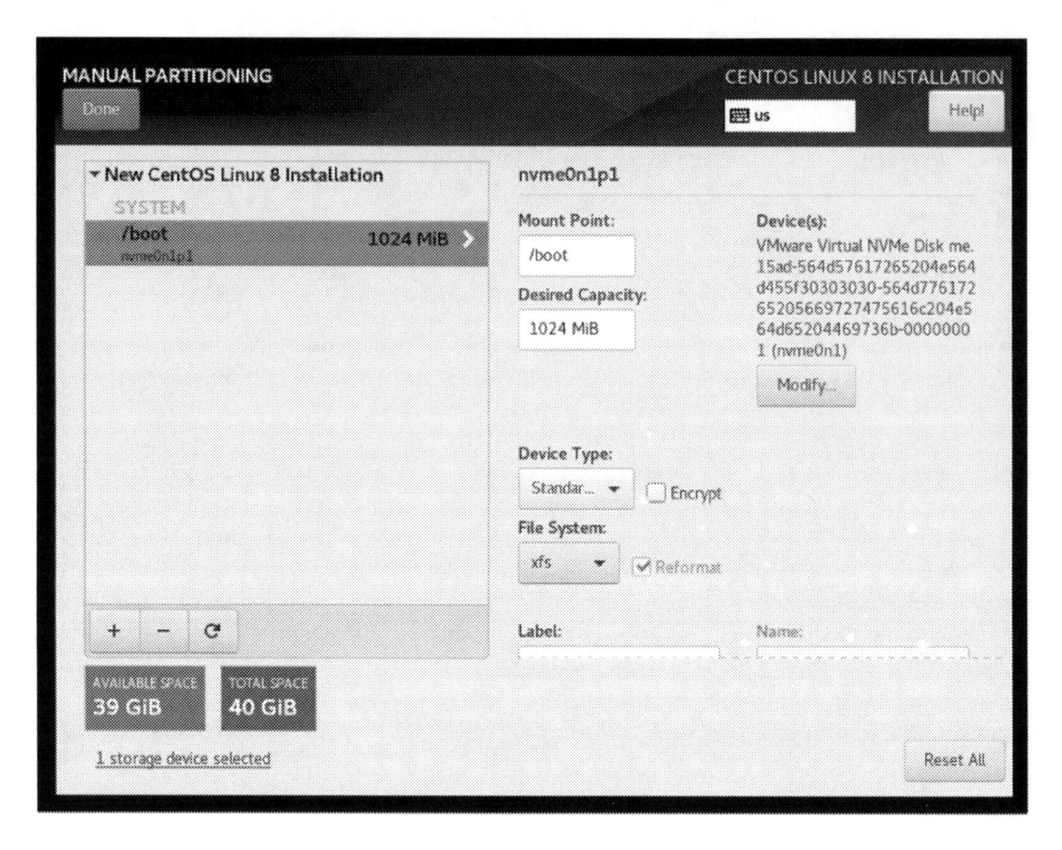

FIGURE 8.11 Manual partitioning windows.

Our first selection is to pick the type of partitioning scheme from a drop-down menu. The choices available for Red Hat 8 are `LVM`, `LVM Thin Provision`, `Standard Partition` and `Btrfs`. In this example, we have selected `Standard Partition` (we look at LVM in a couple of subsections). Now we begin assigning the partitions. We see in Figure 8.11 that one partition has been created, `/boot`.

In the window shown in Figure 8.11, there are three meaningful areas. Near the bottom left we see the total and available space of the device that we are installing onto. This can help us determine if we are specifying reasonable sizes for partitions. Just above these values are two selections denoted as + and –. We use these to add new partitions or remove already specified partitions. Above this is a pane showing all of the currently specified partitions. At this point, we have a single partition, the boot partition mounted at `/boot`. This partition has been assigned 1GB. Clicking on the + brings up a window by which we can add a partition by specifying its mount point, as shown in Figure 8.12, where we are adding the root partition (/).

To add this new partition, we have two decisions to make. What is the mount point for this partition and what size will this partition be? The mount point location is selectable only through a drop-down window. The partition's size is input via a textbox using abbreviations like M or G. Here, we see / is selected with a capacity of 10G. Once a mount point is added, it appears in the main pane.

Selecting any mount point in the main pane displays its details in the right pane of the Manual Partitioning Window. Referring back to Figure 8.11, the `/boot` mount point has been selected. Once specified, we can modify a partition's size and file system type (`xfs` is currently assigned to `/boot`). We can also specify whether the partition should be automatically encrypted. Recommendations for Red Hat 8 partitioning are provided in Table 8.8.

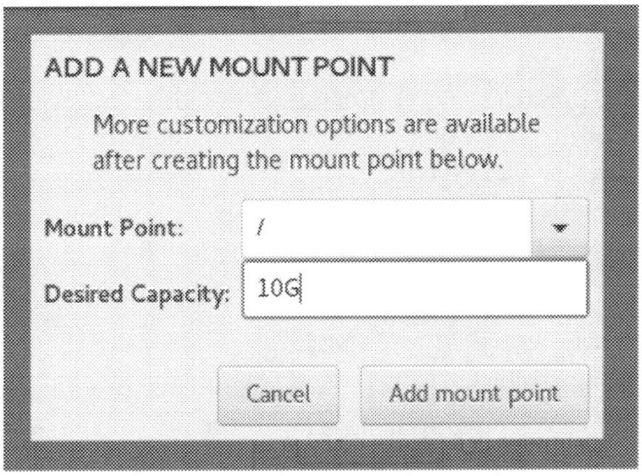

FIGURE 8.12 Add a new mount point window.

TABLE 8.8
Red Hat Recommended Partitions

Partition	Recommended Size	Comments
/	10GB	A "minimal installation" requires at least 5 GB but the minimal installation does not come with a GUI.
/home	1GB or more	Size is dependent on the number of users we expect to use the system; Red Hat recommends encrypting /home.
swap	Between 0.5 and 2 times the size of DRAM	If DRAM<2GB, use twice the size of DRAM; if DRAM is between 2GB and 8GB, use the size of DRAM; if DRAM is between 8GB and 64GB, use ½ size of DRAM; if DRAM is larger, use at least 4GB.
/boot	1GB	This may be more specifically /boot/efi; the reason to separate /boot from / is that /boot is required upon booting while / is not required until after the Linux kernel is running; each kernel stored in /boot requires approximately 56MB divided into initramfs (32MB), kdump initramfs (14MB), system map (3.5MB) and vmlinux (6.6MB).
/var	Varies	The recommendation is to make this part of the root partition; however, when running specialized software like mysql or apache, we might want to have a separate partition and use as much as 3GB.

Upon completing our specifications for partitioning, we select the Done button. Before we complete this, we can make any modifications we like. Before installation begins, we can go back to the Installation Destination window and from there the Manual Partitioning window to make changes if desired.

8.5.4 Repartitioning

What happens if, after OS installation, we find that a partition's size is not sufficient? This can be a challenging problem because we have likely already allocated the full space of the storage device and resizing a partition might wipe out content of another partition. Consider, for instance, that we divided a 1TB hard disk into four partitions giving each one equal size (250GB). Now, we find that /home is nearly full. To provide /home more space, we either have to add hard disk space (e.g., add another hard disk) or we have to repartition. If we repartition, we have to remove space from one of the other partitions. Which one, how much and how do we ensure that doing so does not wipe out content already stored there?

Obviously, this is a non-sensical example because we wouldn't dedicate 250GB to the swap space and so that would be an obvious partition to reduce. But imagine that we have a smaller hard disk to begin with and have stretched it to the limit. Can we remove space from the root partition or the boot partition? Not likely unless we gave those partitions more than needed space. What about a `/var` partition? It depends on what software we might be running.

Assuming we do have space in another partition, then repartitioning is a reasonable solution to adding space as needed to an existing partition. Before attempting to repartition, we need to back up whichever partition(s) we will reduce in size. This ensures that if in encroaching on another partition's disk space it deletes files then we have them available to be restored.

The `parted` program can be used to handle repartitioning. This is a command-line program used to view, alter, add and delete partitions. The program can run with or without user interaction. The non-interactive version requires that we specify the full command on the command line. We will ignore this aspect of `parted` and just concentrate on the interactive version.

To launch `parted` in interactive mode, type `parted` *`device`* where *device* is the name of the device, such as `/dev/sda`. If the device is omitted, `parted` runs on the first block device found (which is likely `/dev/sda`). Running the interactive mode of `parted` places us into the `parted` interpreter with a prompt listed as `(parted)`. From here, we enter (re)partitioning commands.

There are numerous commands available. The more useful ones are listed in Table 8.9. The value *n* refers to the numeric value associated with the partition, which can be obtained through the `print` command. With certain options, `parted` runs without interaction, for instance `-l/--list`) is the same as the `list` command. We are unable to perform tasks like `mkpart`, `mkfs` and `resize` without the interactive interpreter.

Let's step through an example of using `parted`. We issue the command `parted /dev/sda` and are dropped into the `parted` interpreter. From here, we enter `print` which responds with the output shown in Figure 8.13 (introductory output is omitted).

Figure 8.13 shows that there are five partitions where four are of type `xfs` and one is of type `linux-swap`. Among the information reported is the size of each partition and the start and end

TABLE 8.9
`parted` Commands

Command/Syntax	Meaning
`align-check` *`type n`*	*type* is either `minimal` or `optimal`; this command checks if partition *n* aligns to the type specified.
`mklabel` *`type`*	Create an entry in the partition table whose type is *type*; *type* is one of `bsd`, `loop`, `gpt`, `mac`, `msdos`, `pc98` and `sun`.
`mkpart` *`part-type name fs-type start end`*	Create a partition where *part-type* is one of `primary`, `extended` or `logical` and *fs-type* is one of `ext2`, `fat16`, `fat32`, `hfs`, `linux-swap`, `NTFS`, `reiserfs`, `ufs` or `btrfs`; *start* and *end* are block numbers on the disk denoted by block numbers using a notation like 0.0 or 500.10.
`name` *`n name`*	Assigns *name* to partition *n*.
`print [devices | free | list | all |` *`n`*`]`	List the partitions; if one of the parameters is provided then only output the partitions that fit the parameter; `list` and `all` do the same thing.
`rescue` *`start end`*	Rescue a lost partition found near *start*.
`resizepart` *`n end`*	Resize partition *n* to have a new ending block of *end*.
`rm` *`n`*	Delete partition *n*.
`select` *`device`*	Used to specify the device parted will operate on (needed if the device was not specified when parted was started from the command line).

```
Number  Start    End     Size    Type     File system   Flags
1       1049kB   2.1GB   2.1GB   primary  xfs           boot
2       2.1GB    12.1GB  10GB    primary  xfs
3       12.1GB   14.1GB  2GB     logical  linux-swap
4       14.1GB   20.1GB  6GB     primary  xfs
5       20.1GB   32GB    11.9GB  primary  xfs
```

FIGURE 8.13 Output from `parted`'s `print` command.

locations of the partitions. These are given in bytes instead of blocks. We want to reduce the size of the `/var` partition, which is numbered 5, and use the freed-up space to create a new partition. We issue the command `resizepart 5 24GB`. This command informs `parted` to change partition 5's ending block from 32GB to 24GB. This will free up 8GB that we can use for another partition. We would only do this if we knew that the corresponding partition was not using and did not need that extra space.

With 8GB available, starting at 24GB, we can now create a new partition which will contain the `/usr` directory and within it, `/usr/local`. We use `mkpart` to create this partition. As this partition will store `/usr` which itself will contain `/usr/local`, we have a situation where the partition requires its own partition table. For the *part-type*, we will use `extended` instead of `primary` as an extended partition is allowed to have its own partition table. We decide to use `btrfs` for the file system type, and we use 24GB and 32GB respectively for the start and end blocks. Our command is `mkpart extended "/usr/local" btrfs 24GB 32GB`. Reissuing the `print` command shows the same output as in Figure 8.13 except partition 5 now ends at 24 GB, and a sixth entry is listed as `6 24GB 32GB 8GB extended btrfs`.

There are other Linux programs that we can use aside from `parted`. The command `fdisk` is a menu-driven program by which we can create and delete partitions. This command does not format the partition with a file system so we have to also run some command to create the file system. There are numerous options including `mkfs`, `mke2fs`, `mkfs.ext4` and `mkfs.xfs`. Although they involve more work, one advantage of this approach is that we can specify file system options that are not provided in `parted` such as stripe sizes if the file system is being configured for some form of RAID (RAID is covered in Chapter 12), journaling options and inode sizes. We look at using `fdisk` and `mkfs.xfs` in a couple of subsections.

8.5.5 Using a Logical Volume Manager to Partition

Instead of risking the need for repartitioning in the future, another option is to use a *logical volume manager* (LVM). LVM is an approach to disk partitioning where partitions' sizes are not physically imposed at the time of initial disk partitioning. Instead, partitioning is done in a virtual way. Or alternatively, you might think of LVM as a software layer residing over the physical disk where partitioning is handled through this software layer.

LVM views items as physical volumes, physical extents and logical volumes. *Physical volumes* are the collection of storage devices. These are grouped together to form *volume groups*. For instance, we might have two hard disk drives, `/dev/sda` and `/dev/sdb`, that we use to create a single volume group. Now, partitioning places content of a partition wherever space is available on either (or both) of the physical disk drives.

A physical volume is itself made up of *physical extents* (PE). The PEs are fixed in size, much like the page/frame size of virtual memory. A physical volume then is a collection of PEs. The number of PEs for the physical volume is based on the PE size (dictated by the operating system) and the size of the physical volume itself. From our previous paragraph, imagine that `/dev/sda` is 1TB and `/dev/sdb` is 512GB in size. These two physical volumes combine for a single volume group of 1.5TB. Let's further assume that our PE size is 4MB. This gives us more than 393,000 PEs.

We now divide the volume group into partitions, known as *logical volumes*. The logical volume is established with a certain minimum size and an estimated total size. This minimum size dictates the number of initial PEs allocated to the logical volume. The estimated total size is only a suggestion of how many PEs the volume might eventually use.

As the logical volume grows in size, it is allocated more PEs. In this way, a poorly estimated total size would not require physical repartitioning. As long as PEs are available, any of the logical volumes (partitions) can grow. If disk space is freed up from a partition (for instance by deleting unnecessary files), those PEs are handed back to the pool for reuse by any of the logical volumes.

Figure 8.14 provides an illustration of the concepts described here. We see all of the available physical volumes (two disks in this example) united together as a single volume group. Each physical volume has PEs. A logical volume is a collection of PEs. Aside from initially assigning a group of PEs to a logical volume, PEs are allocated on an as-needed basis. Because of this, a logical volume's PEs are not all located within one section of a disk drive but instead can range to include space from all of the physical devices.

There are many reasons to use LVM over physical disk partitioning. The biggest advantage of LVM is that the system administrator will not have to resize partitions in the future if poorly chosen initial sizes were selected. The LVM greatly simplifies any disk maintenance that the administrator may be involved with. Another benefit of LVM is in creating a backup of a partition. This can be done using LVM without interfering with the accessibility of that partition. LVMs are also easily capable of incorporating additional hard disk space as hard disk drives are added to the system without having to repartition or partition anew the added space. In essence, a new hard disk is added to the single volume group providing additional PEs to the available pool. We can also establish redundancy simulating RAID technology.

There are some reasons for not using LVMs. First, the LVM introduces an added layer of indirection in that access to a partition now requires identifying the PEs of the partition. This can lead to challenges in disaster recovery efforts as well as complications during the boot process. Additionally, as a partition may no longer occupy disk blocks in close proximity to each other but instead may be widely distributed because the PEs are provided on an as-needed basis, the efficiency of accessing large portions of the partition at one time decreases. In Figure 8.14, we see that one of the logical volumes (the third from the left) exists on two separate devices.

As booting takes place prior to the mounting of the file systems, the boot sector needs to reside in an expected location. If the `/boot` partition were handled by the LVM, then its placement is not predictable. Therefore, we separate the `/boot` partition from LVM, creating (at least) two physical partitions: one containing `/boot` and the other containing our LVM. The volume group partition contains all of our logical partitions. We would likely have a third partition for swap space.

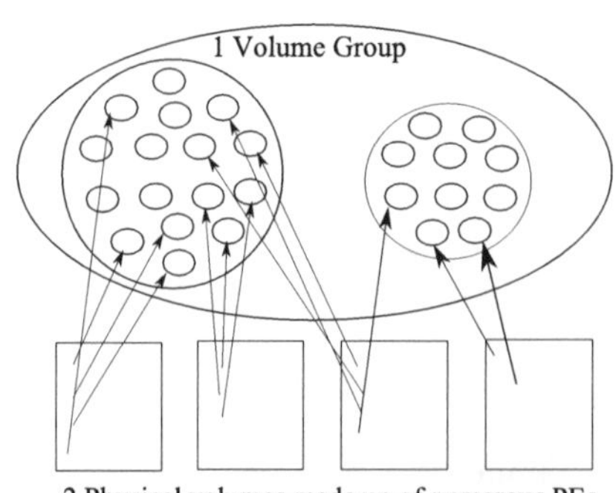

FIGURE 8.14 Implementing logical volumes out of physical volumes using LVM.

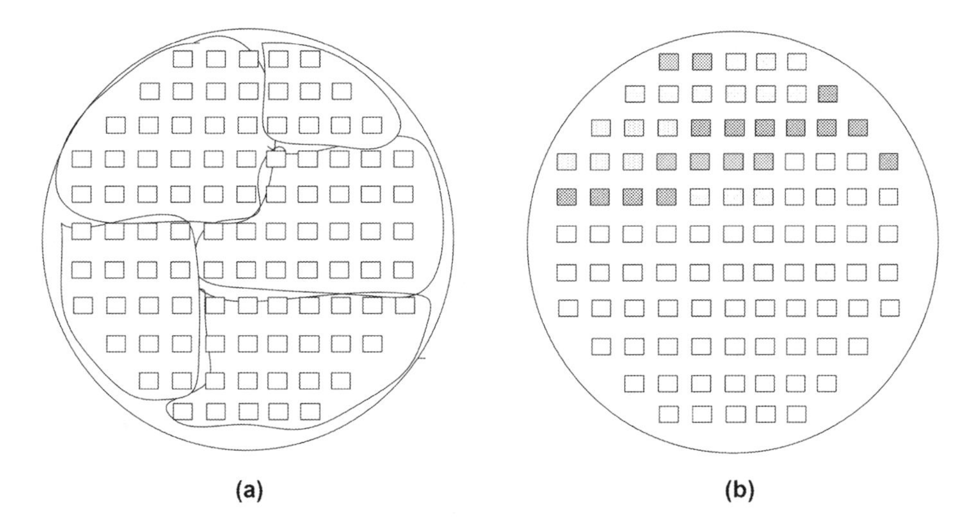

FIGURE 8.15 Physical partitioning (a) vs logical partitioning (b). (From Richard Fox, 2021, *Information Technology. An Introduction for Today's Digital World*, 2nd edition.)

TABLE 8.10
Logical Volume Manager Commands

Command	Meaning
`pvchange, vgchange, lvchange`	Change attributes of a physical volume, volume group, logical volume.
`pvck, vgck`	Check physical volume, volume group.
`pvcreate, vgcreate, lvcreate`	Create a new physical volume, volume group, logical volume.
`pvmove`	Move the PEs of a physical volume.
`pvresize`	Resize a partition.
`vgextend, lvextend`	Add a physical volume to a volume group, extend the size of a logical volume.
`vgmerge`	Merge two volume groups.
`vgreduce`	Remove a physical volume from a volume group.
`lvreduce, lvremove, lvresize`	Reduce the size of a logical volume, delete a logical volume, resize a logical volume.

Figure 8.15 compares a physical partitioning (a) to a logical partitioning (b). The different shades of blocks in the logical partitions are different logical volumes (logical partitions), and the empty blocks are unused. Not shown in the figure is the separate `/boot` partition for the LVM version, which would be a physical partition separate from the LVM.

We establish our disk partitioning using LVM at the time we initially partition our file space. We can later modify the physical and logical volumes through a suite of commands. A few of the more useful commands are provided in Table 8.10. These commands are also available through the interactive, text-based `lvm2` program (which is usually called `lvm` in Linux systems). The `lvm2` program is a front-end or interface for the programs listed in Table 8.10 (which contains just a partial listing of the LVM programs available for creating and maintaining LVM-based partitions).

8.5.6 Adding a Disk Drive

We wrap up this section by looking at the steps to add a disk drive to our computer. We do not simply plug in the device (whether internally or externally). We must also install a file system onto the device and mount the new file system.

We will make the assumption that the new device will be placed internally in our computer's system unit although this does not need to be the case. By making this assumption, we are assured that the device will always be available for mounting (as opposed to an external hard disk that might be plugged in through a USB port and so may or may not be available). Let's further assume that the new drive will be used to extend the user directory space. We will call this new partition `/home2`.

Having inserted the disk drive and rebooted our computer, our first step is to identify the new device's name. It will likely be `sdb` (as our first internal hard disk is named `sda`). We will find this item listed in the `/dev` directory and can identify it by simply performing `ls /dev/sd*`.

We saw earlier that we use `parted` to repartition. One capability of `parted` is to create a new partition. We must do that here to our new device. But instead, we will use `fdisk` for this step to demonstrate this program. As root, we enter the command `fdisk /dev/sdb`. `fdisk` runs and informs us that this drive has no partitions. We are then asked to enter a command, enter `m` for help. Table 8.11 displays the various commands available.

Upon entering a command that asks for additional input (e.g., `d`, `i`, `I`, `p`, `t`), `control+c` will exit from the menu choice but leave us in `fdisk`. This is useful if we happen to enter a command we did not intend, such as `d` to delete a partition. As we have no partitions on our new drive, we want to create one, so we enter `n`. We are first asked if this will be a primary partition (`p`) or extended partition (`e`). We would only select `e` if we are adding space to a physical volume in an LVM. We also specify the partition number, which by default will be 1 greater than the largest current partition. As this is an unpartitioned disk, the default is `1`.

`fdisk` now asks for the starting and ending sectors that will encompass this partition. We are given the next available and last available sectors. As this disk has no other partitions, these will be the first available sector (which is likely to be 2048 instead of 0) and the last sector of the disk respectively. These are listed as defaults, so we do not need to change anything and just press `<enter>` for both. `fdisk` responds that a new partition will be created of type `'Linux'` and

TABLE 8.11
`fdisk` Menu Commands

Command	Meaning	Comments
`d`	Delete a partition.	Asks for partition number.
`F`	List free unpartitioned spaces.	
`g`	Create a new empty GPT partition table.	GPT (GUID Partition Table) is the modern standard.
`G`	Create a new empty SGI partition table.	For SGI/IRIX type machines.
`i`	Print information about a partition.	Asks for partition number.
`I`	Load disk layout from sfdisk script file.	Asks for filename.
`l` (lowercase `L`)	List known partition types.	Lists all available types for various Linux distributions.
`m`	Print the menu of choices.	
`M`	Enter protective/hybrid MBR.	MBR is a DOS-type partition table.
`n`	Add a new partition.	Asks for partition number, first and last sectors.
`o`	Create a new empty DOS partition table.	
`O`	Save disk layout to sfdisk script file.	Asks for filename.
`p`	Print partition table.	
`q`	Quit without saving changes.	
`s`	Create a new empty Sun partition table.	Sun is a BSD-type partition table.
`t`	Change a partition type.	Asks for partition number and new type (can list all available types).
`v`	Verify partition table.	
`w`	Write table to disk and exit.	
`x`	Extra functionality.	Only to be used by `fdisk` experts

provides its size. Now, we write this change to disk by entering w. fdisk responds that the partition table has been altered.

If we exit fdisk and issue ls /dev/sdb*, we will see /dev/sdb and /dev/sdb1. The former is the new disk itself, and the latter is our new partition. As noted earlier, fdisk will create a partition but will not install a file system onto it. We now need to create the file system. The most appropriate file system is xfs, so we run mkfs.xfs.

There are a number of options available for mkfs.xfs that can be used to change default values for such things as the size of blocks, the types of metadata the file system will maintain and the naming options used. We will settle for all of the default values and only add an option to label the file system, which is -L *label*. Our command is mkfs.xfs -L home2 /dev/sdb1. The command creates the file system on the partition and displays information about the result including the block sizes, number of blocks, number of inodes, naming scheme and logging information.

We may now mount this new file system. We cover mounting in the next section so only describe the steps minimally here. We first create a new mount point. This is a new directory. We choose to call it /home2 and use the command mkdir /home2. Next, we issue the mount command, mount /dev/sdb/sdb1 /home2. Finally, we want this new partition mounted automatically and so add it to our file system mount table, /etc/fstab. We edit this file and add the entry /dev/sdb2 /home2 xfs defaults 0 0.

We might want to make one additional change as we are adding this hard disk, in this example, because /home is full and we want to use /home2 for all new users. Thus, we might modify useradd's default to use this new partition with useradd -D -d /home2. In this way, new user accounts will have home directories placed in /home2 on this new disk drive.

There are many other ways we could use our new hard disk. We could have used it to extend the /home partition, requiring that we use parted to repartition /home. If we were using an LVM, we could add the new disk to a physical volume group. Yet another choice is to use an external hard disk that we either physically mount to our computer by USB port or mount over the network. In such a case, we would not be able to use this new drive as an extension to /home or as part of our LVM space as it would not be accessible at all times.

SECTION ACTIVITIES

1. Have you ever partitioned a drive on your computer? It is unlikely that you have. Unix/Linux used to require setting up partitions at Linux installation time but recent distributions have either removed this entirely from the user's control or has left it as an option. It is likely that when you installed Linux you can just use the default. Research reasons why you should do your own non-standard partitioning. What did you find?
2. Windows, by default, places all content on one partition, the C: partition. Other partitions are left for other devices like an optical drive or external hard disk attached via a USB port. In Linux, there are at least three partitions, /boot, swap and the LVM. Which approach is better? See if you can come up with a list of reasons for both philosophies.

8.6 ADMINISTRATIVE FILE SYSTEM TASKS

Now that we've learned about partitions and file systems, we can look at different administrative tasks related to file systems (aside from partitioning/repartitioning). We divide this into four categories: mounting/unmounting, setting up file systems for remote mounting, establishing quotas on a file system and miscellaneous administrative tasks.

8.6.1 Mounting and Unmounting File Systems

The most important activity for a file system is to mount it (and possibly later unmount it). *Mounting* is the process of making a file system accessible. Doing so places the file system at a specified mount point. The mount point must already exist as a directory. Access to the file system then is made via that directory.

Mounting is usually handled automatically at system initialization time. The kernel will mount all of the file systems listed in the `/etc/fstab` file. The system administrator may later unmount any mounted file system and also mount a file system that was not automatically mounted. We use `mount` and `umount` to handle these tasks.

At its simplest form, the `mount` command expects two arguments: the file system to mount and the mount point. The mount point's directory can be either a top-level directory or a subdirectory.

Let's see how to mount a new file system named `/dev/sda3`. We want to place it at `/home2`. First, we create `/home2` using `mkdir /home2`. This is an empty directory. We do not use this directory to create subdirectories but instead mount `/dev/sda3` there. We do so with the command `mount /dev/sda3 /home2`.

If we want to use non-default options when mounting a device, we add `-o` *`option(s)`*. We discussed numerous `mount` options earlier in the chapter. We present these in review as well as others in Table 8.12. Even more options and their descriptions can be found in the man page for `mount`.

As noted in Table 8.12, one option for `mount` is `-a` to mount all partitions in the `/etc/fstab` file. The reason for this is that at runtime some partitions may have been unmounted. With `mount -a`, we can mount everything rather than having to mount each partition one at a time. Another option, `-l`, lists all of the currently mounted partitions. Another option is `-t` *`type`* in which we specify the file system type. This option should not be needed for local partitions but should be used when mounting a remote partition using `nfs` (this is discussed in the next subsection).

Although most partitions will be mounted at the time the Linux kernel is initialized, some partitions will need to be mounted later, or unmounted and remounted. Adding an external storage device such as an external hard disk or a USB drive will require mounting. When we have finished using the device, for safety's sake, we would want to unmount it. Or, we may wish to make a

TABLE 8.12
`mount` Command Options

Option	Meaning
`async/sync`	Access is asynchronous/synchronous; asynchronous access does not guarantee that access will be performed in the order that requests were received; this can lead to unexpected results but is also more efficient in terms of disk access performance.
`auto/noauto`	File system is mounted/not mounted at system initialization time; `auto` also causes mounting to take place with the command `mount -a`.
`context`	Specify SELinux context; useful when the device being mounted is one that may not be trustworthy; related are `defcontext` and `fscontext`.
`defaults`	Use the default options (`rw`, `suid`, `dev`, `exec`, `auto`, `nouser`, `async`, `relatime`).
`dev/nodev`	Treat file system as a character or block special device (as in those stored in `/dev`).
`exec/noexec`	Permit/do not permit execution of executable code on this file system.
`owner/group`	Allow a non-root user to mount the file system if the owner/group is the owner/group of the file system's device.
`relatime/ norelatime`	Update/do not update file system inodes' access times relative to modification times; there are variations called `atime/noatime`, `diratime/nodiratime`, `strictatime/nostrictatime`.
`ro/rw`	Permit read-only/read-write access to the contents of the file system.
`suid/nosuid`	Allow/do not allow SUID/GUID bits to take effect; with `suid` also assign the `suid/guid` number.
`user/nouser`	Allow/forbid a user from mounting/unmounting this file system.

particular partition inaccessible for some time, for instance if moving the physical device or performing a backup of the device's contents. The partition would be unmounted temporarily and then remounted.

The `umount` command unmounts a partition. It is simpler than `mount` with the syntax `umount` *`mount-point`*, as in `umount /home/coolstuff`. There are several useful options for `umount`. If we suspect an unmounting will fail, `-r` changes the file system's status to read-only. Option `-l` unmounts the file system now but cleans up the records of its mounting/unmounting later. Like `mount -a`, `umount -a` can be used to unmount all file systems currently mounted but does not attempt to unmount the root file system. Use `-n` to prevent unmounting from being written to `/etc/mtab`. This causes the file to become out of date. Finally, `-O` *`options`* will unmount all file systems that match the specified option(s). For instance, `umount -O ro` will unmount all read-only file systems.

Although `umount` seems easy to use, it is not without complications. To unmount a file system, two conditions must be true. First, we must be root to issue the command. This differs from `mount` where options like `owner` and `user` can permit mounting by non-root. Second, the file system cannot be busy to be unmounted. Let's take as an example an attempt to unmount `/home`. We receive the error message `umount: /home: device is busy`.

A file system is considered *busy* when there are processes currently accessing the file system. This will be the case when users have logged in. In such a case, their `/home` directories become busy. Even when no users are logged in, running processes might have placed a lock somewhere in `/home`. Note that even `umount -f` (force) will not work on a busy partition.

There are two things we can do about this. The first is to identify the processes using `/home` and kill them. Two programs that can identify file system usage are `fuser` and `lsof`. Both take the file system as a parameter (`/home` in this case).

With `fuser`, the output is a list of all PIDs that are accessing the given file system. The PID is followed by one or more letters indicating that the process is using a directory in the file system as the current working directory (`c`), it is an executable (`e`), it is accessing an open file (`f`), it is writing to an open file (`F`), the directory is the process' root directory (`r`) or the directory is used by a shared file (m). If `fuser` provides no output then there were no processes using that file system.

`lsof` lists open files by process. Again, having provided a file system, `lsof` responds with the processes that have open files within that file system. This output is more detailed. First, the process' name and PID are listed followed by the user. This is followed by a variety of information: the file descriptor by name, the type of entity (`DIR`, `REG`, `BLK`, `CHR`, `FIFO`, `LINK` and `SOCK`, among others), device's device number, size of the file or a file offset, node (a unique identifier that might be an inode number, a device node number or some combination) and the location of the file.

Killing off processes found to be actively using files of a file system can be time consuming. If there are no users logged in and yet the file system is busy, another approach is to change to `rescue` mode (what was *runlevel 1* in systems prior to `systemd`). We can do this using either `systemctl rescue` or `telinit 1` (note that `telinit` is an older program which has been replaced by `systemctl` in `systemd`-versions of Linux so that `telinit` is a symbolic link to `systemctl`). Either instruction will shut down our GUI, turn off networking and start up a text-only single-user (root) mode. We will be asked to log in as root. From here, there should be no processes using files in `/home`.

Having killed off all processes using `/home` or having switched to `rescue` mode, we are able to unmount it. As noted above, the command is simply `umount /home`. Upon completing our work, we want to remount the file system. To mount, we need to know the device name. We would have seen this when using the `df` command. For instance, from figure 8.5 we saw that `/home` was physically on the device `/dev/sda3`. If we did not bother to check this or are unsure, we can also view the `/etc/fstab` file. Assuming `/home` is stored on `/dev/sda3`, then we issue the instruction `mount /dev/sda3 /home`. As a last resort, we can also use `mount -a` to remount all file systems. There is nothing wrong with this, but it is slightly less efficient than remounting only `/home`.

If, prior to unmounting /home, we switched to rescue mode, we need to resume the previous mode. We do so by either entering exit or systemctl default. This resumes the previous mode which offers multiuser login, network access and the GUI. We look at systemctl and the various modes of operation in Chapter 9.

We do not have the ability to mount or unmount the swap space, but the swap partition has its own commands of swapon and swapoff. Normally, we would not shut swap space off. But if we wanted to move the location of our swap space partition, we would use swapoff, establish the new swap space and then issue swapon using the new location. When using these commands, the option -a will turn *all* swap spaces off or on (should we have multiple swap spaces).

8.6.2 Remote File Systems

The ability to mount and unmount file systems extends beyond the partitions of the internal hard disk and attached devices. We can also mount and unmount partitions over a network, referring to these as *remote file systems*. In this subsection, we first look at mounting a remote file system and then setting up our own file systems to be remotely mounted.

Let's take as an example that the remote computer www.someserver.com has a directory /home/stuff set up for us to remotely mount. We decide to mount this directory under our own /home/coolstuff directory. In order to do so, three things must be true. First, we have to create a directory under /home called coolstuff if it does not yet exist. Second, for versions of Linux prior to systemd, the nfs service must be running. Third, we issue the mount command.

Our command will be mount -t nfs www.someserver.com:/home/stuff /home/coolstuff. Here, we add the -t option to specify the file system type (nfs). Notice the notation for specifying the remote device. It is the remote computer's name (or IP address), a colon and the location on that machine of the directory to mount. Our mount point is the newly created /home/coolstuff. We can now cd into /home/coolstuff and see what's there.

Let's consider as an example that coolstuff was made available using the option ro and noexec. Any file we open from coolstuff, for instance into vi, will be read-only. We could modify the file in the editor but we would be forced to save the modified version somewhere outside of coolstuff (for instance, in our home directory). With noexec, any binary files in coolstuff could not be run. If we wanted to run such an executable, we would have to first copy the file to a local location.

Now let's consider that coolstuff had the option rw. What does that allow us to do? We would likely still not be able to write content to files or directories in coolstuff because our user account does not have permission to do so. For instance, imagine /home/stuff (the directory in the remote file system) has permissions 755 and owned by root. Even though we have rw access to the mounted file system, we do not have write access to the directory. What if we were root on our local computer? Even then, we cannot write to the remote directory because we are not root of www.someserver.com.

There are three conditions where we could write to a file or directory. The first is that we have an account on the remote machine. This could be the case if this file system is one shared among the users of our own organization. We could then write content to our own directory on that remote file system.

The second possibility is that the mounted directory, coolstuff (or a subdirectory) is writable by world. In such a case, we are free to write content to that writable directory. This permission is unlikely to be set as it would allow anyone who could mount the partition to write to it.

The third possibility is that /home/stuff was set up for export with the option suid (or sgid). In such a case, remote accessors are assigned the UID or GID as specified in the option.

```
firewall-cmd --add-service mountd
firewall-cmd --add-service rpc-bind
firewall-cmd --add-service nfs
firewall-cmd --add-port=2049/tcp
```

FIGURE 8.16 Commands to open the Linux firewall for `nfs`.

If that UID/GID matched the file's/directory's owner, then the remote user would be considered as the owner.

We now consider how we can take one of our own file systems or directories and make it available for remote mounting. The first step is to specify the access point in the file `/etc/exports`. This file denotes the file systems we want to export for remote mounting. The file should already exist but will be empty.

The entries we add to this file list the local mount points, like the previously mentioned `/home/stuff`, along with restrictions of who can mount it and the mount options. The format is *`local_mount_point network_address(es)(options)`*. Each entry is on a line by itself.

The *network_address(es)* allows us to control who can access this local file system/directory remotely. This can be one or more IP addresses or one or more subnet addresses. We can also use `*` to indicate no limitation on access. The options are as we have previously discussed, many of which are listed in Table 8.12. The entry `/home/stuff *(ro)`would make our `/home/stuff` available as a read-only file system to everyone. If not specified, the default options are used (again, refer to Table 8.12).

We have to next execute the command `exportfs`. This causes any changes to the `/etc/exports` file to take effect. There are a few options of note for `exportfs`. With `-a`, all items in the `/etc/exports` file are exported. This is the default. Adding `-o` allows us to specify a list of export options so that we would not have to specify the options in the `/etc/exports` file itself. If we have made updates to `/etc/exports` after having run `exportfs`, we can flush the runtime export table using `-f` and then rerun `exportfs`.

We also have to make sure the service `rpcbind` is running. If it is currently running, we restart it after running `exportfs` so that `rpcbind` is also aware of the changes to `/etc/exports`. The `rpcbind` service accepts incoming requests and passes them on to the appropriate `nfs` service.

Finally, `nfs` operates over port 2049 (by default). Our firewall is likely not set up to permit requests over this port, so we have to update our firewall. We can do this through the four commands shown in Figure 8.16. `firewall-cmd` is the command line instruction to interface with the Red Hat firewall. We explore the firewall in more detail in Chapter 10. We revisit the steps to export a local directory/file system in more detail in Chapter 9 when we look at services.

8.6.3 Establishing Quotas on a File System

A *disk quota* limits the amount of space available to a user or a group. We establish quotas on a per-partition basis. The only partition that makes any sense to establish quotas on is `/home` as we would only want to place quotas on the user's space. Here, we look at steps to establish disk quotas. Before we start, we need to note that the instructions differ between a partition using the `xfs` file system and an older `ext` file system, like `ext4`. There are multiple ways to establish quotas but for `xfs` we use the following.

The first step is that we have to be logged in as root. Next, we have to enable quotas on the given file system. There are three types of quotas available: those placed on users, those placed on groups and those placed on projects. We look at user quotas. We issue a `mount` command to specify `uquota` (group quotas are `gquota` and project quotas are `pquota`) with the instruction `mount -o uquota /dev/sda3 /home`. Here, we again assume that `/home`'s device is `/dev/sda3`. If `/home` is currently mounted, add `remount` before `uquota` or unmount `/home` and remount it with the above instruction.

The next step is to apply quotas. We use the `xfs_quota` program in expert mode, `-x`. Our commands will follow `-c`. Commands include `report` to output the existing quotas, `limit` to set new quotas, `enable/disable` to enable or disable quota enforcement and `off` to turn off quotas. When disabled, quota accounting remains active so that we can still view whether users have reached their quota or not. Turning off quotas means that we have to repeat the `mount -o uquota` command.

There are two forms of quotas: hard and soft. A *hard quota* cannot be exceeded. Attempting to do so (by saving a file when the user has reached his or her quota) will result in an error. Exceeding a *soft quota* results in the triggering of a *grace period* (which defaults to seven days). At this point, the soft limit becomes a hard limit in that the user is not able to save more to their directory. During the grace period, the user can reduce their directory usage to fall below the soft limit. Warnings are issued during the grace period. There are three types of items we can place quotas on: blocks, inodes and realtime blocks (a feature of xfs file systems which we will not cover).

We can apply quotas in two ways. First, we can establish default quotas to all users. Second, we can issue specific quotas to individual users. We will likely want to either use default quotas or establish default quotas on all users and then change quotas for specific users if we have a reason to do so.

To set a default quota for all users, the instruction is `xfs_quota -x -c 'limit` *`quotalimit`* `-d'` *`filesystem`* where *quotalimit* is the limit we are placing on *filesystem*. To adjust this for a specific user, we would issue a similar command of `xfs_quota -x -c 'limit` *`quotalimit username`*`'` *`filesystem`* where *username* is of the user we are placing the quota on. The file system for our example is `/home`.

Let's be more specific. We have 20 users. We want to limit each user to a soft limit of 500 million blocks and a hard limit of 600 million blocks. However, for two users, foxr and zappaf, we want their soft limits to be 600 million blocks and hard limits to be 750 million and 1 billion respectively. We issue the three instructions shown in Figure 8.17.

We can now issue `xfs_quota -x -c report /home` to view the newly established limits. We use a slight variation of this command, `xfs_quota -x -c 'report -h' /home`. The `-h` option shows values in a human-readable format. Notice that because we had multiple items in the command field, we had to embed them in single quote marks. Figure 8.18 shows an excerpt of this report where we see one of the users who has the default value (andersoni) as well as foxr and zappaf.

Notice in Figure 8.18, zappaf has already exceeded his soft limit. zappaf is given a grace period of seven days before any action takes place. As noted above, because the soft limit has been exceeded, 600M (in blocks) now becomes a hard limit for zappaf. He will not be able to save anything new until he frees up space to move below that limit.

```
xfs_quota -x -c 'limit bsoft=500m bhard=600m -d' /home
xfs_quota -x -c 'limit bsoft=600m bhard=750m foxr' /home
xfs_quota -x -c 'limit bsoft=600m bhard=1g zappaf' /home
```

FIGURE 8.17 Using `xfs_quota` to create or change quotas on users.

```
User quota on /home (/dev/sda3)
                             Blocks
User ID   Used       Soft        Hard             Warn/Grace
------------------------------------------------------------
andersoni  153821    500M        600M             00[--------]
...
foxr       581538    600M        750M             00[--------]
...
zappaf    1247215    600M          1G             00 [7 days]
```

FIGURE 8.18 Viewing established quotas.

8.6.4 Miscellaneous Administrative File System Commands

In this last subsection, we look at several other administrative tasks related to file systems. The system administrator should use the various Linux commands available to inspect and explore the state of the file system. This includes looking at free and used disk space, identifying bad blocks and disk errors, ensuring reasonable permissions are being applied by users and protecting the file system.

To view the file system, we have already seen both the `df` and `stat` commands. For a more specific breakdown of disk usage at the file and directory level, use the `du` (disk usage) command. The command receives one or more files/directories. For files, it outputs the size of the item and its name. For directories, it outputs the size of the contents within the directory and its name. It performs this task recursively so that will report on each subdirectory, and their subdirectories, etc. The `du` command generally is of the form `du [options] file(s)`. Common options for `du` are listed in Table 8.13.

The command `du -s dir` provides a summary of the contents of the given directory and so only outputs the total size used rather than a breakdown of the contents within the directory. If we have not implemented quotas on users, we might use `du` to look at user home directories to see if any users are using more storage space than is desirable. For instance, we might issue the command `du -s /home/*`.

Figure 8.19 provides a script that will iterate through a list of directories supplied as parameters, and for each directory, compute the size of all regular files (`-f` as used in the if statement) and output this summary. Although this is similar to `du -s`, here we are only computing the size of the files in the directories. Notice the use of `awk` in the assignment statement to obtain the size. Remember that `du` responds with both the size and the item's name. We cannot use the name as part of the computation of sum so we ask `awk` to return just the first value.

TABLE 8.13
Common du Options

Option	Meaning
`-a`	Provide details on files and directories.
`--apparent-size`	Provide an estimate (quicker).
`-B size`	Provide sizes in blocks of *size*, e.g., `-B 4K`.
`-b`	Provide sizes in bytes.
`-c`	Provide a total at the end of the listing.
`-h`	Provide human readable output (K, M, G, T).
`-L/-P`	Follow/do not follow symbolic links.
`-s`	Summary only, do not list individual file sizes.
`--time`	Add last modification time/date for each item.

```
#!/bin/bash
for dir in $@; do
   sum=0
   for item in `ls $dir/*`; do
     if [ -f $item ]; then
        size=`du -s $item | awk '{print $1}'`
        sum=$((sum+size))
     fi
   done
echo Size of files in $dir is $sum
done
```

FIGURE 8.19 Script to compute file sizes.

Both du and df can give us statistics on file space and file utilization. The stat command provides specific file details on individual files or file systems including, for instance, size, number of blocks, inode, device number, permissions, last access and modification times and dates.

Linux offers several instructions to examine the file systems for errors. The badblocks instruction calls upon other programs to do much of the work. For badblocks, specify a device to scan. This device will be one of the partitions but specified by device name as in /dev/sda3. We can also supply the first and last blocks to examine. Without these, badblocks will examine all blocks. The order to list these is last block and then first block. If only one is listed, it is assumed to be the last block, not the first. Table 8.14 lists some common options for badblocks.

Why might we use -e in badblocks since the point is to locate all bad blocks? If the file system is truly damaged, we may not care to find all bad blocks only to determine that the file system contains some. After seeing some number, it might convince us to take action. Without this option (or with -e 0), badblocks examines the entire file system.

If bad blocks are found, our next step is to use the fsck program to attempt to repair those bad blocks. fsck, file system check, can attempt to both repair file system problems as well as find them. fsck receives a mount point, file system name or UUID as a parameter. We can specify any number of file systems to check. If the file systems exist on different physical disk drives then they can be checked in parallel. If no file system is specified, fsck will examine every file system, in order, as listed in /etc/fstab.

We can also specify the file system type to be tested using -t *type*. For instance, -t ext4 would inspect all file systems in /etc/fstab that are of type ext4. With this option, we can also accompany it with further options that must match that file system to be searched. The list of options -t ext4 opts=ro,nosync would check file systems of type ext4 which are read-only and asynchronous. Other options in fsck are -C (display progress), -M (do not test mounted

TABLE 8.14
Common Options for badblocks

Option	Meaning
-c *num*	The number of blocks to scan at a time, the default is 64.
-e *num*	Force badblocks to exit after *num* bad blocks have been found.
-i *file*	Scan for bad blocks but skip any block numbers which are listed in the file *file*; this is useful if you already know of bad blocks and you want badblocks to avoid scanning those particular blocks.
-p *num*	Force badblocks to make *num* repeated passes; the default is 0.
-s	Show the progress as a percentage of the file system completed.
-w	Use write-mode to test for badblocks; badblocks writes some pattern to each block and then reads the block to see if it can obtain what was written.

file systems), `-N` (do not repair, only report), `-R` (skip the root file system) and `-a` (repair with no user interaction).

`fsck` is itself a front-end which invokes one of several other `fsck` programs. There are numerous versions, one per type of file system type, known as *filesystem-specific checkers*. Among the specific programs are `e2fsck`, `fsck.ext2`, `fsck.ext3`, `fsck.ext4`, `fsck.fat` and `fsck.xfs`. Some of the above parameters will not work on some of these checkers, so consult the specific program's man page for detail. Note that `fsck.xfs` actually does nothing, returning success; it exists merely as a placeholder to be called by `fsck`. The reason for this is that xfs is a journaling file system which will repair itself prior to being mounted. To explicitly test an xfs file system, use `xfs_repair`.

The system administrator must make sure that the file system is protected so that it is available as needed. Timely backups are one approach to protecting the contents of the file systems. A backup, perhaps stored on tape or to a remotely mounted hard disk, allows the system administrator to replace damaged data files by a simple restoration process. Another precaution to make, if the file system should remain accessible as much as possible, is to employ a form of *redundancy*.

Redundancy is supported using *RAID* (redundant array of independent disks). With RAID, multiple disk drives are employed within a single drive unit. Redundancy data is stored along with the data files so that, should a bad block arise, the remaining data plus the redundancy data can restore the corrupted information. Although not a replacement for backups, it is a wise for any organization that uses a shared file server, database server or webserver to use RAID so that errors do not result in costly downtime.

We examine RAID and backup strategies in Chapter 12. But here, we will concentrate on various Linux programs available to support these endeavors. Two programs are `dump` and `restore` to perform backups and restoration, respectively. `dump` will only operate on entire file systems and so is not useful if we wish to backup individual files and/or directories.

By default, `dump` compares the previous backup created to the current state of the file system using last modification dates of each file to see which files need to be stored in an incremental backup. We specify the degree of a backup using a *dump level*. This value is an integer where 0 indicates a full backup and any higher number is used to indicate that updated files backed up with a lower number should be replaced.

The `dump` program produces backup files in 10KB block sizes. We can override this by specifying `-B` *`blocksize`*. The maximum block size is 64KB. We can also have `dump` compress the backup files as they are being saved. We are also able to alter `dump`'s behavior based on the target location of the backup (e.g., remote hard drive, mounted tape drive).

Option `-u` records data of this backup (the files/directories by name) into the file `/etc/dumpdates`. We might issue a command like `dump 5uf` *`destination source`* where *destination* is the location of the backup and *source* is the file system being backed up. The value `5` controls what is backed up and the `f` indicates destination as the backup location/name. Prior to performing the `dump` command, it would be wise to run `fsck` on the file system (as long as it is not a journaling file system) and change to `rescue` mode (or runlevel 1). The `restore` program is able to restore a full backup, an incremental backup (e.g., files backed up since a particular modification date), a specific file system, directory or even a single file.

Older programs for performing backups of files are `tar` and `cpio`. The `tar`, tape archive, program was originally intended to copy files into an archive to be stored on tape for backup purposes. Today, `tar` is just as commonly used to bundle programs together for convenient transmission over the Internet. The `tar` program expects either `-cf` or `-xf` to specify that a new archive should be created (`c`) or a current archive should be extracted (x). The `-f` option indicates that the target or source is a file. When extracted from an archive, we only specify the archive name but when creating an archive, we specify both the file to be created and the files/directories to be archived.

Let's consider a couple of examples. First, we will `tar` a collection of files using the command `tar -cf archive.tar file1.txt file2.txt file3.txt *.dat`. This creates archive.tar which combines the specified .txt and .dat files. As a second example, we take the contents of a directory and place them into a tar file. This instruction is simpler in that the source is only one item, tar -cf archive2.tar /home/foxr. If we later extract the contents from archive2.tar, it creates a subdirectory of /home/foxr in the current directory.

Untarring (extracting) uses -xf as with tar -xf archive2.tar. As noted above, this would create /home/foxr with a copy of all of its contents placed into this new directory. We would not want to untar archive2.tar in /home as it would overwrite the existing /home/foxr directory (unless we were restoring the directory from backup).

The tar command has numerous options. Table 8.15 describes some of the more common ones. Of particular note are -j, -J and -z to compress/uncompress the tarred file which might be an invaluable way to save space on the backup media.

In order to archive to some externally mounted device such as a tape (or a floppy disk), we specify the device's name in place of the destination file. Let's save the contents of /home/foxr to magnetic tape (/dev/tape0). The instruction is tar -cf /dev/tape0 /home/foxr. If we had chosen to save our backup onto floppy disk (/dev/df0), we might add the option -M to inform tar to expect to use multiple floppy disks, asking the user to exchange disks when the currently inserted disk becomes full. This instruction is tar -cfM /dev/df0 /home/foxr.

tar is not used for backups very often these days because it is outdated, but it is used for bundling software that consists of many files and subdirectories into a single file. We return to tar for the purpose of installing open source software in Chapter 11.

The cpio program is newer than tar and can access either cpio-based or tar-based archives. cpio runs in one of three modes: copy-in, copy-out and copy-pass. The copy-in and copy-out modes are similar to extraction and creation operations in tar respectively. This may seem counter-intuitive in that "in" usually denotes input into the program, but here "in" references movement of data with respect to the file system (from archive into file system) and "out" references movement out of the file system (from file system out to archive). The options for copy-in and copy-out mode are -i and -o, respectively. The copy-pass mode is like using copy-out and copy-in combined. Copy-pass takes files from one location and copies them to another, without creating an archive.

cpio in copy-out mode, unlike tar, is interactive. In this mode, we are placed at a prompt where we enter file names. The instruction for copy-out mode is cpio -o -F *filename* where *filename* is the archive to be created. As we type each file name, that file's contents are copied into the archive. We end our input with control+d. A session to create or add to an archive is shown in Figure 8.20.

Notice that we are not given a cursor when dropped into the cpio interactive mode. At each line, enter the name of the file we want placed into the specified archive (new_archive in Figure 8.20). After typing control+d, cpio archives the files whose names were entered. Once done, cpio responds with the archive's size in blocks and returns us to our Bash prompt.

TABLE 8.15
Common Options for tar

Option	Meaning
-A	Concatenate tar file(s) onto an existing tar file.
-d	Use -diff to compare two archives.
-r	Append files onto an existing tar file.
-t	Output just the list of content (file names, directory names) of a tar file, not the content itself.
-u	Append only files newer than the tar file; used for incremental backup; note that this only adds new files, it does not alter or add modified files.
-j, -J, -z	Compress the tar file using bzip2, xz, gzip.
-I *file*	Compress using the executable compression program *file*.

```
$ cpio -o -F new_archive
file1.txt
file2.txt
file3.txt
<control+d>
1843 blocks
$
```

FIGURE 8.20 Interactive session with `cpio`.

If the archive already exists, it is overwritten. We can append to an existing archive by adding `--append`.

The copy-in mode can be used to either output stored content (restore the content) or output the names of the items in the archive. To extract the files from the archive use `cpio -i -F` *`archive`* and to list the contents use `cpio -i -F` *`archive`* `-t` (or `--list`).

Another form of backup tool is `rsync`. This program, while capable of performing backups, is available as a network-based file copying tool. As such, it is similar to Linux programs `rcp` (remote copy) and `scp` (secure copy). The command that we would issue varies depending on whether we are doing a remote or local copy and in which direction.

The basic syntax is `rsync` *`source destination`*. There are three possibilities: a local copy, copy from a remote machine to our local computer and copy from our local computer to a remote machine. *source* will be a single file, files denoted using wildcards, or a directory and *destination* will be a directory.

When copying from a remote computer the source is indicated using the notation *`username@host:source`*, and when copying to a remote computer the destination is indicated using the notation *`username@host:destination`* again with *source* being one or more files or a directory and *destination* being a directory. *username@* can be omitted if the user account is the same on both remote and local computers. We will be asked to provide *username*'s password. If we omit the destination, `rsync` merely outputs the list of course files and does no copying.

Three examples of `rsync` are shown in Figure 8.21. The first example is the same as performing `cp /home/foxr/* /home/foxr/backup`. The second example copies all files from foxr's home directory on 10.11.12.13 to the current directory. The third example does not perform a copy at all but instead displays the contents of that remote directory.

We noted earlier that RAID can be used to help recover from disk errors. RAID can be implemented physically by having a RAID storage device or in software to simulate RAID. In Linux, `md` and `mdadm` are two programs used to create a *virtual RAID* device out of non-RAID storage. `md` is the multiple device driver while `mdadm` is the `md` administrator program. As `md` is a device driver, we have no direct interaction with the program. Details on `mdadm` are covered in Chapter 12.

Aside from backups and RAID, administrators might want to ensure that users have proper forms of permission placed on their files and directories. We can use the `find` command to search for files whose permissions are unwise and alert users that they should modify them. The `find` command `find / -perm 666 -or -perm 646 -or -perm 446` reports on all files whose permissions are set to be world writable.

Note that symbolic links and sockets are world writable as are many virtual system files found in such directories as `/proc` and `/run`. We might want to restrict this search to `/home` or perhaps

```
rsync /home/foxr/* /home/foxr/backup
--------------------------------------
rsync foxr@10.11.12.13:/home/foxr/* .
--------------------------------------
rsync foxr@10.11.12.13:/home/foxr
```

FIGURE 8.21 Example `rsync` commands.

```
#!/bin/bash
     for file in $(ls -R /); do
          if [ -f $file ]; then
               number=`stat -c "%a" $file`
               if [[ number -eq 666 || number -eq 646 ||
                    number -eq 446 || number -eq 466 ]];
               then
                    echo $file $number >>
                    /root/badfilepermissions.txt
               fi
          fi
     done
```

FIGURE 8.22 Script to search for poor permissions.

`/home` and `/etc`. We can add `-exec chmod 660 {} \;` if we want to take the initiative and modify any such files to no longer be world writable (or readable).

As an alternative, the script in Figure 8.22 does much the same thing by recursively searching all directories beneath / for files that are world writable. Here, the `stat` command is used where the option `-c "%a"` returns the permissions as a 3-digit number. The output is redirected to a file in `/root` although again, we could modify the permissions using `chmod`.

Another approach to protecting data files is encrypting the contents. Encryption can be applied at the file system level (i.e., applied to all contents within a file system) or at the individual file level. To encrypt an entire file system, we have to specify this when we set up the partition. Refer back to Figure 8.11 where there is a checkbox to implement encryption. Unfortunately, if we do not set up the file system to use encryption, we cannot later change this. We could create a new partition that is encrypted and copy the contents of the original file system to the new and then delete the old partition but this would be time consuming.

Another option is to provide encryption tools for the user to select to encrypt and later decrypt their own files. There are a number of such programs available such as `xcrypt` and `gpg` (Open PGP). We can also install encryption software like OpenSSL and TrueCrypt. We examine the concept of encryption and Linux-based encryption software in Chapter 12.

The last form of protection is to ensure that a process cannot accidentally or maliciously impact user or system files. The idea is to *jail* a process within a directory. The process then has no ability to access anything outside of that directory. The command is `chroot`.

Consider for instance a webserver which operates on scripts, password files, log files, error files and the web documents. Let's assume the entire collection of webserver files (including its executable programs) is located under `/usr/local/apache2`. The webserver has no need to access files in `/etc`, `/boot`, `/dev`, `/home` or `/var`. When launching the webserver, if we jail it within `/usr/local/apache2` then it is unable to access anything above this directory. This protects our system in that inadvertent or erroneous code cannot damage our system, nor can a hacker use the webserver to attack our system and damage any part of the file space outside of `/usr/local/apache2`.

The syntax is `chroot [options] directory [command(s)]` where *command(s)* is(are) the process(es) jailed in *directory*. In effect, *directory* becomes the root level of the file space that *command(s)* can access. To start Apache (which has a controlling service called `apachectl`), we use `chroot /usr/local/apache2 apachectl start`. There are only a few options available for `chroot` including `--userspec` and `--groups` to indicate the user and/or groups to use for the root of the isolated file system.

The `chroot` command has other useful applications aside from jailing a process for security purposes. We can use `chroot` to create an isolated file system to test code that we are developing. This is sometimes known as a *sandbox*. This is of use if we are testing the software on a system that has other software and data files.

Another use of chroot arises if we are running software which invokes services, files or programs whose names conflict with system names already installed. Using chroot in this case allows the isolated file space to use the same names without the system confusing which specific files/programs are being requested.

SECTION ACTIVITIES

1. Does your school or workplace have quotas on user accounts? If so, what is the quota placed on your account? Do you have a quota restricting your email? If you are unsure, ask a system administrator.
2. Have you ever had a hard disk fail on you? If so, had you backed up your files (or at least your most important files)? Were you able to scavenge files from the hard disk? Do some research on the failure rate of hard disk drives. Is it more or less than you thought it would be?

8.7 LINUX TOP-LEVEL DIRECTORIES

We wrap up this chapter by examining the Linux top-level directory structure, concentrating on files, subdirectories and uses of the various top-level directories. There are different top-level structures depending on the specific distribution of Linux. Red Hat 7 saw several changes from Red Hat 6 for instance. Figure 8.23 shows the File Browser view of the top-level directories of Red Hat 8 (Red Hat 7 will be similar if not the same). Notice that /bin, /lib, /lib64 and /sbin are symbolic links (the arrow icon denotes this) while /root has an X in it (to denote that the directory is not accessible to the current user).

Many of the top-level directories have already established content, placed there at installation time. Among the content are subdirectories. As we examine many of these, we will talk about the uses of each directory and the content we might expect to find there. We break down our view of these directories by partition.

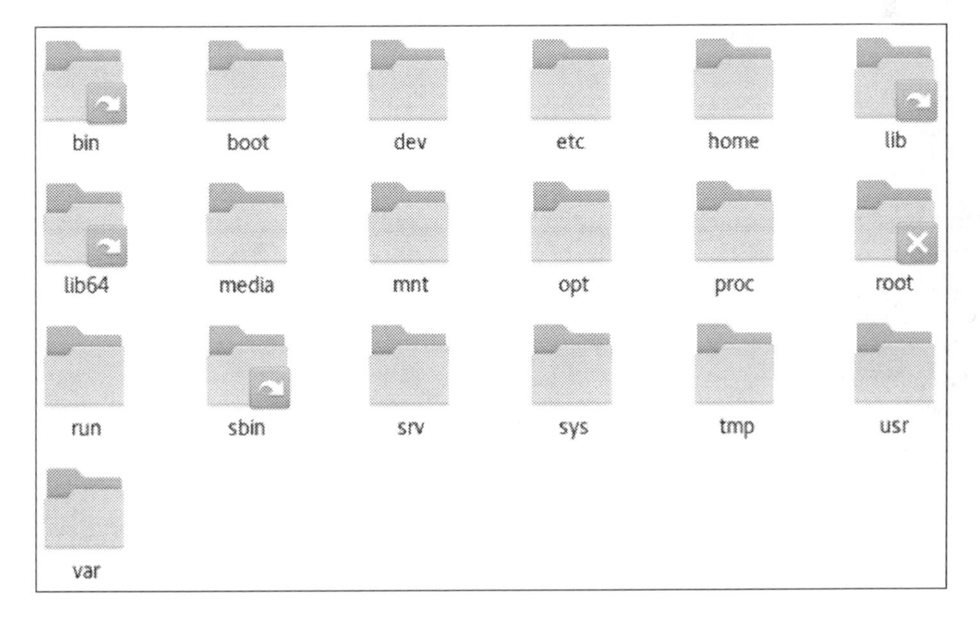

FIGURE 8.23 Top-level directories in Red Hat 8.

8.7.1 Root (/) Partition Directories

The root partition (/, not to be confused with the top-level directory /) typically contains all or most of the operating system's physical files. These files consist of both executable files and data files. We expect to find the top-level directories of /bin, /etc, /lib, /lib64, /lost+found, /mnt, /opt, /root, /sbin, /tmp and possibly /usr and /var in this partition.

The /bin directory stores user programs while /sbin stores administrator programs. Many of the programs in /sbin are executable by normal users but are there more for administrative purposes. There are separate programs found in /usr/bin and /usr/sbin. All of the programs we have discussed throughout the textbook can be found in one of these four directories.

The differentiation of programs found in /bin and /usr/bin, /sbin and /usr/sbin is historical. Unix was a popular platform for workstations in the 1980s and 1990s. In order to keep the cost of these computers down, they often came with no or small hard disk drives. The intention was to have a majority of the software stored on a file server. These programs would be located in /bin and /sbin directories that would be remotely mounted onto the local computer's /usr/bin and /usr/sbin.

Only the essential startup programs were placed locally in /bin and /sbin. The programs expected to be used *after* system initialization were placed in /usr/bin and /usr/sbin. For this reason, /usr may be in the root partition but with /usr/bin and/or /usr/sbin mounted remotely, or if a remote file server were not used then /usr might be in its own partition, separate from the root partition.

The idea of the diskless workstation is outdated and so is this division of programs. In recent Linux distributions, all of the programs previously in /bin have been moved to /usr/bin (or at least those that are still part of the standard installation) and similarly programs from /sbin have been moved to /usr/sbin. Both /bin and /sbin still exist but are now symbolic links to /usr/bin and /usr/sbin, respectively. We would therefore expect /usr to be part of the root partition unless we want a more finely grained separation in our partitions.

In a similar vein, /lib and /lib64 are now symbolic links, and their contents have all been moved into /usr/lib and /usr/lib64, respectively. These directories store shared library files including scripts, configuration files, text files and .so (shared object) files. Also placed into the /usr/lib directory is the systemd directory containing unit files, which we examine in Chapter 9. There was nothing equivalent to do this in earlier versions of Linux. The difference between /lib and /lib64 is that /lib64 is set up specifically to store content for 64-bit applications.

/usr comes with its own subdirectories, somewhat mirroring that of the top-level directory structure. As we've noted, /usr contains its own /bin, /lib, /lib64 and /sbin directories. The top-level directories of these names are now symbolic links to the /usr subdirectories but prior to this change these directories existed both at the top-level and as separate subdirectories of /usr. /usr may also have subdirectories of /games, /include, /libexec, /local, /share, /src and /tmp. The subdirectory local itself has subdirectories that similarly mirror top-level directories just as /usr does.

The /media and /mnt top-level directories serve as temporary mount points. /media is used as the mount point for external storage media like an optical disc drive. /mnt is set aside for the system administrator to mount a remote file system for a short duration.

The /opt top-level directory is set aside for third-party software (which can also be placed in /usr). /root is the home directory for the root account with permissions of r-xr-x--- meaning that only root can cd or ls into it. It is not writable but as root has write access to every file and directory, there is no need to add write-level access to the directory. The reason for these permissions is that root might place password and other secure information here.

/tmp is a directory whose permissions are rwxrwxrwt. Thus, it is writable by all, but the *sticky bit* is set. This directory is made available for software that does not have its own home directory yet needs a place to write data. Most of the contents of this directory are domain sockets so that a process can write to a socket and thus communicate with another process. The /tmp directory is initially empty after every boot/reboot.

We have purposefully skipped discussing two directories. The first is /lost+found. This directory is used to house files that may have been corrupted because they were not closed properly when the system was shut down. Upon booting, Linux examines the contents of this directory and if it finds any files here, it knows that there may have been damaged. It then runs fsck to repair these files.

You might notice that lost+found is missing from Figure 8.23. Once Linux moved from ext4 to xfs as the default file system, this directory became obsolete. Recall that xfs is a journaling file system that ensures that files are properly written to and closed by keeping audit records of every file operation. Should a file not be closed when the computer was shut down, the xfs file system would recover from this mistake at the next system boot without having to resort to examining a lost+found directory.

The last directory is /etc. While /etc is part of the root partition and very important, we dedicate a separate subsection to it due to its complexity.

8.7.2 The /etc Directory

The /etc directory stores configuration files, data files and scripts used by the system. Configuration files are used by root to modify how the system runs. Configuration files and most of the other files in /etc are text files, editable with vi, and many are world readable. The /etc directory also has several subdirectories which contain further groups of configuration and data files.

We have already seen a few of the files in /etc like passwd, group, shadow, fstab, mtab and exports. We will examine yet others in later chapters that deal with such tasks as scheduling, logging and the network.

Table 8.16 lists some of the many files and subdirectories found in /etc. We selected these particular files and subdirectories because we will reference them or we find them to be particularly significant. We have omitted items we have already discussed in this chapter (e.g., exports, fstab, mtab) or Chapter 7 (e.g., passwd, group, shadow, bashrc, profile, skel/). Subdirectories are listed in the table ending with a /. Aside from usage of the item, we list some of its contents. For subdirectories, these include files found within them.

A significant transition took place when Linux moved to systemd. Prior to this time, Linux was based on an older initialization strategy called System V (or later Upstart). Under the System V and Upstart approaches, Linux booted to a specified runlevel. The runlevel defined which services should be running and which services should be stopped.

To control the starting and stopping of services along with other system initialization tasks, Linux relied on a number of scripts which were placed in /etc. Many of these scripts were stored in the subdirectory /etc/init.d, and there were subdirectories like /etc/rc3.d and /etc/rc5.d, one for each of the runlevels. These directories are now empty or nearly empty as the move to systemd and unit files has made them obsolete. We explore the systemd initialization process, with a brief comparison to System V and Upstart, in Chapter 9.

8.7.3 The /boot, /home and /var Directories

We combine these three top-level directories not because they are in any way related but because two of them will take minimal coverage. We also combine them here because they may be in their own partitions, separate from the root partition.

The role of the /boot directory is to store the files needed to boot the operating system. This includes the boot loader program, probably GRUB2, and the Linux kernel, vmlinuz. We omit further detail for now as we will examine the contents of this directory in Chapter 9, Section 9.2.3.

TABLE 8.16
Select Items from /etc

Name	Usage	Types of Content
aliases	File of email aliases.	System user accounts which are aliases of other accounts (primarily root).
anacrontab, crontab	Configuration files for anacron/crond service.	Environment variables and service directives.
at.deny, cron.deny	Usernames who are not allowed to use at/crontab.	Usernames.
alternatives/	Groupings of symbolic links for items moved to /usr.	Varies.
audit/	Subdirectory for auditd.	auditd configuration file and rules.
cockpit/	Cockpit files (if any)	Certificates for secure communication.
cron.d/, cron.daily/, cron.hourly/, cron.monthly/, cron.weekly/	Scripts to be executed by crond or anancron.	Varies but as an example cron.d/ has scripts 0hourly, raid-check, hourly/ has 0anacron and daily/ has logrotate and rhsmd.
cups/	CUPS printer utility files.	Primarily configuration files.
dnsmasq.conf	Configuration file for dnsmasq service.	Configuration directives; the initial file is almost entirely commented out.
default/	Default configuration for grub and useradd.	grub contains directives for GRUB2 and useradd contains default values for useradd.
dnf/	Support files for dnf (replacement to yum).	Configuration files.
firewalld/	Support files for firewalld.	Configuration file and subdirectories containing runtime changes to the firewall.
gdm/	Support files for Gnome Desktop Manager (gdm).	Configuration file and scripts.
hosts	DNS bypass file.	Select list of hostnames and their IP addresses.
logrotate.d/	Configuration files for log files to be rotated.	Each file contains logging directives.
lvm/	LVM configuration files.	Configuration files and subdirectories containing further configuration files and profile files.
modprobe.d/	Configuration files for modprobe (add/remove modules for the kernel).	Various configuration files.
oddjob/	Configuration files for oddjobd service.	oddjob.conf and others.
profile.d/	Startup scripts.	For bash, csh, vi, awk and others.
sysconfig/	Various service configuration files (only found in Red Hat Linux distributions).	Includes configuration files for anaconda, atd, chronyd, crond, firewalld, kdump, rpcbind, rsyslog, samba, snapd, sshd, sysstat and subdirectories containing more.
systemd/	Configuration files to support systemd utilities.	coredump.conf, journald.conf, logind.conf, system.conf, user.conf, among others.
yum.conf, yum/, yum.repos.d/	Support files for yum (even though yum has been replaced by dnf, these directories are still named after yum).	Configuration files and repository files.

The /home directory is the location for all normal user home directories. We say *normal* because root has its own directory (/root) and software accounts either have no home directory or will likely have a home directory in /usr or /var. If we expect to have a lot of users, we might create subdirectories in /home to separate users of different categories into subdirectories, such as faculty and students or divisions within the company. There is little to say about /home aside from what we have already discussed in this chapter about assigning quotas and unmounting the directory to create a backup.

We concentrate the rest of this subsection on the /var directory. This directory contains system and software data files. These data files will change over time, thus the *var* (variable) name. These files are kept separate from the software in /usr for several reasons including the desire to protect the software files in /usr from possibly being accidentally or maliciously written to by running software and to balance out partition sizes. /usr will likely store a lot of content because it stores all of the software while /var is a directory whose content grows (and sometimes shrinks) over time. Having /var on a separate partition allows the contents to increase up until the limit of its size. We would only expect /usr to change when we add or remove software.

As with /usr, /var will most likely have a set structure of top-level subdirectories. These will include account, cache, db, games, lock, log, mail, opt, run, spool, tmp, www and yp. Some of these are worth commenting on. The account directory stores accounting logs (if any). The crash directory stores any recent system crash dumps, if any. The db directory stores various databases of system information. For instance, recent use of sudo will include data here that a user has successfully submitted their password.

The games directory stores any temporary files created by running games. The mail directory stores user email files as used by the default mail program. Variable data created by software running in the /opt third-party software directory is stored in the opt subdirectory. The run directory stores runtime system data since the last boot although starting with Red Hat 7 this directory is now a symbolic link pointing at the new top-level directory /run. We explore /run in the next subsection.

Recall that /tmp (the top-level directory) is used by running software to communicate with each other. But the top-level /tmp directory's contents do not persist across boots. For software that create files that need to persist across boots, there is a tmp subdirectory in /var for this purpose. The yp directory stores network information service (NIS) files, if any.

Now let's focus on some of the remaining subdirectories whose roles are more complex. The /var/cache directory is used to store items generated by various applications that may be needed in the near future. By storing them here, the application does not have to recreate the item. This directory is broken into several subdirectories based on the type of storage. The man subdirectory stores recently accessed man pages. The cups subdirectory stores temporary but "long-lived" files used by cups. fontconfig stores system fonts that have been recently used. The gdm subdirectory stores user GUI configuration settings.

/var/lib contains dozens of subdirectories for the various running applications software and system services. These processes record important runtime data to these subdirectories. These files are not text files and are specific to each running process. Altering the directories or files of this directory will lead to incorrect runtime behavior, and to help protect the system, many of the subdirectories are not world accessible.

The /var/lock subdirectory stores lock files. *Lock files* are the system's way of knowing that a process is running. A lock file's purpose is to ensure that if the same process is running in multiple instances that they do not interfere with each other by attempting to access the same resources. Lock files are empty text files and only exist while the process runs. Starting with Red Hat 7, the /var/lock directory is now a symbolic link pointing to the directory /run/lock.

The /var/log subdirectory stores log files and subdirectories of log files. Log files are usually (but not always) text files storing descriptions of events of note. The files' contents are generated by the running software and managed by one of several services like rsyslog. Log files can be critically important to an administrator in determining what events happened and what events failed. We explore the rsyslog service in Chapter 9 and look at many log files in Chapter 12.

/var/spool stores data awaiting processing. It is divided into subdirectories based on the type of processing. Subdirectories here include anacron, at, cron (scheduling services), cups, lpd (printing), mail and plymouth (system boot splash screen).

8.7.4 Virtual File System Directories

We wrap up our look at top-level directories by examining directories that are not physically stored on disk. These are known as *virtual file systems* and labeled as tmpfs (temporary file systems) when viewed using df or in /etc/mtab. The top-level virtual file systems are /dev, /proc, /run and /sys. These file systems used to be listed in /etc/fstab as file systems to automatically mount but were not listed as having a physical device. Now, these directories do not even appear in /etc/fstab as they will be automatically mounted without needing to be explicitly specified.

The /dev directory stores the interfaces of the devices connected to the computer so that interaction with a device can be treated like a file. The contents of this directory divide into four types of files: character devices (c), block devices (b), subdirectories (d) and symbolic links (l). The subdirectories are groupings of devices, such as block for block devices. Within any subdirectory are symbolic links to the items in the /dev directory. The symbolic links are there because some devices have been renamed and in other cases because they have been moved to either the /proc or /run directory.

The block devices are storage devices (e.g., sda, sdb, etc for entire drives, sda1, sda2, etc for the individual partitions, ht_n for tape drives, fd_n for floppy drives, cdrom for optical disc), LVM device mappers (dm-0, dm-1) and loopback devices (loop0, loop1, etc). The LVM device mappers are assigned to point to each of the logical partitions (if any). A loopback device is an interface for mounting file systems which are not located on block devices. We might use a loopback device to access an ISO file that is not stored on optical disc or to manipulate a file system image without mounting the file system.

Earlier versions of Linux also had numerous ramdisks devices stored in /dev. The *ramdisk* would be stored in RAM (memory) and simulate disk access. The idea was to preload files from disk into memory for quick startup and access. The ramdisk is an outdated idea primarily used in computers of the 1980s and 1990s. If there are ramdisk devices, they will go by names like ram1, ram2, etc.

There are far more character devices than block devices. We describe many of these in Table 8.17. The table lists these by name as stored in /dev. The use of *n* indicates that there are numerous versions of this device where *n* is an integer number. For instance, there may be several printers named lp0, lp1, lp2, etc. and dozens of terminal windows (abbreviated as tty) numbered starting with tty0. The comment field in the table indicates if the item is a program or other useful information.

The /proc directory contains information generated by the Linux kernel about the running processes. Each running process is given its own subdirectory whose name is a number matching the process' PID. An example of this directory is shown in Figure 8.24 where 1, 10, 1041, etc. up through 2225 are process directories. Other directories contain kernel generated information about the system itself such as acpi, bus, driver, fs and irq. The remainder of the content of this directory are files about system processes.

The files in the /proc directory and subdirectories all have sizes of 0 because these are not true files; instead all content is stored in memory. We find that some of the files in /proc are repeated within various subdirectories including cgroups, cmdline, schedstat and stat. The difference is that the files stored directly in /proc pertain to the kernel while the files of the same name in a subdirectory pertain to that particular process. Table 8.18 explores the use of some of the files.

TABLE 8.17
Character Devices in /dev

Name	Use	Comments
autofs	Automatically mounts removable media.	Program
console	Current display (often the same as /dev/tty0).	
fb0	Frame buffer to support video hardware.	
fuse	File system in user space; users can create their own filespaces with fuse.	Software interface
lp*n*	Available printers (possibly connected by network).	
mem	Main memory image.	
null	Linux trash can; used to redirect unwanted output to avoid it being sent to STDOUT, STDERR or elsewhere.	Acts like a file for redirection purposes.
port	I/O port.	
ppp	Bind physical point-to-point protocol device to this file.	
random, urandom	Random number generators; random is considered outdated and only useful for a random number needed during system boot.	Programs; generate binary values not decimal.
sg*n*	Generic SCSI devices.	
tty*n*	Controlling terminal; each is attached to a process.	There are numerous ttys, possibly as many as 64.
vcs*n*, vsca*n*	vcs0 is the memory of the currently displayed virtual terminal; others are virtual console terminals.	

```
1      14    1655  1952  2038  254  8            iomem         partitions
10     1403  1670  1963  2040  26   9            ioports       sched_debug
1041   1404  1672  1965  2041  27   936          irq           schedstat
11     1444  1687  1974  2043  28   937          kallsyms      scsi
1189   1445  17    1975  2044  29   938          kcore         self
12     1484  1757  1977  2045  3    939          keys          slabinfo
1214   15    1763  1979  2047  30   acpi         key-users     softirqs
1260   1566  18    1981  2050  31   buddyinfo    kmsg          stat
1277   1575  1804  1982  2054  32   bus          kpagecount    swaps
1288   1590  1815  1984  2057  348  cgroups      kpageflags    sys
1293   1598  1820  1986  2060  349  cmdline      loadavg       sysrq-trigger
13     16    1830  1995  2086  358  cpuinfo      locks         sysvipc
1300   1606  1839  1997  21    37   crypto       mdstat        timer_list
1301   1617  1847  1998  22    38   devices      meminfo       timer_stats
1318   1630  1848  1999  2225  4    diskstats    misc          tty
13202  1637  19    2     23    40   dma          modules       uptime
13210  1642  1922  20    24    41   driver       mounts        version
13325  1644  1928  2014  243   435  execdomains  mpt           vmallocinfo
13326  1646  1937  2019  245   5    fb           mtd           vmstat
1356   1651  1941  2021  25    6    filesystems  mtrr          zoneinfo
1367   1653  1943  2022  252   7    fs           net
1394   1654  1949  2023  253   71   interrupts   pagetypeinfo
```

FIGURE 8.24 Example contents of the /proc directory.

TABLE 8.18
Select Files Found in /proc and in /proc Subdirectories

Filename	Contents
cmdline	Instruction that launched the process if launched from a command line.
cwd	Link to process's current working directory.
environ	Environment variables of this process (if any).
exe	Link to the process' executable.
fd	File descriptors (if any).
io	I/O utilization information.
limits	Any established limits set on this process.
mounts	Mount information.
root	Link to the root directory of the process.

For the system files, cmdline stores the kernel's startup instruction including all parameters supplied. Other files stored in /proc include information on CPU usage, memory usage, average CPU load, file system usage, mounted partitions and their usage, and virtual memory usage.

Whereas /proc stores information about running processes, /sys stores information about devices and device drivers along with other system configuration data. The /sys directory's file system is called sysfs. In /sys, we find subdirectories for block devices, the bus, firmware, file systems and power management. We also find subdirectories storing crash/core information about the kernel. A quick look at a long listing of /sys shows that, like /proc, the files here have a size of 0.

/sys is a more recent addition to Linux, and it developed from content that had previously been stored in /proc. However, developers felt this content would be better situated in another location. The reason for the change of location is because /proc grew over the years as more and more content was added to it. It reached a point where /proc was felt by some to be a chaotic mess. /sys is an attempt to regain some structure by cleanly separating one category of content (devices/drivers).

Our last directory to examine is also the newest. As noted previously, /var/run has now been made into its own top-level directory, /run. The role of /run is to store runtime system information which describes the state of the system since its last boot. Each user has his or her own /run subdirectory, named using the user's UID. Each running service is denoted with a pid file, storing the service's PID. There is a sudo directory that lists attempts to use sudo. Also, as previously mentioned, the /var/lock directory has been moved to /run/lock. This directory, naturally, contains lock files.

Why was the shift made from /var to /run and how does /run differ from /proc? The answer to the first question is that some of the services and programs that record information in /var need access to the file system early during the initialization process. Since /run is a virtual file system, it is made available before physical mountings take place and so access to /run is available earlier than access to /var. The answer to the second question is that /proc stores statistics about all running processes while /run stores data and only for specific entities (users and system processes). As such, combining the two would make for an inconsistent collection.

We summarize what we have learned here about the top-level directories in Table 8.19. We take a modern view of Linux in this summary by assuming this is the breakdown for the most recent versions of Red Hat (which is mirrored in recent versions of Debian).

TABLE 8.19
Summary of Top-Level Directories

Directory Name	Type of Entity	Role
/bin	Symbolic link.	Content moved to /usr/bin; files are executable programs.
/boot	Physical directory on its own partition.	Boot loader program(s) and configuration; Linux kernel; initramfs images (explored in Chapter 9).
/dev	Virtual file system.	Devices represented as files.
/etc	Physical directory on the root partition.	Configuration files, scripts.
/home	Physical directory on its own partition.	User home directories.
/lib, /lib64	Symbolic links.	Content moved to /usr/lib, /usr/lib64, these directories contain shared library files and systemd unit files (see Chapter 9).
/media, /mnt	Physical directories on the root partition.	Temporary mount points.
/opt	Physical directory on the root partition.	Empty by default, available to store 3rd party software.
/proc	Virtual file system.	Store running process information.
/sbin	Symbolic link.	Content moved to /usr/sbin; files are system administrator executables.
/run	Virtual file system.	Store user and running process data.
swap	Physical file system.	No mount point; contains system swap space.
/sys	Virtual file system.	Store information on devices, drivers and system settings.
/tmp	Physical directory on the root partition.	Used for software communication; world writable; typically stores domain sockets.
/usr	Physical directory; may be on the root partition or its own.	Store all application software (other than any stored in /opt); all content of /bin has been moved to /usr/bin and all content from /sbin has been moved to /usr/sbin.
/var	Physical directory on its own partition.	Variable data generated by running software; /var/lock and /var/run have been moved to the /run directory.

SECTION ACTIVITIES

1. Windows tends to have few top-level directories. Linux has a much more specific breakdown. Now that you have the Linux approach, which do you prefer?
2. As a user, you spend most of your time in your own /home directory and run applications found in /usr/bin and maybe /usr/sbin. As a system administrator, which directory do you expect to use the most? Take a look through the top-level directories that we explored in this chapter to come up with an informed answer. Come up with a list of three reasons why you selected the directory you did.

8.8 CHAPTER REVIEW

Concepts and terms introduced in this chapter:

- B-tree/B+tree – a type of data structure used in many file systems to store disk indexing pointers in a more efficient manner than the outdated file allocation table (FAT).

- Block – a fixed sized unit of storage in the file space; files are typically broken into blocks and distributed across the hard disk surfaces.
- Block device – a type of device, denoted by type `b` in a long listing, that performs input/output on blocks; most storage devices are block devices.
- Block storage – storage at the block level where the operating system must convert file access from files to individual blocks.
- Character device – a type of device, denoted by type `c` in a long listing, that performs input/output on characters; most non-storage devices are character devices, including for instance keyboard, mouse and terminal windows.
- Directory – organizational unit to house files and subdirectories; denoted by `d` in a long listing.
- Domain socket – mechanism to support interprocess communication; denoted by `s` in a long listing.
- ext (extended file system type) – family of file systems supported by Linux; ext is not used but ext2, ext3 and ext4 are all common; Linux switched recently from using ext4 as the default to xfs.
- Extent – a feature of more recent file systems in which disk blocks for a file are allocated in contiguous chunks to prevent the file from being distributed across the disk and thus less efficiently accessible.
- File allocation table (FAT) – used in older Windows (and MS-DOS) operating systems to store the disk block layout to access disk blocks distributed across disk surfaces without requiring numerous disk access; the FAT was loaded from disk into memory upon booting the system or mounting the disk so that searching for a given block was done in memory.
- FIFO – first-in-first-out, an expression used to describe how elements waiting in a queue are serviced; in Linux, a `fifo` is a *named pipe*.
- File descriptor (fd) – designator assigned when a running process opens a file.
- File space – a collection of devices used for storage, typically consisting of an internal hard disk, optical disks and USB drives mounted as needed.
- File storage – a form of storage where files are stored wholly inside of directories; primarily used as network-attached storage rather than internally or externally connected hard disk drives.
- File system – the implementation of a partition that dictates how files will be stored and accessed.
- File type – Linux denotes file types to differentiate between regular files, directories, symbolic links, block devices, character devices, named pipes and domain sockets; the file type is indicated as the first letter of the permissions in a long listed and can also be obtained using the `stat` command.
- Fragment – the portion of a block of a file that is not currently being used.
- Hard limit – when creating a disk quota, the hard limit is the absolute limit which cannot be exceeded; any attempt to store beyond the hard limit results in an error; see also quota and soft limit.
- Hard link – physically stored as the name of the file and its inode number; hard links use the inode number to point to the inode of a file; two files that are hard linked together permit access to the file via either link; see also link count and symbolic link.
- Index – a means of identifying where a file block is to be found; a mapping process is required to convert from a file's block number to the location on disk of that block.
- Indirect block – inodes come with several direct pointers to the first group of disk blocks for the file; the remainder of the disk blocks are pointed to by pointers in indirect blocks; the inode has pointers to indirect blocks, doubly indirect blocks and triply indirect blocks.
- inode – data structure storing file metadata and pointers; the inode is both the indexing mechanism to access the file's disk blocks and a means of maintaining file information.

- Journaling – file system feature in which file operations not yet performed are logged so that, should the access be interrupted, it can be performed once the system resumes to prevent file corruption.
- Link – generic name of a pointer to a file; in Linux, there are hard links and symbolic links.
- Link count – part of the inode metadata that indicates how many hard links are pointing at the inode.
- Lock file – file stored in `/run` to indicate that a program or service is running so that other running instances do not interfere with this one.
- Logical volume – logical volume manager (LVM) manages partitions as logical volumes rather than as physical partitions within the storage device.
- Logical volume manager (LVM) – software means of partition management so that partition sizes can be changed without requiring direct changes to the file system itself.
- Mounting/unmounting – making a file system accessible or inaccessible.
- Mount options – access controls placed on a mounted file system including, for instance read-only versus read/write, synchronized vs asynchronized access, among others.
- Mount point – the logical location of a mounted partition; this will be some directory such as `/opt`, `/mnt` or `/usr/local/mountpoint`.
- Named pipe – file type used to link output of one process to input of another; similar to a Bash pipe but these items are stored in the file system; denoted by `p` in a long listing.
- Network file system (NFS) – a form of file system that permits mounting of partitions over the network.
- Object storage – a newer form of storage where files are objects that contain metadata to support various forms of search for content; primarily used to support cloud storage.
- Partition – a physical division of the file space to protect the contents from other partitions.
- Partitioning – the act of dividing the storage space into independent regions; see also repartitioning and the instruction `parted`.
- Physical extent (PE) – a fixed size unit of storage allocated to a logical volume on demand by the LVM.
- Physical volume – the collection of storage devices pooled together for LVM to manage.
- Pointer to disk blocks – the location on disk of a file's blocks; some indexing scheme is required to maintain these such as FAT, B-tree or inode; in Linux, there are direct pointers, indirect pointers, doubly indirect pointers and triply indirect pointers.
- Quota – limit established by the system administrator on the number of blocks (or inodes) that a given user, group or project is permitted; see also hard limit and soft limit.
- RAID (redundant array of independent disks) – a strategy to protect data by adding redundancy information; RAID is covered in detail in Chapter 12.
- Ramdisk – using memory to store disk files for faster access; ramdisks are used in Linux during system initialization; there may be several ramdisk devices in `/dev`.
- Regular file – Linux treats many entities as files; regular files are true files (data files, executable programs, scripts); indicated by a hyphen (-) in the long listing.
- Remote file system – partition made available over the network.
- Soft limit – a maximum amount of storage permitted in a disk quota before the user is unable to store anything else; upon exceeding the soft limit, the user is given a grace period by which they have to remove content to move below the soft limit.
- Symbolic (soft) link – pointer to a hard link which itself points to an inode; symbolic link takes up less space in a directory and can extend across partitions while hard links cannot; soft links are indicated in a long listing as file type `l` with a name containing the symbolic link's name, an arrow (->) and the location of the file being linked to.
- Top-level Linux directories – standardized directories found in any Linux operating system, all beneath the root level (`/`).

- UUID – device ID number often used to describe partitions, for instance as may be found in /etc/fstab.
- Virtual file system – a type of file system which is stored in memory instead of disk; Linux has several virtual file systems including /dev, /proc, /run and /sys.
- xfs – a more recent type of file system that has become the default for most Linux partitions.

Linux commands covered in this chapter:

- badblocks – locate bad blocks within a particular device or partition.
- chroot – run the given application(s) within the specified directory which becomes the application's top-level directory making directories outside of it inaccessible; a form of security.
- cpio – backup utility.
- df – report on file system usage (amount available, amount used) for all of the mounted or specified partitions.
- du – report on disk usage for given file(s) or directory(ies).
- dump/restore – backup utilities to perform incremental backups and recovery from incremental backups; used in place of cpio for backing up/restoring entire file systems.
- exportfs – export specified file systems to be remotely mounted.
- fdisk – create and delete partitions.
- firewall-cmd – modify Linux firewall (details covered in Chapter 10).
- fsck – file system check; can locate bad blocks and repair files damaged by remaining open at the last system shutdown; this program is a front-end which calls a more specific program based on the type of file system such as fsck.ext4 or e2fsck.
- fuser – file system usage program; displays the PIDs of processes that are using files within the specified file system(s).
- lsof – list open files of the given file system(s).
- lvm2 – handle maintenance on partitions using an LVM.
- mkfifo – create a named pipe.
- mkfs – front-end program which calls specific program like mkfs.ext4 to install a file system on a partition.
- mount/umount – mount specified partition at the specified mount point; unmount partition at the specified mount point.
- parted – utility to display, create, delete partitions and resize existing partitions.
- rsync – remote copy program that can be used to backup file content within a computer or to a remote computer.
- stat – display file statistics (provide details from inodes) or file system statistics.
- swapon/swapoff – turn on/off virtual memory swapping; used to mount/unmount the swap space which has no mount point.
- systemctl – general Linux utility to control system aspects such as services; covered in this chapter to change run mode; see also telinit (systemctl is covered in detail in Chapter 9).
- tar – tape archive, historically used to perform backup to tape but today is more commonly used to create archives of files and directories.
- telinit – change run mode (runlevel); in systemd-versions of Linux, this is now a symbolic link to systemctl.
- xfs_quota – program to establish and manage disk quotas on xfs file systems.

Linux directories and files of note discussed in this chapter:

- /bin/ – location of common binary files (Linux commands and programs); contents moved to /usr/bin and /bin is now a symbolic link to /usr/bin.

- /boot/ – location of boot loader program (e.g., GRUB) and corresponding configuration files, the Linux kernel and any initramfs ramdisks used during system initialization.
- /dev/ – virtual file system and directory storing interfaces to most of the available devices (both physical like hard disk, optical disk, modem and logical like terminal windows (tty), programs like random, and ramdisks).
- /etc/ – stores system configuration files primarily for system administration use; a few notable files and subdirectories are listed below.
- /etc/exports – file system export table for the directories/file systems that can be made available for remote mounting.
- /etc/fstab – file system table specifies mount operations at system initialization time and for `mount -a`.
- /etc/mtab – currently mounted partitions; in recent Linux distributions, this has become a symbolic link to the file in `/proc`.
- /etc/sysconfig/ - subdirectory containing many service configuration files and scripts; found in Red Hat Linux distributions only.
- /home/ – the users' home directory space.
- /lib/, /lib64/ – top-level directories containing runtime libraries and shared object files; `/lib` also contains `systemd` unit files (see Chapter 9); both `/lib` and `/lib64` are now symbolic links with their contents moved to `/usr/lib` and `/usr/lib64`, respectively.
- /lost+found/ – directory used to store file references to files not closed properly at the time the system was shut down; used to restore possibly corrupted files; this directory has been removed with systems using xfs because journaling prevents file corruption.
- /proc/ – virtual file system storing information about all running processes.
- /root/ – the system administrator's home directory.
- /sbin/ – location of system administration executable programs; contents have been moved to `/usr/sbin` leaving `/sbin` as a symbolic link to `/usr/sbin`.
- /usr/ – collection of all software (outside of any stored in `/opt`); previously may have been on its own partition but likely part of the root partition now; `/usr/lib/`, `/usr/lib64/`, `/usr/bin/`, `/usr/sbin/` now combined with the contents formerly in `/lib/`, `/lib64/`, `/bin/`, `/sbin/` respectively.
- /var/ – system data files that persist between reboots; these files typical grow over time; includes application caches, log files, email files, spooler files as stored in such subdirectories as `/var/cache/`, `/var/log/`, `/var/mail/`, `/var/spool/`.

REVIEW PROBLEMS

1. Compare a partition to a file system.
2. Compare the FAT form of indexing to the inode form of indexing.
3. Assume we have the following excerpt of a FAT.

```
Block   500   501   502   503   504   505   506   507   508
Next    613   508   500   505   EOF   501   BAD   781   502
```

   a. One file starts at 501. What are its next three blocks?
   b. What does <EOF> mean?
4. What file systems are available in Linux (note: see Table 8.1).
5. Of the various file systems, which type is used to operate across a network?
6. Which file system is the default used in Red Hat 8?
7. List three types of file systems that use journaling.

8. Match the following Linux file types to the character displayed from an `ls -l` command.

a. Block device	i. -
b. Character device	ii. p
c. Directory	iii. l
d. Domain socket	iv. c
e. Named pipe	v. d
f. Regular file	vi. s
g. Symbolic link	viii. b

9. Provide two differences between a named pipe and a domain socket.
10. Which type of Linux file is used to create a hierarchical structure among the file space?
11. Write a command to create a new named pipe called `pipe1`.
12. Given a named pipe, `pipe1`, you execute `ls -l /etc > pipe1`. What happens?
13. Given a named pipe, `pipe1`, write two commands so that you can run `script1` to fill the pipe and then send its results to `script2` to both process those results and empty the pipe. Assume both scripts are in the current working directory.
14. We have a file called `/home/foxr/foo`. The directory `/home/foxr` is accessible by zappaf. Write two commands to create a hard link to `foo` called `link1` and a symbolic link to `foo` called `link2`, both to be stored in zappaf's home directory.
15. A file's link count is `5`. We delete the file. What impact does this have on the inode? On the file?
16. The file `foo` is pointed to be the symbolic link `mylink`. We delete `foo`. What happens to `mylink`?
17. In Red Hat 6 and earlier, `/sbin/init` was the initial process run once the Linux kernel finished initializing. What is `/sbin/init` now in Red Hat 8?
18. In which directory would you expect to find files whose types are indicated by `b` and `c`?
19. What are the various timestamps found in an inode?
20. Redo Table 8.5 assuming the same layout of pointers in the inode (15 pointers, 12 direct, 1 each of indirect, doubly indirect, triply indirect) whereby a block size is 512 bytes instead of 4096 bytes.
21. True/false: Upon deleting a file, the inode is returned to the file system for reuse and the disk blocks are physically erased.
22. Write a `stat` command to output the file information on the files in `/dev` by file name, type, and time of last access.
23. Write a `stat` command to output the user home directories displaying each directory's name, inode number, owner UID and size in bytes.
24. Write a `stat` command to output for the `/var` file system (assume this is `/dev/sda5`) the total blocks of the file system, total inodes free and the block size.
25. Write a script which obtains the device number for the `/etc/` directory (which will be stored on the root partition) and outputs all top-level directories on the same device. This will inform us which directories are part of the root partition. Hint: to obtain the device number, use `stat` with `%d`.
26. By default, file descriptors are assigned to STDERR, STDIN and STDOUT. What file descriptor numbers are each given?
27. What are the advantages and disadvantages of creating physical disk partitions? Of using an LVM for partitioning?
28. In Figure 8.10, `/dev/sda1` and `/dev/sda6` have values other than `0 0` listed in `/etc/fstab`. What are their values and what do the values represent?
29. In Figure 8.10, why is `/dev/sda2` given a `pri` (priority) value and the other partitions not?
30. What options constitute the `defaults`?

31. How does the file system mounted at /home/coolstuff in Figure 8.10 differ from the other partitions?
32. When do you specify your partitions and their sizes?
33. Your computer has 8GB of DRAM, what is the recommended size of the swap partition as per Red Hat?
34. Of the following top-level directories, which one(s) are recommended to go on partitions outside of the root partition and swap?

    /boot   /etc   /home   /var   /usr

35. You have established partitions but find that the partition for /var is too small. What is the danger in increasing the /var partition's size?
36. In the parted program, what command will you use to display all partitions of the device you are investigating? What command will you use to change the size of an existing partition?
37. If you use an LVM in lieu of partitioning, you will still create two physical partitions.
    a. One of these will store /boot. Why?
    b. What will the other partition store?
38. True/false: /dev/sda is a single hard drive while /dev/sda1, /dev/sda2, and /dev/sda3 are three partitions located on that hard drive.
39. Assume the /home directory is in its own partition and stored on /dev/sda5. Write the command to first unmount /home and then later remount /home.
40. What does the mount option async mean (the answer is not just "asynchronous", explain what this means)? What does the mount option noauto mean? What does the mount option exec mean?
41. You attempt to unmount a file system only to be told it is busy. List two things you can do to make it unbusy.
42. Write a command to mount the remote directory on 10.2.3.15 at /usr/local/stuff, placing it locally at /mnt/other. The mounted partition should be accessible read-only with an SUID of 1000.
43. How does a soft limit differ from a hard limit in a quota?
44. In Red Hat 8, you can place quotas on users, groups and ______.
45. Write two xfs_quota commands to first establish the default that all users have hard limits of 8M inodes and soft limits of 5M inodes and then user foxr has hard and soft limits of 5M/3M, respectively.
46. A user has exceeded his or her disk quota's soft limit. What happens the next time the user tries to save a file if the user has not deleted anything?
47. In the script from Figure 8.19 is the instruction size=`du -s $item | awk '{print $1}'`, explain what this instruction does.
48. What does the fsck program do?
49. If you run fsck.xfs, it always returns success without testing anything. Why?
50. Write a tar command to output the contents of the file /root/repository/stuff.tar without actually extracting the files.
51. What was tar's original intended usage and what is it most used for today?
52. How does rsync differ as a backup utility from tar and cpio? How does dump as a backup utility differ from cpio?
53. Of dump, cpio and tar, which is the easiest to use to perform incremental backups? Of dump, cpio and tar, which is(are) capable of storing individual files, directories and whole file systems in the backup archive?
54. We issue the command chroot /usr/local/mysql mysql. What does this command do?

55. Of the top-level directories in Red Hat 8 (see Figure 8.23), which are actually symbolic links?
56. Of the top-level directories in Red Hat 8 (see Figure 8.23), which are world writable?
57. Of the top-level directories in Red Hat 8 (see Figure 8.23), which are not readable by world?
58. Which top-level directories in Red Hat 8 do you expect to find as part of the root partition?
59. How does /, the root of the top-level directory, differ from /, the root partition?
60. Why had there been a separation of programs between `/bin/` and `/usr/bin/`?
61. What is the main distinction between the programs in `/usr/bin/` and `/usr/sbin/`?
62. Which top-level directory might the system administrator use to store third-party applications software?
63. Starting when xfs became the default for file systems, Linux no longer included the `/lost+found/` top-level directory. Why not?
64. In which top-level directory do you find each of the following?
    a. Most of the system configuration files?
    b. The shared library files?
    c. The runtime data files?
    d. A list of the running processes and information about them?
    e. Domain sockets used for interprocess communication of running processes?
65. Of the top-level directories, which are stored in virtual file systems instead of physical file systems?
66. Between Red Hat 6 and Red Hat 7, `/var/lock/` and `/var/run/` have both changed. What have they changed to and why?
67. What new top-level directory was created starting with Red Hat 7?
68. What is `/var/spool/` used for?
69. What is `/dev/null` and why might you use it?
70. You want to know what instruction was issued at the command line to launch the process with PID of 18531. How do you determine this?
71. What is stored under `/proc/531/exe`?
72. What would you find under the environ file found with every process' data in their `/proc` subdirectories?
73. How does `/sys/` differ from `/proc/`?

9 System Initialization and Services

This chapter's learning objectives are to be able to:

- Explain the boot process and why it is necessary
- Illustrate the Linux kernel initialization process
- Describe the `systemd` startup process and the role of unit files
- Differentiate between Linux service types and roles
- Control services using `systemctl`
- Configure service behavior by modifying configuration files

9.1 INTRODUCTION

The Linux boot and initialization process is well established. It is largely automated requiring little to no system administrator interaction. However, there are reasons for learning the process. If something goes wrong, understanding the process will help you troubleshoot and resolve issues. And, there are some aspects of system initialization that the system administrator may wish to tailor to their needs.

As Linux has become a more complex operating system, so has the startup process. Linux has moved from System V (which dates back to Unix of the 1980s) to Upstart and now `systemd`. As part of our coverage of system initialization, we will make reference to both System V and Upstart in this chapter.

Services are programs run by the operating system in the background to handle on-demand requests of various agents. Services use default configurations that the system administrator can alter. While the operating system is responsible for starting services when needed and stopping them when they are no longer needed, the system administrator can start and stop services as well. Services cover a full range of activities including scheduling, logging, network communication, file system communication, device interaction and power management.

This chapter begins with a look at the boot process for computers. We then focus on the Linux boot process. This is followed by a look at how Linux performs system initialization through `systemd` and the use of unit files. We then turn our attention to services, first looking at the types of services available in Linux and then controlling services. We look at two aspects of controlling services: using `systemctl` to start, stop and query services, and modifying service behavior through configuration files.

SECTION ACTIVITIES

1. If you are a Windows user, open Task Manager and select the Services tab. How many services are there? Can you identify what any of them do?
2. When Linux boots, it outputs messages that describe the steps that it is undergoing. Have you ever looked at this output? What do the Windows and MacOS do during booting? If you have access to a Windows or Mac computer, watch while they boot and compare what you see to the Linux boot process.
3. Have you ever interrupted the boot process before on any computer? If so, why?

DOI: 10.1201/9781003203322-9

9.2 BOOTING THE COMPUTER

We rely on our operating system to present an interface to the user. The operating system is responsible for interpreting user commands, and when those commands involve running programs, the operating system is responsible for locating the executable program within the file space, loading it in memory and starting the program's execution. In order for us then to use our computers, the operating system must be loaded into memory and running.

This presents a paradoxical situation solved by a startup process known as *booting*. The boot process of most computers is similar up to a point. We first look at this paradoxical situation and how the boot program resolves it. In the second subsection, we look at a generic boot process before turning to the Linux boot process in the final subsection of this section.

9.2.1 Volatile and Non-Volatile Memory

The operating system is loaded into random access memory (RAM) memory. RAM is *volatile memory* meaning that it retains its contents only while power is being supplied to it. Shut down the computer and RAM loses its contents resulting in RAM being empty. Turn on the computer and RAM is initially empty.

To run a program, we need the operating system loaded into memory and running. When the computer is first started, memory is empty. How then can we load and run the operating system when we need the operating system in memory and running to load and run programs? If memory were not volatile, we could keep the operating system in memory permanently. But this is not the case.

The reason that RAM memory is volatile is because of the technology we use. There are two forms of RAM memory: SRAM and DRAM. SRAM (*static RAM*) is built out of transistors where several transistors make up each cell (storage location) capable of storing one bit. DRAM (*dynamic RAM*) is made up of one transistor and one capacitor per storage location (bit). SRAM only retains a charge while current is being supplied to it. The capacitor used in DRAM retains the charge but only for a short duration (milliseconds). As the charge dissipates, the DRAM cell must be refreshed with more current. Most modern computers use DDR SDRAM (double data rate synchronous DRAM) chips which require refreshing at a rate of roughly once every 5–20 microseconds. Thus, both SRAM and DRAM lose their contents when power is no longer available.

DRAM is smaller and cheaper than SRAM. Our computers have a large quantity of DRAM storage available and so we use it to build what we call *main memory*. We place DRAM chips on the motherboard. For both economic purposes and room available on a processor chip, we have a much smaller quantity of the faster SRAM. SRAM is used to build *cache* memories and *registers*. See Table 9.1 to see how SRAM, DRAM and ROM differ (we discuss ROM below).

TABLE 9.1
Differences between Memory Types

Type	Volatile or Non-volatile	Typical Amount	Relative Expense	Usage
DRAM	Volatile	4–16GByte	Very cheap	Main memory: stores running program code and data including computer graphics.
SRAM	Volatile	1–64MByte (but usually 8MB or less)	Moderately expensive	Registers and cache memory: stores recently and currently used portions of program code and data.
ROM	Non-volatile	4K or less	Very expensive	Stores unchanging information: the boot program, basic I/O device drivers, microcode.

As noted, both SRAM and DRAM maintain their storage only if there is a continual supply of power from a voltage source. SRAM and DRAM are known as *volatile* forms of memory. Once power is shut off, all contents stored in these forms of memory are lost. Shutting down the computer (or just unplugging it from the power supply) causes memory to become empty. Upon rebooting, memory remains empty and thus we need to first locate and load the operating system.

An alternative form of memory from SRAM and DRAM is called ROM, *read-only memory*. This form of memory is *non-volatile* meaning that when the power is shut off, ROM does not lose its contents. With the power off, ROM is not accessible but once power has been restored, ROM retains the contents that it had previously. To build ROM, the contents (data, program code) are stored permanently. Thus, ROM cannot change or be written to and so we refer to it as *read-only*. We use ROM to help us solve the paradox of restoring the operating system in DRAM upon booting/rebooting the computer. Let's see how ROM is used.

9.2.2 The Boot Process

We need some initialization program which can, upon starting the computer, locate and load the operating system into DRAM. This initialization process is called *booting* (taken from the term bootstrapping). The boot process begins whenever a computer is turned on (cold booting) or the computer is restarted while running (soft or warm booting). We store the boot process (a program) in ROM. However, because ROM is expensive, we actually only store a portion of the boot program in ROM while a majority of it is stored on hard disk.

For any computer, the first step in the boot process is to access the ROM BIOS, the *basic I/O system*. The first task for BIOS is to perform a *Power-On Self-Test* (POST), which examines various pieces of hardware connected to the computer to ensure that they are working properly. Specifically, POST tests the CPU registers, main memory, hardware devices like the interrupt controller, disk controllers and timer, and then identifies all devices currently connected via the system bus (namely the keyboard, mouse, monitor).

Additionally, POST assembles a list of all accessible devices that can be booted from (those devices that could store the operating system). These devices include hard disk(s), floppy disk, optical disk, flash drive and network. The POST step may be skipped during a reboot as these devices are already on and functioning.

Loading the operating system may start automatically. Or, if instructed to by the user, BIOS can present the list of bootable devices and await the user's selection of boot location. Most commonly, the list of bootable locations is pre-enumerated and ordered. The user can later alter this ordered list to change the priority by which the devices are tested. Usually, the network is the last on the list and the hard disk is second to last. In this way, a user can override booting from the default location (which is probably hard disk) if an operating system is present and bootable from USB flash drive, optical disc or floppy disk.

As BIOS is not alterable (because it is stored in ROM), computers come with another form of memory, complementary metal-oxide semiconductor (CMOS). CMOS runs off of a battery placed on the motherboard. The reason to provide this form of memory is to maintain a prioritized boot ordering which can be altered by the user if desired. Thus, upon making a change to the boot order, CMOS remembers the change because ROM cannot. CMOS may also store other alterable boot information and include a clock of the current date and time. The battery ensures that CMOS retains its contents at all times.

9.2.3 The Linux Boot Process

Linux begins its boot operation just like most other computers. Upon receiving power, ROM BIOS is accessed and POST runs. This process is hardware-dependent rather than operating system-dependent. After POST, the *boot loader program* runs. BIOS contains the initial boot loading instructions responsible for locating the remainder of the boot loader program (typically stored on our hard disk).

The very first sector of the hard disk is set aside for the *master boot record* (MBR). The MBR contains this boot loader program (or a portion of it) as well as a partition table and a magic number. The magic number is used for a validation check only. The partition table contains the location on disk of active partitions where bootable operating systems can be found. The MBR is 512 bytes in length. The first part of the MBR consists of 446 bytes that stores a part of the boot loader program. At this point in running the boot loader in the MBR, the boot loader program shifts the boot process from running program code in ROM to running code from hard disk.

For Linux computers, there are three commonly used boot loaders: GRUB and GRUB2, both of which can be used to load different types of operating systems (e.g., Windows and Linux or two versions of Linux) and LILO, which can be used to load different versions of Linux. These three boot loader programs are too large to fit in the 512 bytes available, so the boot loaders are divided into parts generally thought of as the installer and the kernel loader.

LILO (Linux Loader) is the oldest of the boot loaders. It can boot to an operating system stored on either floppy or hard disk, and it can select any of up to 16 different boot images. LILO is file system independent which is both a weakness (GRUB/GRUB2 are able to access the Linux file system to obtain configuration information) and a strength (LILO is simpler). It can also be parameterized based on the boot image selected.

We can alter LILO's behavior (once Linux has booted) by editing the file `/etc/lilo.conf`. This will change how LILO works in future boots. This configuration file stores global options, such as the boot location, the type of installation (e.g., menu or text-based) and the location of file system mapping information. The remainder of the file contains specific configuration information for each of the boot images.

GRUB (GRand Unified Bootloader) is considered obsolete and has largely been abandoned in favor of GRUB2. We compare the two momentarily, but before that let's see how GRUB2 operates. GRUB2 has three stages. The first stage, which executes immediately after POSTfinishes, locates, loads into memory and begins execution of the remainder of the boot loader program. In GRUB2, this program is stored as `boot.img`. This program constitutes stage 1.5. `boot.img` contains device drivers so that GRUB2 can access different types of file systems (EXT, FAT, NTFS).

By providing access to the file system, GRUB2 is able to access the `/boot` partition (which stores the `/boot` top-level directory). Now, GRUB2 loads a default configuration file (stored under `/boot/grub2`) in order to present to the user a list of kernels to select between. The user can also use this interface to drop into a GRUB command line. From here, options include changing kernel parameters, enabling or disabling kernel modules and editing or replacing boot configuration files. Once the user has selected a kernel to boot to, GRUB2 transitions to stage 2. In stage 2, GRUB2 accesses `/boot`. In this directory is the Linux kernel, `vmlinuz`. Stage 2's role is primarily to locate the kernel, load it into memory and begin executing it.

A less common boot loader is called loadlin. This boot loader runs under either DOS or an older Windows operating system. This allows us to boot to Linux from DOS/Windows as opposed to the more traditional boot loader which runs during the boot process. loadlin replaces the current operating system image (of DOS or Windows) with the Linux kernel. As such, loadlin is more useful if we desire booting to Linux occasionally from a fully booted Windows machine. Note that loadlin will not operate with anything newer than Windows ME.

As GRUB2 is the most commonly used boot loader, we will not delve into further details on either LILO or loadlin. But let's compare GRUB and GRUB2. The main difference between the two is that GRUB did not offer easy facilities to change the boot configuration. In GRUB2, this can be done from the GRUB2 interface, as noted above, or by modifying the configuration file, `grub.cfg` (note: the placement of this file varies by distribution but might be found under `/boot/grub2`, `/boot/efi/EFI/redhat` or `/boot/efi/EFI/centos`). The `grub.cfg` file is a shell script which is responsible for setting various environment variables and loading kernel modules using the insmod instruction.

GRUB2 configuration uses a more expressive scripting language than GRUB, permitting an easier and greater degree of configuration modification. It is recommended that rather than modifying grub.cfg directly, any modification be made to the file /etc/default/grub. This is because updates to kernel will replace grub.cfg and any modifications made to that file will be lost. /etc/default/grub is not replaced during a kernel update. This file is accessed by GRUB2, along with grub.cfg during boot. From an already booted Linux, we can also modify GRUB2's boot behavior using the grubby program.

Another difference between the two boot loader versions is that GRUB uses physical and logical disk addresses whereas GRUB2 uses universally unique identifiers (UUIDs). This makes GRUB2's ability to handle initial file system access more reliable. UUIDs also enable GRUB2 to access file systems implemented using an LVM or using RAID software.

Figure 9.1 shows the default version of the configuration specified in /etc/default/grub. The GRUB_TIMEOUT directive specifies for how long the GRUB2 splash screen is displayed. Use 0 for GRUB2 to be "hidden". GRUB_DISTRIBUTOR is used to create a string for output. It is cryptic looking because it uses sed-like notation (refer back to Chapter 5 for a look at sed). GRUB_DEFAULT specifies the default kernel to select. It can either be a number, indicating which entry in the menu of options, or the value saved, indicating that the value specified using GRUB_SAVEDEFAULT should be used. If the latter is not defined, it uses the value specified in the grub.cfg file.

Skipping down the list of directives in Figure 9.1, GRUB_COMMAND_LINUX specifies the options to use in the command to launch the selected kernel. Here, options specify what should happen if the kernel crashes, how to resume from a kernel crash, the name/location of the LVM and entries rhgb (boot to GRUB's GUI) and quiet (disable boot messages). There are dozens of other directives available for the GRUB2 default configuration file. If you ever find yourself needing to modify GRUB2's boot behavior, it would be best to research GRUB2 thoroughly and then create a copy of this default configuration file before attempting to make any changes.

Boot process events are recorded in the *kernel ring buffer*, which provides log storage during system initialization time. Once the system is up and running, boot events are then logged to the file /var/log/boot.log. The buffer also stores later messages generated by the kernel. As the buffer is of a fixed size, once filled new messages cause older messages to be discarded. Therefore, you might find the messages of the ring buffer to diverge from what is stored in /var/log/boot.log.

We can view the ring buffer with the command dmesg. The boot.log file, when it becomes too large, is rotated so that older messages are saved. Also, of note is the /var/log/messages log file, which stores messages generated by the operating system but only after boot. To trace down an event that may have occurred during boot or system initialization, we might need to search both log files and the kernel ring buffer.

```
GRUB_TIMEOUT=5
GRUB_DISTRIBUTOR="$(sed 's, release .*$,,g'
     /etc/system-release)"
GRUB_DEFAULT=saved
GRUB_DISABLE_SUBMENU=true
GRUB_TERMINAL_OUTPUT="console"
GRUB_COMMAND_LINUX="crashkernel=auto
     resume=/dev/mapper/cl-swap rd.lvm.lv=cl/root
     rd.lvm.lv=cl/swap rhgb quiet"
GRUB_DISABLE_RECOVERY="true"
GRUB_ENABLE_BLSCFG=true
```

FIGURE 9.1 The grub.cfg file.

```
[    0.000000] Linux version 4.18.0-257.el8.x86_64 (mockbuild@kbuilder.bsys.centos
.org) (gcc version 8.4.1 20200928 (Red Hat 8.4.1-1) (GCC)) #1 SMP Thu Dec 3 22:16:
23 UTC 2020
[    0.000000] Command line: BOOT_IMAGE=(hd0,msdos1)/vmlinuz-4.18.0-257.el8.x86_64
 root=UUID=d038363f-9014-40fd-87e2-8e4a85e0f8a2 ro crashkernel=auto resume=UUID=85
00fcd8-e934-46c3-b8aa-eb1204c1e984 rhgb quiet
[    0.000000] Disabled fast string operations
[    0.000000] x86/fpu: Supporting XSAVE feature 0x001: 'x87 floating point regist
ers'
[    0.000000] x86/fpu: Supporting XSAVE feature 0x002: 'SSE registers'
[    0.000000] x86/fpu: Supporting XSAVE feature 0x004: 'AVX registers'
[    0.000000] x86/fpu: Supporting XSAVE feature 0x008: 'MPX bounds registers'
[    0.000000] x86/fpu: Supporting XSAVE feature 0x010: 'MPX CSR'
[    0.000000] x86/fpu: xstate_offset[2]:  576, xstate_sizes[2]:  256
[    0.000000] x86/fpu: xstate_offset[3]:  832, xstate_sizes[3]:   64
[    0.000000] x86/fpu: xstate_offset[4]:  896, xstate_sizes[4]:   64
[    0.000000] x86/fpu: Enabled xstate features 0x1f, context size is 960 bytes, u
sing 'compacted' format.
[    0.000000] BIOS-provided physical RAM map:
[    0.000000] BIOS-e820: [mem 0x0000000000000000-0x0000000000097bff] usable
[    0.000000] BIOS-e820: [mem 0x0000000000097c00-0x000000000009ffff] reserved
[    0.000000] BIOS-e820: [mem 0x00000000000ce000-0x00000000000cffff] reserved
[    0.000000] BIOS-e820: [mem 0x00000000000dc000-0x00000000000fffff] reserved
```

```
------------ Wed Jan 27 07:32:40 EST 2021 ------------
[  OK  ] Started Show Plymouth Boot Screen.
[  OK  ] Reached target Paths.
[  OK  ] Started Forward Password Requests to Plymouth Directory Watch.
[  OK  ] Found device VMware Virtual NVMe Disk 5.
[  OK  ] Found device VMware Virtual NVMe Disk 2.
[  OK  ] Reached target Initrd Root Device.
         Starting Resume from hibernation us…cd8-e934-46c3-b8aa-eb1204c1e984...
[  OK  ] Started Resume from hibernation usi…0fcd8-e934-46c3-b8aa-eb1204c
1e984.
[  OK  ] Reached target Local File Systems (Pre).
[  OK  ] Reached target Local File Systems.
         Starting Create Volatile Files and Directories...
[  OK  ] Started Create Volatile Files and Directories.
[  OK  ] Reached target System Initialization.
[  OK  ] Reached target Basic System.
[  OK  ] Started dracut initqueue hook.
[  OK  ] Reached target Remote File Systems (Pre).
[  OK  ] Reached target Remote File Systems.
         Starting File System Check on /dev/disk/by-uuid/d038363f-9014-40fd-87e2-8
e4a85e0f8a2...
[  OK  ] Started File System Check on /dev/disk/by-uuid/d038363f-9014-40f
d-87e2-8e4a85e0f8a2.
         Mounting /sysroot...
```

FIGURE 9.2 Initial output from dmesg and the /var/log/boot.log file.

Figure 9.2 compares the first portion of the kernel ring buffer (via dmesg) with the first contents from /var/log/boot.log. The kernel ring buffer is shown in the top portion of the figure and the boot.log excerpt in the bottom portion (note that these come from two different boots). You can see from these excerpts that both show different types of events. The ring buffer shows such events as running ROM BIOS, testing memory, testing other hardware such as the system bus, cache, I/O interfaces and the system clock. The boot.log file captures the messages that are displayed on the console while booting takes place. These messages include starting the file system and access to the /boot directory (as shown in the figure) but also starting services, sockets and various hardware devices. Note that both of these figures come from CentOS 8.

Before we continue with the boot process, let's take a closer look at the /boot directory's contents. Here, we concentrate on Red Hat 8. The items are shown in Table 9.2.

TABLE 9.2
Contents of the /boot Directory

Item	Explanation
config-...x86_64	Kernel configuration file (the ... is a version number, omitted for brevity here).
efi	Subdirectory containing EFI (extensible firmware interface) programs and the grub.cfg script).
grub, grub2	Subdirectory containing grubenv, a text file storing the GRUB environment block (the grub subdirectory is only available in Red Hat 7, not Red Hat 8).
initramfs images	Disk images of the initramfs file system (explored in the next subsection); the actual images vary by specific distribution but include one file system to be used during rescue operations and one for kdump.
loader	Directory storing boot loader configuration files (not available in Red Hat 7).
System.map-...	The system map is a text file storing the symbol table mapping symbolic names to memory locations of various kernel objects such as cpu_core_map, cpu_info, irq_reqs, numa_node and free_pagetable to list just a few (there are nearly 100,000 entries in the System.map for Red Hat 8); the ... is the kernel version number.
vmlinuz-...	The Linux kernel where ... is the kernel version number; there may be multiple kernels available including a rescue kernel and a normal kernel.

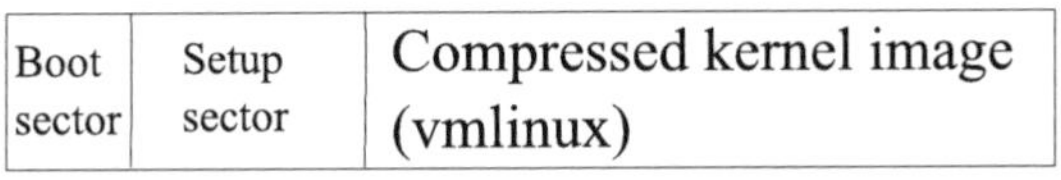

Boot sector	Setup sector	Compressed kernel image (vmlinux)

FIGURE 9.3 vmlinuz file.

At this point, the Linux kernel has been located. The Linux kernel is stored on hard disk as the file vmlinuz (the full file name includes the version number and architecture type, for instance vmlinuz-4.18.0-193.28.1.el8_2.x86_64). The kernel is loaded into memory. The first portion of the kernel image is executable, and it contains two parts: a 512-byte boot sector and a kernel setup program. As the kernel setup portion runs, it performs some basic initial hardware setup. It then uncompresses the latter portion of the vmlinuz file into vmlinux, the rest of the kernel. See Figure 9.3.

It should be noted that the Linux kernel can be stored on a USB drive or optical disk and inserted prior to (or during) the boot process as vmlinuz is bootable from these sources. Once the uncompressed portion of the kernel has completed loading into memory, the process shifts from booting to kernel initialization followed by system initialization.

9.2.4 Loading and Running the Linux Kernel

Kernel initialization begins with additional initialization of hardware beyond that of POST. The kernel searches for and loads an initial ramdisk image and device maps of the hard drives. The ramdisk image, initramfs, is stored in the /boot directory. In Red Hat 8, there are (at least) three different initramfs images available: one version supporting the rescue kernel, one version supporting the kdump service (used if the kernel should crash during initialization) and one version supporting the normal kernel. Also accessed at this time are buses and through the buses the various components connected to the computer (e.g., monitor, keyboard, memory, disk controller(s), timer, plug and play devices). Finally, the kernel sets up interrupt request (IRQ) handling mechanisms.

The reason for initramfs is to set up a file system that the Linux kernel can access while the system initializes because, at this point, the normal file system has yet to be mounted. initramfs is moved from disk to memory for efficient access, thus the use of the ramdisk. `initramfs` contains directories some of which will look familiar because they are similar to the Linux top-level directory structure: `bin`, `dev`, `etc`, `lib`, `loopfs`, `proc`, `sbin`, `sys` and `sysroot`. This allows the kernel to access a file system of programs, modules and data files without having to mount any part of the normal Linux file system. This file system is only stored temporarily and removed from memory midway through kernel initialization.

The `initramfs` directories are minimal, containing only those executables and configuration files needed to finalize the kernel initialization process. The `/dev` directory, for instance, permits the kernel to communicate with the hardware devices needed during hardware initialization. The `/bin` and `/sbin` contain the executable programs necessary for the kernel to start virtual memory and mount the root partition.

Once the root file system has been mounted, the kernel executes a command called `pivot_root`. This alters the root partition from `initramfs` to / and removes `initramfs` from memory. At this point, only the root file system is mounted, consisting of those directories not partitioned separately. The remainder of the file systems, as listed in `/etc/fstab`, are not yet mounted.

To complete the boot process, the kernel finds and executes the `systemd` process. This program is stored in `/usr/lib/systemd`. `systemd` is the latest Linux startup process and differs dramatically from earlier programs that used the System V (SysV) approach. The previous startup process, `/sbin/init`, is still present but is merely a symbolic link to `systemd`. This is available as a fallback should some software call upon `init`. Once `systemd` begins to run, the remainder of the operating system is made ready for users.

SECTION ACTIVITIES (FOR THESE ACTIVITIES, CD TO THE /BOOT DIRECTORY IN LINUX)

1. How many kernel files (these all start with `vmlinuz`) are stored there? How many `initramfs` images are there?
2. Under `grub2/efi/EFI/centos` you will find two grub configuration files, `grub.cfg` and `grubenv`. The latter defines some environment variables. What variables are defined there? The former is a script. Step through this script to see how much you understand from what you learned in Chapter 6. `load_env` is a grub instruction to load environment variables while `save_env` saves environment variables to a file and search is a grub command to search for devices based on options. `insmod` is a Linux command to load a kernel module. All of the set commands are assignment statement.
3. You might notice that significant parts of grub are located in a directory called `efi`. What does EFI stand for and what is its significance with respect to booting?

9.3 INITIALIZATION OF THE LINUX OPERATING SYSTEM

We start this section not with `systemd` but with a look back at the earlier initialization processes, starting with System V's. System V (SysV for short) was the fifth major release of Unix by AT&T, first made available in 1983. SysV launched the program `init` to perform system startup. `init` would launch to a pre-specified runlevel using a configuration entry in the file `/etc/inittab`. The *runlevel* dictated which services would run (and which wouldn't).

Each runlevel established a different type of environment. Runlevel 1 would only launch services to support the system administrator with no other users, no network connectivity and no GUI. Runlevel 2 added multiple users but no network or GUI. Runlevel 3 would add network access. Runlevel 5 would provide full access (including the GUI). Two further runlevels were identified to immediately shutdown the system (runlevel 0) and reboot (runlevel 6). Runlevel 4 was not used, being reserved for "future use".

The starting and stopping of services was controlled by `init`, running various startup scripts. These scripts were stored in `/etc/init.d`. One such script would call upon a runlevel startup directory where the entries were of the form `K##name` and `S##name` where *name* was the name of a service. The `S` denoted that the service should be started, and the `K` denoted that the service should be stopped. The *##* portion of the filename was a 2-digit number used to specify the order that services would be stopped or started.

The directory selected matched the runlevel to initialize to. These directories were located beneath `/etc` and given names like `rc1.d` and `rc5.d` (in actuality, these subdirectories were located beneath `/etc/init.d` where `/etc/rc5.d` and others were symbolic links). The content of each runlevel directory was a set of symbolic links, each pointing to a file in `/etc/init.d`.

The files of `/etc/init.d` were scripts, each used to control a specific service such as `/etc/init.d/atd`, used to control the `atd` service. The scripts contained function definitions, each of which would perform some task on the service like starting it, stopping it or obtaining status information on the service (e.g., like whether it was running or stopped). The remainder of these scripts contained a case statement that, given a parameter, invoked the appropriate function(s). For instance, if a runlevel called upon starting the `nfs` service, the code would invoke the service script and pass it the parameter `start`.

When Linux was implemented, most distributions adopted the SysV approach. But SysV is not without issues. Perhaps the most significant issue was that SysV, through `init`, performed all operations in a pre-specified order. If an operation was slow to start or hung, it paused the initialization process, possibly indefinitely. As an example, today's computers are expected to connect to a network during system initialization. If the network was unreachable at the time of system initialization, then `init` would pause and perhaps never resume. Thus, the system would never complete its initialization process and reach a usable point.

An improved startup routine was implemented called `Upstart`. `Upstart` could operate asynchronously so that tasks may not necessarily start in a proscribed order. Thus, a step that was taking some time to start would not postpone other steps. A system could still be brought up to usability if the network was not responding, as an example. `Upstart` still ran the `init` process and used runlevels. `Upstart` was initially adopted for Ubuntu 6.10 and later adopted for Red Hat 6 and Fedora 9, among others. Debian never adopted `Upstart`.

Upstart has since been abandoned in favor of `systemd`. Fedora 14 was among the first Linux distributions to adopt `systemd`, which was later adopted by Debian 8, Red Hat 7 and Ubuntu 13. The `systemd` initialization process is completely different from SysV/Upstart, for instance eliminating runlevels, `init` and startup scripts. The program `init` (previously found in `/sbin`) has been replaced with a symbolic link from `/usr/sbin/init` to `/usr/lib/systemd/systemd` while most of the former `/etc/init.d` content is missing.

It is `systemd` that is the first process invoked by the Linux kernel rather than `init`. Thus, `systemd` will always have a PID of 1. `systemd` uses unit files to determine how to control system initialization. In this section, we look at the `systemd` startup process and the use unit files. For more information about `Upstart` and the SysV approach, see the supplemental readings on the `init` initialization process.

9.3.1 Target Unit Files

Among `systemd`'s first tasks is to mount all file systems listed in `/etc/fstab`. This will include making a swap space available. Also mounted at this point are the virtual file systems of `/proc`, `/run`, `/sys` and `/tmp` (recall from Chapter 8 that these virtual file systems are no longer listed in `/etc/fstab`). Now `systemd` is responsible for launching a series of services, but which services and in what order?

TABLE 9.3
Unit Types

Type	Usage/Meaning
Automount	Configure a mountpoint to be automatically mounted; each automount unit must have a corresponding `mount` unit; these will include entries in `/etc/fstab`.
Device	Device is managed by `systemd` including any in the `sysfs` file system; a device referenced via `mount` needs to be specified with device unit files.
Mount	Configure a mountpoint for any file system to be managed by `systemd`.
Path	Path to be monitored by `systemd` for activity; predefined paths exist for `/etc/systemd/system.control/*`, `/etc/systemd/system/*`, `/run/systemd/system.control/*`, `/run/systemd/system/*` and `/usr/lib/systemd/system/*`.
Scope	A collection of system processes, known as a *cgroup*, created externally to `systemd` and related to each other; a scope can be managed as a group rather than having to manage individual processes; we explore the `cgroup` later in the text.
Service	Operating system process to handle requests from a number of different sources; we cover service unit files separately in the next subsection and services in Sections 9.4–9.6 of this chapter.
Slice	Resources allocated to a collection of processes arranged in a `cgroup`.
Snapshot	Runtime state of the machine stored for later recall or to be rolled back to.
Socket	A socket unit is assigned to listen to either a predefined domain socket or a network port to help facilitate interprocess communication; every socket has a predefined `.service` file such that the service will be started automatically if activity is seen over the corresponding socket.
Swap	Define a swap space.
Target	Specify startup details; covered in detail throughout this subsection.
Timer	Used by `systemd` to handle scheduled and delayed service events.

SysV's `init` and `Upstart` used runlevels to determine which services should start. `systemd` takes a radically different approach in selecting services to start and stop. This is handled through targets. A target is configured through its own `.target` file. The target is one of several types of unit. *Units* are types of system-level programs and resources. Each specific unit is configured via its own *unit file*. Table 9.3 describes the types of units. Note that a unit file will end with an extension of `.type` where *type* is the unit type, such as `.device`, `.service`, `.socket` or `.target`.

We will look at some of these units in later portions of the chapter, but for now we will concentrate just on the target and target files. What is a target? A *target* describes synchronization points during system startup. Such a point represents a time during the initialization of the operating system whereby certain activities must have taken place and other activities that must take place afterward. The target defines startup dependencies. *Dependencies* specify an ordering by which other targets should run and services need to be started or stopped.

By specifying that target A depends on target B means that for target A to run, target B must first run. If target B depends on service C then service C must be started so that target B can run. Another target, D, to be run after A, may require that C not be running and so C would be started before B and stopped before D. It is through these dependencies that the order that services are started or stopped is decided. Targets replace the older runlevels of `init/Upstart` although the old runlevels map roughly to new targets. We elaborate on this shortly.

Let's start our look at target files by looking at a specific target file. We then explore the contents of target files more generally. Figure 9.4 displays the content of the `graphical.target` file which is, by default, the first target file applied during system initialization. Note that comments from the file have been omitted for brevity. The `After` directive should be on one line but is wrapped around because of the limited width.

```
Description=Graphical Interface
Document=man:systemd.special(7)
Requires=multi-user.target
Wants=display-manager.service
Conflicts=rescue.service rescue.target
After=multi-user.target rescue.service rescue.target
      display-manager.service
AllowIsolate=yes
```

FIGURE 9.4 The `graphical.target` unit file.

`Upstart/init`, the previous initialization programs of Linux, used executable scripts to start and stop services. `systemd`, instead, parses the directives of the target file to determine what needs to be done. In the case of `graphical.target`, `systemd` is told that `multi-user.target` is needed via the `Requires` statement. Further, it specifies that `graphical.target` must start *after* `multi-user.target`. It may be unintuitive that the directive `After` pertains to this target file (`graphical.target`) rather than those listed, but the way to view it is that this target cannot run until those listed in `After` are started.

Notice that listed in `After` are both `rescue.service` and `rescue.target`. Another directive of this target file is `Conflicts`. A target cannot run if the unit files in the `Conflicts` list are currently running. This leads to an interesting situation. Both `rescue.service` and `rescue.target` must run before `graphical.target` but cannot continue to run for `graphical.target` to run. Before `graphical.target` runs, `rescue.service` and `rescue.target` must run and then be stopped. This presents another difference between `systemd` and `Upstart` in that `systemd` is free to start and stop units whereas `Upstart` could only start services.

In the `After` list for `graphical.target` is `display-manager.service`. This is also listed in the `Wants` directive. This service must run and remain running for `graphical.target` to succeed. The other directives of the target file are `Description`, which provides an English description of the target; `Documentation`, which provides the man page to consult for this unit; and `AllowIsolate` (this is a somewhat cryptic directive whose meaning ties in with the `systemctl` instruction which we explore in Section 9.5).

Another way to view `graphical.target` is that before it runs it will run `rescue.service` and `rescue.target`, both of which check to make sure that this initialization of Linux is correct (otherwise, it needs "rescuing"). Once `systemd` knows this initialization does not need rescuing, initialization continues by shutting down both `rescue.service` and `rescue.target`, and then starting both `multi-user.target` and `display-manager.service`. The former is a target file which itself requires other services and targets to run, and the latter is the service that starts and manages the Linux GUI.

`rescue.target` looks similar to `graphical.target` in that it has five of the same seven directives: `Description` (`Rescue Mode`), `Documentation` (the same man page), `Requires` (`sysinit.target`, `rescue.service`), `After` (`sysinit.target`, `rescue.service`) and `AllowIsolate` (`yes`). For `rescue.target` to run, it must first start `sysinit.target` and `rescue.service`. Further, both must remain running for `rescue.target` to succeed.

`sysinit.target` is the next target file applied in the startup process. It is needed by `rescue.target`. The `sysinit.target` file has directives as follows: `Description` (`System Initialization`), `Documentation` (same man page), `Conflicts` (`emergency.service`, `emergency.target`), `Wants` (`local-fs.target`, `swap.target`) and `After` (`local-fs.target`, `swap.target`, `emergency.service`, `emergency.target`). The `sysinit.target` then must run `local-fs.target`, `swap.target`, `emergency.service` and `emergency.target`, but cannot run unless `emergency.service` and `emergency.target` are stopped. Thus, `sysinit.target` causes `emergency.service` and `emergency.target` to run and then stops them both.

Rather than continuing to explore the remaining target files, we will step through what each does and the order they execute. `local-fs.target` corresponds to starting the local file system. It must run after `local-fs-pre.target` and cannot run if `shutdown.target` is running. This unit file contains three directives we haven't yet seen: `DefaultDependencies` (set to no), `OnFailure` (set to `emergency.target`) and `OnFailureJobMode` (set to `replace-irreversibly`).

`local-fs-pre.target` has only one new directive, `RefuseManualStart` (set to `yes`). Before starting the file system, `systemd` notes that the file system cannot be manually started, and if `shutdown.target` is running, the file system cannot start. Now, with the file system started, control resumes with `sysinit.target`.

Unless an error arose that could not be handled, `rescue.service` is shut down, and with it, `rescue.target` ends. This allows `multi-user.target` to run and then `graphical.target`. This completes the target unit files, and with them launching of the needed services. At this point, the system is up and running. The operating system presents a login screen for the user. This is shown in Figure 9.5. The list of users is limited to those who recently used the system. Notice the scroll bar as there are more users not listed. If a user's name is not listed, the user selects `Not listed?` and is given a box to enter their username. After selecting or entering a username, a box appears to enter their password.

Figure 9.6 illustrates the startup process with respect to target, service and other unit files executed. The figure is not complete as some details are omitted including those of `Plymouth-start.service` and `path.target`. Services listed in italics in the figure are started and then stopped as they conflict with targets that started them. Ordering is indicated in a left-to-right manner. For instance, `graphical.target` calls upon `display-manager.service` (which is actually a symbolic link to `gdm.service`), `multi-user.target`, `rescue.service` and `rescue.target`. Both `rescue.target` and `rescue.service` must be stopped before `graphical.target` can run.

All items indented in Figure 9.6 beneath a target or service must start *before* it. There is some repetition that is not shown in the figure. For instance, `multi-user.target` also requires that `rescue.target` and `rescue.service` run and stop before, while `-.slice` and `system.slice` are required by `path.target` and `slice.target`. The list under each item is alphabetized.

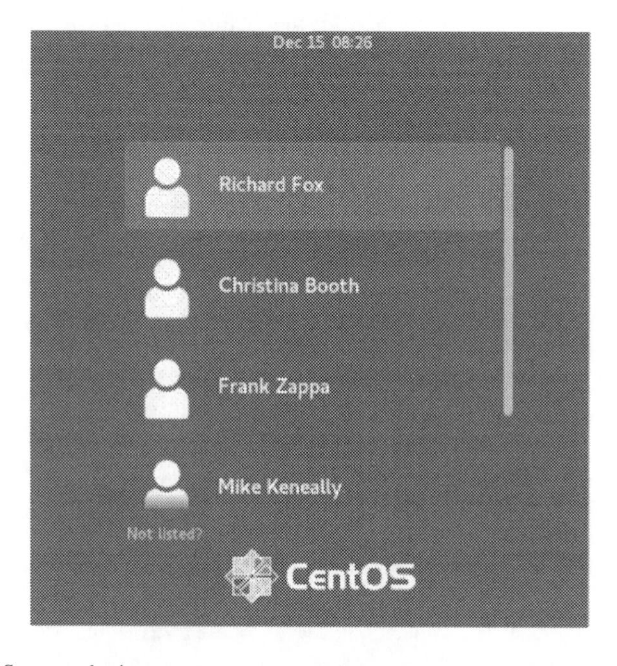

FIGURE 9.5 CentOS stream login screen.

```
graphical.target
   display-manager.service (this is a symbolic link to gdm.service)
      plymouth-quit.service
      rc-local.service
      systemd-user-sessions.sevice
      multi-user.target
        basic.target
           path.target
           slices.target
              -.slice (known as the root slice)
                 system.slice
              sockets.target
              sysinit.target
                 emergency.service
                 shutdown.target (runs after emergency.service)
                 emergency.target
                 local-fs.target
                    local-fs-pre.target
                 swap.target
              timers.target
              tmp.mount
      rescue.service
         plymouth-start.service
      rescue.target
```

FIGURE 9.6 Default order of `.target` and `.service` file execution.

This is not necessarily the order that items run as `systemd` is free to run them in any order. Missing from the figure would be other targets and services that a system administrator may want invoked. Table 9.4 lists some of the other `.target` files that we have not discussed.

While most of the `.target` unit files are located in `/usr/lib/systemd/system` along with other unit files, unit files can also be found in `/usr/lib/systemd/user`. This directory contains a few `.target` files along with numerous `.service` files and some `.socket` files. The difference between the two directories is whether the unit files pertain to the operating system or to user support.

Recall that runlevels no longer exist in `systemd`. However, they can be referenced as the former runlevels now map to various targets. We see these mappings in Table 9.5. There are even runlevel `.target` files which are symbolic links to the target files that they map to. For instance, `Runlevel0.target` is a symbolic link to `poweroff.target`. This is indicated in Table 9.5 under the column `systemd` Target.

TABLE 9.4
Additional `.target` Files

File	Usage
`cryptsetup.target`	For encryption of local file systems.
`getty.target`	Generate getty terminals (login prompts).
`network.target`	Ensure network services are up and running.
`network-online.target`	Test that the network is up and accessible.
`nsf-client.target`	Allow remote mounting.
`nss-user-lookup.target`	Name service lookup for user and group names.
`remote-fs-pre.target`	Like `local-fs-pre.target` but for remotely mounted file systems.
`remote-fs.target`	Like `local-fs.target` but for remotely mounted file systems.
`sshd-keygen.target`	Start `sshd` key generation services.

TABLE 9.5
Mapping Runlevels from SysV/Startup to Targets in system

SysV Runlevel	systemd Target	Description
0	poweroff.target	Halts the system, turns off the power.
1	rescue.target	Single user (root) mode, services stopped (including network), file systems outside of the root partition are not mounted.
2	Not available	Multiuser mode without network or GUI.
3	multi-user.target	Multiuser mode with network but without GUI.
4	Not available	Not used.
5	graphical.target	Multiuser mode with network and GUI, the default mode in most cases.
6	reboot.target	Perform a soft boot.

9.3.2 Service Unit Files

Each type of unit has its own file denoted with the extension *.type* as in .service or .slice. While .target files contain only one section, other unit files have either two or three sections. The first section is usually [Unit]. This is the only section found in .target files. We explored several directives available for the [Unit] section in the last subsection, but there are a few others of note. These additional directives are listed and described in Table 9.6. As these directives are available in the [Unit] section, we can find them in any type of unit file.

TABLE 9.6
List of [Unit] Section Directives

Directive	Meaning
Assert...	See Condition...; with Assert... if the condition(s) tested is(are) untrue, the unit does not start and is indicated as a failure.
Before	The opposite of After, this unit must start before any units listed.
BindsTo	Like Requires except if items listed here stop then this unit must also stop.
Condition...	Numerous directives that, if present, will test a condition before the unit is launched; if the condition is false, the unit will not start but will not be considered a failed start; specific conditions include ConditionArchitecture to test the platform, ConditionVirtualization to test if the system is running in a VM, ConditionHost to test hostname and ConditionKernelVersion to test the kernel version.
OnFailure	Unit(s) to activate should this unit fail; used so that systemd has a course of action when a unit fails.
OnFailureJobMode	Used with OnFailure to specify a mode to execute the listed units; values are fail, replace, replace-irreversibly, isolate, flush, ignore-dependencies and ignore-requirements.
PartOf	Like Requires except when systemd stops or starts any units listed here then this unit must also be stopped or started.
RefuseManualStart, RefuseManualStop	Indicates that this unit cannot be started/stopped manually (using systemctl) but instead can only be invoked by systemd.
Requisite	Like Requires except that if items listed here have not already started then this unit fails.
StopWhenUnneeded	If true, this unit will stop when it is no longer needed.
SuccessAction, FailureAction	Controls what to do if this unit exits successfully or fails; values are none, reboot, reboot-force, reboot-immediate, poweroff, poweroff-force, poweroff-immediate, exit and exit-force.

Aside from .target unit files, the other unit types have a type-specific section. For instance, .service files have a [Service] section, .mount files have a [Mount] section and .timer files have a [Timer] section. Optionally, non-target unit files may also have an [Install] section.

The [Install] section commonly is used to specify directives that serve as inverses to some of the directives available for the [Unit] section. For instance, the WantedBy directive is the inverse of Wants. If a unit lists another unit in its Wants section, then that second unit will list the first unit in its WantedBy section. Similarly, Requires has an inverse of RequiredBy, PartOf has an inverse of ConsistsOf, BindsTo has an inverse of BoundBy, Conflicts has an inverse of ConflictedBy and Requisite has an inverse of RequisiteOf. Any of these directives could appear in an [Install] section. Other [Install] section directives include Alias to specify aliases for this unit and Also which allows units to be grouped together so that if this unit starts or stops the others in the Also list are started/stopped.

.service unit files are called upon to control (start, stop) services. Among the directives needed to control a service are the physical location and startup command for the service and the location of a file that defines environment variables for the service (if any). Let's look at an example. Figure 9.7 displays the atd.service file, which is used to control the atd service (a scheduling service).

In the figure, we see that a .service file has three sections, unlike the .target files. The [Service] section in this figure demonstrates three new directives. EnvironmentFile denotes a file that defines environment variables for the service. ExecStart specifies the startup command, requiring the full path to the service's executable program, as well as any options. In this case, $OPTS is an environment variable that may be defined in the environment file. IgnoreSIGPIPE is set to no to indicate that the service should not ignore a SIGPIPE signal.

Let's focus on the WantedBy directive in the [Install] section. We see that this service is wanted by multi-user.target. For each of the .target files, a subdirectory exists in /etc/systemd/system which contains the services *wanted by* that target. Specifically, the directory /etc/systemd/system/multi-user.target.wants contains symbolic links of all of the items that multi-user.target needs to run. These link to the unit files such as the aforementioned atd.service file. We see the multi-user.target list of wants in Figure 9.8. Notice that this list is not limited to just those unit files that appear directly in the multi-user.target but is a complete list of all unit files that multi-user.target ultimately depends on.

There are a number of additional directives permissible for the [Service] section. Many of these are explained in Table 9.7. ExecStart is required, and if there is a file that defines environment variables, EnvironmentFile will also be present. Other directives are useful but optional.

To better understand the Type directive, let's consider how services are started. As we have discussed, systemd is tasked with starting some services during system initialization and others on-demand. Another way to start a service is from the command line using systemctl. We will refer to the entity that starts the service as the *service manager*.

```
[Unit]
Description=Job spooling tools
After=syslog.target systemd-user-sessions.service

[Service]
EnvironmentFile=/etc/sysconfig/atd
ExecStart=/usr/sbin/atd -f $OPTS
IgnoreSIGPIPE=no

[Install]
WantedBy=multi-user.target
```

FIGURE 9.7 The atd.service file.

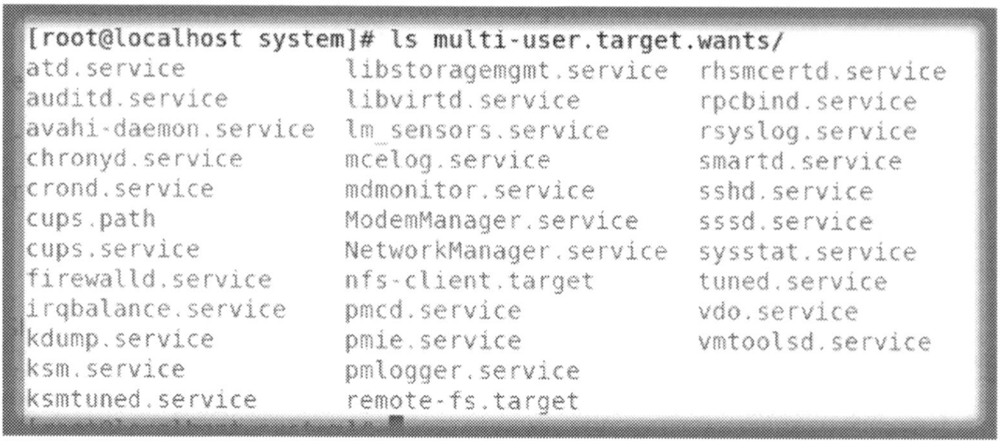

```
[root@localhost system]# ls multi-user.target.wants/
atd.service              libstoragemgmt.service   rhsmcertd.service
auditd.service           libvirtd.service         rpcbind.service
avahi-daemon.service     lm_sensors.service       rsyslog.service
chronyd.service          mcelog.service           smartd.service
crond.service            mdmonitor.service        sshd.service
cups.path                ModemManager.service     sssd.service
cups.service             NetworkManager.service   sysstat.service
firewalld.service        nfs-client.target        tuned.service
irqbalance.service       pmcd.service             vdo.service
kdump.service            pmie.service             vmtoolsd.service
ksm.service              pmlogger.service
ksmtuned.service         remote-fs.target
```

FIGURE 9.8 The list of `multi-user.targets` wants.

TABLE 9.7
Additional Directives for a `[Service]` Section

Directive	Meaning
`BusName`	Used to list the bus name; needed if Type specifies `dbus`; see `Type`.
`ExecReload`	Executable statement used if the service needs to be restarted (similar to `ExecStart`).
`ExecStartPre`, `ExecStartPost`	Executable statement to run before/after the service starts.
`ExecStop`	Executable statement used to stop the service.
`ExecStopPost`	Executable statement to run after the service ends.
`NotifyAccess`	Used if *type* is set to `notify` to indicate the socket that should listen for notifications; values are one of `none`, `main` or `all` to indicate that messages should be ignored, listened only from the main process or listened from all processes in the service's control group respectively.
`PIDFile`	Name and location of service *lock* file, created upon the service's starting; lock files are explored in Section 9.4.
`PrivateTmp`	If `yes` indicates that the service is able to use `/tmp`.
`RemainAfterExit`	Indicates whether the service should or should not remain running after system initialization has completed.
`Restart`	Specifies under what circumstance(s) `systemd` should try to restart the service if it stops; one of `always`, `on-success`, `on-failure`, `on-abnormal`, `on-abort` and `on-watchdog`.
`RestartSec`	If a restart is warranted, the number of seconds `systemd` should wait before attempting the restart.
`RuntimeDirectory`	Specify working directory for the service.
`TimeoutSec`, `TimeoutStartSec`, `TimeoutStopSec`	Specify in seconds how long `systemd` will wait before indicating that a service has failed; the three indicate the timeout duration in attempting to either start or stop the service, start the service only or stop the service only respectively.
`Type`	Indicate how the service will run; one of `dbus`, `exec`, `forking`, `idle`, `notify`, `oneshot`, `simple`; the differences between these is described in the text.
`User`, `Group`	Specify username/groupname service will run under; if not specified the default is to run under `daemon`.

If the Type is simple, exec or oneshot, then the service manager assumes that the service was successfully launched as soon as it issues the fork() system call to invoke the service. The difference between the three is what happens at this point. With simple, the service manager is immediately free to attempt to launch other services as soon as the fork() call is placed. With exec, the service manager waits until the service actually starts. With oneshot, the service manager waits until the service's process exits before launching any follow-up. With forking, it is expected that the process that executed the fork() will itself terminate leaving control with the child service process.

The type dbus is like simple except that the service will communicate over the bus listed under the BusName directive and the service manager will not start a follow-up service until this current service acquires access to the named bus. With notify, the NotifyAccess directive is used to indicate a socket by which this service will send notification messages. Note that if Type=notify and NotifyAccess is missing, it automatically defaults to main. Finally, the type idle indicates that the service's startup is delayed until all other services have been launched.

The simple type is the default if type is omitted. This type is recommended for all long-running services when the service does not require bus or socket communication. The simple type offers the most efficient startup for a service.

Let's take a look at two additional .service unit files. Figure 9.9 shows the contents of the cups.service unit file (top portion) and the cockpit.service unit file (bottom portion, comments omitted). CUPS is the Common Unix Printer Service. Its service file is nearly as simple as that atd.service file but has a few differences. First, this unit file's [Unit] section includes the cupsd man page reference (unlike atd, which did not). The [Service] section contains both Type= and Restart= directives. The [Install] section includes Also to indicate that this service should be treated as a group along with cups.socket and cups.path.

cockpit.service has several additional directives that we haven't seen yet: RuntimeDirectory, Environment (which defines environment variable RUNTIME_DIRECTORY), ExecStartPre, PermissionsStartOnly, User and Group. Here, we see that instead of defining environment variables in a separate file, this .service file is able to define its own environment variable within this file. cockpit.service does not have an [Install] section. Note that some of the directives in cockpit.service had to be wrapped around in the figure due to space restriction.

9.3.3 Other Unit File Types

.target and .service unit files are perhaps the most visible types of unit files because they are used to initialize the operating system. But these unit file types will call upon other types of unit files as well. In this subsection, we look at the remaining unit types.

.socket unit files are used to define domain sockets that will be used in conjunction with services. The reason for having two entities, the service and the socket, is so that the two units can be started in parallel. As services will be stopped when they are not necessary, it is the socket that is responsible for starting up a service when a connection to the service needs to be established.

The .socket unit file has a [Socket] section with a variety of directives. Among the most important are ListenDatagram, ListenStream and ListenFIFO which define attributes of the item to listen to. The former two are used to specify a UDP port address and a TCP port address (see Chapter 10) while the last specifies a buffer by name. Also available for the [Socket] section are directives to specify the socket's user and group owner name (SocketUser, SocketGroup), the permissions that the socket will have (SocketMode) and whether an instance of the socket will be generated for every connection (Accept set to either true or false). The directive Service is used to name the .service unit file should its name not match that of the .socket unit file. Figure 9.10 shows the unit file cups.socket that is used by the cups.service unit file. As the two unit files' names match, the Service directive is omitted.

```
[Unit]
Description=CUPS Scheduler
Documentation=man:cupsd(8)
After=network.target ypbind.service

[Service]
ExecStart=/usr/sbin/cupsd -l
Type=notify
Restart=on-failure

[Install]
Also=cups.socket cups.path
WantedBy=printer.target

-------------------------------------------------------
[Unit]
Description=Cockpit Web Service
Documentation=man:cockpit-ws(8)
Requires=cockpit.socket
Requires=cockpit-wsinstance-http.socket
      cockpit-wsinstance-http-redirect.socket
      cockpit-wsinstance-https-factory.socket
After=cockpit-wsinstance-http.socket
      cockpit-wsinstance-http-redirect.socket
      cockpit-wsinstance-https-factory.socket

[Service]
RuntimeDirectory=cockpit/tls
Environment=RUNTIME_DIRECTORY=/run/cockpit/tls
ExecStartPre=/usr/sbin/remotectl certificate --ensure
      --user=root --group=cockpit-ws
      --selinux-type=etc_t
ExecStart=/usr/libexec/cockpit-tls
PermissionsStartOnly=true
User=cockpit-ws
Group=cockpit-ws
```

FIGURE 9.9 Service unit files for cups and cockpit.

```
[Unit]
Description=CUPS Scheduler
PartOf=cups.service

[Socket]
ListenStream=/var/run/cups/cups.sock

[Install]
WantedBy=sockets.target
```

FIGURE 9.10 The cups.socket unit file.

The .mount and .automount unit files contain [Mount] and [Automount] sections respectively. Automount units require a corresponding mount unit, but mount units do not necessarily require automount units. Because of this dependency, the [Mount] section has many directives while the [Automount] section only allows for two. The [Mount] directives include What (absolute path to the resource being mounted), Where (absolute path to the mount point), Type (type of filesystem to be mounted), Options (any mount options to be applied), SloppyOptions (indicates whether mounting should fail if there are unrecognized options), DirectoryMode (permissions of a parent directory if such a directory needs to be created) and TimeoutSec (amount of time in seconds to wait before the mounting operation fails should the remote device not respond).

```
[Unit]                              [Unit]
...                                 ...
DefaultDependencies=no              DefaultDependencies=no
                                    Before=sysinit.target
[Mount]                             ConditionPathExits=
What=binfmt_misc                         /proc/sys/fs/binfmt_misc/
Where=/proc/sys/fs/binfmt_misc      ConditionPathIsReadWrite=
Type=binfmt_misc                         /proc/sys/

                                     [Automount]
                                     Where=/proc/sys/fs/binfmt_misc
```

FIGURE 9.11 The `proc-sys-fs-binfmt_misc.mount` and `proc-sys-fs-binfmt_misc.automount` files.

The `[Automount]` section of an `.automount` file has the same `Where` and `DirectoryMode` directives.

We look at one pair of mount/automount unit files, for `proc-sys-fs-binfmt_misc`. Both unit files share the same name with different extensions. The important portions of both of these files are shown in Figure 9.11 with the `.mount` unit file on the left and the `.automount` unit file on the right (some of the details are omitted for brevity). Notice that neither file has an `[Install]` section nor do they have a `DirectoryMode` directive. They share the same `DefaultDependencies` and `Where` directives, and the `.automount` file contains two conditional directives.

The path unit is used to monitor a file path for changes. Upon any such modification within the path, the path unit invokes a service whose name matches the name of the path file. For instance, a `.path` unit file named `mypath.path` would have a service unit file named `mypath.service`.

The `.path` unit file may include a number of directives to specify under what circumstance(s) the service should be invoked. These include `PathExists`, `PathExistGlob` (globbing permitted when specifying the path), `PathChanged`, `PathModified` and `DirectoryNotEmpty` each of which tests the given condition such as whether the specified path exists or was changed. `PathModified` triggers if anything in the path was modified such as a file written to or closed. Should the service not be named the same as the path, then a `Unit` directive is required to list the service file's name. `MakeDirectory` can be added so that `systemd` creates the directory to monitor, and if used then `DirectoryMode` can be used to specify the directory's permissions.

To further explore the path unit, we look at `systemd-ask-password-console`, which has both `.path` and `.service` files. The `.path` file is invoked by `plymouth-start.service`. Its condition is `ConditionPathExists=!/run/plymouth/pid where !` in the condition indicates to invert the condition (the `.path` unit triggers if the specified condition does not exist). There are two directives in this unit file's `[Path]` section: `DirectoryNotEmpty=/run/systemd/ask-password` and `MakeDirectory=yes`. These directives inform `systemd` to create the directory and invoke the `ask-password` service should the directory have contents placed into it.

The timer unit is responsible for scheduling tasks to execute either at specified times or after specified events. With timer units, system administrators do not have to rely as much on the `atd` and `crond` scheduling services. A `.timer` unit file has a `[Timer]` section with directives that specify the condition (time or event) and the action scheduled. These directives have the format `OnCondition=`*`value`*. Values vary based on the *condition*. Table 9.8 enumerates some of the more useful condition directives. The action is typically the invocation of a service whose name matches the timer's filename. The unit using the directive can also be specified as `Unit=unitfile` such as with `Unit=dnf-makecache.service`.

The swap unit is used to configure a swap space. The name of the unit file will match that of the swap device (or a swap file). If the swap file system is listed in `/etc/fstab`, then there will be no `.swap` unit file. If there is a `.swap` unit file, the directives for the `[Swap]` section include `What`

TABLE 9.8
Timer Unit File Conditions

Directive	Explanation
OnActiveSec	Specify the number of seconds after the timer activates that the action should execute.
OnBootSec	Specify the number of seconds after system booting that the action should execute.
OnStartupSec	Specify the number of seconds after systemd starts that the action should execute.
OnUnitActiveSec, OnUnitInactiveSec	Specify the number of seconds since the specified unit last activated/stopped that the action should execute.
OnCalendar	Specify a time/date that the action should execute.

(the absolute path to the swap space), TimeoutSec (number of seconds systemd will wait for the swap unit to activate before failing) and Priority and Options (as specified in /etc/fstab).

A control group, called a *cgroup*, is a collection of processes that share the same resources and restrictions on those resources. For instance, a cgroup may have a limitation on the amount of memory they can use. By grouping processes into cgroups, shared limitations, priorities and control can be handled by the kernel at the cgroup level rather than at an individual process level.

We introduce cgroups because they have unit types associated with them. The slice unit is the name applied to a cgroup. Units that manage the processes of a cgroup can then be assigned to the slice unit. Thus, the slice represents the mechanisms by which the cgroup is managed. The other units are typically services and scopes.

A .slice unit file has a [Slice] section but the entries here are not directives specific to slices. Instead, these directives are for resource management, such as TasksMax=200, MemoryHigh=75%, MemoryMax=90%. These specific directives are found in the system-cockpithttps.slice file.

A snapshot is created explicitly using the systemctl instruction and does not have a corresponding unit file. Scopes are similarly not explicitly configured via a unit file but instead created by executing programs. Device units have no specific directives. We omit further detail of these unit types.

9.3.4 Modifying System Initialization

Having explored in detail the systemd initialization process, a natural question is why do we need to learn this? There are three reasons. First, anyone familiar with SysV/Upstart initialization may wonder why the /etc/init.d scripts and subdirectories that support initialization are gone (or have become symbolic links). Knowing that systemd has replaced the init-based startup process will save hours of effort in searching for what no longer exists. Second, if something goes wrong, we now have some idea of where to look for problems. Although the process is complex, we can determine where a unit failed based on its dependencies. But most importantly, we cover the process so that we can make changes if necessary (or to avoid making changes!) to the startup process.

A great deal of thought went into the systemd initialization approach. Although it is possible and even easy to alter system startup, to do so we need a full understanding of the dependencies between the various units. Making arbitrary changes can lead to severe problems. Fortunately, any changes made to the initialization process are recorded and can be viewed using the instruction systemd-delta. This will list every unit file that has been modified. Thus, should we make changes that do not work out, we can view the exact changes to undo them.

With the above warning, let's go ahead and look at reasons for changing the system initialization process, without risk of damaging the systemd startup process. The simplest change we can make is to switch from graphical mode to multiuser mode. This change is made by altering the file default.target, a symbolic link pointing at graphical.target, to point instead at multi-user.target. We do not physically edit graphical.target or multi-user.target. We simply change the symbolic link of default.target. Should we decide to change the default back to graphical.target, we can simply redirect the symbolic link again. The only reason to make this change is if we prefer to boot to a text-based system, perhaps for efficiency purposes.

Let's consider a slightly more elaborate change to our system. In this case, we want to add a service to the startup process other than those already started by default. Specifically, let's assume we have installed the Apache webserver and want to start its process, httpd, upon system initialization. We do so by adding our own .target file with a Wants directive listing the httpd service. We have to indicate when this .target file should start. A good place for this is to start this new .target after network.target has started.

Let's assume our .target file is called httpd.target. Figure 9.12 shows the directives that we would likely place in our httpd.target file. We would also need to set up a subdirectory named httpd.target.wants, placed in /etc/systemd/system. Its contents will be symbolic links to the .service and .target files that the httpd.target wants (that is, that it depends on). In this case, there will be two symbolic links: one to the httpd.service file and one to network.target. We might choose to include a Conflicts directive to ensure that rescue.service and rescue.target are no longer running as we would likely not want to start Apache should the system need rescuing.

We must also add the name of our .target file to one of the other .target files to indicate when this new target should run. We could place it in graphical.target or multi-user.target, adding httpd.target to the list of Requires and Before (as we want to start httpd.target only after graphical.target or multi-user.target runs).

We would also have to add an httpd.service file. The [Unit] section would include a Description and possibly a Documentation directive. The [Install] section would include a WantedBy directive to indicate that this service is wanted by httpd.target.

It is the [Service] section that controls how the httpd service should start. We add an ExecStart directive to specify the command line instruction to launch the service which might be ExecStart=/usr/local/apache/bin/apachectl start. We would include at least a Type statement to indicate how the process should start. As our webserver would not strictly run as a background service but as its own process, we might want to use forking or dbus, as in Type=forking. There may be other directives added to handle restarting the service should it stop and specify the locations of a PID file and Apache's runtime directory, among others.

Another change we might make to the initialization process is to have one or more services start as processes rather than as background service. We noted above that httpd might start this way. By default, all services start in the background, running as daemons. Let's pick one to run as a process. autofs.service starts automount, which runs as a daemon. To start autofs as a process, we make a couple of simple modifications to the autofs.service file. First, the ExecStart statement would need to be altered to indicate that this program should start in the foreground, such as with ExecStart=/usr/sbin/automount $OPTIONS --pid-file /run/autofs.pid. We would add the option -f. Although we could define this as part of the variable OPTIONS, this variable may be used in other locations, so instead we would simply add -f before or after $OPTIONS. The Type directive in this file specifies Type=forking. We would instead want the process to run on its own and not be forked.

```
Requires=network.target httpd.service
After=multi-user.target rescue.service rescue.target
      display-manager.service network.target
```

FIGURE 9.12 Example of an added .target file.

Yet another change we might make to our startup process is to automatically mount some remote file system at system initialization time. In SysV and Upstart, there were scripts that init executed that system administrators could modify. It would have been in one of these scripts that we added the mount command (along with any other necessary commands) to mount the remote file system. Those scripts no longer exist in systemd.

Instead, we want to create both a .mount and an .automount unit file that share the same name. The .mount unit file would specify What and Where to indicate the mount point and the location of the remote file system. It is likely that we would also want to specify mount options using the Options directive. TimeoutSec would be useful so that the mount fails should the remote file system be unavailable. For the .automount unit file, we would use the same Where directive.

Now, we must place the two new file names in a Wants directive. We need to position these correctly. Obviously, we do not want to attempt to mount this remote file system until the network is running and the local file systems have been established. There is also remote-fs.target that should run first. Remote mounting would likely be among the last of the activities performed during initialization. As graphical.target calls upon multi-user.target which calls upon sysinit.target, it would be safe to assume that the new .mount and .automount units should run after sysinit.target runs.

These are just three examples of possible changes we could make to the startup process. None of these examples are complete and have been left a little bit vague to encourage you to further research how to make changes to the startup process should you decide to do so. You might prefer to instead schedule tasks to run or start tasks by hand rather than making changes to the startup process. If you do make changes to existing unit files, make sure you have backups that you can recover from.

SECTION ACTIVITIES

1. Most of the unit files are stored in /usr/lib/systemd/system but you can find others in /usr/lib/systemd/user. Which .target files do you find here and how do they differ in terms of the types of activities they are responsible for (hint: consider the names of the two directories)?
2. If you have experience with older versions of Linux that used SysV or Upstart, you are probably familiar with the directories under /etc/rc.d which contain subdirectories for each of the runlevels. These directories still exist in Red Hat 8. What do they store? Similarly, /etc/init.d used to store a script for each service. These have all been removed in favor of .service files. What is left in the /etc/rc.d/init.d directory?
3. View the README file in /etc/rc.d/init.d. What does it say about the former contents of this directory? What has replaced them? For further reading, which man pages does it point you to?
4. Experiment with modifying the systemd startup process. An example is provided at https://opensource.com/article/20/5/manage-startup-systemd. Read the full article before getting started so you know what to expect. Now, work through the example to implement it. First, implement the hello.sh and hello.service files as per the instructions. Start the service with systemctl (we cover how to do this in Section 9.5) to make sure it works. NOTE: you may need to disable SELinux for this experiment. If so, use setenforce 0 before starting the service. Modify the multi-user.target or graphical.target file and reboot your Linux computer. When done, remove the new entry from the .target file, delete both the hello.sh and hello.service files, and reenable SELinux (setenforce 1).

9.4 LINUX SERVICES

In this section, we explore various services in Linux. In Linux, a service is generally referred to as a daemon (pronounced "demon"). In mythology, a daemon is often conveyed as a guiding spirit that sits in the background, influencing the actions of people. In the case of Linux services, they run in the background, receiving requests from various sources that need some operating system support, and handle those requests. To reinforce their role as daemons, most Linux services' names end with a `d`, as in `atd` which is the "at scheduler daemon".

There are a great variety of services in Linux. The specific services vary by distribution. Red Hat 6 had 66 services. This number increased significantly in Red Hat 7 to 143 while Red Hat 8 increases it again to 167. Some older services have been replaced or superseded by other services. The `network` service, for instance, which was responsible for enabling our network interface, has been replaced by `NetworkManager` starting with Red Hat 7. In Red Hat 7, `network` remained available to serve as a backup to `NetworkManager`. But in Red Hat 8, the `network` service has been removed entirely. Similarly, the service `iptables`, the Linux firewall, was replaced in Red Hat 7 with `firewalld` but in this case, `iptables` has been retained because `firewalld` acts as a front-end to control `iptables`.

In this section, we examine services generically by considering what services are and do. We look at categories of services. We then focus on some of the more significant Linux services found in `systemd`-versions of Linux (and Red Hat 8 in particular). Later in this chapter, we look at how to control services through the `systemctl` instruction (Section 9.5) and how to modify service configuration files (Section 9.6).

9.4.1 What Are Services?

Services are programs that make up part of the operating system separate from the kernel. Each service runs as a collection of one or more processes. The processes run in the background meaning that they only run when needed and have no user interface. Services will not take up processor time unless called upon. Not having a user interface means that the service's behavior is not influenced by runtime interaction but instead proscribed by its configuration, usually found in an editable configuration file. The lack of user interface makes services more secure. Other changes to a service's behavior might be made through its `.service` file.

The role of a service is to provide some type of support to the operating system. But the support is only provided in response to a request. Requests can come from many different sources including users, running applications, the network, other hardware and other operating system services. Some services support a range of requests while other services must combine together to support a request.

Services can be running or stopped, and this status is alterable not only by `systemd` but from the command line by the system administrator. A Linux service will be listed as active or inactive. *Active* services are those that have been loaded into memory. Within this status, services can be running or exited. An *exited* service is one that had been running but has been stopped because it is not currently needed. Any *inactive* service is considered dead. A *disabled* service can be enabled by the system administrator using the `systemctl` instruction or by `systemd`. An enabled service can change state from running to exited by `systemd` when the service is no longer needed or exited to running upon demand. Older versions of Linux did not include an exited state as services were either enabled (running) or disabled (not running).

There are several files related to each service. The most significant is the executable file itself, which is stored in `/usr/sbin`, or in some cases `/usr/bin` (older versions of Linux might also store some services in `/bin` or `/sbin`). Next is the `.service` file which is used to control the starting and stopping of a service. As we noted in the previous section, `.service` files are stored in `/usr/lib/systemd` under the `system` or `user` subdirectory. Older versions of Linux used startup scripts that could be found in `/etc/init.d`.

A service may also have its own environment file, storing environment variable values. A fourth file associated with many services is the service's *configuration file*. This file contains directives and/or service-specific environment variables that will be applied to the running service. Modification of values in this file causes the service's behavior to differ but only when the service is started or restarted.

Service configuration files are stored under `/etc` although the specific placement varies by service and Linux distribution. In Red Hat distributions, many configuration files are found under `/etc/sysconfig`. This directory does not exist in Debian distributions and so configuration files are often found under `/etc/default`. Some services have configuration files stored within service-specific directories such as `/etc/avahi` which stores configuration and related files for the avahi daemon. In some cases, services have multiple configuration files. `logrotate.conf` defines the standard configuration for the `logrotate` service, but there are also files located in the `/etc/logrotate.d` directory for rotation directives for specific applications such as `/etc/logrotate.d/cups`.

Running services may also generate or add content to other files. For instance, running services have events logged to either their own or a collective *log file*. Content generated by a service may be saved to a cache file. Services may generate email, usually sent to root's email file. Some services might generate printer spool data or other types of spool information. Services may generate their own service-specific data to yet other files. Most of these files can be found under various subdirectories of the `/var` directory.

Another form of file created by a running service is its *lock file*. This file is generated by the operating system to note that a service is running. More recent versions of Linux use *pid files* to note the PIDs of running processes. PID files are mostly stored under the `/run` directory but might be found in other locations.

Linux services can be broken into one of several categories such as scheduling, logging, network management and hardware management. Table 9.9 lists some of the many Linux services found in Red Hat 8. The listed services are categorized by a type and listed with a brief description. Services denoted with an `*` in the table are actually a collection of services whose names start with the name listed. For instance, there are eight different services whose names start with `dracut` (e.g., `dracut-pre-pivot`, `dracut-pre-udev`, `dracut-shutdown`).

In the next subsection, we focus on a handful of the more visible services for both Linux users and system administrators. Keep in mind that with well over 100 services, we can't cover them all and so limit our coverage.

9.4.2 An Examination of Significant Linux Services

`anacron` is a scheduling service responsible for running tasks (processes, scripts) based on a specified recurrence but not at a specific time, or for running tasks scheduled at times that the system was not available. `anacron` runs such tasks at the first opportunity after the system has been restarted/resumed. The configuration file for `anacron` is `/etc/anacrontab`. Of the scheduled activities that might be pre-set for `anacrontab` include entries in the directories `/etc/cron.daily`, `/etc/cron.weekly` and `/etc/cron.monthly`.

The `at` daemon, `atd`, is used for *one-time* scheduling of tasks. It runs processes that were scheduled through the `at` and `batch` commands. In `at`, the task is scheduled for a specific time and date. The time/date can be specified using an absolute value such as `today 13:50` (1:50 pm), `tomorrow 9:00 PM` or `12/31/23 2:15 AM`. We can also specify the time relative to the time we submit the command using `now + offset` where *offset* is a numeric value and a time unit, such as one of `5 minutes`, `2 hours` or `1 day`. We can also use such terms as `noon`, `midnight` and `teatime`.

Along with `at` are commands `atq` and `atrm` to view scheduled tasks and remove scheduled tasks. The `batch` program calls upon `atd` to execute the scheduled task once CPU load drops below 80%. In an underused system, the processor load will usually be below 80% and so the

TABLE 9.9
Linux Services

Name	Type	Description
atd	Scheduling	One-time scheduling of a job.
auditd	Logging	Saves audit records to files.
avahi-daemon	Networking	DNS service discovery.
blk-availability	File system	Helper program to unmount/disable LVM.
boltd	Hardware	Manage thunderbolt (hi-res displays/hi-performance data) devices.
chronyd	Networking/hardware	Synchronize system clock with external sources.
cockpit*	Administration support	Web-based administration tool.
colord	Hardware	Generate color profiles to manage color content with I/O devices.
cpupower	Hardware	Set and display CPU power-related values.
crond	Scheduling	Recurring scheduling of tasks.
cupsd	Hardware	Interface between applications and system printers.
dmeventd	Hardware	Event monitor to map device actions.
dnf-makecache	Package management	Support for the dnf package manager (dnf is the next generation of yum).
dracut*	Kernel support	Create and modify initramfs images as used during kernel initialization.
ebtables	Networking	Ethernet bridge table management.
firewalld	Networking/security	Front-end service to Linux firewall.
gdm	User interface	Gnome display manager.
getty@tty1	User interface/security	Prompt user for login, connect to a tty.
gssproxy	Security	Manage and access security credentials.
import-state	Networking	Import network configuration from initramfs.
initrd*	Kernel support	Ramdisk support services.
irqbalance	Kernel support	Distribute hardware interrupts across multiprocessors.
iscsi*	Hardware	Suite of services to support iSCSI management.
kdump	Kernel support/logging	Kernel crash collection (logging) service.
ksm	Kernel support/memory	Copying and merging of shared pages in memory.
ldconfig	Kernel support/memory	Configure runtime bindings of dynamic linked shared objects (.so files).
lvm2*	File system	Suite of services supporting LVM2.
mcelog	Kernel support	Decode the kernel machine-check log.
mdmonitor	File system	Software-based RAID management.
multipathd	Hardware	Manage multipath connections between commonly grouped devices.
NetworkManager	Networking/hardware	Manage network connection via interface device(s).
nfs*	Networking/file system	Support network file system mounting and access.
oddjobd	Miscellany	Used to control other services.
plymouth*	Kernel support/hardware	Provide splash screen and user interaction during system boot.
polkitd	Kernel support/user interface	Provide interprocess communication between unprivileged and privileged programs.
rpcbind	Kernel support	Bind program numbers to absolute addresses for remote procedure calls.
rsyslog	Logging	Map types of and severity of events to logfiles.
smartd	File system/hardware	Monitor SMART ATA/SATA devices.

(Continued)

TABLE 9.9 (*Continued*)
Linux Services

Name	Type	Description
snapd	Package management	Allow for installation and configuration of snaps (third-party software).
sshd	Networking/security	Permit ssh access to the computer.
sysstat*	Logging	Suite of services that collect and store system activity and provide summaries.
systemd-ask-password*	Security	Obtain user password or passphrase.
systemd-fsck*	File system	File system checks on various file systems.
systemd-journal*	Logging	Run, delete or rebuild the journald logging system.
tuned	Kernel support	Dynamically update system settings based on current usage
user*	Kernel support	Services for each logged in user.

batch job would probably execute almost immediately. We explore atd, at, atq and atrm in more detail in Chapter 12.

auditd is the Linux Auditing System daemon responsible for creating log entries of specified activities no matter who performs those actions. There are two programs and three files related to auditd. First is the auditd service itself. Second is auditctl, which the system administrator will use to start and stop the audit daemon. It is auditctl which, when started, will read the auditd configuration rules. And then there are three different configuration files.

The first of the auditd configuration files is auditd.conf, which is a standard configuration file. In this file, we would find directives to specify such aspects of auditd as where the audit log files will be stored, the format for the files, actions to undertake should there be a disk access error, log rotation specifications and so forth. This file does not specify the logging activities. That information is stored as rules in the file audit.rules. The third file, audit-stop.rules, contains rules specifying what auditd should do if the service stops running. All three files, along with any additional auditd configuration data, are stored under /etc/auditd. We explore auditd in more detail in Chapter 12.

Another scheduler is crond. Users use crontab to schedule tasks that crond then runs based on the scheduled time(s). Unlike atd, a user is able to submit only a single crontab job, but each job can contain many scheduled tasks. Also, unlike atd which performs one-time scheduling, crond performs *recurring* scheduling. The recurrence is specified as five values which are either numbers or * (wildcard) to denote that the task should execute on the given minute, hour (in military time), date, month and day of the week. A crontab job contains one line per task to be scheduled where the line encodes the recurrence and the task.

The crontab command permits options of -l to display the crontab job of this user, -r to remove the scheduled crontab job of this user or -e to edit the crontab job of this user (in vi). Otherwise, crontab expects a filename whose contents are the scheduled tasks. We look at crontab in more detail in Chapter 12.

Common Unix Printing System, cups, is a program used to control print jobs on printers connected to a Linux computer. The cups program uses cupsd to schedule print jobs. The printers may be directly connected to the computer or available via a network even if the network consists of computers running different versions of Linux or Unix. When the user issues a printer command (lp, lpr, lpq, lprm), cups takes over to handle the command by accessing the appropriate printer and calls upon cupsd to handle scheduling. The command lpc allows us to alter a printer's properties (this is also available via the Printer Configuration GUI).

dnsmasq is like a mini-domain name system (DNS) server for Linux. The role of dnsmasq is threefold, all of which revolve around caching DNS information. First, it caches responses from

DNS servers so that further repeated accesses can be handled locally while those cached entries remain valid. Second, it uses the /etc/hosts file to handle DNS requests if the mapping information is stored in that file. This file is set up by the system administrator to bypass DNS when IP addresses are known and stable (will not change for years). Third, it responds to requests from DHCP servers regarding hostname to IP address mappings.

With dnsmasq, we reduce the amount of access to the DNS server and thus reduce waiting time for Internet-based communication. dnsmasq has a configuration file of /etc/dnsmasq.conf. The configuration file is used to specify such details as a DNS server to query in case the local cache does not contain the relevant information, what network interface(s) to use and the amount of time that items should remain cached. If DNS is unfamiliar to you, we cover it in some detail in Chapter 10.

dracut is a service used to create initramfs images. An initramfs image is loaded while the Linux kernel is initializing, providing an initial file system for kernel access. By default, any initramfs image is stored in /boot with the name initramfs-....img where ... is a version number and an architecture number (e.g., x86_64). Multiple images can be made available. dracut has both a configuration file and a configuration directory in /etc. It has additional configuration files and a directory of runtime modules in /usr/lib/dracut.

Red Hat 8 comes with several services whose names start with dracut that are used during the initialization process. These include dracut-cmdline.service, dracut-pre-udev.service and dracut-pre-trigger.service which are invoked by sysinit.target, and dracut-pre-mount.service, dracut-mount.serivce and dracut-pre-privot.service, which are invoked by basic.target. Using dracut is quite complex, and we omit further detail. For more information, see dracut's man page as well as dracut.conf's man page.

Older versions of Linux relied on iptables and ip6tables to serve as the front-end to the Linux firewall. The more recent versions of Linux use firewalld or ufw to serve as a more user-friendly front-end to iptables. We examine firewalld and ufw in detail in Chapter 10.

Log files accumulate messages at different rates depending on what source(s) is(are) sending messages to be logged. As a log file fills, it needs to be replaced by an empty file where the current file is set aside. This process is known as *log rotation* and handled by the logrotate service. Log rotation causes the current log file to be renamed and a new, empty, log file to be created. Rotation may add a number to the log file (for instance, messages might become messages.1) or add the date of rotation (messages might be renamed messages-12-16-2023).

There are two ways to configure the log rotate service. First is the logrotate.conf configuration file that contains generic directives for handling log rotation. Second, for several specific log files there are separate configuration files, each with its own directives. All configuration files are stored under the subdirectory /etc/logrotate.d. We look at logrotate's configuration later in this chapter and examine log files in detail in Chapter 12.

To permit the exportation of file systems for remote mounting, we use the Network File System (nfs). The nfs daemon has been replaced with several services named nfs-... including nfs-server, nfs-idmapd and nfs-mountd and we control the nfs suite of services using the Linux command rcp.nfsd. We explored how to mount remote file systems and how to export a local file system for remote mounting in Chapter 8 and cover the specific details of performing file system exportation later in this chapter.

Service requests are made by the user, external messages coming in from the network, other software, hardware or by other parts of the operating system. When requests come from software, it is sometimes the case that the software does not have permission to execute the needed service. The oddjobd service provides a method so that we can specify a mapping of a specific software title to specific services so that the software can invoke the service. The oddjobd configuration file is located in /etc/dbus-1/system.d/oddjob.conf.

Linux offers a mechanism so that the system administrator can define the types of system and software events that should be logged and where those messages should be logged. Originally, this service was called syslogd but is now called rsyslogd. The service has predefined logging rules specified in the configuration file /etc/rsyslog.conf. This file can be edited so that the rules can be modified or new rules added. Rules indicate for various activities, such as authentication attempts, crond events and emergency messages, what should be logged and where. We explore rsyslogd and its configuration file, /etc/rsyslog.conf, later in the chapter. A related service is klogd, which logs kernel messages. klogd is not configurable.

Among the changes brought about by systemd is the inclusion of an automated journaling (logging) facility called journald. journald logs kernel messages, system log messages, structured log messages, standard output and error of services and audit records. journald collects the data from a variety of sources.

The actual service that runs journald is called systemd-journald. Its configuration file is stored at /etc/systemd/journald.conf. systemd-journald uses three socket units of systemd-journald.socket, systemd-journald-dev-log.socket and systemd-journald-audit.socket. With journald running, it means that rsyslogd is of less importance but might still be a preferred service of choice by system administrators.

journald stores log messages under the directory /run/log/journal. Data stored here is in binary and can only be read using the journalctl program (which we present in Chapter 12). journald messages are only temporary as the log file is recreated at each system boot. We can set up a new directory of /var/log/journal for permanent log files, adjusting the configuration file.

Obviously, we have only looked at a few of the many services. Most services will run or stop based on systemd, and there is little we need to do about this. Similarly, we may not ever need to reconfigure services. However, it is important to know about services and familiarize ourselves with the more important ones. For those in the list from Table 9.9 that we did not explore, most have man pages available to further explore the services and configuration files.

It should be noted that there is some inconsistency between the names of services as specified using the .service files and service man pages. As one example, cupsd is a service which is utilized via the cups program. While the service is named cupsd, its unit files are named cups.path, cups.service and cups.socket. Both cups and cupsd have their own man pages. Another example is of rsyslog. While the service is called rsyslogd and the man page is found through man rsyslogd, the service file is rsyslog.service (no d). Viewing the rsyslogd.service file shows that its documentation is man:rsyslogd(8).

SECTION ACTIVITIES

1. If you have access to Red Hat version 6 or earlier, open the Service Configuration tool. This tool enumerates all of the services and allows you to stop or stop them. This GUI tool no longer exists in Red Hat 7 or 8 and so controlling services is handled by systemd or from the command line.
2. If you have access to both Red Hat 6 (or earlier) and Red Hat version 7 or 8, cd to /usr/lib/systemd/system in Red Hat 7/8 and type ls *.service. Compare the list of services in Red Hat 7/8 to those in Red Hat 6. You can view the Red Hat 6 services by looking at the controlling script files in /etc/init.d. How many from Red Hat 6 are still present in 7/8?
3. You can also view service information through Cockpit. Open your web browser and enter as the address *ipaddress*:9090 where *ipaddress* is your machine's IP address. You will be asked to log in, log in using your username and password. Select Services from the menu on the left and you can see all services and their current status. Select any service and it brings up information from its .service unit file like Requires, Before, After. Explore a number of services and see if you can identify the category that it belongs to. You might find that it belongs in some category not listed in Table 9.9.

9.5 USING systemctl

Prior to systemd, the init process was tasked with starting those services that the given runlevel required to run. Such services might have included networking services if the system was launched to runlevel 3 or 5, file system mounting if the system was launched to runlevel 2, 3 or 5, and the graphical user interface if the system was launched to runlevel 5. Once a service started, it remained running even if it was no longer needed. A system administrator could stop, start or restart services using the Linux instruction service.

With systemd capable of stopping services that are no longer needed, there is less need for the system administrator to be concerned with starting or stopping services. But there is still a mechanism to start, stop and restart a service. With systemd, the command is systemctl. systemctl is not just used to controlling services though as it also provides a variety of system-level controls.

Before we get into this in too much detail, let's compare the service and systemctl instructions. service was the command used prior to systemd to start, stop, restart and query services from the command line. Figure 9.13 shows the operations to obtain the status of the rsyslogd service followed by stopping and starting it, all using the service command. Figure 9.14 provides a similar example showing how to obtain the status of the service with systemctl. Notice how much more detailed the status information is. When using systemctl to start or stop a service, unless there is an error, we are provided no feedback (these instructions are not shown but have the format systemctl start/stop *servicename*).

With systemctl, we can issue a number of commands to the service including start, stop and status, just like we could with service. But unlike service, systemctl can be used to control all types of units, not just services. We could similarly obtain the status of, start or stop timers, sockets, slices and so forth.

In Figure 9.14, notice that the command references rsyslog. When referencing any type of unit other than a service, we must include the full unit file's name. We see in Figure 9.14 that we did not have to add .service to rsyslog, but if we wanted to start or stop a timer, we would add .timer as in systemctl stop dnf-makecache.timer. Multiple units can be specified in a single systemctl command by separating the units with spaces as in systemctl stop atd crond rsyslog.

Aside from status, start and stop, there are many other commands available to manipulate a unit. First is restart, which stops and then starts the unit. This can be useful if a configuration file has been modified as the configuration does not take effect until the unit (a service in this case) is started. We could stop the unit and then start it separately, or simply restart it.

Related to restart is try-restart, which stops and starts units that are running but if a unit is not running then it is not started (i.e., stopped units are unaffected). As some units are loaded, a variation of start is reload. This leads to two additional commands of reload-or-restart to reload units that support this and restart the rest, and try-reload-or-restart which is the same except that stopped units are not restarted, much like with try-restart.

Related to status are is-active and is-failed, which will output for the unit(s) whether it is active or is failed, which might be preferred to status when we only want to know if the unit's

```
[root@CentOS6Template init.d]# service rsyslog status
rsyslogd (pid  1169) is running...
[root@CentOS6Template init.d]# service rsyslog stop
Shutting down system logger:                               [  OK  ]
[root@CentOS6Template init.d]# service rsyslog start
Starting system logger:                                    [  OK  ]
```

FIGURE 9.13 Using service in Red Hat 6.

```
[root@localhost system]# systemctl status rsyslog
● rsyslog.service - System Logging Service
   Loaded: loaded (/usr/lib/systemd/system/rsyslog.service; enabled; vendor prese
   Active: active (running) since Sat 2021-05-22 06:56:57 EDT; 1 weeks 6 days ago
     Docs: man:rsyslogd(8)
           https://www.rsyslog.com/doc/
 Main PID: 1307 (rsyslogd)
    Tasks: 3 (limit: 11084)
   Memory: 4.4M
   CGroup: /system.slice/rsyslog.service
           └─1307 /usr/sbin/rsyslogd -n

May 25 18:11:58 localhost.localdomain rsyslogd[1307]: imjournal: journal files ch
May 26 23:37:39 localhost.localdomain rsyslogd[1307]: imjournal: journal files ch
May 28 04:58:47 localhost.localdomain rsyslogd[1307]: imjournal: journal files ch
May 29 09:43:47 localhost.localdomain rsyslogd[1307]: imjournal: journal files ch
May 30 03:21:02 localhost.localdomain rsyslogd[1307]: [origin software="rsyslogd"
May 30 15:28:48 localhost.localdomain rsyslogd[1307]: imjournal: journal files ch
May 31 20:58:48 localhost.localdomain rsyslogd[1307]: imjournal: journal files ch
Jun 02 02:43:49 localhost.localdomain rsyslogd[1307]: imjournal: journal files ch
Jun 03 08:13:49 localhost.localdomain rsyslogd[1307]: imjournal: journal files ch
Jun 04 09:44:07 localhost.localdomain rsyslogd[1307]: imjournal: journal files ch
lines 1-21/21 (END)
```

FIGURE 9.14 Using `systemctl` in Red Hat 8.

specific run status. Another similar command `is-enabled` to determine the enabled status of the unit. `is-enabled` can respond with many different types of status aside from enabled or disabled.

`enabled` and `enabled-runtime` mean that the unit is running where runtime further indicates that the unit was started at runtime rather than a system startup. The values `linked` and `linked-runtime` indicate that access to the unit was made through a symbolic link while `masked` and `masked-runtime` indicate a unit that is not only disabled but will not start (will fail to start if attempted).

`disabled` means that the unit is not currently enabled but can be. `indirect` indicates that the unit is not enabled and can be enabled only indirectly by being wanted by another unit, while `static` is sort of the opposite in that the unit is not enabled but there are no provisions to start the unit in the unit file's `[Install]` section (no other unit wants this unit). Three last possible values are `generated`, `transient` and `bad`, the first two expressing different means for dynamically generating the unit while the last is reserved for unknown units or units that are erroneous.

The `isolate` command starts the unit(s) and all of its(their) dependencies while all other units are stopped. This allows us to run one group of units in isolation. For this to work, `AllowIsolate` must be set to `yes` (for targets) or `IgnoreOnIsolate` must be set to `no` (for all other types of unit) in the unit's unit file.

`systemctl` has a generic syntax of `systemctl [`*`options`*`] `*`command`*` [`*`units`*`]`. As noted, commands include `start`, `stop`, `status`, `restart`, `is-active`, `is-enabled`, among others listed above or not discussed here. We supply the item(s) to be affected by the command as one or more *units*. Again, if the unit is not a service then full unit file name must be provided. Some of the commands we will explore do not require a specified unit. Options vary depending on the command. We explore additional commands, descriptions of those commands and possible options, in Table 9.10.

Let's focus on the first entry of Table 9.10, `list-units`. With no options, `systemctl list-units` lists all known units. We can narrow down the list by specifying some property such as the type of unit or some status of interest. `systemctl list-units -t socket` displays all socket units. We can also specify a type using `--type=`*`type`*. The output of `list-units` is not just the list of units but each unit's status (`loaded` or `unloaded`, `active` or

TABLE 9.10
Other Useful `systemctl` Commands

Command	Meaning	Useful Options
`list-units`	List available units; options limit the units listed based on type or state.	`--type=`*`type`*, `--state=`*`status`*, `--all` (or `-a`)
`list-sockets` `list-timers`	List all units of the given type; note that this is restricted to only these two types.	Same as for `list-units`
`list-dependencies`	List the dependencies of a given unit(s).	`--after`, `--before`
`show`	List properties of the given unit(s) where properties are values of directives.	`--property=`*`prop`*1 `[,prop2,...]`
`set-property`	Change unit's runtime properties; note that the property must be of a changeable type or the instruction fails.	*`PROPERTY=VALUE`* where `PROPERTY` is one of the unit's directives and `VALUE` is a legal value
`edit`	Like `set-property` but drops you in an editor to make changes.	`--full` (copy original unit), `--force` (creates a new unit if the specified unit does not exist), `--runtime` (change only the runtime version of the unit), `--user` (only make changes for this user, not all users)
`help`	Display the unit(s) `man` page(s).	Aside from listing units, we can list PIDs

`enable`
`disable`
`reenable`
	Enable/disable the unit(s); reenable performs disable followed by enable.	

`mask` `unmask`	Mask or unmask the specified units; masked units cause the unit file to map to `/dev/null` so that the unit is no longer available for any operation (except for unmask).	`--now` `--runtime`
`revert`	If any of the units listed have been modified using any of the `systemctl` commands of `edit`, `set-property` and `mask` then revert the unit(s) to the original format.	
`get-default` `set-default`	Returns or changes the startup target by returning or changing what the symbolic link `default.target` points to.	`set-default` requires the unit file of the new default, e.g., `multi-user.target`

`inactive`), and a type-specific status such as `running` or `listening` for sockets, `running` or `exited` for services, `mounted` or `waiting` for mounts/automounts and `running` or `waiting` for timers.

Another useful command is `list-dependencies`. The dependencies are the units that the specified unit depends on. Options are `--before` and `--after`. We demonstrate this with the units that `avahi-daemon.service` must launch before it. The result is shown in Figure 9.15. Notice the hierarchical nature of the output shows not only `avahi-daemon`'s dependencies but what those depend on. We can use this to track dependencies of any unit type.

`systemctl` can query other aspects of our system aside from unit files. The command `list-machines` lists the hostname(s) and status of all running local containers (a consolidated environment of dependent processes). With `list-jobs`, `systemctl` lists all jobs in progress that match the specified units. Use `show-environment` to view all of the environment variables for `systemd` and `set-environment/unset-environment` to change the value of an environment variable where unset will cause the environment variable to have no value.

```
[root@localhost system]# systemctl list-dependencies avahi-daemon.service --before
avahi-daemon.service
● ├─pmcd.service
● ├─multi-user.target
● │ ├─systemd-update-utmp-runlevel.service
● │ ├─graphical.target
● │ │ ├─systemd-update-utmp-runlevel.service
● │ │ └─shutdown.target
● │ └─shutdown.target
● └─shutdown.target
[root@localhost system]#
```

FIGURE 9.15 Dependency output from the systemctl list-dependencies command.

We can control aspects of the operating system through systemctl. With is-system-running, we obtain the status of the operating system as one of running, starting, initializing, degraded (running but with some failures), maintenance, stopping, offline and unknown. We can change the target level of the system by specifying one of default, rescue, emergency, halt, poweroff and reboot (as we noted in Chapter 8 when changing to rescue mode to unmount /home). When changing to default, rescue or emergency, we are launching to the given target level in *isolate* mode. The instruction systemctl default is the same as systemctl isolate default.target. There are other modes which we can explore through systemctl's man page. Every one of those modes has its own target file.

As you learn more about Linux and systemd in particular, you will gain confidence in using systemctl. Early on, perhaps the wisest usages are only of the commands that show you status information (status, list-units, list-dependencies, show). Certainly, you will not want to use systemctl to alter unit's properties. Stopping services can be at least slightly detrimental in that stopping one service may cause others to stop based on dependencies. The good news is that systemd should start a service when needed, including anything it depends on, but this is wasteful of time. On the other hand, usually starting a service that you think should be running should not cause similar problems unless the service conflicts with other running units.

SECTION ACTIVITIES

1. Read the man page for systemctl. Identify five commands that we did not explore in this section and try them (you will have to su to root). Do not try set-property. If you feel the command you entered will mess up your system, use revert to undo your command.
2. You can change the mode of your system using systemctl followed by a new target such as rescue. If no one else is currently logged into your system, use systemctl to change mode to rescue mode. You will be asked to enter the root password again. Upon entering rescue mode, the only allowable user is root. You can exit this mode typing exit.

9.6 CONFIGURATION FILES

In Windows, nearly all aspects of the operating system are controlled by various GUI tools. This is not the same in Linux and may be a source of controversy among users and system administrators. Users and some administrators who are used to Windows might prefer more GUI-based controls. The more purist Linux user will always prefer the command line. One reason to prefer the command-line approach is that an instruction like systemctl gives us greater control and precision in modifying our system. Another argument in favor of the command line is that many changes will take less effort through the command line than in opening a GUI tool, modifying and saving the changes.

It's a moot point in any event because Linux has few GUI tools that support operating system configuration. In Chapter 1, we got our first peak at Cockpit. We also saw various system settings that we can modify in Chapter 1 (and elsewhere). We revisited Cockpit in Chapter 7 when we explored creating user accounts. We will visit the Linux firewall, firewalld, in Chapter 10 and see that we can view and modify its settings through both the command line and a GUI tool. But aside from settings, Cockpit and the firewall GUI, nearly all forms of configuration are left to commands and configuration files. We wrap up this chapter by concentrating on a number of configuration files.

9.6.1 Non-Service Configuration Files

We can divide configuration files into those that proscribe the behavior of a service and those that are data files, used by various services and programs. These data files typically enumerate rules, users or other system objects. Configuration files instead specify directives that control aspects of the service. We start with data files used to configure system performance as these files are generally easier to understand.

The atd and crond scheduling services, by default, are accessible to all users. Linux gives the administrator the ability to blacklist users to disallow them from using these services. An administrator might choose to do this if a given user has abused scheduling privileges by, for instance, continually scheduling tasks at peak usage hours. There are two files, /etc/at.deny and /etc/cron.deny, to create such blacklists. These files are initially empty; root can later add usernames to either or both files. Earlier versions of Unix and Linux also had /etc/at.allow and /etc/cron.allow in case the administrator wanted to disallow all users by default and whitelist those who could use the service(s).

The useradd instruction is not a service but it has its own configuration file of a type. The file /etc/default/useradd stores the useradd default values. These values can be viewed and modified using useradd -D (as discussed in Chapter 7). The specific entries in the file define default values for GROUP (the GID used for user accounts without their own private group), inactive (-1) and expire (none), the default location for the user's home directory, the default shell, the default skel directory and whether an email spool file should be generated for the user.

Other user defaults can be found in /etc/logins.def. These values are primarily used to establish defaults in /etc/shadow for password expiration values. Also in this file are a default umask, whether a user should be allowed to login should the user's home directory be unavailable (CREATE_HOME), and the location of the user's email file (MAIL_DIR). The file also contains minimum and maximum UID values for user accounts and system accounts.

Two files that support network communication and in particular domain name system (DNS) usage are /etc/hosts and /etc/resolv.conf. /etc/hosts allows our computer to bypass DNS by directly mapping hostnames to IP addresses. The contents of this file are a list of entries, each entry on a line by itself where the entry is the host's IP address followed by one or more aliases for that host. We would use this only for well-known and stable (static) IP addresses. /etc/resolv.conf is the hostname resolver file. It provides details for how our computer will use DNS. It can enumerate the options that the resolver should attempt such as to use local cache first, then the /etc/hosts file before attempting to use a DNS name server. The primary purpose of the file is to list the IP addresses of the local name servers. We explore both of these files in Chapter 10.

Another file that supports network communication is called /etc/services. This file enumerates well-known port addresses and the services that use those ports (or that should respond to messages coming in over that port). It also lists the datagram type (tcp or udp) expected. For instance, FTP (file transfer protocol) has four entries, two for ftp data (port 20 tcp, port 20 udp) and

```
# Basic system aliases -- these MUST be present.
mailer-daemon:  postmaster
postmaster:     root

# General redirections for pseudo accounts.
bin:            root
daemon:         root
adm:            root
lp:             root
sync:           root
shutdown:       root
halt:           root
mail:           root
news:           root
uucp:           root
operator:       root
games:          root
gopher:         root
ftp:            root
nobody:         root
usenet:         news
ftpadm:         ftp
ftpadmin:       ftp
ftp-adm:        ftp
ftp-admin:      ftp
www:            webmaster
webmaster:      root
noc:            root
security:       root
hostmaster:     root
info:           postmaster
marketing:      postmaster
sales:          postmaster
support:        postmaster
```

FIGURE 9.16 Excerpts from the /etc/aliases file.

two for ftp operations (port 21 tcp, port 21 udp). Also listed are aliases for the service, in this case fsp and fspd.

The /etc/aliases file allows the administrator to create email aliases. Entries are listed as alias: username where the *alias* is then tied to the *username*. Excerpts from this file are shown in Figure 9.16. Notice that the first two aliases are required. An administrator can create further aliases. Remember that these aliases only tie usernames for email purposes. For instance, any email sent to user daemon or sync is actually sent to root.

The file /etc/exports stores the NFS export table. This file contains the list of local file systems that can be remotely mounted. We explored this in Chapter 8. The file /etc/hostname stores the computer's hostname. This value can be displayed or changed using the command hostname. Similarly, /etc/machine-id stores the machine's ID number. The file /etc/mime.types enumerates mappings between MIME (Multipurpose Internet Mail Extensions) types and file extensions/applications. We explored the use of sudo in Chapter 7. As noted in that chapter, to permit sudo access root must establish the username(s) and command(s) in the /etc/sudoers file.

Now that we've explored files that support non-services, we wrap up this chapter by looking at configuration files for services. We limit our examination because there are dozens of such files found in /etc and various subdirectories. It might be easier to understand configuration files if they all followed the same format but unfortunately this is not the case. Some configuration files simply define environment variables. We see this, for instance, in cpupower, sysstat and rpcbind, all of which are found in /etc/sysconfig (in Red Hat). Other configuration files contain service-specific directives of varying syntaxes. Yet others contain rules although again, in varying types of formats.

Our examination of configuration files will be limited to just a few of the services we had previously described in Section 9.4 and listed in Table 9.11. To learn about other configuration files, consult the service's man page as they will have a CONFIGURATION FILE section to describe the directives available. More information can also be found in the configuration files themselves as most contain comments that describe configuration directives and many have pre-specified directives, possibly commented out. Some configuration files even have their own man pages.

9.6.2 Configuring rsyslog

The rsyslog service uses /etc/rsyslog.conf for its configuration. The file begins with a number of module statements to load needed modules for logging actions. For instance, module(load="imuxsock" SysSock.Use="off") is used to provide support for local logging while turning off message reception over the local socket. The directive module(load="imjournal" StateFile="imjournal.state") provides access by journald using imjournal.state as a file name to store the position of the journal.

The configuration file also stores global directives. These have a similar format to the module directives, such as global(workDirectory="/var/lib/rsyslog") to provide the location of rsyslog's working directory (this is not the same as the location of the log files) and include(file="/etc/rsyslog.d/*.conf" mode="optional") to indicate that other configuration files located in the rsyslog.d directory can be loaded (if any exist).

Much of the remainder of the file consists of rules that instruct rsyslog of the types of events that should be logged and where those messages should be logged. Rules take on the format *source.priority action*, usually separated by one or more tabs. The first pair of items indicates the *source* (type of software) and *log level* of the events that are to be logged. The *action* indicates where the messages should be logged.

Some of the available sources are mail (the system mailer), cron (the crontab service) and authpriv (authentication software). We can also assign specific application software to logging rules using the reserved sources of local0, local1, local2, ..., local7. An * can be used in place of the source indicating all sources. For instance, *.*priority* would indicate an event of the given *priority* level as generated by any system software.

The *priority* dictates which level of event should be logged. There are nine levels of priority: none, debug, info, notice, warning, err, crit, alert and emerg. The meanings of these priorities are given in Table 9.11 (along with the list of possible sources). Each level is progressively more urgent. Using an * for priority indicates that any message sent by the source no matter what level the priority is to be logged.

The *action* is the location of where the log event should be sent. This is typically a file. The log files are text files, and most are stored under /var/log. We can replace the log file with /dev/console to direct log messages to the main console window. Alternatively, the use of * for action causes the messages to be redirected to all open terminal windows. This might be useful for emergency messages such as an impending kernel crash.

Aside from sending log messages to a file, we can redirect the log messages to a named pipe using | *fifo* where *fifo* is the name of the pipe. Another option is to provide a list of users who will receive the log messages. Such messages will be displayed in the user's console window if logged in with an open window.

Yet another option is to forward messages to a remote machine. We first establish forwarding messages in the first portion of the configuration file and follow it with a mapping rule whose action is @*ip_address* as in @192.168.0.1.

TABLE 9.11
Priority Levels and Sources for rsyslog.conf

Priority Level	Meaning	Source	Meaning
none	Do not log messages of the specified source.	auth/authpriv	Security/authentication
debug	Debugging messages; used by programmers/ software testers.	cron	anacron, crond, atd
info	Informational messages capture what the application is doing.	daemon	Other system services
notice	Events worth noting such as opening files, writing to disk and mounting attempts.	kern	Kernel messages
warning	Potential problems.	lpr, mail, news	Printer, email and Usenet news applications
err	Errors that do not cause the program to terminate.	syslog	Issues from rsyslog itself
crit	Errors that will cause the program to terminate.	user	Generic user-level messages
alert	Errors that cause the program to terminate while also possibly presenting problems with other running programs.	uucp	UUCP subsystem
emerg	Errors that could cause the kernel to crash.	local0...local7	Assigned to application software

```
#kern.*                                          /dev/console
*.info;mail.none;authpriv.none;cron.none         /var/log/messages
authpriv.*                                       /var/log/secure
mail.*                                           -/var/log/maillog
cron.*                                           /var/log/cron
*.emerg                                          *
uucp,news.crit                                   /var/log/spooler
local7.*                                         /var/log/boot.log
```

FIGURE 9.17 Example logging rules in /etc/rsyslog.conf.

The initial rules of rsyslog.conf may look like what is shown in Figure 9.17. These rules operate like a nested if-then-else statement. The first rule is commented out in that all kernel messages would be sent to /dev/console. The reason that this rule is commented out is that the klogd daemon is already logging kernel messages for us. If we were to uncomment this rule, not only would klogd log these messages, but they would also be displayed on the administrator's console.

The first uncommented rule causes all informational events except for those originating from mail, authpriv and cron to be logged to /var/log/messages. The reason for an entry of the form *source*.none is to say "exclude messages from this source". The next three rules all pertain to these three sources, sending any events from these three sources to the secure, maillog and cron log files, respectively. The next rule indicates that all *.emerg actions (any software has an emergency which has the potential to crash the kernel) be sent to all open windows (*). You should be able to interpret the final two rules of the figure.

Notice the hyphen appearing before /var/log/maillog. This indicates that writing to this file does not need to be *synchronous*. This means that messages are not necessarily written to the file in the order that they are received. The reason for asynchronous writing to this file is because the maillog may be a very large file and asynchronous access is more efficient than synchronous access.

The bottom of the rsyslog.conf file contains forwarding rules. These rules take on a different format than the mapping rules. Sample forwarding rules appear in this section but are

```
#queue.maxdiskspace="1g"
#action.resumeRetryCount="-1"
#remote_host is: 192.168.0.1:10514
---------------------------------------------------------
daemon.*                        /var/log/daemons
```

FIGURE 9.18 Further directives for the /etc/rsyslog.conf file.

commented out. Three such rules are shown in the top portion of Figure 9.18. The first of these rules specifies that as much as 1GB can be used for the queue that will forward messages. The second rule indicates that if the remote host does not respond, continually retry (-1 indicates no limit). The last rule provides the hostname/IP address of the remote host, which includes a port address (the port is optional).

We would not need to alter /etc/rsyslog.conf unless we find other software that we want to log. If we did add a rule, it might appear like the new rule shown in the bottom portion of Figure 9.18. This rule logs all events originating from any service to the file /var/log/daemons. The problem with this rule is that it logs information messages, debugging messages, warnings, etc., so that this file would grow very quickly. It would be better to limit service-generated events to the more significant issues; we might use one of daemon.warn, daemon.err or daemon.crit instead. With our configuration file modified, we need to save the file and restart the rsyslog service. We would use systemctl to either stop and start or restart the service.

9.6.3 Configuring nfs

The nfs service supports network file sharing by permitting file systems to be remotely accessible over network. Although originally intended for local area network file sharing, file systems can be remotely accessed over any network. We explored this in Chapter 8. Here, we re-examine the details specific to configuring nfs to permit remote access to a local file system.

The first step is to specify the access point in the file /etc/exports. This file will be empty initially. Entries have the format *local_mount_point network_address(options)*. The *local_mount_point* is the local directory. The *network_address* consists of one or more IP addresses or subnet addresses of the allowable clients who can remotely mount this given file system. An * for the IP address indicates that the file system is accessible to everyone. *Options* indicate access (mount) options.

Figure 9.19 shows three examples that we will step through. The first entry offers the local directory /home/coolstuff to anyone on the network 10.11. Access is synchronous and read-only. The second entry allows access of /home/coolstuff to three specific hosts (10.2.1.2, 10.2.3.4 and 10.11.15.21), each with its own set of options. The option rw permits write access for users on 10.2.3.4, but this access is still restricted as discussed below. Access for anyone on 10.2.1.2 and 10.2.3.4 is asynchronous, and the wdelay option for 10.11.15.21 further delays a write if another write is expected to occur shortly. We address no_root_squash momentarily. Note that the second entry should appear on one line of the file but was extended to a second line here due to width limitations.

```
/home/coolstuff      10.11.0.0/16(ro,sync)
-------------------------------------------------------------
/home/coolstuff      10.2.1.2(ro,async) 10.2.3.4(rw,async)
           10.11.15.21(rw,wdelay,no_root_squash)
-------------------------------------------------------------
/home/coolstuff 10.2.3.4(rw,async,anonuid=1005,anongid=1005)
```

FIGURE 9.19 Example /etc/exports entries.

While 10.2.3.4 has write privilege to the /home/coolstuff directory, a user who accesses the directory through the remote mounting has their own permissions. If the directory is not world-writable, then the user is unable to copy or move content into the /home/coolstuff directory. Similarly, if a file is not world-writable, the user on 10.2.3.4 cannot write to or delete an existing file. Thus, rw access is harmless if proper permissions exist on the local directory.

With that said, imagine that the user of 10.2.3.4 has mounted /home/coolstuff and has su'ed to root. What access will root of 10.2.3.4 have on our local file system? The answer is that the local root has the same access rights as other (world) does on 10.2.3.4. This is because local root is not root of the remote computer. nfs assigns the *other* users the username and groupname of nfsnobody by default. This can be overridden by adding anonuid and anongid entries to the options. The third entry in the figure demonstrates this alternative, specifically for 10.2.3.4. In this case, any remote user is given local UID/GID of user 1005.

The second entry in the figure includes the option no_root_squash. With this option in place, if a root user of a local computer saves files to the remote file system, they are not saved under nfsnobody but instead as root. This can be dangerous and is not recommended and so instead anonuid/anongid provide a more secure approach, as shown in the third example.

Once the /etc/exports file has been established, we execute exportfs. If we make changes to already established entries in /etc/exports, it is best to execute exportfs -f to flush the nfs export table and then execute exportfs -a. We then have to restart the service rpcbind (this service accepts incoming requests and passes them on to the appropriate nfs service). We also have to ensure that our firewall is set up to accept requests coming in over the nfs port (by default, 2049). We covered the necessary instructions for this in Chapter 8 but list them again, in Figure 9.20.

NFS requests may use either TCP or UDP packets and so we might want to duplicate the last command in Figure 9.20 but exchange 2049/tcp with 2049/udp so that the firewall accepts both forms of packets. firewall-cmd does not let us add both 2049/tcp and 2049/udp in one command, and thus we would use two separate commands.

The instructions in Figure 9.20 only impact the current (runtime) version of the firewall. We need to add --permanent to each and then issue firewall-cmd --runtime-to-permanent to ensure that these changes are made both now and whenever we restart the firewall. We will cover firewalld and firewall-cmd in more detail in Chapter 10.

There is more to nfs than has been presented here. The program and its suite of services have their own configuration files of note. Most of these services directly or indirectly use the file /etc/nfs.conf, which is divided into sections.

The service nfs-mountd.service uses the [mountd] section of /etc/nfs.conf. The service nfs-server.service calls upon exportfs which itself uses the /etc/exports file, as noted above, and the [exportfs] section of /etc/nfs.conf. nfs-server.service then runs the service rpc.nfsd, which uses the [nfsd] section of /etc/nfs.conf. The service nfs-blkmapd uses the [general] section of /etc/nfs.conf. Aside from nfs.conf, the mount command will use /etc/nfsmount.conf for its configuration and /etc/fstab as mount data for mounting automatically.

As noted above, nfs.conf is divided into sections denoted using [*sectionname*]. While we listed several sections above, there are many others. Most sections have only one or two recognized directives. For instance, [general] is there to specify a directory using pipefs-directory

```
firewall-cmd --add-service mountd
firewall-cmd --add-service rpc-bind
firewall-cmd --add-service nfs
firewall-cmd --add-port=2049/tcp
```

FIGURE 9.20 firewall-cmd instructions to export file system.

and the [exportfs] section only permits a directive to define the debug level. The [lockd] section only permits values for port and udp-port.

The [nfsd] and [mountd] sections permit far more directives. For [nfsd], directives include the debug level, number of available threads, host and port values, grace and lease time (in seconds), whether TCP packets are allowed, and which NSF versions can be responded to. The [mountd] section includes directives for the debug level, number of descriptors available, port to use, number of threads to assign, whether reverse IP lookups are permitted, a default directory path, among others. By default, the entries in /etc/nfs.conf are nearly all commented out. This leaves it up to the administrator to decide if default values should be changed. Consult the nfs.conf man page for more detail.

9.6.4 Configuring logrotate

Log files may grow rapidly. The logrotate service is tasked with rotating log files as they grow and/or age. Log file rotation causes the current log file to be renamed and set aside, and a new file created.

logrotate uses two sets of configuration files. The first configuration file is /etc/logrotate.conf. This file consists of rotation directives used for all software unless specifically overridden with a separate configuration. The directory /etc/logrotate.d contains those software-specific configurations, each in a file whose name matches the software, such as cups, dnf and sssd.

The entries in /etc/logrotate.conf are keyword directives. Some of these require values, others do not include specific values. The directives are largely divided into frequency of rotation, format of the rotated file's name, number of old log files to retain, whether and how to compress log files and actions to take before or after rotation takes place. Table 9.12 provides detail on the directives.

Figure 9.21 demonstrates a setup for /etc/logrotate.conf. We see that the default is to rotate log files weekly, retain four copies, create a new file with each rotation while renaming the current log file by appending the current date to the filename. The newly created file will be given the old file's name. For instance, if the file's name is service_log then it would rotate to something like service_log-20230121 and the new file will be named service_log. Notice that compress is commented out so that files remain uncompressed. The last directive in the figure, include, causes logrotate to also read all configuration files stored in the specified directory.

For log file directives specific to a particular log file/application, there are two options. The first is to include the directives in the logrotate.conf file. Doing so requires preceding the directives with the log file's name followed by the directives in {}, for instance with /var/log/wtmp {#*directives here*#}. The more recent approach is to store the directives in files in an include directory, such as in /etc/logrotate.d, as indicated in Figure 9.21. Figure 9.22 provides two example entries in the file named dnf. Notice that with the two entries, logrotate will be able to rotate two log files generated by dnf.

In comparing the two sets of configurations, notice that both log files generated are treated with identical directives, but that these directives differ from those of the general directives provided in Figure 9.21. The first directive, missingok, informs logrotate that should the current log file be missing do not signal an error but instead move on to the next entry. The next three directives indicate that log rotation should take place weekly but only if the current file is not empty and that four older log files should be retained at a time. The last directive specifies how the new file should be created. Specifically, the new file will have permissions of 600 and its owner and group owner will be root.

In Figure 9.23, we have the configurations for bootlog (top), chrony (middle) and wtmp (bottom). Although the entries shown in Figure 9.23 have many of the same directives that we saw as the defaults from Figure 9.21 and the dnf-specific directives shown in Figure 9.22, we also see several new directives. For instance, boot.log log files are rotated daily with seven old copies retained at a time. copytruncate indicates that rotation will cause the current file to be copied into a new file and then the current file's contents deleted rather than creating a new file.

TABLE 9.12
Various Directive Types and Values for `logrotate.conf`

Directive Type	Possible Values	Meaning/Options
Duration	hourly, daily, weekly, monthly, yearly	Temporal frequency of log rotation.
File size	size, maxsize, minsize, notifempty	Sizes are specified using a notation like 1M or 200K; size indicates that rotation should take place once the current log file exceeds the given size; maxsize causes the file to be rotated even if a duration is provided and the duration has not yet been reached; minsize causes the file to be rotated only once it exceeds both the size and the duration; notifempty will not rotate the current log if it is empty in spite of reaching the duration.
Compression	nocompress, compress, compresscmd, uncompresscmd, compressext, compressoptions	The cmd directives take as an argument the compression/decompression program to use otherwise gzip is used; ext stands for the extension to affix to the file; options are command-line options to pass to the compression program.
Copy	copy, copytruncate, nocopy, nocopytruncate	Copy the current file into a new file retaining both or copy the current file into a new file and then truncate the current file (clear the contents to save from having to create a new empty file).
Create	create, nocreate	Rotate and then create the new file (create is the default); both permit as values the permissions, owner and group of the created file; related is createolddir which creates the default directory (as specified by the olddir directive) if that directory does not yet exist; it also permits permissions, owner and group values.
Limit number of logs saved	rotate, start	Specify a count for the maximum number of rotated logs maintained (not including the new log); once the count has been reached the oldest log file is deleted; default is 0; start includes an integer to indicate the first log file number to use (when not using a date).
Rotated log file's name	dateext, dateformat, datehourago, dateyesterday, nodateext	Specify the naming scheme for the rotated log file; dateext affixes the date as YYYYMMDD whereas dateformat uses a string descriptor like "%m/%d/%Y"; with datehourago and dateyesterday, the date is from 1 hour/1 day ago.
Filename extension	extension, addextension	Both require an extension to affix to the file's name.
Extra script execution	prerotate, postrotate, firstaction, lastaction, endscript	Between prerotate, postrotate, firstaction, lastaction and endscript, we can specify sh script instructions to execute before rotation, after rotation, before prerotation, after postrotation.
Age restrictions	minage, maxage	minage specifies that the log file is not rotated if it is less than the given number of days; maxage indicates an age by a rotated log file must be deleted; both ages are indicated by a number of days (an integer).

For chrony, instead of specifying a filename, the entry specifies a collection of files using a wildcard. These directives apply to every .log file found in the /var/log/chrony directory. The directive nocreate indicates that new log files are not created. This entry contains sharedscripts followed by a script notated between postrotate and endscript. The sharedscripts directive informs logrotate to only run this script once no matter how

```
weekly
rotate 4
create
dateext
#compress
include /etc/logrotate.d
```

FIGURE 9.21 Example /etc/logrotate.conf directives.

```
/var/log/dnf.librepo.log {
     missingok
     notifempty
     rotate 4
     weekly
     create 0600 root root
}

/var/log/hawkey.log {
     missingok
     notifempty
     rotate 4
     weekly
     create 0600 root root
}
```

FIGURE 9.22 Entries in the file /etc/logrotate.d/dnf for dnf log rotation.

```
/var/log/boot.log {
     missingok
     daily
     copytruncate
     rotate 7
     notifempty
}
------------------------------------------------------------
/var/log/chrony/*.log {
     missingok
     nocreate
     sharedscripts
     postrotate
         /usr/bin/chronyc cyclelogs > /dev/null 2>&1 || true
     endscript
}
------------------------------------------------------------
/var/log/wtmp {
     missingok
     monthly
     create 0664 root utmp
     minsize 1M
     rotate 1
}
```

FIGURE 9.23 logrotate configuration for boot.log, chrony and wtmp services.

many files match this entry. The instruction runs chronyc on the cyclelogs file, sending all output to /dev/null. Thus, the command executes but output is ignored.

The last entry is for wtmp and specifies that logrotate will rotate files when both the current log file's size reaches 1M and at least 1 month has elapsed. The new file is given permissions of 664 and owner and group of root and utmp respectively. Only one previous log file is retained at a time. Notice that while these directives are placed in a file named wtmp, this configuration could also have been placed directly in /etc/logrotate.conf in the brackets of /var/log/wtmp {#directives here#}.

Depending on the size of our Linux system, we might find it beneficial to modify some of these configuration files. For instance, if we rarely examine the log files of some software we might prefer to have the rotated log files compressed. If we feel that log files are becoming too large, we might decrease the duration or change from a duration to a size. If we have some data mining application to perform on a particular log file, we would add a `postrotate` section to run the application on the newly rotated log file. We might also have to add a new configuration file if we have installed some software that creates logs.

9.6.5 Configuring auditd

The last of the service configurations that we explore in this chapter is for the `auditd` auditing service. This service has three configuration files and two or three sets of rule files, all located within the `/etc/audit` directory. The primary configuration file is `auditd.conf` which stores the configuration for how `auditd` will run. Included in this file are directives to specify the location and name of the log file generated, the format of the log file, the number of log files to retain, the maximum size of a log file before log file rotation should kick in (or some other action including suspending the daemon until some other action takes place or sending a message to syslog to log a warning) and flushing operations to delete entries from log files. The directives from this configuration file are shown in Figure 9.24.

```
local_events = yes
write_logs = yes
log_file = /var/log/audit/audit.log
log_group = root
log_format = ENRICHED
flush = INCREMENTAL_ASYNC
freq = 50
max_log_file = 8
num_logs = 5
priority_boost = 4
name_format = NONE
##name = mydomain
max_log_file_action = ROTATE
space_left = 75
space_left_action = SYSLOG
verify_email = yes
action_mail_acct = root
admin_space_left = 50
admin_space_left_action = SUSPEND
disk_full_action = SUSPEND
disk_error_action = SUSPEND
use_libwrap = yes
##tcp_listen_port = 60
tcp_listen_queue = 5
tcp_max_per_addr = 1
##tcp_client_ports = 1024-65535
tcp_client_max_idle = 0
transport = TCP
krb5_principal = auditd
##krb5_key_file = /etc/audit/audit.key
distribute_network = no
q_depth = 400
overflow_action = SYSLOG
max_restarts = 10
plugin_dir = /etc/audit/plugins.d
```

FIGURE 9.24 Directives from `/etc/audit/auditd.conf`.

The /etc/audit directory also contains a rules file, audit.rules. See Table 9.13 for a description of the rule format. Rules are specified using options, some of which have parameters. The file /etc/sysconfig/auditd.conf contains directives for extra auditd options. Specifically, this file is used to control how auditd starts and stops as opposed to the configuration of how auditd runs. This file controls the language that auditd uses (defaults to US English) and whether the audit system, including system calls, should be shut down when auditd is not running.

An example of auditd.conf is shown in the left half of Figure 9.25. We see in this example that first any previously defined rules are deleted. Next, auditd will use up to 8192 buffers with a failure flag set to 1 (enable auditing). The entry backlog_wait_time sets the amount of time the kernel will wait should the backlog limit be reached before attempting to queue more messages. The default value is 60 times the clock speed of the CPU if not specified. Since there are no conditions listed, auditd will log all relevant messages.

The file /etc/audit/audit-stop.rules consists of rules to be invoked should auditd stop. We might find this file contains only two entries as shown on the right half of Figure 9.25. The two statements in the /etc/audit/audit-stop.rules file disable auditing and delete all current rules. There may be additional configuration and rule files in subdirectories. Under /etc/audit/plugins.d are configuration files for plugins while under /etc/audit/rules.d may be further rules files.

TABLE 9.13
Rule Directives for auditd

Syntax	Meaning
-D	Delete any previously defined rules.
-c, -i	Continue to load rules in spite of an error or ignore errors.
-b #	Set a maximum number of audit buffers where # is a number.
-e 0, 1 or 2	0 to disable auditing, 1 to enable auditing, 2 to lock configuration (-e 2 should be the last rule in the file).
-f 0, 1, or 2	Set failure flag, 0 for silent, 1 for print failure messages, 2 for panic.
-r #	Limit to # of messages per second (0 for no limit), exceeding this rate with a failure flag of two causes the kernel to get involved.
-R *file*	Read rules from the corresponding file.
-s	Output auditd's status.
-w *directory*/*file*	Log attempts to access the *directory* or *file*.
-w *file* -p [rwxa]*	Log attempts to read *file* (r), write to *file* (w), execute *file* (x) or change *file*'s attributes (a). The * indicates that any combination of the options r, w, x, and a can be listed.
-a *list*,*action*	Append a rule; *list* is one of task, filesystem, exit, user or exclude. The *action* can be one of never or always or a condition specified using any of the following, among other types of tests for rules: -C – compare values like auid, uid, euid, suid, gid, arch (CPU architecture), directory or path, etc. using = and != -F – comparison using numerical values and <, >, !=, =, etc.

```
-D                                  -e 0
-b 8192                             -D
-f 1
--backlog_wait_time 60000
```

FIGURE 9.25 auditd.conf sample entries (left) and /etc/audit/audit-stop.rules (right).

SECTION ACTIVITIES

1. We have only explored a few of the configuration files in /etc. As root, enter `find /etc -name "*.conf"` which will list all .conf files in /etc and subdirectories. Pipe this command to `grep -c` to find out how many there are.
2. Following on from the last activity, look at some of the configuration files you find. How many that you look at are simply files full of comments with no actual directives? For those that have directives, can you understand them? Read the comments if not to see if they give you any more useful information.
3. The configuration file /etc/oddjobd.conf uses XML notation. Do you find this notation to be easier or harder to read?
4. Select several .conf files in /etc and see which ones have man pages. Read a couple of the man pages for these .conf files. Does it give you enough information to understand the role of and type of directives? Do you feel confident that, as a system administrator, you can modify a configuration file when needed?

9.7 CHAPTER REVIEW

Concepts and terms introduced in this chapter:

- After – directive for unit files to indicate that this unit must start after those listed.
- Before – directive for unit files to indicate that this unit must start before those listed.
- Booting – the process of starting a computer; the boot process includes running a power-on self-test, locating bootable devices, loading the operating system kernel and initializing it.
- Boot loader – a program which performs the portion of booting that locates and loads the operating system kernel.
- Configuration file – a file of directives or options that defines how a service will execute; changing the configuration file will alter the service's behavior.
- Conflicts – directive for unit files to indicate which other units this unit conflicts with.
- DRAM – dynamic random access memory; used for main memory; low-cost but volatile.
- ExecStart – directive for service unit files to indicate the command used to start the given service.
- GRUB2 – the current Linux boot loader program.
- initramfs – an image storing a file system loaded into a ramdisk during system booting and used during Linux kernel initialization; one or more initramfs images are stored in /boot.
- Master boot record – a reserved location on the hard disk storing a portion of the boot loader.
- Non-volatile memory – a form of memory whose contents are retained even without power; ROM is a form of non-volatile memory.
- POST – standard routines that any computer will run during the boot process to test available hardware and get the hardware ready to run the boot loader.
- RAM – random access memory; three forms are DRAM, SRAM and ROM (although we typically do not think of ROM as random access); access to any memory location will take the same amount of time as any other.
- Ramdisk – memory used to mimic files so that the operating system can access contents using file commands without the slower interaction with disk files.
- Requires – directive for unit files to indicate which other units this unit depends on.
- ROM – read-only and non-volatile memory; this type of memory has its contents permanently fixed in place so it can be read from but not written to; the primary use of ROM is to store the boot program (or a portion of it).

- ROM BIOS – basic I/O system stored in ROM so that the computer can access standard I/O devices during the boot process.
- Runlevel – the mode that the system is initializing to, numbered 0 through 6; the runlevel dictates the services to start or stop; runlevels have been replaced in `systemd` with targets.
- Service – an operating system program that responds to requests from any number of sources; services are background processes which only execute when called upon.
- SRAM – static random access memory, which is more expensive but faster than DRAM; this form of memory is used to make registers and cache memory; SRAM is volatile.
- System V – a dialect of Unix from the 1980s whose initialization process was adopted by many later Unix and Linux distributions; in Linux, the SysV initialization process was eventually replaced by Upstart and now `systemd`; both SysV and Upstart use runlevels and the `init` process.
- Target – type of unit that is used to group together events needed for system startup; targets rely on other targets and services.
- Unit/unit file – units proscribe initialization events and pertain to such things as startup targets, services, timers, mount points, collections of processes and process resources, among other things; each unit has its own unit file whose name is *`unit.type`* where *unit* is the name of the unit and *type* is the unit's type like service, target, slice, swap or mount.
- Upstart – replaced System V for the Linux startup process; continued to use runlevels but coordinated startup of services in an event-based way to improve performance; replaced by `systemd`.
- Volatile memory – a form of memory that requires a constant power supply to retain its contents; both SRAM (cache, registers) and DRAM (main memory) are forms of volatile memory.
- Wants/WantedBy – directives for unit files that describe what other units this one wants or what other units that want this one; for targets, there is a wants directory that lists all of the units wanted by the given target.

Linux commands covered in this chapter:

- anacron – service for scheduling recurring tasks at non-specific times.
- atd – service for one-time scheduling and called upon by `at` and `batch`.
- crond – service for recurring scheduling and called upon by `crontab`.
- cups (Common Unix Printer System) – service controlling access to system printers providing the ability to print, track print jobs, cancel print jobs and alter printer configuration information; `cups` calls upon `cupsd` to schedule the print jobs.
- dmesg – display the kernel ring (messages generated by the kernel).
- dnsmasq – lightweight DNS service that can launch DNS queries and cache responses.
- dracut – family of services to create and manipulate `initramfs` file systems.
- exportfs – program to refresh the NFS export file systems table.
- firewalld – service that serves as a front-end to the Linux firewall (covered in Chapter 10).
- init (/sbin/init) – System V and Upstart's first process to run after the kernel is loaded and running; responsible for bringing the system to a usable runlevel.
- logrotated – service to rotate log files.
- nfs (Network File System) – suite of programs (including `rpc.bind`, `rpc.nfsd` and various services) to permit remote mounting of file systems and to export file systems for remote mounting.
- pivot_root – kernel-issued command to switch from the `initramfs` file system to the root file system (/).
- rsyslog – configurable logging service.

- systemctl – program to query, manipulate and control units such as starting or stopping units, listing running units and changing unit properties; replaces the `service` instruction in older versions of Linux.
- systemd – first process run by the kernel in modern distributions of Linux; responsible for handling many system duties such as starting and stopping services.
- telinit – outdated command to switch runlevels; replaced by `systemctl`.
- vmlinuz – the Linux kernel, stored in `/boot`, and partially compressed.

Linux files, scripts and directories covered in this chapter:

- /boot/ – top-level directory containing the boot loader program(s), configuration files, Linux kernel and supporting files.
- /boot/grub2/ – directory containing GRUB2 configuration files.
- /boot/initramfs* – disk images storing the `initramfs` file system(s) used by `vmlinuz` during system booting/initialization.
- /boot/vmlinuz* – one or more versions of the Linux kernel; the file is partially compressed so that upon loading it into memory the boot loader must uncompress the remainder before the kernel can execute.
- /etc/ – top-level directory containing various configuration files, data files and scripts as used by Linux programs and especially services.
- /etc/audit/ – subdirectory containing configuration files for the `auditd` service.
- /etc/audit/auditd.conf – `auditd` configuration file.
- /etc/audit/audit.rules – `auditd` rules file.
- /etc/default/useradd – stores default values as used by `useradd`.
- /etc/exports – NFS exports table storing information about local file systems that can be remotely mounted.
- /etc/hosts – file of host name to IP address mappings to bypass DNS.
- /etc/login.defs – configuration information used for `/etc/shadow` password information, among other settings.
- /etc/logrotate.conf – configuration file for `logrotated`.
- /etc/logrotate.d/ – directory containing further configuration files for `logrotated`.
- /etc/resolv.conf – file containing instructions on how to resolve host names using DNS; primarily used to store the IP addresses of local DNS name servers.
- /etc/rsyslog.conf – configuration file for the `rsyslog` logging service.
- /etc/sysconfig/ – subdirectory containing numerous configuration files (Red Hat Linux only).
- /usr/lib/systemd/system/ – subdirectory storing most of the `systemd` unit files.
- /var/log/boot.log – log file storing boot messages.
- /var/log/messages – log file storing system initialization messages.

REVIEW PROBLEMS

1. Use DRAM, SRAM and ROM to answer the following questions.
   a. Which of these forms of memory is non-volatile?
   b. Which of these forms of memory is used to form main memory?
   c. Which of these forms of memory is the fastest?
   d. Which of these forms of memory is the cheapest?
2. List several bootable devices (devices from which the operating system can be booted from).
3. Why might a user want to change the order that bootable devices are tested?
4. You want to set up a dual boot computer between Windows 10 and Red Hat Linux. Which boot loader program should you use?

5. You want to set up a dual boot computer between Ubuntu and Red Hat Linux. Which boot loader program could you use?
6. In your own words, explain why we need a boot process.
7. Order these steps of the Linux boot process from earliest to latest.
    a. Load Linux kernel
    b. Load `initramfs`
    c. Load and run `systemd`
    d. Run boot loader program
    e. Uncompress Linux kernel
    f. Run Linux kernel
    g. Execute `pivot_root`
    h. Locate Linux kernel
    i. Power-on self test
8. To modify GRUB2's behavior from within Linux, you would modify which file(s)?
9. What do the directives `GRUB_TIMEOUT` and `GRUB_DEFAULT` as found in `/etc/default/grub` refer to (see Figure 9.1)?
10. What does the command `dmesg` do?
11. What is the name of the Linux kernel and where would you find it stored?
12. Name three symbols stored in the `System.map` file of the `/boot` directory.
13. What is a ramdisk?
14. True/false: The `initramfs` disk image contains the entire Linux file system.
15. List a specific operating system that uses/used System V/`init` as its startup process. That uses/used Upstart/`init`. That uses/used `systemd`.
16. What does `pivot_root` do?
17. Explain what each of these unit types is used for.
    a. Automount
    b. Path
    c. Service
    d. Socket
    e. Swap
    f. Target
    g. Timer
18. A target file has the following directives. What services and targets does this target need running before it runs? Do any of these need to be stopped before it runs?
    ```
    Requires=abcd.target efgh.target
    Wants=ijkl.service
    Conflicts=mnop.target
    After=abcd.target efgh.target ijkl.service mnop.target
    ```
19. In which directory will you find most of the unit files?
20. Order these targets in terms of the order they will first be called (this does not mean the order that they run). For targets that can run in any order, group them in parentheses.
    a. `basic.target`
    b. `graphical.target`
    c. `local-fs.target`
    d. `local-fs-pre.target`
    e. `multi-user.target`
    f. `rescue.target`
    g. `swap.target`
    h. `sysinit.target`
21. Which target is responsible for testing to see if the network has been started and is accessible? To permit remote mounting of file systems?

22. Which System V runlevel is roughly equivalent to `graphical.target`? To `multi-user.target`?
23. A service, `ijkl.service`, has the following directives in its unit file. In what order do the three services start?

    ```
    After=abcd.service
    Before=efgh.service
    ```

24. What directive is used in a service unit file to specify the service's startup command?
25. What directive is used in a service unit file to specify the name and location of a file that contains environment variables established for this service? Is there a way to establish environment variables for a service without reference to such a file?
26. Which type of unit has a `wants` directory? Where are the `wants` directories stored?
27. Explain each of these service unit file directives.
    a. `ExecStart`
    b. `ExecStartPre`
    c. `ExecstartPost`
    d. `ExecReload`
    e. `ExecStop`
    f. `ExecStopPost`
28. Under what circumstance(s) would you use `Type=notify` for a service?
29. What is the default type for service if the `Type=` directive is omitted?
30. All of the unit files will have a unique section `[Type]` such as `[Service]` or `[Socket]` except for which type?
31. Which of these statements is true?
    a. Every mount unit will have a corresponding automount unit.
    b. Every automount unit will have a corresponding mount unit.
32. Directives like `OnActiveSec`, `OnStartupSec` and `OnCalendar` would be found in a file of which type of unit?
33. You have installed a data mining application that will analyze log files and generate a report daily. You want this application started at system initialization time. The application is controlled by the service `dm.service`. Explain how you would modify the `systemd` startup process to launch this process but not until just before the GUI (`display-manager.service`) is launched.
34. Services run in the foreground, background or could be either?
35. List four entities that a service may receive requests from.
36. List a service that provides each of the following types of activities.
    a. Scheduling
    b. Logging
    c. Kernel support
    d. File system access
    e. Hardware communication
37. Which of these are services and which are programs that use the service? `anacron`, `at`, `atd`, `atq`, `atrm`, `crond`, `crontab`
38. Which service is used to cache responses from DNS requests?
39. What service would you use to create your own `initramfs` image?
40. Specify a `systemctl` command to display the default target unit. Specify a `systemctl` command to modify the default target to `mytarget.target`.
41. Is there any difference between `systemctl status atd` and `systemctl status atd.service`? Between `systemctl status tmp` and `systemctl status tmp.mount`?
42. What is the difference between the `systemctl` command `restart` and `try-restart`?

43. Specify a `systemctl` command for each of the following:
    a. List all of the running mounts
    b. List all of the exited services
    c. List all units
44. True/false: You can display properties of a unit file but not modify them using `systemctl`.
45. Provide a `systemctl` command to list all of the dependent units that must be started before `display-manager.service`.
46. True/false: In Red Hat Linux, all service configuration files are found under `/etc/sysconfig`.
47. True/false: All services have a (or multiple) configure file(s) and no other Linux programs/units have configuration files.
48. You want to disallow zappaf and underwoodr from using `atd` and `crond`. How can you do this?
49. Values used as defaults for users in the `/etc/shadow` file would be found in which configuration file?
50. The command `useradd -D` displays default values for new user accounts as per which configuration file? Using `useradd -D`, can you update the values in this file?
51. Write a rule for `rsyslog.conf` to log all messages from the email service that is of `err` level to the file `/var/log/mail_error.log`.
52. Consider the `rsyslog.conf` rule `*.*          *` What does this rule mean and why would this be a bad rule to add to the file?
53. True/false: You can control log file rotation from `rsyslog.conf`.
54. We want to permit the `/usr/local/apache` directory to be remotely mounted on `10.2.3.0/8` as read-only and `10.11.12.13` as read-write. What entry(entries) would you place in the `/etc/exports` file?
55. Having modified `/etc/exports`, what else do you need to do to permit the specified mount points to be remotely mounted?
56. True/false: The `/etc/logrotate.conf` file contains all log rotation directives.
57. Explain the following specification for log rotation.

```
weekly
minsize 1M
compresscmd /usr/bin/gzip --fast
dateext
prerotate
/usr/bin/datamine /var/log/mylog > /dev/null 2>&1 || true
endscript
```

58. Having modified a service configuration file, when does the updated configuration take effect?

10 Network Configuration

This chapter's learning objectives are to be able to:

- Describe the components of and means to connect to a computer network
- Explain the TCP/IP protocol stack including the protocols that make up TCP/IP, IP addresses and the use of port addresses
- Configure a Linux computer to communicate over network through configuration files and services
- Apply network programs as described in this chapter including `curl`, `dig`, `ftp`, `host`, `ifstat`, `ip`, `nc`, `nmap`, `nslookup`, `ping`, `ss`, `ssh`, `sftp`, `traceroute`, `wget`
- Establish static and dynamic IP addresses
- Use both the `firewalld` GUI and `firewall-cmd` instruction to explore and modify the firewall

10.1 INTRODUCTION

Without computer network access, we are limited to the software that comes with our computer and are unable to access any network resources (printers, file servers, web servers, etc.). Few users would want to work on a computer that is so restricted. With computer networks, we can use resources available across the local network and communicate with other users of the local network, and if our local network connects to the Internet, access information and users across the planet.

In this chapter, we examine how to configure our Linux computer to communicate over network and the Internet. Fortunately, the days of having to perform a number of complex tasks and even write our own code to accomplish network communication are gone. In modern Linux, we rely on Linux services and configuration files to handle many of the tasks.

We begin our look at networks with an examination of TCP/IP (which is optional for students who are already familiar with it). We start with TCP/IP because there are several Linux commands that explore aspects of our computer's network usage based on TCP/IP terminology and concepts. It will also help us understand Linux network communication by knowing more about addressing schemes. We will also have to understand TCP/IP to understand the Linux firewall.

In order for our computer to communicate over network, at a minimum, we need three things. The first and most important requirement for network access is that our computer needs to have a means to connect to the network. This is handled by a specialized piece of hardware in our computers known as a network interface and stored on a network interface card (NIC).

NICs can be wired or wireless. The wired NIC is usually an Ethernet connection where we physically plug our computer into the network through a cable that looks something like a phone jack. The wireless NIC has a small antenna and broadcasts radio signals to a nearby listening device. We make the assumption in this chapter that our connection is through a wired NIC to an Ethernet. Even if you have installed Linux in a VM on a laptop computer that has a wireless NIC, your VM will still emulate that it has a wired connection.

Note that Ethernet is a type of local area network developed in the 1970s and popularized in the 1980s and 1990s. While this is a type of network, with respect to an interface we use the term to indicate how a resource connects to the network over some physical medium.

In our Linux computer, we will usually find two interfaces. If we are using an Ethernet connection, one interface is our Ethernet NIC. In earlier versions of Linux, this was denoted as `eth#`

DOI: 10.1201/9781003203322-10

(usually `eth0`). This has been modified to be `ens#` where the number will vary but is commonly `ens32`. This is the name of our interface to the network.

The other interface is known as the loopback device and is typically labeled as `lo`. The *loopback device* simulates communication over a network but in fact sends any communication back to itself. The reason for having a loopback device is so that running software can send messages to other running software as if it was communicating across a network. Thus, one process will send a message to another over `lo`.

As we have plugged in our computer to the network, or more precisely, we have connected our computer's NIC to the network, our network needs to lead to other computers. The second component then is that our connection to the network itself has a connection to some "next hop". Messages need to go somewhere once they leave our computer. In a local area network, the next device in the network is usually a network switch. In a home network, the next device might be a router connected by phone line or cable to an Internet Service Provider (ISP). Whatever the next hop is, it will be some form of network communication device.

The last component for network connectivity is to run the proper Linux service(s) needed to support network communication. In `systemd`-versions of Linux, this is a single service called `NetworkManager`. `NetworkManager` itself may rely on other services and is responsible for ensuring that such services are running, starting them as needed and stopping those that are no longer needed.

In the next section, we explore some of these concepts in more detail. We then examine the `NetworkManager` and supporting services in the section that follows. Later in the chapter, we focus on Linux commands that use the network, manipulate our network connectivity and/or provide feedback about network communication. We wrap up the chapter by looking at the Linux firewall.

SECTION ACTIVITIES

1. How does your home computer connect to the Internet? Do you have a home network? If so, how many computers connect to it? Is the network wired, wireless or both?
2. Have you ever said "the Internet is down"? This is a popular way to express that you cannot reach the Internet. But the Internet doesn't operate that way. It is almost certainly your computer that cannot connect to the network although it could be a problem with your Internet Service Provider. The irony about this statement is that the Internet, being a distributed, worldwide network, was set up to survive a nuclear war. It would take an awful lot of destruction to bring down the Internet. Research the Internet and see how it is physically implemented.

10.2 COMPUTER NETWORKS AND TCP/IP

To better understand Linux-specific aspects of network configuration, we need to understand TCP/IP. In this section, we first look at the types of devices used to connect networks together. We then look at some specific details of TCP/IP including the various protocols that make up the TCP/IP protocol stack. We separate out some details into their own to ensure that no subsection is overly long.

10.2.1 Network Connection Devices

A *computer network* is a collection of computers, computer resources and network communication devices, all connected together. Devices on a network connect together through some medium such as twisted-wire pair or fiber optic cable. An alternative is for devices to communicate in a wireless fashion such as by radio signals. Computer resources on a network include printers and networked storage devices like file servers and magnetic tape drives.

The network communication devices might more formally be called *network connection devices*. These devices make connections so that the other components can communicate with each other. We categorize these connecting devices as hubs, switches, routers and gateways.

Although outdated, the earliest of the network connection devices is the hub. The *hub* served as a repeater in a bus-style Ethernet network which could only carry a signal along the network medium a certain distance before the signal degraded. The hub would literally repeat or strengthen the signal to carry it along another stretch of cable. Hubs have the behavior that any message that reaches the hub is then transmitted to all connected devices. Thus, the hub can also serve as a device which broadcasts any incoming message to all other devices.

We no longer use hubs because we do not use bus-style networks. The hub has been replaced by more sophisticated devices. The *switch* connects to multiple computers, like the hub, but the switch maintains a table of computer addresses. A message received by the switch has a destination address, and the switch then forwards the message on to the device that has the destination address. The addresses used by the switch are hardware addresses, usually MAC (media access control) addresses.

The *router* uses network addresses, usually IPv4 or IPv6, instead of hardware addresses. The router typically connects to switches and other routers rather than directly to computers. A computer sends a message to a switch which either transmits that message to a computer directly connected to it, or on to a router, which then directs the message along the network to another router or switch. The switch at the endpoint forwards the message to the destination computer. While switches serve as connectors at the endpoints of a network, routers connect networks together.

The *gateway* is much like a router but it has the added capability of converting messages from one protocol into another. Thus, the gateway connects local area networks together that operate on different network protocols, such as an AppleTalk network connecting to a TCP/IP network. A gateway typically appears at the edge of a network, connecting it to another network.

A *local area network* (LAN) is a network where the devices are in close proximity, such as within a building or within a few buildings in one site. We use the term LAN to differentiate a network from one that is spread over much greater distances such as a few miles, hundreds of miles or across the planet. When distances are great, the network is called a *wide area network* (WAN). The Internet is the best example of a WAN.

Figure 10.1 illustrates the layout of a local area network. This network is broken into two smaller wired networks, each with its own network switch. Both switches connect to a router. Also connected to the router is a *wireless access point* (WAP) so that wireless devices can connect. A WAP is a device that receives radio (wireless) signals and passes them on to other wireless devices or to a wired network.

The network itself consists of four computers, a printer and a file server connected to one switch, three computers and a printer connected to another switch, a laptop, a tablet and two smartphones communicating through the WAP to the rest of the network, and the switches and WAP connected to a router. The router either runs a firewall or is connected to a separate security appliance, and from here, the router connects to the Internet. Thus, all of the devices in this network can communicate with each other, or through the router, to devices anywhere on the Internet.

10.2.2 The TCP/IP Protocol Stack

For Internet communication, devices must all utilize the TCP/IP protocol. This four-level protocol is actually a suite of protocols in that each layer of TCP/IP can be implemented by one of many different protocols. TCP/IP provides the rules for how communication must take place between resources on the network. These rules include how messages are broken into packets, how addressing and error handling information is added to the packets, how the packets are treated as they move from location to location, how two-way communication is established, and how received packets are pieced together to make a message. Figure 10.2 illustrates the four layers of the TCP/IP protocol. We examine the details of each layer below.

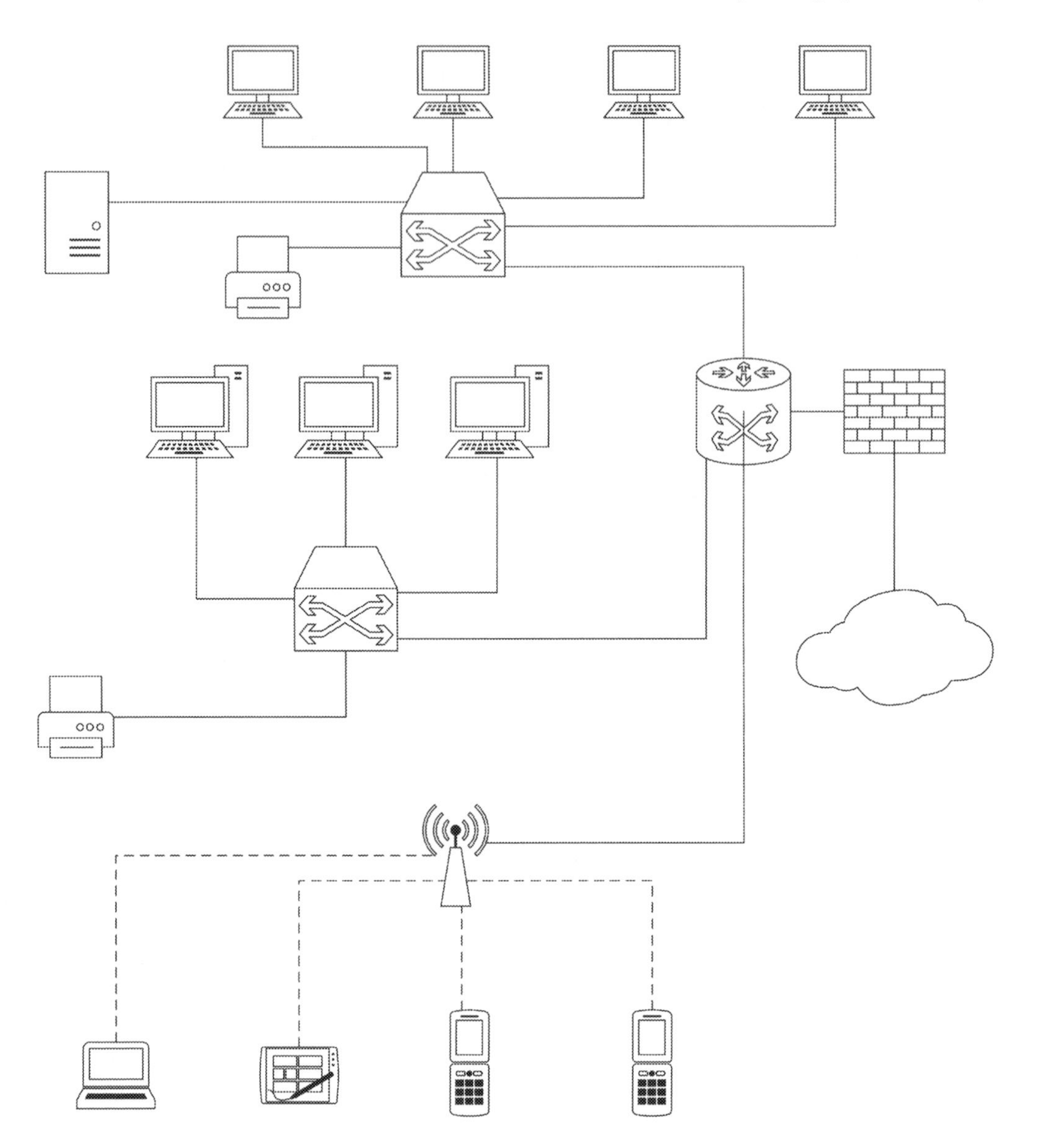

FIGURE 10.1 A local area network. (From Richard Fox, 2021, *Information Technology. An Introduction for Today's Digital World.*)

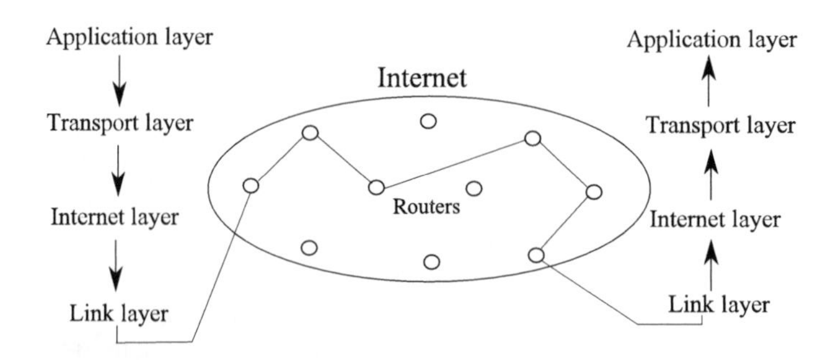

FIGURE 10.2 The TCP/IP protocol stack.

We break the four layers of TCP/IP into two sets. The top set of layers is TCP, the Transmission Control Protocol. It comprises the application and transport layers. The bottom set of layers is IP, the Internet Protocol, which comprises the Internet and link layers. We might think of these two sets of layers as network-neutral layers and network-specific layers. Most network engineers do not consider the physical network itself to be a layer in TCP/IP but that TCP/IP sits on top of the physical network.

Any communication starts with some application software initiating the communication process at the *application layer.* A user might create an email in an email client or select a hyperlink in a web browser, as two examples. The software takes the user's command and produces an initial message to be transmitted. The message is written in one of the many protocols available in the application layer. These include DNS, FTP (and the related SFTP and FTPS), HTTP (and the related HTTPS), IMAP, LDAP, NFS, POP, SSH, SMTP and SNMP to name just a few.

Each protocol has its own syntax and keywords to express how the message should be used at its destination. For instance, HTTP includes a method like `GET`, and an object expressed as a URL. Thus, an HTTP request message might be to `GET` the webpage at a given location on a specified webserver. Other pieces of information that make up the HTTP request can be added to the message, like cookie data and a request that a session with the server be maintained.

The *transport layer* provides a uniform interface between two resources to support host-to-host communication. It is this layer's responsibility to support communication between the two resources no matter what type each device is or what type of network each is operating on. This layer is responsible for separating the initial message as generated in the application layer into distinct *packets* and supplying such information as error correction details (e.g., a checksum) and a count of the number of packets (e.g., packet 6 of 9) so that the recipient's transport layer can determine if an error arose or if the message is incomplete. For a received message, this layer is responsible for identifying the proper application that the message should apply to. This is handled by the message's destination port number. We discuss ports in the next subsection.

There are several different protocols available in the transport layer. The primary protocol is TCP (Transmission Control Protocol), which is confusing because both the protocol to implement the layer and the layer itself are referenced by the same name and acronym. Another protocol used in this layer is UDP (User Datagram Protocol). TCP and UDP dictate different types of packets, also referred to as *datagrams*, particularly with UDP. The TCP packet is larger as it contains data that is used to guarantee delivery. This data comes in two primary forms: a sequence number and an acknowledgment number (plus flags used by routers). The UDP protocol does not guarantee delivery and so the UDP datagram does not contain the same error handling information. This in turn makes UDP datagram smaller than the TCP packet, and so can be delivered more rapidly.

A computer which receives a message using TCP in which some packets are missing (as determined by the sequence number) will ask the source device to resend any of the missing packets. TCP is used for such activities as email and web browsing where every packet is important. UDP facilitates real-time communication (e.g., streaming video) because dropped packets are not resent. Aside from TCP and UDP, other protocols at the transport layer include SCTP, DCCP and UDP Lite.

The remainder of the TCP/IP Protocol stack comprises the IP portion (Internet Protocol). It is at these layers that we see the rules of Internet addressing. Its top layer is the *Internet layer*, responsible for addressing packets and sending them across the Internet. Routers operate at this layer whereby the router receives a packet, examines the destination network address and uses this address to decide how to forward the packet onward.

The most common form of network address used in the Internet layer is the *IPv4* (IP version 4) address. This is a 32-bit number divided into four sets of 8-bit numbers. We can write this number in binary or decimal (in which case each byte is a number between 0 and 255), separating each of the bytes with a period. We refer to each byte, whether written in binary or decimal, as an *octet.*

Historically, networks were assigned to a class which proscribed the octets that encoded the network address and the octets that encoded the host's address within the network. These two parts are in a way like the city/state/zip code/country code and the street address. This allows two computers to have the same host addresses on different networks just as two buildings could conceivably have the same street addresses but in different cities. *Classful networks* were discontinued because they tended to waste addresses. For instance, there were 128 assigned Class A networks, each of which was provided over 16 million unique host addresses. A class A network that did not have 16 million Internet-based resources would waste addresses.

Today, we use a specifier attached to the address to indicate the breakdown of the network address portion versus the host address portion of an IPv4 address. For instance, an address appended with `/19` indicates that the first 19 bits of the 32-bit address constitute the network address portion. The remaining 13 bits would then be the host address. The `/19` is known as a *routing prefix*. By moving from classful to classless networks, it allows a network to be assigned a range of IPv4 addresses that fits its size rather than some arbitrary class.

This leads us to *classless Internet Domain Routing* (CIDR), which is the more common approach to assigning network addresses. The routing prefix is also referred to as *CIDR notation* which in turn dictates the netmask. A *netmask* (also called a subnet mask) is a sequence of 1s followed by 0s, such as 11111111.11111111.11100000.000000000, which would be the netmask for the routing prefix `/19`. Notice that this netmask consists of 19 1s followed by 13 0s. The netmask is ANDed to the IP address, and the result is the network address. The negation (logical NOT) of the netmask can be ANDed to the IP address to obtain the host address.

A subnet is a logical division within a larger network whereby the resources of the subnet share not only the same network address but have similar host addresses. Consider a network with a CIDR notation of `/22`. This leaves 10 bits for the host portion of the address. We might find four subnets then, where the last two bits of the third octet indicate which subnet the host is a part of. For instance, addresses of 00.00000000 through 00.11111111 make up one subnet while 01.00000000 through 01.11111111 make up a second subnet. Subnetworks are typically connected together by router (although can also be combined on the same network switch). We create subnets to better manage our IP addresses.

With 32 bits available for an address, we have over 4 billion unique IPv4 addresses available although many addresses remain reserved for special purposes. For instance, addresses starting with the octet 10 or the octets 192.168 are reserved for private networks (this is why most example IP addresses found in this textbook start with 10). Between having some of the IPv4 addresses go unused and the billions of Internet-capable devices in existence today, we do not have enough IPv4 addresses available to assign to every device. Because of this, and other issues with IPv4, we have moved on to IPv6 (IP version 6) addressing.

An *IPv6 address* is 128 bits long divided into a 64-bit network address and a 64-bit host address. We write IPv6 addresses in hexadecimal notation, dividing each group of 4-hex digits into its own *hextet*, separating hextets with colons. An IPv4 address might appear as 10.11.12.13 while an IPv6 address might be FEA9:A870:0000:0000:0000:0012:3456:789A. There are rules available to remove some of the 0s found in an IPv6 address to write it more succinctly. The above address could be rewritten as FEA9:A870::12:3456:789A.

In Table 10.1, we compare the two types of addresses. Notice how many IPv6 addresses are available. It will be a long time before we run out! We return to IPv6 later in this section.

TABLE 10.1
IPv4 vs IPv6 Addresses

Format	IPv4	IPv6
Bits used for address	32	128
Address divisional unit	Octet (8 bits)	Hextet (4 hexadecimal digits which is 16 bits)
Example address	10.11.12.13	0064:FF9B:0000:0000:0000:0000:0018:5678 or rewritten as 64:FF9B::18:5678
Number of unique addresses	2^{32} = 4,294,967,296	2^{128} which is roughly 340,000,000,000,000,000,000,000,000,000,000,000,000 ($3.403*10^{38}$)

Aside from IPv4 and IPv6, other addressing protocols at the Internet layer include ICMP and ICMPv6, both of which are used to handle network configuration and network error-checking messages. Both the `ping` and `traceroute` programs (which we explore later in this chapter) use ICMP. Another protocol, IGMP, is used for handling multicast communication.

The *link layer* is the lowest layer in the TCP/IP protocol stack and prepares a packet for physical communication across the network media. This layer is responsible for converting messages, stored in binary, to the type of signal that the network medium will handle (e.g., sound, electrical current, pulses of light) and converting the received signals back into binary. Conversion may require such network tasks as modulation and multiplexing. This layer may also add start and stop bits to packets.

At the link layer, message delivery from the network to a specific device is handled by the device's hardware address. We most commonly use MAC (media access control) addresses, which are 48-bit addresses allocated uniquely to every Internet device (although a newer 64-bit scheme is being used in some countries). MAC addresses differ from Internet-based (IPv4, IPv6) addresses because every device is given its own MAC address which it then uses throughout its lifetime. IP addresses can and often do change.

Switches operate at the link layer, unlike routers which operate at the Internet layer. The link layer can differentiate between resources on its local network but not between resources located outside of the local network. To move a message from one network to another, we use routers and the Internet layer. The link layer also handles transmission error detection, bit synchronization and any issues that might arise because of variances in signal strength.

One important protocol of the link layer is the Address Resolution Protocol (ARP). ARP is responsible for mapping IP addresses to MAC addresses so that a switch, upon receiving a message, can identify which specific device it should send the message onto. Another link layer protocol is the Tunnel, which creates a temporary dedicated network connection within a network of another protocol. This is primarily used so that a message that was created within a private network can be sent over a public network (e.g., the Internet). Some other link-layer protocols are PPP (Point-to-Point Protocol), ISDN (Integrated Services Digital Network), DSL (Digital Subscriber Line) and Ethernet.

To transmit a message, the message is processed going down the TCP/IP protocol stack from the application layer to the transport layer to the Internet layer to the link layer. Starting with the application layer, the original message is placed within a protocol-specific message such as an HTTP request. At the transport layer, the message is decomposed into packets with error detection information added to each packet. At the Internet and link layers, yet more information is added in the form of addressing and further error detection data.

The message is then sent across the network as individual packets. When a packet arrives at an intermediate location, a router operating at the Internet layer examines the destination address to decide how to send the packet on its way further across the network. The packets that make up a message may take different routes across the network.

Packets arriving at a switch are then forwarded to the destination device by consulting the destination MAC address and its own table of connected devices. Once a packet arrives at the destination, the destination device sends a receipt packet (if the packet is a TCP packet) to acknowledge that the packet was received. For a UDP message, no acknowledgment is sent.

If the received packet is a UDP datagram, then it is immediately sent up the TCP/IP protocol stack and used by at the application layer. A TCP packet may be one of many and so is set aside until the remainder of the packets that make up the message are received. Upon receiving all portions of the message, the message is then moved up the TCP/IP protocol stack and presented to the proper application software. As a message moves its way up the TCP/IP stack, added information is removed and/or applied to ensure correctness of the data.

No matter what platform of computer is used to create and transmit the message, the recipient need not be using the same platform or even the same application software. The TCP/IP protocol stack presents a device-to-device communication platform so that the two applications can communicate with each other and thus the users at these endpoints can communicate with each other.

As TCP/IP was being developed, the *Open Systems Interconnection* (OSI) model was also being developed. Both were suggested as implementations for Internet communication. It was TCP/IP that was selected, in part because some of the protocols that made up TCP/IP were already in use on the Internet (which at the time was called the ARPANET). In studying TCP/IP, you will find most of its functionality is also specified somewhere in OSI.

There are perhaps three significant differences between the two protocol stacks. First, OSI consists of seven layers, more finely specifying functions to layers. Second, OSI's lowest layer is of the physical network itself, which is omitted from TCP/IP. Third, no single implementation of OSI has ever been produced and instead is used as a model or a standard for new network development.

One last comment with respect to OSI is that many network developers will refer to OSI layers rather than TCP/IP layers. Because of this, switches, which operate in TCP/IP at the link layer, are often referred to as layer 2 devices because they operate in layer 2 of OSI. Routers, which operate in TCP/IP at the Internet layer, are usually referred to as layer 3 devices because they operate in layer 3 of OSI.

10.2.3 Ports

Another form of address is the *port*. This is a 16-bit number that is assigned to a message to denote the application that a message is intended for (whether a network service or application software). There are two port addresses assigned to every packet. The source port is used for return messages so that the transmitting computer knows which message the response pertains to. These are usually randomly generated addresses (likely one greater than the last port number used). Destination port addresses are the ones used to identify the intended application of the message.

There are 65,536 (0–65,535) addresses available in 16 bits. Most destination port addresses are within the first 1024 (0–1023) as these are set aside for well-known applications and known as the *well-known port* addresses. Table 10.2 lists some of the more popular well-known ports. Some of the ports in the well-known port range are currently unassigned or unofficially assigned such as 531 being reserved for the AOL (America On-line) Instant Messenger and 843 for Adobe Flash software communication. Other unofficially reserved ports exist for VMware, Oracle, Cisco, Novell, Symantec and multiplayer computer games. Ports in the range of 1,024–49,151 are known as *registered ports*. Some of these are used as destination ports but most of these, along with any numbered 49,151–65,535, are used by the operating system as source ports.

While the port itself is not a network address, it is used with the IP address to form a more specific destination address that, combined, represents a unique application or service on a destination device. When including a port with an IP address, the port is added at the end after a colon (:) as in 10.11.12.13:22. Together, the two addresses (IP address, port address) can be used by a firewall to safeguard a computer from messages that should be discarded rather than processed.

When an application is listening to a port, it is assigned the port address as an IP address:port address pair, as in 10.11.12.13:22. Once assigned, no other process can be given that same pair. This results in what is called a *port conflict*.

Table 10.2 notes whether a port uses TCP, UDP or both. In fact, this is slightly misleading. TCP has its own set of port addresses and UDP has its own set. They happen to overlap with respect to the numbers assigned to the applications that make up the well-known ports. So, for instance, TCP assigns port 80 to HTTP and port 80 is unassigned for UDP while both TCP and UDP assign port 53 for DNS messages.

10.2.4 IPv6

As described earlier and in Table 10.1, IPv4 addressing offers about 4 billion unique addresses. With the great success of mobile devices and *Internet of Things* (IoT) devices, the number of unique IP addresses needed at any time greatly exceeds 4 billion. In 2011, the last of the IPv4 addresses

TABLE 10.2
Well-Known Port Addresses

Port	Packet Type	Usage
20	Both	FTP data
21	TCP	FTP control
22	Both	ssh (also SCP, SFTP)
23	Both	Telnet
25	TCP	SMTP
43	TCP	WHOIS
53	Both	DNS
57	TCP	Mail transfer protocol
67	UDP	Bootstrap Protocol (used by DHCP)
68	UDP	Bootstrap Protocol (used by DHCP)
70	TCP	Gopher
80	TCP	HTTP
109, 110	TCP	POP2, POP3
118	Both	SQL
123	UDP	Network time protocol
161	UDP	SNMP
194	Both	IRC (Internet relay chat)
443	TCP	HTTPS
514	UDP	Syslog (Linux system logging)
530	Both	RPC
636	Both	LDAP
989	Both	FTPS data (FTP over TLS/SSL)
990	Both	FTPS control (FTP over TLS/SSL)
992	Both	Telnet over TLS/SSL
2049	Both	NFS
3128	TCP	Squid proxy
6660–6669	TCP	IRC
6888–6900	Both	BitTorrent
8008, 8080, 8090	TCP	Alternates for HTTP

were awarded by the Internet Assigned Numbers Authority (IANA). IANA is the organization that assigns IP address ranges to different regions. The last available set of addresses was provided to the Asia-Pacific Network Information Center (APNIC). Having run out of IPv4 addresses to assign is known as *IP Address Exhaustion*. We need to shift to a different form of addressing that provides a greater number of unique addresses. This is the primary motivation for the creation of IPv6.

We noted the difference in address size and notation earlier. Referring back to Table 10.1, we presented an IPv6 address and then the same address in a shortened notation. In order to shorten an IPv6 address, we can apply two shortcut rules. First, multiple hextets that consist solely of zeroes can be replaced by a single :: notation. This rule can only be applied once so we would choose to replace the largest collection of zero hextets.

For example, we can reduce the IPv6 address 1234:5678:9a00:0000:0000:0000:98bc:def0 to 1234:5678:9a00::98bc:def0 by eliminating the three hextets of all zeroes. For the address FE09:90AB:0000:0000:1ABC:0000:1234:4579, we would select the first two zero hextets to replace rather than the single zero hextet that comes later, reducing the address to FE09:90AB::1ABC:0000:1234:4579.

The second shortening rule is to remove all leading 0s within every hextet. Thus, a hextet of 00AB will be reduced to AB and 0009 will become 9. Slightly changing the second address from the previous paragraph to FE09:00AB:0000:0000:0ABC:0000:1234:0579, we would apply both rules to achieve FE09:AB::ABC:0:1234:579. Notice that while we cannot remove the second hextet of all zeroes, we can reduce it to just a single 0.

Let's consider two additional IPv6 addresses. The address of all 0s, 0000:0000:0000:0000:0000:0000:0000:0000 is reduced to ::. The address 0000:0000:0000:0000:0000:0000:0000:0001 is reduced to ::1.

We noted in the last subsection that TCP and UDP have their own port spaces. In fact, with IPv6, we now have four distinct sets of port addresses. TCP has one set of addresses for IPv4 and one set for IPv6 while UDP also has one set for IPv4 and one set for IPv6. Like with IPv4, the IPv6 ports are the same numbers.

As we need to replace IPv4 with IPv6, we need to modify how the Internet works. Unfortunately, IPv4 and IPv6 messages use different formats and so we refer to the two protocols as *non-interoperable*. As a result, until we can completely replace IPv4 with IPv6, we must continue to use both.

To use IPv6, we have to modify both the software that uses TCP/IP and the network hardware. The first part of this replacement process has been completed in that nearly all modern operating systems and network-specific software packages are capable of handling IPv6. The larger issue is that not all networks have been modified to utilize IPv6 addresses. Replacing all of the routers that were put in place prior to the onset of IPv6 has proven to be challenging. As of May 2021, it is estimated that no more than 22% of all existing networks connected to the Internet are IPv6 compliant and only around 30% of the world's top 1000 websites are IPv6 compliant. Until full compliance occurs, our computers and networks will have to continue to use IPv4. In many cases, networks will use both and so your computer may be assigned both an IPv4 and an IPv6 address.

IPv6 is significant for more reasons than the (greatly) enlarged address space. Another feature of IPv6 is that it can make use of security implemented within the protocol itself. TCP/IP with IPv4 lacks any built-in network-based security. For instance, TCP/IP does not include a mechanism for encryption and therefore encryption must be handled at the application layer. With IPv6, there is the *Internet Protocol Security* (IPsec).

Another feature of IPv6 is the ability for a host to automatically configure itself with respect to addressing and locating its router/gateway. Another difference is that IPv6 packets can use headers with optional components. The required portion is simplified over the IPv4 header by discarding seldom used parts. Optional information, or an extension, can be added to specify security options and specifiers that denote such data as abnormally sized packets. All in all, the design of IPv6 is well thought out, having been engineered over a period of years starting in 1998. This is not the case with IPv4 which was designed for an incarnation of the Internet preceding its popularity and in fact preceding the popularity of widespread personal computer usage.

10.2.5 Domains, the Domain Name System and Host Names

Imagine that we want to send a message from our computer to another computer on the Internet. This might be an email which is addressed to a specific user and a mail server, or an HTTP request addressed to a webserver, or a `ping` request to another computer on our local network. In most cases, when we create network messages, as humans, we address them using the destination computer's name. A computer's name consists of a hostname and a domain name. A host name might be mycomputer or zappa or www (commonly used for webservers). Domain names include amazon.com, google.com, nku.edu and wikipedia.org. A *fully qualified domain name* is the combination of the hostname and domain name resulting in a name that unambiguously identifies the resource on the Internet.

Unfortunately, routers cannot handle symbolic names, only numeric addresses (IPv4, IPv6). Unless we plan to memorize hundreds or thousands of IP addresses, we will need some mechanism by which we can translate a name into an IP address to use the Internet. This is handled through a process of *address resolution*.

Address resolution is implemented for us through the *domain name system* (DNS). Across the Internet are DNS name servers whose role is to handle address resolution requests (or DNS requests). A name server receives a request, looks up the appropriate information in a DNS table and returns the requested information to the client in the form of a DNS response message. Although we can send requests directly to name servers, it is generally taken care of for us by either our application software or our operating system (or both in cooperation).

Let's see how this works. First, we, while using the Internet, enter a name, for instance into the address box of a web browser. If our computer maintains its own cache of DNS responses, then this cache is consulted to see if the address can be resolved locally. If not, then our computer contacts our network's *local DNS name server*, which is either maintained by our own organization or by our ISP. The local DNS name server will have its own cache of prior address resolution requests and responses. If found, the IP address is returned to our computer.

If the mapping is not cached by our local DNS name server, then the request must proceed across the Internet. Our local DNS name server forwards our request onward. Where? This is where DNS gets complex. DNS comprises a hierarchy of name servers, each one knowledgeable about the domain it is responsible for. The highest level of this hierarchy is called the DNS *root level*. The name servers at this level are root-level domain name servers. The domains that these servers know of are the *top-level domains*. Given the fully qualified domain name in our request, one of these servers examines it for the top-level domain portion of the address.

There are two categories within the top-level domain. First are the *top-level generic domains* which use names like .edu, .com, .org and .net. Then there are *top-level country domains* like .us, .uk and .ca. The root-level domain name servers know the IP addresses of all of the name servers responsible for each top-level domain. The request that originated from our computer is forwarded on to the appropriate top-level domain name server. For instance, if our request is for somecomputer.someorganization.com, then the root-level name server forwards the request to one of the .com domain name servers.

Just as the root-level domain name servers know the top-level domain servers, the top-level domain servers know the name servers of the second-level domains. *Second-level domains* constitute the resources of a specific organization such as the resources found at nku.edu or amazon.com. In our example, the second-level domain is someorganization.com. The request received by the top-level domain server is forwarded on to the someorganization.com name server for resolution.

Some second-level domains are further subdivided into subdomains. nku.edu might itself be divided into cs.nku.edu and math.nku.edu. The domain of a company, say ourcompany.com, might have subdomains of sales.ourcompany.com, marketing.ourcompany.com and it.ourcompany.com. A request received by a second-level domain name server will either have the address resolution information to handle the request or will forward the request on to a subdomain's name server which will have that information.

Whether the second-level domain name server can handle the request or it is passed on to a subdomain's name server, the requested IP address should be known within the domain so that a response can be created. This step involves consulting a table that stores all of the local resources' name to IP address mappings. For instance, someorganization.com's name server has an entry for somecomputer, and it is this IP address that is placed in the DNS response. The response is then returned our local DNS name server, which might cache the response before forwarding it on to us. Our computer may also cache the response, which is then used to create our TCP/IP message.

The term *domain* conveys a name space whereby the DNS name server(s) for that domain knows about the items found within the domain. The root-level knows about the top-level servers. The top-level servers know about the second-level domains within their domain. The second-level domains know about their resources and/or their subdomain name servers which know of about their resources.

DNS name servers come in two forms: authorities and caches. An *authority* is responsible for its own domain. This means that, for each domain within DNS, there will be at least one authoritative

DNS server that contains information about that domain. The authority contains the records that map its resources from names to addresses.

Caches store responses from authorities, at least for some time. The role of a cache is to respond more quickly. A cache will not be the authority for its cached entries. The response returned from someorganization.com for somecomputer could be cached by our local DNS server for future requests and/or our local computer.

Countless copies of an address might exist in numerous caches across the Internet. *Cache control* is an issue because addresses could be updated at the authority leaving cached copies to be invalid. Thus, cached entries should only persist for a limited amount of time.

DNS name server authorities store information about their local domain in resource records. A *resource record* describes one of the local resources. It can contain IPv4 addresses, IPv6 addresses, canonical names (the true name of the resource rather than an alias like www) and pointers. The pointer is used for what is known as a *reverse IP lookup*. In such an address resolution, we make a request for the name given the IP address. The reason to perform a reverse IP lookup is to ensure that an IP address provided to us is legitimate.

Let's step through a concrete example of address resolution. The user has entered www.nku.edu in her web browser. In order to obtain the web page from the webserver, the name must be converted into its IP address. The user's computer does not have this address in its local cache and so it makes a request and sends it to the user's ISP DNS name server. That server does not have the address cached either and so it forwards the request to the root-level name servers. One of those root-level name servers receives the request and forwards it on to one of the name servers for the .edu domain. The server that receives the request forwards it on to the name server for the nku.edu domain.

It is at the second-level domain that the name server is the authority for the nku domain. The nku name server looks up the computer named www and finds the IP address. It sends back a DNS response which contains this address. The DNS response is received by the user's local DNS name server which caches the entry and forwards it on to the user's computer. That computer pulls out the IP address and adds it to the HTTP request. Finally, the HTTP request can be transmitted. Although we say *finally*, it is likely that this entire process took no more than a couple of seconds. While DNS is complicated, address resolution is transparent to the user.

Figure 10.3 shows the hierarchy structure of DNS (although limits top-level and second-level domains to just a few). The NKU domain has two computers listed, named www and mail. But it also has a subdomain called cs. This subdomain has two entries of www and fs. If the cs subdomain has its own name server then the IP addresses for its resources (www.cs.nku.edu and fs.cs.nku.edu) are stored there, otherwise the name server for nku.edu would have resource records for all four of the named computers. Although two computers at nku.edu are named www, their fully qualified domain names differ because one is www.nku.edu and the other is www.cs.nku.edu.

SECTION ACTIVITIES

1. How much of an understanding of computer networks do you need to be a computer users? A Linux user? A system administrator? Go back through this section and identify which of the sections, if any, would be useful to know for each of the three types of user.
2. If you are a Windows user, open a command prompt and type `ipconfig` to view your IP addresses. If you are a Mac or Linux user, open a terminal window and type either `ifconfig` or `ip addr`. How many interfaces does your computer have? Of them, do they have IPv4 addresses, IPv6 addresses or both?
3. In bygone days, we generally had to memorize telephone numbers of people we called frequently. This is no longer needed as we program our smartphones with

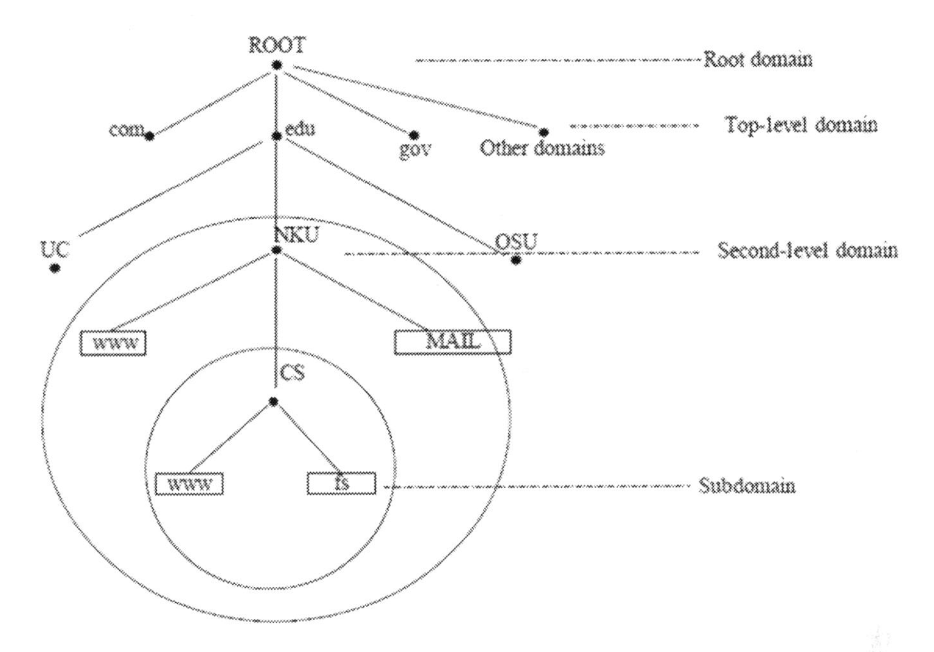

FIGURE 10.3 Example DNS transaction. (Adapted from *Internet Infrastructure. Networking, Web Services and Cloud Computing*, 2018.)

the phone numbers we tend to use. How many phone numbers do you know? Now consider if we didn't have DNS, just how many IPv4 addresses could you memorize? Compare this to the number of names you might use like google.com and amazon.com. Keep in mind that even here we are cheating because the actual name is www.amazon.com (and www is likely not the real name of the webserver, we explore this later in this chapter). If we didn't have DNS, would you be able to cope by memorizing (or saving) IPv4 addresses?

10.3 LINUX `NetworkManager` SERVICE AND RELATED SERVICES AND FILES

Let's now turn to our Linux computer and how to establish network communication. There are several components to Linux's handling of network communication. The most important is the `NetworkManager` service. We spend the first subsection looking at this service and how to control it. We then look at various network files and other services of note.

You might have noted earlier in this chapter and elsewhere in the textbook that we referred to our interface as `ens32` (or `ens33`). In earlier versions of Linux, an Ethernet interface was given the name `eth#` where # was a number, commonly but not always 0. In `systemd`-based versions of Linux, a different naming strategy is employed known as *Predictable Network Interface Names*.

This approach to providing interface names applies one of five schemes, as described in Table 10.3. We will assume that scheme 2 is used and reference all of our interfaces using `ens#`. The number is based on whether a PCI or PCIe slot is used for the interface device. For PCI slots, we typically see 32 or 33 (or a closely related number). For PCIe (PCI Express) slots, the number is a multiple of 32 such as 160 or 192. In the sections that follow, our Red Hat VM has a name of either `ens33` or `ens192`.

TABLE 10.3
Predictable Network Interface Name Schemes

Scheme	Type	Explanation	Example Name
1	Firmware or BIOS index number	For onboard devices if information from the firmware/BIOS is available.	eno1
2	PCI or PCIe slot index	Used as a fall back to the previous scheme, if available.	ens32, ens160
3	Physical location	The location of the connection of the hardware, if available.	enp2s0
4	MAC address	Name based on the MAC address, only used if this is the user's choice.	enx79f3ca42b1a0
5	Kernel naming scheme	Only used if no other choice is available.	enp1s0

10.3.1 NetworkManager

As Linux moved on to systemd, the previously used suite of network services has been replaced by NetworkManager. NetworkManager establishes the interfaces needed to communicate to the local area network. This will likely be an Ethernet NIC or wireless NIC, but may also be a mobile broadband device, a point-to-point connection or some other form of network interface. To create a connection between the computer's interface device and the network, the device must be assigned a network address (e.g., IPv4 and/or IPv6 address) which the NetworkManager oversees. NetworkManager then maintains network information such as the default route(s) to the local router and the IP addresses of the network's local DNS name servers. It makes this information available over the d-bus so that other applications can gain access to it.

NetworkManager runs scripts located in /etc/NetworkManager/dispatcher.d to control or start other network-related services. The actual scripts found in this directory vary by Linux distribution and the type of interface being used. We might find scripts to start dhclient and chrony, for example. dhclient is a service that communicates with a local DHCP server to obtain a dynamic IP address and manage that address' lease. chronyd is a service for synchronizing the computer's system clock with Internet-based time (NTP) servers.

Any scripts found in the dispatcher.d subdirectory are executed by NetworkManager. Each script expects two parameters: the interface device's name (e.g., ens32) and a command. Commands are explored in Table 10.4.

NetworkManager has its own configuration file, NetworkManager.conf. Its location varies by distribution but will likely be in /etc/NetworkManager. The configuration file comprises three specific parts: [main], [ifupdown] and [logging]. There may be other sections dedicated to the keyfile plugin if NetworkManager is using it. The keyfile plugin supports additional network connectivity by maintaining information on various types of network connections, as well as storing network-based passwords and private keys.

[main] is the only required section. It will contain directives to define any plugins, what client will run to obtain a dynamic IP address (dhclient or dhcpcd), a list of interfaces for which NetworkManager will not create a wired connection and whether NetworkManager will use wpa_supplicant for wireless communication. Optionally, an [ifupdown] section will specify if interfaces listed in the interfaces file are handled by NetworkManager or not. Also optional is the [logging] section which indicates the level for logging messages, one of ERR, WARN, INFO or DEBUG, and which type(s) of communication should be logged such as IP4, IP6, DNS, PPP, ETHER.

TABLE 10.4
Commands Accepted by `dispatcher.d` Startup Scripts

Command	Meaning
`dhcp4-change`	Renew or otherwise update or change the lease obtained from the DHCP server for the current IPv4 address.
`dhcp6-change`	Renew or otherwise update or change the lease obtained from the DHCP server for the current IPv6 address.
`down`	Deactivate the interface.
`hostname`	Update the hostname.
`pre-down`	Deactivate but do not disconnect.
`pre-up`	Bring up the interface but do not activate it yet.
`vpn-down`	Deactivate a connection.
`vpn-up`	Activate a connection.
`up`	Activate the interface; this command establishes various network variables including `CONNECTION_UUID` (the UUID of the connection) and `IP4_ADDRESS_N` for the interface's assigned IPv4 address.

Figure 10.4 provides an example of a `NetworkManager.conf` file. The plugin, `ifcfg-rh`, is used in some Red Hat distributions if the older script `/etc/sysconfig/network-scripts/ifcfg-eth0` (or some variation) should be read. The `dhcp` entry indicates that `dhclient` will be used to obtain the dynamic IP address for this computer's interface. Under logging, we see that `WARN` level and higher messages will be logged if generated from any of IPv4, DNS, vpn, firewall, DHCP4 or `systemd`. There is likely little to no reason to modify the `NetworkManager`'s configuration file.

`NetworkManager` is supported by several other programs of note. We first explore `nmcli`, a command-line program which reports on the `NetworkManager`'s state and the state of the devices it governs. The syntax for the command is `nmcli [`*`options`*`]` *`category`* `[`*`command`*`] [`*`arguments`*`]`. The *category* is the only required part and can be omitted if the category is `general`. Each of the categories has its own set of *commands*. We explore some of these in Table 10.5. The various categories can be abbreviated to their first letter.

`nmcli` can provide a great deal of information about our current network connectivity. This could be especially helpful when we have multiple network interface. Modifying existing connections is also available through `nmcli`. To better understand how to use `nmcli`, there are both an extensive man page and a separate man page called `nmcli-examples` (`nmcli-examples` is not a Linux command, just a man page).

We ran `nmcli` with no arguments and obtained the output shown in Figure 10.5. Notice there are multiple interfaces although the Ethernet interface (`ens160`) is the one we use to communicate on the network. The `virbr0` and `virbr0-nic` interfaces are there because this Linux computer is a virtual machine.

```
[main]
plugins=ifcfg-rh
dhcp=dhclient
[logging]
level=WARN
domains=IP4:DNS:VPN:FIREWALL:DHCP4:SYSTEMD
```

FIGURE 10.4 Example `NetworkManager.conf` configuration file.

TABLE 10.5
Categories and Commands for nmcli

Category	Meaning	List of Commands
agent	Run either NetworkManager's secret or polkit agent or both; the secret agent listens for secret requests; polkit is a Linux service which provides authorization for unprivileged programs.	all, polkit, secret
connection	All network configurations are stored as connections; this category lets us control how connections are created, modified and disable connections; an argument follows the command describing the connection by name, UUID or path.	add, clone, delete, down, edit, export, import, load, modify, monitor, reload, show, up
device	Show information about or modify properties of a specified item; most of these commands are followed by an argument which will be an interface's name; wifi has a number of possible arguments.	connect, delete, disconnect, lldp, modify, monitor, reapply, show, status, wifi
general	Show NetworkManager status, permissions, logging level/domain information.	hostname, logging, permissions, status
help	Display 1-page help reference.	None
monitor	Observe and store NetworkManager activity; based on the specified activity we add one or more arguments of the type of activity to monitor such as a connection's name or UUID or the connection path.	add, clone, delete, down, edit, export, import, load, modify, monitor, reload, show, up
networking	Enable or disable network connectivity, or obtain the status of network connectivity (responses are none, portal, limited, full, unknown).	on, off, connectivity
radio	Show status of given wireless connection, or if adding on or off, enable or disable that form of wireless connection.	all, wifi, wwan, all on, all off, wifi on, wifi off, wwan on, wwan off

```
ens192: connected to ens192
        "VMware VMXNET3"
        ethernet (vmxnet3), 00:50:56:B9:47:93, hw, mtu 1500
        ip4 default
        inet4 10.2.56.201/21
        route4 0.0.0.0/0
        route4 10.2.56.0/21
        inet6 fe80::eb01:d2de:de0c:142d/64
        route6 fe80::/64
        route6 ff00::/8

virbr0: disconnected
        "virbr0"
        bridge, 52:54:00:C2:5F:F3, sw, mtu 1500

lo: unmanaged
        "lo"
        loopback (unknown), 00:00:00:00:00:00, sw, mtu 65536

virbr0-nic: unmanaged
        "virbr0-nic"
        tun, 52:54:00:C2:5F:F3, sw, mtu 1500

DNS configuration:
        servers: 172.28.102.11 172.28.102.13 10.11.0.51 10.14.1.10
        domains: hh.nku.edu
        interface: ens192
```

FIGURE 10.5 Output from nmcli.

Another NetworkManager-related program is nmtui. This serves as a menu-driven (text-based) front-end to make changes to the NetworkManager. Upon running nmtui, we are asked which option we wish to pursue: edit a connection, activate a connection, set system hostname or quit.

Through editing a connection, we are shown all of our interfaces. Selecting any interface allows us to edit its name (both the display name and its device name) and modify configuration information (automatic configuration, disabled, enabled, link-local, manual, shared), whether the interface is available to all users and whether the interface is enabled at boot time (automatically connected). We can add a new interface from a list of Bond, Bridge, DSL, Ethernet, InfiniBand, IP tunnel, Team, VLAN and Wi-Fi. We can also delete an interface.

One other notable program is nm-connection-editor, which is a GUI-based program. Running this opens a window listing our available interfaces. From here, we can add or delete interfaces or select an interface and edit it. Figure 10.6 shows an example of editing our VM's Ethernet interface, ens192. We see two views of this interface, General and Ethernet. The other tabs contain no information.

10.3.2 Other Network Services of Note

Many of the previously used network-oriented services have been deprecated as systemd-versions of Linux use NetworkManager for most of the same functionality. However, there are still some network-oriented or supporting services. We first look at those that are found in modern distributions. We then briefly look at some older services. Those services that still exist are placed in /usr/sbin. Those that have configuration files are noted.

The avahi-daemon performs *service discovery* across a network. Through avahi, Linux is able to identify IP addresses of devices on the local network and locate and utilize network services available to clients of the local area network. Services include print and file services. Avahi is a *zero-configuration* service meaning that it can run without user intervention. avahi-daemon has a configuration file and other supporting files located under the /etc/avahi subdirectory.

The rdisc service locates the subnet's router. It does so by using the ICMP (Internet Control Message Protocol) router discovery protocol. Once the router has been identified, this service modifies our computer's router tables to indicate default routes.

dnsmasq is a lightweight network tool that can perform several tasks although it is most commonly used as a DNS caching server for either a single computer or a small network. It can serve as a DHCP server. It can announce the availability of a router on the local network when clients connect to the network. dnsmasq has a configuration file of /etc/dnsmasq.conf. A standard installation provides a version of this configuration file where all directives are commented out so that we would have to edit the file. When used as a DHCP server, dnsmasq stores leasing information under /var/lib/dnsmasq/dnsmasq.leases.

httpd is the Apache web server. It may be preinstalled under /usr/sbin with a configuration file in /etc/httpd/conf/httpd.conf and with a default web space of /var/www. Starting and stopping httpd is usually handled through a program called apachectl. If we choose to install a version of Apache from open source, we can move all of the content of the program to one location, such as /usr/local/apache2. We explore Apache installation and configuration in detail in the supplemental readings.

SSH, secure shell, permits remote access to a computer. The older Telnet protocol did the same but SSH uses public-key encryption. ssh, the program, is used to open, maintain and close secure connections with remote hosts. To permit SSH connections into our computer, we need to set up the sshd service. There are several configuration files for sshd, all under /etc/ssh, including ssh_random_seed, sshd_config, ssh_config, ssh_hot_key and ssh_host_key.pub. In addition, the firewall must be set up to permit SSH messages (which is the default for most Linux distributions).

Linux installations usually provide one of two GUI-based email client programs: Mozilla Thunderbird or Evolution. To use either one, we must specify the location of our email server and

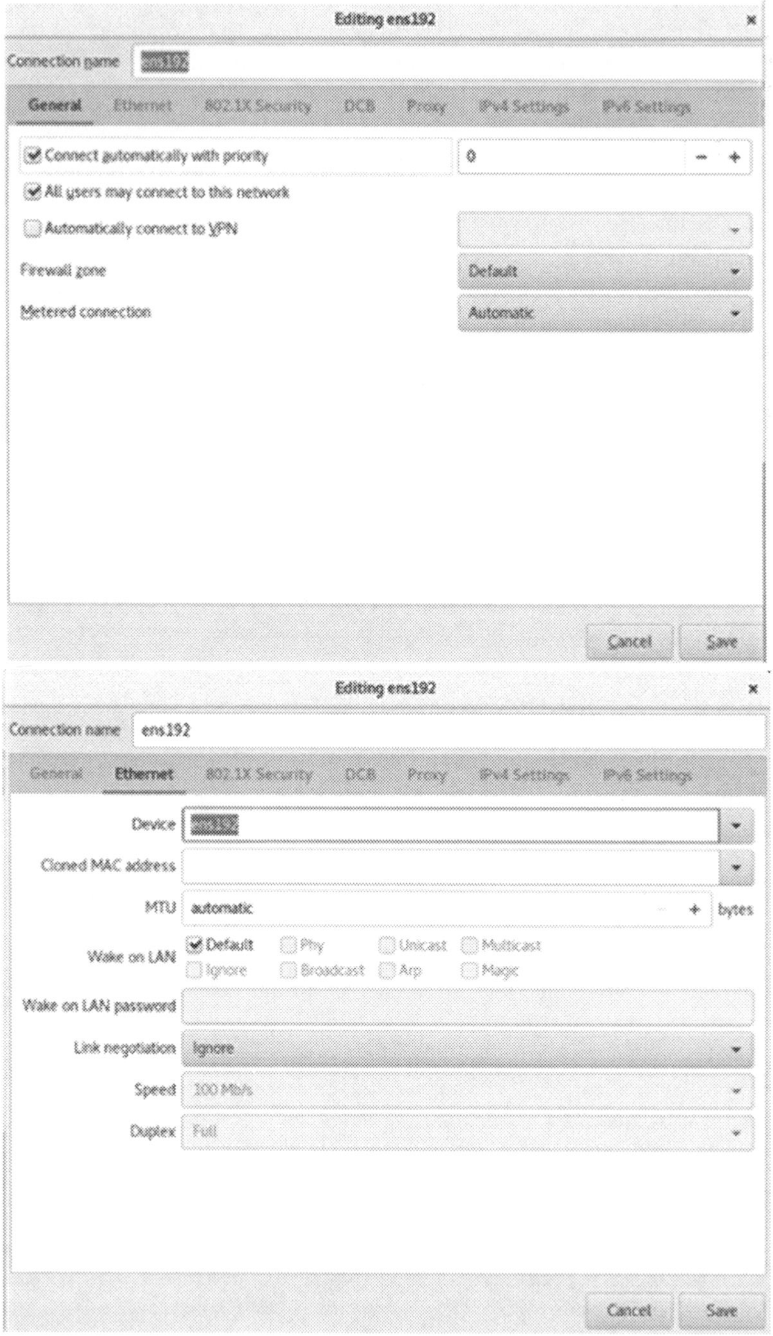

a

FIGURE 10.6 `nm-connection-editor` displaying information on the ethernet interface.

our account. The program, when run, contacts that server to receive our emails and to transmit outgoing emails. Note that a minimal installation would be text-only and so would forego installing either of these software packages.

If we want to run our own email server even if just locally on our Linux computer, we need to set up a mail transfer agent (MTA). Three such programs found in Unix/Linux computers are sendmail, postfix and exim. Depending on your distribution, you may find one or more of these installed, or none. All three use the Simple Mail Transfer Protocol (SMTP). Of the three, sendmail is the oldest and most primitive but also the easiest to configure. It is also the least used of the three MTAs. exim is only available for Debian-based Linux distributions.

postfix has its own directory under /etc that includes a number of both configuration and data files. The main configuration file is called main.cf with additional configuration directives found in master.cf and dynamicmaps.cf. Among the data files are access which stores access control information, canonical for local and nonlocal address mapping, generic for additional but optional address mappings and transport to optionally specify message delivery destinations based on email addresses. As postfix is quite complicated, we do not cover this in any more detail.

To finish this subsection, we look at services which have been likely removed from any recent Linux distribution. snmpd is the service for the SNM (Simple Network Management) Protocol. Its job is to listen for SNMP messages and respond to them. Incoming packets are requests for information from a remote device and commands to alter internal settings. snmpd is primarily used by system administrators to control other network devices, such as servers and routers, across the network.

By default, most network applications use specific ports. The portreserve service does much as the title suggests; it reserves a port for a given application (protocol) while the application is communicating over the network. Additionally, portreserve prevents other programs from utilizing a port which should be reserved for a specific application. Once an application is done with its reserved port, portrelease can be used to release the port so that other applications can use it.

The nfs service provides the functionality of allowing local file systems to be targets for remote mounted. It has been replaced with several nfs-related services including rpc.nfsd to permit remote mounting and mount.nfs/umount.nfs to mount and unmount remote partitions. We covered these services in Chapters 8 and 9.

The xinetd service is a more secure version of the older inetd. Both of these are known as *superservers* in that they are capable of controlling multiple running services. More specifically, these services are configured to select the appropriate application in response to a network message coming in from the network over a specified port. Running xinetd may be preferable over running a number of network-oriented services all the time. We focus on xinetd briefly to get a better idea of how to configure it.

xinetd is invoked when an incoming message arrives. It will first utilize the file /etc/services to map the port address to a service. For instance, if a message arrives over port 22, it will be mapped to ssh. With this information, xinetd will examine its own configuration for how to respond to an ssh message (presumably resulting in the starting of the sshd service).

One way to configure xinetd is through its own configuration file, /etc/xinetd.conf. The configuration file will consist of default directives. These are the directives to apply by default on any incoming message. An example of the defaults is shown in Figure 10.7.

```
defaults {
     instances       = 50
     log_type        = RSYSLOG authpriv
     log_on_success  = PID HOST DURATION EXIT
     log_on_failure  = HOST
     cps             = 25 10
     umask           = 002
}
includedir /etc/xinetd.d
```

FIGURE 10.7 Example /etc/xinetd.conf file.

The `instances` directive dictates the maximum number of simultaneous requests that can be handled. The logging directives specify who is responsible for logging and what to log on a successful versus failed connection attempt. In this case, `rsyslog` is asked to perform logging using `authpriv` as the source where a successful access will log the PID, host IP address, duration and exit status of the communication while the failed attempt will log the host IP address. The directive `cps` establishes the number of connections per second of any given service and the amount of time that a service must wait before it can be restarted respectively.

The `includedir` directive establishes a directory which may store various services' specific configuration files. Each service configuration file contains the directives to be applied for requests for the specified service in place of (or in addition to) the default directives. Figure 10.8 shows an example of one such file for the `rsync` service. We omit further detail.

10.3.3 Establishing DNS Access

The file `/etc/resolv.conf` is used to establish how name resolution should be performed. The default is to store in this file the IP addresses of the local DNS name servers. These values are typically established at the time our computer receives a dynamic IP address from the DHCP server. These entries have the format of `nameserver` *`IPaddress`*. The order of these addresses is the order that our computer will attempt to resolve a host name so if we edit this list by hand, we would want to list the name servers in the order that we want them queried.

Optionally, the `resolv.conf` file may include `search` and `domain` entries. These entries establish the search domain name and the domain name of our computer respectively. The search domain name, by default, is the same as the domain name. We can add to the search entry other domain names. Other directives include `sortlist` to establish an ordering among IP addresses returned by the C function `gethostbyname()`, and `options` to alter default values of variables used by the various C functions that implement name resolution (e.g., `res_init()`, `res_query()`, `res_search()`).

An example `resolv.conf` file is shown in Figure 10.9. This computer's domain name is `somedomain.somecompany.com`, and that the search domain is the same. These entries are followed by the local DNS name servers by their IP addresses. In recent Linux distributions, the contents of this file have been moved to `/run/systemd/resolve/stub-resolv.conf` with `/etc/resolv.conf` being a symbolic link pointing at this new location.

```
service rsync
{
     disable         = yes
     flags           = IPv6
     socket_type     = stream
     wait            = no
     user            = root
     server          = /usr/bin/rsync
     server_args     = --daemon
     log_on_failure += USERID
}
```

FIGURE 10.8 Example `rsync` configuration for `xinetd`.

```
domain somedomain.somecompany.com
search somedomain.somecompany.com
nameserver 10.11.12.13
nameserver 10.11.12.14
nameserver 172.15.183.1
```

FIGURE 10.9 Example `/etc/resolv.conf` file.

If there are machines whose IP addresses are static and which we communicate with often, we may wish to hardcode these addresses locally so that we can bypass DNS entirely and let the mapping be performed by our own computer. Linux provides us with a file to specify such mapping information, `/etc/hosts`. Before any name resolution is attempted, Linux first examines the host's file to see if there are any entries that match the request. Entries in this file are denoted as the IP address followed by host name(s).

Let's consider a local server with a static IP address of 10.11.12.13 called `server1a.internalnet.com`. The organization has aliased this machine to the name `ourserver.internalnet.com`. As we might contact this server frequently and since it has a static IP address, this server seems a useful target for inclusion in the `hosts` file. We place `10.11.12.13 ourserver.internalnet.com server1a.internalnet.com` in the `/etc/hosts` file (this entry should appear on one line). Notice how the entry lists both the name of the server and its aliased name.

The `hosts` file can store other information. We can specify which should be examined first in address resolution, the hosts table or DNS. The two possibilities are labeled as `hosts` and `bind`. If we want to make sure the `/etc/hosts` table is consulted before a name server, we place the entry `order hosts, bind` in the file. The `order` directive can also include `nis` to indicate that name resolution takes place through NIS (network information service).

SECTION ACTIVITIES

1. Open a terminal window in your Linux computer and type `cat /etc/resolv.conf`. How many local DNS name servers are listed? Also type `cat /etc/hosts`. Are there any entries listed? If not, do you know of any servers or computers that you could list there?
2. Start your email client on your home computer (not your Linux computer). Try to identify the name of or IP address of your email server. How did you find it? Now in your Linux computer, try to set up your email client (Thunderbird or Evolution) to receive email from the same server.

10.4 OBTAINING IP ADDRESSES

IP addresses can be assigned in two ways: statically and dynamically. *Static IP addresses* are assigned once by an administrator and persist over time; they may not be permanent addresses but they tend to remain the same for years at a time. Changing a static IP address requires manually modifying DNS resource records and may require other modifications made to network routers and the machine's own interface file(s).

Dynamic IP addresses are assigned upon request (usually at system boot time) from a server responsible for providing IP addresses to clients of its network. Dynamic IP addresses are assigned using a *lease*, which limits the duration that the IP address is available. While many servers and routers are provided static IP addresses, most other Internet-based devices use dynamically allocated IP addresses.

To receive a dynamic IP address, we need some form of dynamic IP address generator. DHCP (Dynamic Host Configuration Protocol) was developed in the early 1990s to replace the outdated Bootstrap Protocol and is the commonly used approach to generating dynamic IP addresses. A DHCP server will supply dynamic IP addresses to clients. The DHCP server is also usually tasked with providing various forms of network configuration information to its clients, such as the addresses of the local network's router and the network's DNS server(s).

In this section, we look at how to set up static and dynamic IP addresses for our computer. The methods for establishing these differ by distribution and have also changed with the transition from init to systemd-based kernels. We concentrate on Red Hat Linux but also note the methods for Debian-based Linux distributions. We then look at configuring our own DHCP server.

10.4.1 Configuring Our Interface Device(s)

Files are set up to specify our interface's configuration, including how it will obtain an IP address, and its address if statically assigned. The location of this directory differs by Linux distribution, being beneath /etc/sysconfig/network-scripts in Red Hat while being found under /etc/network in Debian. The network-scripts directory contains our interface configuration files. Prior to the switch to the NetworkManager service, this directory also contained a number of utilities for bringing up and down various interface devices (which have been removed as unnecessary with NetworkManager). The interface configuration files have the name ifcfg-*interface* such as ifcfg-ens32 (previously, the file might have been named ifcfg-eth0 or some variation).

The ifcfg-ens# configuration file consists of a number of simple directives that assign network-related values. Table 10.6 describes the types of values we might find in this file, their meaning and possible range of values. Some of these entries are required, others are not. The entries we find in this file will differ depending on whether BOOTPROTO is listed as static or dhcp. In the former case, the static IP address is included in the file whereas in the latter case, a DHCP server is used to obtain the IP address. When specifying a static IP address, we provide the address using the directive IPADDR. We also need to include entries for HOSTNAME, NETWORK, NETMASK and BROADCAST. These values are provided to our computer by the DHCP server so are not needed if the IP address is being assigned dynamically.

In Debian, the interface configuration file is /etc/network/interfaces. Entries in this file describe our interfaces. The notation differs dramatically from that of Red Hat's ifcfg files. We see two example entries in Figure 10.10. The first entry defines the loopback device and our Ethernet interface which is given a dynamic address. The lower portion of the figure shows only an Ethernet interface being given a static IP address. If we were to modify the interfaces file, we would have to restart the networking service (networking.service) using systemctl to have the changes take effect.

10.4.2 Setting Up a DHCP Server

In order to set up a DHCP server in Linux, we need the dhcp package. We can obtain this via source code or an executable using dnf or apt-get (as covered in Chapter 11). DHCP comes as both a server named dhcpd, and a client named dhclient. We are interested in the server in this subsection.

The service will be stored as /usr/sbin/dhcpd. The service has several configuration files, each in /etc/dhcp. The configuration files are dhclient.conf, dhcpd.conf and dhcpd6.conf. These files initially contain just a few comments without any directives. To help understand how to configure the DHCP server, a sample configuration is provided in /usr/share/doc/dhcp-server/dhcpd.conf.sample. Table 10.7 lists the directives for the DHCP server's configuration file.

The configuration file begins with global directives which apply to all subnets of the local area network. The domain-name and domain-name-servers directives might be defined here if they cover the entire network. Also defined here are default and maximum duration lease values (in seconds). The configuration file will then contain one or more subnet entries. Each subnet entry includes a netmask and the range of IP addresses that the DHCP will be able to issue to the clients on that subnet. Options include the broadcast IP address and router IP address. Figure 10.11 demonstrates an example of a DHCP server which hosts two subnets, 10.11.0.0 and 10.11.128.0.

In Figure 10.11, we have two subnets specified, each with its own range of IP addresses. The first subnet leases have addresses in the range from 10.11.12.0 to 10.11.12.20 while the second

TABLE 10.6
Contents of `ifcfg-ens#` Configuration File

Directive	Type of Value	Meaning
BOOTPROTO	static, dhcp, none	Source of the IP address (static or via DHCP server or none at all).
BROADCAST	IP address	Broadcast device's address (we usually specify either this value or GATEWAY but not both).
BROWSER_ONLY	yes, no	
DEFROUTE IPV6_DEFROUTE	yes, no	Is this interface the default route to our gateway?
DEVICE	Alphanumeric	Device's name (e.g., ens32, ippp).
DHCP_HOSTNAME	Name	Name of the DHCP server.
DHCP_TIMEOUT	Integer	Number of seconds before timing out when waiting for DHCP server to respond.
DOMAIN	Name	DNS search domain.
DNSN	IP address	IP address of DNS name server.
GATEWAY	IP address	IP address of subnet router/gateway.
HWADDR	Hexadecimal address	MAC address of device.
IPADDR IPV6ADDR	IP address(es)	Set by system administrator for static IP addresses.
IPV4_FAILURE_FATAL IPV6_FAILURE_FATAL	yes, no	Enable/disable device should configuration fail.
IPV6INIT	yes, no	Initialize IPv6 address by default.
IPV6_AUTOCONF	yes, no	
IPV6_ADDR_GEN_MODE	stable-privacy	
NAME	Alphanumeric	Name of device, e.g., ethernet, loopback.
NETMASK	Subnet mask	The netmask used to obtain the local network portion of the IP address.
NETWORK	Network address	IP address of the local network.
NM_CONTROLLED	yes, no	Whether the device is controlled by a network manager program.
ONBOOT	yes, no	Whether to start this interface upon boot or have it manually started.
PREFIX	Number	Number of bits for network portion of IP address (e.g., 19 is equivalent to CIDR /19).
PROXY_METHOD	None	
TYPE	Alphanumeric	Type of device, e.g., Ethernet, PPP.
USERCTL	yes, no	Is user allowed to control this device?
UUID	Hexadecimal address	Address of physical device.

subnet leases have addresses in the range from 10.11.128.129 to 10.11.128.253. Notice that both subnets use the same netmask, where 255.255.128.0 corresponds to the first 17 bits of the 32-bit address. The two subnets have their own broadcast devices and their own routers, so there are two sets of addresses, one per entry. Since both subnets are part of the same organization's network, they share the same domain name and local name servers. They also have equal lease durations, defined in the global section.

When obtaining a dynamic IP address from a DHCP server, a computer is only leased its address from a pool of available addresses. As it is possible (even likely) that there will be more requests

```
auto lo
iface lo inet loopback

auto ens1
iface ens1 inet dhcp
--------------------------------------------
auto ens1
iface ens1 inet static
   address 10.11.12.13
   netmask 255.255.255.0
   gateway 10.11.12.1
   dns-domain ourdomain.net
   dns-nameservers 10.11.12.200 192.168.1.15
```

FIGURE 10.10 Defining interfaces in Debian.

TABLE 10.7
DHCP Directives

Directive	Meaning
`subnet`	DHCP's network (or subnet) address.
`netmask`	DHCP's local network (or subnet) netmask.
`range`	Range of IP addresses available to assign to clients, ranges are indicated by separating IP addresses with a space as in `10.11.12.1 10.11.12.20`.
`routers`	IP address (or alias) of router(s) that this DHCP server responds to.
`domain-name`	Organization (or subnet) domain name.
`domain-name-servers`	IP addresses (or hostnames) of DNS name servers.
`default-lease-time` *`value`*	*value* is the amount of time (in seconds) that an IP address can be made available to a client; once exceeded the IP address expires and the client must renew or request a new one.
`max-lease-time` *`value`*	*value* is the maximum amount of time an IP address can be leased.
`authoritative`	If listed, means this DHCP is the official server for network.
`log-facility` *`level`*	Use *level* listed for `rsyslog` logging (e.g., `local7`).
`group`	Specify parameters that apply to a group of subnets.

for addresses than are available in the pool, the DHCP server limits the amount of time that an IP address is made available. A client with an expired IP address must ask for a new one. The DHCP server may try to assign the same IP address to the client as it had when last leased, but this can only occur if that particular address is currently available.

`dhcpd` stores leasing information in two files, `/var/lib/dhcpd/dhcpd.leases` and `/var/lib/dhcpd/dhcpd6.leases`. These files contain all of the currently leased IP addresses usually by listing the IP address, the client's MAC address, the time and date that the IP address was issued,the time and date the IP address will expire and the computer's hostname (if available).

Aside from building the configuration file, there are a few more tasks to establish a DHCP server. The `dhcpd` service must be started/restarted. DHCP requests must be permitted through the firewall. Typically, DHCP requests come over port 67 under the UDP protocol. We would want to add a rule to our firewall (see Section 10.6) to permit DHCP messages over port 67 using UDP. Third, if necessary, the network's local routers need to know where the DHCP server is located (the machine's IP address).

```
option domain-name somecompany.com;
option domain-name-servers 10.11.1.1 10.11.1.2;

default-lease-time 600;
max-lease-time 7200;

subnet 10.11.0.0 netmask 255.255.128.0 {
     range 10.11.12.0 10.11.12.20;
     option broadcast-address 10.11.12.22;
     option routers 10.11.12.21;
}

subnet 10.11.128.0 netmask 255.255.128.0 {
     range 10.11.128.129 10.11.128.253;
     broadcast-address 10.11.128.255;
     option routers 10.11.128.254;
}
```

FIGURE 10.11 Example entries for DHCP server.

SECTION ACTIVITIES

1. Do a web search on reasons for having static IP addresses. Look at several lists. Do you find any of the reasons to be ones that might fit you? If so, what reason(s)?
2. Research `dhcpd` in more detail on your own. Learn enough to implement a server on a Linux computer.

10.5 NETWORK PROGRAMS

The primary duties of the system administrator with respect to the network are to ensure that the network is accessible and that access to the network is secure. We consider one aspect of security, the firewall, in Section 10.6. Here, we examine some of the available Linux programs that let the system administrator (or the user in many cases) query network status.

10.5.1 THE `ip` PROGRAM

There have been a whole host of programs available for querying network information. Most of these are now obsolete with the same functionality woven into the program `ip`. With `ip`, we can inspect routing tables and rules, view connections between devices, obtain data regarding our interfaces, set routing information and create tunnels, among other tasks.

The basic syntax for ip is `ip [`*`options`*`]` *`object`* `[`*`command`*`]` where *object* is one of `link`, `addr`, `addrlabel`, `route`, `rule`, `neigh`, `tunnel`, `maddr`, `mroute` or `monitor`. The type of *object* dictates the type(s) of *command(s)*, and each *command* might have its own parameters. Options are `-V` for version, `-s` for statistics, `-r` for resolve, `-f` for family which is followed by `inet`, `inet6`, `ipx`, `dnet` or `link`, and `-o` to output on one line. The specified *object* dictates just what `ip` will return or set. Table 10.8 describes each of the options and shows the types of commands available for the given option.

As `ip` controls a great number of network functions, it can be complex to master. At its simplest, we might issue the command `ip addr` to display the values assigned to our interface devices. The response will show MAC addresses, IPv4 address and IPv6 addresses (if any). We see the response from `ip addr` in the top portion of Figure 10.12. Specifically, we see two interface devices, `lo` (the loopback device) and `ens33` (an Ethernet interface).

TABLE 10.8
Objects and Commands in the ip Command

Option	Meaning	Commands
addr	Add, remove or display IP address for one or more devices; add other properties for a device.	add, del, show, flush; del deletes a specified address while flush deletes address(es) based on some specified criteria.
addrlabel	Label an IPv6 address so that later instructions can reference item by label.	add, del, list
link	Status of or alter behavior of an interface.	Varies but includes alias *newname*, allmulticast on/off, arp on/off, mtu value (max transmission unit), up/down.
maddress	Display, add or remove a multicast address.	show, add, delete
monitor	Monitor state of network.	
mroute	Manage the multicast routing cache.	show
neigh	Display, add, change or delete link layer addresses and objects on the subnet (neighbors).	add, change, delete, replace, show
route	Manipulate or inspect a routing table of different types; types are listed in the right column.	add, change, replace, del (delete), show, flush, get Types of routing tables include unicast, unreachable, blackhole, prohibit, local, broadcast, throw, nat, anycast, multicast,
rule	Change the route selection algorithm so that the routing table used in a particular setting is modified.	
tunnel	Add, change or delete a tunnel (connection between two devices).	add, change, delete

```
1: lo: <LOOPBACK,UP,LOWER_UP> mtu 65536 qdisc noqueue state UNKNOWN
group default qlen 1000
   Link/loopback 00:00:00:00:00:00 brd 00:00:00:00:00:00
   inet 127.0.0.1/8 scope host lo
     valid_lft forever preferred_lft forever
   inet6 ::1/128 scope host
     valid_lft forever preferred_lft forever
2: ens33: <BROADCAST,MULTICAST,UP,LOWER_UP> mtu 1500 qdisc noqueue
state UP group default qlen 1000
    link/ether 00:50:56:b9:47:93 brd ff:ff:ff:ff:ff
    inet  10.2.56.201/21  brd  10.2.63.255  scope  global  dynamic
noprefixroute ens33
      valid_lft 564823sec preferred_lft 584823sec
    inet6 f380::eb01:d2d3:de0c:142d/64 scope link noprefixroute
        valid_lft forever preferred_lft forever
----------------------------------------------------------------------
default via 10.2.56.1 dev ens33 proto dhcp metric 100
10.2.56.0/21 dev33 proto kernel scope link src
10.2.56.201
metric 100
```

FIGURE 10.12 Using ip.

For lo, the IPv4 address is 127.0.0.1 (this is always the case for lo), and its IPv6 address is 31 0s followed by 1, which is reduced to ::1. We also see lo has no MAC address (all 0s). This should make sense because lo is not a physical device. The loopback device has no broadcast address (no address from which to broadcast messages) and thus the brd entry is also all 0s. Both the IPv4 and IPv6 addresses are valid forever.

ens33 has a MAC address (00:50:56:b9:47:93), and a broadcast address (ff:ff:ff:ff:ff:ff). This interface has an IPv4 address of 10.2.56.201/21 (where /21 indicates the routing prefix using CIDR notation) and has a broadcast address of 10.2.63.255. We also see that this is a dynamic IP address whose lease is valid for another 564823 seconds (about 6 ½ days). The IPv6 address is shown in the shortened notation, f380::eb01:d2d3:de0c:142d. As IPv6 addresses are typically static, we see that it is valid forever.

The lower portion of Figure 10.12 shows the output from the ip route command, displaying our routing table. Here, we see that the default router for this computer is 10.2.56.1 and that this entry was established at the time we used DHCP to obtain our dynamic IP address. A second router is 10.2.56.201. The label metric 100 found in both listings indicates a priority or preference. In this case, we see that the two routers have equal priority.

The network address for the first router is provided (10.2.56.0/21). The first router was established in our routing table at the time our interface received its IP address (denoted as proto dhcp, where proto stands for protocol) while the second router was established at the time our interface was given its IP address (denoted as proto kernel). Notice that the second entry should be on one line but wraps around.

Aside from viewing the routing table, ip route can be used to modify the routing table by specifying additional routers and establishing or updating the routing policy. A *routing policy* is used to determine which router to select when there are multiple routers available. The *metric* is the simplest form of policy, based strictly on priority and availability. Other policies might factor in packet size, source IP address (instead of destination IP address), number of packets serviced recently and number of hops the packet has already traversed, among others.

ip has superseded older Linux commands of route, ifconfig and iptunnel. route is similar to ip route, displaying router table information. ifconfig is similar to ip addr, displaying interface IP address information. iptunnel has been replaced by ip tunnel to create or change network tunnels. While these programs are simpler to use than ip, the man pages for them report that they have been deprecated in favor of ip. Sometime in the future, these commands are likely to be removed entirely. We have only scratched the surface on what ip can do but we omit further detail and suggest that if you need to know more about ip, consult a network administration text.

10.5.2 Remote Access and File Transfer Programs

One tool that Unix provided from its earlier incarnations was a means to connect to other computers, telnet. With telnet, our terminal window accepts input and sends it to another computer. That other computer receives the input as commands, executes them, and output is sent back across the network to our terminal window to be displayed. In effect, through telnet, we can execute commands on another computer. We must first log in to the remote computer to gain access, so we must have an account.

The telnet program requires as an argument the remote computer's IP address. A number of options are available including -E to prevent escape characters from being interpreted, -a to attempt automatic login, -b to specify a hostname instead of an IP address, -l to specify a username (this option is only useful if we are also using -a) and -x to turn on encryption. We can also specify a port number if we do not wish to use the default port (typically 23).

Communication between the two computers is handled by passing ASCII characters as streams between them. This is unfortunate because it means that all transmitted messages between the two computers are unencrypted, including for instance the password we send to log into the remote computer. Therefore, telnet is not secure and for this reason is unused today.

Another means of communicating between computers is through *r-utilities* (or *r-tools*). This suite of network communication programs is intended to be used on a local network of Unix/Linux computers that share the same authentication server. `rlogin` connects us to a remote computer just as with `telnet`, except that we do not need to log in as we have already logged into the host computer (remember, the two computers share the same authentication server). `rsh` (remote shell) is used to run a command remotely on a networked computer. The command opens a shell, runs the specified command, collects its output and returns the output back to the host computer for display. `rsh` cannot be used to run interactive commands. `rwho` performs a `who` command on the remote computer. Finally, `rcp` is a remote copy command to copy files from the remote computer to the host computer.

Although the r-utilities require that we have an account on the remote computer so that we do not need to authenticate, all interaction is handled in normal text, just as with `telnet`. Thus, any communication using either `telnet` or an r-utility is insecure. In place of `telnet` is `ssh`, the secure shell program. `ssh` is based on the SSH protocol, which uses public-key encryption. When we connect to another computer using `ssh`, our computer is provided a public key to encrypt our messages. The remote computer has a private key used to decrypt messages.

The `ssh` command's syntax is `ssh` *`username@IPaddress`*. If *username@* is omitted then `ssh` attempts to log the user in using their current username. For instance, if we are logged in as foxr and we try to `ssh` to 10.11.12.13, then `ssh` presents to 10.11.12.13 that our username is foxr. If the *IPaddress* is a hostname then DNS will be required to perform address resolution. The first time we login to a remote host using `ssh`, we are asked whether we trust the remote computer's fingerprint (digital certificate). If we answer `yes`, we are logged in and provided the public key via the digital certificate.

Another old technology is FTP (file transfer protocol), developed in 1971. With FTP, a user makes a connection with a remote computer to transfer files. Notice how this differs from `telnet` and `ssh` which both open a shell to the remote computer. With FTP, only files are moved between remote and host computers. Files can be sent to the remote computer, known as *uploading*, and files can be sent from the remote computer to the host computer, known as *downloading*. In order to access the remote computer in this way, we must either have an account on that remote computer or we must log in as an *anonymous* user. If the remote computer permits anonymous logins, this provides the user access only to the public area, perhaps a top-level directory called `/pub`.

The initial FTP command is `ftp` *`address`* where *address* can be the remote computer's IP address or hostname. We are asked to log in. If we are logging in anonymously, we use `anonymous` for the username and our email address for the password. Although the password is not used for authentication, it is captured in a log file for record keeping (thus the login is not truly anonymous). Once connected, either with an account on the remote computer or anonymously, we are given an `ftp>` prompt.

From the `ftp>` prompt, there are several commands available. Perhaps the most important is `get` to download a document. There is also `mget` to retrieve multiple documents (this command can use wildcards). To upload one or more documents, use `put` and `mput`. The commands `ascii` and `binary` change the transfer mode to ASCII text mode (the default) or binary mode (necessary for binary files).

There are `ftp` commands that are virtually the same as Linux commands: `cd`, `ls`, `mkdir`, `pwd` and `rmdir`, which all operate on the remote computer. We can only `cd` and `ls` directories that we have proper permissions for. We would only be able to use `mkdir` and `rmdir` if we had an account on the remote computer and were operating within our own area like our `/home` directory. The command `lcd` performs a `cd` operation on our host computer so that we can change directories locally to upload or download content at this new location. The `delete` command is the same as the Linux `rm` command. Other commands are `open` (to open a new connection to another machine), `close` (to close the current connection) and `quit` (to exit out of the current connection and ftp altogether).

From a closed connection, we can open a new connection using `open` *`address`*. Alternatively, we can launch `ftp` without specifying an address, which places us into the `ftp>` prompt. From there, we open a connection using `open` *`address`*.

ftp, like telnet, is insecure. There are more secure forms of file transfer available. These include *secure FTP* (SFTP) which combines FTP and SSH, *FTP Secure*, which adds TLS security to permit secure file transfer, and SCP (secure copy). Two drawbacks of FTP over SSH are that the remote computer must be able to handle both FTP and SSH and that the user must have an account on the remote machine because SSH does not permit anonymous logins. However, since SFTP is found on most Linux distributions as the program sftp, we will concentrate on it.

sftp is much like ssh in that we specify sftp *username@IPaddress*. If we omit *username@* then sftp attempts to log in us under our current username. In either case, we must authenticate with a password. Like with ssh, the first time we login we are asked whether we trust the remote computer and accept its fingerprint. If we choose yes, we are placed in an sftp> prompt to enter commands. The sftp commands are described in Table 10.9. Notice that there is no equivalent to open as we must specify the remote host in the initial sftp command itself, and no equivalent to ascii or binary.

Today, there are many GUI-based secure FTP programs. These typically have a drag-and-drop feel so that we can copy individual or multiple files very easily. We can change local and remote directories, view contents of local and remote directories, create and delete directories and so forth. FileZilla is one of the more popular GUI-based FTP programs available in Linux.

The newer HTTP (hypertext transfer protocol) and HTTPS (hypertext transfer protocol secure) protocols have largely replaced using FTP/SFTP because most users find it easier to access the world wide web than to use copy commands. HTTP has just a few commands, the most significant are GET (to request a file), HEAD (to request an HTTP response header without the file itself), PUT (to upload a file) and POST (to upload data to a server, such as a post to a discussion board). We don't tend to notice the HTTP commands because we largely access files by clicking on links. But there are mechanisms whereby we can enter HTTP requests from the command line (we explore one approach using netcat later in this subsection).

TABLE 10.9
sftp Commands

SFTP Command	Meaning
bye, exit, quit	Exit sftp.
cd	Change directory on the remote host.
chgrp, chmod, chown	Change the group, permissions or owner of the file/directory on the remote host.
df	Display statistics of the current directory (or the file system if no directory is specified) on the remote host.
get, put	Download/upload specified item(s); wildcards can be used; the instruction can include a local directory to indicate where to store/retrieve the content (otherwise, all items are stored/retrieved to/from the current working directory); options include -a to force a resumption if the transfer is interrupted, -P/-p to copy the items' permissions and -r for a recursive copy.
lcd, lls, lpwd	cd, ls and pwd commands but issued on the local host.
ls, mkdir, pwd, rm, rmdir	Perform ls, mkdir, pwd, rm, rmdir on the specified directory of the remote host.
reget, reput	Same as get and put except that these versions resume a download or upload (like using -a with get/put).
rename	Rename an item on the remote host.
symlink	Create a symbolic link between two items on the remote host.
!	Change from sftp to ssh on the remote host; upon exiting brings the user back to the sftp prompt.
!*command*	Execute *command* in a local shell on the remote host.

HTTPS is a secure form of HTTP using an X.509 digital certificate. This certificate performs two functions. It provides to the client a copy of the public key so that public-key encryption can be used, and, having been signed by a certificate authority, allows the client to know that the website is legitimate.

`wget` is a command line program that allows us to download content from a webserver rather than using a web browser. `wget` is non-interactive. The user enters the complete URL of the file of interest (the URL includes the server's IP address or name). The server responds with the file, which is stored locally on hard disk instead of being displayed in a web browser. In essence, `wget` serves the same purpose as FTP except that the remote computer is a webserver and there is no log in process.

`wget` supports a *recursive* download so that the command can retrieve not only a target file by URL but all of the files that the target file links to, recursively. For instance, if a web page has hyperlinks to three other web pages, then using the recursive download will fetch all four web pages. But then, it operates recursively on those three additional pages. If not done cautiously, a recursive download could result in downloading an entire website! The recursive version of the command is useful for a web spider which attempts to accumulate all pages of a web server.

At its simplest, a `wget` command will look like `wget www.nku.edu/~foxr/index.html` which downloads the index file from the foxr user directory. The recursive option is `-r`. There are GUI-based interactive versions of `wget` available.

Similar to `wget` is `curl`, which is a file transfer program that can communicate with a number of different types of servers using different protocols including FTP, SFTP, SCP, HTTP, HTTPS and email protocols like SMPT and POP3. Files can be uploaded or downloaded, and communication can be encrypted. The command's arguments must include the remote host, directory and filename(s). The syntax for the command varies depending on the particular protocol being used.

We concentrate on using `curl` like we did `wget` above and ignore non-HTTP protocols. To download foxr's index file, we use `curl www.nku.edu/~foxr/index.html`. Multiple files can be specified using globbing as in `curl www.nku.edu/~foxr/file[1-3].html`. We can also use brace expansion to indicate multiple files or multiple servers. Here, we download index files from three different webservers using the one command `curl www.{location1,location2,location3}.com/index.html`. Note that `curl` displays all downloaded content by default. To save to a file, add `-o` *`filename`* to the command.

While `curl` can accomplish the same task as `wget`, `curl` is far more complex. To support the large variety of protocols that `curl` can use, it comes with a lengthy set of options. Most options are specific to a type of protocol, for instance those pertaining to FTP, those pertaining to HTTP and those pertaining to email. There are options for dealing with digital certificates and encryption keys and other options for overriding default session values like timeouts and keepalive values. We omit any further detail and suggest that the reader explore the `curl` man page.

`nc`, or the `netcat` program, has been called the "Swiss army knife" of network programs. Its primary use is to open a network connection with another computer by which we can enter TCP/IP messages directly from the command line. Doing so can be tedious and is generally of value only to someone who is writing network software, learning about TCP/IP or exploring how to accomplish networking tasks. Here, we concentrate on using `nc` to send an HTTP command to a webserver.

The first step in using `nc` is to establish a connection to a remote computer. We specify that computer's IP address/name and the destination port address. For HTTP, we use port 80. We might issue the command `nc www.nku.edu 80`. Upon establishing a connection, our terminal window becomes a message buffer. Anything we enter is sent to the remote computer. As TCP/IP commands may cover multiple lines, pressing `<enter>` does not complete the command and instead requires that we press `<enter>` twice to submit the command.

As noted earlier, HTTP messages contain a command (known as an *HTTP method*) where `GET` is used to download a file from the webserver. Having connected to `www.nku.edu:80`, we might enter the command `GET /index.html HTTP/1.1`. This command requests that the webserver return its index file using HTTP protocol 1.1. To obtain foxr's index page, we issue a slightly different command, `GET /foxr/index.html HTTP/1.1`. Remember that after each of these commands, we press <enter> twice to submit it. What we receive back from the webserver is an HTTP response. With `GET` commands, if the command submitted is valid then the HTTP response includes the entire webpage, which will be written in HTML.

Aside from the HTTP method, URL and protocol, we can add to our HTTP request any number of parameters called *HTTP headers*. Headers might include for instance *content negotiation* values to indicate a selection of file type (e.g., the language we are interested in, or whether and what form of compression should be used). To specify a language preference, we might add the line `Accept-Language: fr en` after the line that contained our `GET` method; this asks for a version of the file in French if available or English if available as a second choice.

Other headers might include the content of a cookie, a request to limit the bytes returned by the webserver and whether a connection should be retained after the response is returned. As we are adding more lines to our message, each time we press <enter> it extends the message but does not submit it until we press <enter> twice.

Let's demonstrate how `nc` works with a simple experiment. Open two terminal windows. In one, `su` to root and type `nc -l` *`port`* (the option is a lowercase L) where *port* is an unused port address, for instance 301. In the other window, type `ip addr` to obtain your IP address and then enter the command `nc` *`IPaddress port`* where *IPaddress* is your computer's IP address, and *port* is the same port as in your first command. These two instructions inform `nc` to listen for messages coming in over the specified port and to open a connection from our computer to our computer using that port.

Upon making a connection, both windows will appear to be suspended but in fact both windows are connected together and waiting for messages. Anything entered in one terminal window appears in the other. Although we are free to type anything we like, a server listens to a port for messages and those messages need to be written in a protocol that the server understands. Thus, the webserver listens for HTTP requests. But `nc` doesn't care what is transmitted.

We can enter text in *either* window and that text is sent to the other window. This shows that our TCP/IP connection creates a two-way communication channel. In a way, this operation can serve as a primitive form of chat in that we can open a connection to someone else's computer and type in text for them to see. The connection remains open unless our computer closes it because we have reached a timeout limit, or one of the two connections is closed by typing `control+c`.

For this experiment to work, we need to ensure that our firewall does not block the selected port. If you try this and are unable to connect from the second window, it is likely that your firewall is blocking messages to the port you selected. We look at modifying our firewall in Section 6. Also make sure that the port selected is not reserved for some specific application because if it is, then upon receiving any message, your computer might pass the text onto that application!

10.5.3 Network Inspection Programs

As a system administrator, user or a network administrator, we might wish to explore the status of the local network or test if a particular networked device is accessible. We can test the accessibility of networked devices with either or both `ping` and `traceroute`. We can use these programs to make sure we can reach our network, that the network is connected to other resources and that a specific device is accessible (responding to network messages).

The `ping` program sends out continual messages to the destination address at one second intervals and reports on responses. For instance, `ping 10.2.3.4` might result in the output shown in Figure 10.13. `ping` continues to send and report on received packets until it is terminated with `control+c`. At this point, `ping` responds with a summary.

As with most Linux commands, `ping` has a number of useful options. We can limit the number of packets and thus the quantity of the output by using `-c` *`n`* where *n* is the number of packets. With option `-f`, `ping` outputs a period for each packet sent rather than the full output. We can specify a different interval for packet transmission using `-i` *`interval`* although intervals of less than 0.2 seconds require system administrator privilege. The option `-R` records the route that the packet took

```
64 bytes from 10.2.3.4: icmp_seq=1 ttl=60 time=0.835 ms
64 bytes from 10.2.3.4: icmp_seq=2 ttl=60 time=0.961 ms
64 bytes from 10.2.3.4: icmp_seq=3 ttl=60 time=1.002 ms
<control+c>
3 packets transmitted, 3 received, 0% packet loss,
time 2798ms rrt min/avg/max/mdev =  0.835/0.933/1.002/0.071
```

FIGURE 10.13 Output from `ping`.

to reach the destination. This is output after the first packet's response. We can also specify our own route if we have knowledge of router addresses that would permit our message to be routed between your computer and the remote computer.

The `traceroute` command is similar to `ping -R`. `traceroute` sends out packets to a remote address and reports statistics on the route (or routes) that the packets took. While `ping` is useful in testing that a remote device is accessible, `traceroute` can be used to determine if parts of the local network are reachable or not, and if any routers in the routes taken are dropping packets (perhaps because of heavy message traffic). A newer version of the program, `traceroute6` (or `traceroute -6`), uses IPv6 addresses rather than IPv4 addresses.

The packets that `traceroute` sends out are known as *probes*. In order to successfully reach the next network location (called a *hop*), three probes must be received by the device at that hop and responses successfully returned to our computer. Probes have a preset *time-to-live* (TTL) value. Should the time in transit exceed the TTL, then whatever receives the packets next drop them rather than forwarding them on to the next hop. The TTL is initially small but if a probe is unsuccessful in reaching the next location, a probe is resent with a larger TTL. If probes are not successful, `traceroute` reports that a particular hop is unreachable by outputting * * *.

Both `ping` and `traceroute` communicate using the ICMP protocol. While both programs can be exceedingly useful, they can also provide a security hole in a network. A hacker could use one or both of these programs to test various IP addresses to locate addresses of resources within someone's local area network. This form of inquiry into a network is called a *reconnaissance attack* and may be a prelude to a larger attack such as a *denial of service* (DOS) *attack* on a server. `traceroute` can permit a hacker to identify router addresses. In order to help protect their network, many network administrators block ICMP packets so that `ping` and `traceroute` do not function.

Several other useful programs, particularly for the system administrator, are `ss`, `nmap`, `tcpdump`, `ifstat` and `lnstat`. There are also older, outdated or deprecated programs of `ifconfig`, `iptunnel`, `netstat`, `route` and `rtstat`. We wrap up this subsection with a brief look at the non-deprecated programs, starting with `ss`. This program displays *socket* usage statistics. Sockets are endpoints in network communication as assigned by applications software. That is, a program creates a socket to be one point of a network message. The socket is assigned a port.

`ss` displays the sockets currently in use. Without options, `ss` displays all open non-listening sockets. We can output both listening and non-listening sockets with the option `-a`. We can narrow down the sockets listed using any or some combination of `-4`, `-6`, `-t` and `-u` to display sockets that are communicating via IPv4, IPv6, TCP and UDP, respectively. Adding `-r` causes `ss` to try to resolve numeric addresses/ports into names, and `-p` shows the process' name that created or is using the socket.

`ss` is intended to replace the older `netstat` which displays current network connections as well as routing tables, interface usage statistics and other information. Figure 10.14 shows first an excerpt of `ss -4 -t -r` followed by `netstat -4 -t`. If you have not yet tried the experiment from the last subsection using `nc`, try it and suspend one of the windows (`control+z`) to enter `ss` to see that the port you selected is currently being listened to.

`nmap` is called a *network exploration tool*. Its role is to send packets onto the network to various network hosts and see what ports are open. In this way, a network administrator can use `nmap` to find potential security holes in the network. Unfortunately, `nmap` can just as easily be used as a tool

```
State Recv-Q Send-Q          Local Address:Port              Peer Address:Port
ESTAB 0      0          localhost.localdomain:ssh      localhost.localdomain:43552
ESTAB 0      0          localhost.localdomain:43552    localhost.localdomain:ssh

Active Internet connections (w/o servers)
Proto Recv-Q Send-Q Local Address           Foreign Address         State
tcp        0      0 localhost.localdoma:ssh localhost.localdo:43552 ESTABLISHED
tcp        0      0 localhost.localdo:43552 localhost.localdoma:ssh ESTABLISHED
```

FIGURE 10.14 Similar output from ss and netstat commands.

for a hacker attempting to find weaknesses in a network. We ran nmap using our own IP address. The results are shown in Figure 10.15. We see that there are four open ports, one each for ssh, rpcbind, nfs and zeus-admin.

The program tcpdump prints out a description of packets that are currently coming into or out of our network interface(s). This is a text-based program similar to the GUI-based Wireshark program that you might have learned about if you have taken a network class. Both programs listen to the specified interface and intercept (capture) packets.

With Wireshark, we can select a particular packet and explore its contents. With tcpdump, by default, all packet contents are displayed. Through various options we can save packets to files and/or limit packets captured through filters such as of a specific IP address or protocol. Wireshark is likely not installed in your Linux system but can be installed using approaches described in Chapter 11. We mention Wireshark because it is easier to use than tcpdump.

The Linux kernel captures network usage statistics, exporting them to files in /proc/net. Two programs that can display such stored information are lnstat and ifstat. The former displays all kernel-saved statistics in a somewhat readable way while ifstat displays statistics about our interface(s) in a readable way.

Of the other programs mentioned, route has been superseded with ip route to manipulate and show our computer's routing table. Similarly, ifconfig has been superseded by ip addr, just as iptunnel, used to create, delete or display network tunnels, has been replaced by ip tunnel. ss has replaced the older netstat. Finally, lnstat has replaced the older rtstat. All of these older programs are still part of Linux distributions, at least for now, so can still be used.

10.5.4 Address Resolution Programs

We explored DNS earlier in the chapter to describe the process involved in name resolution. While Linux handles this process for us, we can also submit our own DNS queries directly from the command line using any of host, dig and nslookup. Before we look at these programs, let's consider in more detail the resource records of an authoritative DNS name server. The first entry is known as a *start of authority* (SOA). A typical entry is shown in Figure 10.16.

```
Starting Nmap 7.70 ( https://nmap.org ) at 2021-03-23 14:15 EDT
Nmap scan report for 10.2.56.201
Host is up (0.00021s latency).
Not shown: 996 closed ports
PORT      STATE SERVICE
22/tcp    open  ssh
111/tcp   open  rpcbind
2049/tcp open   nfs
9090/tcp open   zeus-admin

Nmap done: 1 IP address (1 host up) scanned in 1.63 seconds
```

FIGURE 10.15 Output from nmap.

```
cs.nku.edu.   IN    SOA    ns1.cs.nku.edu.   root.cs.nku.edu.
(
     2021091001;     serial number
     3600;           refresh
     1800;           retry
     604800;         expire
     86500;          minimum TTL
)
```

FIGURE 10.16 Example SOA.

The first entry in the SOA is the *domain name*. If it had previously been listed in the file, the domain name can be replaced with @. The example in Figure 10.16 defines the domain `cs.nku.edu`. The entry `IN` indicates an Internet device. The next entry is a list of server names that can serve as aliases to this name server. We separate multiple aliases with spaces, not commas. The names all end with periods. The list of five values in parentheses specify expiration information. We explore these in more detail in Table 10.10, where they are shown in the order listed in the SOA.

Two terms used in Table 10.10 are master and slave. Many domains utilize multiple authoritative DNS name servers to support what could be heavy request loads. To simplify the process of maintaining multiple servers, we can designate one server as a *master* and the others as *slaves*. When we modify the servers' resource records, we only modify those of the master name server. From time to time, the slave servers query the master server to see if any updates have been made. This is handled through the serial number. If the serial number has not been modified, then no update is waiting.

The first four values in the SOA, as described in Table 10.10, control the slave and master interaction. Upon updating any records in the master, the administrator updates the *serial number*. This might be a simple increment to the number but more commonly includes the date of the update. In Figure 10.16, the serial number is the date of the update followed by a numeric value of the day's update. Ending in 01 means that this was the first update made on the given date.

The slave contacts the master in intervals based on the *refresh rate*. The master responds with the SOA record. If the serial numbers match, no further action is needed by the slave because its records are up to date. However, if the serial number of the master's record is larger, then at least one update has been made. The slave requests the full set of records to update itself. Should the master not respond to the slave's requests, the slave attempts further communication based on the *retry interval*. If the master continues to be inaccessible, the slave must void its own entries after the *expiration period* elapses.

The last entry in the SOA, the *time to live* (TTL), is unrelated to the master/slave interaction. Any DNS response from either the domain's master or slave(s) includes this TTL value. Upon receipt of an authoritative response, a local DNS name server or a client computer may choose to cache the entry. If so, it is cached using the TTL as its expiration date. The entry may remain cached beyond

TABLE 10.10
SOA Values

Value	Meaning
Serial number	A number indicating if resource records have been modified; used to determine if a slave name server is up to date.
Refresh rate	The rate by which a slave will attempt to contact the master name server for updates.
Retry interval	Should the master not respond, this interval is the amount of time before the slaves should try again.
Expiration period	Should the master not respond, this interval is the time until the slave should consider its records out of date and not respond to further requests.
Time to live	This value is attached to all DNS responses to indicate the amount of time the entry is allowed to be cached by a non-authoritative DNS name server before the entry expires.

the TTL but the cache should not respond with the IP address as it would be deemed out of date. The TTL is a form of cache control so that caches across the Internet do not continue to use IP addresses that have expired.

Let's return to the example SOA in Figure 10.16. It has a refresh rate of 3600 (1 hour), a retry interval of 1800 (30 minutes), expiration period of 604800 (1 week) and time to live of 86400 (1 day). An Internet client attempts to resolve www.cs.nku.edu but does not have this information cached locally, nor is it available in the local DNS name server. A query is submitted which arrives at ns1.cs.nku.edu (the cs.nku.edu domain's name server). Let's assume in this case that it is the master DNS name server. It responds and that value is returned to the client, but also cached by the client's local DNS name server with a time to live of 1 day.

Within the next 24 hours, an update is made to the master DNS name server. The slave contacts the master every hour, and after the latest update is made, the slave receives a copy of the master's SOA. The serial number has been modified, so the slave accepts this as an update and requests the entire set of records. All is well at this point.

Hourly, the slave continues to contact the master. At one point over the next week, no response is received. The slave waits 30 minutes and tries again. This continues repeatedly until either the master responds or the expiration period, one week, is reached. Let's assume a week has now elapsed and the slave has not yet heard back from the master.

Returning to our Internet client, it again tries to resolve www.cs.nku.edu. In its attempt to resolve the name, it sends a new request. The local DNS name server does not respond with its cached entry since it is no longer valid and so forwards the request. The request eventually makes its way to the ns1.cs.nku.edu but this name server does not respond to the request because it is inaccessible. Another name server, ns2.cs.nku.edu, is the slave and is available. But it too does not respond because its own entry for www is now considered invalid. To the client, the name goes unresolved and so www.cs.nku.edu is inaccessible (note that if the client knows the IP address, the webserver can still be reached).

The remainder of a name server's entries are the *resource records* of the various named devices found in the domain. When a name server responds to a request, it does so with information found in one or more resource records. Resource records can encode several types of information about a local resource but in most cases, they map from the device's name (its true name or an aliased name) to its IP address. We describe the common types of entries in Table 10.11, although we omit dozens of types that are less frequently used.

TABLE 10.11
Resource Record Types

Record Type	Meaning
A	Resource's IPv4 Address.
AAAA	Resource's IPv6 Address.
CAA	Certificate authority authorization record.
CNAME	A device's canonical name.
MX	One of the domain's mail servers; priority numbers are included to indicate which of the mail servers should be consulted first.
NS	One of the domain's name servers.
PTR	Maps the resource's IPv4 address to its name.
SPF	Sender policy framework lists mail servers which are permitted to send mail from this domain (useful for preventing fraudulent emails).
SVR	Service record maps from a service or server to a domain name in case the domain offers multiple services each handled by a different device.

Most of the entries in Table 10.11 are self-explanatory but let's take a quick look at two. First is CNAME, the device's canonical, or true name. We tend to provide hostnames that are fairly short and easy to remember, such as mycomputer, zappa or hal. But administrators tend to name computers based on some kind of organizing principle such as the office number where the computer is housed, the year the computer was purchased, or a count. For instance, pc17y2022 might be used to indicate that this was the 17th PC purchased in 2022. The CNAME resource record is used to equate a computer's alias to its true name.

The PTR resource record maps the IP address to the resource's name. This entry is used for what is known as a *reverse IP lookup*. This form of lookup is used as a security measure. IP addresses can be spoofed by other computers whereby the user of that computer attempts to intercept messages by claiming to have the IP address of some other computer. To ensure that a computer's IP address is legitimate, we can take the further step of mapping its IP address to its name to see that they match correctly.

Figure 10.17 provides a short example of a domain's resource records. First, we see $ORIGIN defining the subnet for this domain being 10.11 (notice the numbers are written in reverse order, ending with IN-ADDR.ARPA). The records denoted as @ are defined for ORIGIN. These are the name server with the name m15-2022.csc.nku.edu and an alias of the same computer, ns.csc.nku.edu. Next are two resources (computers) named zappa and hackett. For zappa, its IPv4 address is 10.11.12.14 and it has a true name of h1c2, while hackett has both IPv4 and IPv6 addresses defined.

Now that we have some understanding of DNS records, we can explore the DNS query commands. Of the three programs, nslookup is the oldest and most primitive, but this also makes it the easiest to use. nslookup expects as a parameter the hostname of the computer whose IP address we wish to look up. Optionally, we can specify a DNS server's IP address to act as the server to query (without this, our query is sent to our local DNS name server). The format of the instruction is nslookup *hostname* [*DNS_IP_address*] where the *DNS_IP_address* is optional.

The response from nslookup provides all IP addresses known for the specified name and also lists the IP address of the DNS server used. Consider the three nslookup command responses in Figure 10.18. The first example is a request to look up www.nku.edu, sent to a DNS server that is the master for that domain and so the response is authoritative. The second example requests the IP address for www.centos.com and has a non-authoritative response because the response did not come from a centos.com authority. The third is also a non-authoritative response but is of interest because of the number of responses. We receive multiple IP addresses because google has several physical IP addresses to support their servers.

Both the dig (domain information groper) and host programs permit a number of options to obtain more detailed feedback than nslookup. Both commands return portions of the actual resource records of the DNS name server. With dig, the -t option allows us to specify the type of entry we are interested in. We might, for instance, use dig -t MX google.com to request of google.com's authoritative name information about their mail server(s).

A response to the above query might be google.com 460 IN MX 20 *address* (although there will likely be several such entries in the response as Google has numerous mail servers). The value 460 indicates a TTL. The value 20 in this example specifies a priority among the mail servers

```
$ORIGIN 11.10.IN-ADDR.ARPA.
@          IN   NS      m15-2022.csc.nku.edu.
@          IN   PTR     ns.csc.nku.edu.
zappa      IN   A       10.11.12.14
zappa      IN   CNAME   h1c2.csc.nku.edu.
hackett    IN   A       10.11.12.15
hackett    IN   AAAA    e140:f8b0:4010::53e:a1
```

FIGURE 10.17 Example resource records.

```
$ nslookup www.nku.edu
Server:     172.28.102.11
Address:    172.28.102.11#53

www.nku.edu       canonical name = hhilwb6005.hh.nku.edu.
Name: hhilwb6005.hh.nku.edu
Address: 172.28.119.82
-----------------------------------------------------------
$ nslookup www.centos.com
;; Got recursion not available from 172.28.102.11, trying
next server
;; Got recursion not available from 172.28.102.13, trying
next server
Server:     10.11.0.51
Address:    10.11.0.51#53

Non-authoritative answer:
Name: www.centos.com
Address: 87.106.187.200
-----------------------------------------------------------
$ nslookup www.google.com
Server:     172.28.102.11
Address:    172.28.102.11#53

Non-authoritative answer:
www.google.com    canonical name = www.l.google.com.
Name: www.l.google.com
Address: 74.125.227.51
Name: www.l.google.com
Address: 74.125.227.49
Name: www.l.google.com
Address: 74.125.227.48
```

[additional addresses omitted]

FIGURE 10.18 `nslookup` command responses.

for load balancing purposes (other mail services will likely have different values). The command `dig -t NS google.com` to obtain similar information on google.com's name servers gives much the same type of response except that the TTL is very large and there is no load balancing value specified.

The `dig` command responds with several different sections. The first section repeats the command's arguments. It then summarizes the response as a header, a type of operation (Query), status (NOERROR), an ID number, flags and the number of responses received from the DNS server. These are divided into a question (or query) section which in essence repeats the request and then summarizes what we will see in the other sections by listing the number of items we will find in the following ANSWER section, the number of items that appear in the AUTHORITY section and the number of items to appear in the final section called ADDITIONAL. Figure 10.19 demonstrates the result from the query `dig -t NS www.nku.edu`. In this case, we are querying a name server for a specific machine's information so there are fewer entries in the response.

The option `-c` is used to query an entity by type of *class*. All of the classes seen in our examples are `IN`, for Internet. There are other classes but we won't see them when querying Internet resources. Other options include `-p` to specify a port, `-6` to indicate that only IPv6 addresses should be used and `-b` to send the `dig` command to a specified DNS server. We can request responses of multiple machines and/or domain names. If preferred, we can place the request information in a file using `dig -f` *`filename address`*.

The `host` program by default responds with the IP address of the given name. The option `-d` (or `-v`) provides more detail or a verbose response. Like `dig`, `host` will respond with question, answer, authority and additional sections. Unlike `dig`, `host` when supplied with `-d` (or `-v`) responds with information from the SOA record. For instance, the instruction `host -d www.nku.edu` will

```
; <<>> DiG 9.7.3-P3-RedHat-9.7.3-8.P3.el6_2.1 <<>> -t NS www.nku.edu
;; global options: +cmd
;; Got answer:
;; ->>HEADER<<- opcode: QUERY, status: NOERROR, id: 57418
;; flags: qr aa rd ra; QUERY: 1, ANSWER: 1, AUTHORITY: 0,
ADDITIONAL: 0

;; QUESTION SECTION:
;www.nku.edu.              IN    NS

;; ANSWER SECTION:
www.nku.edu.          3600  IN    CNAME hhilwb6005.hh.nku.edu.

;; Query time: 1 msec
;; SERVER: 172.28.102.11#53(172.28.102.11)
;; WHEN: Tue Aug 28 14:23:32 2012
;; MSG SIZE  rcvd: 57
```

FIGURE 10.19 `dig` command response.

provide an AUTHORITY SECTION. Figure 10.20 provides the results of `host -d www.nku.edu`. The information is similar to that from `dig`, but we see that the response includes more than information about the single machine as it includes information about the domain (nku.edu). `host -i` asks the DNS server to perform a reverse IP lookup.

```
Trying "www.nku.edu"
;; ->>HEADER<<- opcode: QUERY, status: NOERROR, id: 2258
;; flags: qr aa rd ra; QUERY: 1, ANSWER: 2, AUTHORITY: 0, ADDITIONAL: 0

;; QUESTION SECTION:
;www.nku.edu.                 IN    A

;; ANSWER SECTION:
www.nku.edu.            3600  IN    CNAME hhilwb6005.hh.nku.edu.
hhilwb6005.hh.nku.edu.  3600  IN    A     172.28.119.82

Received 73 bytes from 172.28.102.11#53 in 2 ms
Trying "hhilwb6005.hh.nku.edu"
;; ->>HEADER<<- opcode: QUERY, status: NOERROR, id: 84
;; flags: qr aa rd ra; QUERY: 1, ANSWER: 0, AUTHORITY: 1, ADDITIONAL: 0

;; QUESTION SECTION:
;hhilwb6005.hh.nku.edu.       IN    AAAA

;; AUTHORITY SECTION:
hh.nku.edu.       3600  IN    SOA   nkuserv1.hh.nku.edu.
postmaster.exchange.nku.edu. 37369675 900 600 86400 3600

Received 104 bytes from 172.28.102.11#53 in 5 ms
Trying "hhilwb6005.hh.nku.edu"
;; ->>HEADER<<- opcode: QUERY, status: NOERROR, id: 15328
;; flags: qr aa rd ra; QUERY: 1, ANSWER: 0, AUTHORITY: 1, ADDITIONAL: 0

;; QUESTION SECTION:
;hhilwb6005.hh.nku.edu.       IN    MX

;; AUTHORITY SECTION:
hh.nku.edu.       3600  IN    SOA   nkuserv1.hh.nku.edu.
postmaster.exchange.nku.edu. 37369675 900 600 86400 3600

Received 104 bytes from 172.28.102.11#53 in 3 ms
```

FIGURE 10.20 `host` command response.

SECTION ACTIVITIES

1. In a terminal window, type `ip addr` and `ifconfig`. Which instruction provides clearer output, if either? Now try `ip route` and `route`.
2. In a terminal window, type `nslookup google.com` and make a note of the server's address. Now type `nslookup www.nku.edu`. Did you get an authoritative or non-authoritative answer? Next type `nslookup www.nku.edu` *`IPaddr`* where *IPaddr* is the address from Google. In this case, you are using one of the Google name servers rather than either your local DNS name server or nku's name server. Did the result differ at all?

10.6 THE LINUX FIREWALL

A *firewall* is a program which examines incoming and outgoing network messages and decides which ones are permitted to be passed along. The firewall itself uses a collection of rules that define attributes of messages and the actions that should be taken on a message that matches those attributes. Typical actions are to allow the message through, reject or drop the message, log information about the message, and in the case of a router, forward the message on to another router or network location. Rules can pertain to incoming messages only, outgoing messages only or both. The criteria tested by the rules can examine source or destination IP addresses (or subnet addresses), source and destination port addresses, protocol(s), message size and the interface device receiving/transmitting the message, to name some of the many attributes.

A *stateful* firewall is able to extend its decisions across groups of messages that make up an ongoing connection. This is useful as a packet is typically just one of many that make up a message. A message may be one of many in an established session between two machines. By maintaining a state, the firewall can reason over more than just the specifics of the one packet. As an example, a firewall can count the number of packets received from a specific IP address, and if this number exceeds some threshold, it can begin blocking any future messages. This could potentially prevent a denial of service attack.

A firewall can be set up to protect a single computer or an entire network. Typically, an organization will employ multiple firewalls. One will be positioned at the Internet point of presence, perhaps as a proxy server (or in addition to a proxy server) or as a dedicated hardware device. Firewalls may run on some or every router. And it is likely that every computer will run its own firewall as part of its operating system.

In Linux, the firewall is called `netfilter`, which is part of the Linux kernel. We do not manipulate `netfilter` directly. Instead, our interface to the firewall is through the programs `iptables` and `ip6tables`. These programs accept a series of rules which specify how to handle incoming and outgoing packets based on attributes of the packets as described earlier. Figure 10.21 lists a few rules defined using the directives found in `iptables`. Comments explain each rule.

We see two forms of actions in Figure 10.21, `ACCEPT` and `DROP`. Other actions are `REJECT`, `FORWARD`, `RETURN` and `LOG`. The difference between rejecting and dropping a message is that

```
# Accept any TCP packets coming in to port 22 (SSH messages)
-A INPUT -p tcp --dport 22 -j ACCEPT
# Drop any packet coming in over port 80 or 8080
# (drop packets intended for a web server)
-A INPUT --dport 80,8080 -j DROP
# Accept any message that is in response to an established
# communication such as a previous outgoing message
-A INPUT -m state --state ESTABLISHED -j ACCEPT
```

FIGURE 10.21 Example `iptables` rules.

rejecting a message still causes a response to be sent to the source so that the sender knows the message was received. Dropping a message ignores the message entirely. Forwarding is primarily used by routers to decide whether a received packet should continue on its way to the next network hop or be dropped. The RETURN action causes the current chain of rules to be discarded and the next set of rules to be examined (see below). Logging can be applied to other actions such as a dropped message, or it can be performed without other actions.

In iptables/ip6tables, rules are cumulative in that if one rule matches, other rules could still be applied. Thus, rules can form chains of logic. One rule might lead to another rule. A chain continues to be examined until either a final action is reached (e.g., ACCEPT or DROP), a RETURN statement is reached or a condition is found to be false. Rules are defined in files that constitute permanent firewall rules (e.g., /etc/sysconfig/iptables) but rules can also be added dynamically from the command line. iptables/ip6tables also utilize their own configuration files.

iptables can be a challenge to use. Most versions of Linux have added a more convenient front-end so that rules do not have to be directly entered from the command line or via one of the iptable rule files. In Red Hat, we use the firewalld service, and in Debian-based versions (including Ubuntu), the front-end is called ufw. In the next three sections, we concentrate on firewalld and its interfaces firewall-config and firewall-cmd. We wrap up this section with a brief look at ufw.

10.6.1 The firewalld Service

With the shift to systemd, Red Hat has added firewalld, a service that serves as a front-end to iptables' configuration. There are two different methods to modify the firewalld settings, and thus the rules that iptables uses. The first is from the command line using firewall-cmd. The second is with a GUI tool called firewall-config. In Red Hat 7, the GUI could be found under the Sundry menu. But with the shift to Red Hat 8, the various menus are no longer present. We might find firewall-config preinstalled. If not, we can install it using dnf (see Chapter 11).

Before we explore firewall-config and firewall-cmd, let's understand what firewalld is. Whereas iptables is divided into rules, firewalld divides network communication into *zones* of trust. There are numerous default zones, described in Table 10.12.

The zones in Table 10.12 are listed in alphabetical order rather than trust. The zone of least trust is the drop zone. This zone drops all packets. If we know of sources that are completely untrusted, we might specify those sources (IP addresses, protocols, ports, etc.) in this zone. The most trusted sources might be placed in the trusted zone, which accepts everything. The other zones are in between these two extremes.

Each zone defines rules and actions. That is, we attach to a zone the various criteria that specify the trust level with respect to ports, IP addresses, protocols, etc. Also attached to a zone is an action such as drop or accept. Although firewalld comes with zones pre-established, we can alter an existing zone by adding new criteria (items) to the zone or removing existing criteria from the zone. We can also define our own zones.

We see in Table 10.12 the predefined items that each zone accepts. ssh messages, for instance, are accepted in many of the zones. If you think about dmz and public, these are zones where security should be increased. Why then should ssh messages be accepted? Because the SSH protocol requires authentication to gain entrance to our computer. Similarly, cockpit requires a login which is why it also appears in numerous zones.

firewalld exists in two different sets of definitions, the current runtime definition and the permanent definition. By default, changes made to the firewall are made to the runtime version. We can save our runtime definition to the permanent firewall. Alternatively, we can specify that a change is to be made to the permanent version. This does not impact the current firewall and requires reloading the firewall from the permanent version. The runtime firewall's definitions are stored in one location and the permanent version in another to ensure that the two versions are kept separate.

TABLE 10.12
`firewalld` Default Zones

Zone	Usage	Preset Values
`block`	Reject incoming packets.	None
`dmz`	For use in a "demilitarized zone" which is, at least in part, publicly accessible; only select items should be accepted.	`ssh`
`drop`	Drop incoming packets.	None
`external`	For use on external networks where some entities of the network are masked to help protect them; these items are considered untrustworthy and only select items should be accepted.	`ssh`, masquerade zone enabled for IP forwarding
`home`	Used in a home network which is protected; items here are largely trusted but only select items should be accepted.	`cockpit, dhcpv6-client, mdns, samba-client, ssh`
`internal`	Used for internal networks where most items are trusted, but only select items should be accepted.	`cockpit, dhcpv6-client, mdns, samba-client, ssh`
`public`	For use in public areas; these items should not be trusted; accept only select items in this zone.	`cockpit, dhcpv6-client, ssh`; Ethernet interface
`trusted`	All items are trusted and should be accepted.	None
`work`	Areas related to or owned by an organization to which we belong (e.g., our company's network, our school's network). Most items in this zone are trustworthy but only select items should be accepted.	`cockpit, dhcpv6-client, ssh`

10.6.2 The Firewall Configuration GUI Tool

The `firewalld` GUI interface is called Firewall Configuration and can be started from the command line with the instruction `firewall-config` (assuming it is installed). Starting the GUI requires being root. Upon running `firewall-config`, the GUI opens with the view shown in Figure 10.22.

The Firewall Configuration GUI has several sets of controls. The default view shows us zones, and within zones, services. These are the Linux services. In the figure, none of the shown services are selected meaning that none of these services are permitted in the given zone, `public`. Scrolling down through the list of services, we would find that `cockpit`, `dhcpv6-client` and `ssh` are all selected for this zone, `public`.

Across the inner set of menus, we can change the view for the selected zone from services to ports, protocols, and source ports assigned to this zone. There is a right arrow at the end of the row. This lets us scroll through further options which are of masquerading, port-forwarding, ICMP filters, rich rules, interfaces, and sources. See Table 10.13 for an explanation of each.

The top of the main pane has a drop-down menu for the configuration type. The default is `run-time` but can also be `permanent`. Depending on which configuration is selected, the GUI tool loads from and writes to a different set of files. Changes made to the permanent firewall do not impact the current version unless reloaded. Changes made to the runtime firewall are not permanently saved unless specified. Along the upper left of the GUI are menus. Under Options is a selection named Runtime to Permanent, which will take any runtime changes and save them permanently.

Beneath the configuration selection is a row of additional menus. Services is the default item selected, displaying all available services. Selecting any service gives us an inner list of menus to see what the service has been assigned in terms of ports, protocols, source ports, modules and destinations. Most services have been assigned ports (including protocols) but nothing else. For instance,

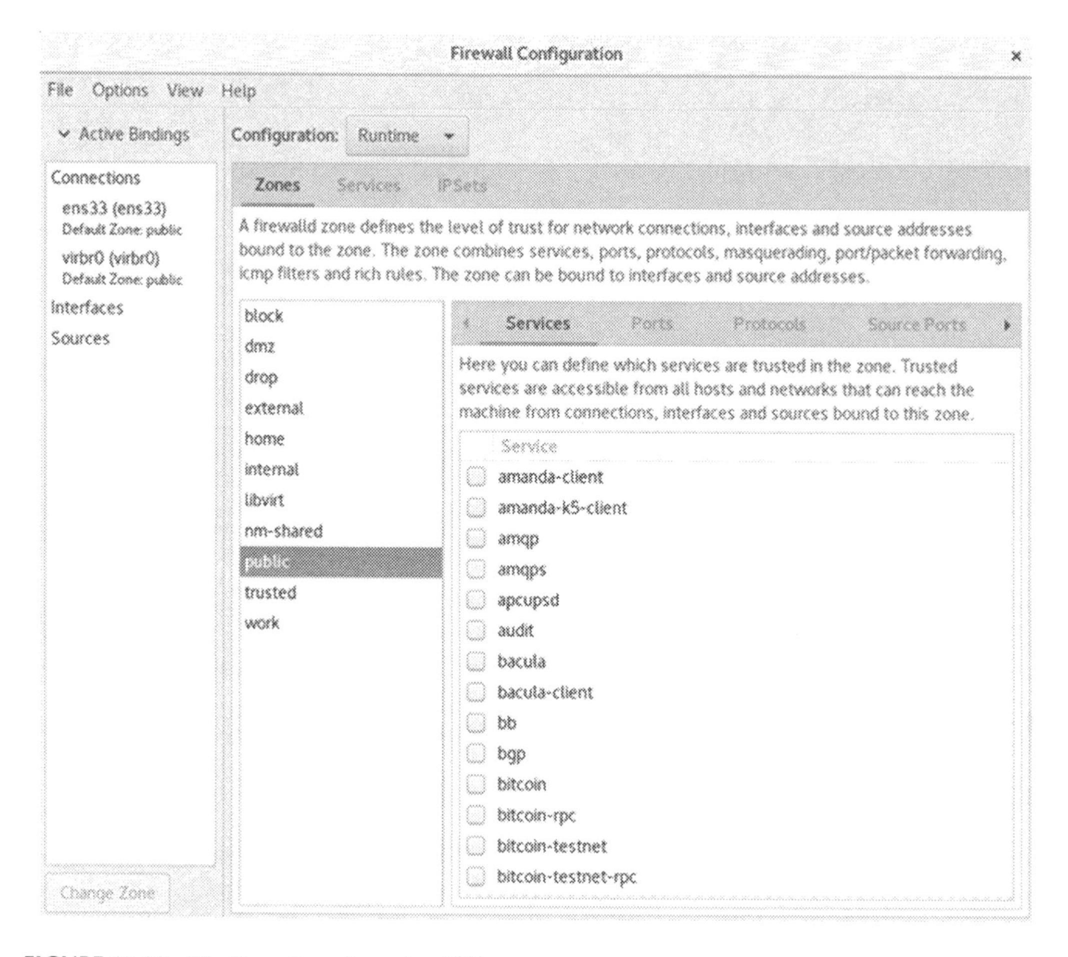

FIGURE 10.22 The firewall configuration GUI.

`ssh` is given the port `22 tcp` but has no entries under protocols, source ports, modules or destinations. Some services have multiple port entries such as `snmp` which is listed as `161 tcp` and `161 udp` while `bacula` has entries of ports `9101`, `9102` and `9103` (all `tcp`).

IPSets lists any defined sets of IP addresses. We must create these. We can place a defined IPSet into one or more zones. This lets us whitelist or blacklist specific IP addresses. New IPSets can only be established under the permanent configuration.

We create an IPSet by first selecting the + found near the bottom of the pane in which IPSets is an option (see the right side of Figure 10.23, near the bottom). A pop-up window appears and asks for a name, description and other information of this IPSet. The left side of Figure 10.23 provides an example where we are creating an IPSet of items at home with static IP addresses. After creating the IPSet, we add specific IP addresses to it. Again, on the right of Figure 10.23 near the bottom of the pane is a button, `Add`. Selecting it allows us to enter an IP address. In the figure, we have added three IP addresses to the IPSet named `Mine`.

Along the left pane (in both Figures 10.22 and 10.23), we see our *connections*. These are our interfaces. Here, we have been operating on our Ethernet interface, `ens33`. Notice that it lists the default zone of public. This means that anything not listed in any other zone such as a service or protocol will use the public zone.

The menus along the upper left have few choices. File only contains a selection to Quit. Options allows us to reload the firewall (from the permanent version), change zones of connections, change

TABLE 10.13
Explanation for Firewall Rule Criteria

Criteria Type	Meaning	Example Values
Service	Assign services to this zone, as described in the text with `ssh` and `cockpit`.	Any of the services.
Port	Message destination port addresses.	`80`, `8080` for a webserver `53` for a DNS name server
Protocol	The protocol of packet.	`TCP`, `UDP`, `ICMP`
Source port	Message source port addresses. We would not normally expect to use this as these addresses are randomly selected by our operating system when sending out messages, but if we are being attacked over an address like 51833, we might add it to the drop zone.	Any of the unreserved port addresses.
Masquerading	Used when we are applying network address translation (NAT).	`yes` (permit NAT), `no`
Port-forwarding	For use when we expect messages to be forwarded.	Initial port, protocol, destination port.
ICMP filters	Specify ICMP message types that are to be blocked (or permitted).	`echo-request`, `echo-reply`, `failed-policy`
Rich rules	Define rich rules through this interface.	See Figure 10.26.
Interfaces	Interface device by name.	`ens32`
Sources	IP addresses or IPSets.	Predefined IPSets.

the default zone, change logging options, obtain help or save any runtime changes to the permanent version. View shows us what items can be viewed and Help brings a window that tells us about the firewall GUI.

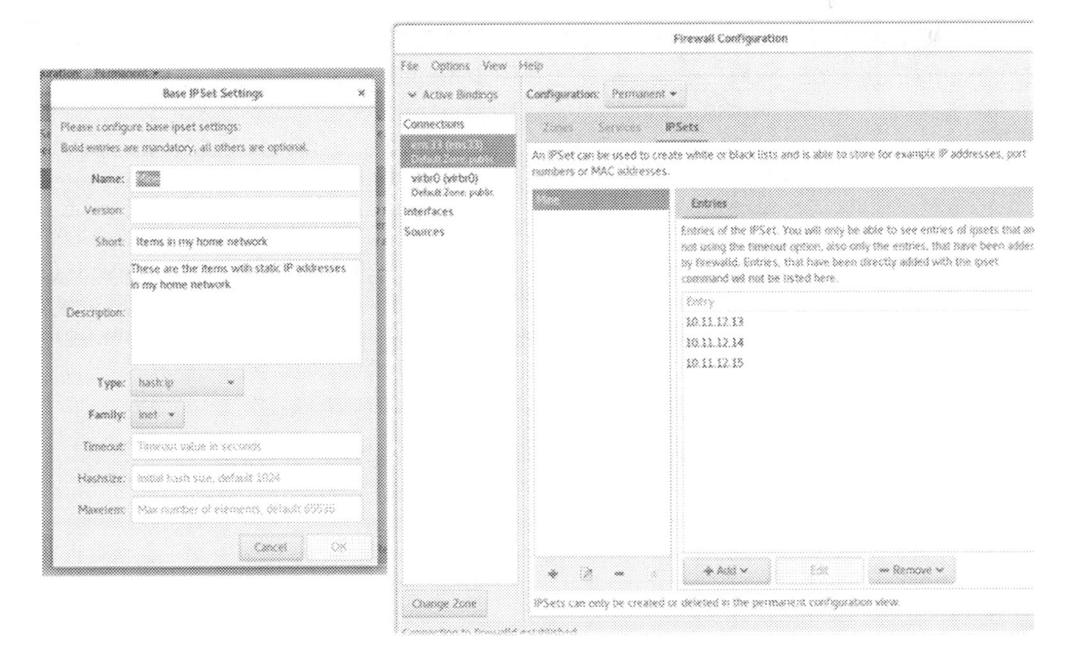

FIGURE 10.23 Adding IPSets with the firewall configuration GUI.

10.6.3 firewall-cmd

While firewall-config is easy to use, it has its own limitations. We can perform more complex and complete modifications to the firewall using the command-line program firewall-cmd. The instruction's syntax is firewall-cmd *command* [*options*] where *options* depend on the *command* specified. Like systemctl and ip, firewall-cmd has a large number of commands, each with its own set of options although some commands do not require or have options.

Table 10.14 presents just a few of the commands. By default, any command issued applies to the runtime version of the firewall. To change this behavior, add --permament to the command so that the command modifies the permanent firewall (but not the runtime firewall). Some commands can only apply to the permanent firewall (noted in the table by stating that the command can only be issued with the specifier --permanent).

Figure 10.24 provides examples of issuing some of the commands listed in Table 10.14. The first two commands list the zones and the default zone. These are simple instructions. The remainder of the commands illustrates how to create our own zone, called http. We issue separate instructions to create the zone and then define its attributes. We separately add ports to the zone as firewall-cmd does not permit us to add both services and ports in one instruction. We finalize the zone by assigning it to a target (the action to take place should a message be placed in this zone). We use the default action although we could have also stated ACCEPT. The last command in the figure outputs the contents of this new zone.

All of the commands regarding this new zone use --permanent to indicate that the changes are to be made to the permanent firewall. These changes do not take effect until we reload the firewall from the permanent version. Note in the figure that some lines extend beyond the margin so they are indented in the line(s) beneath.

In Figure 10.25, we illustrate how to create an IPSet using firewall-cmd. First, we define the IPSet, called hackers. We then add two IP addresses from which known denial of service attacks have originated. We then add the IPSet to the drop zone to ensure any message from one of those IP addresses is dropped. The notation for adding the IPSet to the zone is --add-entry=ipset:*name* where *name* is the name of the IPSet (hackers in this case). We could also specify the IP addresses directly using two --add-entry statements as in --add-entry=10.2.3.4 and --add-entry=10.2.3.5. The last command in Figure 10.25 loads our permanent firewall into the runtime firewall so that we can start using the changes we made.

One entry in Table 10.13 from the previous subsection referenced *rich rules*. Through either the GUI or firewall-cmd, we can specify a mapping of protocols, ports, IP addresses and ports to zones and actions, and we can also define our own rules by hand. We present just a couple of examples of defining rich rules, in Figure 10.26. The first rule accepts FTP messages coming in from the local network (assuming our network is 10.11.12.0/24). The rule logs attempts with the prefix "local FTP access". The second rule rejects any ICMP messages. Neither rule is necessary as we could place 10.11.12.0 in an IPSet and either create a zone that allows the FTP protocol from this IPSet access with an ACCEPT action, or place these in a trusted zone. We can similarly place ICMP into the block zone.

For more information on firewall-cmd, view its man page. You can also find more material on firewalld in its online manual at fedoraproject.org/wiki/Firewalld. This manual includes numerous examples of firewalld settings, commands and rich rules.

10.6.4 ufw

The Uncomplicated Firewall (ufw) presents a different front-end approach to controlling iptables. Depending on the Debian distribution, it and/or the graphical front-end gufw might be preinstalled. A third-party implementation of ufw can be installed in Red Hat.

TABLE 10.14
Some of the Many `firewall-cmd` Commands

Command	Meaning	Comments
`--check-config`	Test the permanent firewall configuration file for errors.	
`--delete-zone=`*zone*	Remove the zone.	Requires the specifier `--permanent`
`--get-active-zones`, `--get-zones`	List all active zones along with the interfaces assigned to each; list all defined zones.	
`--get-default-zone`, `--set-default-zone`	Output or set the default zone.	Default zone is to `public`
`--get-zone-of-interface=`*interface*	Print the zone the interface is bound to.	Outputs no zone if the interface is not bound to a zone
`--info-zone=`*zone*	Display information about *zone*.	
`--ipset=`*ipset*	Create a new IP set, *ipset*.	
`--ipset=`*ipset* `--set-description=`*descr*	View or create a description for *ipset*.	Requires the specifier `--permanent`
`--ipset=`*ipset* `--add-entry=`*entry*, `--remove-entry=`*entry*, `--get-entries`	Add or remove IP addresses to *ipset* or list all defined entries of *ipset*.	
`--reload`	Load the permanent firewall configuration into the runtime firewall (useful if we have changed the permanent firewall).	
`--new-zone=`*zone*	Create a new zone called *zone*.	Once created, we define attributes of the zone using `--zone` (see below).
`--runtime-to-permanent`	Save the active configuration; for use when your `firewalld-cmd` instructions have changed the runtime firewall.	Overwrite the permanent firewall so be careful when using this!
`--zone=`*zone*	Add, modify or delete one or more attributes for this zone; attributes include `service`, `target`, `port`, `description`, `interface`, `rich-rule`.	`Notation is --add-attribute=`*value*, `--change-attribute=`*value*, `--set-attribute=`*value* `and --remove-attribute`
`--zone=`*zone* `--get-description`	Output a description of *zone*.	Requires the specifier `--permanent`
`--zone=`*zone* `--list-all`	List all information about *zone* (if *zone* is omitted, output the default zone).	
`--zone=`*zone* `--get-target`, `--set-target=`*target*	For *zone* output its action or change to a new action.	Actions are one of default, `ACCEPT`, `DROP`, `REJECT` Requires the specifier `--permanent`.

Let's start with a brief look at `gufw`. When first launched, the default view is of the "Home" page. This shows us the standard interface but in place of specific firewall information, it contains help. On the left side of Figure 10.27, we see `gufw` with the Reports tab selected. In the figure, we see that we are viewing the Public profile, which is enabled and by default rejects incoming messages and allows outgoing messages. Reports show us applications (services) whose behavior differs from the default. Selecting one of these and clicking the plus sign at the bottom of the window opens

```
[root@localhost apache2]# firewall-cmd --get-zones
block dmz drop external home internal nm-shared public trusted work
[root@localhost apache2]# firewall-cmd -get-default-zone
public
[root@localhost apache2]# firewall-cmd --new-zone=http --permanent
success
[root@localhost apache2]# firewall-cmd --zone=http --set-
    description="Zone for Apache webserver" --set-service=http --
    set-service=https --permanent
success
[root@localhost apache2]# firewall-cmd --zone=http --set-
    port=80/tcp --set-port=8080/tcp --set-port=443/tcp --permanent
success
[root@localhost apache2]# firewall-cmd --zone=http
    --set-target=default
[root@localhost apache2]# firewall-cmd firewall-cmd
    --info-zone=http --permanent
http
  target: default
  icmp-block-inversion: no
  interfaces:
  sources;
  services: http https
  ports: 80/tcp 8080/tcp 443/tcp
  protocols:
  masquerade: no
  forward-ports:
  source-ports:
  icmp-blocks:
  rich rules:
```

FIGURE 10.24 Example `firewall-cmd` commands.

```
[root@localhost apache2]# firewall-cmd --new-ipset=hackers
    --type=hash:ip --permanent
success
[root@localhost apache2]# firewall-cmd --ipset=hackers
    --add-entry=10.2.3.4 --permanent
success
[root@localhost apache2]# firewall-cmd --ipset=hackers
    --add-entry=10.2.3.5 --permanent
[root@localhost apache2]# firewall-cmd --ipset=hackers
    --get-entries --permanent
10.2.3.4
10.2.3.5
[root@localhost apache2]# firewall-cmd --zone=drop
    --add-source=ipset:hackers --permanent
[root@localhost apache2]# firewall-cmd --reload
success
```

FIGURE 10.25 Another `firewall-cmd` example.

```
firewall-cmd --add-rich-rule='rule family=ipv4 source
address=10.11.12.0/24 service name=ftp log prefix="local FTP
access" level="notice" accept'
----------------------------------------------------------------
firewall-cmd --add-rich-rule='rule protocol value=icmp reject'
```

FIGURE 10.26 Defining rich rules using `firewall-cmd`.

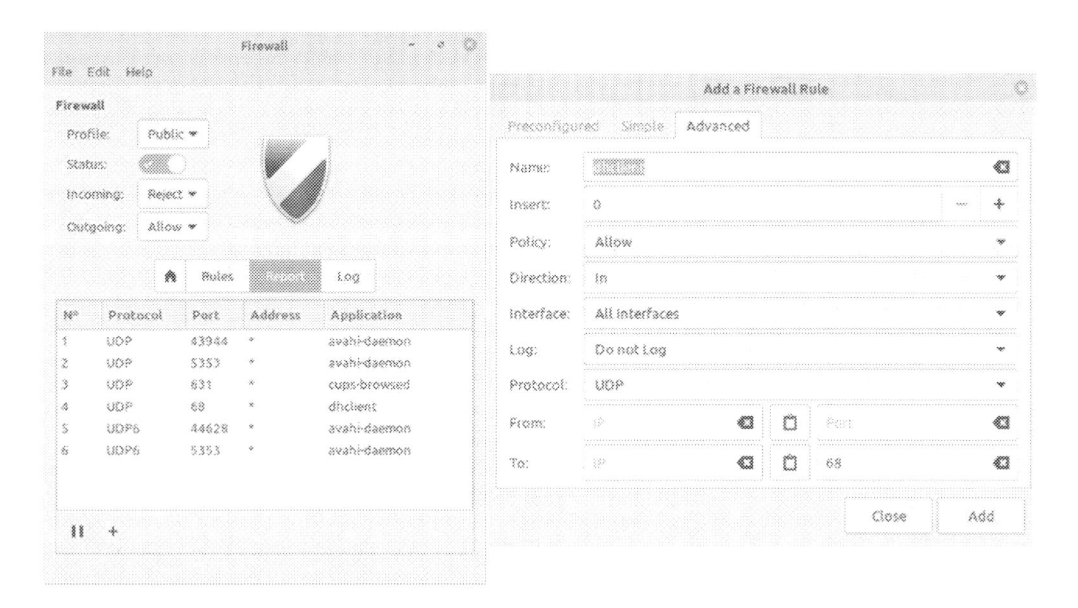

FIGURE 10.27 The `gufw` interface and a rule.

that particular service's rules. We have done so for `dhclient`, whose rule is shown on the right side of the figure, where we have selected the Advanced tab.

We can modify the rule by editing any of the fields. The greyed boxes are controlled through a drop-down menu. For instance, Policy (which we referred to as target or action in previous subsections) permits any of Allow, Deny, Reject or Limit, while Protocol can be TCP, UDP or both. Direction indicates whether the rule pertains to incoming messages (In), outgoing messages (Out) or both. The other tabs in this window show us the preconfigured rule should we want to roll back to the original version, and a simpler interface showing us fewer options for the same rule.

From the `qufw` window, if we select Rules instead of Report, we are able to define our own rules. Adding a rule brings up the same window as the `dhclient` window except that all entries are defaults for us to modify. We might create a new rule, name it http, select a policy of Allow, a direction of In, a protocol of TCP and under Port, enter 80. Notice that unlike `firewall-cmd`, we are limited to one port (or service). To open another port or service, we have to define an additional rule. Having entered our rule, we see our rule displayed, telling us the rule is 80 ALLOW In Anywhere, with a name of `http`.

Lastly, we turn to `ufw`, which is somewhat equivalent to `firewall-cmd` although consists of fewer commands but is equally complex. One nice feature of `ufw` is that it contains an option called `--dry-run`, which does not modify anything but displays the results to show us what would happen should we try the instruction. Table 10.15 describes the commands.

The syntax to specify a rule can be stated simply or more fully. A simple rule is `ufw allow 53` meaning that port 53 is open. A slightly more complex rule is `ufw allow 25/tcp` which says that port 25 is open to TCP packets (but not UDP). Rules can also specify services rather than ports as in `ufw allow http`. We can add `in` or `out` to indicate the direction of packet as in `ufw allow in http`.

Rules can also include a comment field where the comment is in single quote marks. An example is `ufw reject telnet comment 'telnet is unencrypted'`. One last brief example is `ufw allow log 22/tcp`. Here, we are allowing any TCP message over port 22 but any such incidents will be logged.

A rule with full syntax will add words like `proto` for the protocol, `on` for the interface name, `to` for the destination IP address (or network), `from` for the source IP address (or network) and `port`

TABLE 10.15
`ufw` Commands and Options

Command	Meaning	Options (if any) or comments
`disable`, `enable`, `reload`, `reset`	Unload and disable the firewall, reload and enable the firewall, reload the firewall, disable the firewall and reset the firewall to its original settings.	
`default`	Change the default policy to allow, deny or reject; can be specified twice, once per direction, or once for both directions.	`allow`, `deny`, `reject`, and the direction (`incoming`, `outgoing`, `both`)
`logging`	Specify if logging should take place and if so, the log level.	`on`, `off`, `loglevel` (e.g., `warn`, `crit`, `debug`)
`status`	Show firewall status including specific rules.	
`show`	Display information about the runtime firewall based on selected report type.	Type of report
`allow`, `deny`, `limit`, `reject`	Add a rule to allow, deny, limit or reject messages based on criteria specified in the rule.	See the text for a brief description
`insert`, `delete`, `prepend`	Insert new rule given rule number or delete the rule specified; prepend adds the rule at the beginning of the set of rules.	Rule and number (`insert`), Rule or number (`delete`) Rule (`prepend`)

for the destination port address. We omit further detail but the man page for `ufw` has numerous examples of both short and full syntax.

SECTION ACTIVITIES

1. Prior to using Linux, did you ever look at your home computer's firewall? If so, for what purpose? As a home computer user, research reasons why you might want to modify your firewall (for instance because you obtained a new program that by default is blocked by your firewall).
2. If you are a Windows user, open the Windows firewall and compare it to what you have learned about `firewalld/ufw`. Do you find the Windows firewall easier, harder or about equally challenging to use? Which of Windows and `firewalld/ufw` give you more flexibility, if either?
3. As a system administrator of a collection of computers, should you permit users to modify the firewall on their individual work/school computer? What are the advantages and disadvantages of this?

10.7 CHAPTER REVIEW

Concepts and Terms Introduced in This Chapter

- Address resolution – converting a fully qualified domain name (or an alias) into its IP address using a DNS name server, a DNS cache or an entry in `/etc/hosts` (also references ARP, address resolution protocol, to map IP to MAC addresses).
- Anonymous login – used in FTP to permit a client to log into an FTP server and download public files; see File Transfer Protocol.
- Application layer – top layer of TCP/IP protocol stack in which applications generate messages for transmission or receive and display response messages.

- Authoritative name server – DNS name server responsible for its domain.
- Caching name server – DNS name server that caches responses from other name servers and so is non-authoritative.
- Classless Internet domain routing (CIDR) – replacement of classful networks whereby network addresses are assigned not by class but by some smaller division; in CIDR, the network portion of an IPv4 address is indicated using a routing prefix (or CIDR notation), as in `/19`, rather than by the network's class.
- Classful network – older form of assigning IPv4 addresses to networks where the first 8 bits of the IPv4 address dictates the class which in turn determines the portion of the address that makes up the network address.
- Computer network – a collection of computers and computing resources connected together to facilitate communication between resources.
- DHCP server – a device (computer or router/gateway) set up to issue IP addresses dynamically upon request to other devices on the local network.
- DNS name server – a computer designated to resolve host names into IP addresses either directly or by retrieving cached content or forwarding requests on to other name servers.
- Domain name system – the collection of servers, resource records and programs that perform address resolution.
- Dynamic Host Configuration Protocol (DHCP) – service run on a subnet to assign dynamic IPv4 addresses upon request to clients of the subnet.
- Dynamic IP address – an IP address issued to an Internet resource temporarily (for instance, for a few days).
- Ethernet – a technology used to build local area networks (LAN).
- Ethernet interface – a network interface card (NIC) used to connect a computer into an Ethernet-based network.
- File transfer protocol – protocol used to transfer (upload and download) files across the network between two computers; see also anonymous login.
- Firewall – software that helps enforce network security by serving as a filter to allow or prevent messages from coming in from the network or going out onto the network; may also run on dedicated hardware.
- Fully qualified domain name – a symbolic name for a network resource consisting of a host name and a domain name; used in place of IP addresses as the name is easier to remember; address resolution is required to use the name for Internet communication.
- Gateway – a network connection device responsible for joining local area networks of different types together.
- Hextet – a 16-bit portion of an IPv6 address, usually written as four hexadecimal digits such as f31a.
- Hub – a network connection device operating on a subnetwork which, when it receives a message, broadcasts that message to all devices on that subnet; used initially as repeaters to boost messages on bus-style Ethernet networks and are now obsolete.
- Internet domain – a region of IP addresses and names corresponding to one organization and managed by an authoritative DNS name server; top-level domains include .edu, .com, .net, .uk and .ca, and second-level domains are at the organization-level such as amazon.com, google.com and nku.edu.
- Internet layer – the third layer (from the top) of the TCP/IP protocol stack; at this level, IP addresses are used and routers forward messages from one network to another.
- IP address exhaustion – in 2011, IANA issued the last of its available IPv4 addresses; no more IPv4 addresses are available for newer networks; because of this, IPv6 is being used as much as possible.
- IPSet – a list of IP addresses defined in a firewall to create a whitelist (addresses of acceptable hosts) or blacklist (addresses of unacceptable hosts that should be blocked).

- IPv4 address – 32-bit unique address assigned to any Internet resource; usually written as four octets of numbers between 0 and 255 separated by periods, as in 10.11.12.13.
- IPv6 address –128-bit address unique address assigned to an Internet resource; offers a far greater range of addresses than IPv4 addresses; as not all networks are IPv6 compliant, not all Internet resources receive IPv6 addresses.
- Lease – when a dynamic IPv4 address is allocated by a DHCP server, the address is only granted temporarily, known as a lease.
- Link layer – the bottom layer of the TCP/IP protocol; at this layer, messages receive hardware addresses and are directed by the network switch.
- Loopback device – an interface in Linux computers that allows software to communicate to other software on the computer as if the messages were coming over the network.
- Master DNS name server – a DNS name server that is not only an authority for a domain but is the point of contact for all other authoritative name servers (slaves) so that only the one name server needs to be modified.
- Media access control (MAC) address – hardware addresses usually given to specific interfaces like Ethernet cards; these addresses are used by switches in the link layer of TCP/IP; MAC addresses are 48 bits consisting of 12 hexadecimal digits organized in pairs of digits, separated by colons, as in 00:12:34:56:78:9A.
- Network connection device – one of four types of devices which receive network messages from other network resources and transmit the messages to one or more other resources; the devices are hubs, switches, routers and gateways.
- Network protocol – rules by which devices can communicate over a network; the protocol describes the specific syntax, data structures and steps needed to convert a message into one that can be delivered across the network and interpretable by the recipient; example protocols are FTP, HTTP, TCP, UDP, DHCP, DNS and ARP.
- Netmask – a binary number used to AND to an IPv4 address to obtain the network address for the device.
- Network resource – any device connected to a computer network that can send or receive messages such as computers, printers, file servers, switches and routers.
- Octet – one part of an IPv4 address consisting of 8 bits or an integer between 0 and 255.
- Open System Interconnection (OSI) model – a protocol stack describing the functionality of computer networks; never completely implemented but often cited by network architects.
- Port – an address assigned to an application and type of communication protocol and used in conjunction with a network address to specify the source and destination of a message.
- Predictable network interface name – Linux naming strategy to provide a consistent naming scheme for network interfaces.
- Protocol – a set of rules used to describe how entities should interact and/or communicate; see network protocol.
- Resource record – a datum describing a network resource in terms of its IPv4 and/or IPv6 address(es) and name; resource records make up the entries in a DNS name server's tables.
- Reverse IP lookup – mapping an IP address to a name; used to ensure that a message is from a legitimate source and not a spoofed address.
- Router – a network connection device used for forward messages from one network to another; routers operate on IP addresses.
- Routing prefix – the number of bits that make up the network portion of an IPv4 address; also called CIDR notation.
- R-utility – collection of Linux/Unix programs that operate within a local area network whereby user accounts are shared among all computers allowing users to access other computers of the network without having to log in.
- Slave DNS name server – an authoritative DNS name server which requests and receives updates to its resource records from a master DNS name server.

- Start of authority (SOA) record – an entry in an authoritative DNS name server's table that specifies how frequently slaves should communicate with masters while also setting a default time to live for DNS caches.
- Static IP address – an IP address assigned to a computing resource permanently (or at least for a long period of time); changing it requires modifying DNS resource records.
- Subnet – a logical division or subset of a local area network where all computers share not only the same network address but most of the same bits of host addresses.
- Switch – a network connection device operating on a subnetwork; when it receives a message it transmits that message to a single device on the subnetwork based on a table it maintains that maps the devices connected and their MAC address.
- TCP/IP – a network protocol stack and the protocol used to implement the Internet; TCP/IP consists of four layers each of which can be implemented by a number of lesser protocols.
- Time to live – the amount of time that an IP address can be stored in a DNS cache before it expires or the amount of time (or hops) that a packet can be transmitted before the packet should be dropped for not reaching its destination.
- Transport layer – the second layer (from the top) of the TCP/IP protocol stack; at this layer sessions are established between resources and TCP or UDP packets created.
- Tunnel – a temporary dedicated network communication link between two resources that uses a different protocol than the network, such as using an encryption utility to send encrypted data over an unencrypted Internet connection.
- Wireless access point (WAP) – a network connection device used by wireless devices; the WAP transmits any incoming message out to all devices in its region or sends the message to a router via a wired connection.
- Zero-configuration service – a network service which can locate network resources such as a DHCP server without being configured to.
- Zone – the `firewalld` divides message types into zones and assigns zones different targets (actions); zones can be assigned specific services, interfaces, ports, IP addresses and other attributes of incoming and outgoing messages.

Linux Commands Covered in This Chapter

- Avahi – service to discover local area network resources (neighbors).
- chrony – service to synchronize computer clocks with networked resources using NTP (network time protocol).
- curl – non-interactive, text-based program to communicate with servers of various protocols; can accomplish the same tasks as `wget` but is more powerful.
- dhclient – client program to request dynamic IPv4 addresses from DHCP servers.
- dhcpd – DHCP server program to maintain pools of IPv4 addresses and issue them upon request.
- dig – query DNS name servers for resource records; see also `host`.
- dnsmasq – service that can act as a DNS cache and a lightweight DHCP server (for small networks).
- firewall-cmd – command-line program to view settings of and modify the Linux firewall; passes messages on to the `firewalld` service which itself modifies `iptables` data.
- firewall-config – GUI program to view settings of and modify the Linux firewall; see also `firewall-cmd` and `firewalld`.
- firewalld – service that serves as a front-end to the `iptables` Linux firewall.
- ftp – command-line program that uses the File Transfer Protocol (FTP) to upload and download files with another network compute; outdated and largely replaced by secure forms of FTP.

- gufw – graphical front-end to the Uncomplicated Firewall program, `ufw`.
- host – DNS name servers for resource records; see also `dig`.
- httpd – program name for the Apache webserver.
- ifstat – display network interface activity.
- ip – view, modify, add and delete network connectivity data such as IP addresses, routing tables and neighbors; `ip` has replaced older programs like `ifconfig`, `route` and `iptunnel`.
- iptables/ip6tables – Linux firewall; newer front-end services (`firewalld` in Red Hat Linux and `ufw` in Debian) are available to simplify modifications to `iptables/ip6tables`.
- lnstat – display kernel-collected network statistics.
- nc (netcat) – general network communication utility to open network connections with other networked devices and pass messages directly to that device.
- NetworkManager – service in charge of establishing and handling network communications; this newer Linux service has replaced numerous older Linux network services so that network communications can be managed much more easily.
- nmap – network resource discovery program which attempts to locate open ports on networked devices; can be used to secure a network.
- nmcli – command line front-end to modify the `NetworkManager` configuration and obtain network information from the `NetworkManager`.
- nmtui – menu-based front-end to modify the `NetworkManager` configuration and obtain network information from the `NetworkManager`.
- nslookup – submit simple DNS queries.
- ping – send ICMP packets to another network-based resource to test for its availability.
- postfix – Linux mail transfer agent program.
- rdisc – service to discover local area network routers.
- ss – socket investigation program, used to output statistics on current socket usage.
- ssh – remote access program that uses encryption; replaces the insecure and obsolete `telnet`.
- sshd – service to allow a computer to be remotely logged into via `ssh`; requires setting up private and public keys and a digital certificate.
- tcpdump – packet capture and exploration program so that you can view incoming and outgoing TCP/IP packets.
- telnet – obsolete program for remote access to another computer; replaced by `ssh`.
- traceroute – send ICMP packets to some network resource and output the pathways taken (including routers).
- ufw – Uncomplicated Firewall, used as a text-based front-end to `iptables` in Debian versions of Linux.
- wget – text-based non-interactive program to download files from a webserver; similar to `ftp` except that files come from a webserver; see also `curl`.
- xinetd – a superserver capable of invoking appropriate network services based on the ports of incoming messages.

Linux Files and Directories of Note Covered in This Chapter

- /etc/avahi/ - directory storing `Avahi` configuration and other data files.
- /etc/dnsmasq.conf – configuration file for `dnsmasq`.
- /etc/hosts – store name to IP address mapping information for resources that a computer will often communicate with.
- /etc/network/interfaces – used in Debian versions of Linux to define interface data especially when issuing static IP addresses.
- /etc/resolv.conf – store IP addresses of the local name server(s).

- /etc/ssh/ – directory storing data files used by `sshd`, particularly the keys and digital certificate.
- /etc/sysconfig/network-scripts/ifcfg-ens# – used in Red Hat to store configuration information for interface device including whether IPv4 address will be static or dynamic.
- /var/lib/dnsmasq/dnsmasq.leases – information on leased IP addresses if `dnsmasq` is being used as a DHCP server.

REVIEW QUESTIONS

1. Which of a gateway, hub, router and switch would we not expect to find in a modern LAN?
2. How does a gateway differ from a router?
3. Wireless devices can communicate to each other using a _____.
4. List the layers of the TCP/IP protocol stack from top to bottom.
5. Protocols like DNS, HTTP, MIME and SSH are implemented at which layer of the TCP/IP protocol stack?
6. If a TCP packet is dropped, what does the destination computer do? How does this differ from a UDP packet being dropped?
7. To obtain the network address given an IPv4 address, we ____ the address with the netmask. Choose one of AND, OR, XOR, Add, Subtract.
8. We have the following IPv4 address: 192.179.205.144. What is the network address given each of the following router prefixes?
   a. /14
   b. /16
   c. /19
   d. /22
   e. /26
9. Assume your computer's IP address is 10.145.201.12. Compute your network address given each of the following netmasks.
   a. 255.255.255.128
   b. 255.255.255.0
   c. 255.255.192.0
   d. 255.255.128.0
   e. 255.255.0.0
   f. 255.240.0.0
   g. 255.224.0.0
10. With IPv4 addresses, we need a mechanism to divide it between its network address and its host address. Why do we not need such a mechanism for IPv6 addresses?
11. Routers operate at which layer of TCP/IP? Switches operate at which layer of TCP/IP?
12. What type of addresses do switches operate with?
13. What range of addresses are reserved for the well-known ports?
14. Destination port addresses are based on the protocol (application) of the message. How are source ports assigned to a message?
15. True/false: Some protocols can send messages using either (or both) TCP and UDP. For such protocols, the destination port for TCP packets will differ from the destination port for UDP packets.
16. Approximately when did IP address exhaustion occur?
17. Explain the notation 1.2.3.4:55.
18. Let's assume there are 7.7 billion people on the planet. Explain why IPv4 addressing does not provide a sufficient number of IP addresses.
19. Assume there are 7.7 billion people on the planet. Approximately how many different IPv6 addresses could each person be awarded given the 128-bit size for IPv6?

20. Why can we not yet use IPv6 exclusively? That is, why are we still using IPv4?
21. Use the shortcut notation rules to reduce the following IPv6 addresses in size.
    a. 1234:5678:90AB:CDEF:0123:4567:0000:0123
    b. 3A05:9000:0000:0000:0000:0000:0093:06B8
    c. 1020:3040:0506:0708:0000:0000:009A:0BCD
    d. FE34:090C:D804:0012:3404:09A0:00C0:000F
    e. FE00:0000:0000:0001:0000:0001:2345:6789
22. For the following shortened IPv6 addresses, restore them to their full addresses.
    a. A80:EF01::98:11:1234
    b. E0E1:6:9860:153::1
    c. ABCD:EF01:2345:6789:123:4567:789A:BC
    d. 84:AB5::70C4:66F:0:12
23. We use DNS to convert a(n) _____ into a(n) ______.
24. What is the difference between a caching DNS name server and an authoritative DNS name server?
25. What role do the root-level DNS name servers play in address translation?
26. List three top-level generic domains and three top-level country domains.
27. What is the difference between a top-level domain and a second-level domain? What is the difference between a second-level domain and a subdomain?
28. Why should a cached DNS entry expire? What controls the time until it expires?
29. We would expect an interface to be named `ens160` when using which predictable naming scheme?
30. Explain the role of the `/etc/hosts` file.
31. Under what circumstance might you put an entry into the `/etc/hosts` file?
32. The computer `h3dee9.ourorganization.org` is aliased to `ourserver.ourorganization.org` and has an IP address of `10.2.3.15`. We want to bypass DNS to translate this address. What entry would we put in our `/etc/hosts` file?
33. Your `/etc/resolv.conf` file is empty. Does this mean that you cannot access the Internet at all? If not, what restriction(s) might this place on your Internet usage?
34. What type of computer is most likely to receive a static IP address?
35. What are the advantages and disadvantages of using dynamic IP addresses?
36. What single Linux service is responsible for ensuring network communication?
37. Which program(s) might you use to interact with or modify the settings of `NetworkManager`?
38. We run `nmcli` and see the following information. Explain what it means.

```
ens32: connected to ens32
ethernet, 00:48:58:A7:33:21
ip4 default
inet4 10.11.12.13/22
route4 10.11.1.0/22
```

39. Match the service below with its role in supporting the network.

a. `avahi-daemon`	i. Permit SSH connections
b. `dnsmasq`	ii. Discover routers in the network
c. `rdisc`	iii. Map messages to services based on port
d. `sshd`	iv. Serve as a DNS cache or DHCP server
e. `xinetd`	v. Perform zero-configuration network service discovery

40. In Red Hat, what file (including the directory) is used to specify whether an interface uses DHCP or has a static address? In Debian Linux?
41. Refer to Figure 10.11. Why are there two subnet entries? What range of IP addresses are available to this DHCP server?
42. How does `ip addr` differ from `ip route`? How does `ip route` differ from `ip tunnel`?
43. For `ip addr`, what does `del` do? What does `flush` do?
44. List three different things you can use `ip` for.
45. You type `ip addr` and are shown only one interface, `lo`. What can you conclude from this?
46. What are the IPv4 and IPv6 addresses assigned to your loopback (`lo`) interface?
47. You issue `ip addr` and see that your Ethernet interface has an entry for `inet` but not `inet6`. What can you conclude by this?
48. Using `ip route`, you see three different routers with metrics of 1, 50 and 100. Which of the three routers would be the preferred choice?
49. Why do we no longer use `telnet`? What has replaced it?
50. Which r-utility is most closely related to the `telnet` program?
51. What Linux command is most closely related to `rcp`?
52. How do `ssh zappaf@10.11.12.13` and `ssh 10.11.12.13` differ?
53. With `ftp`, you have connected to another computer. You type `mget *.txt`. What does this command do?
54. True/false: Using `ftp`, the command `rmdir` can be used to delete the specified directory on the remote computer.
55. What does `reget` and `reput` do in `sftp`?
56. True/false: Using `sftp`, we can issue shell commands.
57. Which command is `wget` closest to, `ftp`, `ssh` or `nc`?
58. How do these two instructions differ in terms of the response you receive?
    a. `wget www.nku.edu/~foxr/index.html`
    b. `curl www.nku.edu/~foxr/index.html`
59. True/false: `curl` only works when attempting to download files from a webserver.
60. What does the `-l` option do when used with `nc`?
61. Both `ping` and `traceroute` rely on which protocol?
62. You run `traceroute` and receive responses of * * *. What does this tell you?
63. You type `ss -t`. What do you expect to see?
64. Which network tool that we covered in this chapter would be your choice for testing open ports on a computer to see what might need protection?
65. You want to view statistics about your interface. What Linux program should you use?
66. The first value in an SOA's list of numbers is the serial number. How is this used?
67. A slave DNS name server will contact its master DNS name server based on what value as stored in the SOA?
68. Should a master DNS name server not respond to a slave name server's repeated requests and the expiration period elapses, what does the slave do?
69. What is the difference between an `A` and an `AAAA` resource record?
70. An entry in a DNS table indicates that `xxx` has a `CNAME` of `yyy`. What are `xxx` and `yyy`?
71. You use `nslookup` and the response says `Non-authoritative answer`. What can you conclude from this?
72. What Linux command is similar to `host -d`?
73. What is a firewall? What is a stateful firewall?
74. True/false: Firewalls only run on computers.
75. What is the difference between a `DROP` and a `REJECT` action in a firewall?
76. What is the relationship between `firewalld` and `iptables`?

77. What is the default zone used in firewalld?
78. List three services that are already placed in the default zone in firewalld.
79. What action does the block zone automatically perform on all messages that are placed in that zone?
80. True/false: You are unable to add your own zones to firewalld but you can change the existing zones.
81. When you make edits to firewalld, by default does it impact the permanent or runtime version?
82. What is an IPSet and why might you define one?
83. Write a firewall-cmd instruction to display all information on the work zone.
84. Write a firewall-cmd instruction to save the runtime firewall so that it becomes the permanent firewall.
85. You issue the instruction firewall-cmd --zone=block --set-target=ACCEPT. What have you just done?
86. Write a firewall-cmd instruction to modify the block zone by adding all icmp messages (ICMP is a protocol) to it.
87. Write a firewall-cmd instruction to add port 80 using TCP packets to the internal zone.
88. You can establish an IPSet and add it to a zone in the runtime firewall, permanent firewall, both or neither?
89. True/false: ufw can only run in Debian-based Linux distributions.
90. What is the name of the program which runs a graphical interface for ufw?

11 Software Installation and Maintenance

This chapter's learning objectives are to be able to:

- Answer questions about whether and how to install new software
- Use `rpm` or `dpkg` and `dnf/yum` or `apt` to install, upgrade and remove software
- Identify and resolve software dependencies using `rpmfind.net`
- Download and install open-source software using `configure`, `make`, `make install`
- Compile C programs using `gcc` (optional)
- Describe the role of the system administrator with respect to documentation

11.1 INTRODUCTION

Software maintenance, from an administrator's point of view involves selecting, installing, configuring, updating and uninstalling software. There are any number of tools to accomplish the installation, updating and uninstalling steps in Linux, as we will see in this chapter.

Software installation has been greatly simplified in just about all operating systems so that users can install new software with little interaction. In Windows, one downloads an installation program stored as an `msi` file and runs it. The installation program may ask a few basic questions like the destination folder for the executable program and whether a start menu or desktop shortcut icon should be created, but otherwise runs uninterrupted. Even simpler is the "store", a GUI that handles all of the installation steps. We find the "store"-like approach available in Windows, Mac OS and Linux as well as on our smart phones. Linux also has several command-line programs to handle software installation that can range from simple to somewhat more challenging. But also available in Linux is the ability to download the software in its source code format and install from that. We explore all of these approaches in Linux in this chapter.

As a user, how should we decide when and what software to obtain and install? In most cases, a person installs new software because they either have a need for it or a desire to use it. A need for software may be driven by business or personal goals. For instance, if someone is writing a book then that person will need some form of word processor or desktop publishing software. Designing figures would require some kind of drawing and/or image editing software.

The system administrator is often informed that software is needed. It might be up to the administrator to select a software title given a description of the need. This may require some research. Alternatively, the employer may have already selected the software, and it's the administrator's role to purchase and install it.

Let's explore some questions we should ask before attempting to install software. The first question we should ask is whether we need the software. If so, are there alternative titles to explore? What specific features will be needed of the software? Is price a concern?

Do we have similar software already installed? If so, can we make do with the current software, possibly upgrading it rather than obtaining something new? If we are upgrading existing software, is the upgrade free? Are there several versions available to upgrade to and if so, do we know their differences?

Do we or our users already have familiarity with the software. Familiarity reduces the learning curve in mastering the software. Even if we are unfamiliar with a specific title, we might have

DOI: 10.1201/9781003203322-11

experience using other software from the same vendor and the software might have similar features and controls (e.g., the same menu layout).

When it comes to selecting the specific title, we find that the software market is full of competitors for just about every type of software. Let's just focus on one type, drawing software. Table 11.1 compares several different titles along with cost and features. Note that this is not a comprehensive list.

Given a list of software titles, cost is not the only factor in selecting which version we may want to acquire and install. Software has *requirements*. Such requirements might include a minimum amount of memory, a certain type of processor or operating system, enough hard disk capacity to support the software and a need for specific types of input or output devices. Does our computer meet those requirements? If not, our choices are to upgrade our computer or select a different title.

Most Linux software is free but free software comes with the risk of being unsupported. Even with free software, there may be costs involved. Do we need to upgrade hardware to support the software? Does the company or organization that provides the software offer support and if so, is that free? If not, should we purchase support?

How frequently does the software vendor issue updates and upgrades? We differentiate between updates and upgrades in the next section. Are updates free? Are upgrades free? How much support is provided with the software? Does the company producing the software have a reputation for buggy or insecure products?

We might also ask what installation entails. If we are a user of a computer instead of an administrator, do we have access rights to install software? Is installation an arduous task or an easy one? Software used to be purchased and provided in the form of an executable program on one or more floppy disks or optical discs. Today, most software is provided as a download over the Internet and usually in the form of an installation program. The installation program is not the software itself but instead an executable program that, when run, downloads the required software and accompanying files and installs them onto our computer.

As a system administration, we will need to know whether the software is intended to run on a standalone computer, on multiple computers or over a server. This might influence our selection.

TABLE 11.1
Comparing Drawing Software

Title	Cost	Uses	Platform	Comments
Adobe Fresco	Free for Adobe Creative Cloud subscribers, $10 a month otherwise	Illustrations, comics, drawing	Windows, iPad	Familiarity with other Adobe products makes this easier to learn/use.
Clip Studio Paint	$50 or $220 depending on version	Illustration, animation, comics, drawing	Windows, Mac, iPad/iPhone, Android	Can be rented by the month.
Corel Painter	$424	Illustration, comics, drawing	Windows, Mac	Very versatile for drawing/painting.
GIMP2	Free (open source)	Illustrations, photo editing, drawing effects	Windows, Mac, Linux	To fully utilize, may require other computing skills.
ibisPaint	Free (see comments)	Illustration, drawing, comics	iPad/iPhone	Free version is limited; you can also purchase an ad-free version for $8.
Paint Tool SAI	$53	Illustration, drawing	Windows	Fewer features than others.
Photoshop	Less than $100	Illustration, photo editing, custom artwork, drawing effects	Windows, Mac	Similar to GIMP2 but with more features.

For software that will run on multiple computers, does the vendor offer a site license or some form of discount for purchasing multiple copies?

When installing software, aside from installing the executable program, there are likely to be other steps. Environment variables may need to be established and added to the system. The `PATH` variable of the system may need to be modified. Accounts may need to be created. The software may need to be configured (if not required, this step may be useful). In Windows, installation also involves placing one or more entries in the Windows Registry. Are there similar registration steps required in Linux?

SECTION ACTIVITIES

1. What software titles have you installed on your home computer? Create a list. Looking over the list, is it more or less software than you thought?
2. Table 11.1 compares various drawing software. Create your own table for *one* of the following software types where you can compare versions. Your table does not have to be complete but should have several different software titles.
   - Word processor
   - Music composition/sequencer software
   - IDE (integrated development environment) for a programming language of your choice (e.g., Java, C++)
   - Web browser

11.2 SOFTWARE MAINTENANCE TERMINOLOGY

The term software maintenance is often used to describe a part of the software development lifecycle. Having produced a piece of software, some of the developers are tasked with maintaining it. This involves fixing errors and updating the software based on new needs. In the context of this chapter, we refer to software maintenance as the tasks required by the system administrator to select, install, configure, update and, at some point, possibly uninstall software.

Software is the name we use to describe programs. A *program* is the implementation of an algorithm. An *algorithm* is a strategy for solving a problem. A program then is the implementation, in some programming language, of a solution to solve some problem. Solutions are themselves based on ideas and ideas are not tangible (physical). Programs similarly do not exist in any tangible, physical form. Instead, software exists either as current flowing through circuitry in memory and the CPU, as magnetized information stored on disk or tape or encoded through crystals stored on optical disc. Thus, we refer to programs as *soft*ware as opposed to the *hard*ware because the programs cannot be touched or held, unlike the physical components of our computer.

We might further differentiate between a program and software. The program is usually referred to as a single entity. Software on the other hand might comprise a collection of files. These files will include one or more executable programs and supporting files like image files and documentation files.

11.2.1 TYPES OF PROGRAMMING LANGUAGES

Programs must be converted into the computer's native *machine language* to be executed. We can write programs directly in machine language, but this is both challenging because machine language is a low-level language and unnecessary because we have programming language translators that can perform the task for us. Machine languages are written entirely in binary (or hexadecimal) where instructions are encoded into operations (op codes) and operands (values, storage locations).

Operations are limited to very simple tasks such as performing an arithmetic operation on two numbers, performing a bit-level operation (e.g., a shift) on one number, moving a result from a register to memory, comparing two values and deciding whether to branch to another location in the program based on the result of a comparison.

Instead of writing code directly in machine language, programmers today use *language translation software* to convert a program from a higher-level language into machine language. One slightly higher-level language is *assembly language*. The main distinction between machine and assembly language is that assembly code is expressed using short English-like descriptions (called *mnemonics*) for the operations. Examples might include `sub` for subtract, `shl` for left shift and `mov` for data movement. Operands can be referenced as literal values written in decimal (rather than binary) and variables can be referenced using variable names. Most programmers avoid assembly language for the same reason they do not write in machine language: it is challenging and unnecessary.

Today, nearly all programs are written in *high-level languages*. These languages are written using a combination of words (like `for`, `while`, `if`), variable names and numeric literals, arithmetic, relational and Boolean expressions, and data structures like arrays and classes. High-level language instructions can express higher-level ideas than machine or assembly language instructions. Assignments of a variable can be complex as with `x = y * (a - 1) / z`. Decision-making instructions can be written in a more English-like (or mathematical) way such as `if(x > 0) y = sqrt(x)`. In machine or assembly code, both of these two example instructions would take several (perhaps from five to ten) individual operations! In general, a single high-level language instruction might be able to convey what it would take several to dozens of machine or assembly language instructions.

As a computer cannot directly execute a program written in a high-level language, we must translate a program written in one of these languages into the computer's machine language first. We do so by one of two types of language translators: a compiler and an interpreter. The *compiler* translates an entire program into an executable at one time. The executable is then saved or installed and run at a later time. To compile a program, the program must be complete and *somewhat* self-contained (portions of a program may, in some languages, be compiled separately in the form of libraries, a topic we will explore later in this chapter).

The *interpreter* takes one program instruction, translates it into machine language and executes it. The interpreter allows a programmer to build a program in a piecemeal fashion by typing in an instruction to see what it does, and only when successful, the programmer can move on to the next instruction. As you know, Bash contains an interpreter. It interprets Linux commands and Bash scripting language instructions, whether entered one at a time from the command line or presented to the interpreter in the form of a script.

11.2.2 Types of Software

We can classify software in many ways. The broadest division of software is into two categories: applications software and system software. *Applications software* are the programs we run in order to accomplish some task. Examples of applications are listed in Table 11.2. *System software* is the collection of programs that make up the operating system.

There are several forms of operating system software. The most important part of the operating system is the kernel. The kernel is loaded into memory when the computer is booted and remains in memory until the system is shut down. Other types of operating system software are services, device drivers, utility programs and tailorable interfaces such as shells or desktops.

We have already explored the kernel (to some extent), services and shells. Device drivers are programs that serve as interfaces between the operating system and a piece of hardware such as an input or output device. The driver's role is to intercept operating system commands and translate them into specific actions for the device.

TABLE 11.2
Some Types of Applications Example

Type	Usage	Type	Usage
Drawing/image	Art and photo creation/editing	Personal assistant	Address book/contacts, calendar/planner
Education	Classroom management, grading, also productivity software	Productivity software	Database management, presentation spreadsheets, word processing
Enterprise software	Supply chain/inventory, payroll, accounts receivable/payable	Media	Music player, TV/movie player
Entertainment	See media, games	Music/sound	Music and sound creation/editing
Financial software	Accounting, banking, investing, tax	Simulation	Scientific, manufacturing, flight/driving
Games	Entertainment, education	Software development	Compilers, interpreters, editors
Network	Email, web browser	Word processing/desktop	Create, edit and format text documents

Utility programs are similar to services except that it is the user that usually launches the utility and interacts with the utility. As such, utilities are not background processes, and many utility programs provide a GUI-based interface. Examples of utilities include file browsers, disk defragmentation programs, disk formatters and partitioning software, disk backup programs, antiviral software and various forms of diagnostic software.

11.2.3 Types of Software Licenses

Another way to classify software is by the type of *license* that governs the software's usage. Software is written by humans, some of whom produce the software as part of their livelihood. Other software is written by humans who give it away freely to a community of users or to anyone interested in it. Licenses are used to protect the programmers whether the software is sold or given away.

Table 11.3 describes the types of licenses we commonly find applied to software. The column marked *proprietary* indicates whether someone claims ownership of the software. As you can see, all but public domain software is considered proprietary. However, this does not mean that all proprietary software must be purchased. The column *free* indicates those forms of software that typically can be obtained without monetary transaction.

Most Linux software, both operating systems and applications, fall under the open-source branch of software and so are free. They are, however, proprietary software governed by a type of license called the GPL (refer back to Chapter 1, Section 1.4.2). Notice that private software is listed as not free as the company that produces it pays programmers to develop the software. Another word to describe this form of software is *in-house*.

11.2.4 Types of Software Management

Let's explore some terminology specific to software management. The steps involved in software management are installation, configuration, maintenance and removal. *Installation* is the process of obtaining and placing the software on the computer. Installation typically involves more than just copying the executable file from its source location (e.g., optical disc, a webserver on the Internet) to a directory in your computer system.

TABLE 11.3
Classifying Software by License Type

License Type	Meaning	Proprietary	Free
Commercial	Produced for sale; purchase comes with a license to use the software and possibly make copies; a site license gives an organization permission to make copies on internal computers or possibly for employees use at home.	Yes	No
Freeware	Protected by license but freely available.	Yes	Yes
Open source	Protected by license but freely available in source code format; typically, open-source software is also available in executable format for easy installation.	Yes	Yes
Private	Owned by the company that produced it and not available externally.	Yes	No
Public domain	Unlicensed software; no restrictions on its usage or distribution.	No	Yes
Shareware	Protected by license and made available for free in some restricted form such as for a trial period or with limited features; used to entice customers to purchase an unrestricted version.	Yes	Yes

The executable program may require supporting code in the form of libraries. A *library* contains precompiled code that installed software may call upon. The reason that the libraries are separate is that the functions found in a library might be reused by other software. Installing software may require installing one or more libraries.

There may also be helper or support files that the software might use such as image files, help files (e.g., man pages), a welcome splash page, data files and an uninstall program. It is not uncommon for installation to involve storing dozens or hundreds of files. The destination of these files may differ so that they are not placed in the same directory. For instance, the executable may be placed in one directory, the libraries in another, images in a third and help files in yet another.

Once installed, some software may require configuration (not to be confused with software configuration management which is tracking of changes made to software). *Configuration* is the process whereby the administrator of the software selects options and establishes parameters for runtime performance. Additional configuration tasks may involve creating user accounts to use the software and access control lists to provide access to data files. The Apache webserver and the Squid proxy server are both examples of applications that require extensive configuration before usage. We examine both of these open-source products in online supplemental readings.

Another step when preparing software for the users of an organization is to produce *documentation.* Most software comes with either built-in or online help. Yet this may not be sufficient in an organization where the employees need to learn how to use the software quickly. There may be specific steps that users will have to perform to login to the software or set up a printer, for example. Producing documentation is seldom thought of as an administrator's task but it is sometimes the case as no one else is available who either understands the software or understands the systems that support the software. To prepare such documentation, screen captures with step-by-step instructions may be useful.

Once the software is available to be run by the users, the administrator's role changes from installation and configuration to maintenance of the software. *Maintenance* is the step of ensuring that the software remains usable and is kept up to date. For most software, updates are handled automatically by the software itself. Software products are programmed to query a specific webserver to see if updates are available. Upon an update being available, either the software will alert the user for permission to update itself or will update itself without asking for permission, depending on how this is set up.

The *software update* usually provides fixes for known bugs or security issues, or modifies the software in some small way to improve its effectiveness. Such changes may alter one or a few features so that they are more useful, usable or efficient. A software update usually comes in the form of a *patch*. This is a small executable program that modifies some part of the already installed executable program, replacing existing code or possibly adding new code.

Software updates are almost always free for proprietary software as the software license entitles the purchaser to the updates. Companies attempt to make updates when necessary although not too frequently because the task of performing the update may upset users. For instance, some updates require that the computer be rebooted before the update can take effect. Frequent changes to software alienate customers who get used to the software in one way only to find it changing. Additionally, updates must be managed. If a user is asked to update their software and yet missed a previous update, the new update may or may not work correctly.

A more severe form of software modification occurs with a *software upgrade*. Upgrades usually move the software from one major version to another. The new version might have different features, different ways to use the features and a revised *look* (different GUI components). An upgrade may or may not be free depending on the software's license. In many cases, two versions of the same software cannot exist in the same system and so the upgrade replaces the older version. An upgrade's installation may be as involved as the original installation.

Although it is not essential to upgrade software, it is common for a company to discontinue support of the older version. Users might be forced to upgrade to newer versions of software in order to continue receiving support and/or free updates.

Related to a software upgrade is a *software downgrade*. In this case, the user is exchanging a newer version of the software for an older version. This is a rarer event but one that might be necessary if the newer version cannot run on the user's system without hardware upgrade. Alternatively, a user might prefer the older version perhaps because it was easier to use or had a more convenient interface. Some software titles do not make downgrades available as the company no longer wants users to use the older version(s).

As an administrator, part of our task is to troubleshoot our computer system. Troubleshooting is the process of determining why something is not working correctly or efficiently. For the system administrator, troubleshooting includes fixing or correcting hardware, the operating system and applications software. Troubleshooting a malfunctioning program does not mean debugging it as that is the purview of the software developers. Here, our role is to ensure that important files are in their proper place and that they have not been corrupted or deleted. Software may need to write data to the file space. We must make sure that there is available space and that the software has adequate permissions to that location (e.g., write access to `/var/cache`).

In Windows, a common issue is that a computer that has not been rebooted in some time winds up with corruption among its virtual memory space. This causes software to run poorly or incorrectly. This would not be the case in Linux but because of the complex interplay between applications software and the operating system, other issues could arise. To resolve such an issue, one place to look for solutions is the company's (or specific software's) website. Websites have forums for users to post issues and questions that other users or support staff will respond to.

The last step in software maintenance is its removal. We perform *software removal* when the software is no longer of use, or in an extreme case, when we have to free up disk space for other uses. Removing software is not just a matter of deleting the executable program. As noted earlier, most software comes with dozens to hundreds of files which may be located in numerous directories across your hard disk. Locating and removing all associated files may be impossible for the administrator and so a removal program should be used. Even if we know where the files are located, one needs to be careful in removing auxiliary files when uninstalling software because some of those files may be shared with other software. In Windows, most software is installed with a corresponding uninstall program. In Linux, most of the approaches we will examine to install software come with options to uninstall software.

In the next three sections, we examine approaches to install, update/upgrade/downgrade and remove software. We later look at installing software from open source although we do not consider updating/upgrading/downgrading/removing such software.

SECTION ACTIVITIES

1. Have you installed a piece of shareware? If so, did it come with restrictions like a time limit? What restrictions? If there were restrictions, did using the software entice you to buy a copy, did you continue to use the shareware version or did you discontinue using it, perhaps after the time limit expired?
2. There are open-source titles in just about any category of software you can think of. For each commercial product that you have installed on your computer, see if you can find at least one open-source title that is roughly equivalent. Were there any types of software on your computer that you could not find a similar open-source title for?
3. Documentation sometimes is poorly written or does not answer the question you may have. Open a piece of software and select help in whatever form is available. Think of type of problem that you do not know how to solve in the software and search the help available to see if you can find the instructions needed to perform the task. Were you successful? If not, search online to see if someone has posted the instructions.
4. How frequently do you remove software from your computer and for what reason? It is likely that you only remove software that you downloaded for free to try out but remove it after the trial period ends (or if you discover you do not want the software). How many titles have you removed (estimate)? Have you had other reasons for removing software? If so, what are they?

11.3 INSTALLATION AND MAINTENANCE FROM A SOFTWARE STORE

Most operating systems have taken a newer approach for users to easily add, update and remove software using online web portals often referred to as a *software store*. In Windows, there is the Microsoft Store. For Mac and iOS users, it's the Apple Store. Google has its own Chrome Store for Chrome and Android devices. Preceding these stores was the Ubuntu Software Center, which was first made available in 2009. Today, both Red Hat and Debian Linux distributions have their own versions of stores/software centers. In this section, we concentrate on the software GUIs for Red Hat/CentOS, Debian Mint and Ubuntu.

All of these stores operate by connecting to a software repository. A *software repository* is a storehouse of available software along with metadata describing each software title. Such metadata would include the last modification date made to the software so that our store can determine whether an installed piece of software is ready for an update. Quite often, repositories are stored on webservers for easy access.

11.3.1 RED HAT SOFTWARE GUI

Red Hat started offering a "store-based" approach with Red Hat 7, an example of which is shown in Figure 11.1. The main page of this GUI presents the user/administrator with three options: Explore, Installed and Updates. Explore shows software categories. Installed shows the software currently installed. Updates shows the currently installed software that has an update available. In this case, there are seven updates available.

Having selected Explore, we are presented with the categories shown in Figure 11.1. Selecting a category gives us a list of "featured" titles and the full list of available software for this category.

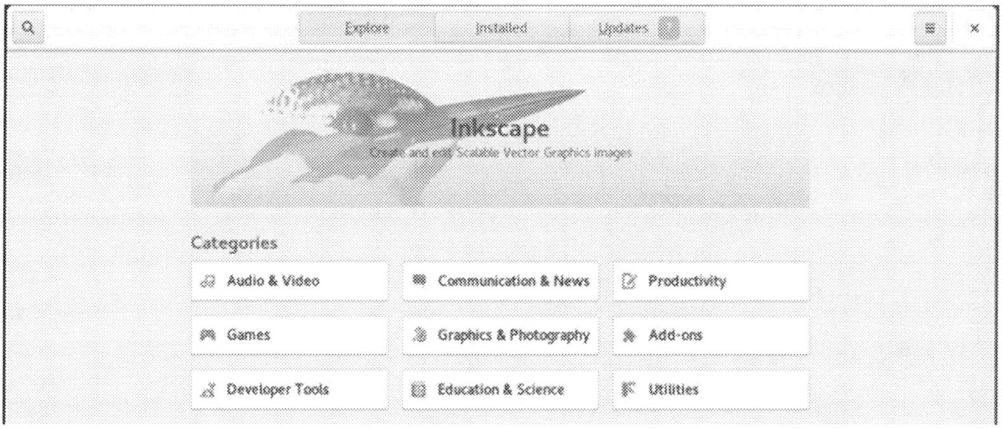

FIGURE 11.1 Explore page in the Red Hat software GUI.

Also, in the Explore window is a search icon that brings up a search box. Any string entered will result in the list of titles that contain the string specified.

Figure 11.2 shows the choices under the Productivity category. Most of these software titles are of LibreOffice, the open-source Office Suite for Linux. The left arrow button in the upper-left corner returns us to the Explore page from Figure 11.1. Notice that Evince has a blue checkmark in its upper-right corner. This indicates that the title is already installed. The titles listed have an average ranking based on user reviews.

Upon selecting a title to install, we are taken to that software title's installation page. On this page is a description of the software, details about the software (current version, license, download size and so forth) and reviews. Some software pages include screenshots of the software. If the software title is not installed our only choice is to install it. If already installed, our choice is to launch it. There is also an option to write a review for any title.

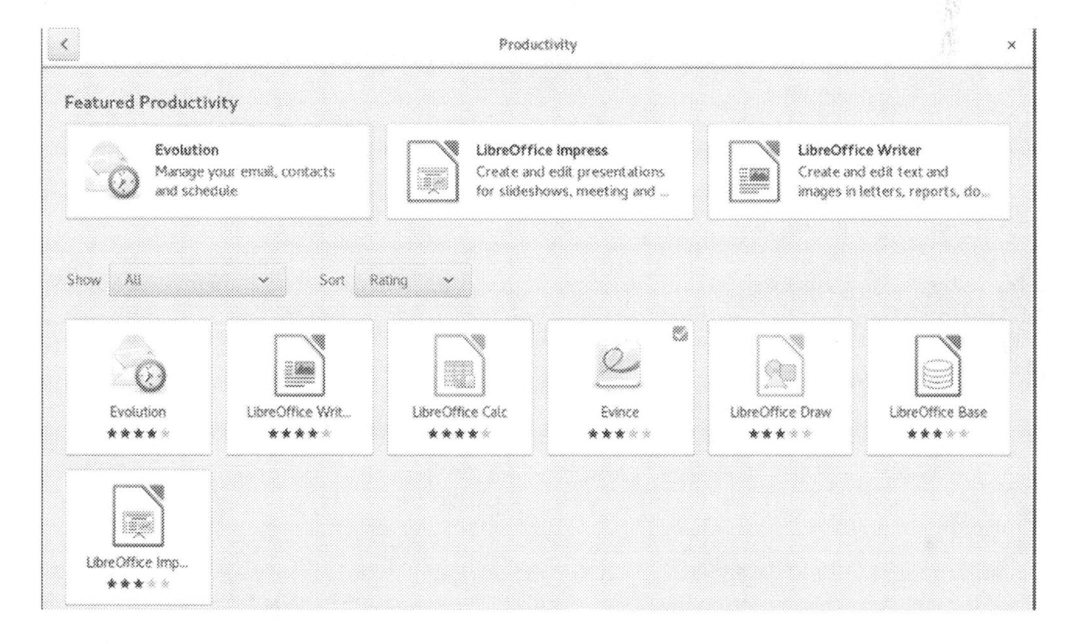

FIGURE 11.2 Productivity software selections.

TABLE 11.4
Red Hat Repositories

Repository	Contents	Examples
AppStream	Applications software, programming language facilities, databases.	`alsa`, `apr`, `CUnit`, `ds-base`, `GConf`, `Judy`, `OpenEXR`, `PackageKit`
BaseOS	Core programs that make up the OS.	Programs found in `/usr/bin` and `/usr/sbin`; C library files
Debuginfo/Devel	Development/debugging tools.	Various C libraries
Extras	Various extra servers and tools.	`ansible`, `ceph`, `nfs`, `openstack`
HighAvailability	Server clustering software.	`clufter`, `corosync`, `pacemaker`
PowerTools	Various applications for media and telecommunications.	`Apache` tools, `avahi`, `evolution`, `Judy`, `Modemmanager`, `OpenEXR`
RealTime/RT	Real-time version of Linux kernel and support packages.	Kernels and kernel tools
GNOME shell extensions	Shells and tools.	Various additional shells and tools to support GNOME.

The Installed tab allows us to remove installed software. However, we are limited to only removing application software. Most of the system software cannot be removed. If we select any installed title, we are taken to that title's install page where the options instead are to launch or remove it.

The Updates tab lists all of the waiting updates. Updates may be to operating system components, application software or both. When updates include the operating system, a restart may be required. The only option here is to update all items listed. If the operating system is listed, selecting it will pop-up a window showing all of the specific components that are ready for updating.

In the upper-right hand corner of the GUI's main page is a button containing three lines. Selecting this brings up a small menu of three choices: Software Repositories, Update Preference and About Software. Software Repositories display the list of available repositories and whether they are enabled or disabled. Any disabled repository can be enabled, and any enabled repository can be disabled, but both actions require submitting the administrator password. If we change the state of any repository, we may see a different list of titles appear when exploring further. Some of the repositories for Red Hat and the types of content stored in each are shown in Table 11.4.

11.3.2 Debian Mint Software GUI

Debian Mint has a similarly styled GUI software manager as Red Hat's. The Debian Mint Software Manager main page, shown in Figure 11.3, provides recommendations for titles displayed as "Editor's Picks". From the main window, we select categories. There is also a search box to list titles that contain the given string.

Selecting a category brings up the list of software titles in that category. Figure 11.4 shows the top portion of the Games selection. When the category contains subcategories, a menu is provided on the left of these subcategories. For Games, we see subcategories consisting of types of games. We will find subcategories in other categories like Graphics, but not all categories (e.g., Accessories and Office).

Similar to Red Hat's software GUI, each title includes an average ranking although in this case we see the number of reviews available. Different from Red Hat is that each of these listed software titles includes a brief description. Already installed titles are indicated with a checkmark. Selecting a title brings us to a page similar to that of Red Hat where there is a longer description and snapshots

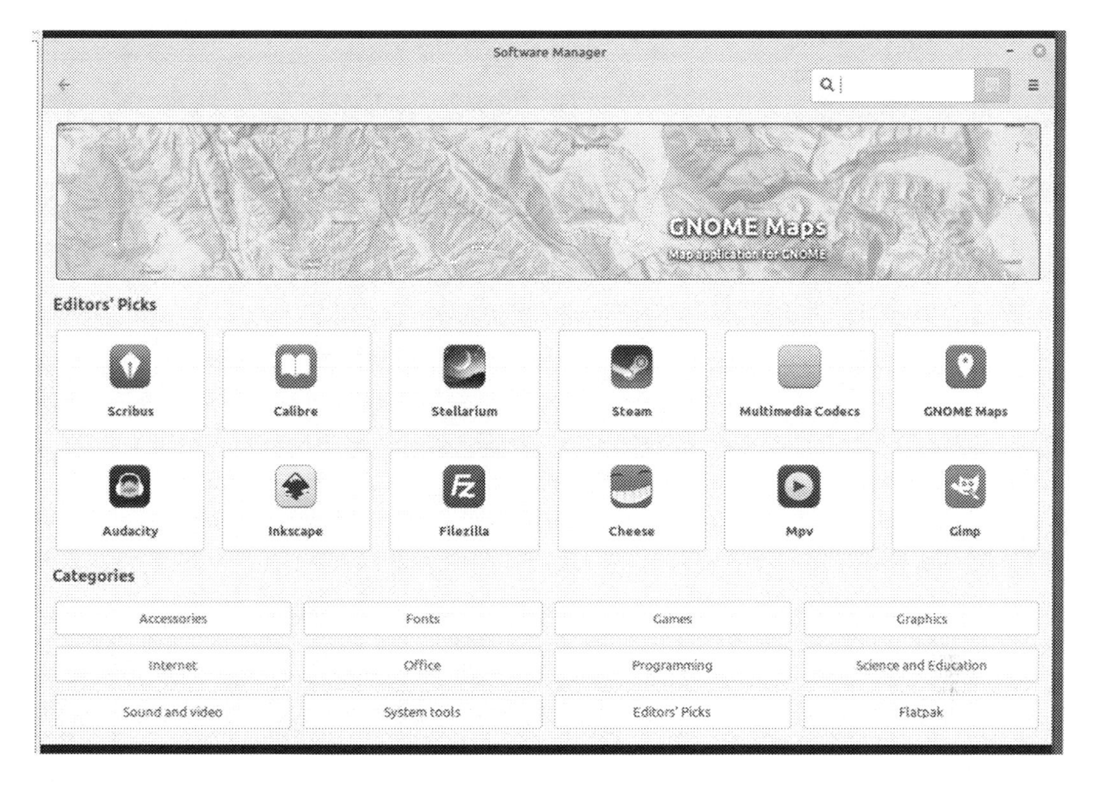

FIGURE 11.3 Debian Mint software manager.

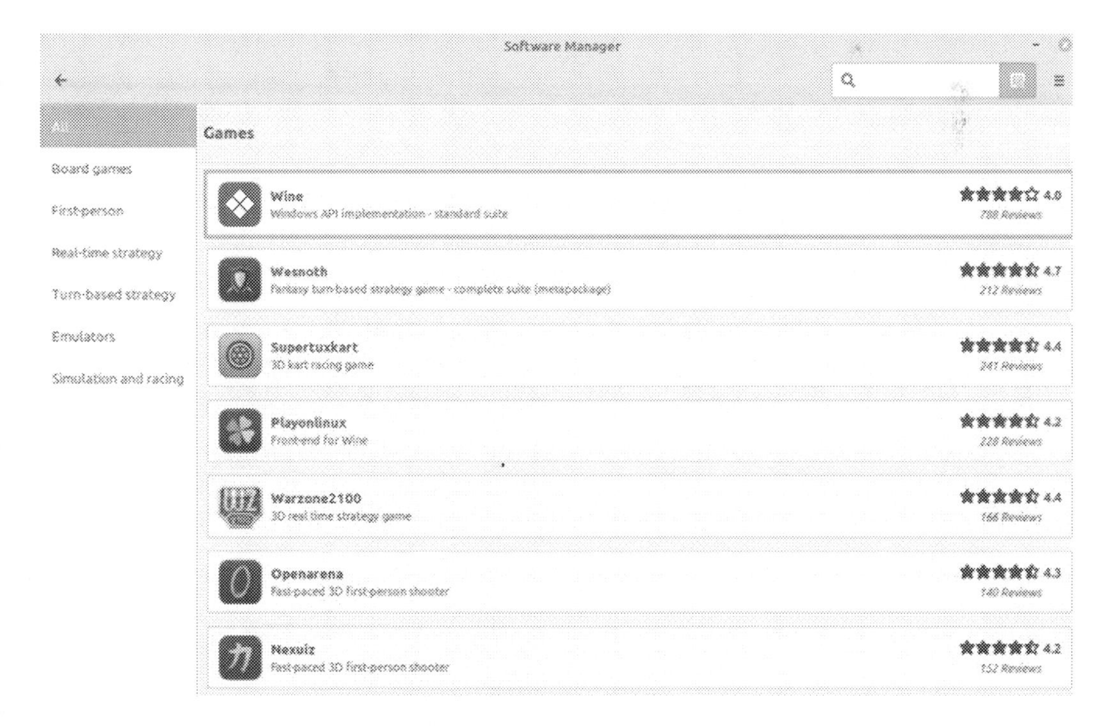

FIGURE 11.4 The games category.

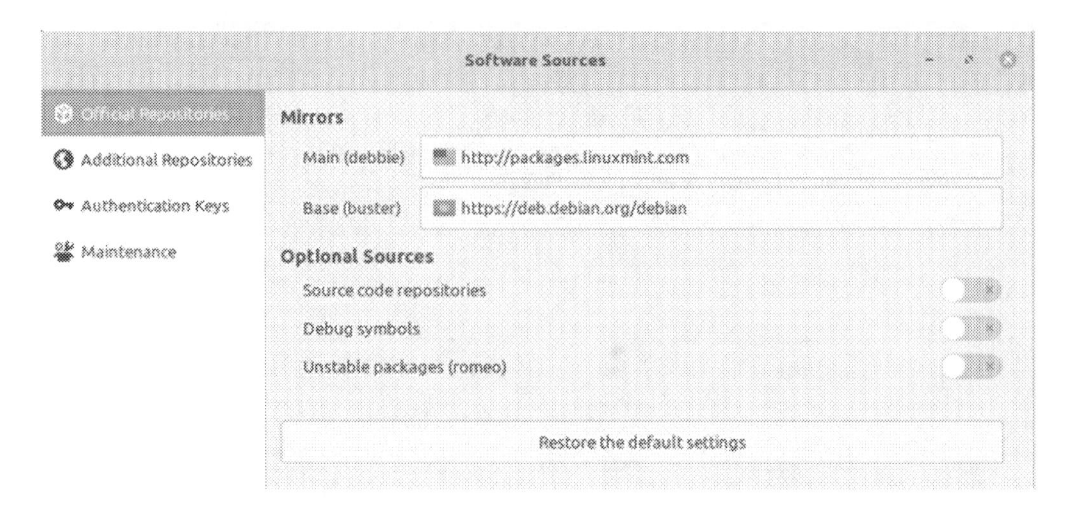

FIGURE 11.5 Selecting a software source (repository).

of the software along with reviews and slightly more information (package, version, size). Our only choice for uninstalled software is to install it. For installed software, we can either launch it or remove it.

Unlike the Red Hat Software GUI where we can update installed packages, Debian Mint has a separate Update Manager. The first time we use this GUI, it presents us with a software sources window (see Figure 11.5) to select either an official repository or add other repositories. We can also bring up this GUI tool separately.

The Update Manager will list any waiting updates and allow us to select any or all of these titles to update. We can also select Preferences which brings up the window shown in Figure 11.6. In the

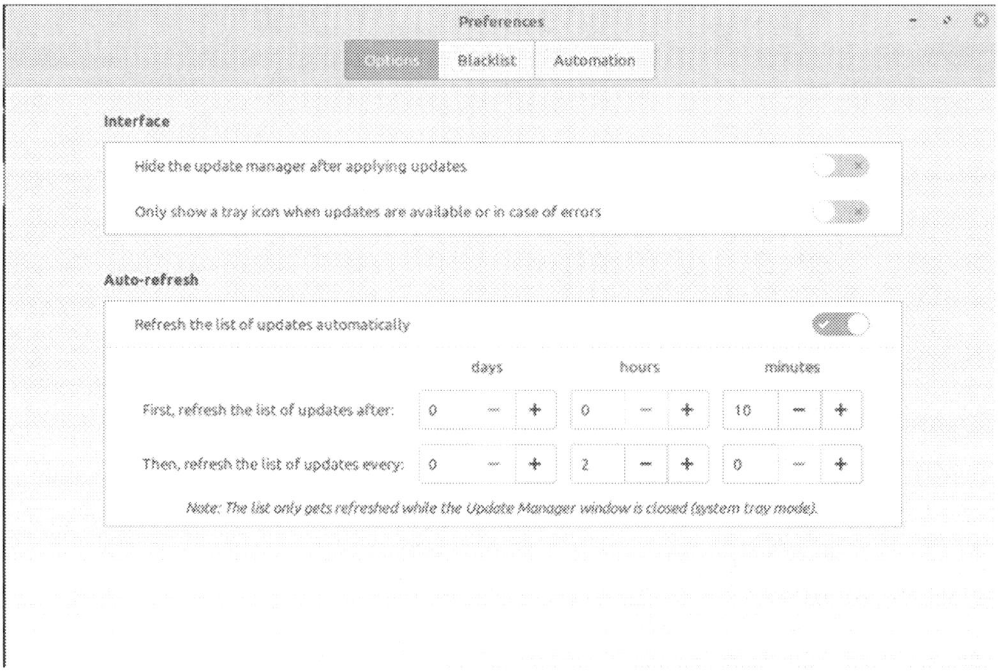

FIGURE 11.6 Update manager preferences.

figure, the Options tab is selected. Blacklist allows us to enumerate packages that should never be updated. We might do this with third-party software that is causing issues with our system. The Automation tab lets us specify whether updates should be handled automatically.

11.3.3 Ubuntu Software Center

Ubuntu was the first of the mainstream Linux distributions to offer a store-like installation program, originally called the Ubuntu Software Center. The current version is shown in Figure 11.7. This GUI is nearly identical to the Red Hat software GUI except that there are more categories, and within categories likely more titles. This is because the Ubuntu Software Center includes software from multiple sources whereas Red Hat only provides access to official Red Hat software. Software for Ubuntu include titles provided by Ubuntu, titles provided by "canonical partners", proprietary software and software provided by independent programmers.

In summary, the software GUIs allow us to install software, remove installed software and update software. In Debian Mint, updating software is handled through a separate tool from installation and removal, which are handled with no need for user interaction.

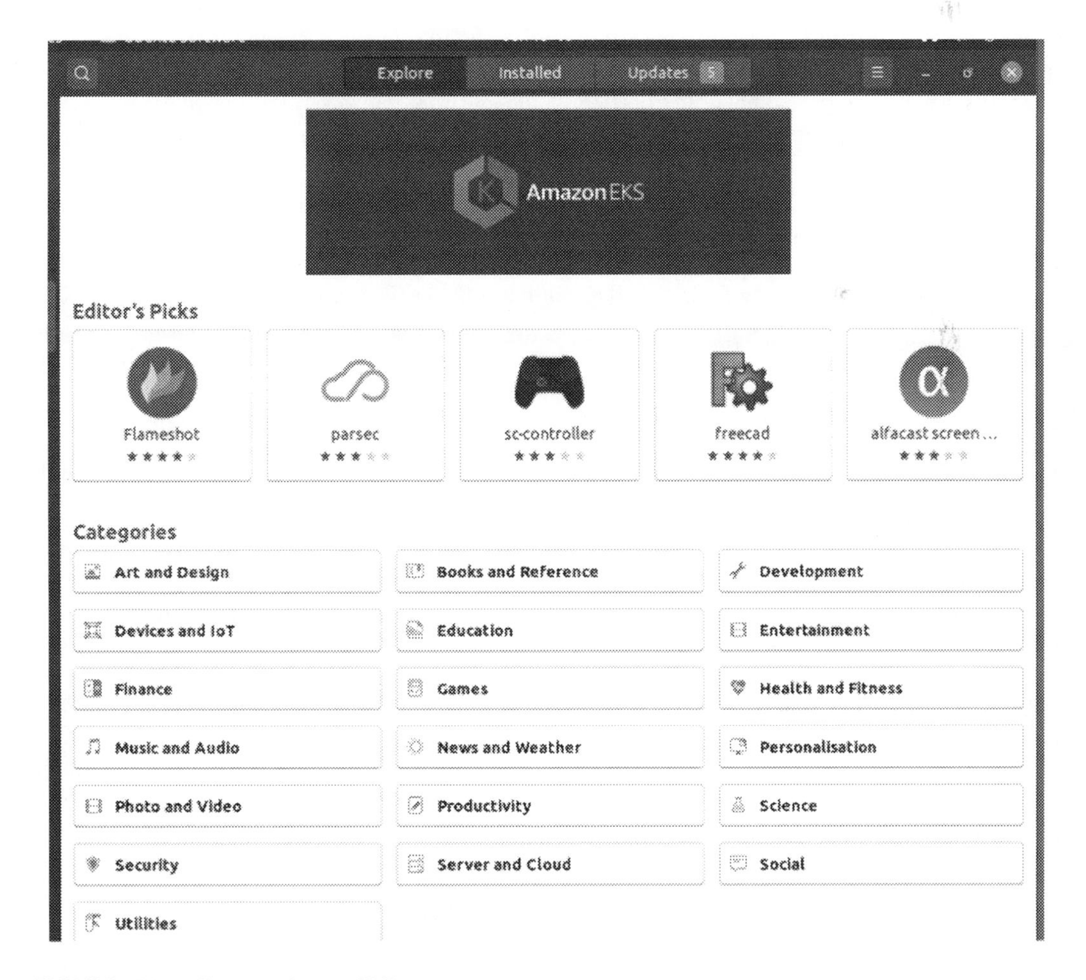

FIGURE 11.7 Ubuntu software GUI.

SECTION ACTIVITIES

1. Have you used the Apple and/or Microsoft Store? Have you used it to purchase music? software? movies? Do you prefer to install from a store-type interface or do you prefer to download an installation program and install from the program? Come up with a reason to support each approach.
2. There is far less software available in the Red Hat store than Debian and Ubuntu. Would this make a good argument for using Debian or Ubuntu Linux instead of Red Hat? Consider the risks of using third-party software in your Linux installation when compared to using software produced by Red Hat software developers for Red Hat.

11.4 `rpm` AND `dpkg`

A *package manager* is a program used by users or system administrators to install, update, upgrade or remove software. The stores described in the last section are GUI-based package managers. As is typical with all things Linux, there are command line approaches to package management as well.

In general, a package manager, or a package management system, provides several different useful functions for both installation and maintenance of software. The package manager operates as an archival tool in that it can unpack and uncompress the files in the package. Since the packages typically consist of executable files, library files and supporting files (e.g., documentation, data files), there is no need to compile any portion of the software. However, the files must be placed correctly. The package manager performs installation by testing files and moving them to their destination directories.

The package manager can be used to verify the correctness of the contents of a package. Correctness means that the files have proper checksum values and the package itself is signed with the proper digital signature. A *checksum* is a number that is derived through some computation of the bytes of the file. A corrupted file (whether maliciously changed or damaged during download) will likely not match its stored checksum and so installation may fail on those grounds.

A *digital signature* encodes the programmer's name to verify that the version we are installing is legitimate. The digital signature is not necessary for software installation but having one lets the package manager know that the software is from the source it claims to be. Having an incorrect signature or no signature is grounds for not trusting the package and therefore perhaps not installing it.

Another form of testing involves checking for dependencies. A *software dependency* arises when a piece of software calls upon functions that are not part of the package itself. Such functions are gathered together into one or more library files. Unmet dependencies because the needed library(ies) is(are) not present can result in a failed installation.

The reason that dependencies exist is that programmers will rely on other, already implemented, software components to simplify their own programming process. Without libraries, programmers would have to implement code that performs tasks whose code has already been implemented by others. Libraries may include, for instance, capabilities like "save as" and "open" windows, virtual memory handling and security mechanisms.

Libraries in Linux consist of `.so` files (*shared object*). These are typically C functions compiled into object files. The `.so` file plays a similar role in Linux as the `.dll` (dynamic linked library) file does in Windows. However, in Linux `.so` files.are versioned meaning that they are stored based on the version number of the Linux system or software that will use them. In Windows, a newer version of the same `.dll` file will replace an older version so that software requiring the given library is forced to use the most recent version. This may or may not complicate matters in running the older software. This leads to a situation that many Windows programmers have dubbed "DLL Hell".

Library files in Linux have names of the form `lib` followed by the library's name followed by `.so.version` where *version* is a version number. For instance, one library your system probably has is gdm whose file is named `libgdm.so.4.0.0` (or some similar version number). Library files are generally stored under `/usr/lib` and `/usr/lib64` (others may be found in other directories such as `/var/lib`). These two `/usr` directories are divided into subdirectories for classes of library files. For instance, security library files are generally found under `/usr/lib/security` and/or `/usr/lib64/security`.

The problem with using libraries is that a programmer cannot guarantee that the needed library(ies) has(have) been installed onto everyone's system. Some package managers will fail when trying to install a package which relies on dependencies that are not currently installed in the system.

Package managers can also be used to update installed software by installing newer versions, patches or additional library files. The package manager can be used to remove software packages if those software titles are no longer of use. This is the antithesis of installation in that the files placed in various directories must be identified, and those that are not involved with dependencies of other packages are deleted.

In this section, we explore the more primitive package managers, `rpm` for Red Hat and `dpkg` for Debian. `rpm` stands for Red Hat Package Manager, and `dpkg` stands for Debian Package. In the next section, we explore more usable managers called `dnf` (or `yum`) for Red Hat and `apt` for Debian.

11.4.1 rpm

To use `rpm`, we first have to download an `rpm` file. The `rpm` file is an archive compressed using `gzip` containing the executable code of the program and any other files required for installation. We can obtain `rpm` files from an *RPM repository*. There are many RPM repositories. A good place to look for `rpm` files is at `www.rpmfind.net/linux/RPM`. We can find other repositories at `rpmfusion.org` (rpm files for Fedora and Red Hat) and `mirror.centos.org/centos` (CentOS).

The `rpm` files are given very expressive names, divided into four sections. The first three sections are separated by hyphens and state the software's name (title), version number and release number. A package name like *title*-2.3.1-3 means that this is *title*'s version 2.3.1, release number 3.

Following these three sections is a period followed by the intended *architecture* type. As the `rpm` file contains compiled software in the form of executable files, the file is intended for a particular platform. Some examples we might see include `i386` for Intel 386 (or later), `i686` for Intel 686 (Pentium II or later), `aarch64` for 64-bit ARM, variations of `arm`, as in `armv7hl` or `armv5tel`, for other ARM processors, `ppc` for PowerPC processors, `s390x` for IBM mainframes and `sparc` for Sun Sparc workstations. The label `noarch` means that the `rpm` file should install correctly on any architecture. The reason `noarch` can install on any platform is that the contents of such an `rpm` file should not contain any executable code but instead shell scripts, text files and man pages.

Having selected and downloaded the `rpm` file(s), the `rpm` instruction is then used to perform the installation. To use `rpm`, we must be root. The command is `rpm [options] file(s)` where the *file(s)* is(are) the `rpm` file(s). There are two categories of options. The first are maintenance tasks (e.g., install, remove). The second category consists of general options that control `rpm`'s behavior.

Table 11.5 lists the maintenance options along with some of the options that are specific to the type of maintenance operation. For instance, `--force` is available to be applied to installing/upgrading/freshening a package. This option attempts to perform the operation even if there are errors. General options include `-v` (verbose), `-vv` (very verbose, outputs debugging information), `--quiet` (output as little information as possible) and `--dbpath PATH` (to specify the RPM database directory rather than the default of `/var/lib/rpm`).

TABLE 11.5
Common rpm Options

Option	Additional options	Meaning
-e, --erase	--allmatches, --nodeps, --noscripts, --notriggers, --test	Remove the software and all corresponding components.
-F, --freshen	--allmatches, --excludedocs, --force, -h/--hash, --ignoresize, --ignorearch,	Same as upgrade (see below) but only if a version is currently installed.
-i, --install	--ignoreos, --includedocs, --nodeps,	Install new package.
-U, --upgrade	--nosignature, --noscripts, --notriggers, --oldpackage, --replacefiles, --replacepkgs, --test	Upgrade or install anew the package except that if an old version exists, it is removed after the new version is installed.
-V, --verify	--nodeps, --nofiles, --scripts, --nosignature, --nosize	Does the package contents match the metadata? Used to ensure reliability of the downloaded rpm file.
-q, --query	--qf/--queryformat	Query the rpm file for information; this option has additional specifiers of the type(s) of information being querie.
--checksig	--nodigest, --nosignature	Verify just the digest/signature information of the rpm file.

Using rpm should be straightforward. An rpm command might be as simple as rpm -i sometitle-1.2.3.noarch.rpm. The additional installation options (which are also used when freshening or upgrading a title) may be of value depending on the situation such as if the downloaded rpm file does not match our architecture. We do not cover these and invite you to read through rpm's man page as needed.

rpm has a complicating factor: dependencies. rpm will fail to install software with unmet dependencies. To resolve this problem, we have to locate the appropriate library(ies), download its(their) rpm file(s) and install it(them) using rpm. If a library has its own unmet dependencies, we have to further resolve those dependencies first before installing the library(ies) that fulfill the original unmet dependencies. This cycle can continue. This situation is referred to as *dependency hell*, or more specifically *RPM hell*, similar to DLL Hell mentioned earlier.

We can test an rpm file for dependencies by adding --test to the install command. For instance, rpm -i --test *somepackage.version*.rpm will test to see if *somepackage* can be installed. Figure 11.8 shows the output of such a report when attempting to install emacs-26.1-5.el8.x86_64.rpm.

The error message from Figure 11.8 informs us that the emacs package has two unmet dependencies. Specifically, we need to include emacs-common, which is a separate software package, and libotf.so.0, a separate library. Resolving dependencies can be a chore. Fortunately, there is a website set up to specifically assist us in locating dependent library files. The site is the RPM database at rpmfind.net.

To resolve the issue from Figure 11.8, we type into the rpmfind.net's search box emacs-common. The site searches for and lists the rpm files that contain the package for this piece of

```
error: Failed dependencies:
     emacs-common = 1:26.1-5.el8 is needed by emacs-26.1-5.el8.x86_64
     libotf.so.0()(64bit) is needed by emacs-26.1-5.el8.x86_64
```

FIGURE 11.8 Dependency error from an rpm command.

software. We similarly search for libotf (without the .so extension). Both searches will result in a lengthy list of .rpm files. We select the two files that match our architecture and operating system (e.g., CentOS 8-stream for x86_64). Once downloaded, we use rpm to try to install both of these dependent packages before we attempt to install emacs again. We can view rpm dependencies with the rpmgraph command.

As noted earlier, rpm may also fail should the rpm file either contain no digital signature (called a key) or have a non-matching digital signature. These two situations are reported as NOKEY and BAD respectively where the latter is an error and the former is a warning. The warning will not prevent the package from being installed/upgraded while the error will.

With all the problems using rpm, you might wonder why we should use it. The answer is that we shouldn't. In the next section, we find better command-line tools. But if you do choose to use rpm, there are additional options worth knowing. First are --ignoresize, --ignorearch and --ignoreos. These three options cause rpm to skip testing if there is adequate disk space to install, that the rpm file is of the appropriate architecture and that the rpm file is of the appropriate operating system. Using any of these options is risky because rpm will try to force an installation resulting in a program that may fail to run.

Similarly, we can avoid the dependency and bad key errors through the options --nodeps (do not test for dependencies) and --nosignature respectively. Again, using these will force rpm to install the software. But if dependencies exist, the software will not run. Installing without testing the signature may result in correctly installed software but is considered a security risk.

11.4.2 dpkg

dpkg is to Debian what rpm is to Red Hat. With dpkg, we first download the installation package which is stored in a .deb file. We then issue the dpkg instruction from the command line. To issue a dpkg instruction, we need to be root. As some Debian installations do not provide us with a root password, the first user account is given administrator-level access via sudo.

The main actions for dpkg are shown in Table 11.6. Most of these are followed by the package's name. Some of these operations permit an option of -R (or --recursive) which then operates not on a package but all packages found in the specified directory.

TABLE 11.6
dpkg Commands

Option	Meaning
-i or --install	Install the package(s).
-l (lower case L) or --list	List installed packages matching the specified pattern.
-L or --listfiles	List files installed from specified package(s).
-p or --print-avail	Print information about specified package(s).
-P or --purge	Remove package(s) including all related files.
-r or --remove	Remove package(s) excluding configuration files.
-s or --status	Print status of specified package(s).
-V or --verify	Verify the integrity of the downloaded package file(s).
--configure	Perform configuration step on an unpacked package(s).
--triggers-only	Process all pending triggers (see Table 11.7).
--unpack	Unpack package(s) but do not configure or install.
--update-avail	Update the dpkg database of available packages (related to this is -A/--record-avail).

TABLE 11.7
dpkg Package and Selection States

Package state	Meaning	Selection state	Meaning
`conf-files`	System only has configuration files.	`deinstall`	Package selected for removal of all components excluding configuration files.
`half-configured`	Configuration started but not completed.	`hold`	Package will not be handled by dpkg.
`half-installed`	Installation started but not completed.	`install`	Package selected to be installed.
`installed`	Installation complete.	`purge`	Package selected for complete removal.
`not-installed`	Package not installed.	`unknown`	Package is in an unknown selection state and may be removed from the database.
`triggers-awaited`	Awaiting installation of another package.		
`triggers-pending`	Another package has been installed triggering this package.		
`unpacked`	Package unpacked but not configured.		

Many of the `dpkg` commands have additional options. We briefly discuss the most significant. As with `rpm`, `dpkg` will fail under a number of circumstances such as dependencies or bad or missing keys (signatures). We can force an installation adding `--force-`*`option`* where *option* can include `all`, `breaks` (install even if the installation harms another installed package), `depends` (turn any dependency errors into warnings), `downgrade` (to install an older version) and `over-write` (install the new version's files over any existing files). We can also specify that dependency checking should be ignored by using `--ignore-depends=`*`packages`* where *packages* are the packages to be installed.

The option `-R/--recursive` can be applied to installation and unpacking in which case we specify a directory which contains the `.deb` files rather than the file(s). The `-G` option causes `dpkg` to skip installation of a package if a newer version is already installed, while `-E` causes `dpkg` to skip installation if the package is the same version of an already installed version.

`dpkg` maintains states for a given package. There are package states and package selection states. These are all described in Table 11.7. There are also two flags set for any given package, `ok` in which the package has a specified state but may not yet be completely installed, and `reinstreq` for a broken package which requires reinstallation.

There are several supporting programs for `dpkg`. `dpkg-deb` can be used to unpack or display information of specified packages. It has various options to build a package, extract from a package, provide information about a package or list the package's contents, among others. `dpkg-query` can be used to query the `dpkg` database. Other helper programs include `dpkg-name`, `dpkg-deb` and `dpkg-split`. `dpkg` invokes these various programs as needed depending on options supplied so that we primarily use `dpkg` and not the helper programs.

We did not explore either `rpm` or `dpkg` in much detail or with numerous examples because, frankly, we don't need to use these programs as there are better programs available which are far easier to use. We cover these, `dnf` and `apt`, in the next section. But should you find yourself needing to use either `dpkg` or `rpm`, it is best to consult the program's man page to explore all of the options.

SECTION ACTIVITIES

1. Visit `rpmfind.net` and search through the various software titles available. Locate three titles that you want to install and download the corresponding `.rpm` files. As root, try to install each one. Do they install or do they have issues like dependencies or bad/missing keys?
2. One of the most significant differences between Red Hat and Debian distributions is the package manager: `rpm` vs `dpkg`. Research the reasons why the two approaches have been taken. In your research, also see if you can find any Linux distributions that can use both package managers.

11.5 `dnf/yum` AND `apt`

There are many challenges in using `rpm/dpkg` but the most frustrating aspect of these programs is the dependency hell that causes an install to fail. To succeed, we must install the dependent packages first. In order to avoid using `rpm` or `dpkg`, Linux provides improved package managers which can resolve dependencies for us. These package managers use `rpm` or `dpkg`, so we refer to these as *front-end package managers*.

In Red Hat, the package manager had been called `yum` (short for Yellowdog Updater Manager). Starting with Red Hat 8, `yum` has been replaced with `dnf` (Dandefied `yum`). In Debian Linux, the program is referred to as `apt` (advanced package tool) although in fact `apt` consists of a suite of programs whose names all start with `apt`. As with `rpm/dpkg`, we must be root (or use `sudo`) to use `dnf/yum/apt`.

Both Red Hat versions of Linux and Debian Linux maintain repository lists. Upon issuing a `dnf` or `apt` command, the program uses the available repository lists to locate the needed `.rpm` or `.deb` file(s), download the file(s) and perform the specified operation, such as installing the packages. In some cases, `.rpm/.deb` files may be preloaded and stored in `/var` to reduce the amount of downloading needed.

We start our look at these two package managers with `dnf`. From a user's point of view, `dnf` is far simpler to use than `rpm` even though it has a great many commands and options. We examine some of the more useful commands in Table 11.8 and then look at some of the most useful options in the text below. Note that `yum` has nearly identical commands although a couple of differences are that `yum` has separate `update` and `upgrade` commands while in `dnf` they do the same thing, and that `resolvedep` does not exist in `dnf`, but `dnf` does have a command `provides` does something similar.

Among the options available, perhaps the most useful is `-y` which causes `dnf` to execute without pausing for permission. The `-b/--best` option, when used with `upgrade`, asks `dnf` to select the best available package that matches the name specified. The option `--skip-broken` causes `dnf` to skip any specified packages that are causing errors because of unresolved dependency issues. `dnf` can also be used to enable or disable a repository with `--enable` and `--disable`, respectively. With these options, we specify the repository(ies) instead of package(s).

As noted earlier, `dnf` does not require that we have already downloaded the needed `rpm` files. Instead, `dnf` maintains a list of repositories and has the ability to search for any needed `rpm` files from one or more of those repositories. Another difference between `dnf` and `rpm` is that the package name(s) specified in a `dnf` command is not the specific `rpm` file but instead the name of the package. For instance, in `rpm` we might install emacs using `rpm -i emacs-27.1-4.3.x86_64.rpm` while in `dnf` it is `dnf install emacs`.

TABLE 11.8
dnf Commands

dnf Command	Meaning
alias	Define, list, manage and remove aliases defined within dnf.
check	Check the dnf database for problems.
check-update	Check if updates to specified packages are available.
clean	Clean up any temporary files regarding repositories.
downgrade	Downgrade an installed package to an earlier version.
history	Display previous dnf operations.
info	Display information about installed and available packages.
install	Install the package and all dependencies.
list	List specific package names based on the string supplied; the string can include wildcards such as *gnome* or gimp[0-9].
remove	Remove the specified software package(s) and all of its dependencies so long as the dependencies are not required by other installed packages; can also be used to remove older installed versions while retaining the most recent version.
repolist	List known enabled/disabled (or all) repositories.
repoquery	Search repositories to respond to a specific query such as listing packages that match a string; requires dnf.plugin.repoquery.
search	Search package(s) for metadata that matches given keywords.
upgrade	Update package(s) to the most recent version; note that the command update does the same thing.

Yet another difference between dnf and rpm, and perhaps the best reason to use it, is that dnf performs dependency handling for us. Specifically, dnf examines the dependencies of an rpm package and tests to see if any of these dependencies are not met by the installed libraries of our system. For every unmet dependency, dnf locates the appropriate rpm file that will fulfill the dependency and downloads and installs it for us. It does this until all dependencies are met.

dnf maintains a cache in the directory /var/cache/dnf. The cache stores already downloaded repository files and repository locations (URLs). dnf has a configuration file of /etc/dnf/dnf.conf. Configuration for the dnf repositories can be found in the directory at /etc/yum.repos.d.

In Figure 11.9, we illustrate output from a dnf command. Here, we are installing the software package emacs, using the previously listed command. We have added the option -y (assume yes) so that dnf does not pause for permission to install the package. Notice in this example that while dnf performed a dependency check, it found no additional packages that required installation. This will certainly not always be the case. Using -y to force installation no matter what may not be wise if there are a number of dependencies because unexpected dependencies will involve using more disk space and more download and installation time.

Aside from installing, upgrading or downgrading a specific software package, we can also update all software awaiting update. This is handled through dnf upgrade (or update). dnf will examine all software titles which have updates available and handle all of these updates for us. This is a convenient way to update our operating system with a single instruction.

Debian Linux uses apt (for Advanced Packaging Tool) as the front-end to dpkg. apt works like dnf in that it contacts one or more repositories that store packages. The file /etc/apt/sources.list stores the URLs of repositories to try (additional files may be stored in the directory /etc/apt/sources.list.d). apt has its own configuration file, /etc/apt/apt.conf. Another file, /etc/apt/preferences, is used to control preferred locations from the sources.list file for different software versions. Retrieved packages and status information are stored in caches under /var/cache/apt/archives and /var/lib/apt/lists respectively.

One of the most complex components in apt is a program which performs *topological sorting* to work out inter-package dependencies to determine the order that packages should be installed. Topological sorting is a graph algorithm often used to organize a sequence of tasks such that any task that has dependencies is executed only after those dependent tasks are executed.

As noted, apt is not a program itself but the name given to a suite of software maintenance tools. The primary program for software maintenance is apt-get, whose syntax is apt-get [*options*] *command package(s)*. The commands available for apt-get are listed in Table 11.9. Options include -d (download only), -f (attempt to fix broken dependencies), -m (ignore missing or corrupt packages), -y (assume yes) and --no-upgrade (when used with install causes apt to install only new packages, not update already installed packages).

While apt-get is the most important of the apt tools, there are three others to note. apt-cache is used to query the apt package cache, which consists of downloaded packages and portions of packages. While this command will not alter the stored data, it can be used to retrieve and summarize information on what has been downloaded. apt-file inspects a package to find out what specific files are included in that package. Finally, apt-secure can be used to ensure the integrity and authenticity of a package through digital signature checking.

```
Loaded plugins: fastestmirror, refresh-packagekit, security
Loading mirror speeds from cached hostfile
 * base: ftp.linux.ncsu.edu
 * extras: mirror.serversurgeon.com
 * updates: mirror.linux.duke.edu

Setting up Install Process
Resolving Dependencies
--> Running transaction check
---> Package emacs.x86_64 1:23.1-21.el6_2.3 will be installed
--> Finished Dependency Resolution

Dependencies Resolved

=============================================================================
 Package       Arch           Version                     Repository    Size
=============================================================================
Installing:
 emacs         x86_64         1:23.1-21.el6_2.3           base          2.2 M

Transaction Summary
=============================================================================
Install       1 Package(s)

Total download size: 2.2 M
Installed size: 11 M

Downloading Packages:
emacs-23.1-21.el6_2.3.x86_64.rpm                        | 2.2 MB     00:00
Running rpm_check_debug
Running Transaction Test
Transaction Test Succeeded
Running Transaction
  Installing : 1:emacs-23.1-21.el6_2.3.x86_64                           1/1
  Verifying  : 1:emacs-23.1-21.el6_2.3.x86_64                           1/1

Installed:
  emacs.x86_64 1:23.1-21.el6_2.3

Complete!
```

FIGURE 11.9 Output from dnf installing emacs.

TABLE 11.9
`apt-get` Commands

Command	Meaning
`build-dep`	Install (or remove) dependencies of the specified package.
`check`	Update package caches and checks for broken dependencies.
`clean` (also `autoclean`)	Remove everything from the local repository (`autoclean` only removes packages that can no longer be downloaded).
`dist-upgrade`	Same as `upgrade` but uses a "smart" conflict resolution tool.
`install`	Install specified packages.
`remove`	Remove specified packages.
`source`	Download but do not install source packages.
`update`	Update the local repository listing of available packages.
`upgrade`	Install the most recent version of all installed packages.

The `apt` suite of programs all operate via the command line. Another front-end tool in Debian distributions is `aptitude`. This program is interactive and uses a menu-based interface (i.e., it is text-based instead of GUI-based but controlled through menus that can be selected using arrow keys). Figure 11.10 illustrates `aptitude`. In `aptitude`, a number of choices are presented to the user to select between: Security Updates, Upgradable Packages, Installed Packages, Not Installed Packages, Virtual Packages and Tasks. Within any one of these categories are subcategories and subsubcategories. Additionally, the menus offer the ability to search for packages, resolve packages and so forth.

We started off our look at Linux package managers with the GUI-based ones. These all serve as front-end interfaces to one of `dnf/yum` or `apt`. Yet another front-end GUI is the Synaptic Package Manager, shown in Figure 11.11. This figure shows upgradable packages. Commands available

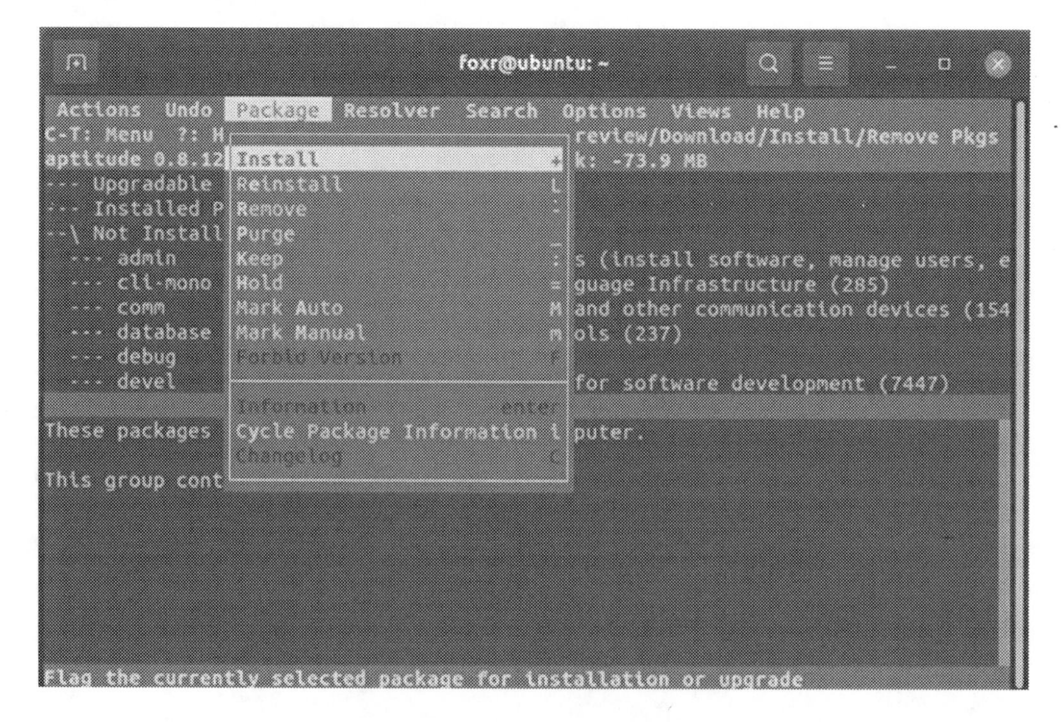

FIGURE 11.10 Debian-based `aptitude` interface.

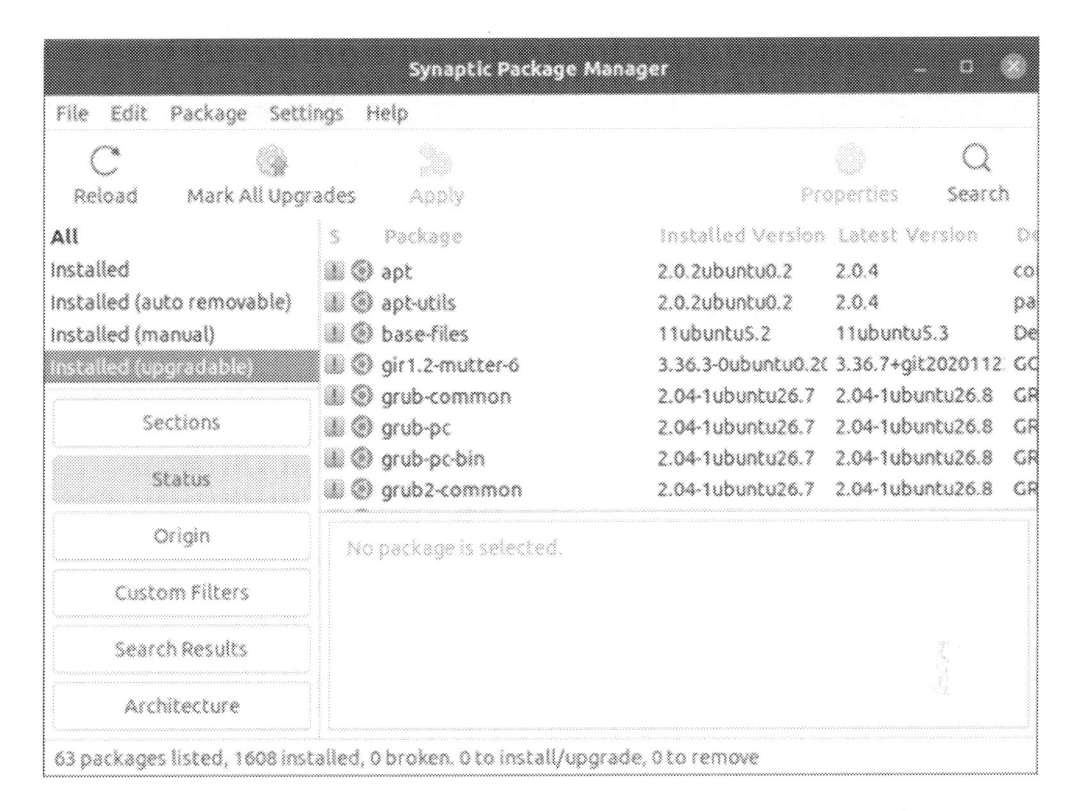

FIGURE 11.11 Debian-based synaptic Package Manager.

include determining a package's status (installed, not installed, upgradable), location (origin), filters (e.g., broken, community maintained, missing recommends) and available architectures. Selecting a package then allows us to install, reinstall, upgrade or remove the package. While Synaptic is available in Debian distributions, it can handle either or both `.deb` packages and `.rpm` packages. One more front-end GUI of note is the `Adept Package Manager`, which can be found in Debian distributions running the `KDE` desktop.

SECTION ACTIVITIES

1. In your Linux system, use the GUI-based software manager available to see if any of your software (whether applications or operating system) has updates waiting. If so, use the appropriate `dnf/yum` or `apt` command to list the waiting updates and then perform the updates. See if you can figure out the necessary commands through the package manager's man page.
2. Read the website `https://linuxconfig.org/comparison-of-major-linux-package-management-systems` which describes the various package managers we covered in this and the previous section. Does it give you a better understanding of the strengths and weaknesses of each approach?

11.6 INSTALLATION OF SOURCE CODE

One of the greatest strengths of Linux is its openness. This comes from the availability of both the operating system and much of the application software in source code format. *Source code* is the original high-level language code, as opposed to a compiled executable. With source code, we can examine, modify and compile the software. By modifying the software, we can improve existing code, add new features or alter available features to perform or appear differently. Of course, any modification to source code requires expertise in programming. One must first understand the source code to figure out how to make the desired changes.

Because of the availability of the software as source code, some administrators will download software packages not in executable form but in the source code form. Such packages are often stored in `.tar` files and compressed using `gzip` or `bzip2`. This allows for easy storage and transport (over network) of the packages. However, because most of the software we are dealing with consists of many files, the compilation and installation process can be very challenging. In this section, we step through the process of installation of source code.

Most open-source software is written in C or C++. To install such open-source software, we need a C/C++ compiler. The expected compiler is the GNU's C/C++ compiler (`gcc`). If you are installing open-source software, first make sure `gcc` is installed by typing `which gcc`. You should find it in `/usr/bin`. If it is not available, install it using `dnf/apt-get`, for instance using `dnf -y install gcc`.

11.6.1 Obtaining Installation Packages

The first step in installing an open-source software package is to obtain the package. Most open-source software has its own website managed by the developers who created the software. This is the most common location where to find the installation packages. For instance, the Apache webserver is available for download at `httpd.apahce.org` while the Squid proxy server is available at `www.squid-cache.org`.

In addition to websites dedicated to the software product, most open-source developers make versions of their software available via the SourceForge site at `sourceforge.net`. This site offers thousands of open-source software titles. Note that SourceForge also contains commercial and shareware forms of software. Another location for open-source software is GitHub, which is more commonly used as a location where programmers share their open-source code as they develop their projects.

The main page of SourceForge provides menus for open-source software, business software and resources. Selecting Open-Source Software takes you to the Browse Open-Source Software page. Along the left side of this page are filters including the operating system (Windows, Linux, Mac, BSD and others) and categories of software such as Internet, Games/Entertainment, Multimedia, Office/Business and Text Editors to name a few. A search box lets you enter keywords. The page also presents the most recent *Top Apps*.

Upon reaching some particular software's page, we will find a variety of information on the software. Among the information are the authors (provided as links to the authors' SourceForge pages), reviews, number of times the software was recently downloaded, a summary of the software's features and screen captures, what support the authors (or the open-source community) offer and links to the software's own page (if any).

Both the software's website and SourceForge offer multiple versions of the software title (both the most recent stable version and older versions) and in both executable and source formats. On the software's website will likely be a link to a Download page. This page will contain any download packages, whether some form of installation program, one or more `.rpm/.deb` files and/or source files packaged in a compressed `.tar` file (and if installing for Windows, files will likely be in `.zip` format). We might also find encrypted versions using one of PGP, MD5, or SHA1 (among others). Moving forward in this section, we consider only source code for Linux, packaged using `tar`.

11.6.2 Extracting from the Archive

Once the installation package has been downloaded, our next step is to extract the files from its archive. We can handle this by issuing a single `tar` command. If, for instance, we are installing the package *somepackage* version 3.1.8 then we will have probably downloaded a file named `somepackage-3.1.8.tar.gz` (or `.bz2`). The command `tar -xzf somepackage-3.1.8.tar.gz` will both uncompress the file (`z`) and extract its contents (`x`). If the file were compressed to `.bz2` format, we change `z` to `j`.

As the package itself most likely contains a directory which itself contains files (and possibly subdirectories), extracting from the package results in a new directory appearing in the current directory. Within this directory will be files and possibly subdirectories. Assuming the file was called `somepackage-3.1.8.tar.gz`, the above command should create a directory called `somepackage-3.1.8`. With the directory created, `cd` into it.

Inside the software's installation directory, we will find a `README` file. This text file contains the instructions for installation, alerting the system administrator to any specific requirements and options for installation. There may be other text files with names like `LICENSE`, `CHANGES`, `NOTIFICATION` (or `NOTICE`), `ABOUT` and `VERSION`.

There will also be at least one script present, `makefile` (alternatively `Makefile` or `Makefile.in`). `makefile/Makefile` is a script that, when run, performs the necessary compilation and installation steps. There may be a `configure` file which is another script that we use to configure the software. Its role is to either create or modify the `makefile`. If there is no `makefile` then the `configure` script uses `Makefile.in` to generate the `makefile`. Among the subdirectories, we will likely find `build` (C program code), `docs` (man page documentation or other forms of documentation), `include` (C header, or `.h`, files) and `modules` (add-on code that the system administrator might wish to add to increase the functionality of the software).

Figure 11.12 comes from untarring the open-source package `apg` (automated password generator). Here, we see most of the source code (the `.c` and `.h` files) is located in the top-level directory. The subdirectories of `bfconvert`, `cast` and `sha` contain additional source code for extra functionality. The `perl` subdirectory contains `perl` code to be used if we want `apg` to run on a server. The files whose names are all in capital letters are instructions, acknowledgments and a to-do list. The two files `install-sh` and `mkinstalldirs`, are scripts used to start and stop `apg` and related services. Finally, `Makefile` contains the compilation and installation operations. The `apg` software was packaged without a `configure` script, so any changes that we want to make to the compilation and installation process would have to be done by hand by altering `Makefile`.

Assuming the package contains all necessary code, and assuming our system has `gcc` installed to compile and install the given package, the next steps should be straightforward. If these steps fail, it is usually because our system is not set up as the installation process expects. If this is the case, we may need to modify the `configure` file, the `makefile` or other files, or we might need

```
apgbfm.c    convert.c  getopt.h         perl        README.CYGWIN  THANKS
apg.c       convert.h  INSTALL          php         restrict.c     TODO
bfconvert   COPYING    INSTALL.CYGWIN   pronpass.c  restrict.h
bloom.c     doc        install-sh       pronpass.h  rnd.c
bloom.h     errors.c   Makefile         randpass.c  rnd.h
cast        errs.h     mkinstalldirs    randpass.h  sha
CHANGES     getopt.c   owntypes.h       README      smbl.h
```

FIGURE 11.12 Contents of untarred open source archive.

to define some needed environment variables or perform other types of installation. It is critical to review the README before attempting installation.

11.6.3 Running the configure Script

The first step may be optional, which is to run the configure script. If there is a makefile present, we are not making modifications to the code and we want to use the default installation, we skip the configure step. We would also skip this step if there is no configure script. If, however, there is no file named makefile (or Makefile), then we *must* run configure. In such a case, we would find a file, probably named Makefile.in. This is not a makefile but instead a partial, or initial, makefile that will be used as a starting point by configure to generate the makefile. If we wish to alter the default installation, we will specify these changes when invoking the configure script as a series of options.

For large pieces of software, either the README file or configure will include instructions on how to use configure. To execute configure, since it is a script, enter the command ./configure. To obtain help, try ./configure -h (or --help). Some of the options available might be to specify installation directories and modules to compile. If we do not provide any options with our ./configure command, our installation will place all components in default directories and the software will be compiled to use the standard features and modules. We view common options in Table 11.10.

The configure step may take seconds or minutes depending on the complexity of the software. The result will not only be a great deal of output sent to our terminal window, but also the creation of a makefile. By default, configure will send a tremendous amount of output to the terminal window, showing every step it is taking. Adding the option -q (or --quiet or --silent) will cause configure to run without displaying this output.

11.6.4 The make Step and the makefile

Once the makefile is successfully generated, our next step is quite easy. Just type make. The make command in Linux causes the makefile/Makefile script to execute. There are numerous options available for make, but perhaps the only significant ones are -i to ignore errors that may occur during compilation, -k to continue to work as much as possible even if errors arise and -I *dir* to specify an *include* directory. If there are include statements, this directory is searched in place of the current directory. As with the configure step, the make step may take seconds or

TABLE 11.10
Common configure Options

Option	Parameter Type	Meaning
--bindir=*dir*	Directory name	Specify destination directory for binary code.
--libdir=*dir*	Directory name	Specify destination directory for library files (object code).
--infodir=*dir*	Directory name	Specify destination directory for documentation.
--prefix=*dir*	Directory name	Specify umbrella directory to store all program components.
--enable-all	None	Enable all modules.
--enable-module=*module(s)*	Module(s)	Enable only modules listed; the list of modules is separated by spaces.
--disable-module=*module(s)*	Module(s)	Disable those modules listed.
--with-*feature*=*value(s)*	Value(s)	*Feature* is the feature name to establish; some features require *values* which are feature-specific.

minutes and will result in numerous messages being displayed. Notice that we do not execute the makefile directly through ./makefile.

Most makefiles are written in parts. These parts may include a compilation section, an installation section, a clean-up section and an archiving section, among others. To run the compilation section, just issue the make command by itself. To perform the installation separately (which is usually required), use the command make install. If we have already attempted to compile and/or install the software and it failed, before trying again, issue make clean. Finally, if we want to take the files and wrap them up (or back up) into an archive, use make tar. The command make all should perform a make clean, make and make install all in one. Not all makefiles will have all of these sections and we can find out which sections are available in the README file.

The makefile itself consists of variable definitions, called *macros*, and commands. Variables store program names, gcc options and flags, and destination directories. The top portion of Figure 11.13 provides an example of some commonly used macros. We see, among other things, that CC defines the name of the compiler, FLAGS specify a gcc option (in this case -Wall to enable all warnings) and LIBS and LIBM define library options. The directories specify destination locations; these directories were established by using options in the ./configure step. Notice how these macros are assignment statements but unlike those that we issue in Bash (from the command line or a Bash script), we can have blank spaces around the equal sign.

Variables storing program names often store lists of programs. These lists are used during different compilation steps. We might divide programs into source code files, header files and already-compiled (object) files. The variable PROGRAM_NAME might be used to store an English description of the program being installed. The source files will be C/C++ programs, usually ending with either a .c or .cpp extension. Header files use a .h extension and object files a .o extension. We might find, for instance, the two variable definitions shown in the bottom of Figure 11.13 to define the source and object files.

After variables are defined, we then find the various sections of the makefile. Most sections are named with the terms that make will respond to, such as an install section and a clean section. One section will be given the name of the software, or possibly defined as $(PROGRAM_NAME). This will be the set of instructions executed by make when issued without a section name.

We see a partial example of the sections portion of a makefile in Figure 11.14. We assume the variables from Figure 11.13 had been defined earlier in this file. If the command provided is make all, then the all section indicates that the *name* section should be invoked (where *name* is the name of the software like apg). The *name* section issues the command ${CC} which, as defined in Figure 11.13, invokes gcc. It uses ${FLAGS}, or -Wall, and then compiles the files listed under ${SOURCES} using the ${LIBS} and ${LIBM} options, delivering as a result an executable program named ${PROGRAM_NAME} (the -o option in gcc specifies the executable file's name).

```
CC = gcc
FLAGS = -Wall
LIBS = -lcrypt
LIBM = -lm

INSTALL_PREFIX = /usr/local
BIN_DIR = /bin
MAN_DIR = /man/man1
BIN_DIR_D = /sbin
--------------------------------------------
SOURCES = file1.cpp file2.cpp file3.cpp
OBJECTS = file4.o file5.o file6.o
```

FIGURE 11.13 makefile macros (variables).

```
all:  name

name:
     ${CC} ${FLAGS} -o ${PROGRAM_NAME} ${SOURCES}
               ${LIBS} ${LIBM}

install:
     mv ${PROGRAM_NAME} /usr/bin

clean:
     rm -f ${PROGRAM_NAME} ${OBJECTS} *core*
```

FIGURE 11.14 Example makefile.

The install section simply moves the executable to the directory denoted by $INSTALL_PREFIX. Note that we would not need to be root to issue the ./configure and make steps, but make install will likely fail if we are not root because our normal user account would not have write access to /usr/bin. The clean section removes any object files, executable and core files that might have been generated in a previous attempt to build the program.

Makefiles can be far more complex than this brief example. We limit our discussion to what we have already covered. Interested readers should consult an online manual on makefiles such as the one at https://www.gnu.org/software/make/.

11.6.5 The make install Step

We noted in the last subsection that the makefile may contain an install section. If so, we are not done with our open-source installation. Once the make step has completed, we run make install. The install section is primarily comprised of Linux operations to move the files to their destination. Figure 11.14 showed the absolute minimal install section. It is likely that more components need to be moved and it is possible that they will be moved to differing destination directories.

The destination locations are stored in the previously defined variables, which are either default values or are values specified when issuing the ./configure command. The default destination locations tend to place files in numerous locations. We might find, for instance, the executable program placed in /usr/bin as /usr/bin/*name*, service programs placed in /usr/sbin/*name* (or perhaps /usr/sbin/*named*) and documentation in /usr/share/man. Many complex pieces of software have configuration files. It is likely that such a file will be placed somewhere in /etc, perhaps as /etc/*name*.conf or /etc/*name*.d/*name*.conf. These destination locations will be overridden if we had specified our own destinations individually (e.g., with --bin-dir=) or using --prefix=.

Instead of using mv or cp instructions, the make install section will probably use the Linux install command. This command has the ability to not only move files but also change the permissions of those files. In this way, the files generated by the make command are not limited to the permissions of the gcc compiler (or the user who issued the make command(s)). Nor will the author of the makefile have to include chmod commands. We might see an instruction like install -m 0755 ${PROGRAM_NAME} ${BIN_DIR}/${PROGRAM_NAME}. This instruction causes the compiled program to be moved to the location specified as the destination directory changing the permissions to 755.

Let's consider a more complex installation section than the one presented in Figure 11.14. Assume the make command resulted in the creation of an executable to be placed in $BIN_DIR, a script file to control the executable to be placed in $SCRIPT_DIR, a man page to be placed in $MAN_DIR, a configuration file to be placed in $INSTALL_PREFIX/conf/ and a group of icons (.png files) to be placed in $INSTALL_PREFIX/icons/. The install section of the

```
test -d $INSTALL_PREFIX || mkdir $INSTALL_PREFIX
test -d $INSTALL_PREFIX/conf || mkdir $INSTALL_PREFIX/conf
test -d $INSTALL_PREFIX/icons ||
         mkdir $INSTALL_PREFIX/icons
install -m 0700 a.out $BIN_DIR/$PROGRAM_NAME
install -m 0700 $PROGRAM_NAME.sh $SCRIPT/$PROGRAM_NAME
install -m 664 $PROGRAM_NAME.man $MAN_DIR
install -m 644 $PROGRAM_NAME.conf $INSTALL_PREFIX/conf
install -m 644 *.png $INSTALL_PREFIX/icons
```

FIGURE 11.15 Example `install` section for a `makefile`.

`makefile` might then look like the code shown in Figure 11.15. The `test` commands are used to determine if a given directory exists and if not, create it. This is necessary before we attempt to move files into those directories via the `install` commands. Note that `a.out` is the name of the executable file created by `gcc` if the name is not overridden, and this file is being renamed to the value stored in `$PROGRAM_NAME`.

The last step in the installation process, if desired, is to issue `make clean`. We would do this if either an error arose during compilation or installation, or to finalize the process so that any temporary files created during this process can be cleaned up (deleted). If the `makefile` does not have a specific `clean` section, then we might have to perform the removal operation by hand.

If there is a `tar` section in the `makefile`, we might issue `make tar` to automatically package up the important files produced during the compilation process. We could then delete the entire set of files used to compile and install the software. We would `cd` up one directory from where we used the `make` commands and type `rm -rf` *`dir`* as in `rm -rf somepackage-3.1.8`. This will remove the entire directory and all of its contents. We might either keep the original `tar.gz` file or take the files created by `make tar` and move it/them to some location to archive the installation package. In this way, should we ever need to reinstall the package, we can bypass the original download. Of course, by the time we might need to reinstall, there could be newer versions available for us to download.

SECTION ACTIVITIES

1. Search `sourceforge.net` for ten open-source software titles that might be of interest to you. Of those you found, how many have executable installation programs available? Of those, which are available for your Linux version?
2. We noted github but didn't talk much about it. Research github and compare its use to that of sourceforge. How do they differ? Which of these sites would you more commonly look at to install open-source software?
3. Find a piece of open-source software and try to install it. List any issues you had and share it with others to see if they have had similar issues.

11.7 THE gcc COMPILER

This optional section is provided for readers who are interested in learning more about the GNU C/C++ compiler (`gcc`). `gcc` was originally a C compiler, but later versions were updated to compile C and C++ programs.

`gcc` is an essential tool in Linux because most open-source software is written in C or C++. Although we can install a good deal of Linux software from executable code, to modify software we will need to obtain the source code, modify it and compile it. Thus, we will need `gcc`. As noted in the last section, if `gcc` is not already installed, we can install it easily using `dnf/apt-get`.

To use gcc, we will need to have some C or C++ source code. The gcc program then allows us to perform preprocessing, compilation, assembly and linking of the gcc files. Compilation is the process of converting source code of a program into executable code. Compilation has multiple, distinct phases. Compilers handle these in order and usually if one phase cannot be completed because of an error, the compiler abandons the attempt and provides the programmer with syntax error messages.

11.7.1 Preprocessing

In C/C++, *preprocessing* occurs before compilation. This step involves the compiler executing *compiler directives*. These instructions are indicated with a pound (#) symbol. Collectively, compiler directives are placed at the top of a file prior to any C/C++ code. They can be placed in .c, .cpp or .h files. Table 11.11 describes some of the more common compiler directives. Note that the second example for #define in Table 11.11 should appear on one line but is placed on two lines due to space restriction.

11.7.2 Lexical Analysis and Syntactic Parsing

The first step in the compilation process is *lexical analysis*, whereby the program is broken into component parts. These parts are referred to as *lexemes*. Each lexeme is then categorized using a *token*. In English, we have tokens that describe the grammatical category of words, such as "noun", "verb" and "adjective". In C/C++, tokens include "identifier", "literal", "for-operator", "semicolon", "arithmetic-operator" and so forth. The "for-operator" is literally the word for, used to denote the beginning of a for loop. An arithmetic-operator is any legal arithmetic operator in C/C++ such as +, -, *, /, % (mod), << (shift) or & (bitwise AND). The token literal may be more precisely defined as string-literal, char-literal, int-literal and float-literal. A literal is a specific value such as 0, 1.234 or "Frank", hard-coded in the source code. The identifier token is applied to named entities, which consist of variables (and constants), functions/methods, classes and user-defined types.

Lexical analysis may result in errors if any particular lexeme does not have a corresponding category. For instance, words that do not match any reserved word and yet are not identifiers result

TABLE 11.11
Common C/C++ Compiler Directives

Directive	Meaning	Example
#define	Macro substitution definition to replace a given string with another string.	#define MAX 1000 #define getmax(a,b) ((a>b)?(a):(b))
#include	Include specified header file's contents into this file.	#include <stdio.h> #include "myheader.h"
#ifdef... #endif	If a parameter has been defined, perform the following task.	#ifdef MAX int a[MAX]; #endif
#ifndef...#endif	If a parameter is not defined, perform the following task.	#ifndef MAX #define MAX 1000 #endif
#if... #else... #endif	Test given parameter and perform associated action(s) based on results; #else clause is optional; #elif clauses are also permitted.	#if MAX>500 int a[500]; #else int a[MAX]; #endif

in lexical errors. Two examples are For (capitalization) and whle (misspelled). Other lexical errors might arise if some punctuation is used that is not recognized in C/C++. Should a lexical error be found, compilation ceases and all such errors are reported. Otherwise, compilation continues.

With lexemes identified and labeled, *syntactic parsing* groups lexemes together into higher-level categories. These categories are not necessarily full instructions but identifiable constructs within instructions. An assignment statement, for instance, will consist of an identifier, the equal-operator and an expression. The expression's form can vary based on the type of identifier (it might be a string expression, an arithmetic expression, a Boolean expression, etc.). One example is an identifier, an arithmetic-operator and an int-literal, as found in the expression x + 5. Thus, an assignment-statement might be composed of identifier equal-operator expression where the expression is composed of identifier arithmetic-operator int-literal.

Figure 11.16 provides an example C if-statement which we will examine to better understand syntactic analysis. The if-statement consists of if-operator (the word if), condition (in parentheses) and block. The block itself is decomposed into open-block-operator, assignment-statement, assignment-statement and close-block-operator. The first assignment statement consists of identifier, assignment-operator, arithmetic-expression and semicolon-operator. The arithmetic-expression is identifier, arithmetic-operator and float-value. The second assignment statement uses the postfix increment and might be categorized as identifier, postfix-operator and semicolon.

The compiler creates what is known as a *parse tree* to represent the syntactic structure of the code. Figure 11.17 illustrates the parse tree for the C code from Figure 11.16. Creation of the parse

```
if(x > y - z) {
      x = x + 1.5;
      z++;
}
```

FIGURE 11.16 Example C if statement to demonstrate parsing.

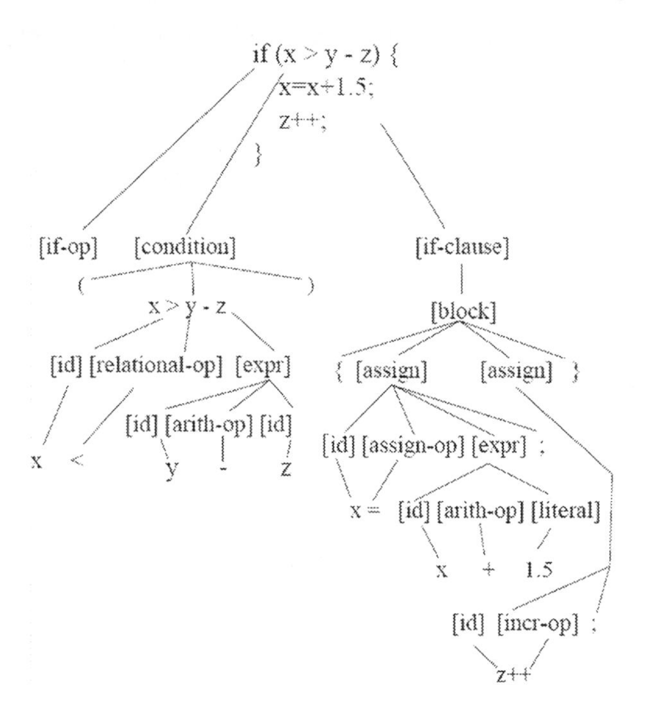

FIGURE 11.17 Sample parse tree for example C if statement.

tree requires analyzing the code recursively by continuing to group together tokens into larger categories. For instance, `x + 1.5;` makes up an arithmetic expression which itself is part of the right-hand side of an assignment statement which is part of a block which is the if clause which is part of the if-statement.

Syntactic parsing may discover errors and so parsing would be aborted at that point with all such syntax errors reported. Slight variations of the code in Figure 11.16 could result in errors such as omitting the first semicolon. Such an omission would lead the compiler to view the if clause as being `x = x + 1.5 z++;` which is lacking an arithmetic operator between 1.5 and `z++`. Alternatively, omitting one of the plus signs in `z++;` would similarly be flagged as an error. Another error would arise if either the open or close parenthesis or curly brace was omitted.

11.7.3 Semantic Analysis, Compilation and Optimization

Semantic analysis is next performed. This step ensures that instructions are being used appropriately based on the context of the instruction. For instance, does the comparison in the `if` statement compare items that can be compared? This will be the case if the operator applies to the data types of the two values being compared and that the data types are compatible with each other. Two examples whereby this `if` statement's condition would cause semantic errors are comparing strings using < or > and subtracting a string from a string. That is, should `x`, `y` or `z` be a string, then a semantic error will have been found.

The `if` statement's block could also have semantic errors. For instance, if `x` is an int variable, then there is an error because we cannot add 1.5 to an int and store the result in an int. Or, if `x` is an array or string, the error arises because we cannot add 1.5 to a string or an entire array. Similarly, if `z` is a type that cannot be incremented through ++, it is another semantic error.

While lexical and syntactic errors are misuses of the programming language itself, semantic errors are *misapplications* of the programming language. Misuse means that the syntax is incorrect while misapplication means that the statement is syntactically correct but used in an inappropriate way or context.

If the semantic analysis completes without error, the code can now be translated. Typically, as an intermediate stage, the code is first converted to assembly language; thus, we have an assembly component to compilation. Next, the assembly code is converted by an assembler within `gcc` to machine language code (executable code).

Even here, `gcc` is not finished. Optionally, we can ask `gcc` to perform optimization on the code. *Optimization* asks the compiler to rearrange the machine code in such a way that the instructions can take the most advantage of the parallel hardware available. Optimization may require several additional passes through the code. The result of compilation (assuming no errors arose) is an executable program (or multiple executable programs).

11.7.4 Linking

Earlier in this chapter, we examined the role of the library file and the need to track down dependencies for a successful installation. The C/C++ programming languages use library files extensively. This allows us to utilize pieces of precompiled code without having to write our own versions. Libraries in both C and C++ are extensive and include such necessary functions as input and output operations, string and character operations, random number generation, dynamic memory allocation functions, math functions and type conversion operations to name just a few.

In order to utilize a function written in an external file, we must include the function's *prototype*. This is a definition of the function, similar to a function header. For instance, the function prototype for the `printf` output function is `int printf(const char *format, ...);` We can interpret this particular prototype as defining the function `printf` which returns an `int` value and expects to receive at least one parameter, a char pointer (string). The ... indicates optional parameters may also be passed to the function.

To simplify the use of library functions, *header files* collect together information that the compiler will need in order to use library functions. A header file will include at a minimum the function prototypes of the functions defined in the given library. A header file may also include definitions for constants and data types. Header files may include their own precompiler directives, particularly #include and #define statements to include yet other header files and define constants.

If we are to place such definitions into one or more header files, we then need to indicate that our C program needs to include that or those header file(s). Header files are labeled with a .h extension. In our program, we provide statements of the form #include <somelibrary.h> and #include "myheader.h". The use of <> is for header files that are part of standard C libraries while those indicated in "" are our own header files.

If our C code uses library file functions, then we require another step in the compilation process known as *linking*. Linking can take place either during compilation, in which case it is known as *static linking*, or at run-time, in which case it is known as *dynamic linking*. Linking takes already-compiled functions and links the components together.

If linking is done statically, then our program will be completely translated into an executable during the compilation process. The advantage here is that the execution of the program is not slowed by run-time linking. Our program can also be completely optimized at compile time. The advantage of dynamic linking is that there are some, perhaps many, functions that a program might never use. Why link to those functions unnecessarily?

Another process is called *loading*. At the time the program starts, its executable is loaded into memory. If linking was used, then there is a question of when those linked functions are loaded. If statically linked, the executable is the entire program, including linked functions, and so loading takes place all at one time. If dynamic linking is being used, then external functions are linked only if they are called upon and loaded at that time. This process is called *dynamic linking and loading* and gives rise to .dll files in Windows. Dynamic linking and loading requires run-time overhead, slowing down the program's execution.

Figure 11.18 illustrates the benefit of using dynamic linking and loading. Here, the code includes several function calls (foo1, foo2, foo3, foo4, foo5). These functions are only invoked if x is less than 0. If this condition is rarely true, it is better to dynamically link and load these five functions because statically linking and loading them may be a waste of time and effort.

In C/C++, functions default to being dynamically linked but this can be overridden by using the reserved word static. When compiling our program, statically linked files use the .a extension while dynamically linked files use the .o extension. As noted here, linking is a separate compilation step. In the next subsection, we examine gcc specifically and see that we can include linking with compilation or perform linking separately.

```
void foo(int x) {
   // other code goes here
     if (x < 0) {
          foo1();
          foo2();
          foo3();
          foo4();
          foo5();
     }
     // other code may go here
}
```

FIGURE 11.18 Example benefitted by dynamic linking and loading.

11.7.5 Using gcc

The gcc command can range from simple to convoluted depending upon the operations that the user wishes to perform. At its simplest, we specify gcc [*compilation-option*] *filename*. Compilation options are one of -c, -S or -E where -c performs compilation or assembly but no linking, -S performs compilation but no assembly and -E forces the compiler to stop after preprocessing. If no compilation option is given, then compilation performs preprocessing, compilation, linking and assembly combined.

By using -c, the output produced by the compiler is an object file (with a .o extension). We would use -c to create a library file that other programs might link to. By itself, the .o file is not executable and can only be used by the compiler to produce other compiled programs. Without the -c option, the default is to create an executable file named a.out in the current directory. To alter the output file name, add the option -o *filename*. Thus, gcc foo.c produces a.out while gcc foo.c -o foo produces foo.

There are a number of different options available to specify the type of language that gcc is to compile. These include -ansi for ANSI (standard) C, adding -std=*standard* to specify the standardized version desired, and -x *language* where *language* specifies which language is included with the C code. It is possible, for instance, to include Ada or FORTRAN code in a C program (among other languages).

Although not important from gcc's point of view, there are a number of naming conventions used for file extensions. The .c and .cpp extensions are typical for C and C++ source code respectively. A .i or .ii file contains C and C++ source code that should not be preprocessed. The .m extension is used with objective-C source code. The .h and .o files are C/C++/Objective-C header and object files, respectively. There are extensions for FORTRAN source code including .f, .for and .ftn, while .f90 and .f95 represent FORTRAN 90 and FORTRAN 95, respectively. Ada code is placed in .ads or .adb files. Assembly code is commonly placed in .s or .S files.

As gcc is expected to compile very large files, compiler warnings can be crucial in discovering potential problems or flaws with the source code. There are a great number of different warning options available. Each warning option is preceded by -W. These options include the aforementioned -Wall which informs the compiler to use all warning options. More specific options include -Wcast-align, -Wfatal-errors, -Wno-overflow, -Wnonnull, -Wpointer-arith, -Wsign-conversion, -Wtype-limits, -Wunused-value and -Wunused-variable to mention but a few. Similarly, there are a number of debugging options available. These are best explored by programmers who are looking for specific debugging assistance.

gcc provides a large number of optimization choices. These are specified using -O options and include -O, -O1, -O2, -O3, -Os, -O0, -Og, -Ofast. Aside from the general optimizations, we can also specify a machine-dependent option to indicate the intended destination architecture. Such architectures include mainframe computers produced by DEC, Intel-based architectures like i386 and x86-64, and ARM processors often found in handheld devices. Table 11.12 provides example gcc commands with explanations. All commands should be on a single line.

One last comment about gcc. There are alternate programs available that also compile C/C++ programs. First is g++, a variant of gcc which will treat any C file as a C++ file instead. Although it is a subtle difference, many C++ programmers prefer to use g++ as it is specifically tuned to compile C++ rather than C. This is especially true if the C++ program links to the standard libraries in C++ (which gcc is not capable of handling). GNAT is a version of gcc which specifically will compile Ada programs. While gcc can handle multiple languages, GNAT is specific to Ada. There is also gdb, the GNUs Debugger, which is not a replacement for gcc but an addition to gcc.

TABLE 11.12
Example gcc Commands

Command	Explanation
gcc -Wall hello.c -o helloworld	Compile the program hello.c into the executable helloworld using the -Wall warning option.
gcc -Wall -O -mpower hello.c -o helloworld	Same as above but optimize executable for the PowerPC processor.
gcc -Wall hello1.c hello2.c hello3.c -o helloworld	Compile three different .c files into one executable file called helloworld.
gcc -Wall -g hello.c -o helloworld	Same as the first example except that this version includes debugging help.
gcc -Wall -I ../lib hello.c -o helloworld	Same as the first example except that hello.c uses header files found in the directory ../lib.
gcc -c hello.c -o hello.o	Compile hello.c into the object file hello.o rather than an executable; the object file could then be used as a library for other programs.

SECTION ACTIVITIES

1. Download and install gcc (use dnf/yum or apt-get). Write a hello world program (see below) saved as hello.c. Try to compile it. List any difficulties you had.

```
#include <stdio.h>
int main() {
    printf("hello world!\n\n");
    return 0;
}
```

2. As an IT person, should you learn C? Why do you feel that way? If you learn C and are a Windows or Mac user, is there any reason to learn gcc? Why or why not? Consult the IT people you know to see how many of them know C or some other high-level language and see how much programming they do as part of their job.

11.8 SOFTWARE DOCUMENTATION

We noted early in this chapter that one aspect of software maintenance is the production of documentation. There are many forms of documentation including information intended for the system administrator and information intended for the user. In this section, we look at both forms and discuss how the system administrator might need to use the former and write the latter.

As system administrators, we will need to research the software we are installing to ensure that it fulfills the need it was selected for, that our system has adequate resources to support the software and to see if there are special installation instructions. If we are installing from a GUI, an installation executable, or dnf/apt-get, we may not need to explore installation any further. If we are installing using rpm/dpkg, we might want to explore the chain of dependencies before performing the installation. If we are installing from source code, we should explore the various help text files (e.g., README, INSTALLATION) that come with the software package.

After installation, we might need to configure the software. The Apache webserver, Squid proxy server and BIND DNS name server are examples of software that require specific types of configuration instructions. Software that requires adjustments to configuration usually come with instructions for modifying the configuration file(s). These instructions might be bundled with the software although it is also common to find such instructions on the project's website (or alternatively, in books that either come with the software or that you can purchase). Additionally, configuration files often are commented to explain how to modify the files based on your needs. In many cases, Linux offers man pages for configuration files as well.

Once installed and configured, the administrator may need further instruction on how to maintain the software. Those developers of the software product may have ideas for the frequency of updates or specific instructions that need to be performed prior to an update. For instance, it would not be uncommon to require that users discontinue accessing the software during the update. Another maintenance task is *bug reporting* whereby we, as the system administrator, may be tasked with compiling issues that arise while our users use the software and submit those issues.

There may also be documentation available to support the software's use. If not, now it falls on us, the system administrators, to produce this documentation. There are three forms of such documentation. The first are instructions for how to run the software. In our system, we might need users to set up environment variables, have a separate software-specific login or establish other settings before running the software. We would learn about the settings needed as we prepare the software for general use. It is best to document our work so that we can return to it later to produce the user's manual for our users.

The second form is a true *user's manual* to explain how to use the software. This might include instructions to operate the software's features. It may be a step-by-step document showing how to create, modify and save artifacts. Likely, we would include screen captures to illustrate how users perform the various actions. Documentation may only be required for more advanced operations of the software depending on the users' familiarity with this or related software titles.

The third and final form is errata. While we might compile such problems for bug reports, we would also collect the issues to warn other users. For instance, if a user has discovered a sequence of operations causes the software to crash then it might be our responsibility to report this to all other users so that they do not have the same issue.

In most cases, the system administrator is not tasked with producing the user's manual. One should be available as any good software product should have such documentation available. However, we will likely be tasked with the first and third form of documentation if necessary. And of course, if the system administrator learns the software to produce documentation, this makes the administrator a go-to person for questions from the users of the software. To write useful documentation, learning how to perform technical report writing could be advantageous.

SECTION ACTIVITIES

1. Select any piece of software that you have on your Windows or Mac computer and run it. Examine the available form(s) of help. Now in your Linux computer, do the same. Compare the forms of help (e.g., GUI-based, takes you to a website, man pages). Which of the platforms offers the better form(s) of help?
2. Have you ever written a description of how to accomplish a task step-by-step? For this exercise, write one. It doesn't matter what the task is. It could be how to run a piece of software, how to cook a meal, how to wash clothes or how to pick a song to listen to. See how well you can do in writing your "guide to" solving the given task. If possible, add pictures to your description.

11.9 CHAPTER REVIEW

Concepts and terms introduced in this chapter:

- Algorithm – the description of a solution to a problem.
- Application software – the category of software that users use to solve a problem.
- Architecture – in the context of this chapter, architecture refers to the type of processor; `.rpm` and `.deb` files are executables and so are processor-specific.
- Assembly language – low-level programming language that is slightly more readable and usable than machine language.
- BAD error – an RPM error indicating that the digital signature does not match the expected signature.
- Checksum – a number added to a file used as an error-checking mechanism to ensure the software package has not been corrupted or tampered with.
- Compiler – language translation program that translates a program in a specific high-level language (e.g., C) to a specific machine language (e.g., Intel x86).
- Deb file – software package containing the executable code and auxiliary files needed to install and use a piece of software; `.deb` files are used by `dpkg` and `apt` in Debian Linux.
- Dependency – a piece of software which uses already-compiled functions stored in one or more library files; if the library is not installed, either it needs to be installed or the installation will fail.
- Dependency hell – the name given to a situation where you are forced to track down libraries that the software to be installed requires only to find that those libraries have dependencies of other libraries.
- Digital signature – a file included in software packages to ensure the integrity of the software product; a bad or missing signature might result in a failed installation attempt.
- Dynamic linking and loading – the process of linking to an already-compiled function and loading it only when it is needed at run-time.
- Freeware – the category of software that is freely available in an executable form but restricted by a license in terms of how the software can be used.
- Front-end manager – a program that serves as an interface between the user and another program.
- GNU C compiler (`gcc`) – the standard C/C++ compiler used in Linux; required to install open-source software.
- High-level language – a programming language in which instructions are at a higher-level and so easier to use; most programming uses high-level languages rather than low-level assembly or machine language.
- Language translation software – a program that takes a program in one language and creates an equivalent program in a different language; the most commonly used language translators are compilers and interpreters, but assemblers also fall into this category.
- Lexeme – a collection of characters that are grouped together to comprise a component of a program such as a numeric literal value, the name of a variable or a reserved word.
- Lexical analysis – a step in the compilation process whereby the compiler breaks the program into lexemes and categorizes each as a category (token).
- Library file – a collection of already-compiled functions that can be used by multiple programs.
- License – an agreement between a user and a programmer or company describing the rights assigned to the user to use the software; licenses had previously only come with purchased software but are now common for most software.

- Linking – the step in the compilation process whereby subroutine calls are linked to already-compiled subroutines; linking can be done at compile time (static) or while the program executes (dynamic).
- Loading – the process of loading the executable code of a program into memory; this is handled prior to the beginning of program execution unless dynamically linked functions are involved in which case those functions are loaded when first called.
- Machine language – the language native to a processor and the only language of which programs can run on a computer; machine language is written in binary and most instructions are simple/low-level operations.
- Makefile/makefile – a script called upon by the `make` command to compile and install open-source software.
- NOKEY error – an RPM error if there is no digital signature in the package to be installed.
- Open-source software – software produced by the open-source community and made available both for free and in source code format (most open-source software is also available for free in executable format).
- Optimization – the process of rearranging program instructions to take advantage of hardware to improve execution performance; optimization is an optional step of some compilers.
- Package manager – a program that helps facilitate the installation, upgrading and removal of software.
- Parse tree – the product of syntactic parsing; the parse tree shows the breakdown of the input (instruction, sentence) into tokens and then lexemes.
- Patch – a small program that serves to update an installed piece of software; patches usually fix errors or resolve security issues by making minor modifications to existing code.
- Preprocessing – a step in C/C++ program compilation whereby the compiler adds to or changes some of the source code; C preprocessing steps are indicated using a # before the command as in `#define` and `#include`.
- Program – a list of instructions written in some programming language that encodes an algorithm to solve a problem.
- Proprietary software – a category of software that is sold by the programmer (or company) to individuals and organizations for use; aside from software produced "in-house", proprietary software has the most restrictions in its licensing agreement.
- Public domain software – a category of software in which ownership of the software no longer exists and users are free to use the software with no restrictions.
- README file – when installing open-source software, a text file of this name is usually available to describe to the administrator or user how to install the software.
- Repository – a central location that contains a collection of available software; there are many Linux repositories of different kinds to support the different Linux versions.
- RPM File – a file used by the `rpm` (and `dnf`/`yum`) programs to install, remove and upgrade Red Hat Linux software.
- rpmfind.net – a website that stores `rpm` files.
- Semantic analysis – a step in the compilation process whereby instructions are examined to make sure they are being used correctly such as that the values being multiplied are numeric or that the result of a numeric operation is being stored in a numeric variable.
- Shareware – a category of software in which the software is available for free but with restrictions (like a limited number of uses or limited features); offered for free to entice users to purchase the full version of the software.
- Software – term to describe programs as opposed to physical devices (hardware); while a program is considered a singular item, software typically includes one or more executable programs, library files, help files data files and other auxiliary files.

- Software configuration – the step of establishing or modifying software settings to tailor how the software runs.
- Software downgrade – changing installed software from the current version to an older version.
- Software installation – the process of saving onto your system the executable software and all related files so that the software can run.
- Software maintenance – the steps involved in installing, configuring, troubleshooting and removing software.
- Software package – a bundle of the files necessary to perform the installation of software onto a computer.
- Software removal – the step of removing software from a system; removal involves removing the executable file(s) along with auxiliary files related to the software and deleting directories and making other modifications to the system.
- Software requirements – the hardware needed for a piece of software to run, or to run effectively, such as a minimum amount of DRAM, hard disk space, processor speed and type and needed I/O devices; requirements will also include a type of operating system.
- Software store – a website or GUI portal that presents to the user available software and software upgrades; in Linux, these GUI-based package managers are front-ends to one of `apt or dnf/yum`.
- Software update – installing and running a patch to fix bugs/security holes or modify existing features in software; purchasing software usually gives you access to free updates.
- Software upgrade – installing a new version of the software which may include new features/functionality; purchasing software may or may not give you access to free upgrades.
- Source code – program code written in a high-level language (e.g., C/C++, Java) which cannot be directly executed on a computer and must be translated into machine language by a compiler or interpreter; software developers work on the source code while most users only interact with the executable code.
- sourceforge.net – website storing thousands of primarily open-source software packages; the open-source software is available in both source code and executable forms.
- Syntactic parsing – a step in the compilation process whereby the tokens identified in lexical analysis are combined into meaningful instructions; this step creates a *parse tree*.
- System software – categorization of any software that is part of the operating system; this includes the kernel, services, device drivers, utilities and shells.
- Software version – a specific release of some software package, typically identified by a naming and numbering scheme such as *sometitle*.5.3 meaning the fifth major version and the third minor release of the fifth major version.
- Token – the category assigned to a lexeme such as `integer-literal`, `identifier` or `semicolon-operator`.
- Topological sorting – a graph problem whose solution is used by `apt` to determine the order that dependencies should be met.
- Update manager – in some Linux distributions, updating repositories is handled separately from the package manager; if so, it is the update manager's responsibility.
- User's manual – a document that describes how to use the software.

Linux commands and files covered in this chapter:

- apt – an umbrella term for the `apt` command line package manager tools for Debian Linux.
- apt-get – the specific program of `apt` used to install, upgrade, update and remove software; front-end to `dpkg`.
- aptitude – a text-based, menu-driven version of `apt`.

- configure – a script written using Linux instructions to generate a makefile as part of the process of installing open-source software.
- dnf – Dandefied yum (yum's replacement), front-end tool to rpm for software management; unlike rpm, dnf will track down and install dependent packages.
- dpkg – the Debian package manager program which operates on .deb files; as this is a more primitive tool that does not perform dependency handling, apt is preferred.
- gcc – the GNU's C/C++ compiler which is a requisite program for installing most open-source software.
- install – similar to mv but allows you to change a file's permissions before moving the it to its destination location.
- make – a command to execute the makefile script which is in charge of compiling large pieces of open-source software.
- make all – typically this command handles all of the parts of the make operation (make clean, make, make install and possibly make tar).
- make clean – when a make command fails, use make clean to clean up any temporary or partially created files before trying make again.
- make install – this portion of the make process is used to move the produced executable and supporting files to their destination directories.
- make tar – some makefiles contain a portion that allows you to package the result of the compilation process into a single file.
- rpm – the RedHat package manager command line program which can install software from an .rpm file.
- tar – the tape archive program used to package together and unpackage files and directories; while it is used for backup purposes, it is commonly used by the open-source community to create software packages for easy transport over the Internet.
- test – used to test if a directory or file already exists.

REVIEW QUESTIONS

1. What is an msi file and for what operating system(s) would you find it?
2. Examine Table 11.1. Given the expense of some of these titles, provide an argument for selecting an open-source or free version and an argument for selecting a commercial product.
3. Before installing software, what questions should you ask?
4. List three types of three requirements of software that you should know about your system before you attempt to install some software.
5. Make an argument for why already familiar software should be favored over unfamiliar software, particularly for an organization. Can you think of any drawbacks of basing a choice on familiarity? If so, what?
6. Compare an algorithm and a program.
7. What are the types of language translators?
8. What are the two categories of software?
9. The operating system is not just the kernel. List three other components of the Linux operating system.
10. What is the difference between the following?
    a. Freeware and shareware
    b. Freeware and open-source software
    c. Freeware and public domain software
11. Of the various classifications of software licenses listed in Table 11.3, which form is the only one where software is available in a form that a person can modify the code?

12. To install software, do you have to do anything more than store the executable program somewhere? If so, what other steps are required?
13. List three types of documentation a system administrator might be required to develop for employees.
14. Compare a software update and a software upgrade. Which is the more significant form of installation? A patch is associated with which of the two?
15. Open the Software GUI in your Linux distribution. Step through all of the categories to see the list of available titles and answer the following.
    a. Which category has the most titles in it?
    b. How many titles are available in all and how many are already installed?
16. What is a package dependency?
17. What is an `.so` file and what does one store?
18. Where are you likely to find most of the `.so` files of your Linux computer?
19. What is the relationship between `dnf` and `rpm`? Between `apt` and `dpkg`?
20. A `.rpm` file has the name `mysoftware-3.1.5-2.i386.rpm`. What is the version number of this software package? What is the release number?
21. What does `noarch` mean when found in a `.rpm` file's title?
22. What is the difference between upgrading and freshening an `rpm` package?
23. What does the `--ignorearch` option mean when attempting to install or upgrade software in `rpm`?
24. What does the `--nodeps` option mean when attempting to install software in `rpm`?
25. You want to install a piece of software using `rpm` but do not want to install any documentation (e.g., man pages). What option should you add to your command?
26. What happens if, when using `rpm` to install a package, there are dependencies that are unmet?
27. What option can you add to an `rpm` instruction to attempt an installation even if there are unmet dependencies?
28. What information can you find at `rpmfind.net`?
29. What happens if, when using `rpm` to install a package, there is no digital signature?
30. What happens if, when using `rpm` to install a package, the digital signature is not recognized?
31. How can you test for an `rpm` file's digital signature before trying to install it?
32. What is the difference between the message `BAD` and `NOKEY` when reported by `rpm`?
33. What does the option `--force-depends` do when used with `dpkg`?
34. What do the options `-G` and `-E` do when used with `dpkg`?
35. What is a package state and a selection state in `dpkg`?
36. What is the relationship between `dnf` and `yum`?
37. What is a software repository? Of `apt`, `dnf`, `dpkg` and `rpm`, which maintain a list of repositories?
38. What does the instruction `dnf list *` do?
39. What does `dnf`'s `downgrade` option do?
40. What is the difference between `dnf install somesoftware` and `dnf -y install somesoftware`?
41. You specify `dnf upgrade`. What do you expect to happen?
42. Which specific command from the `apt` suite of tools do you use to install software?
43. The GUI-based `Synaptic` program is a front-end to which package manager program?
44. Why might you want to install open-source software from source code rather than an already created executable or installation program?
45. What website is a common location for software developers to post their open-source software?
46. If software is available via open source, is it likely to only be available in that format and not available as an already compiled executable to be installed by an installation program?

47. What program will you probably need to have in your Linux system to successfully install open-source software from its source code?
48. You have downloaded the file `mysoftware.tar.gz`. What `tar` command should you issue to untar this file?
49. Having untarred an open-source software package, which file should you review to see installation instructions and other comments that might help you install the software?
50. Examine Figure 11.12. What are the `.c` files? The `.h` files? `doc` is a subdirectory, what do you expect to find there?
51. Why might you want to run `./configure` when installing open-source software?
52. What types of options might you specify when running the configure script?
53. You want to install the components of a piece of software under `/usr/local/piece-of-software`. What configure option would you use?
54. What is the difference between `make` and `makefile`?
55. When might you run `make clean`?
56. What is the difference between `make` and `make install`?
57. What does the `test` instruction do and why might we expect to find it used in a `makefile`'s install section?
58. How does the `install` instruction differ from `mv`?
59. Explain what the following C compiler directives will do.

```
#ifndef MAX
#define MAX 1000
#endif
```

60. Draw a parse tree for the following C instruction.

```
x=y*z++;
```

61. Draw a parse tree for the following C instruction:

```
while(x>=0)
      x=x-y;
```

62. Assume in the following C code that x and y are both int values and z is an int array. For each of the following instructions, will they cause a lexical error, syntactic parsing error, semantic parsing error or no error?
    a. `if(z[x]>z[y]) x++; else y++;`
    b. `if(x>y) x++; else y+;`
    c. `if(x>y) x++ else y++;`
    d. `if(x>y) x=x+1.5;`
63. What does optimization mean with respect to compilation?
64. In `gcc`, what does `-Wall` mean?
65. Why might you use the `-c` option in `gcc` so that compilation does not include linking?
66. What is the difference between using the `-o` option in `gcc` and not using it?
67. If you were asked to produce documentation on how to use a piece of installed software, what information might you include?

12 Maintaining and Troubleshooting Linux

This chapter's learning objectives are to be able to:

- Explain the importance of backups and how to implement a backup strategy
- Differentiate between RAID levels and be able to compute parity bits and parity bytes
- Describe encryption/decryption algorithms and use Linux encryption software
- Schedule tasks using `at`, `batch` and `crontab`
- Select appropriate Linux performance tools and apply them to understand the state of the system
- Examine log files in order to determine what events took place
- Perform operating system troubleshooting

12.1 INTRODUCTION

So, you've successfully installed Linux, created accounts, mounted file systems, automated the initialization process, configured your network and installed software. There is nothing more to do and you can coast for the rest of your career, right? Wrong. These steps have established your Linux system, giving users the ability to log in and use your Linux system. But this is only half of your job. Now you have to *maintain* the system.

Maintaining a computer system requires that we examine the usage of system resources to ensure that our users have adequate resources. We have to monitor for external threats over the network. We have to keep our software updated. We need to ensure the integrity of the data files in the system by protecting files through encryption and safeguarding files through backups. We should examine log files at regular intervals to explore if something has happened or gone wrong so that we can correct the issue and/or prevent it from happening again.

We have already examined how to update software, so in this chapter we examine four other activities that all relate to system maintenance and troubleshooting. First, we look at maintaining file system integrity. This examination explores how to perform backups, what RAID can do for us and how to encrypt files. We then turn to task scheduling. We saw briefly the services `atd` and `crond` in Chapter 9, but here we look at how to use these in more detail. We next explore numerous Linux tools for system monitoring, some of which we introduced in previous chapters. Finally, we turn our attention to log files and the content we can expect to find in them. We put all of these ideas together by covering some troubleshooting scenarios: problems, ways to discover the cause(s) and solutions.

SECTION ACTIVITIES

1. Have you troubleshooted your own PC? If so, what did you have to do? In Windows, how often do you use the Task Manager to see what's going on in your computer and possibly kill processes? Have you used other OS utilities such as the Device Manager and Computer Manager? Make a list of some of the tools you have used. As you work through this chapter, compare the Linux tools to similar tools in Windows (or Mac).

DOI: 10.1201/9781003203322-12

2. We troubleshoot all the time. Why did the light not turn on? Why does the car not start? Why is the milk suddenly spoiled? There is mud in the entryway, who left it? Think of three situations where you had to troubleshoot (other than of your computer) and list what you did to solve the problem in each case.

12.2 FILE SYSTEM INTEGRITY: BACKUPS, RAID AND ENCRYPTION

In Chapter 8, we explored file system administration. Although we discussed encryption and backups in that chapter, we did not go into detail. In this section we look at three forms of ensuring the integrity of a file system: through timely backups and a sound backup strategy, using some form of RAID storage and encrypting the files of the file system.

12.2.1 Backups: Why, How and When

Although this may sound unintuitive, it is the hard disk that is both the most important component of our computer and the one most liable to failure. The reason that the hard disk is so important is that it is where we store our files. Our files might store anything from important work to school assignments to financial records to sensitive personal data to photographs/keepsakes. The value of the information in the files might range from days' worth of effort to priceless.

Why is the hard disk more valuable than the CPU or memory (components necessary to run our programs)? Quite simply because the CPU and memory, along with most other hardware in the computer are replaceable. Should one fail, although it may cost money, it can be replaced. The hard disk can also be replaced, but the hard disk's contents cannot.

Why is the hard disk most liable to fail? Surprisingly, it is one of the only components of our computer with moving parts. The disk platters spin and the read/write heads move across the surface of those platters. A motor drives the entire unit. As the disk platters are permanently sealed inside the disk drive unit, should the drive fail then the information on those platters becomes inaccessible.

For these reasons, it is important to back up the contents of a computer's hard disk onto some other form of storage. Most people back up their hard disk to another hard disk, either a local but external hard drive or to the cloud, accessed over the Internet.

The advantage of the former approach is that external hard disks are readily available and inexpensive. Additionally, any confidential data never leaves the user's possession. Disadvantages are that external hard disks are portable and so could be stolen (although then so could a person's computer) and because people have a tendency to keep the external hard disks near their computer; should the computer be damaged by fire, flood or similar disaster, it is likely that the external hard disk would be similarly damaged.

The advantages of cloud storage are that the cloud provides remote access so that damage to both a computer and the cloud storage is unlikely, and that the cloud company is charged with ensuring its security. Remote access also makes the files available anywhere there is Internet access, as opposed to an external hard disk that can only be accessed when it is in the same place as the computer. Security provided by the cloud company should ensure that no one can access the storage and that the storage is backed up in a timely fashion. The disadvantages of the cloud are that confidential data is now being held by a third party and that it costs money to store data in the cloud (although many cloud companies offer small amounts of storage for free).

For the sake of this subsection, we will concentrate solely on maintaining our own backups. The destination might be magnetic tape, another partition of our computer, or a remotely mounted file server. We concentrate on the how and when of backups, starting with when.

In Chapter 8, we examined possible partitions for our system. If we explicitly created our own partitions, rather than using an LVM, it is easier to back up our content, partition-by-partition.

TABLE 12.1
Partitions and Backups

Partition	Usage	Frequency of backup
/	Core components of the operating system outside of the kernel; includes directories of /etc, /root, /bin, /lib, /sbin, possibly /usr.	It might be helpful to back up any updated configuration files in /etc and important data files in /root, but otherwise backing up / can be done infrequently, perhaps when system updates are made, and possibly never as the system can be reinstalled if needed.
/boot	Boot sector and Linux kernel.	Only if changes were made for instance to grub configuration or a new initramfs created.
/home	All user files.	Depends on number of users but at least weekly if not more frequently.
/var	System data files including log files.	Depends on system load but probably at least once a week to ensure log files and other data file are backed up.
/usr	All software and support files (e.g., configuration files).	Only when updates are made to the software or support files, if ever.
swap	Virtual memory, refreshes with each boot.	Never as this partition does not contain permanent data storage.
Virtual file systems	Various system files stored in memory.	Never as these file systems are stored in memory.

Table 12.1 describes the common partitions, their uses and an indication of the frequency for performing backups. The partition storing /home is by far the most important to back up with the /var partition being the second most important to back up.

We need to determine the frequency that each partition should be backed up: daily, weekly, monthly, less frequently or never. The /home directory will be the partition that requires the most frequent backups. If we are dealing with a single workstation used by only one or a few users, we might be able to perform weekly backups. If we have more users, say dozens, who use a single workstation, the frequency of backups is increased, possibly to daily backups. Since backing up /home will likely require unmounting it, we should perform our backups during off-hours.

If we have many workstations, say one per user, then backing up the hard disks becomes more of a burden because we have to perform backups individually, perhaps by working on each computer or by remotely logging in to each computer. Centralizing all of the /home directories on a networked file server would be far easier. But this in itself would require more frequent backups because the amount of content to back up would be greater and change more often. Fortunately, backing up a single server is easier than backing up several or dozens of individual workstations.

The /var directory would be the next most common file system in need of backing up. As with /home, when /var is backed up, it is best to unmount it. We have the same issue when backing up /var as we would with individual /home directories, having to work on individual computers. While /home might be centrally mounted, it is more likely that each computer would have its own /var partition.

What content on /var needs to be backed up? Certainly, the log files will be significant as we might need to explore older logs to track down the root cause of issues. Log rotation ensures that there is adequate disk space available for logs but log rotation also deletes older log files, so we need to backup these older files before we might lose them. If we have implemented mail, then it becomes critical to back up the /mail directory in /var. This directory might grow rapidly but we can institute disk quotas on user files to force users to move a majority of their emails to their /home directories or delete them. Alternatively, email might be implemented on a separate server.

Many of the other /var directories are not as critical although this will vary based on just what software we are running. The Apache webserver, for instance, has a default subdirectory under /var to store the website. If placed there, it could be modified frequently such that it should be backed up frequently. Therefore, the frequency of backups for /var will have to be determined by the computer's usage, but backing up /var once per week might be a reasonable goal.

As noted in Table 12.1, the other partitions require backups less often, if at all. Although it is reasonable to think that / should be backed up frequently, it is likely that the only content that changes among the operating system files are configuration files (mostly in /etc) and anything the administrator places in /root. Thus, the root partition could be backed up only when significant changes have been made and we need only back up those changed files.

Once we've decided to perform backups, we need to explore how long a backup should be saved. Imagine that we back up the entire contents of our hard disk drive every day, storing its entire contents onto an external hard disk. Let's assume we only have one external hard disk so we reuse it daily. We modify a file a day later and then modify it again another day later. Now, two days after the file was first modified, we discover that the file needs to be restored to its original version. The problem is that the backup of the unmodified file is now deleted because we erased the backup of two days ago when performing a backup yesterday. It might be wise to not erase our backup for a new backup. This requires retaining multiple backups.

The problem with retaining multiple backups, in this scenario, is that each backup is of our full hard disk and that means we need one equivalently sized external hard disk for each backup we retain. That could be a lot of hard disks! It may not be practical for a user or a company to have several or dozens of backups per computer. Fortunately, this scenario is unrealistic for two reasons. As discussed above, we do not need to regularly back up the full hard disk, only one or two partitions (/home and /var). Additionally, we do not need to resort to creating a new backup every day. Instead, we can merely save those files that have been updated. This is the difference between a full backup and an incremental backup.

A *full backup*, as the name implies, is a backup of the entire contents of the partition. An *incremental backup* is a backup of only files that are newly created or modified since the last backup. Incremental backups are preferable because they take up less storage space and take less time to perform the backup. Full backups are still necessary from time to time, perhaps weekly if we are performing daily incremental backups.

Consider the following strategy where we back up /home to magnetic tape. Once per week, we perform a full backup onto tape 1. During that week, we perform incremental daily backups. As these are not nearly as sizeable, we can store several such incrementable backups on a single tape. We see in Table 12.2 a pattern of storing incrementable backups until the tape is full and then moving onto the next tape. Assuming that six days of incrementable backups will take up two tapes, we can see that we use three tapes per week. At the beginning of each week, we perform a full backup onto one tape. Over the next three days, we perform incrementable backups on a second tape. For the last three days of the week, we perform incremental backups on a third tape. Should we only have six total tapes, we would have to start reusing tapes beginning in week 3. Tapes actually store far more than a single hard disk and so it is likely that two or three full backups and several incremental backups (say 2 weeks' worth) could be stored on a single tape.

Let's look at how we restore files from the backups, referencing the tape numbers listed in Table 12.2. Imagine that a user's file, myfile1.txt, was backed up on tape 1 as part of the week 1 full backup. The file was modified at the end of the week and so was placed on an incremental backup (tape 3). During week 2, the file was again backed up during the full backup, and so now exists also on tape 4. The user modified this file early in the week and was also backed up during one of the nightly incremental backups onto tape 5. The user accidentally deletes myfile1.txt and wants it restored from backup. Of the four versions backed up onto tape, which version should we restore?

TABLE 12.2
Possible Daily Backup Strategy

Tape Number	Week	Usage
1	1	Full backup
2	1	Daily incremental backup
2	1	Daily incremental backup
2	1	Daily incremental backup
3	1	Daily incremental backup
3	1	Daily incremental backup
3	1	Daily incremental backup
4	2	Full backup
5	2	Daily incremental backup
5	2	Daily incremental backup
5	2	Daily incremental backup
6	2	Daily incremental backup
6	2	Daily incremental backup
6	2	Daily incremental backup
1	3	Full backup

We do not have to consider tapes 1 or 3 because tape 4 has the file in a full backup. So, the question is whether we should restore from tape 4 or tape 5. Keep in mind that we do not retain a list of all saved content so we will have to search tapes without knowing if or where an item is stored. But we do know that anything on tape 1 will also appear on tape 4 but perhaps as a newer version.

We do not want to restore the file from tape 4 because that version may be out of date. We do not know if the file has been modified, but if it has, it will be on tape 5 or 6 (or both). We work backward starting with tape 6. As we do not find the file, we move to tape 5. Here, we find a copy and so it is the most recent version and the one we restore. We did not have to search tape 4. Had the file not been modified during this week, we would not have found it on any incremental backup (tape 5 or 6) but would have found it on the week's full backup, tape 4.

With this example, it seems odd to retain 2 weeks' worth of full and incremental backups. Why do we even need tapes 1–3 once we have tapes 4–6? Assume that the user did not delete `myfile1.txt` but instead has come to realize that the modifications made over the past 2 weeks are incorrect and wants the file restored to its original version. Again, there are four different versions of the file on backup tapes. The modification made most recently is on tape 5. The previous modification was placed onto tape 3 and also appears on tape 4 during the week's full backup. The unmodified file would be found only on tape 1. Thus, retaining two full weeks' worth of backups allows us to restore the file. If the user knows when the first incorrect modification was made, we can move to the full backup tape prior to that time and so we can restore from tape 1 without having to search other tapes.

What if the file had been modified prior to the full backup onto tape 1? Now we have an issue because we have not retained enough backups. The question is how long will it take a user to discover they need to restore a previous version? In this example, it was within 2 weeks. What if a user modifies a file and discovers that the modification was incorrect but not for months? Should we retain backups going back months? This may not be practical. Yet, as we will explore shortly, we are likely to be able to retain more than 2 weeks' worth of backups if we are using magnetic tape.

Now let's consider that the file system failed completely and we need to restore the entire contents of the `/home` directory from backup. We need to work our way forward instead of backward through the tapes. We start with the most recent full backup (in this case, tape 4) and copy its contents into a newly created version of the /home partition. Next, we work through each incremental backup, replacing versions of files that were just restored with the newer versions. Restoration of an entire file system will likely take far more time than restoration of a single file. Why did we work our way forward? Had we started with the most recent incremental backup, we would wind up erasing one version of a file for an older version in other incremental backups or in the full backup.

We retain the contents of our hard disk (or at least /home) to create archives. While backups are kept so that we can restore files should damage or accident occur, *archives* are stored for *permanence* (or at least for years). The archive is a form of security to ensure an electronic copy of the organization's (or individual's) files exists in case we need to refer back to earlier files. For personal records, we might archive our taxes and expenses. A company might archive all transactions of a fiscal year. It is possible that a company might keep archival data indefinitely, years at least. Some companies retain data for decades.

Our example in Table 12.2 is not particularly realistic because incremental backups will likely not be very large. Additionally, if we store our backups onto tape, we should be able to store far more onto a single tape than the contents of a hard disk. Let's use some numbers to see how long it might take to fill up our backup medium.

Assume we are only backing up /home and that this partition is on a shared file server, comprising 1TB of storage. Let's further assume that no more than 10% of this partition changes in any one day. A daily incremental backup will be 100GB. A full backup plus six incremental backups would take up roughly 1.6TB of storage space for a week. If we were backing these up to a 1TB external hard disk, we could store one full backup or 10 incremental backups. If we were backing up to a 15TB tape, we could store over 9 weeks' worth of backup content on the one tape! Returning to our previous question of how long should we retain a backup, using tape might allow us to retain months of backups on just a few tapes.

Home computer users will likely not use magnetic tape for backups and so may need a more reasonable backup strategy to ensure that contents are backed up in a timely fashion but also that significant files are retained going back multiple versions. Organizations would be best served with magnetic tape because they provide a cost-effective way to back up a lot of content.

We shift now to the *how* of backups. The tar program has several benefits over other approaches. We can run tar without unmounting the file system being backed up, leaving it accessible during the backup. It is easy to use and the contents can be compressed during archiving.

tar is also capable of incremental backups although not as easily controlled as through dump. One way to perform an incremental backup is to specify in a separate file, *filename*, all of the files (and directories) that have been modified or created since the last backup using find -cnewer and -anewer. With this file created, we can then add -g *filename* or --listed-incremental=*filename* to our tar command so that the files archived are those listed in *filename*. There is also a --newer=*date* option available in tar.

Newer archiving programs are cpio and dump. Both cpio and tar have an advantage over dump in that we can select specific files or directories or both to archive. dump operates only on the level of file systems and so is unable to create an archive in a piecemeal fashion. On the other hand, dump provides an easy facility for incremental backups by using a *level*. We denote a full backup as level 0. Our next backup would be indicated as level 1. Any files created or modified since the level 0 backup are saved on the level 1 backup. We continue to increase the level with each new incremental backup. Our next full backup would again use level 0.

dump's partner, restore, lets us interactively find *and* restore specific files. So, while we perform a backup of all files in the file system (full backup) or those new or modified files (incremental backup), we can locate and restore any single files we need. cpio is in some ways a compromise between tar and dump in that it has a slightly improved incremental backup facility over tar while allowing us to save and restore individual files as well as file systems.

12.2.2 RAID for File System Integrity

When performing a backup, we may have to make the partition inaccessible. Certainly, if restoring files from a backup, those files are inaccessible until the restoration is complete. For a single user, the impact of downtime during backup and restoration may not be particularly significant. For a company, inaccessible data might be a loss in productivity, revenue and reputation. It is important in such cases that such files are always available.

It seems impossible that a file can always be accessible when that file may be damaged or destroyed. One way to mitigate the problem is to store not just the file but *redundancy* data. In doing so, a damaged file might still be accessible by restoring the inaccessible data from the portions of the file that are not damaged and redundancy data. This permits continuous access to the file without having to restore it from backup. RAID technology provides this form of accessibility.

RAID, redundant array of independent disks, provides a storage unit that consists not of one disk drive but several disk drives. Each drive can be accessed independently. Independent access can improve disk performance. By providing extra disks, we can store more data. We dedicate (or sacrifice) part of this additional space to store redundancy data.

The advantages of RAID are twofold. Because of redundancy, it is possible, even likely, that a data file can be accessed even when a portion of the file has been damaged. Because of having independent disk drives, if we are clever with how data is distributed across disk surfaces, we can improve the performance of disk access. Combined, these ideas constitute the R and I of RAID. The disadvantage of RAID is its expense because of the need for increased storage combined with the hardware needed to control the RAID storage device. To offset the increased expense, early forms of RAID offered inexpensive disk drive units in the RAID cabinet, and thus RAID could also stand for *redundant array of inexpensive disks*.

There are several different forms of RAID. These are known as *RAID levels*. The lowest level is RAID 0. As we increase levels we are not building upon the previous level. Some levels are related while others are not. There are seven commonly cited RAID levels (although not all are used) along with some hybrid levels. There are also proprietary RAID levels which we will not comment on. The seven RAID levels and two of the hybrid levels are described in Table 12.3. We concentrate on levels 0, 1, 3, 5 and 6 below.

We can break down the forms of redundancy into four categories. RAID 0 has no redundancy and so does not count. RAID 1 uses a *mirror*, duplicating all data from one set of disks onto another. RAID 2 uses Hamming Distance Codes. RAIDs 3–6 use parity bytes. RAID 6 uses both parity bytes and Reed-Solomon redundancy.

RAID 1's mirror has an obvious advantage in that it maintains a complete copy of the data, offering 100% redundancy. Perhaps less obvious is that with a mirror two disk reads can be performed simultaneously, one to the original and one to the mirror. As a write must be made to both the original and mirror simultaneously, a write cannot be accomplished simultaneously with a read or a second write. The disadvantage of RAID 1 is the expense as 50% of the storage space is dedicated to the mirror.

In Hamming Distance Codes, a byte of data is used to generate error detection and correction information. This information is encoded through four additional bits. Although this is not as expensive as RAID 1, it still requires sacrificing a good deal of storage space to redundancy. With these codes, we can detect and correct an error, or detect up to three errors. Unfortunately, should there be multiple errors, we cannot recover from them. The Hamming Distance Codes were proposed for RAID 2 but in practice are not used because Hamming Distance Codes take some time to compute, require more complex hardware and require more disk space than other forms of RAID (outside of RAID 1).

RAID levels 3–6 all use the parity byte to encode redundancy. This requires a brief explanation. A *parity bit* is a simple error detection mechanism. The bit is added to any byte of data to encode the evenness or oddness of the byte. Let's assume we are using even parity. We want to ensure that

TABLE 12.3
RAID Levels

Level	Description	Advantages/Disadvantages	Usage
0	Striping at block level, no redundancy	A: Improved disk performance over standard disk drive. D: No redundancy.	For superior disk performance without redundancy; there is an increased cost but no sacrificed storage space for redundancy.
1	Complete mirror	A: Provides 100% redundancy and improves disk access for parallel reads. D: Most costly form of RAID.	Safest form of RAID; best choice if cost is not a factor.
2	Striping at bit level, redundancy through Hamming codes	A: Fast access for single-disk operations. D: Hamming codes are time consuming to compute.	Not used in practice because of the complexity in using Hamming codes.
3	Striping at byte level, parity byte redundancy	A: Fast access for single-disk operation; compromise between expense and redundancy. D: All drives active for any single access so cannot accommodate parallel accesses.	Useful for single-user systems.
4	Striping at block level, single parity disk	A: Larger stripes accommodate parallel accesses (like RAID 0) while adding redundancy. D: Single parity disk is a bottleneck defeating advantage gained by striping.	Not used in practice because the parity disk is a bottleneck.
5	Striping at block level, parity distributed across disks	A: Same as 4. D: None.	Useful for multiuser systems (e.g. file servers).
6	Striping at block level, two forms of redundancy recorded (parity and Reed Solomon codes)	A: Same as 4 and 5 except that with double the redundancy data provides a greater degree of security. D: More expensive than RAIDs 3–5 because of the extra storage dedicated to redundancy.	Same as 5.
10	Striping at block level, complete mirror (RAID 0 and RAID 1 combined)	A: Same as 0 and 1 combined. D: Requires twice the disk drive capacity as RAID 1.	For file servers that require both parallel disk access and redundancy.
53	Extra disks to support RAID 3 and RAID 5 striping	A: Best overall access as one can access the RAID 3 or the RAID 5 set of disks. D: Requires more disks so is more expensive than 3 or 5.	Useful for multiuser systems (e.g. file servers).

the byte and the parity bit combined contain an even number of 1 bits. To compute the parity bit, we add a 0 if the byte currently has an even number of 1s and a 1 if the byte currently has an odd number of 1s. The result is that the nine bits combined will always contain an even number of 1s. Should the byte and parity bit be received somewhere in our computer such that there are an odd number of 1s then the receiving unit in the computer has detected an error and can ask that the nine bits be resent.

The parity bit is simple to compute and an error can be easily detected. Thus, this approach is far more attractive than the Hamming codes. Unfortunately, the parity bit does not let us figure out where the error arose, nor can we use it to handle multiple errors. Let's consider a brief example before we move on to the parity byte.

In Figure 12.1, we have an original datum (byte) of 11100101. With an odd number of 1 bits, the parity bit is computed as a 1. This gives us 11100101, 1. These nine bits are transmitted from one location in the computer to another. We see four versions of this nine-bit datum being received in

Original datum: 11100101 Parity bit: 1

Transmit 11100101 1

Receive 10100101 1	11100101 0	11000111 1	11101101 1
Result: error, resend	error, resend	error but undetected, do not resend	no error

FIGURE 12.1 Using the parity bit.

the figure. In the first, we get 10100101, 1, which has an odd number of 1 bits so an error arose. In the second, we receive 11100101, 0, an odd number of 1 bits so an error arose. In the third, the nine received bits have even parity and so it looks like the datum has no error. In fact, there are two errors, but this goes undetected. In the fourth example, we receive the nine bits correctly.

If you look carefully, the first of the errors presented in Figure 12.1 shows that one of the bits in the original byte changed. In the second example, it was the parity bit itself that had changed. In the third example, two bits of the original byte changed. Figure 12.2 demonstrates how to compute even parity using XORs.

The parity bit is a simple solution to detecting an error and used in all computers. The parity bit is only enough extra data to detect that one error arose but not enough to either correct it or detect multiple errors. Fortunately, as a single bit error is very unlikely to arise, two errors in the same 9-bit sequence arising are even more unlikely. The reason we can recover from a transmission error with the parity bit is that we ask the sending unit to resend the original datum. The error arose in transmission, not storage. To recover from a data storage error, we have to use a different approach.

A *parity byte* is computed by computing the parity bit of the same bit in a collection of bytes, bit-by-bit. For instance, if we have four bytes, we compute the parity bit of the collection of bytes' leftmost bit, and then the parity of bit for the second column, and then the third, and so forth until we have 8 parity bits. That group of 8 bits makes up our parity byte.

The number of bytes we start with can be any number from two bytes upward although if we have too many bytes, the chances of multiple errors become more likely. Figure 12.3 presents an example where we compute the parity byte of four bytes of data. In order to compute each parity bit, we XOR the bits of each column. For instance, the first column consists of 0, 1, 1 and 0. We would perform (0 XOR 1) XOR (1 XOR 0) which is 1 XOR 1 which is 0, thus the first column's parity bit is 0. The second column is (0 XOR 1) XOR (0 XOR 0) which is 1.

Given a byte: XOR pairs of bits, XOR those results, XOR those results
Result: Even parity bit

Example: 11100101

```
1 XOR 1   XOR   1 XOR 0   XOR   0 XOR 1   XOR   0 XOR 1

   0      XOR      1      XOR      1      XOR      1

          1                XOR             0

                            1
```

The even parity bit of 11100101 is 1

NOTE: to obtain the parity bit using odd parity, NOT the result

FIGURE 12.2 Computing an even parity bit.

Byte 1:	00000011
Byte 2:	11110010
Byte 3:	10101011
Byte 4:	00101111
Parity byte:	01110101

FIGURE 12.3 Computing a parity byte.

In RAIDs 3–6, we compute the parity byte of some number of bytes. The parity byte becomes our redundancy data. If an error arises such that one of the four bytes plus parity byte becomes inaccessible, we can recover from it. If we have five disk drives in our RAID unit, we might then have four bytes of data and the fifth being the parity, as demonstrated in Figure 12.3.

Let's see how we can use the parity byte. In Figure 12.4, we see our four bytes and the parity byte from Figure 12.3, but in this case the block storing byte 3 has become corrupted, at least in that location. As we cannot access the third byte of data, we must restore it from the other bytes and the parity byte. We can compute this missing byte by performing the same XOR operations on the remaining three bytes plus the parity byte. The figure steps through each bit's computation, resulting in the missing value for byte 3.

Depending on the RAID level, we distribute the bytes of a disk block across different surfaces. Figure 12.5 illustrates a RAID cabinet of five drives, each with four surfaces. The three black boxes indicating inaccessible blocks of data. Since each is in a different location, we can compute the missing data by using the corresponding bytes from the other four disks. Had two or more disk surfaces become inaccessible in the same location, we might have a problem in that more than one byte of the five is no longer accessible and so we may not be able to restore the missing or damaged data.

Byte 1:	00000011	bit 0 = ((0 XOR 1) XOR (0 XOR 0)) = 1
Byte 2:	11110010	bit 1 = ((0 XOR 1) XOR (0 XOR 1)) = 0
Byte 3:	xxxxxxxx	bit 2 = ((0 XOR 1) XOR (1 XOR 1)) = 1
Byte 4:	00101111	bit 3 = ((0 XOR 1) XOR (0 XOR 1)) = 0
Parity byte:	01110101	bit 4 = ((0 XOR 0) XOR (1 XOR 0)) = 1
		bit 5 = ((0 XOR 0) XOR (1 XOR 1)) = 0
		bit 6 = ((1 XOR 1) XOR (1 XOR 0)) = 1
		bit 7 = ((1 XOR 0) XOR (1 XOR 1)) = 1

Restored byte (byte 3): 10101011

FIGURE 12.4 Restoring a byte.

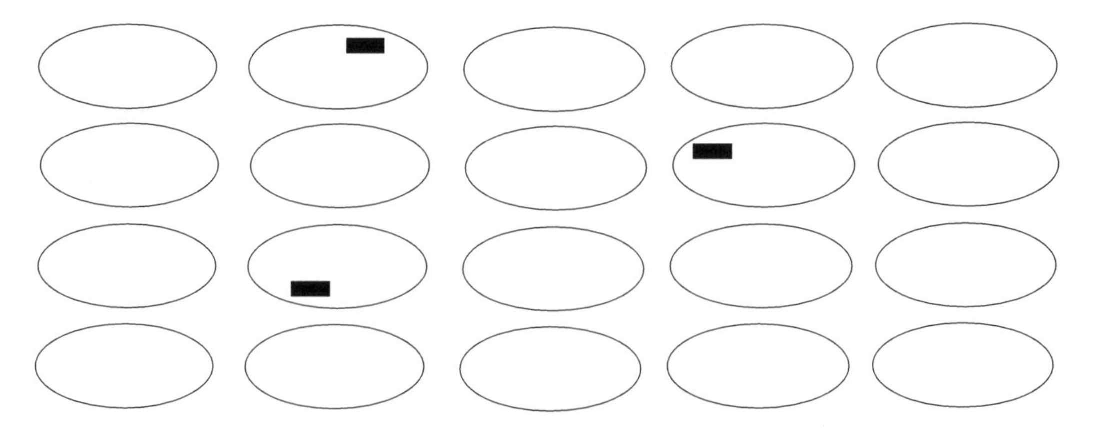

FIGURE 12.5 Bad blocks in RAID storage.

RAID 6 adds a further form of redundancy, Reed-Solomon error correction code. We will not cover this form of code. But with RAID 6, we need additional storage to accommodate the additional form of redundancy data. On the positive side, having an extra set of redundancy allows us to potentially recovery from multiple errors.

Aside from the type of redundancy, the other difference between the RAID levels is the strategy of distributing data (both original and redundancy) across the disks to take advantage of the disk independence. In RAIDs 0, 4, 5 and 6, a disk block is decomposed into smaller parts, called *stripes*, which are then distributed across disk surfaces. For instance, a block might be divided into two stripes, locating the first half of the block on drive 0 and the second half in the same location on drive 1.

With large stripes, a block will consist of only a few stripes while small stripe sizes cause a block to be stored broken into more stripes and so are distributed across more disk surfaces. With large stripes, we can potentially have simultaneous disk accesses of different blocks. Consider, for instance from Figure 12.5 that one block is located on disks 1 and 2 and a second block on disks 3 and 4. If two processes need to access these two blocks at the same time, both could be accessed because each of the four drives can act independently of each other.

If our stripe size was smaller, a block might be spread across too many disk surfaces to permit parallel access. Assume a block is placed on four surfaces. Thus, accessing a block employs four of the five disk drives. Since another block would also employ four of the five drives, only one access could take place at a time.

Figure 12.6 illustrates the differences between RAIDs 4, 5 and 6. In RAID 4, no matter what the stripe size is, all parity information is recorded on one drive. In RAID 5, the parity information is interleaved (distributed) across all drives. RAID 6 is the same as RAID 5 except that there are two forms of redundancy data gathered and these are interleaved across all drives.

RAID 4, because of a single drive being dedicated as a parity drive, is unused. The reason is that, even with large stripe sizes, we could never accommodate two parallel accesses because both would require access to the parity drive. The parity drive is a bottleneck. RAIDs 5 and 6 have the potential for parallel disk accesses depending on the size of the stripes and the location of the stripes of the blocks that are to be accessed. As RAID 6 will have an additional disk over RAID 5, it is likely that RAID 6 could accommodate parallel disk accesses more frequently than RAID 5.

In RAIDs 2 and 3, the idea of small stripes is taken to the extreme. Here, disk blocks are broken into bits and distributed across *every* disk surface. In this way, any disk access involves every surface. This provides for a faster single-disk access. If there are five disks of three platters each, the RAID cabinet then comprises a total of 30 surfaces (remember that we use both the top and bottom surfaces of every platter). The actual time it takes to access a block (not including seek time, rotation latency or transfer time) is as much as 1/30th the time it would take to access the file in a non-RAID storage unit.

As noted in Table 12.3 and discussed above, RAID 2 is not used because of the time-consuming nature of computing and using Hamming distance codes. In RAID 4, having all parity data on one drive creates a bottleneck. In theory, RAID 3 offers the maximum speedup available for a single-disk access which would be beneficial for a computer being used by a single person. In practice, though, RAID 3 is seldom used because RAID is primarily intended for organizations and file servers and so we want to accommodate parallel accesses (via RAID 5 or 6) rather than fast single accesses

There are two ways to utilize RAID. The more traditional approach is through RAID hardware. A RAID cabinet appears to the computer as a single drive unit. However, internally, striping, redundancy and independence are handled through hardware controllers. Between the extra hardware from controllers and the greater storage capacity needed to encode redundancy (if not using RAID 0), RAID hardware tends to be more expensive than non-RAID hard disk drives, but not outrageously so.

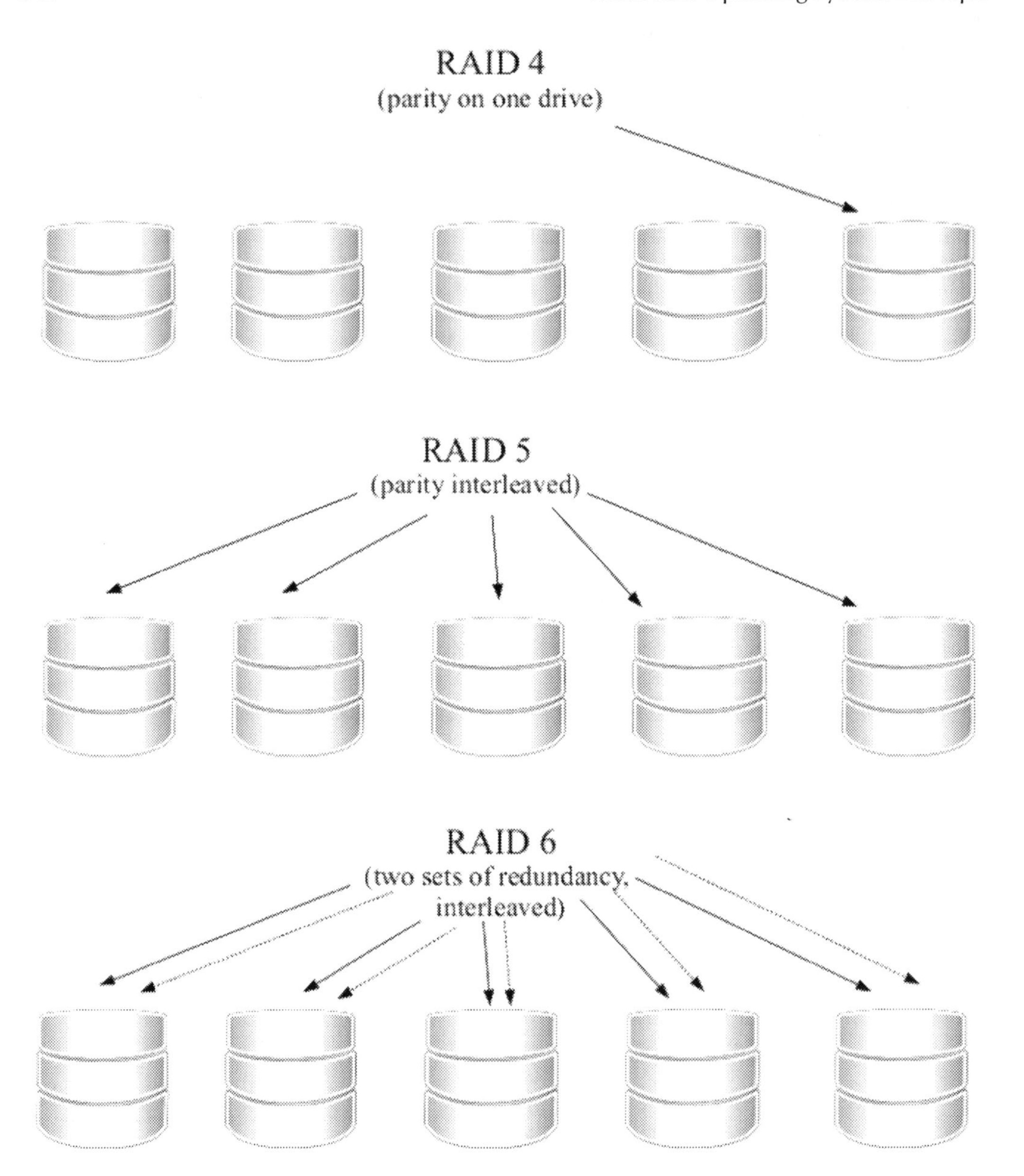

FIGURE 12.6 Comparing RAID 4, RAID 5 and RAID 6. (From Richard Fox, 2021, *Information Technology. An Introduction for Today's Digital World, 2nd edition.*)

The other approach in using RAID is to simulate RAID through software. With only one internal hard disk, software-based RAID cannot improve disk access performance because the single drive has a single set of read/write heads and so there is no independent access available. Software-based RAID can provide forms of redundancy though. Alternatively, if we have two internal hard disks, we can employ software-based RAID and gain some disk independence that might improve disk performance.

In Linux, software-based RAID is offered through the `mdadm` (multiple device administration) program. `mdadm` itself calls upon the `md` device driver. To create a RAID drive, specify `mdadm` *`mode virtual-device(s)`* `[`*`options`*`]` *`physical-device(s)`*. As seen in the instruction's syntax, there are two forms of devices: virtual and physical. The physical hard disk(s) is(are) denoted

TABLE 12.4
`mdadm` Forms of RAID

Form of RAID	Usage	RAID Form?
CONTAINER	Combine multiple physical devices into a single logical device so that RAID can be later applied.	No
FAULTY	Employ algorithms to handle disk faults (without using RAID); only applicable to single physical devices.	No
LINEAR	Combine multiple physical devices into a single logical device.	No
MULTIPATH	Utilize multiple physical devices.	No
RAID0	Striping but no redundancy; only available with multiple physical disks.	Yes
RAID1	Establish full mirroring.	Yes
RAID4	Striping with a dedicated parity drive.	Yes
RAID5	Striping with parity interleaved.	Yes
RAID6	Striping with two types of parity interleaved.	Yes
RAID10	Stripes several RAID 1s to get both mirroring and striping.	Yes

by name as with `/dev/sda` and/or `/dev/sdb`, etc. Even with a single hard disk, we can apply `mdadm`, in which case we use `mdadm` to employ RAID on one or more partitions such as `/dev/sda1` and `/dev/sda3`. The virtual devices will be denoted as *arrays*.

The *mode* portion of the instruction indicates what `mdadm` will do with the specified devices. `Assemble` collects previously created arrays and forms them into an active array. `Incremental assembly` adds one array to an existing active array. `Build` and `create` generate new arrays to then be assembled. The difference between the two is that `build` does not generate metadata and so is of more restricted use. `Grow` increases or decreases the size of an array. `Manage` manages the components that make up an array.

Some of the options are mode-specific such as `-n` (or `--raid-devices`) to indicate the specific device for `create`, `build` and `grow` and `-l` to specify the type of RAID to apply (again to `create`, `build` and `grow`). The types of RAID available in the `mdadm` command are described in Table 12.4. Notice that several of these types are not true forms of RAID.

Let's build a RAID 5 device out of three existing internal hard disk drives, `/dev/sda`, `/dev/sdb` and `/dev/sdc`. We will refer to the RAID device as `/dev/md0`. The command is `mdadm --create /dev/md0 --level=5 --read-devices=3 /dev/sda /dev/sdb /dev/sdc`. In this case, having three physical drives lets us use any of the modes listed in Table 12.4. For instance, we could use `--level=0` or `--level=faulty` instead of choosing level 5.

With `mdadm`, we can apply RAID to a partition, multiple partitions, a disk drive or multiple disk drives. Notice though that since `mdadm` is run by the kernel and the kernel is not loaded until system initialization time, there is an issue in having the `/boot` partition included in the RAID distribution. We can get around this problem by including `mdadm` as part of the `initramfs` used during system initialization or by not having `/boot` as part of our software-based RAID configuration. If we select the latter approach, we need to have `/boot` stored separately from the disk drives we use in `mdadm` or at least on a partition not included in the RAID configuration.

12.2.3 Encryption and Encryption Programs

Encryption is an important part of telecommunications because the Internet is an insecure way to transmit data. Encryption is perhaps of less importance in a file system because files stored on our own computer should not be accessible to anyone else. However, there are two important reasons to encrypt files. First, if the computer is stolen, then that data can fall into others' hands. Even with password protected accounts, it may not be hard for a thief to break that password protection. If the thief can gain root access, then all files become accessible. Second, in spite of firewalls and passwords, it is possible for a hacker to remotely log into a computer and gain access to files (we often refer to such a person as a *cracker* as they need to crack password protection to do so).

Recall from Chapter 8 (and Figure 8.11 specifically) that we can employ encryption on an entire file system. Most types of file systems available in Linux support encryption. We might choose to encrypt all of `/home`, for instance, so that all user files are encrypted. Unfortunately, this option is only available at the time we create the partition and the option, if chosen, impacts the entire partition.

An alternative approach is to let users decide if and which files to encrypt. By using encryption at the file level instead of the file system level, any encrypted file would have to be unencrypted before it could be used (e.g., opened for editing). When encryption is employed at the file system level, it is the file system that performs encryption and decryption for us but when it is employed at the file level, it is the user's responsibility to encrypt and decrypt files manually.

Let's explore what encryption is before we move on to look at available forms of encryption in Linux. *Encryption* is the process of taking a message and encoded it into a form that makes it difficult for someone else to read. Once encrypted, while the message's contents might be accessible, it would likely look like gibberish until it is decrypted. Commonly, encryption substitutes one group of characters for another group of characters. *Decryption* is the process of taking an encrypted message and restoring it to its original form. The processes of encryption and decryption revolve around one or more *keys*, which proscribe how the characters of the file will be encoded and later decoded.

Encryption applies some form of code, or *cipher*. A simple cipher is a rotation cipher whereby each character is rotated by some amount to another. For instance, rotation +1 would cause each character to be encoded as the next character in the ASCII table. Such a simple cipher would be easy to crack and thus would not be very safe. Modern encryption uses complicated ciphers that might employ many different strategies to make it nearly impossible to crack.

There are two forms of encryption that we utilize: public-key encryption (or asymmetric encryption) and private-key encryption (or symmetric encryption). *Public-key encryption* uses two different keys: a *public key* to encrypt messages and a *private key* to decrypt messages. The public key can be provided to anyone who might want to send encrypted messages while the private key is held securely so that only the recipient can decrypt such messages. In *private-key encryption*, there is a single key to both encrypt and decrypt messages. In this form of encryption, the key is always kept secure. Private-key encryption will be used to encrypt files and file systems.

There are many programs available in Linux to perform both public-key and private-key encryption. Some of these are available in Linux without installation, and others must be installed. Table 12.5 describes a few of the software titles.

TABLE 12.5
Encryption Programs Available in Linux

Encryption Title	Capable of	Pre-Installed?
7-zip	Data compression and archiving; can encrypt archives using private-key encryption.	No
bcrypt	Primarily intended to encrypt passwords in Linux.	Some distributions
ccrypt, mcrypt	Private-key file encryption; replaced the older UNIX `crypt` program.	No
Gnu Privacy Guard (gpg)	Same functionality as PGP.	Yes
OpenSSL	Same functionality as PGP.	Yes
Pretty Good Privacy (PGP)	Symmetric and asymmetric encryption; hashing, data compression, generating private keys, generating public keys from private keys, generating digital certificates and fingerprints, automated key management.	No
zip	Same as 7-zip except that it is typically pre-installed in most Linux distributions.	Yes

PGP is a proprietary piece of software. OpenPGP was developed from PGP to be an open-source version. OpenPGP is not the same as PGP, nor does it have all of the same capabilities and features, but there is a lot of overlap between them. GPG is based on OpenPGP but is a third implementation. All three pieces of software can generate private keys, public keys, digital certificates, digital signatures and digital fingerprints. Given a generated key, we can apply the key to encrypt and decrypt files. A file encrypted by any of these three programs can be decrypted by either of the other two programs as long as the key is available.

We will take a brief but specific look at another software title that offers the same functionality, OpenSSL, which is a command-line tool. One task that OpenSSL can perform is to generate a private key. Given a private key, OpenSSL can then generate a public key and a digital certificate. Given a private key, we can encrypt and decrypt a file. We can also add a passphrase to the encryption/decryption process in an attempt to increase the security. Figure 12.7 steps us through some sample interactions with OpenSSL. Comments have been interspersed among the instructions. Note that the command-line prompt has been shortened to `$` to save space.

Let's look at the commands and responses shown in Figure 12.7. The first instruction lists the available ciphers in the installed version of `openssl`. There will likely be dozens. The list is displayed on one long line with each cipher separated from the next by a colon. Adding the option `-v` provides a more readable output by displaying the ciphers on separate lines and providing more detail for each. We choose the cipher AES256 for the remaining instructions.

The next two instructions encrypt the file `f1.txt` using `AES256` and then decrypt the file. The encrypted file is stored in the file `f1.out`. The encryption command pauses to ask for a passphrase which must be verified. Decrypting the file uses the added option `-d`. As we did not specify an output file, the decrypted version is displayed in the window. We are asked for the passphrase but we do not verify it.

The final four instructions in the figure demonstrate the sequence of operations to generate keys and an x.509 certificate. *x.509 certificates* are used to share public keys with other users, particularly in protocols like HTTPS and SSH. The first two instructions in this block generate a private key. The first does not include a passphrase while the second does. Both private keys generated are 2048 bits. By specifying `-aes256`, we are asking `openssl` to use our passphrase as part of the

```
$ openssl ciphers
   Displays the list of all available ciphers
$ openssl enc -aes256 -in f1.txt -out f1.enc
enter aes-256-cbs encryption password:
Verifying - enter aes-256-cbs encryption password:
   Passwords will not be displayed as they are entered
$ openssl enc -aes256 -d -in f1.enc
enter aes-256-cbs encryption password:
   Displays decrypted version of file
------------------------------------------------------------
$ openssl genrsa -out mykey.key 2048
Generating RSA private key, 2048 bits
   Also output are some .....++++ characters and some data on the key
$ openssl genrsa -aes256 -out mykey.key 2048
enter pass phrase for mykey.key:
Verifying - Enter pass phrase for mykey.key:
   Also output are some .....++++ characters and some data on the key, again, passwords
   do not appear
$ openssl rsa -in mykey.key -pubout > mykey.pub
Enter pass phrase for mykey.key:
writing RSA key
$ openssl req -x509 -new -key mykey.key -days 365
     -out mycert.pem
   User is queried for data for the certificate: country, state/province, city,
       organization name, unit name, common name and email address
```

FIGURE 12.7 Interactions with OpenSSL.

generated key. We issue the genrsa command twice, once without the passphrase. We use the same output file, mykey.key, so the key generated by the first genrsa command is overwritten by the second version of the instruction.

The next instruction in the figure uses the rsa command to apply our private key in order to generate a public key, stored in the file mykey.pub. Had we not issued the second genrsa command, we would not have needed to input a passphrase when generating the public key. The final command shown in Figure 12.7 generates an x.509 certificate by using the previously generated private key.

x.509 certificates have an expiration date. For ours, we specify that the certificate will be valid for 365 days. After entering the command, we are asked by openssl for information about our organization. This information is placed into the certificate and includes the organization's location (country, state/province, city), name, subunit name (if any), a common name and an email address for the person who will maintain the certificate.

The result of this final command is a certificate that has no signature. We next should either generate a certificate request so that a *certificate authority* can sign our certificate and thus validate it or sign it ourselves creating a *self-signed certificate* (which web browsers are built to distrust). To sign it ourselves, we add req to the instruction (prior to -x509). To generate a certificate request, we add the option -x509toreq to the command, or openssl -x509 -x509toreq -in mycert.pem -out mycert.csr -signkey mykey.key.

In the few commands from Figure 12.7 and the one in the previous paragraph, we have used several different file extensions. We made up our own extension of .enc to indicate an encrypted text file and .pub to indicate a public key. Standard extensions when dealing with keys and x.509 certificates are .key (for keys), pem (used for certificates), .crt (or .cer or .cert, also used for certificates) and .csr (for certificate signing requests).

PEM is short for *Privacy Enhanced Mail* which is a form of encoding that uses Base64 ASCII to represent, as text, sequences of bits. Specifically, Base64 ASCII stores 6-bit binary sequences as an equivalent ASCII character using the upper and lowercase letters, digits and the + and / characters. Figure 12.8 shows a public key in PEM format generated from a private key using the RSA cipher. The key itself is a sequence of bits, but stored in PEM format, it is a series of Base64 ASCII characters.

We have only touched on a few of the many uses of programs like gpg and openssl. More detail is beyond the scope of this textbook. If you are interested in learning more, first study the man pages of the given program. For more information, consult both software titles' web pages. You can also find numerous examples on the web stepping you through their usages. Information on modern encryption techniques can be found in both cryptography texts and computer security texts.

SECTION ACTIVITIES

1. Do you back up the important files on your home computer? If so, to where? Have you explored other choices? Research your choices for backups, which include external hard disks, USB flash drives, optical discs, magnetic tape and cloud storage. Create a list and enumerate each one's advantages and disadvantages.
2. As a home computer user, do you use RAID? Is there a need for using RAID as a home computer user? Review the material on RAID in this section and decide how you might go about it (software-based, hardware-based) and which level you might want to apply.
3. Have you ever encrypted a file? If so, for what purpose(s) and how? When encrypting a file, you specify a passphrase. Did you use strong password concepts for the passphrase?

```
-----BEGIN PUBLIC KEY-----
MIGfMA0GCSqGSIb3DQEBAQUAA4GNADCBiQKBgQDKiy3wq4QnkYP2/81MJGsLP7gD
NeGulHwURDIdY+xHuD8tRi4iSdtdJ03Q52peL10Ho38HHWwfo1kHZhHWgZRmPBHt
u5UGjsIV1c4PL/OH76dPFL2jQjbH54IWvcGz+MxLIg2QXFr2ICgaEkO1GIVbX8q4
J3jsmajF2nK/sllyWwIDAQAB
-----END PUBLIC KEY-----
```

FIGURE 12.8 Example encryption key in PEM format.

12.3 TASK SCHEDULING

We had previously looked at one-time scheduling with `at` through the `atd` service and recurring scheduling with `crontab` through the `crond` service (Chapter 9). The reason to schedule tasks is to increase the amount of automation in our system so that we are not burdened with remembering to execute operations and then having to enter the commands. We take a closer look at these services and how to use them to schedule tasks. We first look at the `at` program and then at `crontab` separately.

12.3.1 at AND atd

`at` is a program that the user uses to submit one-time scheduling requests. These requests are handled by the `atd` service. We specify the time that the event should take place and the command(s) to be executed. The time includes both the date and the time on that date. Time and date are either absolute values or are relative to when the command is issued.

For an absolute specifier, the time format is `HH:MM`, optionally followed by `AM` or `PM`. If `AM/PM` is omitted, then the time is interpreted as being in *military notation*. Thus, 12:00 PM and 12:00 are treated the same but 1:00 PM would be noted as 13:00 in military time (1:00 would be interpreted as 1:00 AM). The military time for 12:00 AM is 0:00. If we use `AM` or `PM`, it can be listed using either uppercase or lowercase letters. For convenience, three prespecified times are available as `noon`, `midnight` and `teatime` (4 PM).

The date is specified using one of four formats: `MMDDYY`, `MM/DD/YY`, `DD.MM.YY` or `YYYY-MM-DD`. If no date is provided, then the event is scheduled to take place at the next occurrence of the listed time. For instance, if it is currently 3:35 PM and we schedule a task for 14:30 (2:30 PM), then the event is scheduled to take place tomorrow at 2:30 PM. We can also use the words `today` and `tomorrow` for the date.

To specify a relative time, we use the notation `now + ` *`count unit`* where *count* is an integer and *unit* is a temporal unit such as `minutes`, `hours` or `days`. For instance, `now + 5 minutes` will launch the task 5 minutes after being scheduled and `now + 4 weeks` will launch the task in exactly 4 weeks. We are not allowed to combine multiple units as in `now + 1 day 2 hours`. The result would be a `Garbled time` message.

We can combine both absolute and relative values but doing so may lead to confusion. `tomorrow + 2 hours` will schedule the task to run 26 hours from now, and `1:00 PM + 2 days` will schedule the task to run 2 days from now but at 1:00 PM. But consider `tomorrow + 2 days`. While `at` will accept this, the task will be scheduled for 3 days from now.

The event to be scheduled is commonly specified in a script and added to the `at` command using `-f` *filename* as in `at tomorrow 5:00AM -f ./my_event.sh`. Without the `-f` option, `at` places us into an interactive buffer allowing us to enter one or more commands. The prompt provided is `at>` and the input ends when we type `control+d`.

Once we have scheduled events, we can examine all of our scheduled events using `atq`. The response is a list of each waiting event scheduled through either the `at` or `batch` program for the current user. Each event is given its own job number. We can delete a scheduled item using `atrm` followed by the job number, as in `atrm 2` to remove the second item in the scheduled list.

The batch command uses atd but rather than scheduling the event for a specific date/time, the event executes as soon as system load drops below 80%. The batch command does not include a time/date specifier. The event to schedule can either be specified using -f *filename* or through the same at> prompt as we had with at in which case we enter commands one per line, ending with control+d.

The atd service executes all tasks scheduled using at and batch. These tasks are executed using the /bin/sh shell rather than the user's default shell. So be warned in case you submit a script expecting it to be executed by Bash. Upon execution, unless redirected to a file, any output from the scheduled at command is sent to the user's email. Errors generated by the execution of the command(s) are also sent to the user's email.

Figure 12.9 demonstrates an interaction with at. We see several commands scheduled including a couple through the at> prompt. We also see the use of atq and atrm.

Looking at the interaction shown in Figure 12.9, let's focus on the first atq command. We see jobs listed using numbers 2 and 3. The first job had already executed by the time we ran atq. Job numbers are not reused from already completed jobs so we do not regain a job number 1. Jobs are listed by job number, date and time that the job is scheduled to execute, the username of the job's scheduler and a letter. This letter, a in the figure, indicates the job queue storing these jobs.

The fourth at command did not include -f so we are dropped into an at> prompt to enter the commands. After entering the third command, we type control+d which causes at to output <EOT> (end of task) and process the at request, which is numbered job 5. We are then returned to our normal Bash prompt. We later issue a batch command and are placed into an at> prompt again.

By default, at is set to be used by all users. We may wish to blacklist some users from using it, for instance if users in the past have abused the scheduling privilege by scheduling tasks at the busiest times of day. The file /etc/at.deny is initially empty, but we can add usernames to it to prevent those users from using at.

```
[foxr@localhost ~]$ at now + 1 minute -f ./script1.sh
job 1 at Tue Feb 16 10:47:00 2021
[foxr@localhost ~]$ at tomorrow -f ./script2.sh
job 2 at Wed Feb 17 10:48:00 2021
[foxr@localhost ~]$ at 11:45 09202022 -f ./script3.sh
job 3 at Tue Sep 20 11:45:00 2022
[foxr@localhost ~]$ atq
2    Wed Feb 17 10:48:00 2021 a foxr
3    Tue Sep 20 11:45:00 2022 a foxr
[foxr@localhost ~]$ at now + 10 minutes
at> ./script4.sh
at> ./script5.sh
at> ./script6.sh
at> <EOT>
job 4 at Tue Feb 16 11:00:00 2021
[foxr@localhost ~]$ batch
at> ./script7.sh
at> <EOT>
job 5 at Tue Feb 16 10:55:00 2021
[foxr@localhost ~]$ atq
2    Wed Feb 17 10:48:00 2021 a foxr
3    Tue Sep 20 11:45:00 2022 a foxr
4    Tue Feb 16 11:00:00 2021 a foxr
[foxr@localhost ~]$ atrm 2
[foxr@localhost ~]$ atq
3    Tue Sep 20 11:45:00 2022 a foxr
4    Tue Feb 16 11:00:00 2021 a foxr
```

FIGURE 12.9 Example at/batch commands.

12.3.2 crontab AND crond

at and batch perform one-time scheduling. crontab is used to issue *recurring* scheduling. crontab is used to schedule tasks, and the service crond is responsible for executing those scheduled tasks. Unlike at/batch which allows a user to submit any number of jobs, crontab permits up to one job per user. However, that job can consist of any number of scheduled tasks.

The syntax for running crontab is crontab *filename* where *filename* contains the list of recurring tasks to run. Once submitted, we can view our list of tasks using crontab -l, remove our crontab job with crontab -r and edit our list of scheduled tasks using crontab -e. In the latter case, all scheduled tasks are displayed in a vi editor so that we can modify, delete or add to the list. We can issue crontab -e without first issuing crontab *filename* whereby we enter our tasks in the editor directly.

The listing of tasks in the file, or entered in the editor, is one task per line. The line specifies the recurrence and the task to execute. Because each line must be self-contained, the event(s) scheduled will need to be limited in size to fit on the one line. To avoid this restriction, we might place the commands to be executed in a script and then place the script name as the event to execute.

Specifying recurrences is where crontab becomes complicated. Every task's recurrence is indicated with five values. These values are listed in a row of the file, separated by spaces. The five values indicate the minute of the hour (0–59), hour of the day (0–23, using military time), date of the month (1–31), month of the year (1–12) and day of the week (0–7 where both 0 and 7 represent Sunday and 1–6 are Monday through Saturday). Each of these values can be a specific integer or the wildcard *, which indicates *every* unit of the time period. For instance, 30 8 * * 1 means 8:30 AM on Monday for every date (the first *) and every month (the second *). The notation 0 * 1 1 * means January 1st on the hour (0 minutes) for every hour (the first *) no matter the day of the week (second *).

We can specify multiple times within a specified interval by listing the time units separated by commas. For instance, 0,30 for the minute would mean that the task should execute on the hour and at 30 past the hour. The notation 0,30 * * * * means to execute the task twice an hour, on the hour and at 30 minutes past the hour, every day. If there are several recurrences to specify, we can use the notation *time/x* to indicate the given *time* at intervals of *x* units. The notation 0/10 12 1 1 * specifies that the given task will run at 12:00, 12:10, 12:20, 12:30, 12:40 and 12:50 PM on January 1st.

Table 12.6 presents numerous examples. Each line omits the actual command being scheduled (for brevity). If we wanted to execute myscript.sh, the first entry in its entirety would appear as 30 1 15 * * ./myscript.sh. See if you can determine the recurrence specified in each of the examples before reading the meaning column, which explains the recurrences.

TABLE 12.6
Example crontab Recurrences

Recurrence					Meaning
30	1	15	*	*	1:30 AM, the 15th of every month.
0	12	*	*	6	12:00 PM every Saturday.
0	0	13	*	5	Every Friday the 13th at midnight.
15	3	14,28	*	2	3:15 AM on the 14th and 28th of each month if the day falls on a Tuesday.
0,15,30,45	19	*	*	0	7:00, 7:15, 7:30 and 7:45 PM every Sunday.
0/5	9	1	*	*	Every 5 minutes from 9:00 AM through 9:55 AM on the first of every month.
45	18	*	*	1,4	Every Monday and Thursday at 6:45 PM.
0/10	0,12	*	*	*	Every 10 minutes starting from midnight to 12:50 AM and again from noon to 12:50 PM daily.
59	23	31	12	*	11:59 PM on December 31.

Root can view and control other users' crontab jobs by adding -u *username* to the instruction. In this way, root can view (using -l), edit (-e) and delete (-r) other users' jobs. However, should we, as root, delete or modify another user's crontab job we should inform the user(s).

When atd runs a scheduled job, it uses /bin/sh to run the instructions. When crond executes scheduled commands, it has a few environment variables set. These are HOME, LOGNAME, SHELL and a limited PATH variable which consists solely of the directories /sbin, /bin, /usr/sbin and /usr/bin. Thus, crond will run scheduled tasks under the user's username, using that user's default SHELL and that user's HOME directory. The PATH used may or may not include all of the necessary directories. In order to establish our PATH when crond executes one of our jobs, we might include source $HOME/.bashrc as the first executable instruction in the script we specify. This runs our .bashrc file so that we can ensure the PATH is as we want it to be.

Similar to atd, there is /etc/cron.deny to blacklist users so that they cannot issue crontab jobs. crond also has five directories set up in /etc. Four of these directories are named cron.daily, cron.hourly, cron.monthly and cron.weekly. Placing a script in one of these directories causes crond to execute that script based on the recurrence of the directory (e.g., daily or hourly). We might place a script in one of these directories if we have a task that we need to be executed in that recurrence, such as monthly, but not at a specific time. Some scripts are likely already placed in some of these directories, such as the script logrotate in /etc/cron.daily. logrotate executes under /bin/sh and runs the service /usr/sbin/logrotate on the data file /etc/logrotate.conf.

There is another scheduler named anacron. Its role is like crond but it does not assume that the computer is running at a time when a scheduled task is to take place and so executes the task when it can. anacron runs tasks no more than daily. We do not schedule anacron from the command line, unlike crontab and at. Instead, we would modify the anacron configuration file, /etc/anacrontab. Recurrences are specified by an interval of days (e.g., 1 for daily, 7 for weekly) and a delay in minutes to start after the system boots.

We will likely find the three lines shown in Figure 12.10 already in the configuration file. The command run-parts invokes the scripts and executables located in the given directory. In the figure, we see that the /etc/cron-daily directory's contents should be executed by run-parts daily, with a delay of 5 minutes after system initialization; cron.weekly every week after a delay of up to 25 minutes; and cron.monthly every month with a delay of up to 45 minutes. Missing from the figure is the establishment of environment variables defined in the anacrontab file such as those for SHELL and PATH.

SECTION ACTIVITIES

1. List five activities that you might want to schedule on your home computer. If you struggle coming up with enough activities, do a web search to get some ideas.
2. If you are a Windows user, open the Task Scheduler (right-click on the Start button, select Computer Management, and in the left pane, expand System Tools and click on Task Scheduler). In the right pane, select Create Basic Task and from the pop-up window, type in a name and then select Next so you can see the options for Trigger, Action and Finish. Cancel the task. Compare this tool versus using at and crontab. Which is easier? Which is more flexible?

```
1         5    cron.daily    nice run-parts /etc/cron.daily
7         25   cron.weekly   nice run-parts /etc/cron.weekly
@monthly  45   cron.monthly  nice run-parts /etc/cron.monthly
```

FIGURE 12.10 /etc/anacrontab file contents.

12.4 SYSTEM MONITORING

We need to ensure that our Linux systems are running effectively. We have a suite of tools available to inspect resource utilization so that we can determine what system resources are being stressed. We have already explored a number of programs for system monitoring including `top`, `ps`, the System Monitor, `vmstat` and `free`. In this section, we look at these and other programs. Before we do so, let's gain a better understanding of a reasonably running system by examining some operating system concepts.

12.4.1 Operating System Issues That Degrade Performance

The idea that a process is making progress toward completion is known as *liveness*. To promote liveness, resources that the process needs during its execution must be made available such that the process is not forced to wait an indefinite period of time. The opposite of liveness is *starvation*, which means that the resources that the process needs are continually being withheld from the process resulting in the process not being able to proceed toward completion.

Resources include access to the operating system, a sufficient amount of main memory, access to the file system and specific files within the file system, network access and most especially execution cycles from the CPU. The reason that starvation may arise is subtle but has to do with concurrent processing in operating systems (refer back to Chapter 4). Let's examine why.

Imagine that we have two running processes, P0 and P1, both of which need to operate on a shared file, F0, which stores the single value 0. P0 is going to read the file and update the datum by adding 3 to it. P1 is going to read the file and update the datum by subtracting 2 from it. Both processes will write the modified datum back to the file. No matter which order P0 and P1 execute, the resulting value stored in the file after both accesses complete should be the 1 (0+3 – 2=1).

If the operating system is multitasking, then it might alternate between the two processes, running some number of instructions of P0 followed by some instructions of P1 before resuming with P0. The actual pattern is not possible to predetermine as a process may suspend itself, another process may enter the system and one or both process priorities might be changed during execution. Figure 12.11 illustrates one possible path through the sequence of steps that the two processes are taking.

The issue arising from the pattern we see in Figure 12.11 is that P0 reads 0 from the file into X and then P1 reads 0 from the file into Y. Unfortunately, three copies of the datum now exist: X, Y and what is stored in the file. As time continues, the values stored in X and Y diverge from the original value, 0. Based on the order that X and Y are written back to the file, we have different possible results.

We can avoid this problem if we either move step 5 from Figure 12.11 up to follow step 1b or move step 1 below step 3c. These changes resolve our issue because the two processes are now accessing the datum not in an overlapped fashion but in a sequential order. However, the order of

1. P0 begins executing
   a. P0 reads the datum from the file and stores the datum in a local variable, X.
   b. P0 adds 3 to X. X is now 3 (the file is still storing 0).
2. The CPU is interrupted by the timer and the operating system performs a context switch to P1.
3. P1 begins executing
   a. P1 reads the datum from the file and stores the datum in a local variable, Y.
   b. P1 subtracts 2 from Y. Y is now -2 (the file is still storing 0).
   c. P1 writes Y back to the file (the file now stores -2).
4. The CPU is interrupted by the timer and the operating system performs a context switch back to P0.
5. P0 writes X (3) back to the file (the file now stores 3).

FIGURE 12.11 One possible sequence of operations between multitasking processes.

the steps shown in Figure 12.11 is not predeterminable because of multitasking and so we run the risk of the sequence causing an overlapped access.

We refer to the situation in Figure 12.11 as a *race condition*. The problem is that two processes are racing each other to store their differing results back into the shared datum. If we allow the race condition to occur, then we could have *data corruption*. As seen in Figure 12.11, the shared datum winds up storing 3 instead of 1. Should the order of 3c and 5 be swapped, we would have had yet another result.

In order to prevent race conditions, we need a mechanism to enforce *mutually exclusive* access to the shared datum in the file. This means that once one process has started to access the datum, we lock access to the datum. This causes any other process that requests access to the datum to be *blocked*. Blocking should result in that process having to wait until the process using the datum frees the datum. In Figure 12.11, we should lock access to the datum once P0 reads the datum into X (step 1a). When P1 reaches step 3a, it finds that access to the file is blocked and so P1 must wait.

Operating systems implement mutually exclusive access to a shared datum through some form of *synchronization*. This permits one process to request and gain access to a datum and then any other requesting process that requests the datum is blocked. A blocked process is forced to wait for access which results in the process not moving forward toward execution.

While synchronization is a reasonable way to avoid data corruption, it can lead to two significant problems: starvation and deadlock. *Starvation* arises when a process is continually forced to wait for access to a shared resource. *Deadlock* occurs when two or more processes are holding resources that the other process(es) need so that neither process can make progress.

Returning to Figure 12.11, imagine that there are three processes in the system where process P2 has no interest in the shared data. P0 begins (steps 1 and 2 in the figure) and a context switch starts P1. P1 requests the shared datum and is blocked, so waits. The operating system switches to P2 to let it run for a while. Upon the next context switch, the operating system resumes P0, which writes its shared datum back to the file, thus freeing it up.

Before the next context switch takes place, P0 again asks for access to the file. As the file is currently available, access is granted to P0, which again holds the file. The next context switch resumes P1 but it must wait again, so a context switch moves on to P2. After P2 runs for a while, a context switch resumes P0. P0 again writes its result to the file, runs a while, requests access to the file and is granted access, just before another content switch causes the CPU to switch to P1. If this cycle continues indefinitely, then P1 starves because every time it gains access to the CPU it is still in a blocked state.

If programmers are careless about implementing synchronization, starvation may be the result. The reason why is that our example from the previous paragraph shows our implementation of synchronization uses an unfair policy. We can achieve, or at least attempt to achieve, a fairer policy by placing P1 in a waiting queue upon suspending it initially. When P0 frees its hold on the shared file, the operating system can look to see if any other process is waiting for it. As P1 is in the wait queue for that file, it can be resumed at this point so that it can gain access to the file. This prevents the previously described scenario where P0 gets to access the file again before P1 gets its next attempt.

Deadlock is a more complicated situation, so let's take a closer look at it. Figure 12.12 continues the demonstration of two processes, P0 and P1, but now there are two files being accessed, F0 and F1. Both processes run concurrently using multitasking. Both processes request access to one of the two files. The issue arises in step 5a when one process requests access to the file currently held by the other process.

By the time the situation in Figure 12.12 reaches step 7c, we have both P0 and P1 waiting. What are they waiting for? For the other process to release its hold on the file it has been allocated. When will each process release that file? Unfortunately, P0 cannot continue executing and so will not reach a point where it can free P0. Similarly, P1 cannot continue executing and so will not reach

1. P0 begins execution.
   a. P0 requests access to F0.
   b. As no other process is using F0, the operating system grants P0 access.
2. A context switch forces the CPU to switch to P1.
3. P1 begins executing.
   a. P1 request access to F1.
   b. As no other process is using F1, the operating system grants P1 access.
4. A context switch forces the CPU to switch to P0.
5. P0 resumes executing.
   a. While holding onto F0, P0 requests access to F1.
   b. The operating system denies access to P0 as P1 is holding onto F1.
   c. P0 enters a waiting state
6. With P0 waiting, the CPU performs a context switch to P1.
7. P1 resumes executing.
   a. While holding onto F1, P1 requests access to F0.
   b. The operating system denies access to P1 as P0 is holding onto F0.
   c. P1 enters a waiting state.

FIGURE 12.12 Multitasking situation resulting in deadlock.

a point where it can free P1. Both processes are in a waiting state meaning that neither can move forward. As neither moves forward, neither releases its held file leaving the other process waiting. This could virtually go on forever. *Deadlock.*

The situation might be worse than described. Imagine the system starts process P2 and it requests file F2. No process is using F2 and so the access is granted. Later, P2 also requests F1. As F1 is held, P2 cannot be granted access so it is forced to wait. Process P3 requests access to F0. Process P4 requests access to F2. All five processes are now deadlocked. As deadlocked processes make no progress toward completion, all five processes are starving.

Operating systems handle deadlock in differing ways. Some aggressively prevent deadlock from arising by keeping processes that might cause a deadlock waiting before they can begin execution. Known as deadlock prevention, it might result in our system only running one process at a time (single tasking) since running multiple processes might cause a deadlock.

Another approach is to run processes as they are requested but not to allocate a requested resource to a process if the action *might* result in a deadlock. In this strategy, called deadlock avoidance, we are multitasking but possibly forcing processes to wait as soon as they request any resources.

Another strategy is to ignore the possibility of deadlock when granting resource requests and to look for deadlocks from time to time (such as every few minutes). If a deadlock is detected, the deadlock is resolved by arbitrarily killing off some of the processes involved. This is known as deadlock detection and handling.

In fact, the most common approach to deadlock is to ignore deadlock entirely. Instead, deadlocks can arise and it is up to the user (or administrator) to detect the issue and handle it. To handle it, we might need to kill one or more of the deadlock processes and restart them later. This (lack of) approach to deadlock has been called the *Ostrich algorithm*, a sarcastic name alluding to the operating system "burying its head in the sand".

Linux handles deadlock in two ways. For kernel processes, deadlocks are avoided because kernel processes are granted access to resources in a specific order. A particular resource will only be granted to a kernel process when it is that process' turn to access any or all resources. For user processes, Linux uses the Ostrich algorithm allowing deadlocks to arise for any non-kernel processes and letting the users deal with the consequences!

Another situation leading to poor system performance is known as thrashing. *Thrashing* occurs when the operating system spends more time performing page swapping than processing. Recall that virtual memory stores portions of our running processes on hard disk. When a particular portion (page) is needed that is not in memory, the operating system performs page swapping, locating the page on disk and swapping it with a page currently in memory.

As disk access is slow, paging slows down processing. When memory is full, the operating system must select a page to discard. If it selects poorly, the discarded page may be needed in the near future requiring another round of paging. Thrashing can arise if the operating system continually selects pages to discard from memory that are still needed and thus has to move them back into memory.

Let's consider a simple but unrealistic situation. Memory consists of 16 frames, each of which stores one program page. Currently, there are four processes running, each of which is allocated four frames. Process P0 is currently executing; it accesses a portion of the process not currently in memory. The needed page, page 7224, is only going to be accessed for a few cycles before the process requires another page, 4813. When the operating system retrieves page 7224 from disk, it discards page 4813 as this page hasn't been accessed recently. Page 7224 is loaded into memory, accessed, and then the process needs to access 4813 again. Discarding 4813 was a poor choice. In this case, page 7224 is discarded to make room for the return of 4813. But after only a few cycles, page 7224 is needed again. Assume this situation continues for several iterations. The operating system is paging more frequently than the CPU is executing instructions. As paging is far more time consuming than executing program instructions, performance degrades, badly. Again, this is unrealistic as any process is most likely granted dozens or hundreds of frames in memory but thrashing does happen.

12.4.2 Processor and Process System Monitoring Tools

The Linux System Monitor provides us with both an overview and detail of what is going on with many system resources. Specifically, we can see the amount of CPU usage, memory usage, network usage, virtual memory usage and file system usage. For CPU, memory, swap space and network utilization, the tool shows us the changes taking place over the past minute.

As a system administrator, a quick look at this tool can indicate that our system is running smoothly. We would hope to see that CPU load is not approaching or at 100% for any length of time and that memory utilization is not at or close to 100%. If we see memory utilization running high, then we would probably see swapping occurring more frequently. If we find frequent swapping, it means that our system is doing a lot of very inefficient paging.

Through the System Monitor or `top`, we can inspect the running processes and see the status, CPU and memory usage of the processes. We might use this program to see if either there are too many processes running or if a handful of processes are trying to monopolize the CPU. We can then look to suspend or kill processes. Are there root-owned processes that can be postponed? Are any user processes running in the background that can be postponed? Can we modify some running user processes by lowering their priority (raising their niceness)? Are there scheduled tasks that are running that can be rescheduled for a later time?

If memory utilization is approaching capacity causing a lot of virtual memory usage, we might select some of the processes to kill off and schedule to run at a later time. Lowering process priority will likely have no impact on memory usage so may not be an adequate solution to this problem. We might decide that we have too many users for the amount of memory in our computer. This might argue for decreasing the number of users (impractical), buying more computers or increasing the size of memory.

See Chapter 4 for more details on both the System Monitor and `top`. Other tools for querying the system are command-line programs. These include `ps`, `mpstat`, `sar`, `pidstat`, `uptime`, `vmstat`, `free`, `stat`, `df`, `du`, `netstat`, `nmap`, `ip`, `ss`, `strace`, `iostat`, `lnstat` and `lpstat`. We have already looked at some of these programs in Chapters 3, 4, 8 and 10. We examine the remainder here and re-examine `mpstat`. Recall that some of these commands may not be installed in your version of Linux; to install `mpstat`, `iostat` and others, install the package `sysstat`.

As noted in Chapter 4, mpstat reports processor-related statistics for each of the system's processors. It is implied that mpstat is used on a multiprocessor system, but it can also be used for a single processor computer. mpstat provides for each CPU the percentage of time it spends on user activities, system activities, hardware interrupts, software interrupts, guest activities, ideal time, cycle stealing and periods of input or output when it is forced to wait. The user activity is broken into two different values: ordinary user activity and user activity with niced processes. Guest activities are actions that the CPU takes to run a virtual processor. We describe cycle stealing in the next subsection.

Similar to mpstat, sar reports on CPU utilization. The primary difference is that mpstat reports an average over some duration of time. sar, on the other hand, outputs a series of snapshots of processor utilization over a period of time. The output shows, row-by-row, the processor's utilization at each timed interval. By default, the interval is every 10 minutes. This information is stored in log files under the directory /var/log/sa where these files have names of sa# or sar# (# is a number).

For each interval, sar provides us with the CPU number (or statistics on all CPUs), percentage of time the CPU was used in user and nice modes, system usage, wait time, cycle stealing time and ideal. We are also given averages over the entire recorded time period. We can limit the intervals by specifying a different start and/or end time for the report, or by altering the interval. We can also ask sar to specify different types of statistics than CPU usage. These include I/O transfer rate statistics, memory usage statistics, paging statistics network statistics and block device usage.

Another command that reports on the processor's performance is pidstat. This program provides CPU statistics for each running process. These statistics are broken into user, system, guest and CPU usage where CPU usage is the sum total of user, system and guest. We are also told which processor is running the particular process (for a multiprocessor system). So, while mpstat reports on each processor, pidstat shows similar information broken down at the process level.

pidstat has a number of options that allow us to specify what type of information we want reported. -d shows I/O usage including the amount of reading and writing to hard disk. -r provides virtual memory information (number of minor and major page faults occurring per second, virtual memory size and resident size). -w displays task switching activity.

The uptime program reports on the duration that the system has been running (since the last boot/reboot). The output is presented in a single line that states the current time, the number of days, hours and minutes of uptime, the number of current users (logged in) and the average load over the past 1, 5 and 15 minutes.

strace may or may not be useful as a troubleshooting tool, but it is an interesting program. Applications software and portions of the operating system invoke the kernel through system calls. strace maps the number and nature of system calls of a given Linux command. As an example, using strace on pwd results in a listing of the system calls made by pwd. The output includes the parameters passed to those functions. pwd calls, in order, brk, mmap, access, open, fstat, mmap, close, open, read, fstat, mmap, mprotect, mmap (four times), arch_prctl, mprotect (twice), munmap, brk (twice), open, fstat, mmap, close, getcwd, fstat, mmap, write, close, munmap, close and exit_group.

strace invokes the Linux command that we specify and traces what the command does while it executes. It intercepts and records the system calls. strace is primarily used as a debugging tool for those who are writing Linux commands but can be used by system administrators to troubleshoot problems within the operating system itself. Chances are, if we have installed a stable version of Linux and not modified it, we would not need to resort to strace.

12.4.3 Memory System Monitoring Tools

We examined vmstat and free in Chapter 4. Let's take another look at them to better understand the information it reports to us. Figure 12.13 (which is the same as Figure 4.21) shows the results of free (top half of the figure), giving us the sizes of free and used memory and swap space currently in use. The information output is the current utilization, unlike vmstat which reports on an average over a period of time. The output provides for us amounts for Mem (memory) and Swap (swap space) broken into amount used, amount free and total amount.

For memory, we also see the amount dedicated to cache, buffers and shared memory. The line indicated as -/+ buffers/cached describes the amount of main memory that is allocated for a buffer in support of some application(s) or as disk cache. In this example, notice that swap space is only one-half a GB. This is unusual in that swap space (virtual memory) is usually at least the size of main memory.

The bottom half of Figure 12.13 provides an example of vmstat's output. The default report from vmstat is broken into six sections showing process information, memory usage, swap space usage, I/O usage, system usage and CPU usage. These values are all averages, as computed since the last time the system was booted. See Table 12.7 for a description of the abbreviations from vmstat's output.

Let's explore the meaning behind some of the terms in Table 12.7. Linux is a multitasking operating system where, at any time, it is likely that there are multiple processes in some state of execution.

```
[foxr@localhost ~]$ free
              total        used        free      shared  buff/cache   available
Mem:        1832412     1216820      171304       30680      444288      420452
Swap:       1048572      284160      764412
[foxr@localhost ~]$ vmstat
procs -----------memory---------- ---swap-- -----io---- -system-- ------cpu-----
 r  b   swpd   free   buff  cache   si   so    bi    bo   in   cs us sy id wa st
 4  0 284160 171304     56 444256    0    1     8    17   35    8 98  2  0  0  0
[foxr@localhost ~]$
```

FIGURE 12.13 Sample free and vmstat output.

TABLE 12.7
vmstat Abbreviations

Header	Meaning
r	Number of processes waiting for run time.
b	Number of processes in uninterruptible sleep.
swpd	Amount of virtual memory used (in KB).
free	Amount of free RAM (in KB).
buff	Amount of RAM used as buffers (in KB).
cache	Amount of RAM used to cache hard disk data (in KB).
si	Average KB/second of data swapped into memory from disk.
so	Average KB/second of data swapped out of memory to disk.
bi	Average blocks/second of data swapped out from memory to disk.
bo	Average blocks/second of data swapped into memory from disk.
in	Number of interrupts/second.
cs	Number of context switches/second.
us	User time (non-kernel).
sy	System (kernel) time.
id	Idle time.
wa	Wait time.
st	Cycle stealing time.

The CPU switches off between processes, giving each a little bit of CPU time. The number of waiting processes (r) are those that are active (loaded in memory) but not being executed at that moment by the CPU. Some processes had been executing but have suspended themselves and so are sleeping. There are two categories of sleeping processes: interruptible and uninterruptible. Some processes are uninterruptible because interrupting them could damage data. Portions of the kernel run in uninterruptible sleep. vmstat reports on those uninterruptible processes have put themselves to sleep (b).

Most context switches between running processes occur because of the time limit imposed by process niceness. Other switches occur because of interrupts. The number of context switches per second (cs) can give us an indication of how readily our processor is executing our processes. Other values of note are us, sy, id and wa which are the amounts of time that the processor is spending on user processes, system processes, idling (no process executing) and waiting on I/O respectively.

Cycle stealing (st) is a situation where memory needs to be accessed by both the CPU and an I/O device. Governing the I/O device's communication with memory is a direct memory access (DMA) controller. Cycle stealing arises when the CPU's access is postponed to give priority to the DMA controller. The CPU is not usually idle during a stolen cycle but instead turns its attention to a non-memory operation before access can be regained.

The snapshot of the vmstat command shown in the bottom part of Figure 12.13 indicates a system with a light load. There are few processes running, and a more than sufficient amount of main memory so that disk swapping is not needed at the moment and no significant cycle stealing has arisen. A majority of the CPU time is spent idling rather than either running the user processes or system processes.

12.4.4 I/O System Monitoring Tools

There are a number of programs that report on I/O performance. These are iostat to report on the file system, lpstat to report on the printer, network programs like ip, ss, nstat, lnstat, rtacct and nmap and file system commands stat, df and du. We explored the file system commands and some of the network programs earlier in the textbook. Here, we will concentrate on those we haven't seen. Remember we can also view resource usage in files stored in /proc.

The iostat command gives us the flexibility to obtain three types of reports: CPU utilization, device utilization and network filesystem utilization. CPU utilization provides the percentage of CPU time used for ordinary and niced user applications, system operations, cycle stealing time, I/O wait time and idle time. The device report lists each type of connected block device. This report displays the average number of transfers per second, blocks read per second, blocks written per second and total number of blocks read and written.

If there are any mounted file systems (including USB or CD-ROM mounted devices), iostat also provides a report for each of these by listing the filesystem name, number of blocks read and written per second and total number of blocks read and written. We can obtain more detailed information on any one device by using -x *devicename*, as in iostat -x /dev/sda5. This report includes the number of seçtors read/written per second, the average size of requests and the average wait time.

Two network programs we have not yet explored are nstat and rtacct. Both provide network interface statistics. These are useful if we want to examine the types of messages that we have received or sent. Messages are specified in categories like IcmpInErrors (ICMP incoming message errors), IpOutRequests (outgoing message requests), TcpActiveOpens (TCP active ports) and Ip6InDelivers (incoming IPv6 messages). The lnstat program provides network routing cache statistics and additional network statistics from the Linux kernel, as stored in the directory /proc/net/stat.

```
Nmap scan report for 10.11.12.13
Host is up (0.000120s latency).
Not shown:  993 closed ports
PORT        STATE      SERVICE
22/tcp      open       ssh
25/tcp      closed     smtp
80/tcp      open       http
111/tcp     open       rpcbind
113/tcp     closed     auth
2049/tcp    open       nfs
----------------------------------------------------------
22/tcp      open       ssh  OpenSSH 5.3 (protocol 2.0)
| ssh-hostkey: 1024 … (DSA)
| 2048 ... (RSA)
80/tcp      open       http Apache httpd 2.2.4 ((CentOS))
```

FIGURE 12.14 Output from nmap.

We already examined nmap briefly in Chapter 10 but let's take another look at it. nmap (which may or may not be pre-installed) scans network hosts to see what ports are available. It does this by sending out IP packets and examining the responses. By determining port access, it can report on the services that the given computer offers (e.g., ssh, http). Additionally, it can report on firewall activity, operating system type and numerous other features of a given computer. With nmap, we can investigate a network's security (or the individual computer's network access security) as well as monitor network components. We can also use nmap to accumulate network statistics.

The only argument required for nmap is the address to be investigated. This can be a specific IP address, a host name or a range of network addresses. For instance, nmap 10.11.12.5-205 would investigate all devices in the range from 10.11.12.5 through 10.11.12.205. We can also specify a subnet such as nmap 10.11.12 to scan all devices within the network 10.11.12. nmap responds with a report of accessible ports for each device contacted. The list of ports includes their status (open, closed), the service implemented on that port and the version (if available). For instance, the command nmap 10.11.12.13 might receive a report like the one shown in the top half of Figure 12.14. Using option -A, nmap provides more detail on version types for the various services and the operating system. We see a partially expanded output in the lower half of Figure 12.14 specific to the ports used by ssh (20) and http (80). The … in the figure are specifications of keys, omitted for space.

One last program that we have used at points in the book is who. Although this merely lists the logged users of the system, we can use this program to see how many users are currently and commonly logged in as well as where they have logged in from. We can also obtain this information by viewing log files (see the next section) that store login attempts.

With all of these programs, we have tools that provide us several different views of our system. We can see current statistics of resource usage to see what resources are or have been overly burdened with heavy demand. We can determine peak usage time perhaps to generate policies on scheduling tasks. We can view current processor load to see if we need to remove processes (suspend or kill). We can view averages over time to see if we have insufficient resources. We can gage if there is performance degradation over time. We might also determine if and how often we need to reboot our system.

Table 12.8 provides a summary of many of the commands we have explored in this section and throughout the book. For each command, we note the type(s) of information it provides us. Note that VM stands for virtual memory in the table.

TABLE 12.8
Comparing System Administration Monitoring Programs

Name	Processor Info	Process Info	Memory Info	VM Info	File System Info	I/O Info	Network Info	Comments
df					*			
du					*			
free			*	*				
iostat	*			*	*	*		
lnstat							*	
lpstat						*		
mpstat	*							
netstat							*	Deprecated, replaced by ss
nmap							*	
nstat							*	
pidstat	*	*		*				
ps	*	*	*	*		*		
rtacct							*	
System monitor	*	*	*	*	*	*	*	Graphical, persistent
sar	*	*	*	*	*	*	*	
ss							*	
stat					*			
strace	*	*						
top	*	*	*	*				Persistent
uptime	*							System uptime
vmstat	*	*	*	*				
who							*	Lists logged in users

SECTION ACTIVITIES

1. There are topics from the first subsection of this section that are covered in computer science operating system courses. If you are taking an IT course, how important is it that you should learn these topics?
2. The System Monitor looks like a poor man's Windows Task Manager. If you are a Windows user, compare the two in terms of features both have and features one has that the other does not.
3. Of the Linux tools described in this section, which have you already had experience with? Of those, which do you regularly use? Do you use them for administrative purposes or other reasons?

12.5 LOG FILES

Logging is an automated process in Linux. It is up to the system administrator to review the log files for evidence of problems that need to be resolved. Log files usually contain different types of event information that, collectively, can give us a picture of how our system is running. We will find, among other pieces of information, who is logged in and who has tried to authenticate, what services have successfully started or have stopped, what software has been successfully installed or updated, any errors generated because of hardware issues and what scheduled tasks have run successfully. In this section, we concentrate on a few specific log files to better understand the types of information we will find there. We also explore the frequency by which we might examine these logs.

12.5.1 rsyslogd-Created Log Files

Recall the rsyslog.conf file proscribes the types of events rsyslogd will log and where those messages will be logged. We repeat the default list of this file in Figure 12.15. This informs us that the rsyslogd-created files are all stored in /var/log with the names of messages, secure, maillog, cron, spooler and bootlog. Inspecting these log files will show that they store different types of information but they all contain the date and time of the event, the hostname of the computer, the name of the program that caused the event to be generated and a short description of the event. In most logs, entries are one line of text or less. If the process that generated the message is not rsyslogd or klogd, then that process' PID is included.

The messages file stores events from a variety of sources. Among entries stored here will be non-kernel errors generated during booting and system initialization and messages generated from some of the services that support user applications. Some messages might overlap entries found in the kernel ring buffer, which can be viewed through the dmesg command. In Debian-based Linux, these events are recorded in the file /var/log/syslog.

The secure log (which is called auth.log in Debian distributions) stores information generated by authorization and authentication software. These events occur when users attempt to log in to the system or must otherwise provide a password (e.g., when using su, sudo or passwd). An example of some secure log entries is shown in Figure 12.16. The first three messages pertain to successful authentication events while the bottom half consists of a single attempt that results in unsuccessful authentication.

The first logged message from Figure 12.16 is of a user attempting to log into mycomputer using ssh over port 22. The second message indicates an attempt to enter a password through the GUI by foxr. uid=0 means that root has handled the log in attempt. The third entry line denotes that foxr (UID of 1000) attempted to su to root. The last three items show foxr attempting to su to root and failing. All three entries are of a single event. While the successful authentication attempts may be useful, it is likely that unsuccessful attempts will inform us of more important events such as a potential hacker repeatedly trying to log in.

Figure 12.17 contains excerpts from the log file /var/log/cron of events generated by anacron and crond. For the crond entries, we see that the user who submitted the crontab

```
*.info              /var/log/messages
mail.none           /var/log/messages
authpriv.none       /var/log/messages
cron.none           /var/log/messages
authpriv.*          /var/log/secure
mail.*              /var/log/maillog
cron.*              /var/log/cron
uucp,news.crit      /var/log/spooler
local7.*            /var/log/boot.log
```

FIGURE 12.15 Default rsyslog.conf log file mappings.

```
Nov 23 10:29:16 mycomputer sshd[1781]: Server listening
on 0.0.0.0 port 22.

Nov 23 10:29:41 mycomputer pam: gdm-password[2041]:
pam_unix(gdm-password:session): session opened for user
foxr by (uid=0)

Nov 23 10:32:18 mycomputer su: pam_unix(su:session):
session opened for user root by foxr(uid=1000)
-------------------------------------------------------
Nov 23 11:37:20 mycomputer su: pam_unix(su:session):
session opened for user root by foxr(uid=1000)

Nov 23 11:37:27 mycomputer unix_chkpwd[4993]: password
check failed for user (root)

Nov 23 11:37:27 mycomputer su: pam_unix(su:auth):
authentication failure; logname=foxr uid=1000 euid=0
tty=ptrs/1 ruser=foxr rhost= user=root
```

FIGURE 12.16 Excerpts from the `secure` log file.

```
Nov 20 11:01:01 mycomputer anacron[5013]: Anacron started on
2012-11-20

Nov 20 15:10:01 mycomputer CROND[5042]: (foxr) CMD
(./my_scheduled_script >> output.txt)

Nov 20 16:43:01 mycomputer CROND[5311]: (foxr) CMD (echo "did
this work?")

Nov 18 03:49:01 mycomputer run-parts(/etc/cron.daily)[14361]:
starting logrotate

Nov 18 03:49:01 mycomputer run-parts(/etc/cron.daily)[14364]:
finishing logrotate
```

FIGURE 12.17 Excerpts from `/var/log/cron`.

jobs was foxr. The second of the entries in the figure shows that foxr scheduled the script `my_script` to run and send output to `output.txt` while the third entry shows that `crontab` performed an `echo` instruction as the scheduled task. The first, fourth and fifth entries were of events generated by `anacron`. The first shows that `anacron` started. The fourth and fifth are of `anacron` running `run-parts` to execute the `logrotate` script stored in `/etc/cron.daily`. The fourth entry records when the script started, and the fifth entry records when the script ended.

The `boot.log` file contains logged entries of events that took place during the last boot and system initialization attempt. These entries, unlike the previous log file entries, are of a single event, a system boot, and so only contains the date/time once (Dec 15, 2020 in this case). The events are segmented into groups based on the phase of the initialization step. Figure 12.18 shows several excerpts, each part separated by a dashed line. The first of the excerpts shows steps starting the boot process. Later is the message "Starting Switch Root". Steps that follow include system initialization steps starting (and stopping) various services as target files are utilized. The figure only shows a small portion of a `boot.log` file.

12.5.2 `auditd` Logs

A subdirectory of `/var/log` is `audit` which is used by the `auditd` service to store its log information. `auditd` is responsible for logging general activities of the users. A few of the types of

```
------------ Tue Dec 15 08:21:44 EST 2020 ------------
[[0;32m  OK  [0m] Mounted /sysroot.
[[0;32m  OK  [0m] Reached target Initrd Root File System.
         Starting Reload Configuration from the Real Root...
[[0;32m  OK  [0m] Started Reload Configuration from the Real Root.
[[0;32m  OK  [0m] Reached target Initrd File Systems.
[[0;32m  OK  [0m] Reached target Initrd Default Target.
         Starting dracut pre-pivot and cleanup hook...
[[0;32m  OK  [0m] Started dracut pre-pivot and cleanup hook.
         Starting Cleaning Up and Shutting Down Daemons...
[[0;32m  OK  [0m] Stopped dracut pre-pivot and cleanup hook.
[[0;32m  OK  [0m] Stopped target Remote File Systems.
[[0;32m  OK  [0m] Stopped target Remote File Systems (Pre).
[[0;32m  OK  [0m] Stopped target Initrd Default Target.
[[0;32m  OK  [0m] Stopped target Initrd Root Device.
--------------------------------------------------------------------------
         Starting Switch Root...
--------------------------------------------------------------------------
[[0;32m  OK  [0m] Started Monitoring of LVM2 mirrors, snapshots etc.
using dmeventd or progress polling.
[[0;32m  OK  [0m] Started LVM event activation on device 8:3.
[[0;32m  OK  [0m] Started udev Wait for Complete Device Initialization.
[[0;32m  OK  [0m] Reached target Local File Systems (Pre).
         Mounting Mount unit for gtk-common-themes, revision 1514...
         Mounting Mount unit for gnome-3-34-1804, revision 60...
         Starting File System Check on /dev/disk/by-uuid/7AE9-808C...
--------------------------------------------------------------------------
[[0;32m  OK  [0m] Mounted /boot.
         Mounting /boot/efi...
[[0;32m  OK  [0m] Mounted /boot/efi.
[[0;32m  OK  [0m] Reached target Local File Systems.
         Starting Tell Plymouth To Write Out Runtime Data...
         Starting Import network configuration from initramfs...
         Starting Restore /run/initramfs on shutdown...
[[0;32m  OK  [0m] Started Restore /run/initramfs on shutdown.
[[0;32m  OK  [0m] Started Tell Plymouth To Write Out Runtime Data.
[[0;32m  OK  [0m] Started Import network configuration from initramfs.
         Starting Create Volatile Files and Directories...
[[0;32m  OK  [0m] Started Create Volatile Files and Directories.
--------------------------------------------------------------------------
[[0;32m  OK  [0m] Started Login Service.
[[0;32m  OK  [0m] Started firewalld - dynamic firewall daemon.
[[0;32m  OK  [0m] Reached target Network (Pre).
         Starting Network Manager...
[[0;32m  OK  [0m] Started Network Manager.
         Starting Network Manager Wait Online...
[[0;32m  OK  [0m] Reached target Network.
         Starting OpenSSH server daemon...
```

FIGURE 12.18 Excerpts from the `boot.log`.

activities that we can find logged include successful and failed login attempts, opening of terminal windows, opening, closing and saving files, launching of executable programs, successful and failed system calls of user processes and cryptographic events including the generation of keys. The audit log files are quite large so we will view their content using two handy programs, `aureport`, which summarizes events, and `ausearch`, which root can use to query the audit log files for specific types of information.

For `ausearch`, we specify a type of event that we are interested in. In some cases, we also include a specific value to match against. For instance, we might include a UID to search for entries of a particular user or a PID for entries generated by a particular process. Tables 12.9 and 12.10 provide options for `aureport` and `ausearch`, respectively. The `ausearch` program is more complex but gives us more detailed feedback as it outputs specific audit entries rather than summaries.

TABLE 12.9
aureport Options

Option	Meaning
-au	Authentication attempts
-c	Configuration changes
-e	Events
-f	File operations
-i	Convert numeric (UID, GID, etc) entries into text
-l	Login attempts
-m	Modification of user accounts
-n	Anomalous events
-p	Process initiated events
-s	System calls
-u	User initiated events
-x	Processes executed

TABLE 12.10
ausearch Options

Option	Meaning
-a *EID*	All entries for event # *EID*.
-gi *GID*	All entries of processes owned by group *GID*.
-i	Convert numeric (UID, GID, etc) entries into text.
-k *string*	All entries that contain *string*.
-m *type*	All entries whose message type is listed in *type*.
-p *PID*	All entries generated by process *PID*.
-pp *PID*	All entries generated by process whose parent is *PID*.
-sc *name*	All entries generated by the system call *name* (name may either be a string or number).
-ui *UID*	All entries generated by user *UID*.
-x *name*	All entries generated by the executable program *name*.

Running aureport with no options gives the output shown in top portion of Figure 12.19. We can obtain more information about specific events by using one of the options shown in Table 12.9. For instance, aureport -au provides a summary of the authentication events. Three such events are listed in the middle portion of Figure 12.19. The information reported consists of the authentication event number, the date and time, the user account responsible for the event, the host and terminal window from which the event was generated, the executable program responsible for the event, whether it was successful or not, and the overall event number.

An excerpt from aureport -e (all events) is shown in the bottom portion of Figure 12.19. Here, we see events from the morning of March 19, 2013. These entries indicate user 1000 (foxr) attempted to log in. Using -i in place of -e replaces the UID with the username in the report. These three sections of the figure are separated by a dashed line.

To obtain more detail on an event, we use ausearch. The last event in Figure 12.19 has an event number of 35610. We issue the instruction ausearch -a 35610 and receive the output shown in the top half of Figure 12.20. Note that the formatting will differ in that the entire entry has no line breaks, unlike what we have shown here.

Comparing the last entry in Figure 12.19 to the top entry in 12.20, we see the same event but in more detail. The outputs from both aureport and ausearch provide the date, type of event,

```
Summary Report
======================
Range of time in logs: 12/02/2020 10:11:02.774 -
02/18/2021 10:21:15.081
Selected time for report: 12/02/2020 10:11:02 -
02/18/2021 10:21:15.081
Number of changes in configuration: 18
Number of changes to accounts, groups, or roles: 47
Number of logins: 35
Number of failed logins: 9
Number of authentications: 271
Number of failed authentications: 11
Number of users: 4
Number of terminals: 16
Number of host names: 2
Number of executables: 21
Number of commands: 6
Number of files: 2
Number of AVC's: 0
Number of MAC events: 20
Number of failed syscalls: 0
Number of anomaly events: 3
Number of responses to anomaly events: 0
Number of crypto events: 68
Number of keys: 0
Number of process IDs: 4787
Number of events: 29145
-----------------------------------------------------------------
2. 02/14/2021 10:11:16 foxr ? :0 /usr/libexec/gdm-session-worker
yes 35602
3. 02/14/2021 10:15:59 foxr ? ? /usr/sbin/userhelper yes 35609
4. 02/14/2021 10:22:35 root ? pts/0 /bin/su yes 35617
-----------------------------------------------------------------
10. 02/14/2021 10:11:16 35607 USER_START 1000 yes
11. 02/14/2021 10:11:16 35608 USER_LOGIN 1000 yes
12. 02/14/2021 10:15:59 35609 USER_AUTH 1000 yes
13. 02/14/2021 10:15:59 35610 USER_ACCT 1000 yes
```

FIGURE 12.19 Sample outputs from `aureport` on `auditd` logged information.

event number, UID of the user and PID corresponding to the process that generated the event. But with `ausearch`, we get details that include the user's SELinux context, session and terminal window in which the event occurred and the full path to the executable of the process that generated the event (`gdm-session-worker`). As a point of comparison, a second event is listed in the bottom half of Figure 12.20 that reports on a failed `su` attempt.

```
time->Sun Feb 14 10:11:16 2021
type=USER_START msg=audit(1363702276.674:35607):
user pid=511948 uid=0 auid=1000 ses=2
subj=system_u:system_r:xdm_t:s0-s0:c0.c1023
msg='op=PAM:session_open acct="foxr"
exe="/usr/libexec/gdm-session-worker" hostname=?
addr=? terminal=:0 res=success'
----------------------------------------------
time->Sun Feb 14 11:15:48 2021
type=USER_AUTH msg=audit(1366125348.228:13171):
user pid=521909 uid=1000 auid=1000 ses=1
subj=unconfined_u:unconfined_r:unconfined_t:s0-
s0:c0.c1023 msg='op=PAM:authentication
acct="root" exe="/bin/su" hostname=? addr=?
terminal=pts/0 res=failed'
```

FIGURE 12.20 Sample output from `ausearch` on `auditd` logged information.

12.5.3 Examining the Log Files

The previously examined log files are likely the most significant. But there are others that we need to know about. Several notable log files, not yet examined, are described in Table 12.11.

We would likely only examine files listed in Table 12.11 under specific circumstances that led us to question the behavior of a particular event like issues opening windows, issues with the firewall or issues with installing Linux itself. We would never bother with the httpd logs if we were not running Apache or maillog if we were not running an email server. lastlog, btmp, utmp and wtmp are only worth exploring if we want to see patterns of logins from our users or attempts to hack into our system.

The more significant log files are messages, secure, boot.log and the auditd log files. How frequently should we examine these files? What should we look for? The answers to these questions depend on how many users use our Linux system and whether the system is running well or poorly. Certainly, the messages and secure files should be examined at least weekly if not more frequently. Others may be examined on an as-needed basis. We would normally not need to examine the cron log file unless we suspect scheduled tasks were not being executed. Table 12.12 provides an indication of why and how often we might examine some of the log files presented in this section.

TABLE 12.11
Other Log Files of Note

File (Including Location)	Type of Content	Usage
/var/log/anaconda/	Directory containing logs of events generated during system installation.	Useful if our system does not run correctly after first installing it.
/var/log/dnf.log (or yum.log)	dnf/yum operations.	Did expected updates and installations occur without error?
/var/log/faillog, /var/log/btmp	Failed login attempts.	Search these and secure for evidence of hacking attempts.
/var/log/firewalld	Updates made to the firewall.	Is the firewall set up as expected? Inspect changes made.
/var/log/httpd/	Directory storing two log files from the Apache web server.	Use these if running the Apache server; Apache messages are stored in error_log and HTTP requests in access_log.
/var/log/kern.log	Events logged by the kernel.	Useful if we have modified the kernel to debug our modifications.
/var/log/lastlog	Last login attempts by every user.	Stored in binary, use the lastlog command to view the contents.
/var/log/maillog (or mail.log)	Mail server-related events.	Is our mail server running correctly? This log can also be viewed to see if users are receiving spam.
/var/log/sa/	Directory storing sa log files for sar.	Refer back to Section 12.4.2 for details on sar.
/var/log/spooler	Events generated by USENET.	If we are not running a USENET service, this will be empty.
/var/log/utmp (or /run/utmp) /var/log/wtmp	Currently logged in and previously logged in users.	See who is currently logged in or has logged in and out; both files are stored in binary; use lslogins to view login information from these files.
/var/log/Xorg	Events generated by X Windows.	See what GUI-based events occurred including failures.

TABLE 12.12
Reasons to Examine Log Files

Log file/Program	Reason	Frequency
anaconda	System did not install correctly.	Only if installation is not running properly.
aureport	See a summary of events.	Daily to weekly.
ausearch	View specific events in detail.	If something from aureport caught our attention.
boot.log	Last reboot/initialization resulted in errors or poorly running system.	After each boot or after each boot in which the system is not running well.
cron	See if scheduled events did not run; see what anacron has done.	As needed or weekly at most.
dnf.log/yum.log	Installation/update of software did not work or take place as expected.	Only after failed installation/update attempts.
kern.log	We have modified the kernel.	Never unless we have modified our kernel's source code and recompiled.
messages	Software not running properly or we suspect a problem.	Weekly, more frequently if problems have arisen.
sar/sa logs	We have received reports from users about failed operations or we want to inspect operations that have taken place.	Weekly, possibly more frequently if system is running poorly.
secure	We suspect unauthorized access (even by our own users).	At least weekly; possibly also look at btmp/faillog, utmp/wtmp/lastlog.

12.5.4 journald

systemd-based versions of Linux introduced a new service to oversee all logging called journald. You might wonder why another logging mechanism was needed when there are already so many log files and with services like rsyslogd and auditd. The problem with the various log files is that, taken collectively, they are inconsistent. Different programs log different types of information using different formats. Information is spread out across many log files so that a system administrator will have to search multiple files to get a full perspective of the system.

What journald provides us is a more structured form of logging. journald takes the events generated by other logging services and stores them in a uniform way no matter the source of the event. It then provides a more convenient way to view this collection of log entries through the journald interface, journalctl. journalctl offers a search facility which might allow us to locate useful information faster and more easily than manually or through ausearch.

The journald log files are stored under /run/log/journal. The logs created by journald are formatted and non-text, so the only way to examine their contents is through journalctl. As journald is a service, it has a configuration file, /etc/systemd/journald.conf. We explore some of its directives in Table 12.13.

The journalctl instruction is the administrator's interface with the journald log files. If invoked with no arguments, the entire log file(s) is displayed, which is too much information to be useful. Instead, the command should be issued with some filter(s) to limit the output. Filters are based on fields that the log files store. We list some of the many fields in Table 12.14. The possible values vary based on the field. For instance, _EXE will be a command name, _PID should be a PID and _SYSTEMD_UNIT is a specific unit file (including its type as in dbus.socket, system.slice or gdm.service).

We use a filter with the notation *name=value* where *name* is the filter's name and *value* is the value we want to restrict our search to. For instance, we might issue the instruction journalctl

TABLE 12.13
Directives of the `journald.conf` Configuration File

Directive	Meaning	Possible Values
`Audit`	Whether kernel auditing should run or not.	`true, false`
`Compress`	Should large entries be compressed before writing to the file system?	`true, false`
`RateLimitBurst, RateLimitIntervalSec`	Specifies the maximum number of items that can be logged; overflow messages are dropped.	Numeric value and time unit, defaults to 10,000 messages per 30 seconds.
`RuntimeKeepFree, RuntimeMaxFile, RuntimeMaxFileSize, RuntimeMaxUse`	Size specifications for memory (volatile) dictate the amount of free space that should be maintained, max size of files, total amount of combined space available and number of files that can be used.	Size units are in bytes (M, G, T, etc. can be used); the number of files is an integer with a default of 100.
`Seal`	Whether forward secure sealing is permissible (requires a sealing key).	`true, false`
`SplitMode`	How to divide journal files.	`uid` (split by user), `none` (do not split)
`Storage`	Where to store journal data such as memory or disk.	`auto, none, persistent, volatile`
`SystemKeepFree, SystemMaxFiles, SystemMaxFileSize, SystemMaxUse`	Size limits on contents stored on disk (persistent) dictate the amount of free space that should be maintained, max size of files, total amount of combined space available and number of files that can be used.	Size units are in bytes (M, G, T, etc. can be used); the number of files is an integer with a default of 100.

`_UID=1000` to obtain all logged events pertaining to the user whose UID is 1000. Specifying multiple filters acts like a logical AND operation in that an entry must fulfill all of the filters specified. For instance, `journalctl _TRANSPORT=syslog _UID=1000` will output only those entries that have both fields of `_TRANSPORT=syslog` and `_UID=1000`.

The fields/filters listed in Table 12.14 all start with an underscore (_). These denote *trusted* fields. Other fields may be added by specific software (in which case they do not start with an underscore or are addresses of field entries which begin with two underscores). We omit these non-trusted fields from the table.

By default, `journalctl` outputs the date/time, hostname, process generating the event and its PID and a brief message. The option `-o` (or `--option`) allows us to specify how much information should be displayed, defaulting to `short`. Several options are available to modify the timestamp.

Timestamp options include `short-full` to provide more detail in the timestamp and `short-iso` to use a wallclock timestamp. `short-iso-precise` adds microseconds to the `short-iso` timestamp while `short-precise` uses a normal timestamp but with microseconds added. `short-monotonic` specifies a monotonic timestamp. `short-unix` provides the timestamp in seconds since the epoch.

Other formats for `-o` include `verbose` to provide full detail, `export` for a verbose output but with some of the identifiers expressed numerically in hexadecimal notation, `json`, `json-pretty` and `json-sse` to output verbosely but using JSON data structures, `cat` for a terse output and `with-unit` which is the same as `short-full` but with unit prefixes added to names.

TABLE 12.14
journald Fields and journalctl Filters

Field	Value	Use
_BOOT_ID	Kernel ID	Messages generated by the boot process of the specified kernel.
_CMDLINE	Command line (full command)	Messages generated by the specified command.
_COMM	Executable name	Messages generated by the specified executable.
_EXE	Executable path	Same as _COMM but includes the full path.
_GID, _UID	GID, UID	Messages generated from any process running under the given GID or UID.
_PID	PID	Messages generated by the process specified by PID.
_SELINUX_CONTEXT	SELinux context	Messages generated by any process running under the given SELinux context.
_SYSTEMD_CGROUP, _SYSTEMD_UNIT	Control group path or systemd unit	Messages generated by any process running in the specified cgroup or of the specified unit.
_TRANSPORT	One of audit, driver, syslog, journal, stdout, kernel	Messages received by journald using the specified service or software type.

We can also control the fields to output using `--output-fields=` and list each field, separated by a comma.

Several other options are worth noting. `-n` *n* (or `--lines=`*n*) outputs the last *n* events in the journal. `-r` reverses the output to appear from most recent to most distance. `-g` *regex* (or `--grep=`*regex*) further filters `journalctl` as if we piped the result of our `journalctl` instruction to `egrep`. If `journalctl` is drawing from multiple journals, `-m` merges the contents in an interleaved fashion. `--no-hostname` causes output to appear without the hostname.

We can apply some of the filters more easily through options. `_BOOT_ID` can be replaced with `-b` *ID*. `PRIORITY` (which is not a trusted field) can be replaced with `-p` *priority* or `--priority=`*priority*. `-u` *unit* or `--unit=`*unit* can be used in place of `_SYSTEMD_UNIT=`*unit*. We can also add or use `-S` (or `--since=`) and/or `-U` (or `--until=`) to specify a time/date that we want to filter from or to. Times and dates are listed as `yyyy-mm-dd hh:mm:ss`. If the time is omitted, it is assumed to be midnight of the specified date. If the date is omitted, it is assumed to be today. Dates of `yesterday`, `today` and `tomorrow` can be used instead of specific dates.

As `journald` stores its journal(s) under */run*, the journal only exists until we shut down or reboot the system. We can rotate the journal from `/run/log/journal` into `/var/log/journal` for permanent storage using `--flush`, and we can force a journal stored in `/var/log/journal` to be rotated using `--rotate`. The option `--list-boots` causes `journalctl` to only list the IDs of available boots as stored in the journal. `-k` (or `--dmesg`) displays kernel messages only. `-F` *field* (or `--field=`*field*) lists the values found in the specified field throughout the journal. Finally, `-N` (or `--fields`) lists all available fields.

Advantages of using `journalctl` are that `journald` presents a uniform interface to examine log entries and `journalctl` gives us a single instruction to access the combined log entries. On the other hand, there are reasons to explore an individual log file such as being able to view all authentication events in `secure` without having to guess at proper filters to provide such data.

Having a one-size-fits-all tool like `journalctl` might also make `journalctl` challenging to master. By combining all log entries into one file, looking for a specific incident might require more thought than scanning a single log file. For this and other reasons, the other logs remain available and should be used if you are more comfortable with them over `journald`.

SECTION ACTIVITIES

1. In Windows, you can view events with the Event Viewer tool. Run it and take a look. Can you find anything useful in the logged messages?
2. Windows uses the Event Viewer, thus consolidating all logging under one tool. As we see in Linux, logging is handled in a variety of ways. `journald` is an attempt to provide one mechanism to view logs. Which approach do you prefer, a single Event Viewer which logs and displays log messages, the older Linux approach where logs are creating by different software and viewed by different software, or the `journald` approach where there are still different logs being created but `journald` consolidates log entries and `journalctl` lets you view messages with a single tool?
3. Look at the log files and subdirectories under `/var/log` to see which are readable to non-root users. Why do you suppose these specific log files are set this way and the others are not?

12.6 TROUBLESHOOTING

We have now presented the tools available to troubleshoot a system. Which tools should we apply and when? This section examines a number of technical problems and discusses system administration efforts to resolve those problems. For each problem, we look at steps to further identify the cause of the problem followed by easy or short-term solutions and then more involved or long-term solutions. The solutions described in these scenarios are not intended to be complete but should illustrate some of the types of efforts that you, the system administrator, should take when faced with similar problems.

Problem 1: System is running ineffectively.

Description: Simple tasks are taking too long to execute. Login is taking more time than expected. There is a delay between issuing a command and seeing its result.

Steps to determine the problem:

1. Use one or more of `top`, `ps`, the system monitor, `mpstat`, `sar` and `pidstat` to view the running processes.
   a. Is CPU load heavy, approaching 100%?
   b. Are there processes that are taking most of the CPU time or are there many processes taking little time but combined cause a heavy load?
2. Use `vmstat` and `free` to examine main memory and swap space utilization.
   a. Is main memory full?
   b. Is the system spending a lot of time swapping?
   c. Are there too many processes in memory?
3. Use `who` to determine if there are currently a lot of users logged in.
   a. Are there more users than expected?
4. Use `uptime` to see how long the system has been running without a reboot. While Linux seldom needs to be rebooted, a reboot may resolve some problems.

Short-term solutions: Identify system processes that can be halted and rerun later. Possibly schedule some of these. Identify processes whose priorities can be lowered through renice, or processes which could be moved to the background. Contact users of some of the resource-heavy processes and ask them to discontinue their processes and/or log off. Ask them to schedule their processes at times you consider as non-peak times. Reboot the computer if the above steps do not solve the problem.

Long-term solutions: Purchase more main memory, increase the size of the swap partition (possibly add a second hard disk to contain more of the swap partition) possibly purchase a more powerful processor (or additional processors). Separate users across multiple Linux computers if absolutely necessary.

Problem 2: Service not responding.

Description: Some service fails to respond such that users are receiving error messages when they attempt to use it. For instance, HTTP requests to a webserver result in server not found.

Steps to determine the problem:

1. Identify the service in question (from the users), and if this information is not readily available, consult `/var/log/messages` for evidence of a service stopping.
2. See if the service is running (e.g., `systemctl status` *`servicename`*).
3. See if the service's supporting programs are running.
4. View the configuration file for the service to see if it is configured correctly.
5. View the log file(s) generated as a response to the service.
6. If the service is one that listens over the network, make sure the port it listens to is open and that the port is open through the firewall.

Short-term solutions: Restart the service. Restart supporting services. Make sure the network is running correctly (restart NetworkManager if needed, look at the firewall). If absolutely necessary, reboot (softboot) the computer.

Long-term solution: If the problem is related to a server (not service) which is not functioning correctly, reinstall the server or view the server's websites to see if there are solutions to the given problem. If the service failed for a known reason, investigate that reason. It may require reconfiguring the service through the service's configuration file.

Problem 3: Inadequate hard disk space.

Description: One or more of the file systems is filling up or has become full. Users cannot save files. Or, swap space is commonly low on available space.

Steps to determine the problem:

1. Use `df` to view how full each file system is. To identify specific directories' sizes, use `du`.
2. Use `find /home -size +`*`c`* to search for inordinately large user files (using a value of *c* that you feel is too large for files) if `/home` is low on space. Look at the log files' sizes in `/var/log` to see if any need rotating and rotate them by hand. Look at the files used in other `/var` directories including `cache` (including `dnf/yum`), `mail`, `spool`.
3. If swap space is low, examine swap history using `vmstat`, `sar` and `pidstat`.

Short-term solutions: Back up the file system which is running out of space to ensure that everything is available should the system fail. Delete (or at least compress) overly large files and warn the user/owner of the files (e.g., "I have removed several core dumps found in your directory"). Ask users to clean up their file space. Implement disk quotas, if necessary, to prevent user spaces from filling up in the future. Initiate mail quotas. Move large log files to an archive and/or modify log rotation behavior (e.g., start compressing rotated files).

Long-term solutions: Back up *all* file systems. Purchase additional hard disks and either divide up the previous file system that was filling up across the disks (e.g., divide user home directories across the disks) or repartition the file systems to use the added disk space and enlarge the file systems filling up.

Problem 4: Suspicious system behavior.

Description: Services or programs are not working as they should. System might be too slow. Files might have disappeared. Processes are being run by users that shouldn't be running them.

Steps to determine the problem:

1. Examine the log files, particularly `secure`, `lastlog` and `btmp`, to look for unusual patterns of logins. Examine the `sa` files using `sar` or `auditd` files using `ausearch` and look for activities of file deletions that should not occur.
2. Look for running processes with peculiar ownership.
3. Use `ausearch` to look at authentication events, particularly failed ones.
4. Look for evidence of computer virus or Trojan horse.

Short-term solutions: Kill any suspicious processes (with apologies to any users who own those processes). Run antiviral software. Reboot the computer if needed. Examine the firewall to make sure it is running.

Long-term solutions: Implement a more secure authentication system and a more secure firewall. Implement an intrusion detection system. Discuss account protection with the users. Require all users to change passwords at the next log in, increasing the strength of passwords. Delete any suspicious user accounts.

Problem 5: Scheduled event did not take place.

Description: An event scheduled through `crontab`, `at`, `batch` or `anacron` did not take place as expected.

Steps to determine the problem:

1. Examine the `cron` log file to see if other events did not take place as planned.
2. See if system downtime occurred when the event was scheduled.
3. Check to see if the event took place at the next boot (via `anacron`).
4. Check to see if the event failed because the scheduler did not have adequate permission (e.g., the scheduler's user name appears in `/etc/cron.deny` or `/etc/at.deny`, or if the user attempted to run a program that the user does not have access to).
5. See if the event took place but generated an error, through the `audit` or `cron` log.

Short-term solution: Re-schedule the event.

Long-term solution: Move the event from `cron` to one of the `/etc/cron` directories (e.g., `cron.daily`). Change the scheduled event date and time. Remove the user from the blacklist file if the user has been blacklisted or schedule the event yourself in lieu of the blacklisted user.

Problem 6: Network not responding.

Description: Users are unable to reach other computers via web browser or other network tools.

Steps to determine the problem:

1. See if the `NetworkManager` is running.
2. Check the physical connection to the network to see if there is something wrong with the cable or port.
3. Use `ip` to check the status of your interface device(s) to make sure it is up and running and has an IP address. Use `ip route` to make sure there is a connection to a gateway (router). Test various network resources using `ping` and/or `traceroute`. There might be network access but not access outside of the local area, or the computer that the user is trying to reach may be inaccessible.

4. Make sure there is access to a DNS name server. Make sure `/etc/resolv.conf` has IP addresses of your DNS name servers. Test a DNS lookup using `nslookup`, `host` or `dig`. If the local DNS name server is inaccessible, use a public one like google's or try to connect to the computer of interest using the IP address rather than a hostname.
5. See if other users are also unable to communicate via the network.

Short-term solution: Restart the `NetworkManager` service. Examine the interface configuration file (e.g., `ifcfg-ens33`) and correct any mistakes found (for instance, if a netmask is listed and is inaccurate, fix this). If using DHCP, make sure the DHCP server is responding, possibly asking the network administrator to restart it. Test the local name servers to make sure they are responding. If they are unavailable, network and Internet access may still be available but only through IP addresses. If there is an issue with the local name servers, you might change to using a public one such as those available from Google). Add address resolution information to your `/etc/hosts` file. Ask the network administrator to reboot the local switch and/or routers if any are not responding. If nothing else works, try to reboot your computer.

Long-term solutions: Reconfigure the network itself by replacing the DHCP server and/or hardware (switches, routers, gateway). Test all network cables (not just the one leading to your computer). Try an alternate network interface device (e.g., replace your Ethernet card with a new one).

Problem 7: Software not running correctly.
Description: Some software that was previously running is no longer running correctly.
Steps to determine the problem:

1. View the `messages` and `audit` logs to see if the software is causing errors.
2. Check the `dnf` (`yum`) log to see if the software is in need of an update.
3. Examine the software's configuration and/or rules file(s) (if any) and see if they have been inappropriately modified.
4. View the software's website to see if known errors have been found recently.

Short-term solution: Using `dnf/yum`, upgrade the software. If this fails, attempt to remove the software and install it anew, possibly moving to a newer version. If the software was recently updated, see if rolling it back to the previous version resolves the issue.

Long-term solution: If the software comes with support, contact the support team and detail the problem. If there is a website that permits problem uploads, upload the problem there. Explain what has been tried but failed, and any modifications made to the software since the installation. Usually tech support in these websites requires very detailed information about the computer, operating system and specific steps that already tried and the results.

Problem 8: Device failure.
Description: Hardware device is not functioning.
Steps to determine the problem:

1. Check the device's physical connection(s) to the computer.
2. Make sure the device is turned on, or if the device has its own power source, make sure it is plugged in or that the battery is adequate.
3. Make sure the device's driver software is installed in your system.
4. Check the `boot.log` and `messages` log file to see if an error message was generated at the last reboot (or if this is a plug and play device, when the device was connected).
5. If the device uses a service, see if the service is running.

Short-term solution: Restart any necessary service(s). Disconnect and reconnect the device and check the power. Reboot the device if it is bootable. Reboot the computer.

Long-term solution: Connect the device to another computer to see if it works from there. If so, check the port/connection on the computer where it was not working. Replace the device driver on that computer. If the device still does not work on different computers, then replace the device.

Problem 9: Cannot access file system.
Description: Root file system available but other file system(s) unavailable.
Steps to determine the problem:

1. Use `df` and examine `/etc/mtab` to see if the partition is mounted and `/etc/fstab` to make sure the file system is listed as one that should be mounted.
2. Check the mode of the computer to see if you are in rescue mode or a state where file systems are not mounted.
3. Run `fsck` on the file system to see if it is damaged.
4. If the file system is to be remotely mounted, make sure `rpc.mountd` (the NFS mount daemon) is running and that the network is accessible.

Short-term solution: Use `systemctl` to change modes to multiuser or graphical mode if in rescue mode. Attempt to mount the file system by hand using `mount`. Update `/etc/fstab` if the entry is missing or erroneous. If the file system is not on the internal hard drive, make sure the external device is connected and accessible over network. Restart `NetworkManager`. The problem may be with your network connectivity.

Long-term solution: Repartition to create a new version of the file system (if it is damaged); add hard disk space as needed. If the file system is being remotely mounted, contact the system administrator of the remote system and make sure the file system is available.

Problem 10: System does not initialize correctly.
Description: Upon boot/reboot, the operating system does not come up in a usable mode.
Steps to determine the problem:

1. Check `dmesg` for errors during system boot.
2. If the system initializes to Linux, see what target the system initialized to.
3. Does `/usr/sbin/systemd` exist?
4. Is the root file system being mounted?
5. Examine GRUB2 to make sure it is configured correctly.
6. Examine `/boot` to make sure there is at least one usable `vmlinuz` image available.

Short-term solution: If errors arose during boot (from `dmesg`), try to diagnose the cause of those errors (bad device, bad kernel image), fix the issue and reboot. If the system came up in the wrong target level, use `systemctl` to initialize to the desired level and then use `systemctl` to change the default target. See if any of your partitions are full requiring repartitioning of one or more of your file systems. If the system does not come up at all, interrupt the boot process to reach the `GRUB2` command line and make sure that GRUB2 is configured and can find and access the Linux kernel. Another issue might be that GRUB2 cannot locate `/boot`. Test to make sure `/boot` is accessible and contains both the proper `initramfs` and `vmlinuz` images.

Long-term solution: Reinstall the OS.

Although the above ten troubleshooting scenarios cover serious problems, they are only a few possible situations that you may encounter. Hopefully in reading this text, you know enough now to explore problems and find solutions. Remember that there are tens of thousands of Linux users and administrators who regularly contribute to the community. Don't be afraid to ask your questions on a Linux website.

SECTION ACTIVITIES

1. Of the scenarios explored in this section, have you had similar incidents occur on your home computer? If so, what tools did you use to identify the problem and solve it?
2. There are dozens of websites for people to post questions seeking help for problems with their operating system. Have you visited any of these? If so, for what reason? If not, do a search on some issue your computer is having and see what sites you find.

12.7 CHAPTER REVIEW

Concepts and Terms Introduced in This Chapter

- Archive – moving files deemed no longer needed to long-term/permanent storage; companies typically archive transactions, tax records, etc., from previous years.
- Backup – storing files from a hard disk onto a separate media in case damage occurs to the hard disk.
- Backup media – physical storage media used to store backup files (or archive); commonly, a remote or external hard disk or magnetic tape but may also be USB flash drive or optical disc.
- Backup strategy – plan that specifies the partitions to backup and the frequency of those backups, a duration that backup media is retained, and the pattern of full versus incremental backups.
- Cycle stealing – situation where the CPU waits for another device before it is allowed to access memory.
- Certificate authority – organization trusted to sign an x.509 digital certificate to ensure the certificate's authenticity.
- Data corruption – result of a race condition in which a datum's value is modified incorrectly due to overlapping access by multiple processes.
- Deadlock – two or more processes holding resources that the other processes have requested resulting in the processes involved all remaining indefinitely in a blocked state.
- Digital certificate – used in telecommunications to both provide users with a public key and to show that the organization can be trusted to facilitate monetary transactions; specifically we use x.509 certificates to support public key encryption in HTTPS and other Internet protocols (e.g., SSH).
- Encryption – the process of encoding a message or file into a form which is unintelligible if read.
- Full backup – storing the entire contents of a file system onto a backup medium.
- Hamming code – mathematical process encoding error detection and correction bits; proposed for use in RAID storage but not used due to the complexity of the mathematical process involved; used in some forms of telecommunication.
- Incremental backup – backing up only those files in the file system that are new or have been modified since the last full or incremental backup.
- Liveness – the state of a process which is making progress toward eventual completion and termination.
- Log file – a collection of messages that describe events that may be of use to the system administrator; log files are generated automatically by logging software.
- Mirror – using two sets of disk drives so that any file is stored on both sets providing 100% redundancy on all storage; this form of redundancy is used in RAID 1.

- Mutually exclusive access – disallowing any process from accessing a datum while another process is accessing the datum; this form of access removes potential race condition; typically implemented using some form of synchronization.
- One-time scheduling – scheduling an event to run one time at a specific time and date; the `atd` service runs one-time scheduled tasks as scheduled by `at` and/or `batch`.
- Ostrich algorithm – ignoring the possibility of deadlock; most operating systems use this approach (or lack of approach) in dealing with deadlocks.
- Parity – the use of an extra bit to facilitate error detection; even parity requires that the number of 1 bits is even.
- Parity bit – bit added to a byte to encode evenness or oddness of the number of 1 bits in the byte; the parity bit is used for simple error detection in computers.
- Parity byte – encoding a parity bit for each bit of a group of bytes; the parity byte is a common form of redundancy used in many RAID levels.
- Privacy Enhanced Mail (PEM) – format used to store encryption keys and digital certificates taking the binary data and storing each 6-bit sequence as one of 64 ASCII characters.
- Private key – a key used to both encrypt (in private-key encryption) and decrypt (in public-key and private-key encryption) files/network messages; the key is kept secret to ensure the integrity of the files/messages.
- Private-key encryption – a type of encryption algorithm that applies a private key using a passphrase so that the user can encrypt and decrypt files and messages.
- Public key – a key that can be shared publicly so that users can encrypt files and messages; the public key cannot be used to decrypt messages encrypted by a public key; public keys are generated from private keys.
- Public-key encryption – type of encryption algorithm that uses two keys, a publicly available public key for encryption and a secretly held private key for decryption.
- Race condition – two or more multitasking processes accessing a shared datum in an overlapped fashion, whereby each process has a different value to write back to the shared storage unit and thus the pattern of writes corrupts the datum.
- RAID – redundant array of independent (or inexpensive) disks to provide file storage integrity if some contents become corrupted; redundancy encodes extra data so that corrupted data can be recovered while independent disk accesses support more efficient disk accesses; there are several different forms of RAID technology known as levels.
- RAID level – a type of strategy used to provide redundancy and independence of disk accesses; there are seven commonly cited levels (although not all of them are used) along with hybrid and proprietary forms.
- Recurrent scheduling – scheduling of tasks to be performed in a recurring pattern such as at a specific time every day or a specific time and day every week; users can schedule such tasks using `crontab` with services `crond` and `anacron` being responsible for executing such scheduled jobs.
- Restoration – a process of copying damaged, destroyed or lost files from a backup to a computer's hard disk.
- Self-signed certificate – a type of digital certificate that was signed by the owner rather than a certificate authority and so may not be trustworthy.
- Starvation – a situation where a process is not able to progress toward completion because resources that it needs are continually being held by other processes.
- Stripe – a strategy in RAID storage in which a file's block is further subdivided and distributed across multiple disks to promote faster disk access.
- Synchronization – controlling access to shared resources so that once a process is holding a resource, all other processes requesting that resource are blocked from access and forced into a waiting state.

- Thrashing – a situation in which the operating system is spending too much time paging between main memory and virtual memory and so little to no processing is accomplished.
- X.509 digital certificate – a format of digital certificate used to support encrypted forms of telecommunication in HTTPS and SSH.

Linux Commands Covered in This Chapter

- anacron – service to execute scheduled events in a recurring pattern but not at a specific time.
- at – program to schedule one-time events at a specific date and time.
- atd – service to run one-time scheduled tasks submitting using `at` and `batch`.
- atq – list a user's `at/batch` scheduled tasks.
- atrm – delete `at/batch`-based scheduled tasks.
- auditd – service which logs software events.
- aureport – query `auditd` log files for summaries of types of events.
- ausearch – query `auditd` log files for details of types of events.
- batch – schedule tasks when system load drops below 80%.
- cpio – backup utility that can store individual files, directories or partitions; contains a crude incremental backup facility.
- crond – service to run recurring scheduled events as scheduled by `crontab`.
- crontab – schedule recurring tasks; unlike `at` in which a user can schedule any number of events, each user is able to submit a single `crontab` job which can include many different schedules and tasks.
- df – display file system usage statistics.
- du – display amount of disk usage for files and directories.
- dump – backup utility which only operates on file systems but can perform full and incremental backups.
- free – report on the amount of used and available memory.
- iostat – display processor, I/O, virtual memory and file system usage statistics.
- klogd – the kernel logging daemon.
- journalctl – interface with `journald` log files.
- journald – `systemd` service used to consolidate all logging.
- lnstat – display network statistics including routing cache information.
- lpstat – display printer statistics.
- mdadm – set up RAID storage for Linux; can be used to simulate some RAID levels on a single drive or group multiple drives into a logical RAID unit.
- mpstat – report on processor utilization, primarily intended for multiprocessor systems.
- nmap – scan a network host's available ports.
- openssl – open-source program used for encryption, decryption, generation of public and private keys and generation of digital certificates.
- pidstat – report on processor utilization broken down by process.
- ps – process snapshot program to display the currently running processes and their resource utilization.
- run-parts – used by the `anacron` schedule to run scripts.
- sar – report on statistics compiled into the `sa` log files; logged events are snapshots of the system at specified intervals (usually every 10 minutes).
- strace – trace through the system calls made by executing a Linux program; useful for debugging the kernel.
- syslogd – logging daemon for non-kernel system events.

- System monitor – GUI program to display running processes and their resource utilization as well as system-wide resource utilization.
- tar – tape archive program for creating and extracting archives; originally intended for backups but primarily used today as a package manager utility.
- top – text-based program to display running processes and their resource utilization as well as system-wide resource utilization.
- uptime – report on the time since the last boot.
- vmstat – report on the amount of memory and virtual memory utilization during the current uptime.
- who – output usernames who are currently logged in.

Linux Log Files under /var/log Covered in This Chapter

- audit/ – directory storing the `auditd` log files.
- boot.log – events related to the last boot.
- btmp – failed login attempts.
- cron – scheduled events executed by `anacron` and `crontab`.
- dnf.log – may also be `yum.log`, repository of `dnf/yum` update and installation actions.
- faillog – failed login attempts.
- lastlog – users' last login information.
- maillog – messages related to the mail program.
- messages – system events as logged by `rsyslog` and `klogd`.
- sa/ - directory storing snapshots of the system in intervals (usually 10 minutes) as used by `sar`.
- secure – authentication events as generated by any authentication program.
- utmp – up-to-date status of all login/logout events and related system events since the last boot.
- wtmp – compilation of `utmp` logs.
- Xorg – X Windows operations of different console windows.

REVIEW QUESTIONS

1. Why is the hard disk the component most liable to fail in a computer?
2. Assume you back up your hard disk every week. Is there a reason to retain a week-old backup when you replace it with the next week's backup? If so, why?
3. Rank these partitions in terms of the frequency of backups from most to least: /, `/boot`, `/home`, `/var`, `/usr`, `swap`
4. What is the advantage of using incremental backups rather than full backups? What is the disadvantage of using incremental backups?
5. You perform a full backup of `/home` the first day of the month and incremental backups every 3 days afterward (9 incremental backups). Assuming `/home` stores 512MB and that approximately 5% of /home changes every 3 days, what is the total size of the backups for the month?
6. For a standalone Linux workstation, how often should you back up `/home`? `/var`?
7. For a file server storing dozens of users' home directories, how often should you back up `/home`?
8. What are the advantages of using `tar` over `dump` for backups? What are the advantages of using `dump` over `tar` for backups?

9. You perform a full backup of /home on the first day of the month (refer to this as backup 0) and an incremental backup each week afterward (assume three total incremental backups, referred to as backups 1, 2 and 3, respectively).
   a. A user reports a corrupted file. In what order are the backups searched for a copy of the file?
   b. A user reports that he mistakenly changed a file sometime in the last month but can't remember when. In what order will you search the backups to find an unchanged version of the file?
   c. Will your answer to part b change if the user says he updated it several times during the month?
   d. /home has been damaged and you need to restore dozens of files. In what order will you restore the files from the backups?
10. You have three tapes to store backups on. You plan to put a full backup on one tape and then incremental backups on the other two tapes doing the incremental backups every other day. You estimate that you can place 4 incremental backups on each tape. How many days will it be before you run out of tapes and must overwrite the full backup on the first tape?
11. How does RAID 1 differ from RAID 0?
12. Why is RAID 2 not used?
13. How does RAID 4 differ from RAID 5? What aspect of RAID 5 alleviates the problem that RAID 4 presents?
14. RAID 1 is the most expensive because it uses 50% of the storage space for redundancy. Which level would be the second most expensive? Which level is the cheapest?
15. If you use RAID 1, does this mean you don't need to back up your hard disk?
16. Compute the parity byte for the six bytes listed here: 00000001, 11111111, 01010101, 11110000, 01110101, 01100111.
17. In Figure 12.4, if the parity byte were 01101011 instead of 01110101, what should byte 3 be?
18. True/false: Using parity byte redundancy, you can recover a missing byte given the other bytes and the parity byte but you would not be able to recover two missing bytes given the other bytes and the parity byte.
19. Why would you not want to implement RAID 0 on a single disk?
20. Which of these programs that can encrypt files would you expect to find pre-installed in a Linux distribution? `7-zip`, `ccrypt`, `gpg`, `pgp`, `zip`
21. Which of a public key and a private key do you use to decrypt files in public-key encryption?
22. In private-key encryption, we use a private key. Do we also use a public key?
23. Which of these is true?
   a. We use a public key to generate a private key.
   b. We use a private key to generate a public key.
   c. We generate the public and private keys independently.
24. In Figure 12.7, the last command generates an X.509 digital certificate. Explain what each of the options in the command refers to.
25. Given the following three strings, use a Base64 table (for instance, at https://en.wikipedia.org/wiki/Base64) to convert each character to its equivalent binary code (6 bits) and then group the binary into 8-bit sequences and convert each 8 bits to a character using an ASCII table (you can find an ASCII table at http://www.asciitable.com/).
   a. `QUJDREVGR0g`
   b. `Rlogcm9ja3Mh`
   c. `SXMgTGludXggZnVuPz8/`

26. It is currently 6:15 PM on 12/25/23. When will each of the following commands execute?
    a. `at teatime -f foo.sh`
    b. `at tomorrow -f foo.sh`
    c. `at 6:16 -f foo.sh`
    d. `at now + 15 hours -f foo.sh`
    e. `at 22:22 020224 -f foo.sh`
27. You have submitted five `at` jobs and realize that you want to cancel the third one. What steps do you take to accomplish this?
28. What happens when you submit an `at` command without using the `-f` option?
29. As the system administrator, you have decided to disallow user zappaf from using `at`. How do you enforce this?
30. Specify the following occurrences using the five values in a `crontab` job. Remember to use * when needed.
    a. Every day at 10 AM.
    b. The first Monday of every month at 12:01 AM.
    c. The 1st and 15th of January and June at 2 PM.
    d. Every 5 minutes between 12:00 PM and 1:00 PM (inclusive) on August 10.
    e. Every December 31st at 11:59 PM.
31. What is the difference between `crontab -e` and `crontab filename` where *filename* is a file storing crontab scheduling information?
32. What is the recurrence that each of the following specifies for a `crontab` job?

a.	*	*	1	*	5
b.	0	1,13	*	*	*
c.	30	19	*	6	*
d.	*/5	23	*	*	*
e.	59	11	15	3	*

33. True/false: When `crond` runs a scheduled job, it runs under the user's permissions and environment variables.
34. What type of item would you place in the directory `/etc/cron.daily` and why would you do this?
35. Are starvation and deadlock the same thing? If not, how do they differ?
36. Similar to Figure 12.11, assume we have two processes, P0 and P1, sharing a datum X, which is initially 1. P0 will add 3 to the datum and P1 should then double the resulting datum. If synchronization is not properly employed, what possible results values could result in the shared datum X after both P0 and P1 run?
37. Can we have a race condition in a single-tasking system? Why or why not?
38. What approach does Linux take with deadlock of user processes?
39. What is cycle stealing?
40. True/false: `iostat` reports I/O statistics but no other useful information.
41. When viewing the report from `vmstat`, there are fields listed as `us`, `sy`, `id`, `wa`. What do these four items tell us?
42. Which of these Linux programs provides information about network usage? `mpstat`, `netstat`, `nmap`, `nstat`, `pidstat`, `sar`, `strace`
43. True/false: There is no reason to use `mpstat` if your computer has a single processor.
44. You want to see how much space is still available in a partition. Should you use `df`, `du`, `stat` or `vmstat`?
45. What does the program `uptime` tell you?
46. You suspect a scheduled process did not run correctly. What log file(s) might you examine?
47. By default, what are the intervals of data collected by `sar`? Where is the data stored?

48. You suspect that a user has been trying to su to root. What log file(s) should you examine for such attempts?
49. During the last boot, you suspect a piece of hardware failed. What log file should you look at to see if a failure event was logged?
50. The cron log file(s) stores information on crond events. Does it also store information on atd events? anacron events?
51. What command would you enter to find the number of login attempts since your operating system was last booted? The number of system calls?
52. Which of ausearch and aureport provide more detail about a single event?
53. Provide a command to display all auditd log messages from the user whose UID is 1501.
54. The aureport -m option displays user modification events. What is a user modification event and what programs can you think of that would generate such an event?
55. What would ausearch -sc open do?
56. Explain the auditd log entry shown in the bottom half of Figure 12.20. What was the event and what user (by UID) caused the event to happen?
57. Given the following audit log entry, explain each part that is underlined.

```
time->Wed Jul 17 12:20:10 2014
type=CRED_ACQ msg=audit(1374078010.023:59603): user pid=3102 uid=501
auid=501 ses=1 subj=unconfined_u:unconfined_r:unconfined_t:s0-s0:c0.
c1023 msg='op=PAM:setcred acct="root" exe="/bin/su" hostname=? addr=?
terminal=pts/2 res=success'
```

58. Which of the log files stored under /var are stored in a non-text format?
59. True/false: journald stores its log information in memory rather than on disk.
60. What is the difference between the journald.conf directive RuntimeKeepFree and SystemKeepFree?
61. Specify a command to obtain all of the journald log events of the user whose UID is 1501.
62. Specify a command to obtain all of the journald log events of the process whose PID is 123456 and which was logged by rsyslogd.
63. Provide a reason for using journalctl to examine log entries rather than searching logs by hand. Provide a reason to still look at logs by hand rather than using journalctl.
64. Your Linux computer boots to rescue mode. How will you troubleshoot the situation? Provide an analysis similar to the 10 troubleshooting scenarios presented in Section 12.6.
65. A user is not able to log into his or her account even though other users can. How will you troubleshoot the situation? Provide an analysis similar to the 10 troubleshooting scenarios presented in Section 12.6.
66. A remotely mounted file system is no longer accessible. How will you troubleshoot the situation? Provide an analysis similar to the 10 troubleshooting scenarios presented in Section 12.6.
67. Too much disk swapping seems to be taking place. How will you troubleshoot the situation? Provide an analysis similar to the 10 troubleshooting scenarios presented in Section 12.6.

Bibliography

Abrams, M., LaPadula, L., Eggers, K., and Olson, I. A Generalized Framework for Access Control: An Informal Description, *Proceedings of the 13th National Computer Security Conference*, pp. 135–143, 1990.

Accetta, M., Baron, R., Bolosky, W., Golub, D., Rashid, R., Tevanian, A., and Young, M. Mach: A New Kernel Foundation for Unix Development, *Proceedings of the Summer 1986 USENIX Conference*, pp. 93–112, 1986.

Adelstein, T., and Lubanovic, B. *Linux System Administration*, Sebastopol, CA: O'Reilly, 2007.

Aho, A., Kernighan, B., and Weinberger, P. *The AWK Programming Language*, Boston, MA: Addison-Wesley, 1988.

Al-Hafeedh, A., Crochemore, M., Ilie, L., Kopylova, E., Smyth, W., Tishcler, G., and Yusufu, M. A Comparison of index-based Lempel-Ziv LZ77 factorization algorithms, *ACM Computing Surveys*, Vol. 45, Issue 1, 5:9–5:17, Article 5, 2012.

Allen, N. *Network Maintenance and Troubleshooting Guide: Field Tested Solutions for Everyday Problems*, Boston, MA: Addison-Wesley, 2009.

Almesberger, W. Booting Linux: The History and the Future, *Proceedings of the Ottawa Linux Symposium*, 2000.

Almgren, M., Debar, H., and Dacier, M. A Lightweight Tool for Detecting Web Server Attacks, *Networks and Distributed System Security Symposium*, pp. 157–170, 2000.

Anderson, K. Convergence: A holistic approach to risk management, *Network Security*, Vol. 2007, Issue 5, pp. 4–7, Amsterdam: Elsevier Science Publishers, 2007.

Aulds, C. *Linux Apache Web Server Administration*, Laguna Beach, CA: SYBEX, 2000.

Bach, M. *The Design of the Unix Operating System*, Hoboken, NJ: Prentice-Hall, 1999.

Bagherzadeh, M., Kahani, N., Bezemer, C., Hassan, A., Dingel, J., and Cordy, J. Analyzing a decade of Linux system calls. *Empirical Software Engineering*, Vol. 23, Issue 3, pp. 1519–1551, New York: Springer, 2018.

Balasubramanian, K., and Johnson, D. Linux Memory Management Overview, *The Linux Kernel Hacker's Guide*, http://www.redhat.com:8080/HyperNews/get/memory/memory.html, 1993.

Bar, M. *Linux File Systems*, New York: McGraw-Hill, 2001.

Barrett, D., Silverman, R., and Byrnes, R. *Linux Security Cookbook*, Newton, MA: O'Reilly, 2003.

Bembenek, J., and Klus, A. *Grep Pocket Reference*, Newton, MA: O'Reilly, 2009.

Bennett, H. CD-E: Call It Erasable, Call It Rewritable, but Will It Fly? *CD-ROM Professional*, 1996.

Benvenuti, C. *Understanding Linux Network Internals*, Newton, MA: O'Reilly, 2006.

Berry, D. *Copy, Rip, Burn: The Politics of Copyleft and Open Source*, London: Pluto Press, 2008.

Billimoria, K. *Linux Kernel Programming: A Comprehensive Guide to Kernel Internals, Writing Kernel Modules, and Kernel Synchronization*, Birmingham: Packt, 2021.

Black, D., Golub, D., Julin, D., Rashid, R., Draves, R., Dean, R., Forin, A., Barrera, J., Tokadu, H., Malan, G., and Bohman, D. Microkernel Operating System Architecture and Mach, *Proceedings of the USENIX Workshop on Micro-Kernels and Other Kernel Architectures*, pp. 11–30, 1992.

Blum, R., and Bresnahan, C. *Linux Command Line and Shell Scripting Bible*, Hoboken, NJ: Wiley & Sons, 2021.

Bonaccorsi, A., and Rossi, C. Why open source software can succeed, *Research Policy*, Vol. 32, Issue 7, pp. 1149–1292, Elsevier, 2003.

Both, D. *systemd, Using and Administering Linux*, Vol. 2, pp. 379–410, Berkeley, CA: Apress, 2020.

Bourne, S.R. The UNIX Shell, *The Bell System Technical Journal*, NJ: American Telephone and Telegraph Company, Vol. 57, Issue 6, pp. 1971–1990, 1978.

Bovet, D. *Understanding the Linux Kernel*, Newton, MA: O'Reilly, 2005.

Bradley, D. The Divergent Anarcho-Utopian discourses of the open source software movement, *Canadian Journal of Communication*, Vol. 30, pp. 585–611, 2005.

Bridger, R. *Introduction to Human Factors and Ergonomics*, Orlando, FL: CSC Press, 2017.

Brinch, P. *Classic Operating Systems: From Batch Processing to Distributed Systems*, New York: Springer-Verlag, 2001.

Brookshear, J. *Computer Science: An Overview*, Boston, MA: Pearson, 2018.

Brown, F. *Boolean Reasoning: The Logic of Boolean Equations*, New York: Dover, 2012.

Brown, G. *zOS JCL (Job Control Language)*, Hoboken, NJ: John Wiley & Sons, 2002.

Burtch, K. *Linux Shell Scripting with Bash*, Boston, MA: Pearson, 2004.

Calderon, P. *Nmap: Network Exploration and Security Auditing Cookbook*, Birmingham: Packt Publishing, 2017.

Callaghan, B. *NFS Illustrated*, Boston, MA: Addison-Wesley, 2000.

Ceruzzi, P. *A History of Modern Computing*, Cambridge, MA: The MIT Press, 2003.

Chau, P. Selection of packaged software in small business, *European Journal of Information Systems*, Vol. 3, Issue 4, pp. 292–302, 1994.

Chen, P., Lee, E., Gibson, G., Katz, R., and Patterson, D. RAID: High-performance, reliable secondary storage, *ACM Computing Surveys*, Vol. 26, Issue 2, pp. 145–185, 1994.

Chervenak, A., Vellanki, V., and Kurmas, Z. Protecting File Systems: A Survey of Backup Techniques, *Proceedings of the Joint NASA and IEEE Mass Storage Conference*, 1998.

Cheung, W., and Loong, A. Exploring issues of operating systems structuring: From microkernel to extensible systems, *Operating Systems Review*, Vol. 29, pp. 4–16, 1995.

Ciampa, M. *Security+ Guide to Network Security Fundamentals*, Boston, MA: Thomson Course Technologies, 2011.

Clements, A. *The Principles of Computer Hardware*, New York: Oxford, 2006.

Cole, E. *Network Security Bible*, Hoboken, NJ: Wiley and Sons, 2009.

Comer, D. *Internetworking with TCP/IP*, Boston, MA: Addison-Wesley, 2013.

Corner, D. *Computer Networks and Internets*, Boston, MA: Pearson, 2014.

Cox, R., Muthitacharoen, A., and Morris, R. Serving DNS using a peer-to-peer lookup service, *Lecture Notes in Computer Science*, Vol. 2429, pp. 155–165, New York: Springer, 2002.

Cramer, R., and Shoup, V. Design and analysis of practical public-key encryption schemes secure against adaptive chosen ciphertext attack, *SIAM Journal on Computing*, Vol. 33, pp. 167–226, 2001.

Crawley, D. *The Accidental Administrator: Linux Server Step-by-Step Configuration Guide*, Seattle, WA: CreateSpace, 2010.

Dada, E., Bassi, J., Chiroma, H., Adetunmbi, A. and Ajibuwa, O. Machine learning for Email spam filtering: Review, approaches and open research problems, *Heliyon*, Vol. 5, Issue 6, pp. e01802, 2019.

Dallheimer, M., and Welsh, M. *Running Linux*, Newton, MA: O'Reilly, 2005.

Davis, D., and Swick, R. Network security via private-key certificates, *ACM SIGOPS Operating Systems Review*, Vol. 24, Issue 4, pp. 64–67, 1990.

De Goyeneche, J. Loadable kernel modules, *IEEE Software*, Vol. 16, Issue 1, pp. 65–71, 1999.

Denning, P. Thrashing: Its Causes and Prevention, *Proceedings of the December 9–11 1968 AFIPS Fall Join Computer Conference*, pp. 915–922, New York: ACM, 1968.

Denning, P. Virtual memory, *ACM Computing Surveys*, Vol. 2, Issue 3, pp. 153–189, 1970.

Denning, P. A Short Theory of Multiprogramming, *Proceedings of the 3rd International Workshop on Modeling, Analysis and Simulation of Computer and Telecommunication Systems*, pp. 2–7, 1995.

Dijkstra, E. The structure of the THE multiprogramming system, *Communications of the ACM*, Vol. 11, Issue 5, pp. 341–346, New York: ACM, 1968.

Doeppner, T. *Operating Systems in Depth: Design and Programming*, Hoboken, NJ: Wiley and Sons, 2010.

Droms, R. Automated configuration of TCP/IP with DHCP, *IEEE Internet Computing*, Vol. 3, Issue 4, pp. 45–53, 1999.

Durumeric, Z., Li, F., Kasten, J., Amann, J., Beekman, J., Payer, M., Weaver, N., Adrian, D., Paxson, V., Bailey, M. and Halderman, J. The Matter of Heartbleed, *Proceedings of the 2014 Conference on Internet Measurement Conference*, pp. 475–488, New York: ACM, 2014.

Easttom, W. *Computer Security Fundamentals*, Boston, MA: Pearson, 2019.

Ebrahim, M. and Mallett, A. *Mastering Linux Shell Scripting: A Practical Guide to Linux Command-line, Bash Scripting and Shell Programming*, Birmingham: Packt, 2018.

Economides, N., and Katsamakas, E. Linux vs. windows: a comparison of application and platform innovation incentives for open source and proprietary software platforms, *The Economics of Open Software Development*, J. Bitzer and P. Schroder (eds), Amsterdam: Elsevier, 2006.

Elmasri, R., Carrick, A., and Levine, D. *Operating Systems: A Spiral Approach*, New York: McGraw-Hill, 2009.

Fedora DOCS, https://docs.fedoraproject.org/en-US/docs/.

Fenwick, P. The burrows-wheeler transform for block sorting text compression: Principles and improvements, *Computer Journal*, Vol. 39, Issue 9, pp. 731–740, New York: Oxford University Press, 1996.

Fitzgerald, B. The transformation of open source software, *MIS Quarterly*, Vol. 30, Issue 30, pp. 587–598, 2006.

Forouzan, B. *TCP/IP Protocol Suite*, New York: McGraw-Hill, 2010.

Forouzan, B. *Data Communications and Networking*, New York: McGraw-Hill, 2012.

Fox, R. *Information Technology: An Introduction for Today's Digital World*, Boca Raton, FL: CRC Press, 2021.
Fox, R., and Hao, W. *Internet Infrastructure: Networking, Web Services and Cloud Computing*, Boca Raton, FL: CRC Press, 2017.
Friedl, J. *Mastering Regular Expressions*, Newton, MA: O'Reilly, 2006.
Frisch, E. *Essential System Administration*, Newton, MA: O'Reilly, 2002.
Gajewska, H., Manasse, M., and McCormack, J. Why X is not our ideal window system, *Software—Practice & Experience*, Vol. 20, Issue S2, 1990.
Galov, N. 111+ Linux Statistics and Facts – Linux Rocks!, Hosting Tribunal blog, https://hostingtribunal.com/blog/linux-statistics/.
Gancarz, M. *Linux and the Unix Philosophy*, Daytona Beach, FL: Digital Press, 2003.
Garfinkel, S., Spafford, G., and Schwartz, A. *Practical Unix and Internet Security*, Newton, MA: O'Reilly, 2003.
Garrels, M. *Bash Guide for Beginners*, CA: Fultus Corporation, 2010.
Garrido, J., and Schlesinger, R. *Principles of Modern Operating Systems*, Burlington, MA: Jones and Bartlett, 2011.
Gay, J., Stallman, R., and Lessig, L. *Free Software, Free Society: Selected Essays of Richard M. Stallman*, Seattle, WA: CreateSpace, 2009.
Geisshirt, K. *Pluggable Authentication Modules*, Birmingham: Packt, 2007.
Gillay, C. *Linux User's Guide: Using the Command Line and Gnome Red Hat Linux*, OR: Franklin, Beedle & Associates, 2003.
GNU/Free Software Foundation, *General Public License*, http://www.gnu.org/copyleft/gpl.html.
GNU/Free Software Foundation, *GNU Make*, https://www.gnu.org/software/make/
Goldman, D., and Bonzini, P. *Definitive Guide to Sed: Tutorial and Reference*, Renton, WA: EHDP Press, 2013.
Gomberg, M., Evard, R., and Stacey, C. A Comparison of Large-Scale Software Installation Methods on NT and Unix, *Proceedings of USENIX Technical Program on Large Installation System Administration of Windows NT Conference*, pp. 37–48, 1998.
Goodheart, B., and Cox, J. *The Magic Garden Explained: The Internals of UNIX System V Release 4 An Open Systems Design*, Hoboken, NJ: Prentice-Hall, 1994.
Goyal, V., Horman, N., Ohmichi, K., Soni, M., and Garg, A. Kdump: Smarter, Easier, Trustier, *Ottawa Linux Symposium*, 2007.
Goyvaerts, J., and Levithan, S. *Regular Expressions Cookbook*, Newton, MA: O'Reilly, 2012.
Grampp, F., and Morris, R. UNIX operating-system security, *AT&T Bell Laboratories Technical Journal*, Vol. 63, pp. 1649–1672, 1984.
Green, R., Baird, A., and Davies, J. Designing a fast, on-line backup system for a log-structured file system, *Digital Technical Journal of Digital Equipment Corporation*, Vol. 8, Issue 2, pp. 32–45, 1986.
Gregg, J. *Ones and Zeros: Understanding Boolean Algebra, Digital Circuits, and the Logic of Sets*, Hoboken, NJ: Wiley-IEEE Press, 1998.
Groom, F. The structure and software of the internet, *Annual Review of Communications*, Vol. 50, pp. 695–707, 1997.
Guttman, E. Service location protocol: automatic discovery of IP network services, *IEEE Internet Computing*, Vol. 3, Issue 4, pp. 71–80, 1999.
Haff, G. *How Open Source Ate Software: Understanding the Open Source Movement and So Much More*, Hoboken, NJ: Apress, 2018.
Hagen, S. *IPv6 Essentials*, Newton, MA: O'Reilly, 2014.
Halsall, F. *Data Communications, Computer Networks, and Open Systems*, Boston, MA: Addison-Wesley, 1996.
Hamming, R. Error detecting and error correcting codes, *Bell System Technical Journal*, Vol. 29, Issue 2, pp. 147–160, 1950.
Hansen, P. (editor). *Classic Operating Systems: From Batch Processing to Distributed Systems*, New York: Springer, 2010.
Harker, J., Brede, D., Pattison, R., Santana, G., and Taft, L. A quarter century of disk file innovation, *IBM Journal of Research and Development*, Vol. 25, Issue 5, pp. 677–689, 1981.
Harkins, M. *Managing Risk and Information Security: Protect to Enable*, Hoboken, NJ: Apress, 2016.
Hartig, H., Hohmuth, M., Liedtke, J., Schonberg, S., and Wolter, J. The Performance of Micro-Kernel-Based Systems, *Proceedings of the 16th ACM Symposium on Operating Systems Principles*, 1997.
Hecker, F. Setting up shop: the business of open-source software, *IEEE Software*, Vol. 16, Issue 1, pp. 45–51, 1999.
Helmke, M. *Ubuntu Unleashed*, Boston, MA: Addison-Wesley, 2020.

Hennessy, J. and Patterson, D. *Computer Architecture: A Quantitative Approach*, San Francisco, CA: Morgan Kaufman, 2017.

Hicks, B., Rueda, S., St. Clair, L., Jaeger, T., and McDaniel, P. A logical specification and analysis for SELinux MLS policy, *ACM Transactions on Information and System Security*, Vol. 13, Issue 3, pp. 26–31, Article 26, 2010.

Hill, B., Burger, C., Jesse, J., and Bacon, J. *The Official Ubuntu Book*, Boston, MA: Pearson, 2016.

Holcombe, C., and Holcombe, J. *Survey of Operating Systems*, New York: McGraw-Hill Education, 2019.

Howser, G. *Computer Networks and the Internet: A Hands-On Approach*, New York: Springer, 2019.

Hunt, C. *TCP/IP Network Administration*, Newton, MA: O'Reilly, 2002.

Jacob, B., and Mudge, T. Virtual memory in contemporary multiprocessors, *IEEE Micro Magazine*, Vol. 18, pp. 60–75, 1998.

Jacob, B., and Wang, D. *Memory Systems: Cache, DRAM, Disk*, San Francisco, CA: Morgan Kaufmann, 2007.

Jang, M. *Security Strategies in Linux Platforms and Applications*, Burlington, MA: Jones and Bartlett, 2015.

Kalkhanda, S. *CentOS Quick Start Guide: Get Up and Running with CentOS Server Administration*, Birmingham: Packt, 2018.

Kalkhanda, S., *Learning AWK Programming: A Fast and Simple Cutting-edge Utility for Text-Processing on the Unix-like Environment*, Birmingham: Packt, 2018.

Kamel, M., Keast, J., and Pal, C. Concrete Architecture of the Linux Kernel, technical report, University of Waterloo, Waterloo, Ontario, N2L 3G1. Department of Electrical and Computer Engineering, Department of Computer Science, 1998.

Katz, J., and Lindell, Y. *Introduction to Modern Cryptography*, Boca Raton, FL: CRC Press, 2020.

Kernighan, B. *UNIX: A History and a Memoir*, Independently Published, 2019.

Kernighan, B., and Pike, R. *The Unix Programming Environment*, Boston, MA: Pearson, 2015.

Kernighan, B., and Ritchie, D. *The C Programming Language*, Hoboken, NJ: Prentice-Hall, 1988.

Kiddle, O., Stephenson, P., and Peek, J. *From Bash to Z Shell: Conquering the Command Line*, Hoboken, NJ: Apress Media LLC, 2004.

Kirkbride, P. *systemd, Basic Linux Terminal Tips and Tricks*, pp. 221–234, Berkeley, CA: Apress, 2020.

Kirtch, O. *Linux Network Administrator's Guide*, Newton, MA: O'Reilly, 1995.

Knuth, D. Dynamic huffman coding, *Journal of Algorithms*, Vol. 6, Issue 2, pp. 163–180, Amsterdam: Elsevier, 1985.

Krishnamurthy, B., and Rexford, J. *Web Protocols and Practice: HTTP/1.1, Networking Protocols, Caching and Traffic Measure*, Boston, MA: Addison-Wesley, 2001.

Kumar, A. Migration from Microsoft to Linux on Servers and Desktops, *Proceedings of the 20th Annual Conference of the National Advisory Committee on Computing Qualifications*, S. Mann and N. Bridgeman (eds), pp. 117–123, 2007.

Langfeldt, N. *The Concise Guide to DNS and Bind*, Indianapolis, IN: Que Corp, 2000.

Laurent, A. *Understanding Open Source and Free Software Licensing*, Newton, MA: O'Reilly, 2004.

Leach, R. *Advanced Topics in UNIX: Processes, Files and Systems*, Hoboken, NJ: Wiley and Sons, 1994.

LeBlanc, D., and Yates, I. *Linux Install and Configuration Little Black Book: The Must-Have Troubleshooting Guide to Installing and Configuring Linux*, Scottsdale, AZ: Coriolis Open Press, 1999.

Ledin, J. *Modern Computer Architecture and Organization: Learn x86, ARM and RISC-V Architectures and the Design of Smartphones, PCs, and Cloud Servers*, Birmingham: Packt, 2020.

Lempel, A., and Ziv, J. On the complexity of finite sequences, *IEEE Transactions on Information Theory*, Vol. 22, Issue 1, pp. 75–81, 1976.

Lerner J., Pathak P. A., and Tirole, J. The dynamics of open source contributors, *American Economic Review*, Vol. 96, Issue 2, pp. 114–118, 2006.

Lewine, D. *POSIX Programmer's Guide: Writing Portable Unix Programs*, Newton, MA: O'Reilly, 1992.

Li, Y., Li, W., and Jiang, C. A Survey of Virtual Machine Systems: Current Technology and Future Trends, *Proceedings of the Third International Symposium on Electronic Commerce and Security*, 2010.

Liang, Y. and Liang, Y., *Introduction to Java Programming and Data Structures*, Boston, MA: Pearson, 2017.

Liedtke, J. On Micro-Kernel Construction, *Proceedings of the Fifteenth ACM Symposium on Operating Systems Principles*, 1995.

Liedtke, J. Toward real microkernels, *Communications of the ACM*, Vol. 39, Issue 9, pp. 70–79, 1996.

Limoncelli, T., Hogan, C., and Chalup, S. *The Practice of System and Network Administration*, Boston, MA: Addison-Wesley, 2016.

Lin, Y., Hwang, R., and Baker, F. *Computer Networks: An Open Source Approach*, New York: McGraw-Hill, 2011.

Liu, C. *DNS and BIND on IPv6: DNS for the Next-Generation Internet*, Newton, MA: O'Reilly, 2011.

Loscocco, P., and Smalley, S. Integrating Flexible Support for Security Policies into the Linux Operating System, *Proceedings of the FREENIX Track: 2001 USENIX Annual Technical Conference, The USENIX Association*, pp. 29–42, CA: USENIX Association, 2001.

Love, R. *Linux Kernel Development*, Hoboken, NJ: Addison-Wesley, 2010.

Lu, L., Arpaci-Dusseau, A, Arpaci-Dusseau, R., and Lu, S. A study of Linux file system evolution, *ACM Transactions on Storage,* Vol. 10, Issue 1, pp. 1–32, 2014.

Luotonen, A., and Altis, K. World-Wide Web Proxies, *Computer Networks and ISDN Systems*, Vol. 27, Issue 2, pp. 147–154, Amsterdam: Elsevier, 1994.

Mallett, A. *CentOS System Administration Essentials*, Birmingham: Packt, 2014.

Mann, S., and Mitchell, E. *Linux System Security: The Administrator's Guide to Open Source Security Tools*, Hoboken, NJ: Prentice-Hall, 2002.

Mansfield, K., and Antonakos, J. *Computer Networking for LANs to WANs: Hardware, Software and Security*, Boston, MA: Thomson Course Technology, 2009.

Markatos, E., Katevenis, M., Pnevmatikatos, D., and Flouris, M. Secondary Storage Management for Web Proxies, *Proceedings of USITS 99: The 2nd Conference on USENIX Symposium on Internet Technologies and Systems*, Vol. 2, pp. 93–114, CA: USENIX, 1999.

Matotek, D., Turnbull, J. and Lieverdink, P. *Pro Linux System Administration: Learn to Build Systems for Your Business Using Free and Open Source Software*, Hoboken, NJ: Apress, 2017.

Matthews, J. *Computer Networking: Internet Protocols in Action*, Hoboken, NJ: Apress, 2005.

Mauelshagen, H. Logical Volume Manager (LVM2). *Red Hat Magazine*, 2004.

Mauerer, W. *Professional Linux Kernel Architecture*, Hoboken, NJ: Wrox, 2008.

Meeker, H. *The Open Source Alternative: Understanding Risks and Leveraging Opportunities*, Hoboken, NJ: Wiley & Sons, 2008.

Melve, I., Slettjord, L., Bekker, H., and Verschuren, T. Building a Web Caching System—Architectural Considerations, *Proceedings of the 1997 NLANR Web Cache Workshop*, CA: National Laboratory for Applied Network Research, 1997.

Menezes, A., Oorschot, P., and Vanstone, S. *Handbook of Applied Cryptography (Discrete Mathematics and Its Applications)*, Boca Raton, FL: CRC Press, 1996.

Miles, S. Linux Closing in on Microsoft Market Share, Study Says, CENT News.com, http://news. com. om/2100-1001-243527.html?tag=mainstry, 2000.

Mockapetris, P., and Dunlap, K. Development of the Domain Name System, *SIGCOMM 88 Symposium Proceedings on Communications Architectures and Protocols*, pp. 123–133, New York: ACM, 1988.

Moody, G. *Rebel Code: Linux and the Open Source Revolution*, New York: Basic Books, 2009.

Morris, R., and Thompson, K., Password security: a case history, *Communications of the ACM*, Vol. 22, Issue 11, pp. 594–597, New York: ACM, 1979.

Moshe, B. *Linux File Systems*, New York: McGraw-Hill, 2001.

Mustonen, M. Copyleft—the economics of Linux and other open source software, *Information Economics and Policy*, Vol. 15, Issue 1, pp. 99–121, Amsterdam: Elsevier, 2003.

Negus, C., and Boronczyk, T. *CentOS Bible*, Hoboken, NJ: Wiley and Sons, 2009.

Negus, C., and Bresnahan, C. *Linux Bible*, Hoboken, NJ: Wiley and Sons, 2020.

Nemeth, E., Snyder, G., Hein, T., Whaley, B., and Mackin, D. *Unix and Linux System Administration Handbook*, Boston, MA: Addison-Wesley, 2017.

Newham, C. *Learning the Bash Shell: Unix Shell Programming*, Newton, MA: O'Reilly, 2005.

Null, L., and Lobur, J. *The Essentials of Computer Organization and Architecture*, Burlington, MA: Jones and Bartlett, 2018.

Odom, W. *Computer Networking First-Step*, Indianapolis, IN: Cisco Press, 2004.

O'Regan, G. *Introduction to the History of Computing*, New York: Springer, 2016.

Otero, A. *Information Technology Control and Audit*, Boca Raton, FL: Auerbach Publications, 2020.

Paar, C., and Pelzl, J. *Understanding Cryptography*, New York: Springer. 2010.

Pate, S. *UNIX Filesystems: Evolution, Design and Implementation*, Hoboken, NJ: Wiley and Sons, 2003.

Patterson, D., Gibson, G., and Katz, R. A Case for Redundant Arrays of Inexpensive Disks (RAID), *Proceedings of SIGMOD '88*, pp. 109–116, New York: ACM, 1988.

Patterson, D., and Hennessy, J. *Computer Organization and Design: The Hardware/Software Interface*, San Francisco, CA: Morgan Kaufmann, 2011.

Peek, J., Powers, S., O'Reilly, T., and Loukides, M. *Unix Power Tools*, Newton, MA: O'Reilly, 2007.

Peterson, L., and Davie, B. *Computer Networks: A Systems Approach*, San Francisco, CA: Morgan Kaufmann, 2021.

Petersen, R. *Linux: The Complete Reference*, New York: McGraw-Hill, 2007.

Pfleeger, C., and Pfleeger, S. *Security in Computing*, Boston, MA: Pearson, 2018.
Plank, J. A Tutorial on reed-solomon coding for fault-tolerance in RAID-like systems, *Software Practice and Experience*, Vol. 27, Issue 9, pp. 995–1012, Hoboken, NJ: Wiley and Sons, 1997.
Portnoy, M. *Virtualization Essentials*, Hoboken, NJ: Sybex, 2016.
Prabhakaran, V., Arpaci-Dusseau, A., and Arpaci-Dusseau, R. Analysis and Evolution of Journaling File Systems, *USENIX '05 Online Proceedings*, pp. 105–120, 2005.
Prasse, P., Sawade, C., Landwehr, N. and Scheffer, T. Learning to identify concise regular expressions that describe email campaigns, *Journal of Machine Learning Research*, Vol. 16, Issue 1, pp. 3687–3720, 2015.
Puryear, D. *Best Practices for Managing Linux and Unix Servers*, New York: Penton, 2006.
Quarterman, J., and Hoskins, H. Notable computer networks, *Communications of the ACM*, Vol. 29, Issue 10, pp. 932–971, New York: ACM, 1986.
Quigley, E. *UNIX Shells by Example*, Hoboken, NJ: Prentice-Hall, 2001.
Ramey, C., and Fox, B. Bash Reference Manual, http://www.gnu.org/software/bash/manual, Network Theory LTD, 2012.
Rash, M. *Linux Firewalls: Attack Detection and Response with iptables, psad and fwsnort*, San Fransico, CA: No Starch Press, 2007.
Raymond, E. *The Cathedral and the Bazaar: Musings on Linux and Open Source by an Accidental Revolutionary*, Newton, MA: O'Reilly, 2001.
Red Hat Inc, Product Documentation for Red Hat Enterprise Linux 8, https://access.redhat.com/documentation/en-US/Red_Hat_Enterprise_Linux/8/, 2021.
Refsdal, A., Solhaug, B. and Stolen, K. *Cyber-Risk Management*, New York: Springer, 2015.
Rescorla, E. An introduction to openssl programming, *Linux Journal*, Vol. 2001, Issue 89, p. 3, Article 3, 2001.
Reynolds, G. *Ethics in Information Technology*, Boston, MA: Cengage, 2018.
Riehle, D. The economic motivation of open source: stakeholder perspectives, *IEEE Computer*, Vol. 40, Issue 4, pp. 25–32, 2007.
Ritchie, D. *The Evolution of the UNIX Time-Sharing System, Language Design and Programming Methodology, Lecture Notes on Computer Science*, Vol. 79, Berlin: Springer-Verlag, 1979.
Ritchie, D., and Thompson, K. The UNIX time-sharing system, *Communications of the ACM*, Vol. 17, Issue 7, pp. 365–375, 1974.
Robbins, A. *VI Editor Pocket Reference*, Newton, MA: O'Reilly, 1999.
Robbins, A. *Sed and Awk: A Pocket Reference*, Newton, MA: O'Reilly, 2002.
Robbins, A. *Bash Pocket Reference*, Newton, MA: O'Reilly, 2016.
Robbins, A., and Beebe, N. *Classic Shell Scripting*, Newton, MA: O'Reilly, 2005.
Rodeh, O., Bacik, J. and Mason, C. BTRFS: the Linux B-tree filesystem. *ACM Transactions on Storage*, Vol. 9, Issue 3, pp. 1–32, 2013.
Rosenfeld, L. and Downey, A. *Think Perl 6: How to Think Like a Computer Scientist*, Newton, MA: O'Reilly, 2017.
Ross, K. and Kurose, J. *Computer Networking: A Top-Down Approach*, Boston, MA: Pearson, 2017.
Royon, Y., and Frenot, S. A Survey of Unix Init Schemes, Technical report arXiv:0706.2748v2, Cornell University Library, 2007.
Rusen, C. *Networking Your Computers & Devices Step by Step*, Redmond, WA: Microsoft Press, 2011.
Rusling, D. The Linux Kernel, http://sunsite.unc.edu/Linux/LDP/tlk/tlk.html, 2001.
Ryan, P. Linux Market Share Set to Surpass Win 98, OS X Still Ahead of Vista, http://arstechnica. com/apple/2007/09/linux-marketshare-set-to-surpass-windows-98, 2007.
Saini, K. *Squid Proxy Server 3.1: Beginner's Guide*, Birmingham: Packt Publishing, 2011.
Salas, P. *The Daemon, the GNU and the Penguin*, Virginia Beach, VA: Reed Media Services, 2008.
Salus, P. (ed). *A Quarter Century of Unix*, Boston, MA: Addison-Wesley, 1994.
Samar, V. Unified Login with Pluggable Authentication Modules, *Proceedings of the 3rd ACM Conference on Computer and Communications Security*, pp. 1–10, ACM Press, 1996.
Sandberg, R., Goldberg, D., Kleiman, S., Walsh, D., and Lyon, B. Design and Implementation of the Sun Network File System, *Proceedings of the 1985 USENIX Summer Conference*, pp. 119–130, 1985.
Sandhu, R., and Samarati, P. Access control: principles and practice, *IEEE Communications*, Vol. 32, pp. 40–48, 1994.
Sarwar, S., and Koretsky, R. *Unix: The Textbook*, Boca Raton, FL: CRC Press, 2016.
Sawicki, E. *Guide to Apache*, Boston, MA: Thomson, 2008.
Schach, S., Jin, B., Wright, D., Heller, G., and Offutt, A. Maintainability of the Linux Kernel, *IEE Proceedings—Software*, Vol. 149, Issue 1, pp. 18–23, London: Institute of Engineering and Technology, 2002.

Scheifler, R., and Gettys, J. *X Window System: Core and Extension Protocols: X Version 11, Releases 6 and 6.1*, Daytona Beach, FL: Digital Press, 1996.

Schneier, B. *Applied Cryptography: Protocols, Algorithms and Source Code in C*, Hoboken, NJ: Wiley & Sons, 2015.

Schwartz, M. Linux Job Scheduling, *Linux Journal*, Vol. 2000, Issue 77, Article 8, 2000.

Sebesta, R. *Concepts of Programming Languages*, Boston, MA: Addison-Wesley, 2015.

Shoch, J., Dalal, Y., Redell, D., and Crane, R. Evolution of the ethernet local computer network, *Computer*, Vol. 15, Issue 8, pp.10–27, 1982.

Shotts Jr., W. *The Linux Command Line: A Complete Introduction*, San Fransico, CA: No Starch Press, 2019.

Siever, E., Weber, A., Figgins, S., Love, R., and Robbins, A. *Linux in a Nutshell: A Desktop Quick Reference*, Newton, MA: Riley, 2009.

Silberschatz, A., Galvin, P., and Gagne, G. *Operating System Concepts*, Hoboken, NJ: Wiley & Sons, 2021.

Silva, S. *Web Server Administration*, Boston, MA: Thomson Course Technology, 2008.

Sloan, J. *Network Troubleshooting Tools*, Newton, MA: O'Reilly, 2001.

Smalley, S., and Fraser, T. A Security Policy Configuration for the Security-Enhanced Linux, *NAI Labs Technical Report*, 2001.

Smyth, N. *Red Hat Enterprise Linux 8 Essentials: Learn to Install, Administer and Deploy RHEL 8 Systems*, Cary, NC: Payload, 2019.

Sobell, M. *A Practical Guide to Ubuntu Linux*, Boston, MA: Pearson, 2015.

Sobell, M. *A Practical Guide to Linux Commands, Editors, and Shell Programming*, Boston, MA: Addison-Wesley, 2017.

Spafford, E. The internet worm: crisis and aftermath, *Communications of the ACM*, Vol. 32, Issue 6, pp. 678–687, 1989.

St. Laurent, A. *Understanding Open Source and Free Software Licensing*, Newton, MA: O'Reilly, 2004.

Stallings, W. *Data and Computer Communications*, Boston, MA: Pearson, 2013.

Stallings, W. *Computer Organization and Architecture*, Boston, MA: Pearson, 2015.

Stallings, W. *Cryptography and Network Security: Principles and Practices*, Boston, MA: Pearson, 2016.

Stallings, W. *Operating Systems: Internals and Design Principles*, Boston, MA: Pearson, 2017.

Stallings, W., and Brown, L. *Computer Security: Principles and Practices*, Boston, MA: Pearson, 2017.

Stallman, R. The GNU Operating System and the Free Software Movement, *Open Sources: Voices from the Open Source Revolution*, C. DiBona, S. Ockman and M. Stone (eds), Sebastopol, CA: O'Reilly, 1999.

Stallman, R. Why 'Free Software' Is Better than 'Open Source', http://www.gnu.org/philosophy/free-software-for-freedom.html, 2007.

Stallman, R., and Gay, J. *Free Software, Free Society: Selected Essays of Richard M. Stallman*, New York: SoHo Books, 2002.

Stankovic, J. Software communication mechanisms: procedure calls versus messages, *Computer*, Vol. 15, Issue 4, pp. 19–25, 1982.

Statcounter, Operating System Market Share Worldwide, https://gs.statcounter.com/os-market-share.

Stewart, J. and Kinsey, D. *Network Security, Firewalls, and VPNs*, Burlington, MA: Jones and Bartlett, 2020.

Stubblebine, T. *Regular Expression Pocket Reference*, Newton, MA: O'Reilly, 2003.

Tam, L., Glassman, M. and Vandenwauver, M. The psychology of password management: a tradeoff between security and convenience. *Behaviour & Information Technology*, Vol. 29, Issue 3, pp. 233–244, 2010.

Tanenbaum, A. *Structured Computer Organization*, Boston, MA: Pearson, 2016.

Tanenbaum, A. *Modern Operating Systems*, Boston, MA: Pearson, 2016.

Tanenbaum, A. Feamster, N. and Wetherall, D. *Computer Networks*, Boston, MA: Pearson, 2020.

Tanenbaum, A., Herder, J., and Bos, H. Can we make operating systems reliable and secure? *Computer*, Vol. 39, Issue 5, pp. 44–51, 2006.

Tevanian, A., Rashid, R., Golub, D., Black, Dl, Cooper, E., and Young, M. Mach Threads and the UNIX Kernel: The Battle for Control, *Proceedings of the Summer 1987 USENIX Conference*, 1987.

Tirgari, V. Information technology policies and procedures against unstructured data: a phenomenological study of information technology professionals. *Journal of Management Information and Decision Sciences*, Vol. 15, Issue 2, pp. 87–106, 2012.

Toigo, J. *Disaster Recovery Planning: Preparing for the Unthinkable*, Hoboken, NJ: Prentice-Hall, 2002.

Tominaga, A., Nakamura, O., Teraoka, F., and Murai, J. Problems and Solutions of DHCP, *Proceedings of INET 95*, 1995.

Toxen, B. *Real World Linux Security: Intrusion Prevention, Detection, and Recovery*, Hoboken, NJ: Prentice-Hall, 2003.

Ts'o, T., and Tweedie, S. Planned Extensions to the Linux Ext2/Ext3 Filesystem, *Proceedings of the FREENIX Track: 2002 USENIX Annual Technical Conference*, pp. 235–243, 2002.

Ubuntu Documentation Team, Official Ubuntu Documentation, https://help.ubuntu.com/.

Uti, N. Real Time Mobile Video Streaming Using Wavelet Transformation and Run-Length Coding, *Proceedings of the 2009 International Conference on Wireless Networks*, 2 *(ICWN '09)*, pp. 368– 373, 2009.

Vacca, R. *Computer and Information Security Handbook*, San Francisco, CA: Morgan Kaufmann, 2017.

Van Winkle, L. *Hands-On Network Programming with C*, Birmingham: Packt, 2019.

Vance, N. and Polik, W. Understanding firewalld in Multi-zone Configurations, *Linux Journal*, Vol. 269, pp. 80–88, 2016.

Vermeulen, S. *SELinux Cookbook*. Birmingham: Packt, 2014.

Viega, J., Messier, M. and Chandra, P. *Network Security with openSSL: Cryptography for Secure Communications*, Newton, MA: O'Reilly, 2002.

Warford, J. *Computer Systems*, Burlington, MA: Jones and Bartlett, 2016.

Watt, A. *Beginning Regular Expressions (Programmer to Programmer)*, Hoboken, NJ: Wrox, 2005.

Welch, T. A., A technique for high performance data compression, *IEEE Computer*, Vol. 17, Issue 6, pp. 8–19, 1984.

Wells, N. *The Complete Guide to Linux System Administration*, Boston, MA: Thomson Course Technology, 2005.

Wessels, D. *Squid: The Definitive Guide*, Newton, MA: O'Reilly, 2009.

West, J., Dean, T. and Andrews, J. *Network+Guide to Networks*, Boston, MA: Cengage, 2018.

Whitesitt, J. *Boolean Algebra and Its Applications*, New York: Dover, 2010.

Williams, S. *Free as in Freedom: Richard Stallman's Crusade for Free Software*, Seattle, WA: CreateSpace, 2009.

Wirzenius, L. (ed). Linux System Administrator's Guide, Linux Documentation Project, http://www.tldp.org/LDP/sag/html/index.html.

Wright, C. Linux Security Module Framework, *Proceedings of the Linux Symposium*, pp. 604–610, 2002.

Wu, C., Gerlach, J., and Young, C. An empirical analysis of open source software developers' motivations and continuance intentions, *Information and Management*, Vol. 44, pp. 253–262, Amsterdam: Elsevier, 2007.

Ziv, J., and Lempel, A. Compression of individual sequences via variable-rate coding, *IEEE Transactions on Information Theory*, Vol. 24, Issue 5, pp. 530–536, CA: IEEE, 1978.

Index

Note: **Bold** page numbers refer to tables; *italic* page numbers refer to figures.